WORLD ATLAS

COMPACT

Previously published as *Compact Atlas of the World*

DK | Penguin Random House

FOR THE SIXTH EDITION

SENIOR CARTOGRAPHIC EDITOR
Simon Mumford

PRODUCTION CONTROLLER
Rita Sinha

SENIOR PRODUCER
Luca Frassinetti

PUBLISHER
Andrew Macintyre

PUBLISHING DIRECTOR
Jonathan Metcalf

ASSOCIATE PUBLISHING DIRECTOR
Liz Wheeler

ART DIRECTOR
Philip Ormerod

First American edition 2001
Published in the United States by DK Publishing,
345 Hudson Street, New York, New York 10014

Reprinted with revisions 2002. Second edition 2003. Reprinted with revisions 2004.
Third edition 2005. Fourth edition 2009. Fifth edition 2012. Sixth edition 2015.
Previously published as the Compact Atlas of the World.

Copyright © 2001, 2002, 2003, 2004, 2005, 2009, 2012, 2015
Dorling Kindersley Limited
A Penguin Random House Company

15 16 17 18 19 10 9 8 7 6 5 4 3 2 1
001–265178–May/2015

A catalog record for this book is available from the Library of Congress.
ISBN 978-1-4654-2991-9

DK books are available at special discounts when purchased in bulk for sales promotions, premiums,
fund-raising, or educational use. For details, contact: DK Publishing Special Markets,
345 Hudson Street, New York, New York 10014 or SpecialSales@dk.com

Printed and bound in Hong Kong

A WORLD OF IDEAS:
SEE ALL THERE IS TO KNOW

www.dk.com

Key to map symbols

Physical features

Elevation

6000m/19,686ft

4000m/13,124ft

3000m/9843ft

2000m/6562ft

1,000m/3281ft

500m/1640ft

250m/820ft

0

Below sea level

△ Mountain

▽ Depression

◬ Volcano

)(Pass/tunnel

Sandy desert

Drainage features

Major perennial river

Minor perennial river

– – – Seasonal river

Canal

| Waterfall

Perennial lake

Seasonal lake

Wetland

Ice features

Permanent ice cap/ice shelf

Winter limit of pack ice

Summer limit of pack ice

Borders

━ ━ ━ Full international border

– – – – Disputed de facto border

· · · · · Territorial claim border

x x x Cease-fire line

– – – Undefined boundary

Internal administrative boundary

Communications

Major road

Minor road

Railway

✈ International airport

Settlements

◉ Above 500,000

◉ 100,000 to 500,000

○ 50,000 to 100,000

○ Below 50,000

● National capital

● Internal administrative capital

Miscellaneous features

+ Site of interest

⌐⌐⌐⌐ Ancient wall

Graticule features

Line of latitude/longitude/Equator

– – – Tropic/Polar circle

25° Degrees of latitude/longitude

Names

Physical features

Andes

Sahara | Landscape features

Ardennes

Land's End | Headland

Mont Blanc 4,807m | Elevation/volcano/pass

Blue Nile | River/canal/waterfall

Ross Ice Shelf | Ice feature

PACIFIC OCEAN

Sulu Sea | Sea features

Palk Strait

Chile Rise | Undersea feature

Regions

FRANCE | Country

BERMUDA (to UK) | Dependent territory

KANSAS | Administrative region

Dordogne | Cultural region

Settlements

PARIS | Capital city

SAN JUAN | Dependent territory capital city

Chicago

Kettering | Other settlements

Burke

Inset map symbols

Urban area

City

Park

▪ Place of interest

□ Suburb/district

WORLD ATLAS

COMPACT

Contents

The World's Regions

North & Central America

South America

Africa

Europe

North & West Asia

South & East Asia

Australasia & Oceania

Index – Gazetteer

The Political World

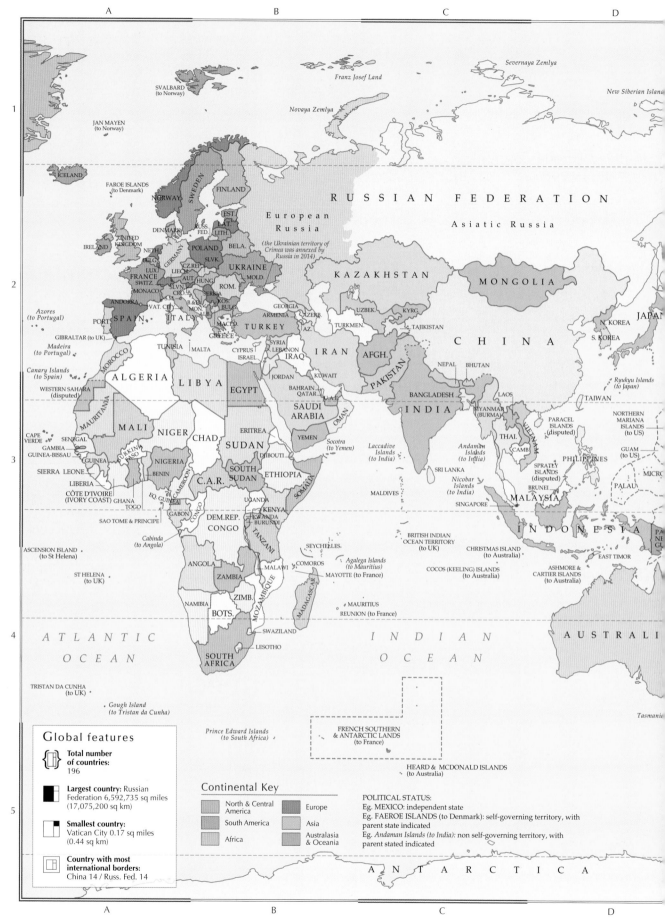

A | B | C | D

1

SVALBARD
(to Norway)

JAN MAYEN
(to Norway)

Franz Josef Land

Severnaya Zemlya

New Siberian Island

Novaya Zemlya

ICELAND

FAROE ISLANDS
(to Denmark)

NORWAY

SWEDEN

FINLAND

RUSSIAN FEDERATION

European
Russia

Asiatic Russia

DENMARK

RUSS.
FED.

EST.
LAT.
LITH.

(the Ukrainian territory of
Crimea was annexed by
Russia in 2014)

2

UNITED
KINGDOM

IRELAND

NETH.
BELG.
LUX.
FRANCE
SWITZ.
MONACO

GERMANY

POLAND

BELA.

SLVK.
CZ.REP.
LIECH.
AUT.
SLVN.
CRO.

UKRAINE

HUNG.
ROM.
MOLD.

KAZAKHSTAN

MONGOLIA

JAPAN

N. KOREA
S. KOREA

ANDORRA

Azores
(to Portugal)

PORT.

SPAIN

VAT. CITY
S.M.
B.&H.
MON.

ITALY

SERBIA
KOS.
BUL.
MACED.
GREECE

GEORGIA

ARMENIA
AZERB.
AZ.

TURKMEN.

UZBEK.

KYRG.

TAJIKISTAN

CHINA

TURKEY

GIBRALTAR (to UK)

MOROCCO

TUNISIA

MALTA

CYPRUS
ISRAEL

SYRIA
LEBANON

IRAQ

IRAN

AFGH.

PAKISTAN

NEPAL

BHUTAN

Ryukyu Islands
(to Japan)

Madeira
(to Portugal)

Canary Islands
(to Spain)

WESTERN SAHARA
(disputed)

ALGERIA

LIBYA

EGYPT

JORDAN

KUWAIT

BAHRAIN
QATAR

U.A.E.

SAUDI
ARABIA

OMAN

BANGLADESH

INDIA

MYANMAR
(BURMA)

LAOS

TAIWAN

NORTHERN
MARIANA
ISLANDS
(to US)

PARACEL
ISLANDS
(disputed)

CAPE
VERDE

MAURITANIA

MALI

NIGER

CHAD

SUDAN

ERITREA

YEMEN

Socotra
(to Yemen)

Laccadive
Islands
(to India)

Andaman
Islands
(to India)

THAI.

CAMB.

SPRATLY
ISLANDS
(disputed)

VIETNAM

GUAM
(to US)

PHILIPPINES

MICRO

SENEGAL

GAMBIA

GUINEA-BISSAU

GUINEA

SIERRA LEONE

LIBERIA

CÔTE D'IVOIRE
(IVORY COAST)

BURKINA
FASO

NIGERIA

BENIN

GHANA
TOGO

CAMEROON

EQ. GUINEA

SAO TOME & PRINCIPE

DJIBOUTI

SOUTH
SUDAN

C.A.R.

ETHIOPIA

SOMALIA

UGANDA

KENYA

SRI LANKA

Nicobar
Islands
(to India)

MALDIVES

SINGAPORE

BRUNEI

PALAU

MALAYSIA

3

GABON

CONGO

DEM.REP.
CONGO

RWANDA
BURUNDI

TANZANI

INDONESIA

PA
NE
GU

Cabinda
(to Angola)

SEYCHELLES

BRITISH INDIAN
OCEAN TERRITORY
(to UK)

CHRISTMAS ISLAND
(to Australia)

EAST TIMOR

ASCENSION ISLAND
(to St Helena)

ANGOLA

ZAMBIA

MALAWI

COMOROS

Agalega Islands
(to Mauritius)

COCOS (KEELING) ISLANDS
(to Australia)

ASHMORE &
CARTIER ISLANDS
(to Australia)

ST HELENA
(to UK)

ZIMB.

MOZAMBIQUE

MADAGASCAR

MAYOTTE (to France)

MAURITIUS

REUNION (to France)

NAMIBIA

BOTS.

SWAZILAND

LESOTHO

ATLANTIC

OCEAN

INDIAN

OCEAN

AUSTRALI

4

SOUTH
AFRICA

TRISTAN DA CUNHA
(to UK)

Gough Island
(to Tristan da Cunha)

Prince Edward Islands
(to South Africa)

FRENCH SOUTHERN
& ANTARCTIC LANDS
(to France)

Tasmanie

HEARD & MCDONALD ISLANDS
(to Australia)

Global features

Total number
of countries:
196

Largest country: Russian
Federation 6,592,735 sq miles
(17,075,200 sq km)

Smallest country:
Vatican City 0.17 sq miles
(0.44 sq km)

Country with most
international borders:
China 14 / Russ. Fed. 14

Continental Key

North & Central
America

South America

Africa

Europe

Asia

Australasia
& Oceania

POLITICAL STATUS:
Eg. MEXICO: independent state
Eg. FAEROE ISLANDS (to Denmark): self-governing territory, with
parent state indicated
Eg. Andaman Islands (to India): non self-governing territory, with
parent stated indicated

ANTARCTICA

A | B | C | D

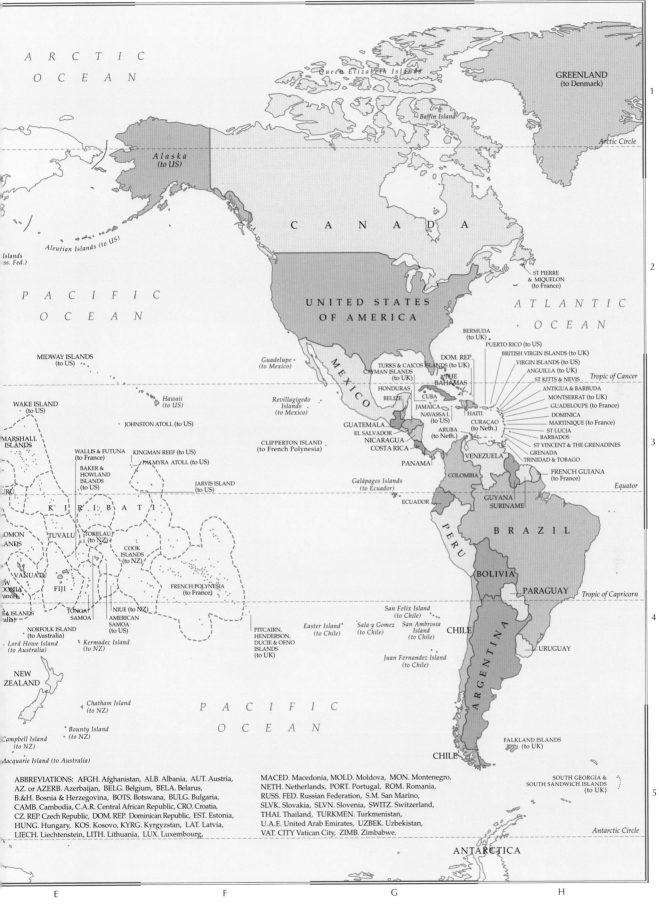

ABBREVIATIONS: AFGH. Afghanistan, ALB. Albania, AUT. Austria, AZ. or AZERB. Azerbaijan, BELG. Belgium, BELA. Belarus, B.&H. Bosnia & Herzegovina, BOTS. Botswana, BULG. Bulgaria, CAMB. Cambodia, C.A.R. Central African Republic, CRO. Croatia, CZ. REP. Czech Republic, DOM. REP. Dominican Republic, EST. Estonia, HUNG. Hungary, KOS. Kosovo, KYRG. Kyrgyzstan, LAT. Latvia, LIECH. Liechtenstein, LITH. Lithuania, LUX. Luxembourg, MACED. Macedonia, MOLD. Moldova, MON. Montenegro, NETH. Netherlands, PORT. Portugal, ROM. Romania, RUSS. FED. Russian Federation, S.M. San Marino, SLVK. Slovakia, SLVN. Slovenia, SWITZ. Switzerland, THAI. Thailand, TURKMEN. Turkmenistan, U.A.E. United Arab Emirates, UZBEK. Uzbekistan, VAT. CITY Vatican City, ZIMB. Zimbabwe.

The Physical World

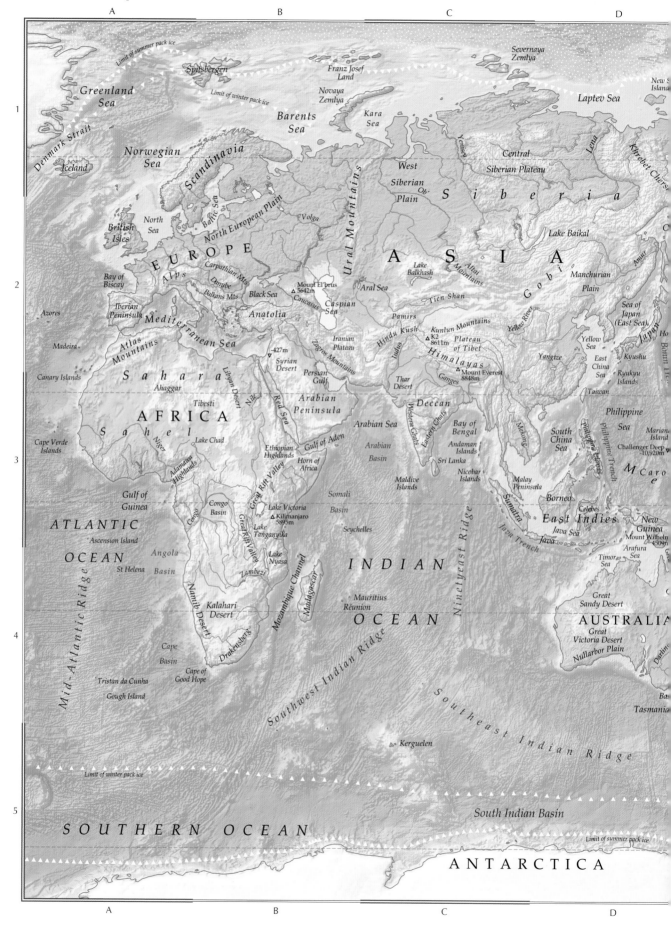

Limit of summer pack ice
Spitsbergen
Franz Josef Land
Severnaya Zemlya
Greenland Sea
Limit of winter pack ice
New Siberian Islands
Barents Sea
Novaya Zemlya
Kara Sea
Laptev Sea
Norwegian Sea
Scandinavia
West Siberian Plain
Central Siberian Plateau
Lena
Khrebet Cherski
Iceland
Denmark Strait
Ob'
Siberia
British Isles
North Sea
Baltic Sea
EUROPE
North European Plain
Volga
Ural Mountains
Yenisey
Lake Baikal
ASIA
Amur
Bay of Biscay
Alps
Carpathian Mts
Danube
Balkans Mts
Mount El'brus △5642m
Caucasus
Black Sea
Caspian Sea
Aral Sea
Lake Balkhash
Altai Mountains
Gobi
Manchurian Plain
Azores
Iberian Peninsula
Mediterranean Sea
Anatolia
Tien Shan
Pamirs
Sea of Japan (East Sea)
Japan
Madeira
Atlas Mountains
Syrian Desert
△ -427m
Iranian Plateau
Zagros Mountains
Persian Gulf
Hindu Kush
Kunlun Mountains
K2 8611m △
Plateau of Tibet
Yellow River
Yellow Sea
Kyushu
Canary Islands
Sahara
Ahaggar
Libyan Desert
Nile
Red Sea
Arabian Peninsula
Indus
Himalayas
Ganges
△ Mount Everest 8848m
Thar Desert
Deccan
Yangtze
East China Sea
Ryukyu Islands
Cape Verde Islands
AFRICA
Sahel
Tibesti
Niger
Lake Chad
Ethiopian Highlands
Gulf of Aden
Horn of Africa
Arabian Sea
Western Ghats
Eastern Ghats
Bay of Bengal
Andaman Islands
Sri Lanka
Philippine Sea
Philippine Islands
Philippine Trench
Mariana Islands
Challenger Deep -10,920m
Caro
Adamawa Highlands
Great Rift Valley
Somali Basin
Arabian Basin
Maldive Islands
Nicobar Islands
Malay Peninsula
South China Sea
Borneo
Celebes
East Indies
New Guinea
Gulf of Guinea
Congo
Congo Basin
Lake Victoria
△ Kilimanjaro 5895m
Lake Tanganyika
Seychelles
Ninetyeast Ridge
Sumatra
Java Trench
Java Sea
Java
Mount Wilhelm 4509m
Arafura Sea
ATLANTIC
Ascension Island
Angola Basin
St Helena
Lake Nyasa
Zambezi
Mozambique Channel
Madagascar
INDIAN
Timor Sea
OCEAN
Mid-Atlantic Ridge
Namib Desert
Kalahari Desert
Mauritius
Réunion
OCEAN
Great Sandy Desert
AUSTRALIA
Cape Basin
Drakensberg
Great Victoria Desert
Tristan da Cunha
Cape of Good Hope
Nullarbor Plain
Gough Island
Southwest Indian Ridge
Southeast Indian Ridge
Tasmania
Kerguelen
Limit of winter pack ice
South Indian Basin
SOUTHERN OCEAN
Limit of summer pack ice
ANTARCTICA

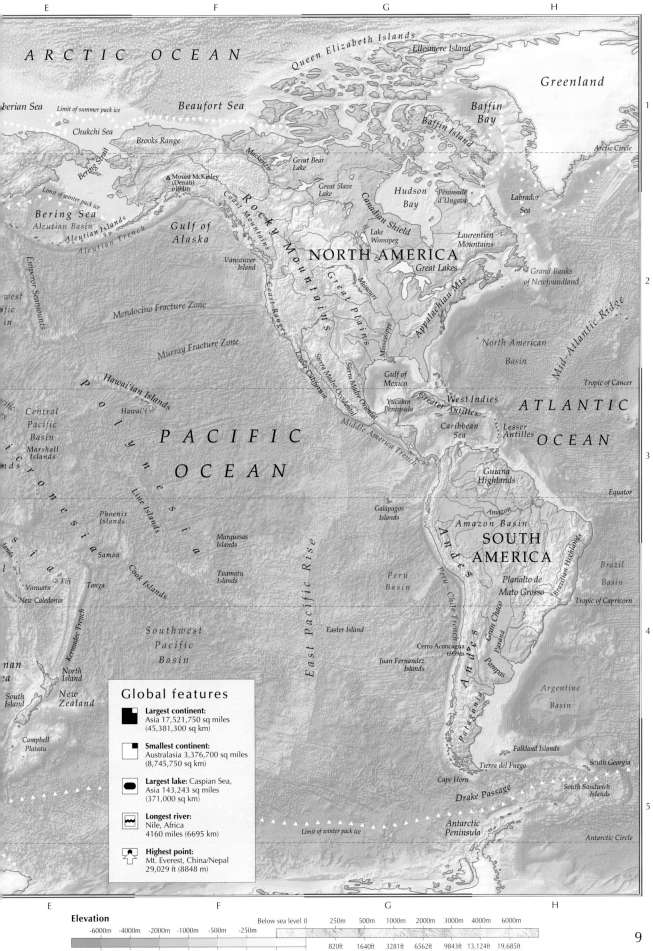

ARCTIC OCEAN

Queen Elizabeth Islands

Ellesmere Island

Greenland

...berian Sea

Limit of summer pack ice

Beaufort Sea

Baffin
Bay

1

Chukchi Sea

Brooks Range

Baffin Island

Arctic Circle

Mackenzie

Great Bear
Lake

△ Mount McKinley
(Denali)
6194m

Limit of winter pack ice

Great Slave
Lake

Hudson
Bay

Péninsule
d'Ungava

Labrador
Sea

Bering Strait

Bering Sea

Aleutian Basin

Aleutian Islands

Aleutian Trench

Gulf of
Alaska

Coast Mountains

Rocky Mountains

Canadian Shield

Lake
Winnipeg

NORTH AMERICA

Laurentian
Mountains

...west
...ific
...in

Emperor Seamounts

Vancouver
Island

Coast Ranges

Great Plains

Missouri

Great Lakes

Grand Banks
of Newfoundland

2

Mendocino Fracture Zone

Appalachian Mts

North American
Basin

Mid-Atlantic Ridge

Murray Fracture Zone

Sierra Madre Occidental

Lower California

Mississippi

Sierra Madre Oriental

Hawaiian Islands

Hawai'i

Gulf of
Mexico

Yucatán
Peninsula

West Indies

Greater
Antilles

Tropic of Cancer

ATLANTIC

Central
Pacific
Basin

Polynesia

PACIFIC

Middle America Trench

Caribbean
Sea

Lesser
Antilles

OCEAN

3

Marshall
Islands

OCEAN

Micronesia

...nds

Guiana
Highlands

Equator

Phoenix
Islands

Line Islands

Galápagos
Islands

Amazon

Amazon Basin

SOUTH
AMERICA

Marquesas
Islands

Andes

Brazilian Highlands

Brazil
Basin

Samoa

Tuamotu
Islands

Peru
Basin

Planalto de
Mato Grosso

Tropic of Capricorn

...l

Vanuatu

Fiji

Tonga

Cook Islands

East Pacific Rise

Easter Island

Peru-Chile Trench

Gran Chaco

Paraná

New Caledonia

4

...nan
...ea

Kermadec Trench

North
Island

Southwest
Pacific
Basin

Juan Fernandez
Islands

Cerro Aconcagua
6959m

Andes

Pampas

Argentine
Basin

South
Island

New
Zealand

Patagonia

Campbell
Plateau

Global features

▪ **Largest continent:**
Asia 17,521,750 sq miles
(45,381,300 sq km)

▫ **Smallest continent:**
Australasia 3,376,700 sq miles
(8,745,750 sq km)

⬬ **Largest lake:** Caspian Sea,
Asia 143,243 sq miles
(371,000 sq km)

〰 **Longest river:**
Nile, Africa
4160 miles (6695 km)

⛰ **Highest point:**
Mt. Everest, China/Nepal
29,029 ft (8848 m)

Falkland Islands

South Georgia

Tierra del Fuego

South Sandwich
Islands

Cape Horn

Drake Passage

5

Limit of winter pack ice

Antarctic
Peninsula

Antarctic Circle

Elevation

| -6000m | -4000m | -2000m | -1000m | -500m | -250m | | Below sea level 0 | 250m | 500m | 1000m | 2000m | 3000m | 4000m | 6000m |

-19,658ft -13,124ft -6562ft -3281ft -1640ft -820ft -328ft/-100m 0

820ft 1640ft 3281ft 6562ft 9843ft 13,124ft 19,685ft

Standard Time Zones

The numbers at the top of the map indicate how many hours each time zone is ahead or behind Coordinated Universal Time (UTC). The row of clocks indicate the time in each zone when it is 12:00 noon UTC.

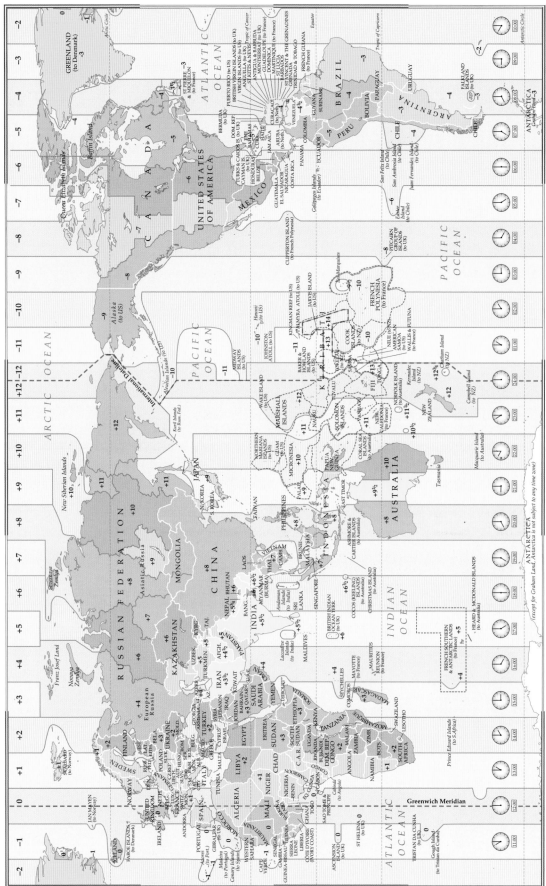

TIME ZONES

Because Earth is a rotating sphere, the Sun shines on only half of its surface at any one time. Thus, it is simultaneously morning, evening, and night time in different parts of the world. Because of these disparities, each country or part of a country adheres to a local time. A region of the Earth's surface within which a single local time is used is called a time zone.

COORDINATED UNIVERSAL TIME (UTC)

Coordinated Universal Time (UTC) is a reference by which the local time in each time zone is set. UTC is a successor to, and closely approximates, Greenwich Mean Time (GMT). However, UTC is based on an atomic clock, whereas GMT is determined by the Sun's position in the sky relative to the 0° longitudinal meridian, which runs through Greenwich, UK.

THE INTERNATIONAL DATELINE

The International Dateline is an imaginary line from pole to pole that roughly corresponds to the 180° longitudinal meridian. It is an arbitrary marker between calendar days. The dateline is needed because of the use of local times around the world rather than a single universal time.

The
WORLD
ATLAS

THE MAPS IN THIS ATLAS ARE ARRANGED CONTINENT BY CONTINENT, STARTING FROM THE INTERNATIONAL DATE LINE, AND MOVING EASTWARD. THE MAPS PROVIDE A UNIQUE VIEW OF TODAY'S WORLD, COMBINING TRADITIONAL CARTOGRAPHIC TECHNIQUES WITH THE LATEST REMOTE-SENSED AND DIGITAL TECHNOLOGY.

North & Central America

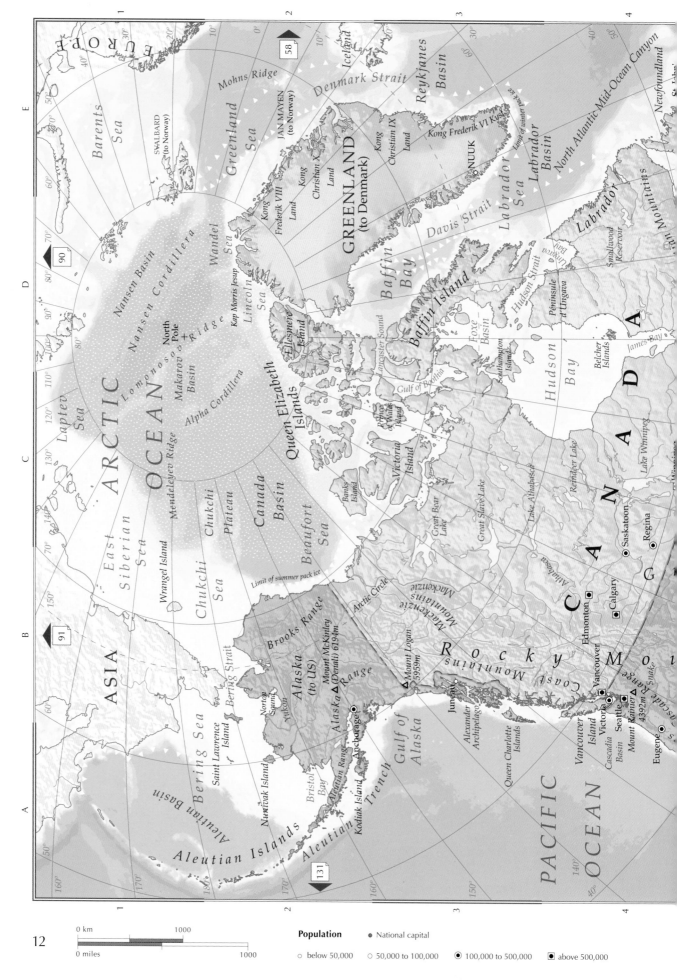

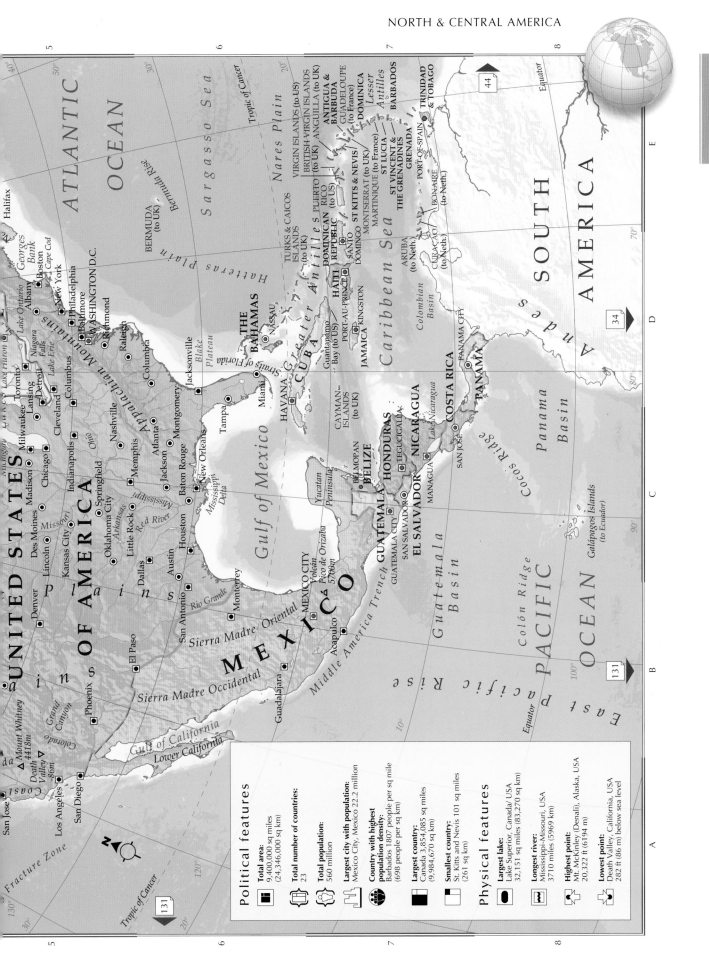

Political features

Total area:
9,400,000 sq miles
(24,346,000 sq km)

Total number of countries:
23

Total population:
560 million

Largest city with population:
Mexico City, Mexico 22.2 million

**Country with highest
population density:**
Barbados 1807 people per sq mile
(698 people per sq km)

Largest country:
Canada 3,854,085 sq miles
(9,984,670 sq km)

Smallest country:
St. Kitts and Nevis 101 sq miles
(261 sq km)

Physical features

Largest lake:
Lake Superior, Canada/ USA
32,151 sq miles (83,270 sq km)

Longest river:
Mississippi-Missouri, USA
3710 miles (5969 km)

Highest point:
Mt. McKinley (Denali), Alaska, USA
20,322 ft (6194 m)

Lowest point:
Death Valley, California, USA
282 ft (86 m) below sea level

Western Canada & Alaska

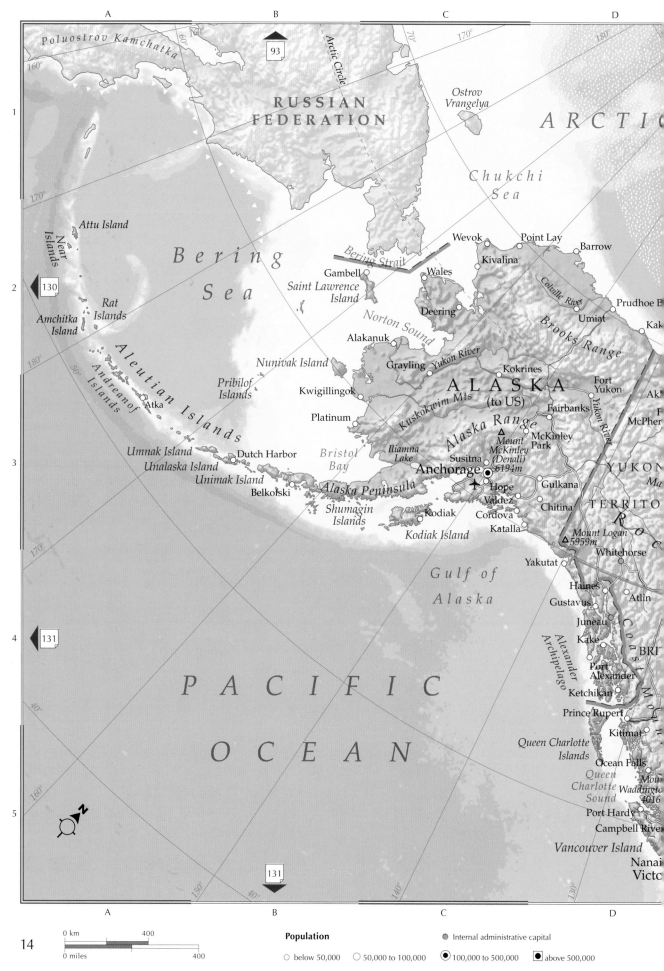

Poluostrov Kamchatka

RUSSIAN
FEDERATION

ARCTIC

Ostrov
Vrangelya

Arctic Circle

Chukchi
Sea

Attu Island

Near
Islands

Bering
Sea

Rat
Islands

Amchitka
Island

Saint Lawrence
Island

Norton Sound

Aleutian Islands

Andreanof
Islands

Atka

Pribilof
Islands

Nunivak Island

Kwigillingok

Platinum

Bristol
Bay

Umnak Island
Unalaska Island
Unimak Island

Dutch Harbor

Belkofski

Shumagin
Islands

Alaska Peninsula

Kodiak

Kodiak Island

Wevok
Kivalina
Point Lay
Barrow

Wales

Gambell

Bering Strait

Deering

Alakanuk

Grayling
Yukon River
Kokrines

ALASKA
(to US)

Kuskokwim Mts

Iliamna
Lake

Alaska Range

Mount
McKinley
(Denali)
6194m

Susitna

Anchorage
Hope
Valdez
Cordova
Katalla

Gulkana
Chitina

McKinley
Park

Fairbanks

Fort
Yukon

Yukon River

Prudhoe B
Umiat
Kak

Colville River

Brooks Range

YUKON

TERRITO

Ak
McPher

Ma

Mount Logan
5959m

Whitehorse

Yakutat

Gulf of
Alaska

Haines

Atlin

Gustavus

Juneau

Kake

Alexander
Archipelago

Port
Alexander

Ketchikan

Prince Rupert

Kitimat

Queen Charlotte
Islands

Ocean Falls

Queen
Charlotte
Sound

Waddingto
4016

Mou

BRI

Coast

Port Hardy

Campbell River

Vancouver Island

Nanai

Victo

PACIFIC

OCEAN

0 km   400
0 miles   400

Population

○ below 50,000   ○ 50,000 to 100,000   ◉ 100,000 to 500,000   ● above 500,000

● Internal administrative capital

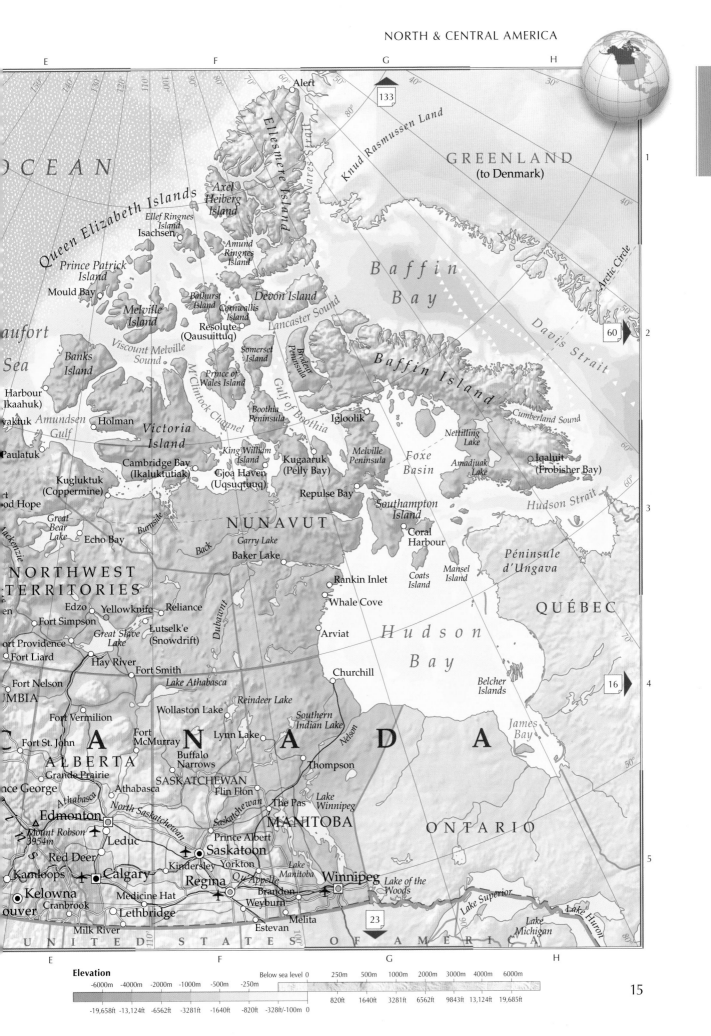

E F G H

1500 140° 130° 120° 110° 100° 90° 80° 70° 60° 50° 40° 30°

Alert

133

80°

OCEAN

Queen Elizabeth Islands

Axel Heiberg Island

Ellef Ringnes Island

Isachsen

Ellesmere Island

Naris Strait

Knud Rasmussen Land

GREENLAND
(to Denmark)

1

Prince Patrick Island

Amund Ringnes Island

Arctic Circle

40°

Mould Bay

Melville Island

Bathurst Island

Devon Island

Cornwallis Island

Lancaster Sound

Baffin Bay

50°

60

2

Resolute (Qausuittuq)

aufort

Banks Island

Viscount Melville Sound

Somerset Island

Prince of Wales Island

MʳClintock Channel

Brodeur Peninsula

Gulf of Boothia

Baffin Island

Davis Strait

Sea

Harbour (Ikaahuk)

yaktuk *Amundsen Gulf*

Holman

Victoria Island

King William Island

Boothia Peninsula

Igloolik

Melville Peninsula

Cumberland Sound

60°

Paulatuk

Cambridge Bay (Ikaluktutiak)

Gjoa Haven (Uqsuqtuuq)

Kugaaruk (Pelly Bay)

Foxe Basin

Nettilling Lake

Amadjuak Lake

Iqaluit (Frobisher Bay)

60°

od Hope

Kugluktuk (Coppermine)

Burnside

Repulse Bay

Southampton Island

Hudson Strait

3

Great Bear Lake

Echo Bay

Back

Garry Lake

Baker Lake

Coral Harbour

Coats Island

Mansel Island

Péninsule d'Ungava

Mackenzie

NUNAVUT

Rankin Inlet

QUÉBEC

70°

NORTHWEST TERRITORIES

en

Edzo **Yellowknife** **Reliance**

Whale Cove

rt Providence

Fort Simpson

Great Slave Lake

Lutselk'e (Snowdrift)

Dubawnt

Arviat

Hudson Bay

50°

Fort Liard

Hay River

Fort Smith

Churchill

16

4

Fort Nelson

Lake Athabasca

Belcher Islands

James Bay

MBIA

Fort Vermilion

Reindeer Lake

Wollaston Lake

Southern Indian Lake

Nelson

A N A D A A

Fort St. John

Fort McMurray

Lynn Lake

Thompson

ONTARIO

ALBERTA

Grande Prairie

Buffalo Narrows

SASKATCHEWAN

Flin Flon

The Pas

Lake Winnipeg

ce George

Athabasca

Athabasca

North Saskatchewan

Saskatchewan

Edmonton

△ Mount Robson 3954m

Leduc

Prince Albert

MANITOBA

5

Red Deer

Saskatoon

Kindersley **Yorkton**

Lake Manitoba

Winnipeg

Lake of the Woods

Kamloops

Calgary

Regina

Qu'Appelle

Lake Superior

Lake Huron

uver

Kelowna

Cranbrook

Medicine Hat

Brandon

Weyburn

Lake Michigan

Lethbridge

Milk River

Estevan

Melita

23

U N I T E D S T A T E S O F A M E R I C A

110° 100° 90° 80°

E F G H

Elevation

-6000m -4000m -2000m -1000m -500m -250m   Below sea level 0   250m 500m 1000m 2000m 3000m 4000m 6000m

-19,658ft -13,124ft -6562ft -3281ft -1640ft -820ft -328ft/-100m 0   820ft 1640ft 3281ft 6562ft 9843ft 13,124ft 19,685ft

Eastern Canada

NORTHWEST TERRITORIES

NUNAVUT

SASKATCHEWAN

MANITOBA

Churchill

Southern Indian Lake

Nelson

Hayes

Cedar Lake

Lake Winnipeg

Lake Winnipegosis

Lake Manitoba

Sandy Lake

C A N

O N T A R I O

Severn

Fort Severn

Peawanuk

Winisk

Attawapiskat

Attawapiskat

Albany

Fort Albany

Moosonee

Moose

Harricana

HUDSON Bay

Ottawa Islands

Coats Island

Mansel Island

Inukjuak (Port Harrison)

Belcher Islands

James Bay

Akimiski Island

Ivujivik

Charles Island

Péninsule d' Ungava

Lac Mi

Rivi

QU

Eastmain

Rivière de Rupert

Mistass

Chibougama

Réservo Gouin

Lac Seul

Kenora

Dryden

Armstrong

Lake of the Woods

Lake Nipigon

Longlac

Hearst

Kapuskasing

Cochrane

Amos

Rouyn-Noranda

Val-d'Or

Fort Frances

Atikokan

Nipigon

Marathon

Tip Top Mountain △ 640m

Timmins

Foleyet

Kirkland Lake

Rainy Lake

Thunder Bay

Wawa

NORTH DAKOTA

MINNESOTA

SOUTH DAKOTA

UNITED STATES

Red River

100°

45°

40°

Lake Superior

M I C H I G A N

Sault Ste.Marie

Sudbury

North Bay

Pembroke

Gatineau

Hull

OTTAWA

NEBRASKA

OF AMERICA

IOWA

WISCONSIN

Mississippi River

ILLINOIS

INDIANA

OHIO

Lake Michigan

Manitoulin Island

Georgian Bay

Lake Huron

Midland

Peterborough

Brampton

Kitchener

Hamilton

Sarnia

London

Windsor

Leamington

Oshawa

Toronto

St.Catharines

Niagara Falls

King

Lake Ont

NEW YOR

PENNSYLVANIA

Lake Erie

15

15

23

18

0 km   300
0 miles   300

Population

National capital

Internal administrative capital

○ below 50,000   ○ 50,000 to 100,000   ◉ 100,000 to 500,000   ◼ above 500,000

E 65° 60° 55° F 60° 50° 45° G 55° H

Baffin Island
Resolution Island
Strait
Akpatok Island
Ungava Bay
uaq
Rivière à la Baleine
Caniapiscau
Scheffervile
Gagnon
Réservoir Manicouagan
voir de piscau
Button Islands

Labrador Sea

Nain
Hopedale
Makkovik
Cape Harrison
Cartwright

NEWFOUNDLAND & LABRADOR
Smallwood Reservoir
Lake Melville
Churchill
St.Anthony

Strait of Belle Isle

E C D A
Laurentian Mountains
Havre-St-Pierre
Sept-Îles
Baie-Comeau
Île d'Anticosti
Corner Brook

Newfoundland

Gander
Grand Falls
St.John's
Channel-Port aux Basques
Cape Race

St.Lawrence
Matane
Rimouski
Chicoutimi
Rivière-du-Loup
Edmundston
Charlesbourg
Québec
us-vières
St-Georges
ummondville
éal
Sherbrooke

Gaspé
Péninsule de Gaspé
Gulf of St. Lawrence
Îles de la Madeleine
Cabot Strait

PRINCE EDWARD ISLAND
Bathurst
Sydney
Glace Bay
NEW BRUNSWICK
Charlottetown
Moncton
Amherst
Oromocto
New Glasgow
Fredericton
Truro
Saint John
NOVA SCOTIA
Dartmouth
Halifax
Bay of Fundy
Liverpool
Yarmouth

ST PIERRE & MIQUELON
(to France)

Cape Breton Island

Sable Island

MAINE

NEW HAMPSHIRE

ATLANTIC

ACHUSETTS
Cape Cod

NECTICUT RHODE ISLAND
70°

OCEAN

60
44
44
1
2
3
4
5

50°
45°
50°
40°

65° 40° 60° 55°

N

Elevation

-6000m	-4000m	-2000m	-1000m	-500m	-250m	Below sea level 0	250m	500m	1000m	2000m	3000m	4000m	6000m
-19,658ft	-13,124ft	-6562ft	-3281ft	-1640ft	-820ft	-328ft/-100m 0	820ft	1640ft	3281ft	6562ft	9843ft	13,124ft	19,685ft

USA: The Northeast

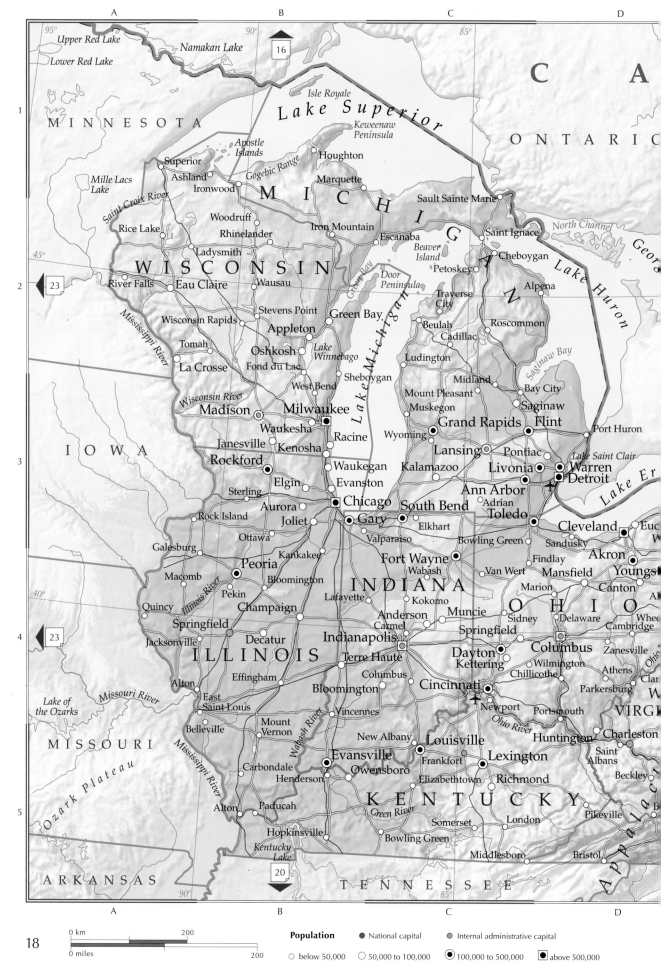

Upper Red Lake
Lower Red Lake
Namakan Lake
16
Lake Superior
Isle Royale
Keweenaw Peninsula

MINNESOTA
ONTARIO
CANA

Apostle Islands
Houghton
Superior
Ashland
Ironwood
Marquette
Mille Lacs Lake
Gogebic Range
Sault Sainte Marie
MICHIGAN
Saint Ignace
North Channel
Georg

Saint Croix River
Woodruff
Rhinelander
Iron Mountain
Escanaba
Cheboygan
Rice Lake
Ladysmith
Beaver Island
Petoskey
Alpena
Lake Huron

WISCONSIN
River Falls
Eau Claire
Wausau
Stevens Point
Traverse City
Roscommon
23

Wisconsin Rapids
Beulah
Cadillac
Tomah
Oshkosh
Lake Winnebago
Ludington
La Crosse
Fond du Lac
Door Peninsula
Green Bay
Appleton
Sheboygan
Midland
Bay City
Mount Pleasant
Saginaw Bay

Wisconsin River
West Bend
Muskegon
Saginaw
Madison
Milwaukee
Grand Rapids
Flint
Port Huron
Waukesha
Wyoming
Racine
Lansing
Pontiac
Lake Saint Clair

Janesville
Kenosha
Kalamazoo
Livonia
Warren
IOWA
Rockford
Waukegan
Evanston
Ann Arbor
Detroit
Lake Er

Elgin
South Bend
Adrian
Sterling
Chicago
Gary
Toledo
Cleveland
Euc
W
Aurora
Elkhart
Rock Island
Joliet
Valparaiso
Bowling Green
Sandusky
Akron
Ottawa
Findlay
Galesburg
Kankakee
Fort Wayne
Mansfield
Youngs
Peoria
Bloomington
Wabash
Van Wert
Marion
Canton
Macomb
INDIANA
Kokomo
OHIO
A
Pekin
Lafayette
Sidney
Delaware
Quincy
Champaign
Anderson
Muncie
Springfield
Cambridge
Whee

Springfield
Carmel
Indianapolis
Dayton
Columbus
Zanesville
Jacksonville
Decatur
Kettering
Wilmington
Athens
ILLINOIS
Terre Haute
Columbus
Chillicothe
Parkersburg
Clar
Alton
Effingham
Bloomington
Cincinnati
Portsmouth
VIRGI
East Saint Louis
Vincennes
Newport
Ohio River
Huntington
Charleston
Mount Vernon
New Albany
Louisville
Saint Albans
Belleville
Wabash River
Frankfort
Lexington
Beckley
Carbondale
Evansville
Owensboro
Richmond
Henderson
Elizabethtown
MISSOURI
Lake of the Ozarks
Missouri River
Alton
Paducah
KENTUCKY
Pikeville
Somerset
London
Hopkinsville
Green River
Bowling Green
Middlesboro
Bristol
Kentucky Lake
20
Ozark Plateau
Mississippi River
ARKANSAS
TENNESSEE
Appalac

18

0 km   200
0 miles   200

Population ● National capital ● Internal administrative capital
○ below 50,000 ○ 50,000 to 100,000 ● 100,000 to 500,000 ■ above 500,000

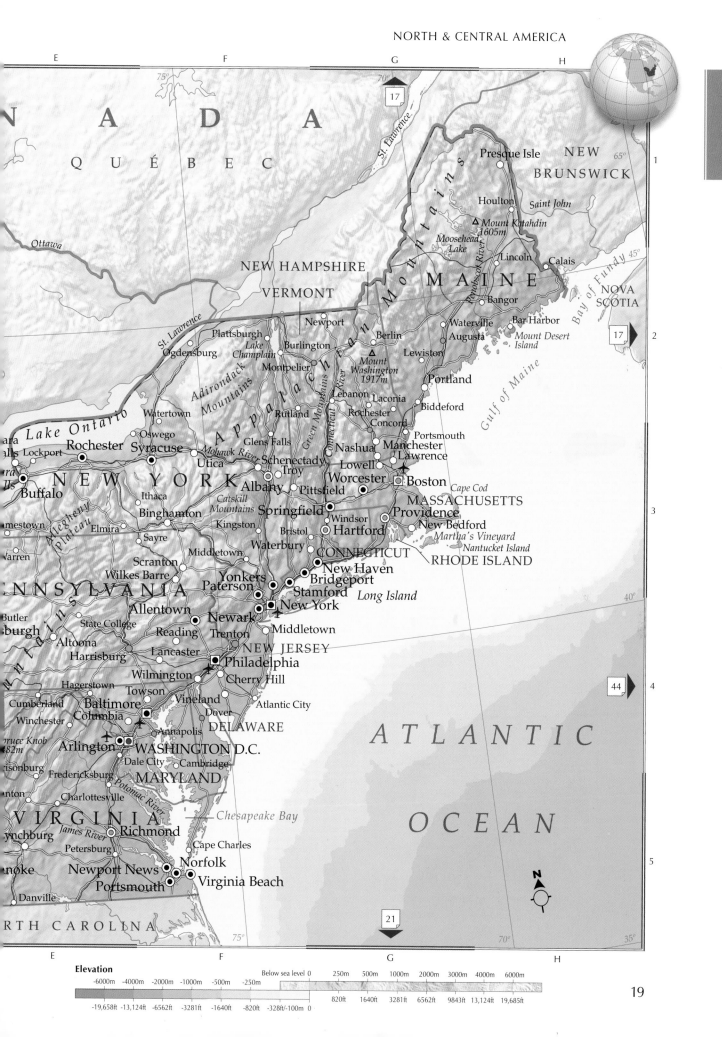

Elevation

| Below sea level 0 | | | | | | | 250m | 500m | 1000m | 2000m | 3000m | 4000m | 6000m |

-6000m -4000m -2000m -1000m -500m -250m

-19,658ft -13,124ft -6562ft -3281ft -1640ft -820ft -328ft/-100m 0   820ft 1640ft 3281ft 6562ft 9843ft 13,124ft 19,685ft

USA: The Southeast

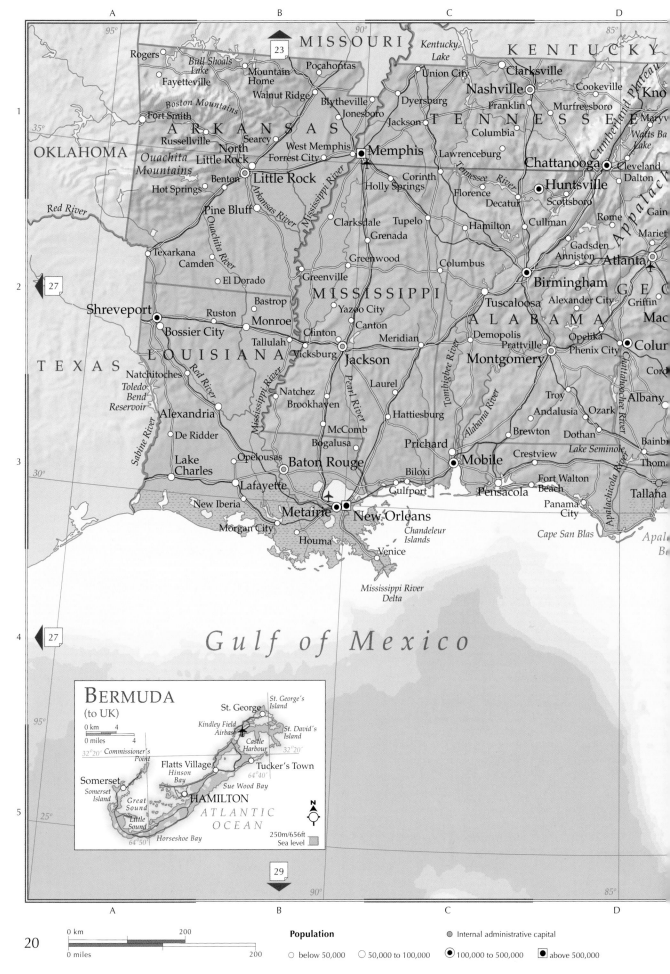

MISSOURI

KENTUCKY

Kentucky Lake

Rogers
Bull Shoals Lake
Fayetteville
Mountain Home
Walnut Ridge
Pocahontas
Blytheville
Jonesboro
Union City
Clarksville
Nashville
Cookeville
Kno

Fort Smith
Boston Mountains
ARKANSAS
Jackson
Dyersburg
Franklin
Murfreesboro
Maryv
Watts Ba
Lake

OKLAHOMA
Russellville
Searcy
West Memphis
Forrest City
Memphis
TENNESSEE
Columbia
Lawrenceburg
Chattanooga
Cleveland

North Little Rock
Little Rock
Holly Springs
Corinth
Florence
Decatur
Scottsboro
Huntsville
Dalton

Ouachita Mountains
Benton
Hot Springs
Clarksdale
Tupelo
Hamilton
Cullman
Rome
Gain

Pine Bluff
Ouachita River
Arkansas River
Mississippi River
Grenada
Greenwood
Columbus
Gadsden
Anniston
Atlanta
Mariet

Texarkana
Camden
El Dorado
Greenville
MISSISSIPPI
Birmingham
GEO

Red River
Bastrop
Yazoo City
Canton
Tuscaloosa
Alexander City
Griffin
Mac

Shreveport
Ruston
Monroe
Clinton
Meridian
ALABAMA
Opelika
Phenix City
Colur

Bossier City
Tallulah
Vicksburg
Jackson
Demopolis
Prattville
Montgomery
Cord

TEXAS
LOUISIANA
Natchitoches
Natchez
Brookhaven
Laurel
Tombigbee River
Troy
Andalusia
Ozark
Albany

Toledo Bend Reservoir
Red River
Alexandria
McComb
Hattiesburg
Alabama River
Brewton
Dothan
Bainbr

De Ridder
Bogalusa
Prichard
Crestview
Lake Seminole
Thom

Lake Charles
Opelousas
Baton Rouge
Biloxi
Mobile
Fort Walton Beach
Pensacola
Panama City
Apalachicola River
Tallaha

Sabine River
Lafayette
Gulfport
Apal
Ba

New Iberia
Metairie
New Orleans

Morgan City
Houma
Chandeleur Islands
Cape San Blas

Venice

Mississippi River Delta

Gulf of Mexico

BERMUDA
(to UK)

0 km 4
0 miles 4

St. George's Island
St. George
Kindley Field Airbase
St. David's Island
Castle Harbour
Commissioner's Point
Flatts Village
Tucker's Town
Somerset
Hinson Bay
Sue Wood Bay
Somerset Island
HAMILTON
Great Sound
Little Sound
Horseshoe Bay

ATLANTIC OCEAN

N

250m/656ft
Sea level

0 km 200
0 miles 200

Population

○ below 50,000   ○ 50,000 to 100,000   ◉ 100,000 to 500,000   ● above 500,000

● Internal administrative capital

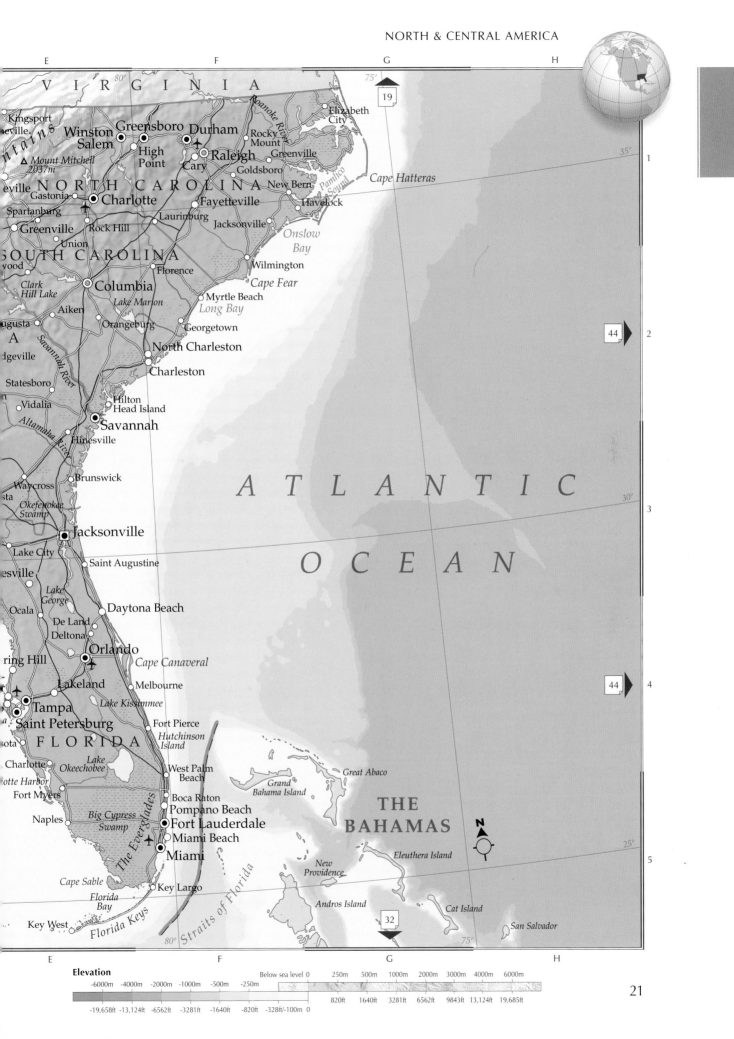

E   F   G   H

VIRGINIA

80°

Kingsport
eville
Mount Mitchell
2037m

Winston
Salem
NORTH CAROLINA
eville
Gastonia
Charlotte
Spartanburg
Greenville
Rock Hill
Union

Greensboro
High
Point
Cary
Durham
Raleigh
Goldsboro
New Bern
Fayetteville
Laurinburg
Jacksonville

Roanoke River
Rocky
Mount
Greenville
Havelock

Elizabeth
City
75°
19
35° 1

Cape Hatteras

Pamlico
Sound

SOUTH CAROLINA
wood
Augusta
dgeville
A
Statesboro
Vidalia
Waycross
sta

Clark
Hill Lake
Aiken
Columbia
Lake Marion
Orangeburg
Florence
North Charleston
Charleston

Wilmington
Cape Fear
Myrtle Beach
Long Bay
Georgetown

Onslow
Bay

44 2

Savannah River

Hilton
Head Island
Savannah
Hinesville
Brunswick
Okefenokee
Swamp
Altamaha River

A T L A N T I C

30° 3

O C E A N

Jacksonville
Lake City
esville
Ocala
De Land
Deltona
Saint Augustine

Lake
George
Daytona Beach

ring Hill
Orlando
Cape Canaveral
Lakeland
Melbourne
Lake Kissimmee
44 4

Tampa
Saint Petersburg
FLORIDA
a
ota

Fort Pierce
Hutchinson
Island

Charlotte
Lake
Okeechobee
otte Harbor
Fort Myers
Naples
Big Cypress
Swamp

West Palm
Beach
Boca Raton
Pompano Beach
Fort Lauderdale
Miami Beach
Miami

The Everglades

Great Abaco

Grand
Bahama Island

**THE
BAHAMAS**

Eleuthera Island

N

25° 5

Cape Sable
Florida
Bay
Key Largo
New
Providence

Key West
Florida Keys
80°
Straits of Florida
Andros Island
32
75°
Cat Island
San Salvador

E   F   G   H

Elevation

	Below sea level 0	250m	500m	1000m	2000m	3000m	4000m	6000m
-6000m	-4000m	-2000m	-1000m	-500m	-250m			
-19,658ft	-13,124ft	-6562ft	-3281ft	-1640ft	-820ft	-328ft/-100m 0		
					820ft	1640ft	3281ft	6562ft	9843ft	13,124ft	19,685ft

USA: Central States

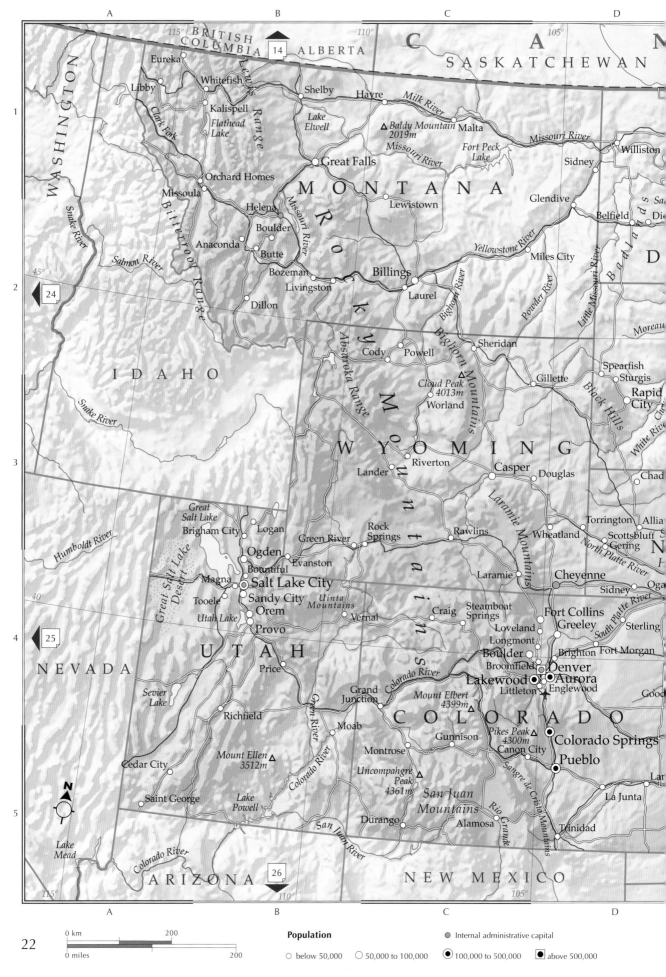

0 km   200

0 miles   200

Population

○ below 50,000   ○ 50,000 to 100,000   ◉ 100,000 to 500,000   ◼ above 500,000

● Internal administrative capital

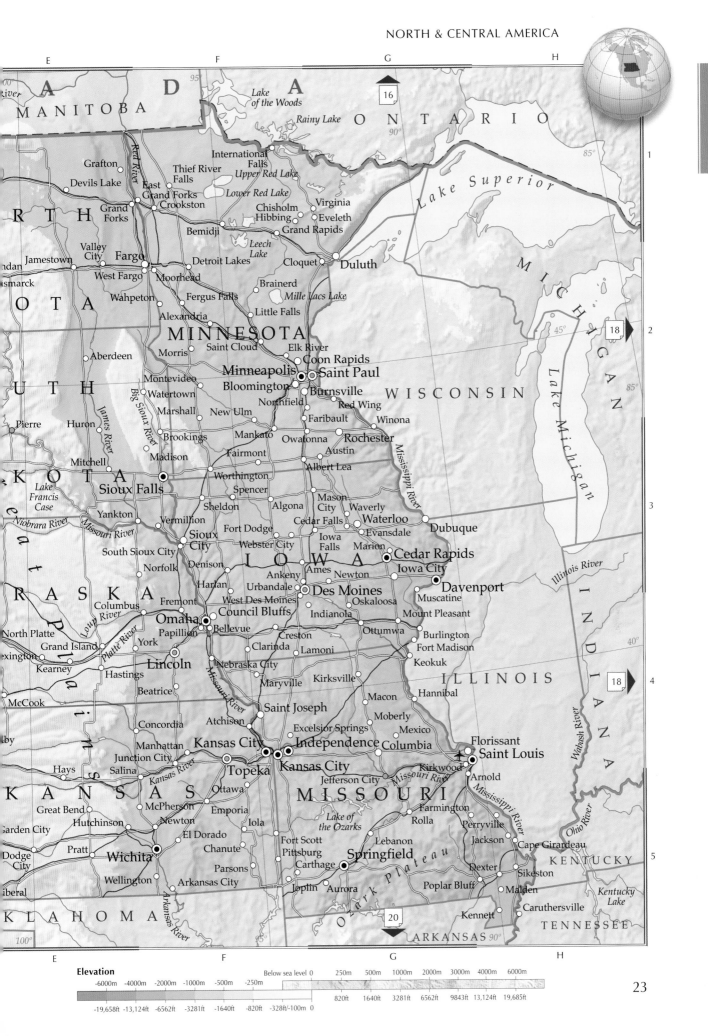

E   D   F   G   H

CANADA

MANITOBA

ONTARIO

Lake of the Woods

Rainy Lake

Lake Superior

MICHIGAN

Lake Michigan

95°

90°

85°

45°

85°

16

18

Grafton

Devils Lake

East Grand Forks

Grand Forks

Crookston

International Falls

Thief River Falls

Upper Red Lake

Lower Red Lake

Bemidji

Chisholm

Hibbing

Virginia

Eveleth

Grand Rapids

Leech Lake

Valley City

Jamestown

Fargo

West Fargo

Moorhead

Detroit Lakes

Fergus Falls

Cloquet

Duluth

Wahpeton

Brainerd

Mille Lacs Lake

Red River

smarck

Bismarck

NORTH

DAKOTA

Aberdeen

Alexandria

Little Falls

MINNESOTA

Saint Cloud

Morris

Elk River

Coon Rapids

Minneapolis

Saint Paul

WISCONSIN

SOUTH

DAKOTA

Montevideo

Watertown

Bloomington

Burnsville

Northfield

Red Wing

Marshall

New Ulm

Faribault

Winona

Pierre

Huron

Brookings

Madison

Mankato

Owatonna

Rochester

Austin

Albert Lea

Big Sioux River

James River

Lake Francis Case

Mitchell

Sioux Falls

Worthington

Spencer

Fairmont

Sheldon

Algona

Mason City

Waverly

Cedar Falls

Waterloo

Evansdale

Dubuque

Yankton

Vermillion

Fort Dodge

Iowa Falls

Marion

Niobrara River

Missouri River

South Sioux City

Sioux City

Webster City

Ames

Cedar Rapids

Iowa City

Davenport

Norfolk

Denison

IOWA

Newton

Muscatine

NEBRASKA

Columbus

Fremont

Harlan

Urbandale

Ankeny

Des Moines

Oskaloosa

Mount Pleasant

Loup River

Platte River

Council Bluffs

West Des Moines

Indianola

Burlington

North Platte

Omaha

Papillion

Bellevue

Creston

Ottumwa

Fort Madison

Grand Island

Columbus River

York

Clarinda

Lamoni

Keokuk

exington

Lexington

Lincoln

Nebraska City

Maryville

Kirksville

ILLINOIS

Illinois River

INDIANA

40°

18

Kearney

Hastings

Beatrice

McCook

Saint Joseph

Macon

Hannibal

Concordia

Atchison

Excelsior Springs

Moberly

Mexico

Wabash River

Hays

Manhattan

Junction City

Salina

Kansas City

Kansas City

Independence

Columbia

Florissant

Saint Louis

Topeka

Kirkwood

Arnold

Ottawa

Jefferson City

MISSOURI

Missouri River

Mississippi River

KANSAS

Great Bend

McPherson

Newton

Emporia

Iola

Lake of the Ozarks

Farmington

Rolla

Perryville

Jackson

Ohio River

arden City

Garden City

Hutchinson

El Dorado

Chanute

Fort Scott

Lebanon

Cape Girardeau

KENTUCKY

Dodge City

Pratt

Wichita

Parsons

Pittsburg

Carthage

Springfield

Ozark Plateau

Dexter

Sikeston

Kentucky Lake

iberal

Liberal

Wellington

Arkansas City

Joplin

Aurora

Poplar Bluff

Malden

Arkansas River

OKLAHOMA

ARKANSAS

Kennett

Caruthersville

TENNESSEE

100°

90°

20

E   F   G   H

Elevation

Below sea level 0   250m   500m   1000m   2000m   3000m   4000m   6000m

-6000m   -4000m   -2000m   -1000m   -500m   -250m

820ft   1640ft   3281ft   6562ft   9843ft   13,124ft   19,685ft

-19,658ft   -13,124ft   -6562ft   -3281ft   -1640ft   -820ft   -328ft/-100m   0

USA: The West

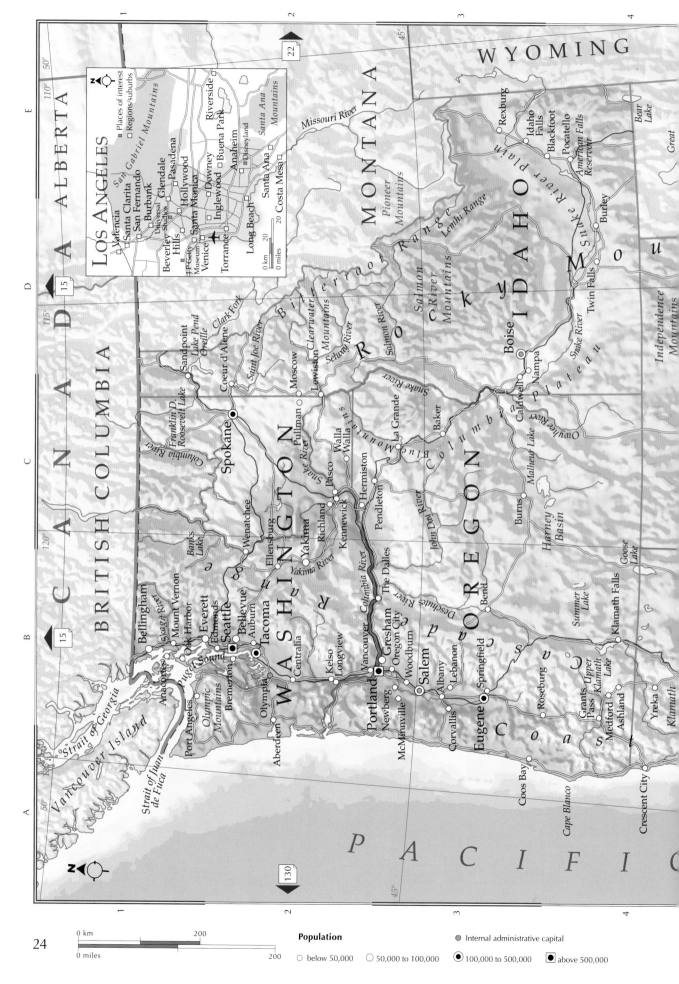

LOS ANGELES

- Places of interest
- Regions/suburbs

Valencia
Santa Clarita
San Fernando
San Gabriel Mountains
Burbank
Glendale
Pasadena
Universal Studios
Beverley Hills
Hollywood
Getty Museum
Santa Monica
Venice
Inglewood
Downey
Disneyland
Anaheim
Torrance
Santa Ana
Buena Park
Santa Ana Mountains
Long Beach
Costa Mesa

0 km 20
0 miles 20

WYOMING

MONTANA

IDAHO

ROCKY Mountains

Pioneer Mountains

Salmon River Mountains

Lemhi Range

Snake River Plain

Rexburg
Idaho Falls
Blackfoot
Pocatello
American Falls Reservoir
Twin Falls
Burley
Bear Lake
Great

Boise
Nampa
Caldwell
Owyhee River
Columbia Plateau
Malheur Lake
Independence Mountains

CANADA
ALBERTA
BRITISH COLUMBIA

Missouri River

Bitterroot Range
Clearwater Mountains
Selway River
Salmon River

Clark Fork
Saint Joe River
Lake Pend Oreille
Sandpoint
Coeur d'Alene
Moscow
Pullman
Lewiston
Walla Walla
La Grande
Baker
Snake River

Franklin D. Roosevelt Lake
Columbia River
Spokane

WASHINGTON
Wenatchee
Ellensburg
Yakima
Yakima River
Richland
Pasco
Kennewick
Hermiston
Pendleton
Blue Mountains
Harney Basin
Burns

Banks Lake

Bellingham
Skagit River
Mount Vernon
Oak Harbor
Everett
Edmonds
Seattle
Bellevue
Auburn
Tacoma
Olympia
Centralia
Kelso
Longview

Anacortes
Puget Sound
Bremerton
Port Angeles
Olympic Mountains
Aberdeen

Strait of Georgia
Vancouver Island
Strait of Juan de Fuca

Vancouver
Gresham
Oregon City
Woodburn
Portland
Newberg
McMinnville

OREGON
Columbia River
Deschutes River
John Day River
The Dalles
Bend
Salem
Albany
Lebanon
Springfield
Eugene
Corvallis
Roseburg
Grants Pass
Medford
Ashland
Yreka
Klamath
Klamath Falls
Upper Klamath Lake
Summer Lake
Goose Lake

Coast Range

Coos Bay
Cape Blanco
Crescent City

PACIFIC

MONTANA

50°
110°
115°
120°

45°

45°

22
15
15
130

200

0 km 200
0 miles 200

Population

Internal administrative capital

○ below 50,000   ○ 50,000 to 100,000   ◉ 100,000 to 500,000   ◼ above 500,000

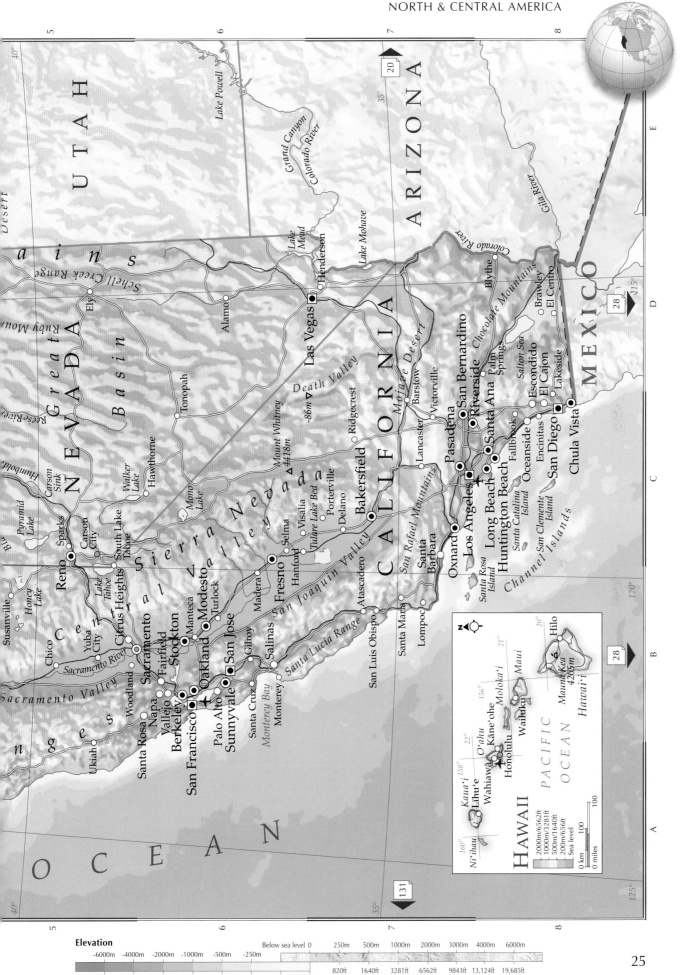

UTAH

Desert

NEVADA

ains

Schell Creek Range

Ruby Mou

Great

Basin

Reese River

Humbold

Carson Sink

Pyramid Lake

Honey Lake

Susanville

Bir

Ely

Alamo

Tonopah

Hawthorne

Walker Lake

Mono Lake

South Lake Tahoe

Lake Tahoe

Sparks

Reno

Carson City

Chico

Yuba City

Citrus Heights

Woodland

Ukiah

Sacramento River

Sacramento Valley

Santa Rosa

Napa

Vallejo

Berkeley

San Francisco

Oakland

Palo Alto

Sunnyvale

San Jose

Santa Cruz

Monterey Bay

Monterey

Manteca

Modesto

Turlock

Stockton

Fairfield

Gilroy

Salinas

Madera

Fresno

Selma

Hanford

Visalia

Porterville

Delano

Sierra Nevada

Central Valley

San Joaquin Valley

Bakersfield

Santa Lucia Range

San Luis Obispo

Santa Maria

Lompoc

Mount Whitney △ 4418m

Ridgecrest

Tulare Lake Bed

Atascadero

CALIFORNIA

Death Valley

-86m ▽

Mojave Desert

Lancaster

Barstow

Victorville

San Rafael Mountains

Santa Barbara

Oxnard

Los Angeles

Pasadena

San Bernardino

Riverside

Santa Ana

Palm Springs

Long Beach

Huntington Beach

Fallbrook

Oceanside

Encinitas

Escondido

El Cajon

Lakeside

San Diego

Chula Vista

Salton Sea

Brawley

El Centro

Chocolate Mountains

Gila River

Colorado River

Blythe

ARIZONA

MEXICO

Lake Powell

Grand Canyon

Colorado River

Lake Mead

Henderson

Las Vegas

Lake Mohave

Santa Rosa Island

Santa Catalina Island

San Clemente Island

Channel Islands

OCEAN

PACIFIC OCEAN

20

28

131

28

HAWAII

Kaua'i 158°

Ni'ihau 160°

Lihu'e

O'ahu 22°

Wahiawā

Kāne'ohe 156°

Honolulu

Waialua

Moloka'i 21°

Maui 20°

Hilo

Mauna Kea 4205m

Hawai'i

PACIFIC OCEAN

N

z

2000m/6562ft
1000m/3281ft
500m/1640ft
200m/656ft
Sea level

0 km   100
0 miles   100

Elevation

							Below sea level 0	250m	500m	1000m	2000m	3000m	4000m	6000m
-6000m	-4000m	-2000m	-1000m	-500m	-250m									
-19,658ft	-13,124ft	-6562ft	-3281ft	-1640ft	-820ft	-328ft/-100m	0		820ft	1640ft	3281ft	6562ft	9843ft	13,124ft	19,685ft

USA: The Southwest

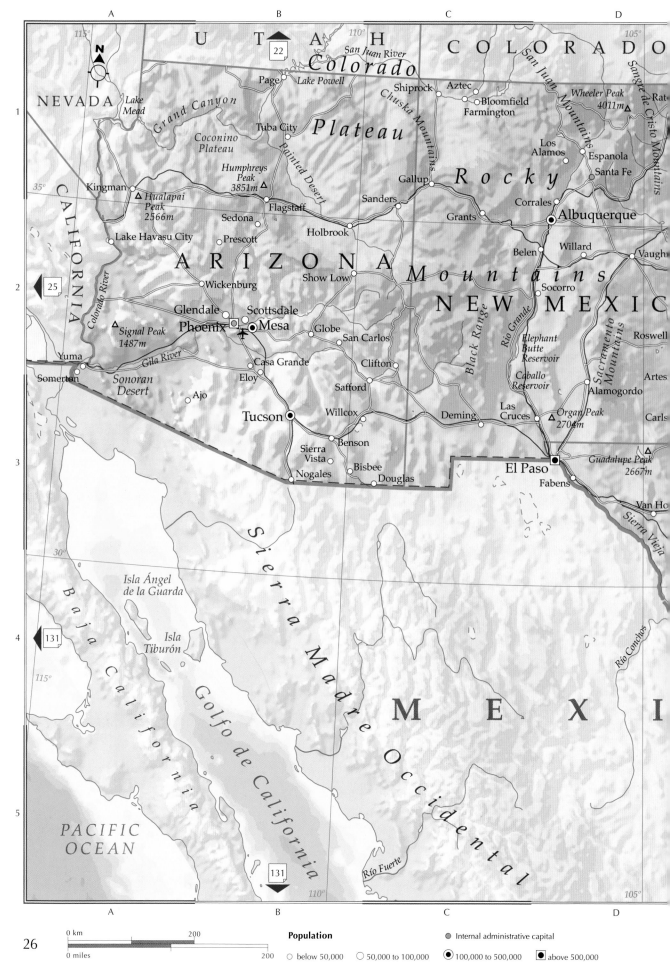

0 km   200

0 miles   200

Population

● Internal administrative capital

○ below 50,000   ◯ 50,000 to 100,000   ◉ 100,000 to 500,000   ■ above 500,000

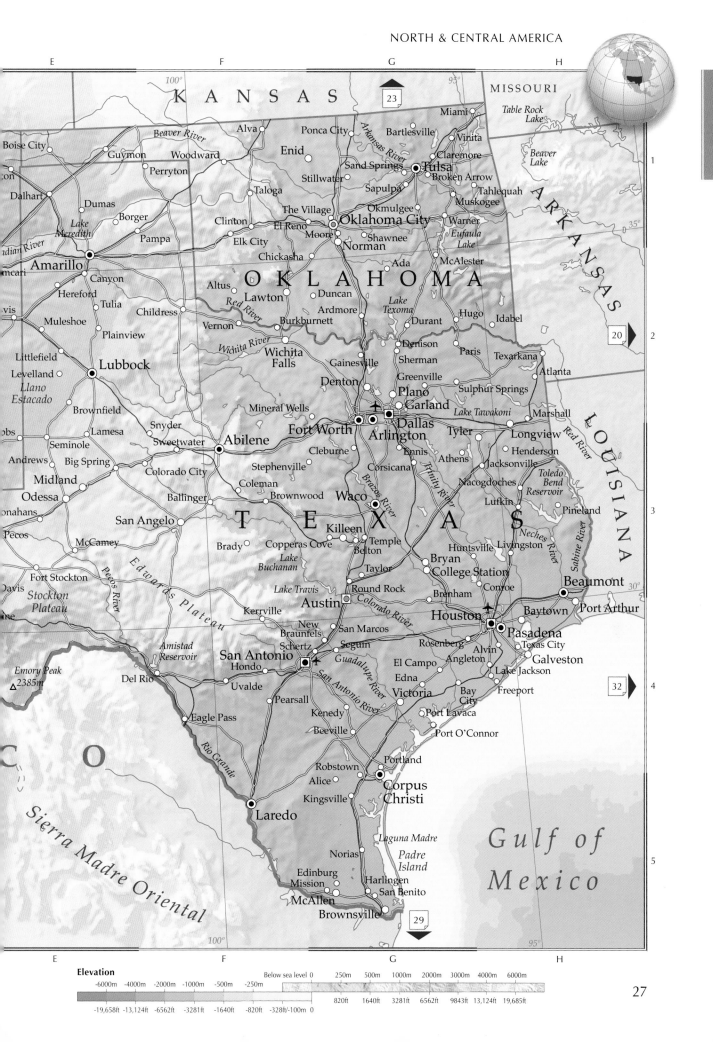

KANSAS

MISSOURI

23

E F G H

Boise City
Guymon
Woodward
Beaver River
Alva
Ponca City
Bartlesville
Miami
Table Rock Lake
Vinita
Claremore
Beaver Lake
1
Dumas
Borger
Perryton
Dalhart
Enid
Sand Springs
Stillwater
Tulsa
Broken Arrow
Tahlequah
Muskogee
Lake Meredith
Pampa
Taloga
The Village
Okmulgee
Warner
Eufaula Lake
35°
Amarillo
Canyon
Clinton
El Reno
Moore
Oklahoma City
Shawnee
Norman
Ada
McAlester
Hereford
Tulia
Altus
Lawton
Duncan
OKLAHOMA
Indian River
Childress
Vernon
Ardmore
Lake Texoma
Durant
Hugo
Idabel
2
Muleshoe
Plainview
Red River
Burkburnett
Wichita River
Wichita Falls
Denison
Sherman
Paris
Texarkana
20
Littlefield
Levelland
Llano Estacado
Gainesville
Greenville
Sulphur Springs
Atlanta
Lubbock
Brownfield
Denton
Plano
Garland
Marshall
Lake Tawakoni
Snyder
Mineral Wells
Fort Worth
Dallas
Arlington
Tyler
Longview
Henderson
Red River
Lamesa
Sweetwater
Abilene
Cleburne
Ennis
Athens
Jacksonville
Nacogdoches
Toledo Bend Reservoir
Seminole
Big Spring
Colorado City
Stephenville
Corsicana
Trinity River
LOUISIANA
3
Andrews
Midland
Coleman
Brownwood
Waco
Brazos River
Lufkin
Pineland
Odessa
Ballinger
TEXAS
Killeen
Huntsville
Livingston
Neches River
Sabine River
San Angelo
Copperas Cove
Temple
Bryan
Pecos
McCamey
Brady
Lake Buchanan
Belton
College Station
Conroe
Beaumont
30°
Fort Stockton
Pecos River
Edwards Plateau
Taylor
Brenham
Port Arthur
Stockton Plateau
Lake Travis
Round Rock
Colorado River
Houston
Baytown
Davis
Kerrville
Austin
San Marcos
Rosenberg
Pasadena
Texas City
Emory Peak 2385m
New Braunfels
Schertz
Seguin
Alvin
Galveston
San Antonio
Hondo
Guadalupe River
El Campo
Angleton
Lake Jackson
4
Amistad Reservoir
Del Rio
Uvalde
Pearsall
San Antonio River
Edna
Victoria
Bay City
Freeport
32
Eagle Pass
Kenedy
Port Lavaca
Port O'Connor
Rio Grande
Beeville
Sierra Madre Oriental
Robstown
Portland
Alice
Corpus Christi
Laredo
Kingsville
Laguna Madre
Padre Island
Gulf of Mexico
5
Norias
Edinburg
Mission
Harlingen
San Benito
McAllen
Brownsville
29

O

E F G H

Elevation

| Below sea level | 0 | 250m | 500m | 1000m | 2000m | 3000m | 4000m | 6000m |

-6000m -4000m -2000m -1000m -500m -250m

-19,658ft -13,124ft -6562ft -3281ft -1640ft -820ft -328ft/-100m 0

820ft 1640ft 3281ft 6562ft 9843ft 13,124ft 19,685ft

Mexico

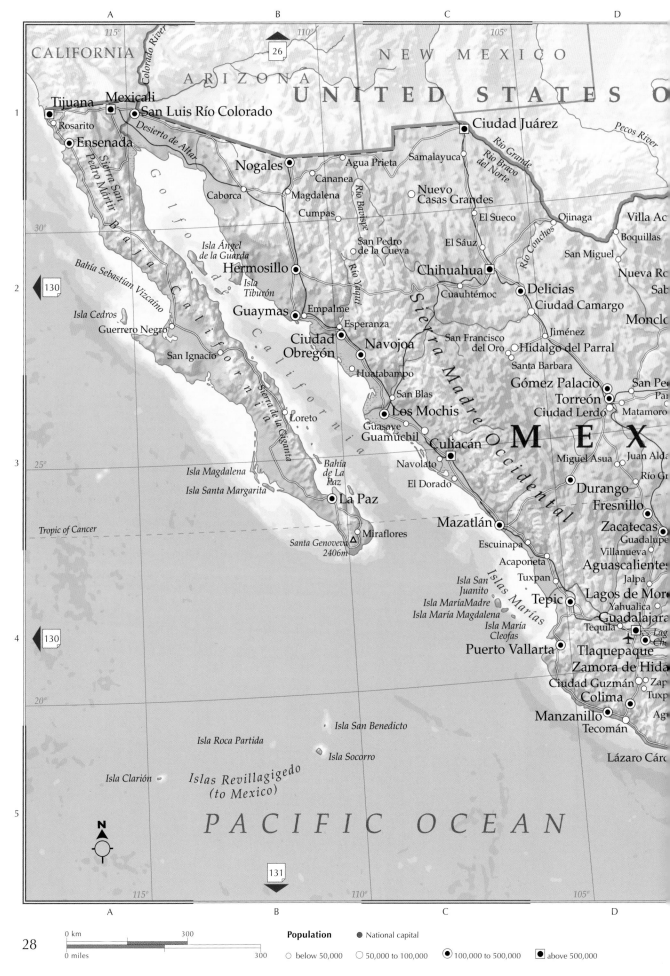

CALIFORNIA

ARIZONA

NEW MEXICO

UNITED STATES O

Colorado River

Pecos River

Tijuana
Rosarito
Mexicali
San Luis Río Colorado
Ensenada

Desierto de Altar

Ciudad Juárez

Nogales
Agua Prieta
Samalayuca

Río Grande
Río Bravo del Norte

Cananea

Caborca
Magdalena
Cumpas

Nuevo
Casas Grandes

El Sueco
Ojinaga
Villa Ac

Río Bavispe

San Pedro
de la Cueva

El Sáuz
Boquillas

San Miguel

Bahía Sebastián Vizcaíno

*Isla Ángel
de la Guarda*

Hermosillo

Chihuahua
Delicias
Nueva Ro
Sab

*Isla
Tiburón*

Cuauhtémoc

Ciudad Camargo
Moncl

Isla Cedros

Guaymas
Empalme
Esperanza

Jiménez

San Francisco
del Oro
Hidalgo del Parral

Sab

Río Yaqui

Guerrero Negro

Ciudad
Obregón
Navojoa

San Ignacio

Huatabampo

Santa Barbara

Gómez Palacio
San Pe

Sierra San Pedro Mártir

San Blas

Torreón
Par

Los Mochis
Ciudad Lerdo
Matamoro

Sierra de la Giganta

Loreto

Guasave
Guamúchil

M E X

Baja California

Golfo de California

Culiacán
Miguel Asua
Juan Alda

Isla Magdalena

Navolato

Río Gr

*Bahía
de La
Paz*

Durango
Fresnillo

Isla Santa Margarita

El Dorado

La Paz

Zacatecas
Guadalupe

Mazatlán

Tropic of Cancer

Miraflores
Escuinapa

Villanueva

Aguascalientes

Santa Genoveva
2406m

Acaponeta

Jalpa

Tuxpan

Lagos de More

*Isla San
Juanito*

Tepic

Yahualica

Guadalajara

Isla MaríaMadre

Islas Marías

Tequila

Lag
Che

Isla María Magdalena

*Isla María
Cleofas*

Puerto Vallarta
Tlaquepaque

Zamora de Hida

Ciudad Guzmán
Zap

Colima
Tuxp

Isla San Benedicto

Manzanillo
Ag

Tecomán

Isla Roca Partida

Lázaro Cárd

Isla Socorro

Isla Clarión

*Islas Revillagigedo
(to Mexico)*

PACIFIC OCEAN

N

0 km   300

0 miles   300

Population   ● National capital

○ below 50,000   ○ 50,000 to 100,000   ◉ 100,000 to 500,000   ◼ above 500,000

E F G H

95° 90° 85°

ALABAMA

FLORIDA

20

MISSISSIPPI

30°

LOUISIANA

1

Brazos River

Red River

Sabine River

Mississippi River

Mississippi River Delta

AMERICA

TEXAS

Colorado River

dras Negras

Grande

Nuevo Laredo

25°

44

2

Sabinas Hidalgo

Ciudad Miguel Alemán

Padre Island

Gulf of

85°

Reynosa

Río Bravo

Matamoros

Laguna Madre

illo

Monterrey

Mexico

Tropic of Cancer

Montemorelos

Linares

O

Sierra Madre Oriental

Ciudad Victoria

Yucatan Channel

3

Ciudad Mante

Rio Lagartos

Tizimín

Cancún

Ciudad Madero

Progreso

Motul

Isla Cozumel

Luis

Pánuco

Tampico

Mérida

Ciudad Valles

Umán

Ticul

Valladolid

20°

Verde

Laguna de Tamiahua

Peto

Dolores Hidalgo

Tamazunchale

Oxkutzcab

Tekax

Guanajuato

Túxpan

Bahía de Campeche

Campeche

Felipe Carrillo Puerto

Poza Rica

Yucatan Peninsula

Querétaro

Papantla

Champotón

apuato

Pachuca

Tulancingo

Teziutlán

Chetumal

elia

Xalapa

Laguna de Términos

Fransisco Escárcega

30

4

MÉXICO (MEXICO CITY)

Perote

Veracruz

Frontera

Carmen

Toluca

Tlaxcala

Alvarado

Comalcalco

Gulf of Honduras

Cuernavaca

Puebla

Córdoba

Coatzacoalcos

Villahermosa

Zacatepec

Popocatépetl 5452m

Tehuacán

San Andrés Tuxtla

Macuspana

BELIZE

Taxco

Cuautla

Tuxtepec

Minatitlán

Teapa

Río Usumacinta

Iguala

Sierra Madre del Sur

Huajuapan

Istmo de Tehuantepec

Túxtla

San Cristóbal de Las Casas

alsas

Chilpancingo

Oaxaca

Ocozocuautla

Chiapa de Corzo

Comitán

Tecpan

Ixtepec

Matías Romero

15°

Pinotepa Nacional

Tehuantepec

Juchitán

Arriaga

Presa de la Angostura

Acapulco

Miahuatlán

Salina Cruz

Pijijiapán

GUATEMALA

HONDURAS

Puerto Escondido

Puerto Angel

Golfo de Tehuantepec

Escuintla

Huixtla

5

Tapachula

Ciudad Hidalgo

EL SALVADOR

131

E F G H

100° 95° 90°

Elevation

| Below sea level 0 | 250m | 500m | 1000m | 2000m | 3000m | 4000m | 6000m |

-6000m -4000m -2000m -1000m -500m -250m

-19,658ft -13,124ft -6562ft -3281ft -1640ft -820ft -328ft/-100m 0

820ft 1640ft 3281ft 6562ft 9843ft 13,124ft 19,685ft

Central America

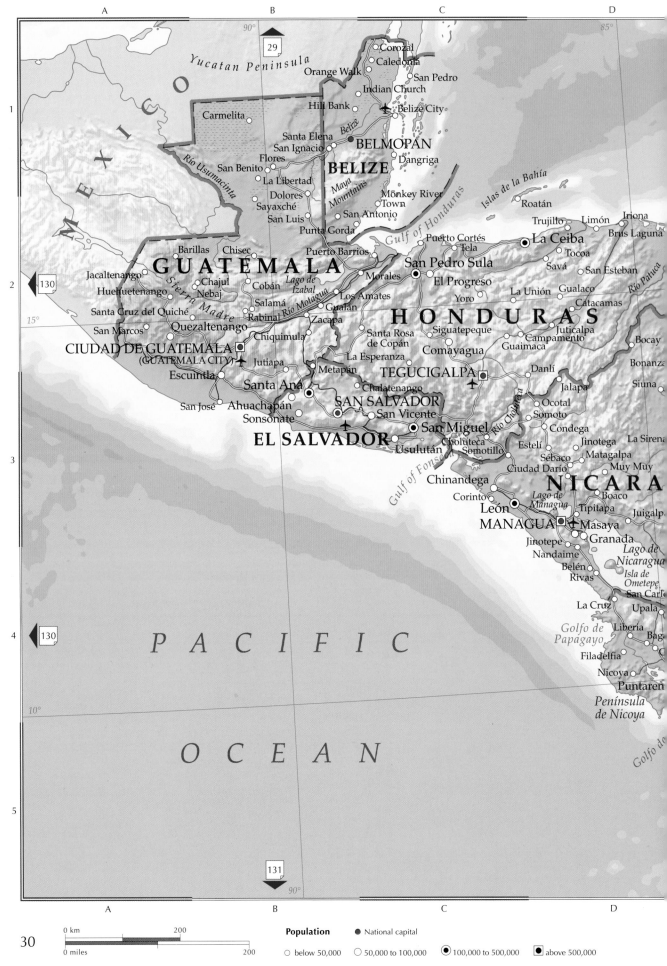

MEXICO

Yucatan Peninsula

Carmelita

Corozal
Caledonia
Orange Walk
San Pedro
Indian Church
Hill Bank
Belize City
Santa Elena
San Ignacio
BELMOPAN
Flores
Dangriga
San Benito
Río Usumacinta
La Libertad
Maya
Dolores
Mountains
Sayaxché
Monkey River
San Luis
Town
San Antonio
Punta Gorda

BELIZE

GUATEMALA

Barillas
Chisec
Puerto Barrios
Gulf of Honduras
Islas de la Bahía
Roatán
Trujillo
Limón
Iriona
Jacaltenango
Chajul
Coban
Lago de
Izabal
Morales
Puerto Cortés
La Ceiba
Brus Laguna
Nebaj
Salamá
Rabinal
Río Motagua
Los Amates
San Pedro Sula
Tela
Savá
Tocoa
San Esteban
Sierra Madre
Huehuetenango
El Progreso
Gualán
Santa Cruz del Quiché
Yoro
La Unión
Gualaco
Río Patuca
San Marcos
Quezaltenango
Chiquimula
Zacapa
HONDURAS
Cátacamas
Bocay
Siguatepeque
Juticalpa
Santa Rosa
Campamento
Bonanza
CIUDAD DE GUATEMALA
de Copán
Comayagua
Guaimaca
(GUATEMALA CITY)
Jutiapa
La Esperanza
TEGUCIGALPA
Danlí
Siuna
Escuintla
Metapán
Jalapa
Santa Ana
Chalatenango
Ocotal
San José
Ahuachapán
SAN SALVADOR
San Vicente
Somoto
Condega
Sonsonate
San Miguel
Río Choluteca
Estelí
Jinotega
La Sirena
EL SALVADOR
Usulután
Choluteca
Somotillo
Sébaco
Matagalpa
Gulf of Fonseca
Ciudad Darío
Muy Muy
Chinandega
Boaco
NICARA
Corinto
Lago de
Tipitapa
Juigalp
León
Managua
MANAGUA
Masaya
Jinotepe
Granada
Nandaime
Lago de
Belén
Nicaragua
Rivas
Isla de
Ometepe
San Carlo
La Cruz
Upala
Golfo de
Liberia
Bag
Papagayo
Filadelfia
Nicoya
Puntaren
Península
de Nicoya
Golfo de

PACIFIC

OCEAN

Population

○ below 50,000   ○ 50,000 to 100,000   ◉ 100,000 to 500,000   ◼ above 500,000

● National capital

0 km ——— 200

0 miles ——— 200

30

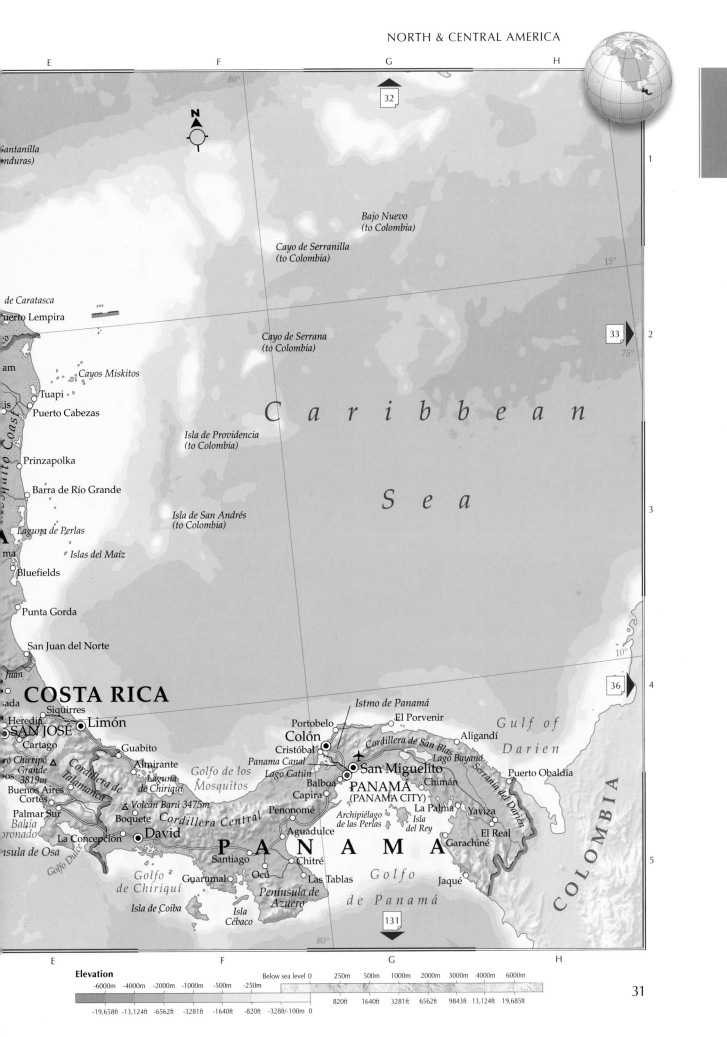

E   F   G   H

32

N

80°

1

Bajo Nuevo
(to Colombia)

Cayo de Serranilla
(to Colombia)

15°

Santanilla
nduras)

de Caratasca

uerto Lempira

Cayo de Serrana
(to Colombia)

33

75°

2

am

Cayos Miskitos

Tuapi

Puerto Cabezas

C a r i b b e a n

Prinzapolka

Isla de Providencia
(to Colombia)

Barra de Río Grande

S e a

3

is

Mosquito Coast

Laguna de Perlas

ma

Islas del Maíz

Isla de San Andrés
(to Colombia)

Bluefields

Punta Gorda

San Juan del Norte

10°

Juan

36

4

sada

COSTA RICA

Siquirres

Heredia

SAN JOSÉ   Limón

Cartago

ro Chirripó
Grande
3819m
os

Guabito

Almirante

Laguna
de Chiriquí

Buenos Aires
Cortés

Palmar Sur

Bahía
oronado

La Concepción

David

Boquete

△ Volcán Barú 3475m

Cordillera Central

*Cordillera de
Talamanca*

P A N A M A

Istmo de Panamá

El Porvenir

Portobelo

Colón

Cristóbal

Panama Canal

Lago Gatún

Balboa

Capira

Penonomé

Aguadulce

Santiago

Chitré

Guarumal

Ocú

Las Tablas

Cordillera de San Blas

Aligandí

*Golfo de los
Mosquitos*

San Miguelito

PANAMÁ
(PANAMA CITY)

Chimán

La Palma

*Archipiélago
de las Perlas*

Isla
del Rey

Golf of

D a r i e n

Lago Bayano

Serranía del Darién

Puerto Obaldía

Yaviza

El Real

Garachiné

Jaqué

*Golfo
de Panamá*

C O L O M B I A

*Golfo
de Chiriquí*

*Península de
Azuero*

Isla de Coiba

Isla
Cébaco

sula de Osa

Golfo Dulce

131

80°

5

E   F   G   H

Elevation

-6000m   -4000m   -2000m   -1000m   -500m   -250m

Below sea level 0   250m   500m   1000m   2000m   3000m   4000m   6000m

-19,658ft   -13,124ft   -6562ft   -3281ft   -1640ft   -820ft   -328ft/-100m   0

820ft   1640ft   3281ft   6562ft   9843ft   13,124ft   19,685ft

The Caribbean

UNITED STATES OF AMERICA

Gulf of Mexico

The Everglades

Florida Keys

Straits of Florida

Tropic of Cancer

Grand Bahama Island

Freeport

Marsh Harbour

Great Abaco

Bimini Islands

Berry Islands

Northeast Providence Channel

Nicholls Town

NASSAU

New Providence

Eleuthera Island

Rock Sound

Cat Island

Andros Town

Andros Island

Exuma Cays

Exuma Sound

San Salvador

THE BAHAMAS

George Town

Great Exuma Island

Long Island

Rum Cay

Clarence Town

Anguilla Cays

Cay Sal

Archipiélago de Camagüey

Ragged Island Range

Crooked Island Passage

Crooked Island

Acklins Island

Mayaguana Passage

Mayag

Caicos Pass

Little Inagua

Lake Rosa

Matthew Town

Great In

LA HABANA (HAVANA)

Guanabacoa

Artemisa

Cárdenas

Matanzas

Sagua la Grande

Pinar del Río

Consolación del Sur

Santa Clara

La Fé

Cienfuegos

Placetas

Yucatan Channel

Nueva Gerona

Sancti Spíritus

Morón

Ciego de Ávila

Isla de la Juventud

Cayo Largo

Archipiélago de los Canarreos

Bahía de Cochinos

C U B A

Camagüey

Nuevitas

Las Tunas

Holguín

Archipiélago de los Jardines de la Reina

Manzanillo

Bayamo

Guantánamo

Palma Soriano

Santiago de Cuba

Guantánamo Bay (to US)

Windward Passage

Ha

Haiti

Gonaïves

Little Cayman

Cayman Brac

GEORGE TOWN

Grand Cayman

CAYMAN ISLANDS (to UK)

G

r

e

a

t

e

r

NAVASSA ISLAND (to US)

Île de la Gonâve

HA

Jérémie

PORT-AU-PRINCE

Montego Bay

Spanish Town

Portmore

KINGSTON

Jamaica Channel

Cayes

Jacr

JAMAICA

Pedro Cays

C

a

r

i

b

b

e

a

n

HONDURAS

NICARAGUA

COSTA RICA

COLOMBI

JAMAICA

Caribbean Sea

Montego Bay

Lucea

Falmouth

Discovery Bay

St Ann's Bay

Ocho Rios

The Cockpit Country

Cambridge

Christiana

Ewarton

Annotto Bay

Buff Bay

Port Antonio

Savanna-La-Mar

Mandeville

Spanish Town

Blue Mountain Peak △2258m

Black River

May Pen

Old Harbour

Portmore

KINGSTON

Morant Bay

Portland Bight

Caribbean Sea

| 2000m/6562ft |
| 1000m/3281ft |
| 500m/1640ft |
| 200m/656ft |
| Sea level |

0 km 20
0 miles 20

0 km 200
0 miles 200

Population

● National capital

○ below 50,000   ○ 50,000 to 100,000   ◉ 100,000 to 500,000   ▣ above 500,000

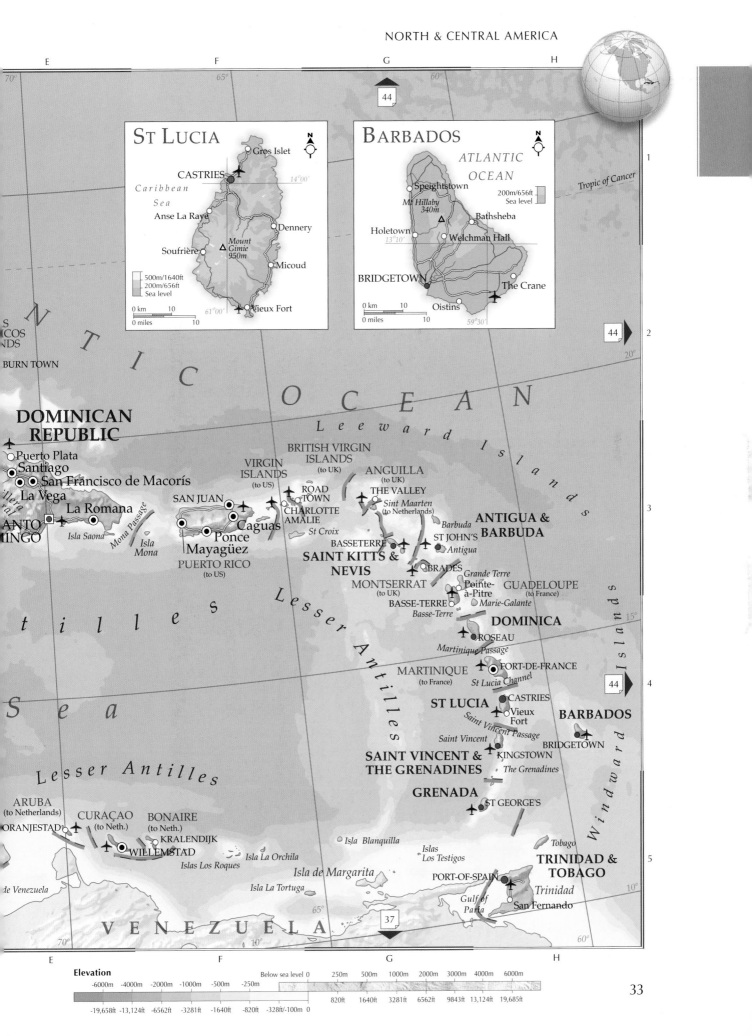

South America

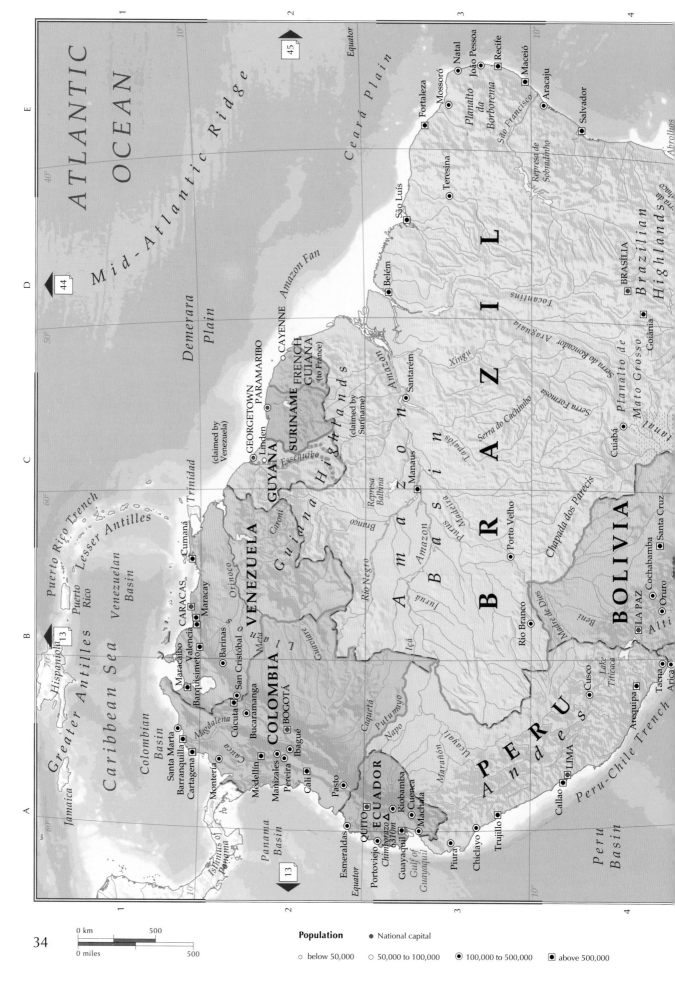

ATLANTIC OCEAN

Mid-Atlantic Ridge

Equator

Ceará Plain

Natal
Mossoró
João Pessoa
Recife
Maceió
Aracaju
Fortaleza
Planalto da Borborema
São Francisco
Salvador
Abrolhos

Teresina
Represa de Sobradinho

São Luís

BRASÍLIA
Brazilian Highlands

Belém

Tocantins
Araguaia
B R A Z I L
Goiânia
Planalto de Mato Grosso
Serra do Roncador
Iauví

Demerara Plain

Amazon Fan

CAYENNE
PARAMARIBO
SURINAME FRENCH GUIANA (to France)
(claimed by Venezuela)
GEORGETOWN
Linden
GUYANA
(claimed by Suriname)
Guiana Highlands
Essequibo

Santarém
Amazon
Xingu
Serra do Cachimbo
Tapajós
Chapada dos Parecis

Lesser Antilles
Puerto Rico Trench
Puerto Rico
Venezuelan Basin
Trinidad
Cumaná

CARACAS
Maracay
Valencia
Maracaibo
Barquisimeto
VENEZUELA
Barinas
San Cristóbal
Cúcuta

Caroní
Orinoco
Meta
Guaviare
Río Negro
Branco
Represa Balbina
Manaus

A m a z o n   B a s i n

Madeira
Purus
Amazon
Porto Velho

BOLIVIA
Santa Cruz
Cochabamba
LA PAZ
Oruro
Alti

Caribbean Sea
Greater Antilles
Jamaica
Hispaniola
Colombian Basin
Santa Marta
Barranquilla
Cartagena
Montería
Medellín
Magdalena
Cauca
Manizales
Pereira
Bucaramanga
COLOMBIA
BOGOTÁ
Ibagué
Cali
Pasto

Caquetá
Putumayo
Napo
Juruá
Içá
Marañón
Ucayali

P E R U
A n d e s
Cusco
Lake Titicaca
Tacna
Arica
Arequipa
Callao
LIMA
Trujillo
Peru-Chile Trench
Peru Basin

ECUADOR
QUITO
Portoviejo
Chimborazo 6310m
Guayaquil
Gulf of Guayaquil
Riobamba
Cuenca
Machala
Piura
Chiclayo

Esmeraldas
Equator

Panama Basin
Isthmus of Panama

34

0 km   500
0 miles   500

Population   ● National capital

○ below 50,000   ○ 50,000 to 100,000   ◉ 100,000 to 500,000   ◼ above 500,000

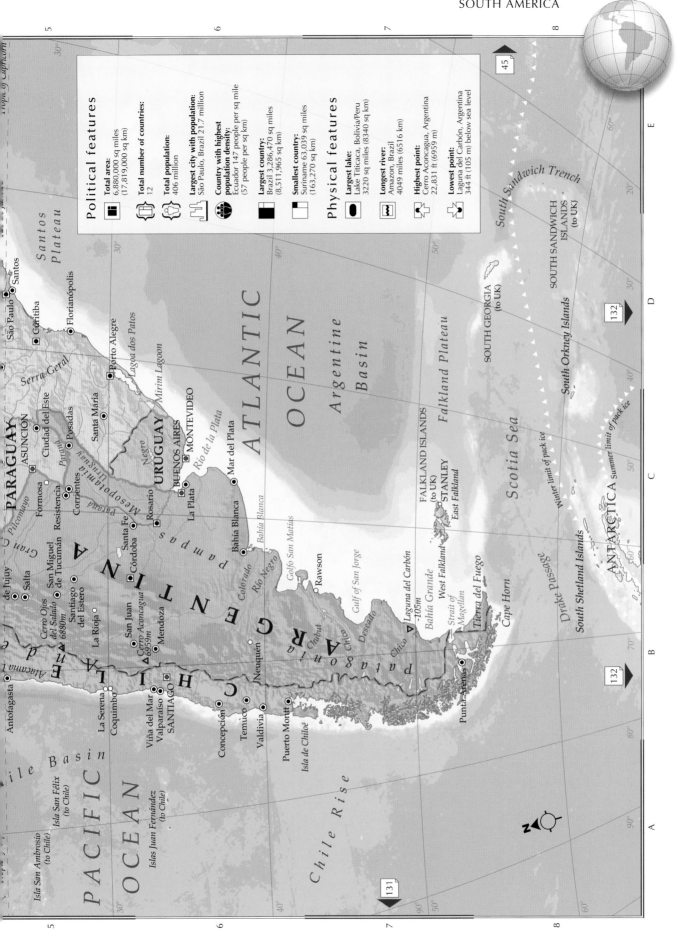

Northern South America

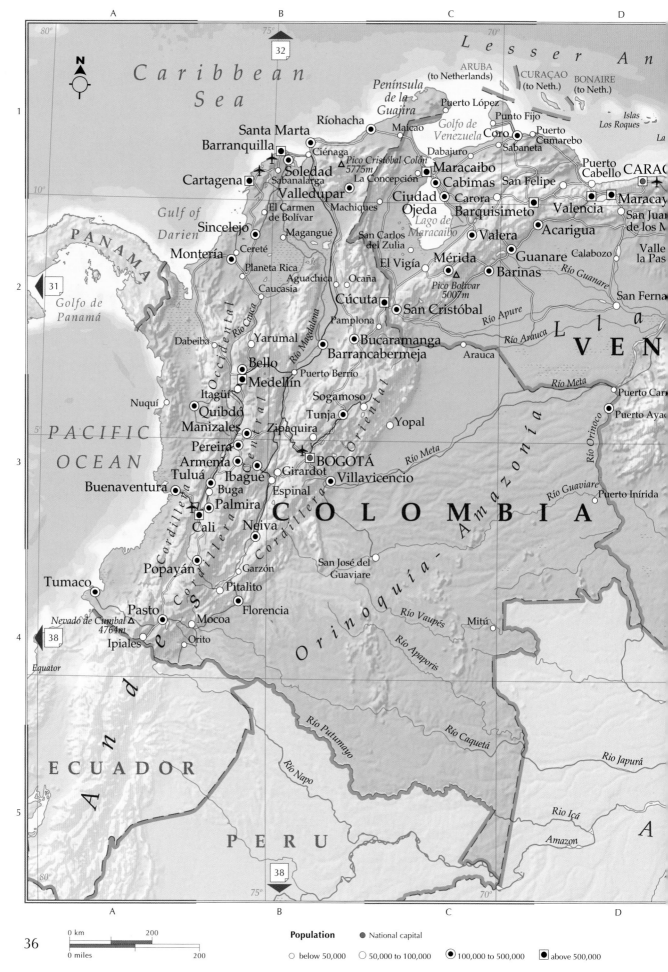

A   B   C   D

Caribbean Sea

Lesser An

ARUBA
(to Netherlands)

CURAÇAO
(to Neth.)

BONAIRE
(to Neth.)

Península
de la
Guajira

Golfo de
Venezuela

*Islas
Los Roques*

La

Ríohacha

Puerto López

Punto Fijo

Coro

Puerto
Cumarebo

Santa Marta

Maicao

Sabaneta

Barranquilla

Ciénaga

Dabajuro

Puerto
Cabello

CARAC

Cartagena

Soledad

Pico Cristóbal Colón
5775m

Maracaibo

San Felipe

Valledupar

Sabanalarga

La Concepción

Cabimas

Valencia

Maracay

El Carmen
de Bolívar

Machiques

Ciudad
Ojeda

Carora

Barquisimeto

San Juan
de los M

Sincelejo

Magangué

San Carlos
del Zulia

Lago de
Maracaibo

Valera

Acarigua

Gulf of
Darien

Montería

Cereté

El Vigía

Mérida

Guanare

Calabozo

Valle
la Pas

Planeta Rica

Ojeda

Barinas

Río Guanare

Aguachica

Ocaña

Pico Bolívar
5007m

Caucasia

Cúcuta

San Cristóbal

Río Apure

San Fernа

Dabeiba

Pamplona

Río Arauca

V E N

Yarumal

Bucaramanga

Arauca

L

l
a

Río Cauca

Barrancabermeja

Río Meta

Puerto Car

PANAMA

Golfo de
Panamá

Bello

Puerto Berrío

Medellín

Río Magdalena

Puerto Ayac

Nuquí

Itagüí

Sogamoso

Río Orinoco

Quibdó

Tunja

Yopal

Manizales

Zipaquira

**PACIFIC
OCEAN**

Pereira

BOGOTÁ

Río Meta

Armenia

Girardot

Villavicencio

Puerto Inírida

Tuluá

Ibagué

Río Guaviare

Buenaventura

Buga

Espinal

C O L O M B I A

Palmira

Cali

Neiva

San José del
Guaviare

Popayán

Garzón

Pitalito

Río Vaupés

Mitú

Tumaco

Florencia

Mocoa

Río Apaporis

Nevado de Cumbal
4764m

Pasto

Orito

Orinoquía-

Amazonía

Ipiales

Cordillera Occidental

Cordillera Central

Cordillera Oriental

Andes

Equator

ECUADOR

Río Putumayo

Río Napo

Río Caquetá

Río Japurá

PERU

Río Içá

Amazon

A

80°   75°   70°

A   B   C   D

0 km   200

0 miles   200

36

Population   ● National capital

○ below 50,000   ◯ 50,000 to 100,000   ◉ 100,000 to 500,000   ■ above 500,000

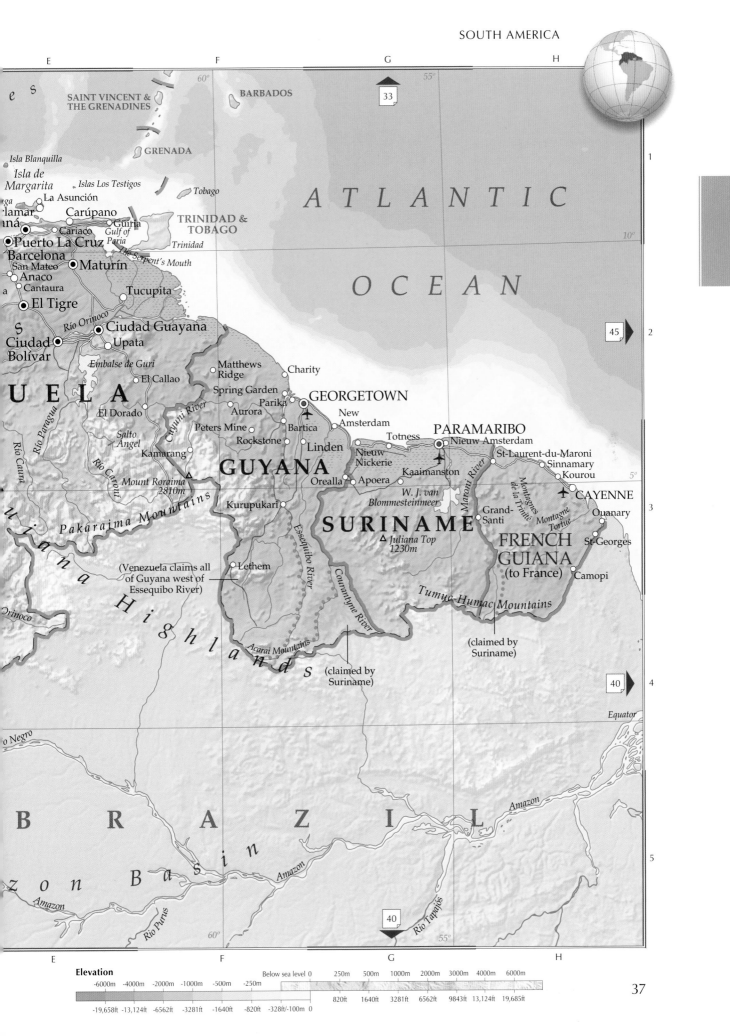

E   60°   F   G   55°   H

SAINT VINCENT &
THE GRENADINES

BARBADOS

33

1

GRENADA

Isla Blanquilla

Isla de
Margarita
ga   Islas Los Testigos

La Asunción   Tobago

lamar   Carúpano
iná   Güiria   TRINIDAD &
Cariaco   Gulf of   TOBAGO
Paria   Trinidad
Puerto La Cruz   The Serpent's Mouth
Barcelona
San Mateo
Anaco   Maturín
Cantaura
El Tigre   Tucupita

S   Río Orinoco   Ciudad Guayana   **2**
Ciudad   Upata   45
Bolívar
Embalse de Guri
El Callao   Charity
UELA   Matthews
Ridge
Río Paragua   Spring Garden   GEORGETOWN
El Dorado   Cuyuni River   Aurora   Parika
Peters Mine   Bartica   New
Salto   Rockstone   Amsterdam   PARAMARIBO
Ángel   Río Caroní   Linden   Totness   Nieuw Amsterdam
Kamarang   Nieuw   St-Laurent-du-Maroni
Nickerie   Sinnamary
Mount Roraima   **GUYANA**   Orealla   Kaaimanston   Kourou   5°
2810m   Apoera   St-Georges   CAYENNE
Pakaraima Mountains   W. J. van   Maroni River   **3**
Blommesteinmeer   Montagnes   Ouanary
Kurupukari   de la Trinité
iana   **SURINAME**   Grand-   Montagne   FRENCH
(Venezuela claims all   Santi   Tortue   GUIANA   St-Georges
of Guyana west of   Essequibo River   △ Juliana Top   (to France)
Essequibo River)   1230m   Camopi
Lethem
Orinoco   *Highlands*   Courantyne River   Tumuc-Humac Mountains   (claimed by
Suriname)
Acarai Mountains   (claimed by   40   **4**
Suriname)
Equator
o Negro
B   R   A   Z   I   L
Amazon
5
zon   Basin   Amazon
Amazon
Rio Purus   Rio Tapajós   40
60°   55°

E   F   G   H

Elevation
Below sea level 0   250m   500m   1000m   2000m   3000m   4000m   6000m
-6000m   -4000m   -2000m   -1000m   -500m   -250m
820ft   1640ft   3281ft   6562ft   9843ft   13,124ft   19,685ft
-19,658ft -13,124ft -6562ft   -3281ft   -1640ft   -820ft -328ft/-100m 0

37

Western South America

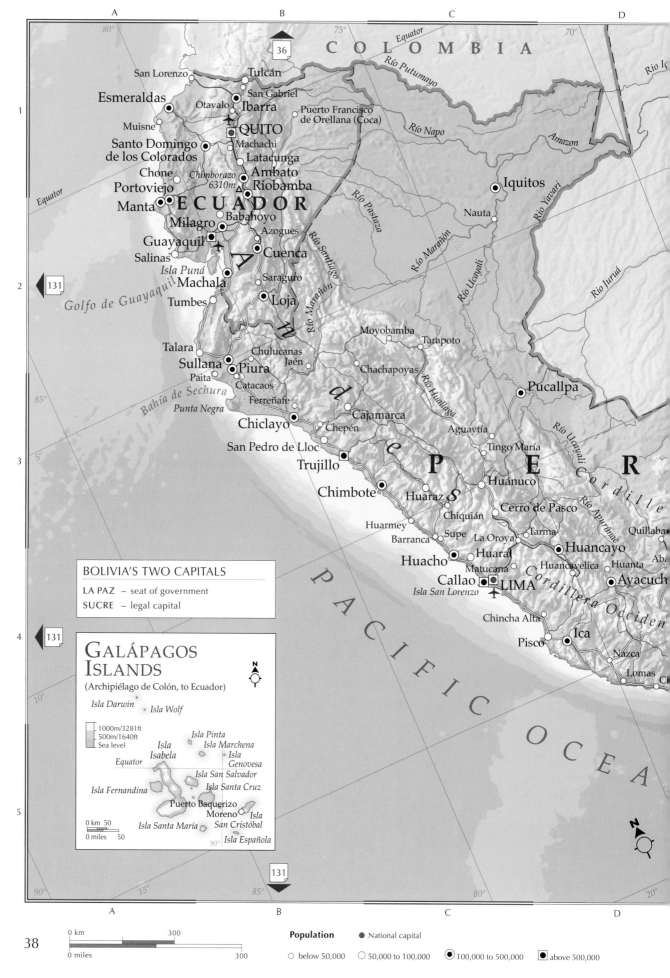

COLOMBIA

ECUADOR

San Lorenzo
Tulcán
Esmeraldas
San Gabriel
Muisne
Otavalo
Ibarra
Puerto Francisco de Orellana (Coca)
QUITO
Machachi
Santo Domingo de los Colorados
Latacunga
Chone
Chimborazo 6310m
Ambato
Riobamba
Portoviejo
Manta
Milagro
Babahoyo
Azogues
Guayaquil
Cuenca
Salinas
Isla Puná
Saraguro
Machala
Loja
Tumbes

Río Putumayo
Equator
Río Içá
Río Napo
Amazon
Río Pastaza
Iquitos
Nauta
Río Marañón
Río Santiago
Río Ucayali
Río Yavari
Río Jurua
Río Marañón

Moyobamba
Tarapoto
Talara
Chulucanas
Jaén
Chachapoyas
Sullana
Piura
Paita
Catacaos
Ferreñafe
Bahía de Sechura
Punta Negra
Cajamarca
Chiclayo
Chepén
Río Huallaga
Pucallpa
San Pedro de Lloc
Trujillo
Aguaytía
PERU
Río Ucayali
Cordillera
Chimbote
Huaraz
Tingo María
Huánuco
Chiquián
Cerro de Pasco
Huarmey
Río Apurímac
Quillaba
Barranca
Supe
La Oroya
Tarma
Huaral
Huancayo
Huanta
Huacho
Matucana
Huancavelica
Ayacuch
Callao
LIMA
Cordillera Occiden
Aba
Isla San Lorenzo
Chincha Alta
Ica
Pisco
Nazca
Lomas

Golfo de Guayaquil

PACIFIC OCEA

BOLIVIA'S TWO CAPITALS

LA PAZ – seat of government

SUCRE – legal capital

GALÁPAGOS ISLANDS

(Archipiélago de Colón, to Ecuador)

Isla Darwin
Isla Wolf

1000m/3281ft
500m/1640ft
Sea level

Isla Pinta
Isla Marchena
Isla Isabela
Isla Genovesa
Equator
Isla San Salvador
Isla Fernandina
Isla Santa Cruz
Puerto Baquerizo Moreno
Isla
Isla Santa María
San Cristóbal
Isla Española

0 km 50
0 miles 50

0 km 300
0 miles 300

Population ● National capital

○ below 50,000   ○ 50,000 to 100,000   ◉ 100,000 to 500,000   ■ above 500,000

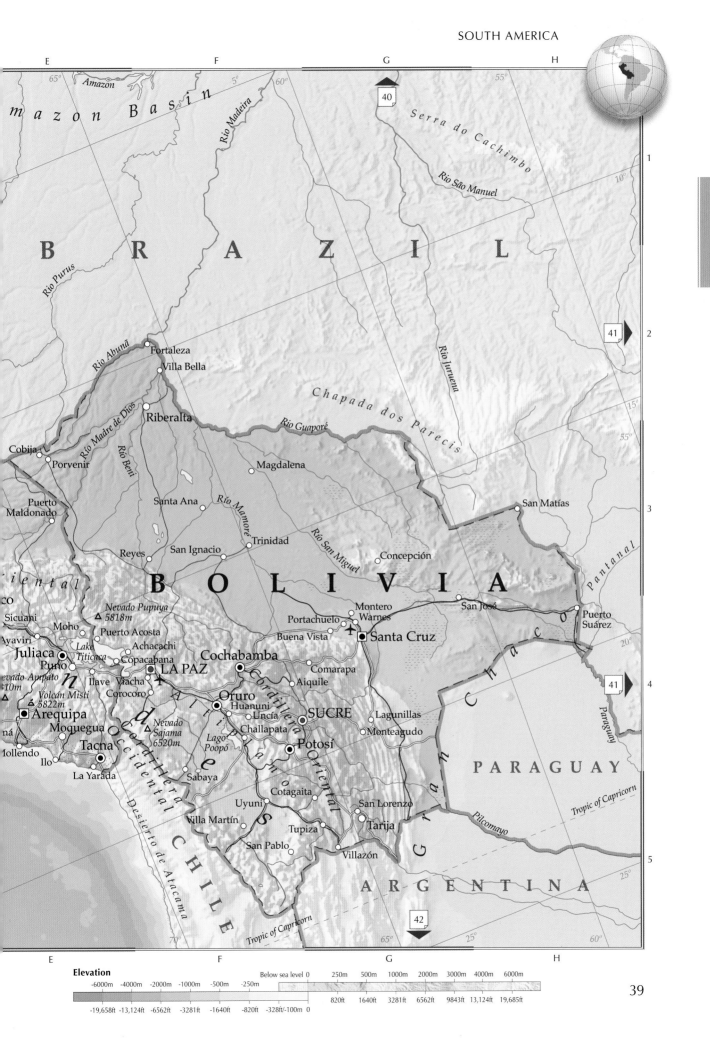

E   F   G   H

65°   5°   60°   55°

Amazon

mazon Basin

Rio Purus   *Rio Madeira*

B   R   A   Z   I   L

10°   1

Serra do Cachimbo

Rio São Manuel

40

Rio Abunā

Fortaleza

Villa Bella

41   2

Rio Jurnena

Chapada dos Parecis   15°

Rio Madre de Dios

Riberalta   *Rio Guaporé*   55°

Cobija

Porvenir   *Rio Beni*

Magdalena

Santa Ana   *Rio Mamoré*

Puerto
Maldonado   San Matías   3

Reyes   San Ignacio   Trinidad   *Rio San Miguel*   Concepción

riental   B   O   L   I   V   I   A   *Pantanal*

co   Montero   San José   Puerto
Suárez

Sicuani   *Nevado Pupuya*   Portachuelo   Warnes

Moho   △ *5818m*   Puerto Acosta   Buena Vista   Santa Cruz   20°

Ayaviri   Achacachi

Juliaca   *Lake*   Copacabana   Cochabamba   Comarapa
Titicaca
Puno   LA PAZ   Aiquile
evado Ampato   Ilave   Viacha

10m   Corocoro   Oruro   San José   Lagunillas
△   *Volcán Misti*   Huanuni   SUCRE   41   4
5822m   Uncía   Monteagudo

Arequipa   *Nevado*   Challapata
△ *Sajama*
Moquegua   *6520m*   Potosí   P A R A G U A Y
Lago
Tacna   *Poopó*

ná   Sabaya   *Gran Chaco*

Mollendo   Cotagaita   *Tropic of Capricorn*

Ilo   Uyuni   San Lorenzo
La Yarada   Villa Martín   Tupiza   Tarija   *Pilcomayo*

San Pablo   Villazón   5   25°

A R G E N T I N A

42

Tropic of Capricorn   25°

70°   65°   60°

E   F   G   H

Elevation

-6000m	-4000m	-2000m	-1000m	-500m	-250m	Below sea level 0	250m	500m	1000m	2000m	3000m	4000m	6000m

-19,658ft -13,124ft -6562ft -3281ft -1640ft -820ft -328ft/-100m 0   820ft 1640ft 3281ft 6562ft 9843ft 13,124ft 19,685ft

39

Brazil

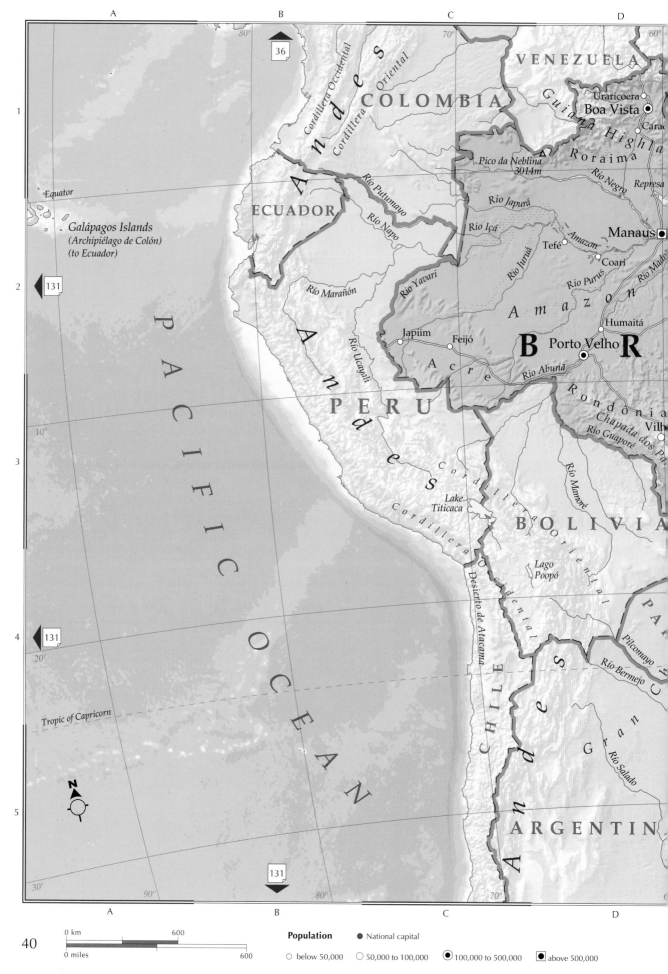

A

B

C

D

80°

70°

60°

36

VENEZUELA

COLOMBIA

Guiana Highla

Uraricoera

Boa Vista

1

Carac

Cordillera Occidental

Cordillera Oriental

A n d e s

Pico da Neblina
3014m

Río Negro

Roraima

Represa

ECUADOR

Río Putumayo

Río Japurá

Río Içá

Tefé

Río Juruá

Amazon

Manaus

Galápagos Islands
(Archipiélago de Colón)
(to Ecuador)

Equator

Río Napo

Coari

Río Purus

Río Made

131

Río Marañón

Río Yavari

A m a z o n

A n d e s

Humaitá

2

Río Ucayali

Japiim

Feijó

B

Porto Velho

R

Acre

Río Abunã

Rondônia

P A C I F I C   O C E A N

P E R U

Chapada dos Pa

Vilh

Río Guaporé

10°

Río Mamoré

3

Cordillera

Lake
Titicaca

Cordillera   Occidental

Cordillera   Oriental

B O O R I E N T A L I V I A

Lago
Poopó

P A

131

Desierto de Atacama

Pilcomayo

4

20°

Río Bermejo

C

A n d e s

CHILE

G r a n   C h

Tropic of Capricorn

Río Salado

N

5

A R G E N T I N

131

80°

90°

30°

A

B

C

D

40

0 km   600

0 miles   600

Population   ● National capital

○ below 50,000   ○ 50,000 to 100,000   ◉ 100,000 to 500,000   ◼ above 500,000

ATLANTIC OCEAN

FRENCH
GUIANA
(to France)

Tumuc-Humac
Mountains

Amapá

Macapá

Ilha Caviana de Fora

Ilha
de Marajó

Baía de Marajó

Belém

Mouths of the Amazon

Baía de São Marcos

São Luís

Parnaíba

Camocim

Equator

Amazon

quer

Santarém

Altamira

Bacabal

Piripiri

Teresina

Fortaleza

Atol das Rocas

San Fernando de Noronha
(to Brazil)

Itaituba

Represa de
Tucuruí

Imperatriz

Maranhão

Mossoró

Assu

Cabo de São Roque

45

Rio Xingu

Marabá

Carolina

Floriano

Ceará

Juazeiro do Norte

Rio Grande do Norte

Natal

Pará

Balsas

Picos

Paraíba

João Pessoa

Campina Grande

do Cachimbo

Z I L

Piauí

Pernambuco

Recife

Represa de Sobradinho

Alagoas

Maceió

Palmas do
Tocantins

Juazeiro

Chapada
Diamantina

Serra Formosa

Serra dos Gradaús

Tocantins

Rio São Fransisco

Aracaju

Estância

10°

Grosso

Taguatinga

Goiás

Bahia

Feira de Santana

Salvador

Baía de Todos os Santos

Cuiabá

Planalto
Central

BRASÍLIA

Janaúba

Itabuna

Rio Araguaia

3

Anápolis

Vitória da Conquista

Canavieiras

onópolis

Goiânia

Jataí

Minas

Montes Claros

Araçuai

Mato Grosso
do Sul

Araguari

Gerais

Governador Valadares

do Sul

Uberlândia

Uberaba

Espírito
Santo

Campo Grande

Belo Horizonte

Divinópolis

Vitória

idauana

Ribeirão Preto

Juiz de Fora

Campos dos Goytacazes

20°

45

dente Prudente

Marília

Campinas

Londrina

São Paulo

Nova
Iguaçu

Rio de Janeiro

Maringá

Paraná

Santos

Tropic of Capricorn

Represa
de Itaipú

Ponta Grossa

Curitiba

Saltos do Rio Iguaçu
Iguaçu

Joinville

raná

Santa Catarina

Blumenau

Florianópolis

Passo Fundo

Rio Grande

a Maria

Canoas

do Sul

Porto Alegre

5

Negro

Bagé

Lagoa dos Patos

Rio Grande

UGUAY

Mirim Lagoon

30°

ATLANTIC OCEAN

44

45

Elevation

| -6000m | -4000m | -2000m | -1000m | -500m | -250m | Below sea level 0 | 250m | 500m | 1000m | 2000m | 3000m | 4000m | 6000m |

| -19,658ft | -13,124ft | -6562ft | -3281ft | -1640ft | -820ft | -328ft/-100m 0 | | 820ft | 1640ft | 3281ft | 6562ft | 9843ft | 13,124ft | 19,685ft |

Southern South America

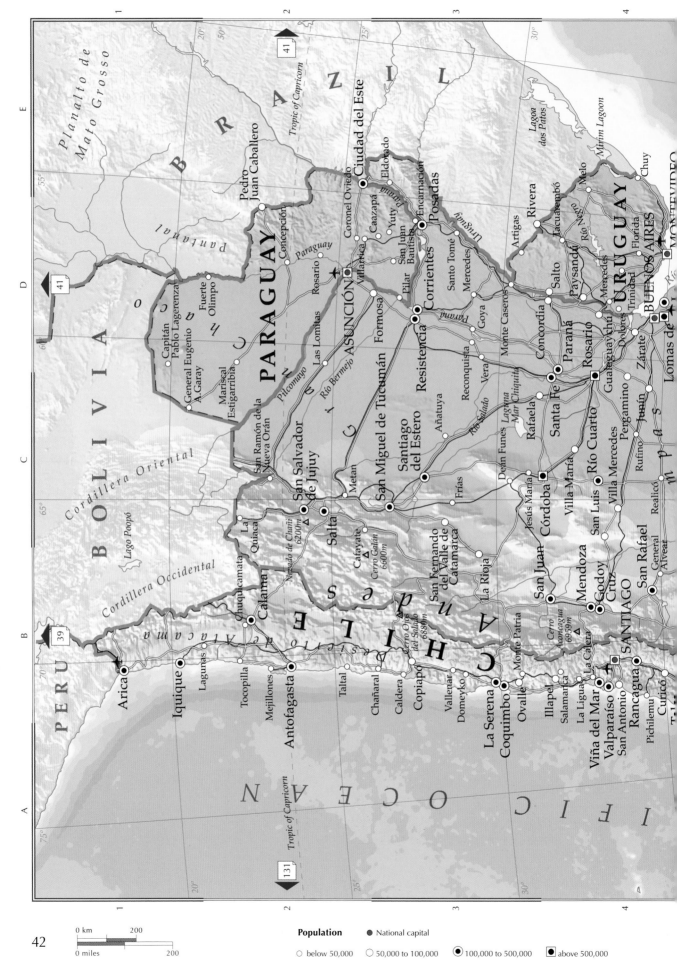

0 km   200
0 miles   200

Population   National capital

○ below 50,000   ○ 50,000 to 100,000   ◉ 100,000 to 500,000   ■ above 500,000

Labels on map

BRAZIL

Planalto de Mato Grosso

Pantanal

BOLIVIA

PERU

Cordillera Oriental

Cordillera Occidental

Lago Poopó

Desierto de Atacama

CHILE

PARAGUAY

URUGUAY

Lagoa dos Patos

Mirim Lagoon

ANDES

PAMPAS

PACIFIC OCEAN

Tropic of Capricorn

Cities and places

Pedro Juan Caballero
Concepción
Capitán Pablo Lagerenza
Fuerte Olimpo
General Eugenio A. Garay
Mariscal Estigarribia
Las Lomitas
Rosario
Paraguay
Pilcomayo
Río Bermejo
San Ramón de la Nueva Orán
La Quiaca
San Salvador de Jujuy
Nevado de Chañi 6200m
Salta
Calama
Chuquicamata
Metán
Cerro Galán 6600m
Catavate
Cerro Ojos del Salado 6880m
Cerro Aconcagua 6959m
ASUNCIÓN
Villarrica
Coronel Oviedo
Caazapá
Yuty
San Juan Bautista
Ciudad del Este
Eldorado
Encarnación
Posadas
Pilar
Formosa
Corrientes
Resistencia
Santo Tomé
Mercedes
Artigas
Rivera
Tacuarembó
Río Negro
Melo
Paraná
Uruguay
Salto
Paysandú
Concordia
Goya
Monte Caseros
Reconquista
Vera
Río Salado
Añatuya
San Miguel de Tucumán
Santiago del Estero
Frías
Laguna Mar Chiquita
Rafaela
Santa Fe
Paraná
Rosario
Gualeguaychú
Dolores
Trinidad
Florida
Mercedes
Zárate
Lomas de
BUENOS AIRES
MONTEVIDEO
Chuy
Deán Funes
Jesús María
Villa María
Córdoba
Río Cuarto
Villa Mercedes
Pergamino
Junín
Rufino
Realicó
San Fernando del Valle de Catamarca
La Rioja
San Luis
San Juan
Mendoza
Godoy Cruz
San Rafael
General Alvear
Monte Patria
Illapel
Salamarca
La Ligua
La Calera
Valparaíso
Viña del Mar
San Antonio
SANTIAGO
Rancagua
Pichilemu
Curicó
Arica
Iquique
Lagunas
Tocopilla
Mejillones
Antofagasta
Taltal
Chañaral
Caldera
Copiapó
Vallenar
Domeyko
La Serena
Coquimbo
Ovalle
San Juan

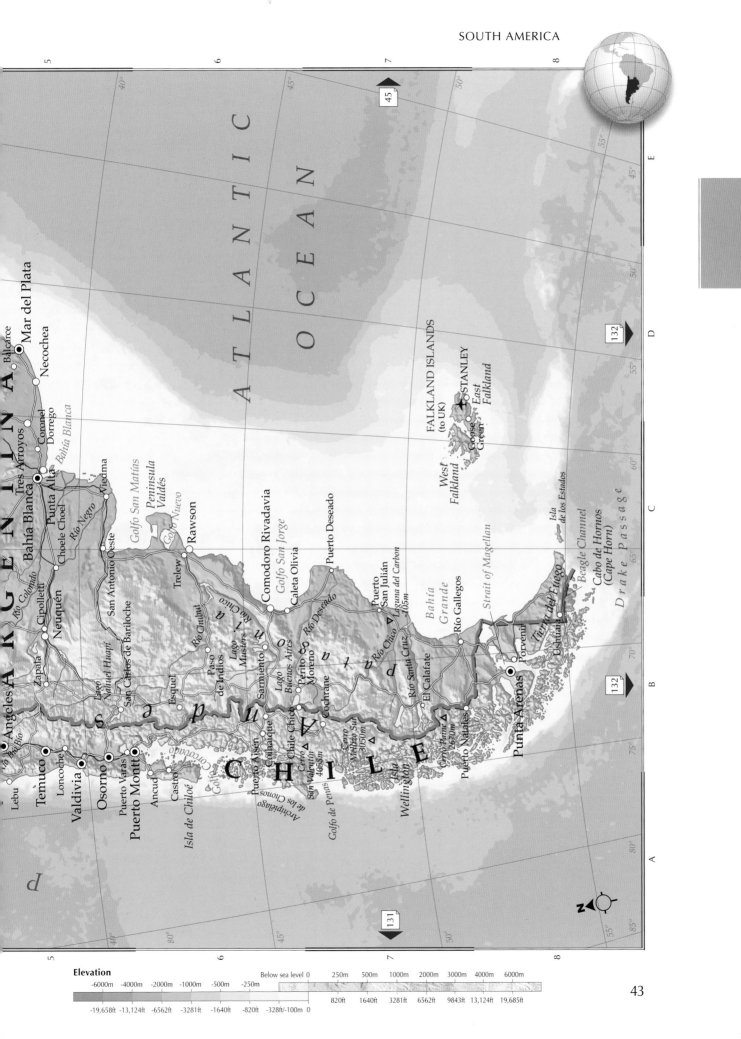

Elevation

-6000m	-4000m	-2000m	-1000m	-500m	-250m
-19,658ft	-13,124ft	-6562ft	-3281ft	-1640ft	-820ft	-328ft/-100m 0

Below sea level 0   250m   500m   1000m   2000m   3000m   4000m   6000m

820ft   1640ft   3281ft   6562ft   9843ft   13,124ft   19,685ft

The Atlantic Ocean

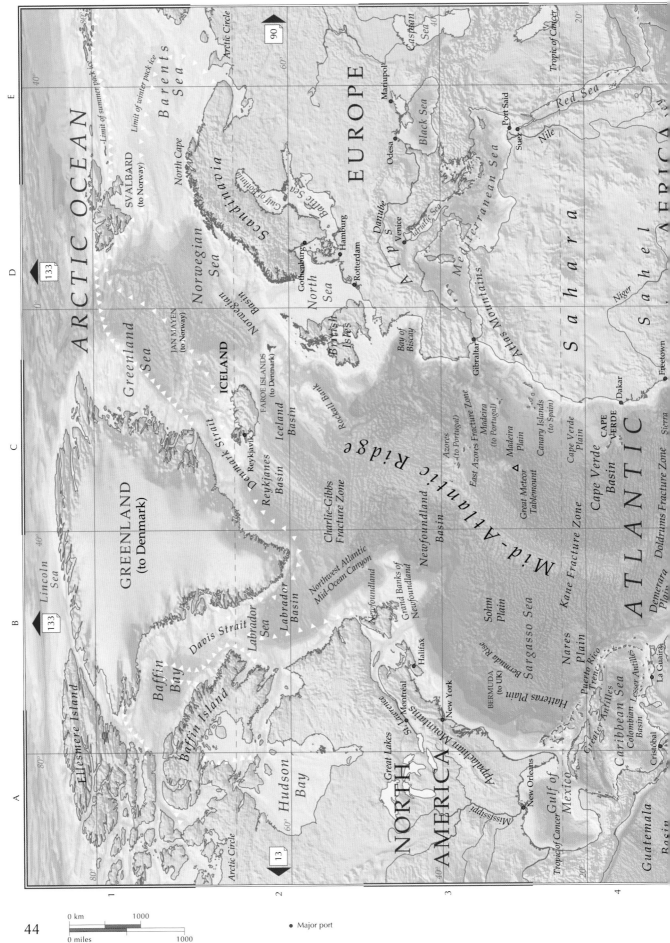

ARCTIC OCEAN

EUROPE

NORTH AMERICA

ATLANTIC

Limit of summer pack ice
Limit of winter pack ice
Arctic Circle

Barents Sea

SVALBARD
(to Norway)

North Cape

Scandinavia

Gulf of Bothnia

Baltic Sea

Mariupol
Odesa

Black Sea

Caspian Sea

Port Said
Suez
Nile
Red Sea

Sahara

Sahel

AFRICA

Niger

Danube
Venice
Adriatic Sea

Alps

Atlas Mountains

Mediterranean Sea

Hamburg
Rotterdam

North Sea

Gothenburg

Norwegian Sea

Norwegian Basin

JAN MAYEN
(to Norway)

Greenland Sea

ICELAND

FAROE ISLANDS
(to Denmark)

British Isles

Bay of Biscay

Gibraltar

Reykjavik
Denmark Strait

Reykjanes Basin

Iceland Basin

Rockall Bank

Mid-Atlantic Ridge

Azores
(to Portugal)

East Azores Fracture Zone

Madeira
(to Portugal)

Madeira Plain

Great Meteor Tablemount

Canary Islands
(to Spain)

Cape Verde Plain

CAPE VERDE

Dakar

Freetown

Cape Verde Basin

Sierra

Doldrums Fracture Zone

GREENLAND
(to Denmark)

Lincoln Sea

Charlie-Gibbs Fracture Zone

Northwest Atlantic Mid-Ocean Canyon

Labrador Basin

Labrador Sea

Davis Strait

Baffin Bay

Baffin Island

Ellesmere Island

Hudson Bay

Newfoundland

Newfoundland Basin

Grand Banks of Newfoundland

Halifax

Montreal
St. Lawrence
Great Lakes

New York

BERMUDA
(to UK)

Bermuda Rise

Hatteras Plain

Sohm Plain

Sargasso Sea

Kane Fracture Zone

Nares Plain

Puerto Rico Trench

Demerara Plain

Appalachian Mountains

New Orleans

Mississippi

Tropic of Cancer

Gulf of Mexico

Greater Antilles

Lesser Antilles

Caribbean Sea

Colombian Basin

Cristóbal

La Guaira

Guatemala Basin

Tropic of Cancer

0 km   1000
0 miles   1000

● Major port

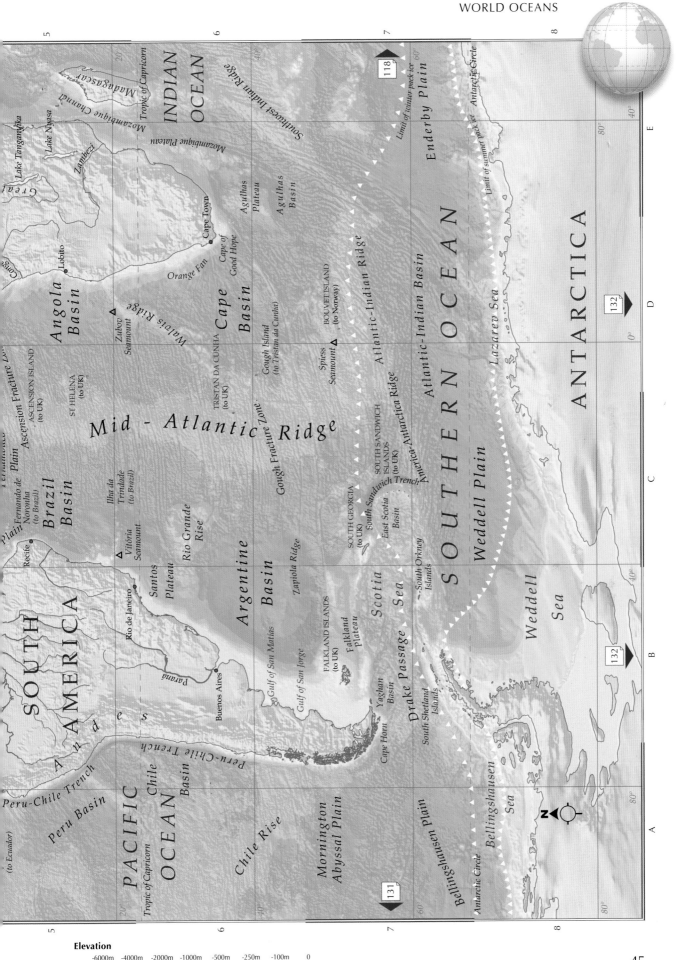

Elevation

| -6000m | -4000m | -2000m | -1000m | -500m | -250m | -100m | 0 |

| -19,658ft | -13,124ft | -6562ft | -3281ft | -1640ft | -820ft | -328ft/-100m | 0 |

Africa

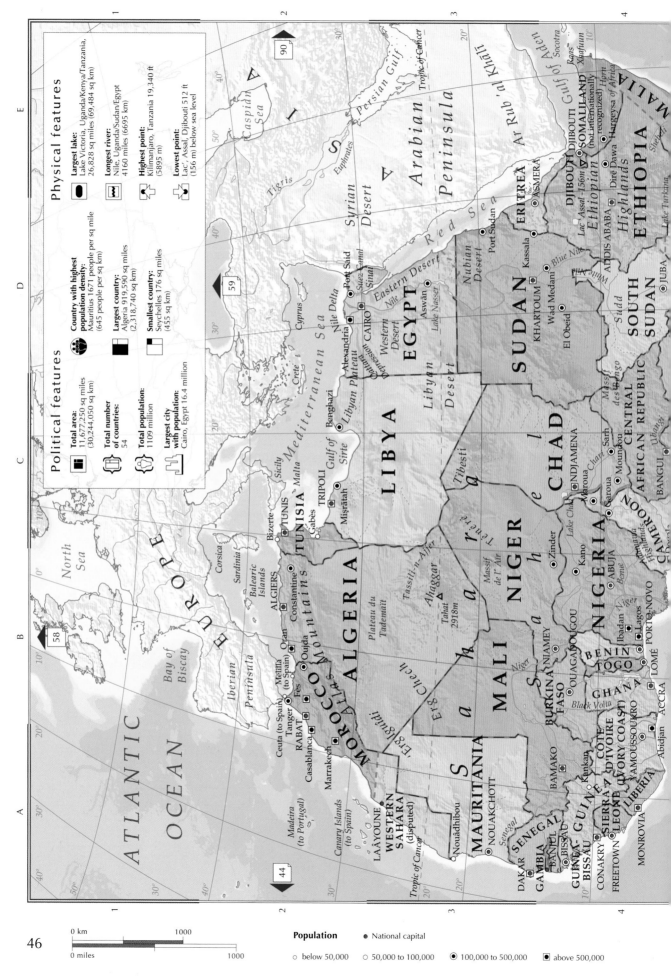

Political features

Total area:
11,677,250 sq miles
(30,244,050 sq km)

Total number of countries:
54

Total population:
1109 million

Largest city with population:
Cairo, Egypt 16.4 million

Physical features

Largest lake:
Lake Victoria, Uganda/Kenya/Tanzania,
26,828 sq miles (69,484 sq km)

Longest river:
Nile, Uganda/Sudan/Egypt
4160 miles (6695 km)

Highest point:
Kilimanjaro, Tanzania 19,340 ft
(5895 m)

Lowest point:
Lac' Assal, Djibouti 512 ft
(156 m) below sea level

Country with highest population density:
Mauritius 1671 people per sq mile
(645 people per sq km)

Largest country:
Algeria 919,590 sq miles
(2,318,740 sq km)

Smallest country:
Seychelles 176 sq miles
(455 sq km)

Population
○ below 50,000   ○ 50,000 to 100,000   ◉ 100,000 to 500,000   ◼ above 500,000
● National capital

0 km _____ 1000
0 miles _____ 1000

46

Somali Basin
Basin

Aldabra Group

Kilimanjaro 5895m▲
Mombasa
Tanga
Pemba
Zanzibar
Dar es Salaam
Masai Steppe

COMOROS
MORONI
MAYOTTE (to France)
Nacala
Mahajanga

ANTANANARIVO
Fianarantsoa
Toliara

MADAGASCAR

Madagascar Basin

Tropic of Capricorn

Madagascar Plateau

INDIAN

OCEAN

Southwest Indian Ridge

Crozet Plateau

Prince Edward Islands (to South Africa)

RWANDA
BUJUMBURA
BURUNDI
DODOMA
Lake Tanganyika
Lake Rukwa
Lake Nyasa
TANZANIA
MALAWI
LILONGWE
Luangwa
Lurio
Nampula

Rovuma

Mozambique Channel

MOZAMBIQUE

DEM. REP.
CONGO
Bukavu

Great Rift Valley
Kalemie
Lake Mweru
Lubumbashi
Ndola
Kitwe
Lualaba
Luvua

Blantyre
Beira
Zambezi
Lake Kariba
HARARE
Bulawayo

Mozambique Plateau

ZAMBIA
LUSAKA

ZIMBABWE

MAPUTO
MBABANE
SWAZILAND
PRETORIA
Limpopo
Durban
East London
Port Elizabeth

Ilebo
Kasai
KINSHASA
BRAZZAVILLE
Matadi
Cabinda (to Angola)
KanDanga

Victoria Falls
Cuando
Okavango Delta
Francistown
GABORONE
Johannesburg
BOTSWANA
Kalahari Desert

LESOTHO
MASERU
BLOEMFONTEIN
SOUTH AFRICA
Great Karoo
Drakensberg
Vaal

Agulhas Plateau

Agulhas Basin

ANGOLA
LUANDA
Cuanza
Bié Plateau
Môco 2619m▲
Huambo
Lubango
Namibe
Cuango
Cubango

NAMIBIA
WINDHOEK
Etosha Pan
Namib Desert
Nossob
Orange River
Cunene

CAPE TOWN
Cape of Good Hope

Orange Fan

Cape Basin

SAINT HELENA (to UK)

Ascension Fracture Zone
. ASCENSION ISLAND (to UK)

Angola Basin

ATLANTIC

OCEAN

Walvis Ridge

TRISTAN DA CUNHA (to UK)

Gough Island (to Tristan da Cunha)

Mid-Atlantic Ridge

Atlantic-Indian Ridge

Atlantic-Indian Ridge

Winter limit of pack ice

Tropic of Capricorn

Basin

N

119
132
132
45

Northwest Africa

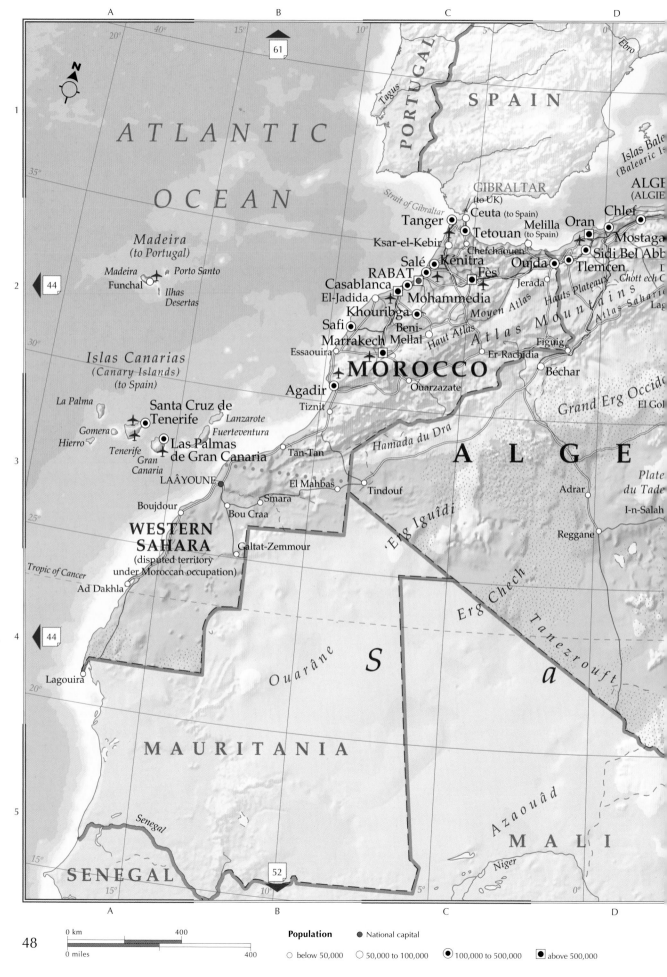

ATLANTIC

OCEAN

PORTUGAL

SPAIN

Islas Bale
(Balearic Is

Tagus

Ebro

GIBRALTAR
(to UK)

Ceuta (to Spain)

ALGE
(ALGIE

Chlef

Tanger

Tetouan

Melilla
(to Spain)

Oran

Mostaga

Strait of Gibraltar

Ksar-el-Kebir

Chefchaouen

Sidi Bel Abb

Salé

Kénitra

Oujda

Tlemcen

Madeira
(to Portugal)

RABAT

Fès

Jerada

Chott ech C

Madeira

Porto Santo

Funchal

Ilhas
Desertas

Casablanca

El-Jadida

Mohammedia

Hauts Plateaux

Atlas Sahari

Lag

Khouribga

Moyen Atlas

Figuig

Safi

Beni-
Mellal

Haut Atlas

Atlas Mountains

Er-Rachidia

Islas Canarias
(Canary Islands)
(to Spain)

Marrakech

Essaouira

MOROCCO

Béchar

La Palma

Ouarzazate

Grand Erg Occide

El Go

Agadir

Gomera

Santa Cruz de
Tenerife

Lanzarote

Tiznit

ALGE

Hierro

Fuerteventura

Tenerife

Gran
Canaria

Las Palmas
de Gran Canaria

Tan-Tan

Hamada du Dra

Plate
du Tade

Adrar

I-n-Salah

LAÂYOUNE

El Mahbas

Tindouf

Boujdour

Smara

'Erg Iguîdi

Reggane

Bou Craa

WESTERN
SAHARA

Galtat-Zemmour

(disputed territory
under Moroccan occupation)

Erg Chech

Tanezrouft

Tropic of Cancer

Ad Dakhla

S

Lagouira

Ouarâne

a

MAURITANIA

Azaouâd

MALI

Senegal

Niger

SENEGAL

0 km   400

0 miles   400

Population   ● National capital

○ below 50,000   ○ 50,000 to 100,000   ◉ 100,000 to 500,000   ■ above 500,000

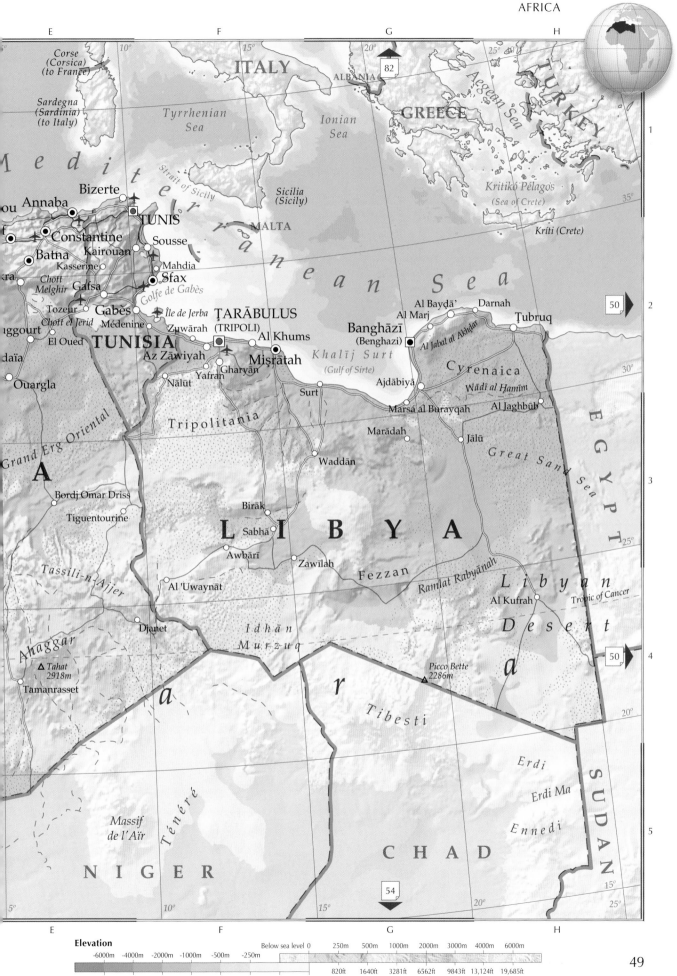

AFRICA

E · 10° · F · 15° · G · 20° · H · 25°

ITALY

ALBANIA

82

GREECE

TURKEY

Corse
(Corsica)
(to France)

Sardegna
(Sardinia)
(to Italy)

Tyrrhenian
Sea

Ionian
Sea

Aegean Sea

Kritikó Pélagos
(Sea of Crete)

1

Bizerte

Annaba

Strait of Sicily

Sicilia
(Sicily)

MALTA

Kríti (Crete)

35°

TUNIS

Constantine

Sousse

Batna

Kairouan

Mahdia

M e d i t e r r a n e a n   S e a

50

Kasserine

Chott
Melghir

Gafsa

Sfax

Golfe de Gabès

2

Tozeur

Gabès

Chott el Jerid

Médenine

Île de Jerba

ṬARĀBULUS
(TRIPOLI)

Al Khums

Banghāzī
(Benghazi)

Al Bayḍā'

Darnah

Al Marj

Ṭubruq

El Oued

Zuwārah

Al Jabal al Akhḍar

daïa

TUNISIA

Az Zāwiyah

Miṣrātah

Khalīj Surt
(Gulf of Sirte)

Cyrenaica

30°

Ouargla

Yafran

Gharyān

Wādī al Ḥamīm

Nālūt

Surt

Ajdābiyā

Al Jaghbūb

Tripolitania

Marsá al Burayqah

Grand Erg Oriental

Marādah

Jālū

Great Sand Sea

E G Y P T

A

Bordj Omar Driss

Waddān

3

Tiguentourine

Birāk

L I B Y A

25°

Sabhā

Awbārī

Zawīlah

Fezzan

Ramlat Rabyānah

Tassili-n-Ajjer

Al 'Uwaynāt

L i b y a n

Al Kufrah

Tropic of Cancer

Ahaggar

D e s e r t

Djanet

Idhān

Picco Bette
2286m

Murzuq

4

▲Tahat
2918m

Tamanrasset

a

r

50

20°

Tibesti

a

Erdi

Erdi Ma

S U D A N

Ténéré

Massif
de l'Air

Ennedi

5

N I G E R

C H A D

54

5° · 10° · 15° · 20° · 25°

E · F · G · H

Elevation

Below sea level 0   250m 500m 1000m 2000m 3000m 4000m 6000m

-6000m -4000m -2000m -1000m -500m -250m

820ft 1640ft 3281ft 6562ft 9843ft 13,124ft 19,685ft

-19,658ft -13,124ft -6562ft -3281ft -1640ft -820ft -328ft/-100m 0

49

Northeast Africa

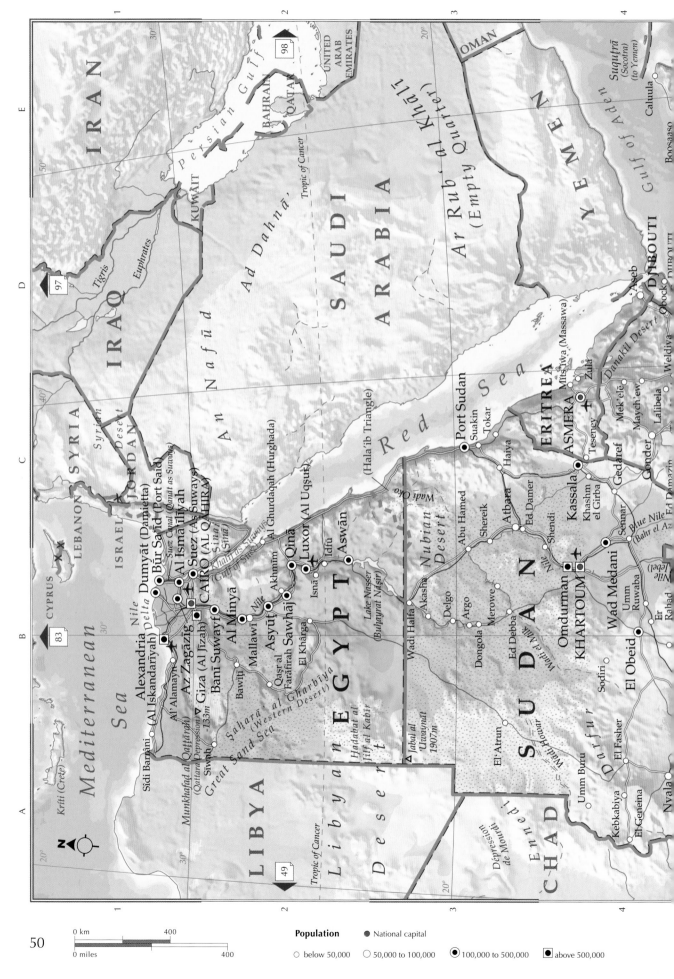

IRAN

IRAQ

SYRIA

LEBANON

ISRAEL

JORDAN

CYPRUS

KUWAIT

BAHRAIN

QATAR

UNITED ARAB EMIRATES

OMAN

SAUDI ARABIA

YEMEN

Persian Gulf

Tigris

Euphrates

Syrian Desert

An Nafūd

Ad Dahnā'

Ar Rub' al Khālī (Empty Quarter)

Rub' al Khālī

Gulf of Aden

Suquṭrā (Socotra) (to Yemen)

Caluula

Boosaaso

DJIBOUTI

Aseb

Obock

Weldiya

Tropic of Cancer

Red Sea

Port Sudan

Suakin

Tokar

Haiya

Mits'iwa (Massawa)

Zula

Danakil Desert

ERITREA

ASMERA

Teseney

Mek'elē

Maych'ew

Lalibela

Gedaref

Gonder

Kassala

Khashm el Girba

Ed Damer

Atbara

Shereik

Abu Hamed

Akasha

Delgo

Argo

Merowe

Dongola

Ed Debba

Wadi Halfa

Nubian Desert

Wadi Oko

Shendi

Omdurman

KHARTOUM

Wad Medani

Sennar

Umm Ruwaba

Blue Nile (Bahr el Jebel)

Nile (Bahr el Azra)

El Obeid

Er Rahad

Sodiri

SUDAN

Darfur

Ennedi

El Fasher

Umm Buru

Kebkabiya

El Geneina

Nyala

Nile

Lake Nasser (Buḥayrat Nāṣir)

Jabal al 'Uwaynāt 1907m

El 'Atrun

Wadi el Howar

CHAD

Dépression de Mourdi

Mediterranean Sea

Kriti (Crete)

Sidi Barrani

Al'Alamayn

Alexandria (Al Iskandariyah)

Nile Delta

Dumyāt (Damietta)

Būr Saʿīd (Port Said)

Al Ismāʿīlīyah

Suez (As Suways)

Suez Canal (Qanāt as Suways)

CAIRO (AL QĀHIRA)

Az Zagāzig

Giza (Al Jīzah)

Banī Suwayf

Al Minyā

Mallawī

Asyūt

Sawhāj

Akhmīm

Qinā

Luxor (Al Uqṣur)

Idfū

Isnā

Aswān

Qaṣr al Farāfirah

El Khārga

Bawīṭī

Siwah

Munkhafaḍ al Qaṭṭārah (Qattara Depression) -133m

Ṣaḥarā' al Gharbīya (Western Desert)

Great Sand Sea

Hadabat al Jilf al Kabīr

EGYPT

Libyan Desert

Sīnā (Sinai)

Gulf of Suez

Khalīj as Suways

Al Ghurdaqah (Hurghada)

(Hala'ib Triangle)

LIBYA

Tropic of Cancer

Gulf of Suez

N

83

49

97

98

50

Population

- ○ below 50,000
- ○ 50,000 to 100,000
- ◉ 100,000 to 500,000
- ■ above 500,000

● National capital

0 km   400

0 miles   400

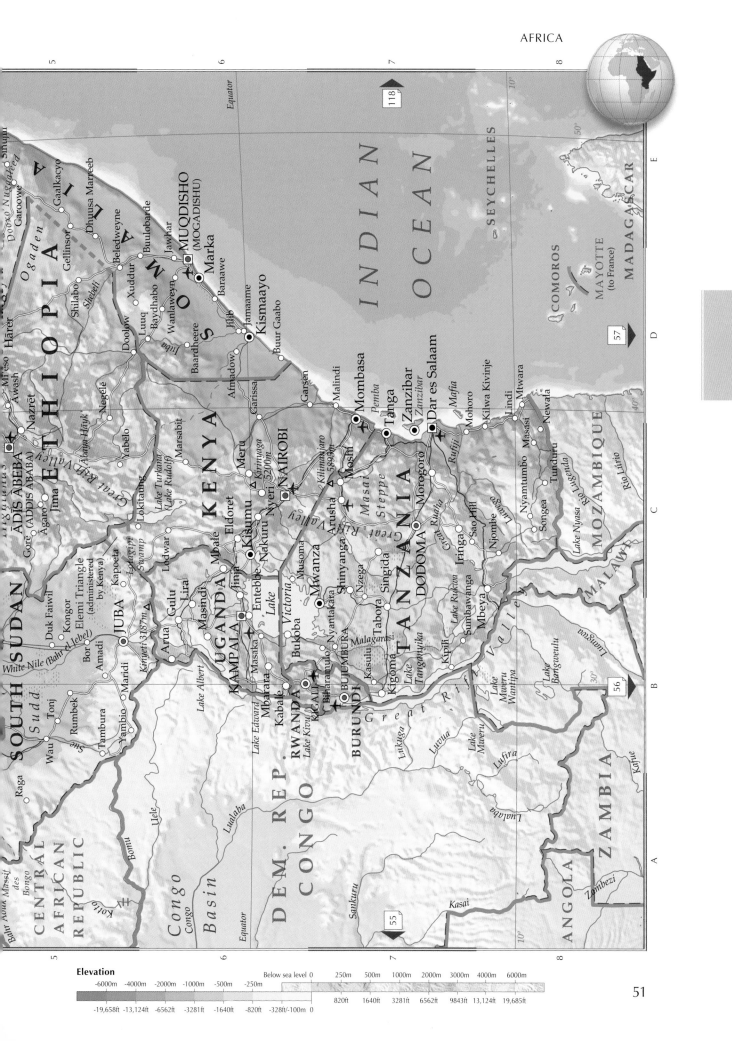

Equator

INDIAN

OCEAN

SEYCHELLES

COMOROS

MAYOTTE
(to France)

MADAGASCAR

E

118

57

D

SOMALIA

Dooxo Nugaaleed

Sinujif
Garoowe
Gaalkacyo
Dhuusa Marreeb

Shinbir
Ogaden

ETHIOPIA

Harer
Mi'ēso
Awash
Nazrēt
ADIS ABEBA
(ADDIS ABABA)
Gorē
Agaro
Jima

Great Rift Valley

Abaya Hāyk'

Negēlē

Yabēlo

Beledweyne
Buulobarde
Jawhar
MUQDISHO
(MOGADISHU)
Marka
Baraawe

Gellinsor
Xuddur
Baydhabo
Wanlaweyn
Jilib
Jamaame
Kismaayo
Buur Gaabo

Shilabo
Shebeli

Doolow
Luuq
Juba
Baardheere

Afmadow

Shabeelle

C

Garissa
Garsen
Malindi
Mombasa
Pemba
Tanga
Zanzibar
Zanzibar
Dar es Salaam
Mafia
Mohoro
Kilwa Kivinje
Lindi
Mtwara
Newala
Masasi

MOZAMBIQUE

Rio Lúrio

MALAWI

Lake Nyasa/Lago Niassa

KENYA

Marsabit
Meru
Kirinyaga
5200m
NAIROBI
Kilimanjaro
5895m
Moshi
Arusha

Masai
Steppe

Great Ruaha
Morogoro
Rufiji

Sao Hill
Iringa

Njombe

Nyamtumbo
Masasi
Songea
Tunduru

Ruvuma

SOUTH
SUDAN

Raga

CENTRAL
AFRICAN
REPUBLIC

Bahr Aouk Massif
des
Bongo

Kotto

Kongor
Duk Faiwil
Bor
Amadi
JUBA
Maridi

White Nile (Bahr el Jebel)

Sudd

Wau
Tonj
Rumbek
Tambura
Yambio

Sue

Uele

Lualaba

Congo
Basin

Congo

Equator

DEM. REP.

CONGO

Sankuru

Kasai

Lualaba

Lokitaung
Lake Turkana/
Lake Rudolf

Lodwar
Liral

Sagiri
Saamp

Kapoeta
Elemi Triangle
(administered
by Kenya)

Gulu
Arua
Masindi

UGANDA

Jinja
Entebbe
KAMPALA

Masaka
Mbarara
Kabale
Bukoba
Biharamulo

RWANDA
KIGALI
Lake Kivu
BUJUMBURA
BURUNDI

Lake Albert

Lake Edward

Kenyatti 3187m

Eldoret
Kisumu
Nakuru
Nyeri

Musoma
Mwanza
Nyantakara

Shinyanga
Nzega
Singida

Tabora

DODOMA

TANZANIA

Mbeya

Lake Victoria

Malagarasi

Kasulu
Kigoma

Lake
Tanganyika

Kipili
Sumbawanga

Lake Rukwa

Lake Mweru
Wantipa

Lake Mweru

Great Rift Valley

Lukuga
Luvua
Lufira

Lake
Bangweulu

Luangwa

ANGOLA

ZAMBIA

Zambezi

Kafue

B

A

56

55

Elevation

-6000m	-4000m	-2000m	-1000m	-500m	-250m

Below sea level 0   250m   500m   1000m   2000m   3000m   4000m   6000m

-19,658ft -13,124ft -6562ft -3281ft -1640ft -820ft -328ft/-100m 0

820ft   1640ft   3281ft   6562ft   9843ft 13,124ft 19,685ft

West Africa

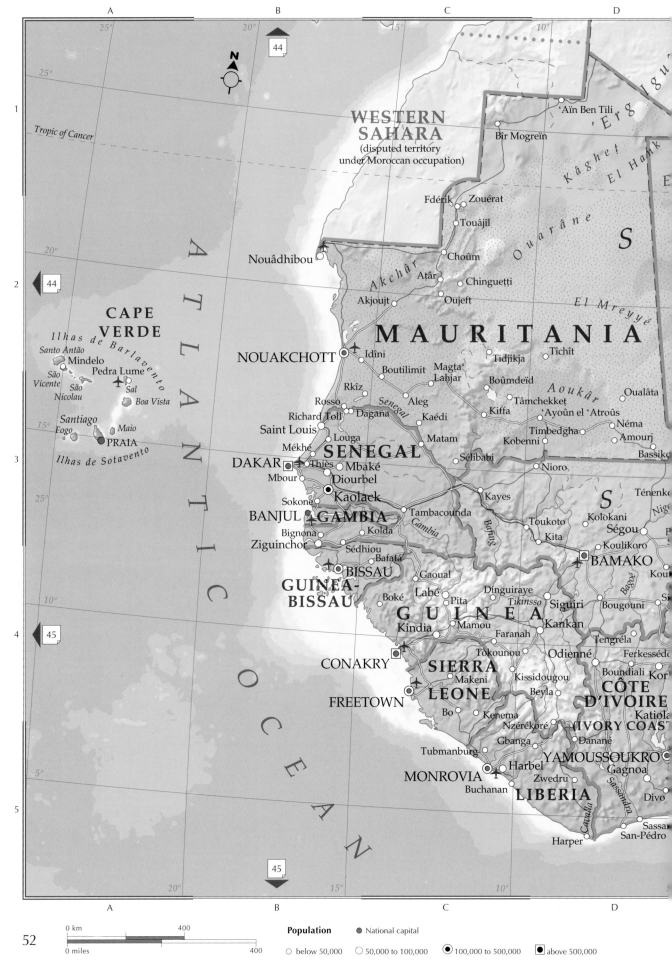

A
B
C
D

1

N

Tropic of Cancer

25°

25°

WESTERN SAHARA
(disputed territory
under Moroccan occupation)

'Aïn Ben Tili

Bîr Mogreïn

'Erg Igu

Kâghet

El Hamk

Fdérik • Zouérat

Touâjîl

Ouarâne

S

20°

Choûm

Nouâdhibou

Akchâr

Atâr

Chinguetti

El Mreyyé

Akjoujt

Oujeft

2

MAURITANIA

**CAPE
VERDE**

Ilhas de Barlavento

Santo Antão

Mindelo

NOUAKCHOTT

Idîni

Boutilimit

Tidjikja

Tîchît

São
Vicente

São
Nicolau

Pedra Lume

Sal

Boa Vista

Rkîz

Magta
Lahjar

Boûmdeïd

Aoukâr

Oualâta

Rosso

Aleg

Tâmchekket

'Ayoûn el 'Atroûs

Néma

Santiago

15°

Richard Toll

Senegal

Kaédi

Kiffa

Timbedgha

Amourj

Fogo

Maio

Dagana

Saint Louis

Matam

Kobenni

Bassik

PRAIA

Louga

Nioro

Ilhas de Sotavento

Mékhé

SENEGAL

Sélibabi

3

DAKAR

Thiès • Mbaké

Kayes

S

Ténenk

25°

Mbour

Diourbel

Toukoto

Kolokani

Niger

Sokone

Kaolack

Kita

Séguou

BANJUL

GAMBIA

Tambacounda

Koulikoro

Bignona

Kolda

Gambia

Bafing

BAMAKO

Kou

Ziguinchor

Sédhiou

Bafatá

Gaoual

Dinguiraye

Bagoé

10°

BISSAU

Boké

Labé

Tikinsso

Siguiri

Bougouni

**GUINEA-
BISSAU**

Pita

Mamou

Kankan

G U I N E A

Kindia

Faranah

Tengréla

4

Tokounou

Odienné

Ferkessédo

CONAKRY

**SIERRA
LEONE**

Makeni

Kissidougou

Boundiali

Kor

**CÔTE
D'IVOIRE**

FREETOWN

Beyla

Bo

Kenema

Nzérékoré

Katiola

(IVORY COAST

Gbanga

Đanané

Gagnoa

5°

Tubmanburg

YAMOUSSOUKRO

MONROVIA

Harbel

Zwedru

Divo

Buchanan

LIBERIA

Sassa

Harper

San-Pédro

5

20°

15°

10°

A
B
C
D

A T L A N T I C   O C E A N

0 km   400

0 miles   400

Population   ● National capital

○ below 50,000   ○ 50,000 to 100,000   ◉ 100,000 to 500,000   ◼ above 500,000

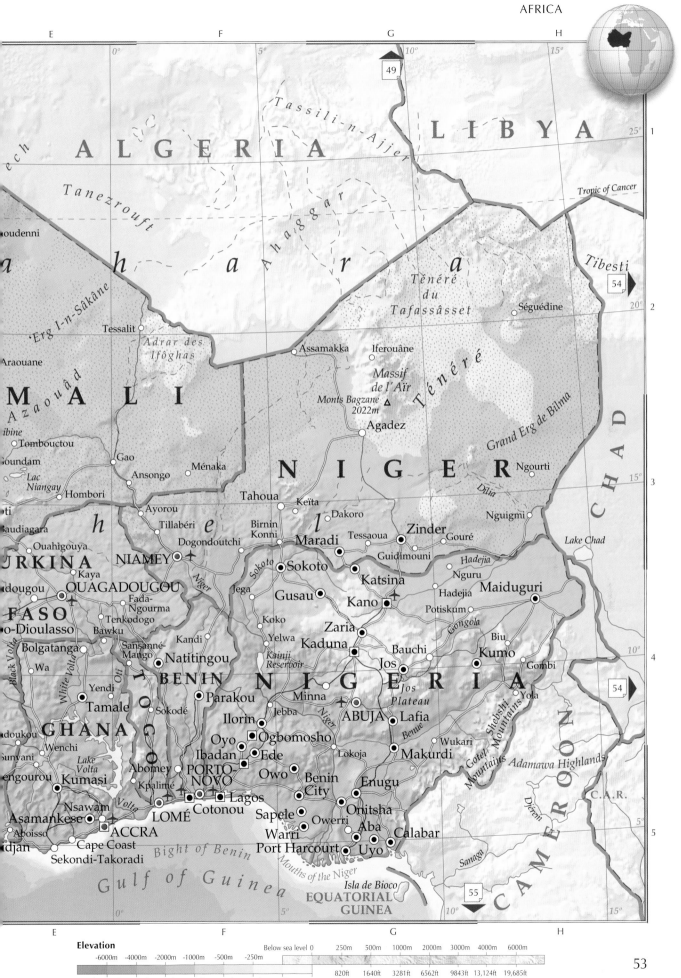

ALGERIA LIBYA

Tassili-n-Ajjer

Tanezrouft *Ahaggar*

Tropic of Cancer

oudenni

Erg In-Sâkane Tessalit *Adrar des Ifôghas* Assamakka Iferouâne *Tibesti* 54

Araouane *Ténéré du Tafassâsset* Séguédine

a h a r a

Massif de l'Aïr *Ténéré*

ibine *Monts Bagzane* △ 2022m

Tombouctou Agadez *Grand Erg de Bilma*

Goundam *Azaouâd*

Lac Niangay MALI Gao Ménaka N I G E R Ngourti

ti Ansongo *Dïla* CHAD

audiagara Hombori Tahoua Keïta Ngourti

ti Ayorou Dakoro *Lake Chad*

h Tillabéri *e* Birnin Konni Tessaoua Zinder Gouré

RKINA Dogondoutchi *l* Maradi Guidimouni *Hadejia* Nguigmi

dougou Kaya NIAMEY Sokoto Katsina *Hadejia* Nguru

OUAGADOUGOU Fada-Ngourma *Niger* Jega Gusau Kano Hadejia Maiduguri

FASO Tenkodogo Koko Zaria Potiskum *Gongola* Biu

o-Dioulasso Bawku Yelwa Kaduna Bauchi Kumo Gombi

Bolgatanga Sansanné-Mango Kandi *Kainji Reservoir* Jos Yola 54

Wa Natitingou *Jos Plateau* *Shebshi Mountains*

Yendi BENIN N I G E R I A *Adamawa Highlands*

Tamale Parakou Minna Lafia Wukari *Gotel Mountains* C.A.R.

doukou Sokodé Jebba ABUJA Benue Makurdi *Djérem*

GHANA Ilorin Oyo Ogbomosho Lokoja CAMEROON

unyani Wenchi Ibadan Ede Owo

engourou Kumasi Abomey PORTO-NOVO Benin City Enugu

Nsawam Kpalimé Owo Onitsha

Asamankese LOMÉ Cotonou Lagos Sapele Owerri Aba Calabar

Aboisso ACCRA Warri Port Harcourt Uyo

idjan Cape Coast *Bight of Benin* *Mouths of the Niger* *Sanaga*

Sekondi-Takoradi

Gulf of Guinea *Isla de Bioco* 55

EQUATORIAL GUINEA

Elevation

-6000m	-4000m	-2000m	-1000m	-500m	-250m	Below sea level 0	250m	500m	1000m	2000m	3000m	4000m	6000m

-19,658ft -13,124ft -6562ft -3281ft -1640ft -820ft -328ft/-100m 0   820ft 1640ft 3281ft 6562ft 9843ft 13,124ft 19,685ft

Central Africa

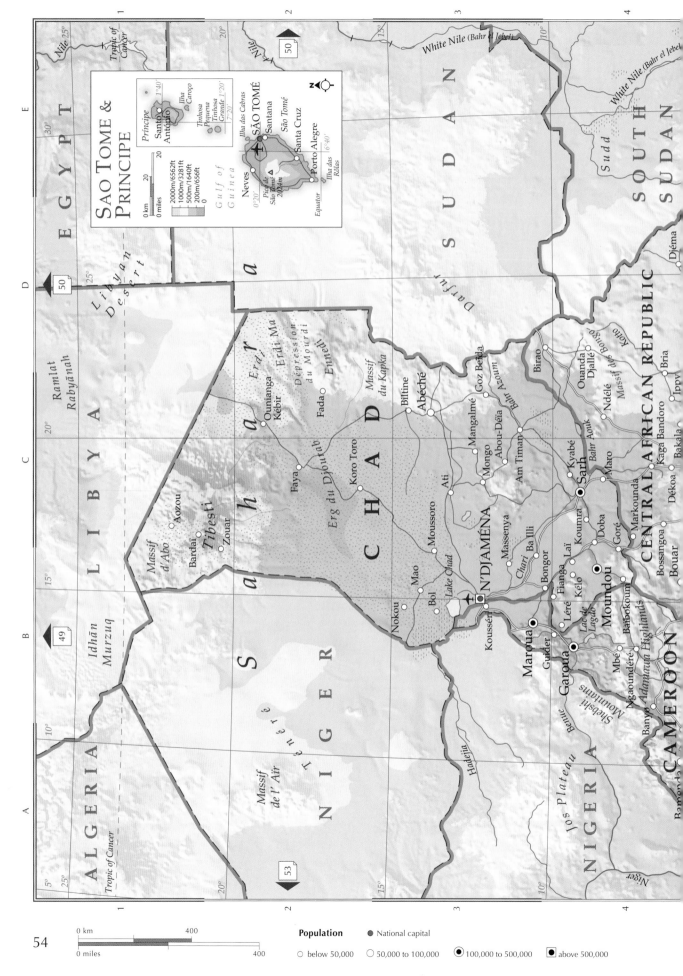

Population   ● National capital

○ below 50,000   ○ 50,000 to 100,000   ◎ 100,000 to 500,000   ■ above 500,000

0 km   400

0 miles   400

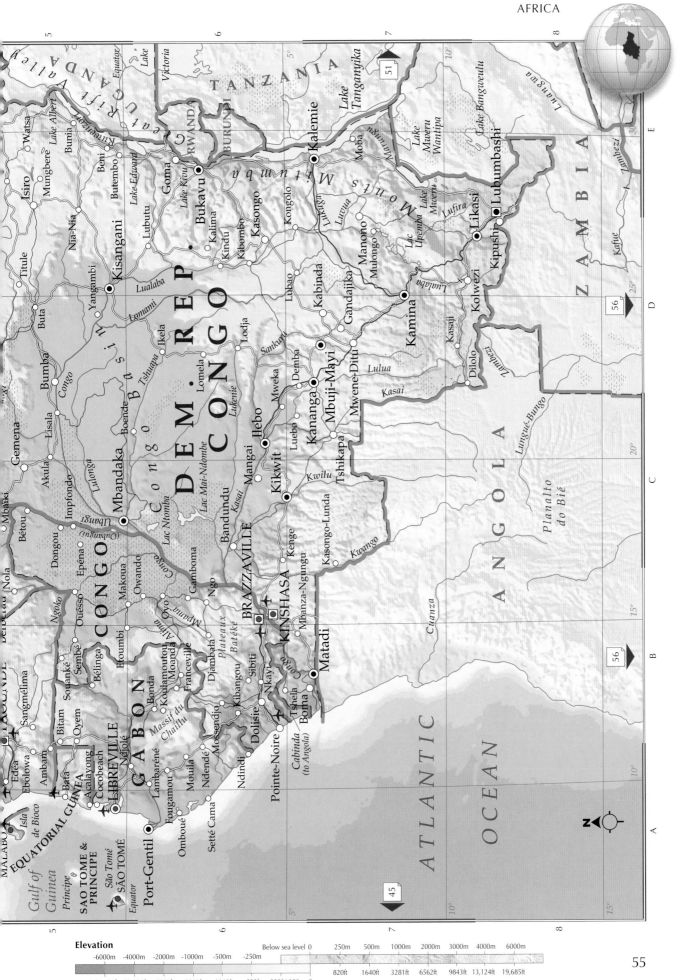

Elevation

| Below sea level 0 | 250m | 500m | 1000m | 2000m | 3000m | 4000m | 6000m |

| -6000m | -4000m | -2000m | -1000m | -500m | -250m |

| -19,658ft | -13,124ft | -6562ft | -3281ft | -1640ft | -820ft | -328ft/-100m 0 |

| 820ft | 1640ft | 3281ft | 6562ft | 9843ft | 13,124ft | 19,685ft |

Southern Africa

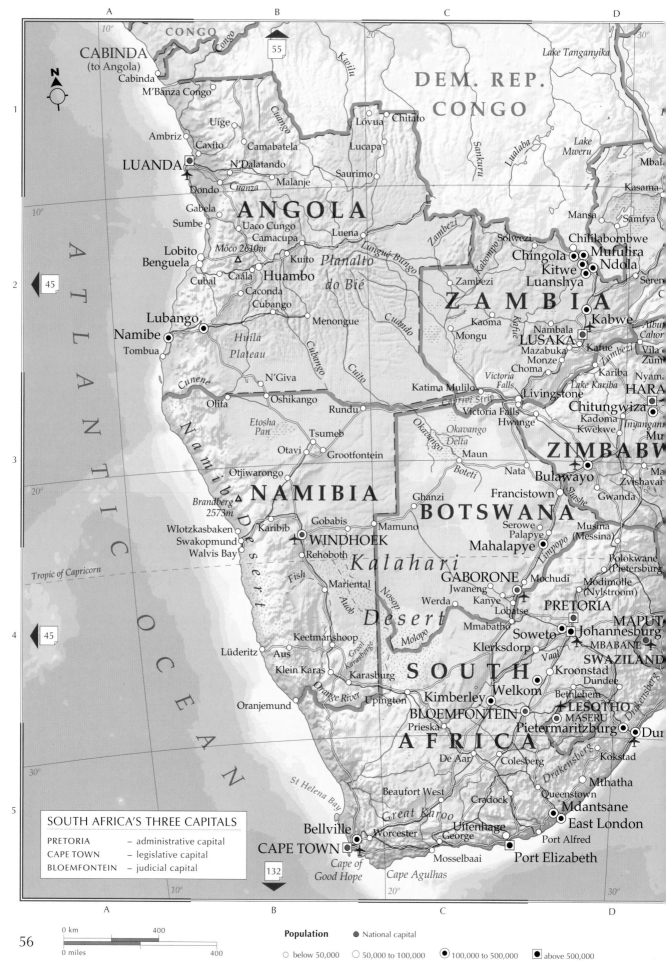

CABINDA
(to Angola)
Cabinda
M'Banza Congo

CONGO

Congo

Kwilu

DEM. REP.
CONGO

Lake Tanganyika

Uíge
Ambriz
Caxito
Camabatela
Lovua
Chitato

Lucapa

Lake
Mweru

Mbal

LUANDA
N'Dalatando
Dondo
Cuanza
Malanje
Saurimo
Kasama

Gabela
Sumbe
Uaco Cungo
Camacupa
Luena
Planalto
do Bié

Zambezi

Solwezi
Mansa
Samfya

Chililabombwe
Mufulira
Lobito
Benguela
Môco 2610m
Kuito
Luena
Lunguê-Bungo
Chingola
Kitwe
Ndola
Seren

Cubal
Caála
Huambo
Zambezi
Luanshya

Caconda
Cubango

Kabompo
Kafue
ZAMBIA
Kabwe

Menongue

Kaoma
Nambala
Albu

Lubango
Namibe
Cubango
Cuito
Mongu
LUSAKA
Cahor

Tombua
Huíla
Plateau
Mazabuka
Monze
Choma
Zambezi
Vila
Zum

N'Giva
Cuando
Kariba
Nyama

Cunene
Katima Mulilo
Victoria
Falls
Lake Kariba
HARA

Olifa
Oshikango
Rundu
Caprivi Strip
Livingstone
Victoria Falls
Chitungwiza

Etosha
Pan
Tsumeb
Okavango
Delta
Hwange
Kadoma
Inyangani
Mu

Otavi
Grootfontein
Okavango
Maun
Kwekwe

ZIMBABW

Otjiwarongo
Boteti
Nata
Bulawayo

NAMIBIA
Ghanzi
Francistown
Zvishavar
Ma

Brandberg
2573m
Gobabis
Mamuno
BOTSWANA
Gwanda

Wlotzkasbaken
Swakopmund
Walvis Bay
Karibib
Serowe
Palapye
Musina
(Messina)

WINDHOEK
Rehoboth
Kalahari
Mahalapye
Limpopo
Polokwane
(Pietersburg)

Fish
Mariental
Nosop
Desert
GABORONE
Mochudi
Modimolle
(Nylstroom)

Tropic of Capricorn
Jwaneng
Kanye
Lobatse
PRETORIA
MAPUT

Keetmanshoop
Auob
Werda
Mmabatho
Soweto
Johannesburg
MBABANE

Lüderitz
Aus
Groot
Karasberge
Molopo
Klerksdorp
Vaal
SWAZILAND

Klein Karas
Karasburg
SOUTH
Kroonstad
Dundee

Oranjemund
Orange River
Upington
Kimberley
Welkom
Bethlehem
Drakensberg

BLOEMFONTEIN
Kroonstad
LESOTHO
Dun

Prieska
Pietermaritzburg
MASERU

De Aar
Colesberg
AFRICA
Kokstad

Beaufort West
Cradock
Mthatha

St Helena Bay
Great Karoo
Queenstown
Mdantsane

Bellville
Worcester
Uitenhage
George
East London

CAPE TOWN
Mosselbaai
Port Alfred

Cape of
Good Hope
Cape Agulhas
Port Elizabeth

SOUTH AFRICA'S THREE CAPITALS

PRETORIA — administrative capital
CAPE TOWN — legislative capital
BLOEMFONTEIN — judicial capital

0 km   400

0 miles   400

Population   ● National capital

○ below 50,000   ○ 50,000 to 100,000   ◉ 100,000 to 500,000   ▣ above 500,000

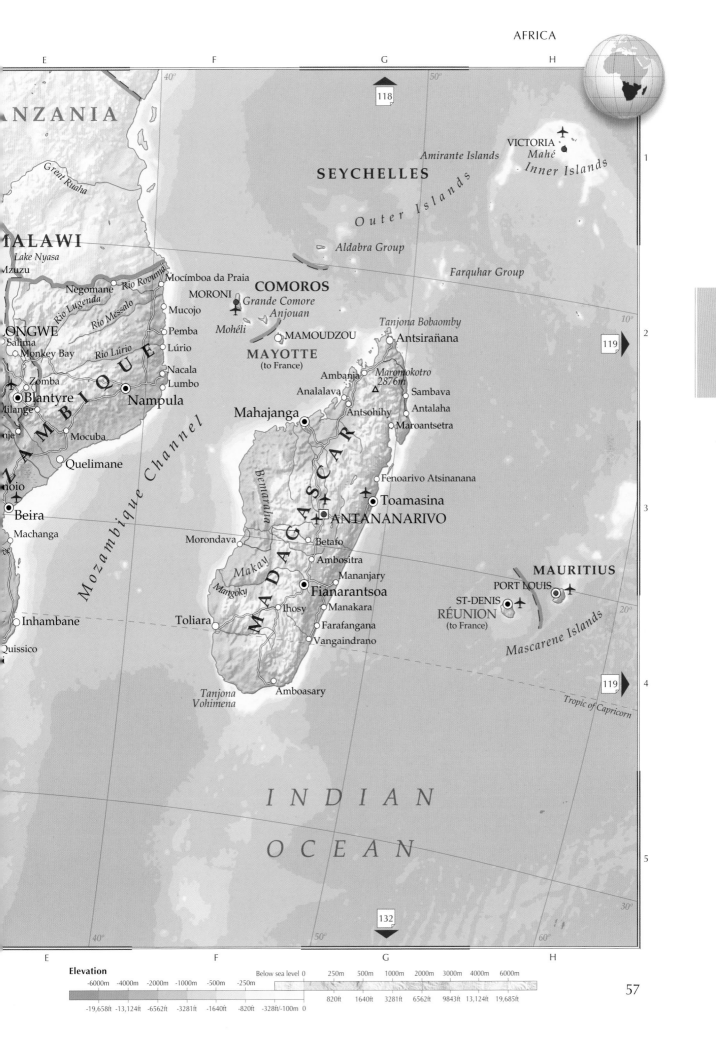

E   F   G   H

40°   50°

118

ANZANIA

Great Ruaha

Amirante Islands

VICTORIA
Mahé
Inner Islands

1

SEYCHELLES

O u t e r   I s l a n d s

10°

MALAWI

Lake Nyasa

Mzuzu

Negomane   *Rio Rovuma*

Mocímboa da Praia

Aldabra Group

Farquhar Group

ONGWE

Rio Lugenda

Mucojo

COMOROS

MORONI

Grande Comore
Anjouan

Tanjona Bobaomby

Salima

Rio Messalo

Pemba

Mohéli

MAMOUDZOU

Antsirañana

2

119

Monkey Bay

Rio Lúrio

Lúrio

MAYOTTE
(to France)

Ambanja *Maromokotro*
2876m

Zomba

Nacala

Analalava

Sambava

Milange

Lumbo

Antsohihy

Antalaha

Blantyre

Nampula

Mahajanga

Maroantsetra

nje

Mocuba

M o z a m b i q u e   C h a n n e l

B e m a r a h a

Quelimane

Fenoarivo Atsinanana

Beira

Toamasina

3

Machanga

ANTANANARIVO

Morondava

Betafo

Ambositra

MAURITIUS

Makay

Mananjary

PORT LOUIS

e

Mangoky

Fianarantsoa

ST-DENIS

RÉUNION
(to France)

Inhambane

Ihosy

Manakara

20°

Quissico

Toliara

Farafangana

M A D A G A S C A R

Vangaindrano

Mascarene Islands

119

Tanjona
Vohimena

Amboasary

Tropic of Capricorn

4

I N D I A N

O C E A N

5

30°

132

40°   50°   60°

E   F   G   H

Elevation

Below sea level 0   250m   500m   1000m   2000m   3000m   4000m   6000m

-6000m   -4000m   -2000m   -1000m   -500m   -250m

820ft   1640ft   3281ft   6562ft   9843ft   13,124ft   19,685ft

-19,658ft   -13,124ft   -6562ft   -3281ft   -1640ft   -820ft   -328ft/-100m   0

Europe

Political features

Total area:
4,809,200 sq miles
(12,456,000 sq km)

Total number of countries:
44

Total population:
721 million

Largest city with population:
Moscow, European Russia 16.7 million

Country with highest population density:
Monaco 48,181 people per sq mile
(18,531 people per sq km)

Largest country:
European Russia 1,527,341 sq miles
(3,955,818 sq km)

Smallest country:
Vatican City, Italy 0.17 sq miles
(0.44 sq km)

Physical features

Largest lake:
Lake Lagoda, European Russia
7,100 sq miles (18,390 sq km)

Longest river:
Volga, European Russia
2,290 miles (3,688 km)

Highest point:
El'brus, Caucasus, European Russia
18,510ft (5,642 m)

Lowest point:
Volga Delta, Caspian Sea, European
Russia 92 ft (28m) below sea level

0 km 500
0 miles 500

Population   ● National capital
○ below 50,000   ◎ 50,000 to 100,000   ◉ 100,000 to 500,000   ◼ above 500,000

Barents Sea

North Cape

Ostrov Kolguyev

Arctic Circle

● Murmansk

Kola
Peninsula

White
Sea

Ural Mountains

Ob'

Irtysh

133

70°

80°

R U S S I A N

FINLAND

● Archangel

Northern Dvina

Perm' ■

90

● Tampere

● Turku
■ HELSINKI

Lake Onega

F E D E R A T I O N

70°

50°

KHOLM
■ TALLINN

Lake Ladoga

■ Saint Petersburg

● Vologda

Ufa ■

ESTONIA

● Yaroslavl'

Kazan' ■

LATVIA

■ RIGA

● Nizhniy
Novgorod

Ul'yanovsk ■

Orenburg ■

European Plain

MOSCOW
●

● Samara

Ural

ITHUANIA
rad ● Kaunas

● Vitsyebsk

Volga Uplands

Aral Sea

Syr Darya

VILNIUS

Central
Russian
Upland

Volga

KALININGRAD
(to Russ.Fed).
cz

■ MINSK

Amu Darya

● Babruysk

● Homyel'

● Voronezh

WARSAW
BELARUS

60°

40°

● Brest

Pripet
Marshes

Ural

Bug

Dnieper Lowlands

Don

LAND

L'viv ●

■ KIEV

Kharkiv ●

Volgograd ■

Dnieper

UKRAINE

Dnipropetrovs'k ■

Astrakhan' ●

Volga Delta
-28m

Caspian Sea

AKIA

Dniester

Donets'k ■

Chernivtsi ●

● Rostov-na-Donu

GARY

MOLDOVA

Sea of
Azov

● Stavropol'

90

Cluj-Napoca ●

■ CHIŞINĂU

● Odesa

Crimea

Caucasus

ROMANIA

Simferopol' ●

El'brus 5642m △

LGRADE

Braşov ●

(the Ukrainian territory
of Crimea was annexed
by Russia in 2014)

I A

■ BUCHAREST

Danube

Black
Sea

BIA

BULGARIA
● Varna

OVO
puted)

Balkan Mountains

● Burgas

S

RIGA ● SOFIA

SKOPJE

ACED

TURKEY

NIA

Aegean
Sea

Anatolia

Zagros Mountains

GREECE
● ATHENS

50°

● Piraeus

30°

oponnese

Irákleio ●

Cyprus

96

Tigris

Euphrates

a

Crete

30°

40°

The North Atlantic

Devon Island
Ellesmere Island
Gulf of Boothia
Nares Strait
Qaanaaq
Knud Rasmusse
Innaanganeq
Savissivik
NUNAVUT
Hudson Bay
Southampton Island
Foxe Basin
Qimusseriarsuaq
Baffin Bay
Kullorsuaq
CANADA
Upernavik
Baffin Island
Limit of summer pack ice
Péninsule d'Ungava
Uummannaq
Qeqertarsuaq
Qeqertarsuaq
QUÉBEC
Hudson Strait
Cumberland Sound
Davis Strait
Qeqertarsuup Tunua
Qasigianguit
Arnaud
GREENLAND
(to Denmark)
Frobisher Bay
Sisimiut
Kong Frederik IX Land
Ungava Bay
Maniitsoq
George
NUUK
Kong Christian IX Land
Gunnbjø
Mont Forel 3360m
NEWFOUNDLAND & LABRADOR
Paamiut
Ammassalik
Ivittuut
Kong Frederik VI Kyst
Denm
Labrador Sea
Qaqortoq
Nanortalik
Reykjanes Basin
Nunap Isua (Kap Farvel)
Limit of winter pack ice
ATLANTIC
OCEAN

Arctic Circle

16
16
17
44

0 km   400
0 miles   400

Population   ● National capital

○ below 50,000   ○ 50,000 to 100,000   ◉ 100,000 to 500,000   ■ above 500,000

ARCTIC OCEAN

Kap Morris Jesup

Wandel Sea

ncoln Sea

Independence Fjord

Nord

SVALBARD
(to Norway)

Kvitøya

Zemlya Frantsa-Iosifa

Nordaustlandet

Kong Karls Land

Spitsbergen

Barentsøya

Edgeøya

LONGYEARBYEN
Barentsburg

Storfjorden

Barents Sea

Novaya Zemlya

88

Kong Frederik VIII Land

Greenland Sea

Limit of winter pack ice

Christian X Land

Limit of summer pack ice

Daneborg

Petermann Bjerg
2940m

Bjørnøya
(to Norway)

Mohns Ridge

Nordkapp
(North Cape)

Kong Oscar Fjord

Ittoqqortoormiit

Kangertittivaq

Kangikajik

JAN MAYEN
(to Norway)

Norwegian Sea

Norwegian Basin

FINLAND

Vestfjorden

Arctic Circle

rait

62

ICELAND

ungarvík

Siglufjörður
our
Húsavík
Akureyri
Stykkishólmur
Seyðisfjörður
REYKJAVÍK
Neskaupstaður
Selfoss *Vatnajökull*
Djúpivogur
lákshöfn *Hvannadalshnúkur*
2119m
tsey
Vestmannaeyjar

Raufarhöfn

S W E D E N

Gulf of Bothnia

FAROE ISLANDS
(to Denmark)

TÓRSHAVN

N O R W A Y

63

Shetland Islands

N

Elevation

| -6000m | -4000m | -2000m | -1000m | -500m | -250m | Below sea level 0 | 250m | 500m | 1000m | 2000m | 3000m | 4000m | 6000m |

| -19,658ft | -13,124ft | -6562ft | -3281ft | -1640ft | -820ft | -328ft/-100m 0 | 820ft | 1640ft | 3281ft | 6562ft | 9843ft | 13,124ft | 19,685ft |

Scandinavia & Finland

RUSSIAN FEDERATION

Barents Sea

ARCTIC OCEAN

Nordkapp (North Cape)

Varangerhalvøya

Varangerfjorden

Porsangenfjorden

Magerøya

Sørøya

Ringvassøya

Kvaløya

Senja

Andøya

Vesterålen

Lofoten

Vestfjorden

Norwegian Sea

Kirkenes

Tana Bru

Deatnu

Lakselv

Alta

Talvik

Tromsø

Harstad

Narvik

Finnsnes

Finnmarksvidda

Válljohka

Karigasniemi

Inarijärvi

Kaamanen

Ivalo

Saariselkä

Kaaresuvanto

Torneträsk

Kiruna

Kebnekaise
2117m

Malmberget

Skalka

Jokkmokk

Fauske

Bodø

Mo i Rana

Mosjøen

Vega

Namsos

Steinkjer

Lapland

Kittilen

Sattanen

Sodankylä

Kemijärvi

Rovaniemi

Muonio

Kolari

Ounasjoki

Tornionjoki

Muonionjoki

Visttasjohka

Gällivare

Luleälven

Boden

Kalix

Haparanda

Tornio

Kemi

Luleå

Piteå

Skellefteå

Skellefteälven

Arvidsjaur

Storuman

Storman

Vilhelmina

Lycksele

Angerman

Dorotea

Borgefjell

Kvarnbergs-
vattnet

Kutsamo

Kemijärvi

Kemijoki

Oulujoki

Iijoki

Pudasjärvi

Kajaani

Oulujärvi

Oulu

Oulujoki

Hailuoto

Kempele

Raahe

Kokkola

FINLAND

Suomussalmi

Kuhmo

Sotkamo

Nordkapp

Arctic Circle

Arctic Circle

Population

● National capital

○ below 50,000   ○ 50,000 to 100,000   ◉ 100,000 to 500,000   ▣ above 500,000

0 km   200
0 miles   200

62

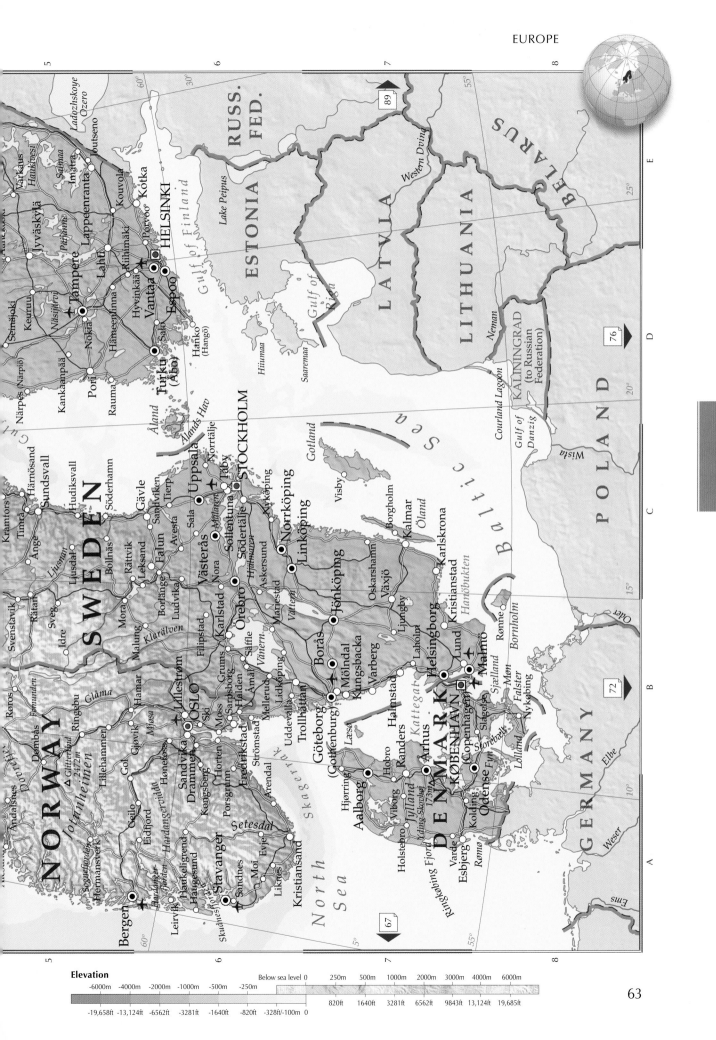

Elevation

-6000m	-4000m	-2000m	-1000m	-500m	-250m

Below sea level 0   250m   500m   1000m   2000m   3000m   4000m   6000m

-19,658ft	-13,124ft	-6562ft	-3281ft	-1640ft	-820ft	-328ft/-100m	0

820ft   1640ft   3281ft   6562ft   9843ft   13,124ft   19,685ft

The Low Countries

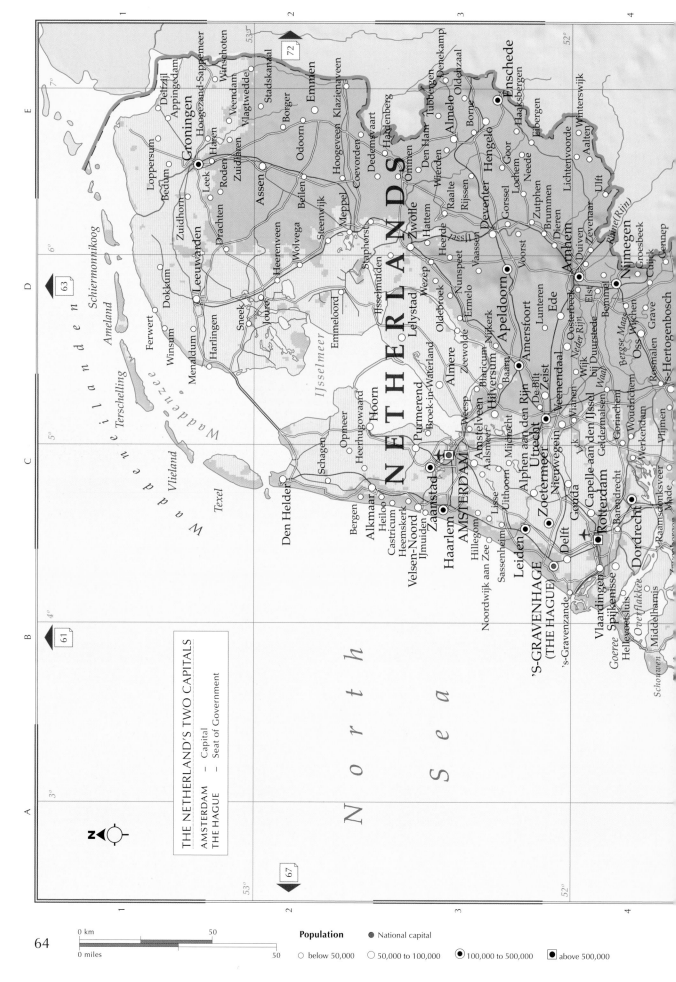

THE NETHERLAND'S TWO CAPITALS

AMSTERDAM – Capital
THE HAGUE – Seat of Government

Population   ● National capital

○ below 50,000   ○ 50,000 to 100,000   ◉ 100,000 to 500,000   ■ above 500,000

0 km   50
0 miles   50

64

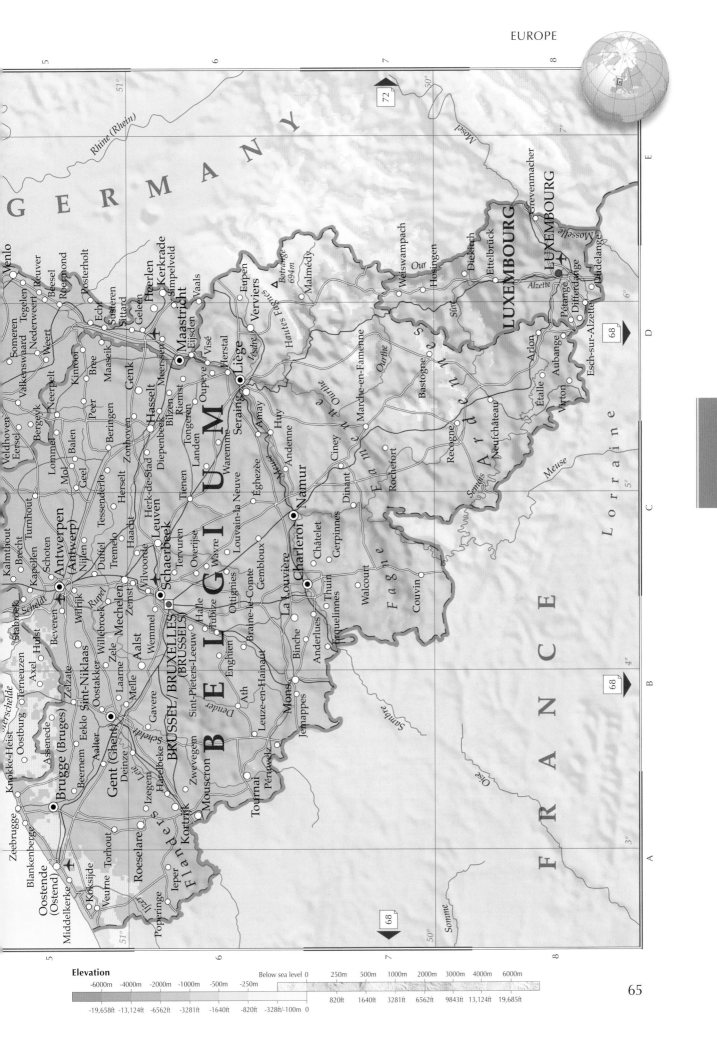

GERMANY

BELGIUM

LUXEMBOURG

FRANCE

Rhine (Rhein)

Mosel

Moselle

Grevenmacher
LUXEMBOURG
Ettelbrück
Diekirch
Alzette
Pétange
Differdange
Dudelange
Esch-sur-Alzette
Arlon
Aubange
Étalle
Virton
Neufchâteau

Weiswampach
Hosingen
Our
Sûre
Bastogne
Recogne
Marche-en-Famenne
Rochefort
Ciney
Dinant
Meuse
Andenne
Huy
Amay
Seraing
Liège
Herstal
Visé
Oupeye
Eijsden
Vaals
Verviers
Eupen
Malmédy
Botrange
694m
Hautes Fagnes
Vesdre
Ourthe

Ardenne
Semois
Stôr

Maastricht
Meerssen
Riemst
Tongeren
Bilzen
Genk
Diepenbeek
Hasselt
Zonhoven
Beringen
Herk-de-Stad
Landen
Waremme
Eghezée
Gembloux
Namur
Charleroi
Châtelet
Gerpinnes
Walcourt
Couvin
Fagne
Thuin
Binche
Anderlues
Frameries
Mons
Jemappes
La Louvière
Braine-le-Comte
Enghien
Ath
Leuze-en-Hainaut
Péruwelz
Tournai
Mouscron
Kortrijk

Menlo
Roermond
Posterholt
Echt
Susteren
Sittard
Geleen
Heerlen
Kerkrade
Simpelveld
Beesel
Reuver
Tegelen
Nederweert
Weert
Maaseik
Bree
Kinrooi
Peer
Bergeyk
Valkenswaard
Eersel
Veldhoven
Someren
Neerpelt
Overpelt
Lommel
Mol
Balen
Geel
Herentals
Turnhout
Brecht
Kapellen
Schoten
Nijlen
Duffel
Tremelo
Haacht
Leuven
Tienen
Louvain-la-Neuve
Wavre
Overijse
Tervuren
Ottignies
Vilvoorde
Zemst
Mechelen
Lier
Antwerpen
(Antwerp)
Wijnegem
Wilrijk
Rupel
Schelde
Beveren
Willebroek
Zele
Laarne
Wetteren
Aalst
Denderleeuw
Ninove
Halle
Tubize
Zottegem
Ronse
Oudenaarde
Zwevegem
Izegem
Roeselare
Ieper
Poperinge
Diksmuide
Veurne
Koksijde
Nieuwpoort
Middelkerke
Oostende
(Ostend)
Blankenberge
Knokke-Heist
Zeebrugge
Brugge (Bruges)
Oostkamp
Torhout
Tielt
Deinze
Gavere
Melle
Gent (Ghent)
Eeklo
Beernem
Aalter
Maldegem
Assenede
Zelzate
Sint-Niklaas
Temse
Hamme
Dendermonde
Lokeren
Stekene
Hulst
Axel
Terneuzen
Kalmthout
Stabroek
Westerschelde

BRUSSEL/BRUXELLES
(BRUSSELS)
Schaerbeek

Dender
Scheldt
Leie
IJzer
Sambre
Oise
Somme
Lorraine
Flanders

72
68
68
68

Elevation

-6000m	-4000m	-2000m	-1000m	-500m	-250m	Below sea level 0	250m	500m	1000m 2000m 3000m 4000m 6000m

-19,658ft -13,124ft -6562ft -3281ft -1640ft -820ft -328ft/-100m 0   820ft 1640ft 3281ft 6562ft 9843ft 13,124ft 19,685ft

The British Isles

North Sea

ATLANTIC OCEAN

Shetland Islands
Unst
Yell
Fetlar
Mainland
Lerwick

Fair Isle

Orkney Islands
Sanday
Kirkwall
Mainland
Hoy
John o'Groats

Thurso
Ben Hope 927m ▲

North West Highlands
Ullapool
Inverness
Loch Ness
Aviemore
Elgin
Spey
Moray Firth

Fraserburgh
Peterhead
Aberdeen
Dee
Montrose
Arbroath
Forfar
St Andrews
Dundee
Firth of Forth
Edinburgh
Dunfermline
Perth
Tay
Grampian Mountains
SCOTLAND
Berwick-upon-Tweed
Galashiels
Hawick

Stromeferry
Mallaig
Fort William
Ben Nevis 1343m ▲
Loch Lomond
Stirling
Forth
Glasgow
Greenock
Paisley
Hamilton
Clyde
East Kilbride
Kilmarnock
Prestwick
Isle of Ayr
Oban
Firth of Lorn

The Minch
Isle of Lewis
Stornoway
Harris
The Little Minch
Isle of Skye
Rhum
Eigg
Coll
Tiree
Isle of Mull
Jura
Islay
Kintyre
Arran
Inner Hebrides

North Uist
South Uist
Barra
Outer Hebrides

St Kilda

N

0 km — 100
0 miles — 100

Population ● National capital   ● Internal administrative capital

○ below 50,000   ○ 50,000 to 100,000   ◉ 100,000 to 500,000   ◼ above 500,000

FRANCE

Seine

E N G L I S H   C h a n n e l

English Channel

Celtic Sea

Irish Sea

St George's Channel

Cardigan Bay

Bristol Channel

UNITED KINGDOM

ENGLAND

WALES

IRELAND

Isle of Man
(British Crown Dependency)

GUERNSEY
(British Crown Dependency)

ST PETER PORT

Sark

ST HELIER

JERSEY
(British Crown Dependency)

Channel Islands

Alderney

Scarborough
Whitby
Northallerton
Harrogate
York
Leeds
Bradford
Kingston upon Hull
Beverley
Bridlington
Grimsby
Louth
Skegness
Lincoln
Doncaster
Sheffield
Huddersfield
Castleford
Manchester
Bolton
Preston
Blackpool
Lancaster
Kendal
Barrow-in-Furness
Liverpool
Birkenhead
Bangor
Holyhead
Anglesey
Chester
Crewe
Stoke-on-Trent
Stafford
Shrewsbury
Wolverhampton
Birmingham
Kidderminster
Worcester
Gloucester
Cheltenham
Cotswold Hills
Swindon
Reading
Oxford
Luton
Milton Keynes
Stevenage
Bedford
Northampton
Coventry
Leicester
Nuneaton
Derby
Nottingham
Boston
King's Lynn
The Wash
The Fens
Peterborough
Kettering
Cambridge
Newmarket
Ipswich
Colchester
Harwich
Felixstowe
Lowestoft
Great Yarmouth
Norwich
Southend-on-Sea
Canterbury
Margate
Dover
Channel Tunnel
Folkestone
Hastings
Eastbourne
Brighton
Hove
Crawley
Woking
Guildford
Winchester
Havant
Portsmouth
Isle of Wight
Newport
Bournemouth
Poole
Southampton
Eastleigh
Andover
Salisbury
Yeovil
Weymouth
Lyme Bay
Bridport
Torquay
Exeter
Exmoor
Exmouth
Dartmoor
Saltash
Plymouth
Truro
Falmouth
St Austell
Bodmin
Newquay
Penzance
Land's End
Isles of Scilly
Tiverton
Taunton
Bridgwater
Barnstaple
Ilfracombe
Bideford
Weston-super-Mare
Bristol
Bath
Newport
Cardiff
Swansea
Port Talbot
Brecon Beacons
Llanelli
Carmarthen
Milford Haven
Haverfordwest
Fishguard
Aberystwyth
Tywyn
Barmouth
Cambrian Mountains
Snowdonia
Watford
St Albans
Harlow
Windsor
LONDON
Croydon
Maidstone

Dublin
Dún Laoghaire
Lucan
Drogheda
Dundalk
Newry
Downpatrick
Armagh
Portadown
Omagh
Enniskillen
Sligo
Castlebar
Galway
Galway Bay
Ennis
Tralee
Dingle Bay
Killarney
Bantry Bay
Cork
Munster
Connaught
Leinster
Longford
Athlone
Newbridge
Port Laoise
Kilkenny
Carlow
Clonmel
Limerick
Waterford
Wexford
Barrow
Wicklow Mts
Lough Corrib
Lough Derg
Shannon
Blackwater
Lough Erne
Lower Lough Erne
Upper Lough Erne
Lough Neagh

Ouse
Mersey
Ribble
Trent
Wye
Severn
Thames
Teme
Tamar
Exe

DOUGLAS
Downpatrick

64

68

68

44

Elevation

| -6000m | -4000m | -2000m | -1000m | -500m | -250m | Below sea level 0 | 250m | 500m | 1000m | 2000m | 3000m | 4000m | 6000m |

-19,658ft -13,124ft -6562ft -3281ft -1640ft -820ft -328ft/-100m 0   820ft 1640ft 3281ft 6562ft 9843ft 13,124ft 19,685ft

France, Andorra & Monaco

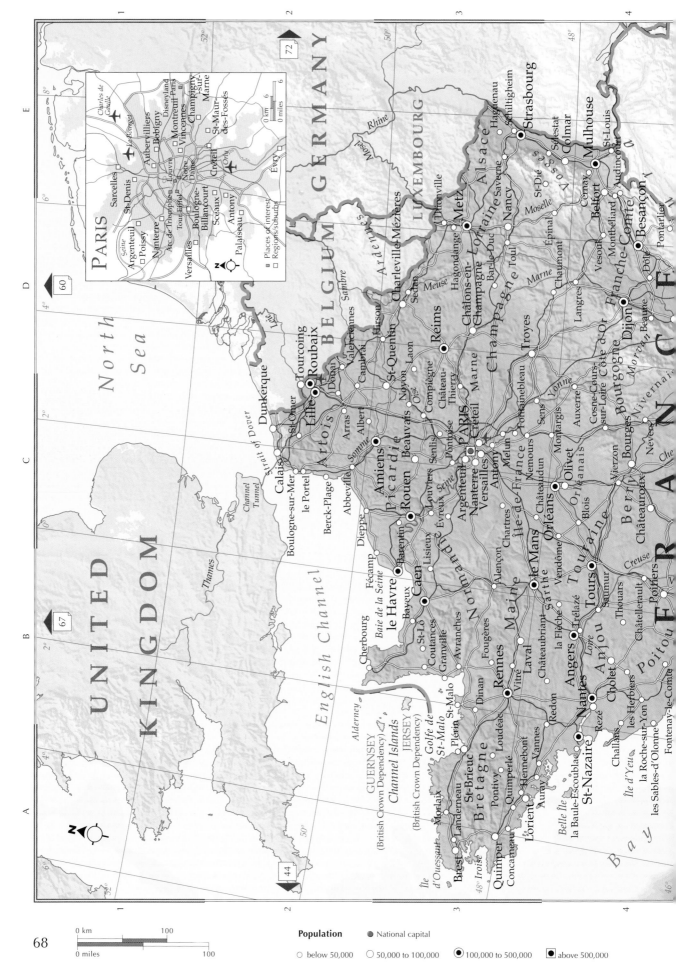

PARIS

■ Places of interest
□ Regions/suburbs

Charles de Gaulle
le Bourget
Sarcelles
St-Denis
Aubervilliers
Bobigny
Nanterre
Arc de Triomphe
Tour Eiffel
Louvre
Montreuil-Paris
Vincennes
Disneyland
Champigny-sur-Marne
St-Maur-des-Fossés
Seine
Argenteuil
Poissy
Versailles
Boulogne-Billancourt
Notre Dame
Créteil
Orly
Évry
Sceaux
Antony
Palaiseau

North Sea

UNITED KINGDOM

BELGIUM

GERMANY

LUXEMBOURG

Rhine
Mosel
Meuse
Sambre
Lys
Escaut
Ardennes
Moselle
Marne
Yonne
Seine
Che
Creuse

Dunkerque
Tourcoing
Roubaix
Calais
le Portel
Boulogne-sur-Mer
Berck-Plage
St-Omer
Lille
Denain
Valenciennes
Cambrai
St-Quentin
Noyon
Laon
Hirson
Charleville-Mézières
Sedan
Thionville
Hagondange
Metz
Haguenau
Schiltigheim
Strasbourg
Sélestat
Colmar
Mulhouse
St-Louis
Audincourt
Belfort
Cernay
Montbéliard
St-Dié
Épinal
Vesoul
Besançon
Pontarlier
Dole
Beaune
Côte-d'Or
Dijon
Langres
Chaumont
Troyes
Châlons-en-Champagne
Bar-le-Duc
Toul
Nancy
Saverne
Reims
Compiègne
Beauvais
Senlis
Pontoise
Château-Thierry
Sens
Montargis
Nemours
Fontainebleau
Auxerre
Cosne-Cours-sur-Loire
Nevers
Bourges
Vierzon
Châteauroux
Issoudun
Blois
Vendôme
Orléans
Olivet
Chartres
Dreux
Évreux
Louviers
Rouen
Barentin
Dieppe
Fécamp
le Havre
Caen
Bayeux
Cherbourg
St-Lô
Coutances
Granville
Avranches
Fougères
Lisieux
Alençon
le Mans
la Flèche
Saumur
Tours
Châtellerault
Poitiers
Thouars
Fontenay-le-Comte
Niort
les Sables-d'Olonne
la Roche-sur-Yon
Challans
les Herbiers
Cholet
Trélazé
Angers
Nantes
Rezé
St-Nazaire
la Baule-Escoublac
Redon
Châteaubriant
Laval
Vitré
Rennes
Dinan
St-Malo
St-Brieuc
Plérin
Loudéac
Pontivy
Hennebont
Lorient
Quimperlé
Concarneau
Quimper
Auray
Vannes
Landerneau
Morlaix
Brest

île d'Ouessant
île d'Yeu
Belle Île

GUERNSEY (British Crown Dependency)
Channel Islands
JERSEY (British Crown Dependency)
Alderney
Golfe de St-Malo
Baie de la Seine

Strait of Dover
English Channel
Channel Tunnel
Thames

Artois
Picardie
Somme
Albert
Arras
Amiens
Abbeville
Île-de-France
Melun
Nanterre
Versailles
Antony
Créteil
PARIS
Normandie
Maine
Anjou
Touraine
Sarthe
Loire
Bretagne
Berry
Orléanais
Poitou
Nivernais
Bourgogne
Morvan
Franche-Comté
Champagne
Lorraine
Alsace
Vosges
Sarthe

Bay of Biscay

FRANCE

Population
● National capital
○ below 50,000
○ 50,000 to 100,000
◉ 100,000 to 500,000
■ above 500,000

0 km 100
0 miles 100

68

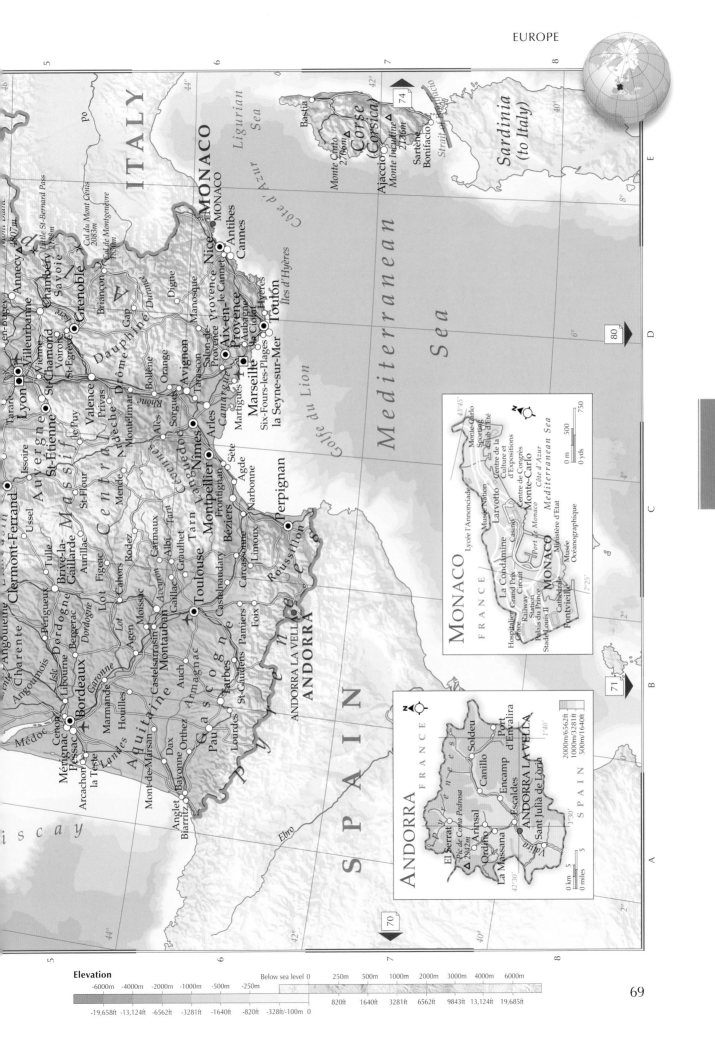

Spain & Portugal

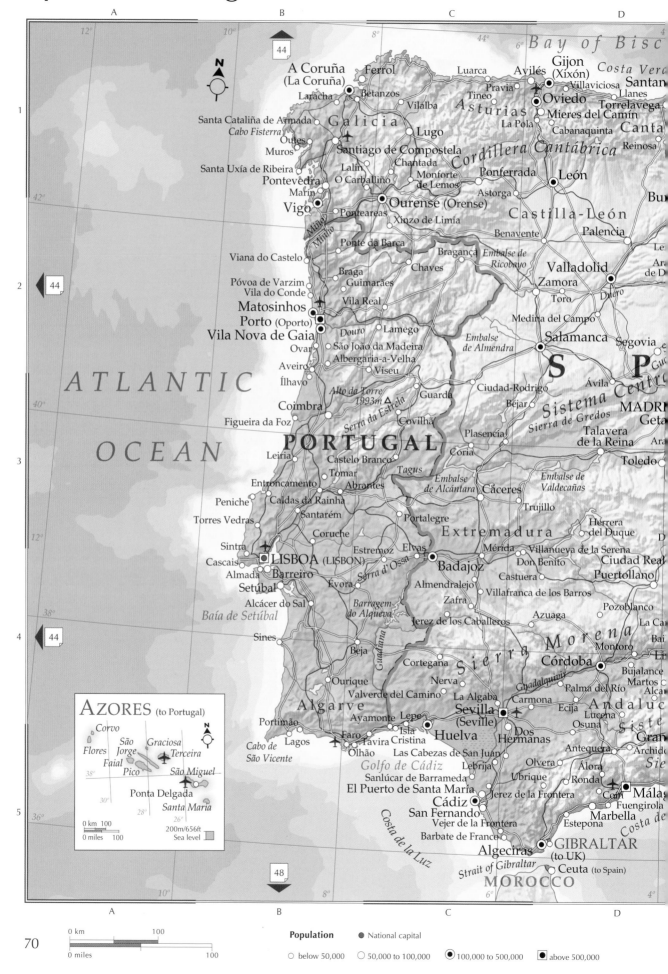

A Coruña (La Coruña)
Ferrol
Luarca
Avilés
Gijon (Xixón)
Costa Verde
Bay of Biscay
Santan
Larracha
Betanzos
Vilalba
Tineo
Pravia
Villaviciosa
Santa
Santa Cataliña de Armada
Cabo Fisterra
Lugo
Asturias
La Pola
Oviedo
Torrelavega
Llanes
Cabanaquinta
Canta
Outes
Santiago de Compostela
Chantada
Monforte de Lemos
Cordillera Cantábrica
Ponferrada
León
Reinosa
Galicia
Muros
Lalín
O Carballiño
Astorga
Bu
Santa Uxía de Ribeira
Pontevedra
Marín
Ourense (Orense)
Benavente
Castilla-León
Palencia
Le
Vigo
Ponteareas
Xinzo de Limia
Zamora
Toro
Duero
Ar
de D
Viana do Castelo
Ponte da Barca
Bragança
Embalse de Ricobayo
Valladolid
Medina del Campo
Braga
Chaves
Guimarães
Vila Real
Salamanca
Segovia
Póvoa de Varzim
Vila do Conde
Matosinhos
Porto (Oporto)
Vila Nova de Gaia
Ovar
São João da Madeira
Albergaria-a-Velha
Lamego
Douro
Embalse de Almendra
S P
Aveiro
Viseu
Ciudad-Rodrigo
Ávila
MADR
Geta
Ílhavo
Alto da Torre 1993m
Guarda
Béjar
Sistema Central
Coimbra
Serra da Estrela
Covilhã
Sierra de Gredos
Talavera de la Reina
Ara
Figueira da Foz
Plasencia
PORTUGAL
Coria
Toledo
Leiria
Castelo Branco
Tagus
Embalse de Alcántara
Cáceres
Embalse de Valdecañas
Entroncamento
Tomar
Abrantes
Trujillo
Peniche
Caldas da Rainha
Santarém
Portalegre
Herrera del Duque
Torres Vedras
Coruche
Extremadura
Mérida
Villanueva de la Serena
Sintra
Estremoz
Elvas
Badajoz
Ciudad Real
Cascais
LISBOA (LISBON)
Serra d'Ossa
Don Benito
Puertollano
Almada
Barreiro
Évora
Almendralejo
Villafranca de los Barros
Setúbal
Castuera
Baía de Setúbal
Alcácer do Sal
Barragem do Alqueva
Zafra
Pozoblanco
La Ca
Jerez de los Caballeros
Azuaga
Morena
Bai
Sines
Beja
Montoro
Córdoba
Li
Cortegana
Sierra
Guadalquivir
Bujalance
Martos
Alca
Ourique
Nerva
Palma del Río
Algarve
Valverde del Camino
La Algaba
Carmona
Ecija
Andaluc
Portimão
Ayamonte
Lepe
Sevilla (Seville)
Lucena
Osuna
Siste
Lagos
Faro
Isla Cristina
Huelva
Dos Hermanas
Antequera
Gran
Sie
Cabo de São Vicente
Olhão
Tavira
Golfo de Cádiz
Lebrija
Olvera
Álora
Archido
Las Cabezas de San Juan
Ubrique
Ronda
Málaga
Sanlúcar de Barrameda
El Puerto de Santa María
Fuengirola
El Puerto de Santa María
Jerez de la Frontera
Marbella
Cádiz
Estepona
Costa de
San Fernando
Vejer de la Frontera
Costa de la Luz
Barbate de Franco
Algeciras
GIBRALTAR (to UK)
Ceuta (to Spain)
Strait of Gibraltar
MOROCCO

ATLANTIC
OCEAN

AZORES (to Portugal)

Corvo
Flores
São Jorge
Graciosa
Faial
Pico
Terceira
São Miguel
Ponta Delgada
Santa Maria

0 km 100
0 miles 100
200m/656ft
Sea level

Population
● National capital
○ below 50,000
○ 50,000 to 100,000
◉ 100,000 to 500,000
■ above 500,000

0 km 100
0 miles 100

E   F   G   H

2°   0°   44°   2°   4°

68

F R A N C E

Golfe du Lion

Bermeo
Zarautz
Eibar
Donostia/San Sebastián
Irun
Tolosa
Bergara
ís Vasco
oria-Gasteiz
Miranda
de Ebro
goño
Arnedo
La Rioja
Tarazona
Soria
Calatayud
Medinaceli
Daroca
Alcañiz
Pamplona
(Iruña)
Estella
Calahorra
Tudela
Ejea de
los Caballeros
Huesca
Barbastro
Monzón
Balaguer
Cervera
Fraga
Jaca
Monte Perdido
3348m
La Seu d'Urgell
Berga
ANDORRA
Ripoll
Banyoles
Manlleu
Vic
Figueres
Girona
(Gerona)
Palafrugell
Palamós
Blanes
Arenys de Mar
Costa Brava

42°

74

Navarra
Cataluña
Zaragoza
Lleida
(Lérida)
Tàrrega
Sabadell
Terrassa
Mataró
Barcelona
L'Hospitalet de Llobregat
Sitges
Aragón
Vilafranca del Penedès
Valls
Reus
El Vendrell
Tarragona

I   N

ema   Ibérico

uadalajara
lá de Henares
ón de Ardoz
gus

Teruel
Javalambre
2020m △
Cuenca

Tortosa
Amposta
Sant Carles de la Ràpita
Vinaròs

Ciutadella
Menorca
(Minorca)
Maó

40°

País Valenciano

Castellón de la Plana
Borriana
Vall d' Uxó
Sagunto
(Sagunt)
Burjassot
Onda

Pollença
Sa Pobla
Palma
Manacor
Felanitx
Llucmajor
Mallorca
(Majorca)

Golfo de
Valencia

Costa del Azahar

3

stilla-La Mancha
Mota del Cuervo
ampo de Criptana
Socuéllamos
La Roda
Tomelloso
anares
Solana
eñas
Villanueva de los Infantes
Hellín
Segura
Beas de Segura
Moratalla
Villacarrillo
a
Cazorla
éticos
Huéscar
Baza
uadix
hacén
1m
vada
Berja
Adra

Valencia
Torrent
Catarroja
Sueca
Cullera
Gandia
Oliva
Dénia
Benidorm
Villajoyosa (La Vila Joíosa)
Sant Joan d'Alacant
Alicante (Alacant)
Callosa de Segura
Orihuela
Murcia
La Unión
Cartagena
Aguilas
Mojácar
Almería
Algemesí
Xàtiva
Júcar
Almansa
Ontinyent
Alcoy
Villena
Elda
Jumilla
Monóvar
Elche
(Elx)
Cieza
Mula
Murcia
Totana
Lorca

Ibiza
Eivissa (Ibiza)
Formentera

Illa de
Cabrera

Islas Baleares
(Balearic Islands)

Albacete

Costa Blanca

Mediterranean Sea

A L G E R I A

49

GIBRALTAR (to UK)

N

5°21'
SPAIN
Gibraltar
Airport
North Mole
Gibraltar
Harbour
Catalan Bay
Bay of Gibraltar
Rosia
Rosia
Bay
Little
Bay
The Rock
Catalan
Bay
36°8'
Sandy
Bay
Summit
426m △
Buena Vista
Europa Point
Strait of Gibraltar

200m/656ft
Sea level
0 km   1
0 mile   1

75

38°

4°

2°   0°   2°   4°

36°

E   F   G   H

Elevation

-6000m -4000m -2000m -1000m -500m -250m   Below sea level 0   250m 500m 1000m 2000m 3000m 4000m 6000m

-19,658ft -13,124ft -6562ft -3281ft -1640ft -820ft -328ft/-100m 0

820ft 1640ft 3281ft 6562ft 9843ft 13,124ft 19,685ft

Germany & The Alpine States

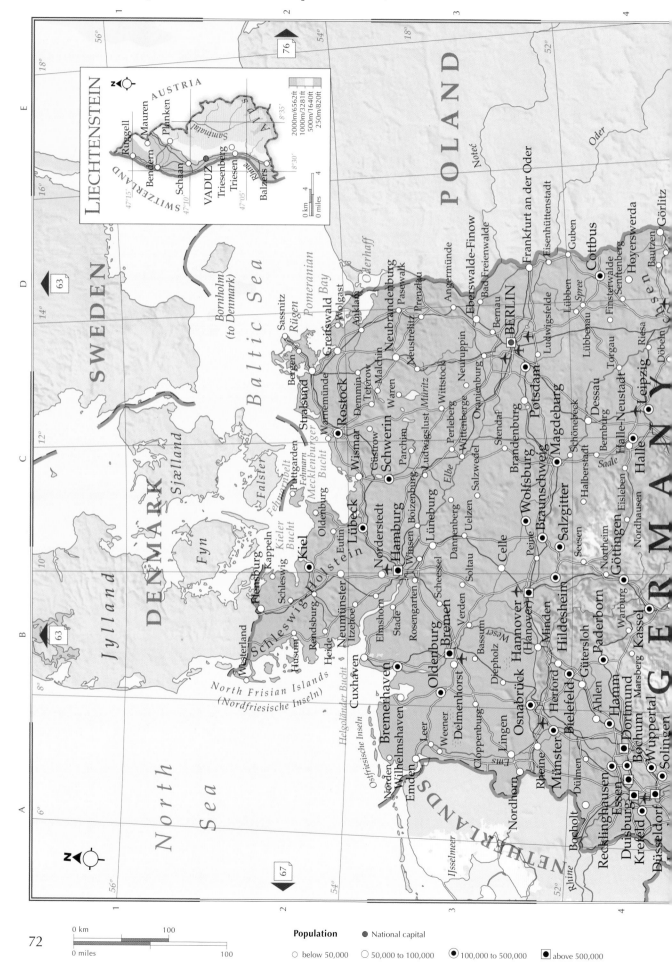

LIECHTENSTEIN

AUSTRIA

SWITZERLAND

Ruggell
Mauren
Schellenberg
Bendern
Planken
Schaan
VADUZ
Triesenberg
Triesen
Balzers

Samínatal
Rhine

2000m/6562ft
1000m/3281ft
500m/1640ft
250m/820ft

0 km   4
0 miles   4

SWEDEN

DENMARK

Jylland

Sjælland

Fyn

Falster

Bornholm
(to Denmark)

Baltic Sea

Rügen
Sassnitz
Bergen

North Sea

North Frisian Islands
(Nordfriesische Inseln)

Ostfriesische Inseln

POLAND

Oder
Noteć

GERMANY

NETHERLANDS

IJsselmeer

Rhine
Ems
Weser
Elbe
Saale
Spree

Cities

Flensburg
Kappeln
Schleswig
Husum
Westerland
Rendsburg
Heide
Itzehoe
Kiel
Eutin
Neumünster
Plön
Oldenburg
Fehmarn
Puttgarden
Lübeck
Norderstedt
Hamburg
Winsen
Lüneburg
Stade
Scheessel
Rosengarten
Elmshorn
Cuxhaven
Bremerhaven
Wilhelmshaven
Norden
Emden
Leer
Weener
Delmenhorst
Oldenburg
Cloppenburg
Lingen
Nordhorn
Rheine
Osnabrück
Bremen
Bassum
Diepholz
Verden
Dannenberg
Uelzen
Soltau
Celle
Salzwedel
Stendal
Hannover
(Hanover)
Peine
Minden
Hildesheim
Hameln
Herford
Bielefeld
Gütersloh
Paderborn
Warburg
Northeim
Göttingen
Nordhausen
Kassel
Marsberg
Ahlen
Hamm
Dortmund
Bochum
Wuppertal
Solingen
Münster
Dülmen
Recklinghausen
Essen
Duisburg
Krefeld
Düsseldorf
Bocholt

Rostock
Warnemünde
Stralsund
Wismar
Güstrow
Teterow
Demmin
Malchin
Waren
Schwerin
Parchim
Ludwigslust
Boizenburg
Wittstock
Perleberg
Wittenberge
Neuruppin
Neustrelitz
Müritz
Oranienburg
Bernau
Brandenburg
Potsdam
BERLIN
Magdeburg
Schönebeck
Halberstadt
Seesen
Wolfsburg
Braunschweig
Salzgitter
Bernburg
Dessau
Bitterfeld
Halle
Halle-Neustadt
Eisleben
Leipzig
Döbeln
Riesa
Bautzen
Görlitz

Pasewalk
Prenzlau
Anklam
Wolgast
Greifswald Bay
Pomeranian Bay
Oderhaff
Angermünde
Eberswalde-Finow
Bad Freienwalde
Frankfurt an der Oder
Eisenhüttenstadt
Guben
Cottbus
Senftenberg
Hoyerswerda
Finsterwalde
Lübbenau
Lübben
Ludwigsfelde
Torgau

Kieler Bucht
Mecklenburger Bucht
Fehmarnbelt
Helgoländer Bucht

63
63
67
76

72

Population ● National capital

○ below 50,000   ○ 50,000 to 100,000   ◉ 100,000 to 500,000   ■ above 500,000

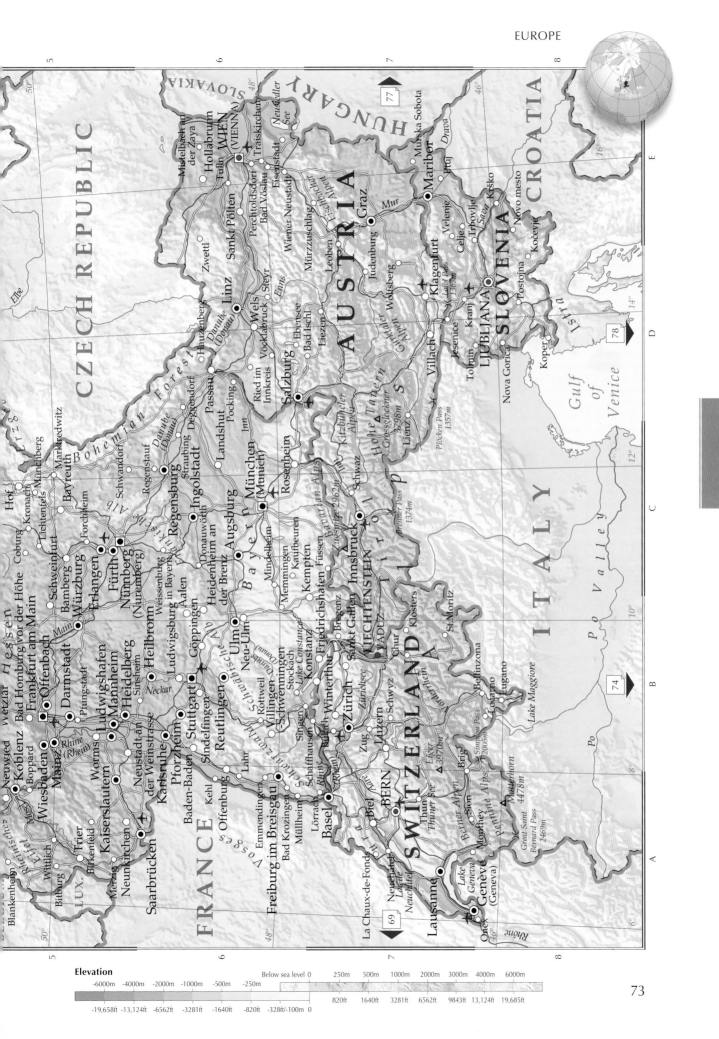

Italy

74

Population   ● National capital

○ below 50,000   ○ 50,000 to 100,000   ◉ 100,000 to 500,000   ■ above 500,000

0 km   100
0 miles   100

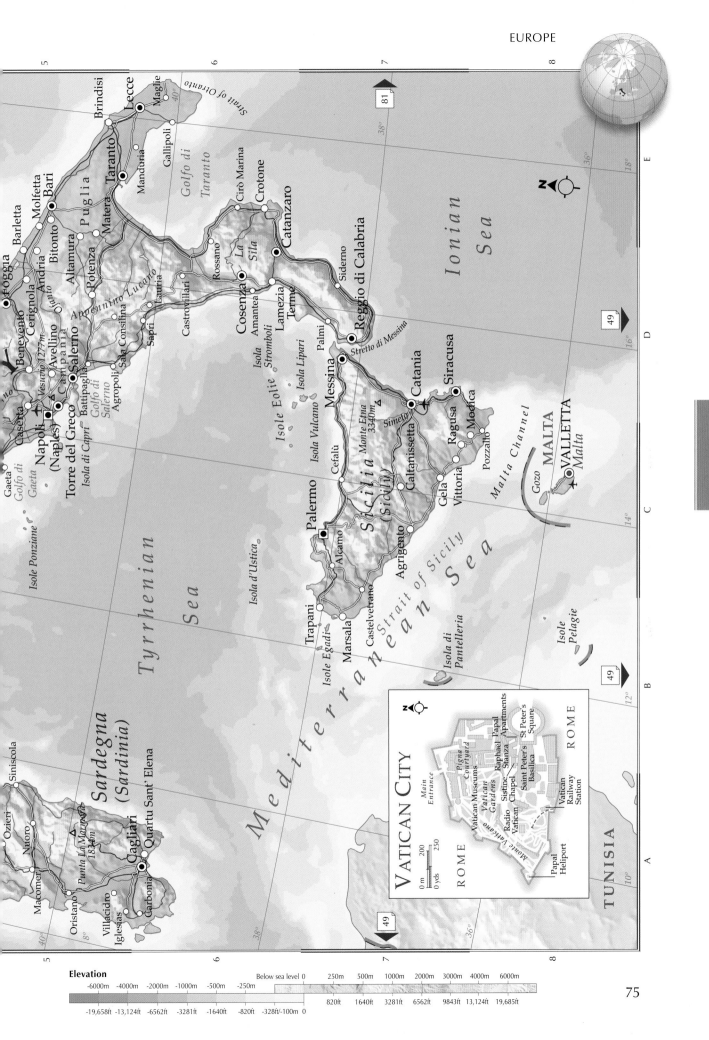

5   6   7   8

Brindisi
Lecce
Maglie
Gallipoli
Strait of Otranto
40°
81
38°
36°
18°
E
Bari
Molfetta
Barletta
Andria
Bitonto
Foggia
Cerignola
Benevento
Avellino
Vesuvio 1277m
Salerno
Torre del Greco
Napoli (Naples)
Caserta
Gaeta
Golfo di Gaeta
Isole Ponziane
Altamura
Matera
Taranto
Manduria
Golfo di Taranto
Puglia
Potenza
Sala Consilina
Appennino Lucano
Lauria
Agropoli
Campania
Battipaglia
Golfo di Salerno
Isola di Capri
Sapri
Castrovillari
Rossano
La Sila
Ciró Marina
Crotone
Catanzaro
Cosenza
Amantea
Lamezia Terme
Siderno
Reggio di Calabria
Calabria
Palmi
Stretto di Messina
Isola Stromboli
Isola Lipari
Isole Eolie
Isola Vulcano
Messina
Cefalù
Catania
Monte Etna 3340m
Sineto
Siracusa
Modica
Ragusa
Caltanissetta
Gela
Vittoria
Pozzallo
Palermo
Alcamo
Sicilia (Sicily)
Agrigento
Trapani
Isole Egadi
Marsala
Castelvetrano
Strait of Sicily
Isola di Pantelleria
Isola d'Ustica

Ionian Sea

MALTA
VALLETTA
Malta
Gozo
Malta Channel
36°
14°
D
49
16°

Tyrrhenian Sea

Mediterranean Sea

Isole Pelagie
12°
B
49

Sardegna (Sardinia)
Siniscola
Ozieri
Nuoro
Macomer
Oristano
Villacidro
Iglesias
Carbonia
Punta La Marmora 1834m
Cagliari
Quartu Sant'Elena
Sant'Elena
8°
40°
38°
36°
10°
A

TUNISIA

49

Elevation

-6000m   -4000m   -2000m   -1000m   -500m   -250m   Below sea level 0   250m   500m   1000m   2000m   3000m   4000m   6000m

-19,658ft -13,124ft -6562ft -3281ft -1640ft -820ft   -328ft/-100m 0   820ft   1640ft   3281ft   6562ft   9843ft   13,124ft   19,685ft

Central Europe

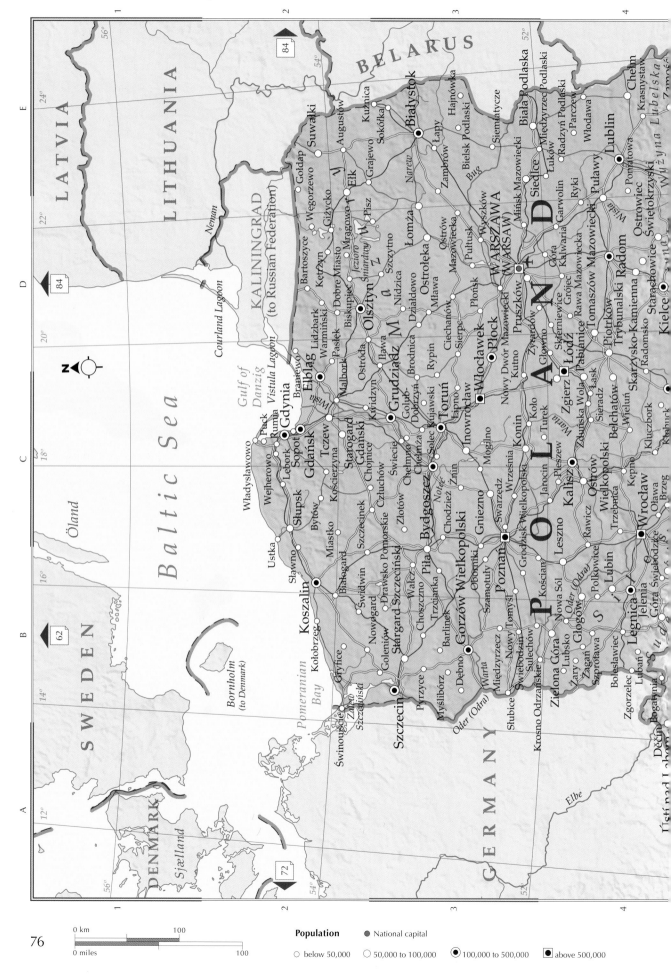

LATVIA

LITHUANIA

SWEDEN

DENMARK

GERMANY

BELARUS

KALININGRAD
(to Russian Federation)

P O L A N D

Baltic Sea

Öland

Bornholm
(to Denmark)

Sjælland

Pomeranian Bay

Gulf of Danzig

Courland Lagoon

Vistula Lagoon

Neman

Wisła

Noteć

Warta

Oder (Odra)

Oder (Odra)

Elbe

Świnoujście
Zalew Szczeciński
Szczecin
Pyrzyce
Myślibórz
Gryfice
Kołobrzeg
Koszalin
Ustka
Sławno
Słupsk
Bytów
Miastko
Białogard
Nowogard
Goleniów
Świdwin
Szczecinek
Czaplinek
Drawsko Pomorskie
Wałcz
Choszczno
Trzcianka
Piła
Złotów
Człuchów
Chojnice
Kościerzyna
Władysławowo
Wejherowo
Rumia
Puck
Lębork
Sopot
Gdynia
Gdańsk
Tczew
Starogard Gdański
Świecie
Chełmno
Chełmża
Grudziądz
Kwidzyn
Malbork
Elbląg
Braniewo
Pasłek
Ostróda
Iława
Lidzbark Warmiński
Bartoszyce
Biskupiec
Olsztyn
Dobre Miasto
Kętrzyn
Mrągowo
Giżycko
Węgorzewo
Gołdap
Suwałki
Augustów
Grajewo
Ełk
Pisz
Szczytno
Nidzica
Działdowo
Mława
Ciechanów
Płońsk
Nowy Dwór Mazowiecki
Pułtusk
Ostrów Mazowiecka
Wyszków
Zambrów
Łomża
Ostrołęka
Kolno
Grajewo
Kuźnica
Sokółka
Białystok
Łapy
Bielsk Podlaski
Hajnówka
Siemiatycze
Międzyrzec Podlaski
Biała Podlaska
Radzyń Podlaski
Parczew
Włodawa
Chełm
Krasnystaw
Lublin
Poniatowa
Puławy
Dęblin
Ryki
Garwolin
Kalwaria
Łuków
Siedlce
Mińsk Mazowiecki
WARSZAWA (WARSAW)
Pruszków
Żyrardów
Grodzisk Mazowiecki
Skierniewice
Rawa Mazowiecka
Tomaszów Mazowiecki
Piotrków Trybunalski
Radomsko
Radom
Starachowice
Ostrowiec Świętokrzyski
Skarżysko-Kamienna
Kielce
Stąporków
Bełchatów
Zduńska Wola
Łask
Sieradz
Zgierz
Łódź
Pabianice
Turek
Koło
Konin
Września
Jarocin
Pleszew
Kalisz
Ostrów Wielkopolski
Kępno
Kluczbork
Brzeg
Oława
Wrocław
Legnica
Trzebnica
Rawicz
Leszno
Kościan
Góra
Głogów
Polkowice
Lubin
Szprotawa
Żagań
Żary
Lubsko
Nowa Sól
Zielona Góra
Świebodzin
Nowy Tomyśl
Grodzisk Wielkopolski
Szamotuły
Poznań
Oborniki
Wągrowiec
Gniezno
Mogilno
Żnin
Inowrocław
Toruń
Solec Kujawski
Bydgoszcz
Chodzież
Barlinek
Gorzów Wielkopolski
Debno
Sulechów
Sulechów
Świebodzin
Międzyrzecz
Słubice
Krosno Odrzańskie
Gubin
Zielona Góra
Bogatynia
Zgorzelec
Lubań
Bolesławiec
Jelenia Góra
Góra Świebodzka
Děčín
Ústí nad Labem
Golub-Dobrzyń
Brodnica
Rypin
Sierpc
Lipno
Włocławek
Płock
Kutno
Łowicz
Głowno
Zgierz
Wieluń

Population

0 km   100
0 miles   100

● National capital

○ below 50,000   ○ 50,000 to 100,000   ◉ 100,000 to 500,000   ◼ above 500,000

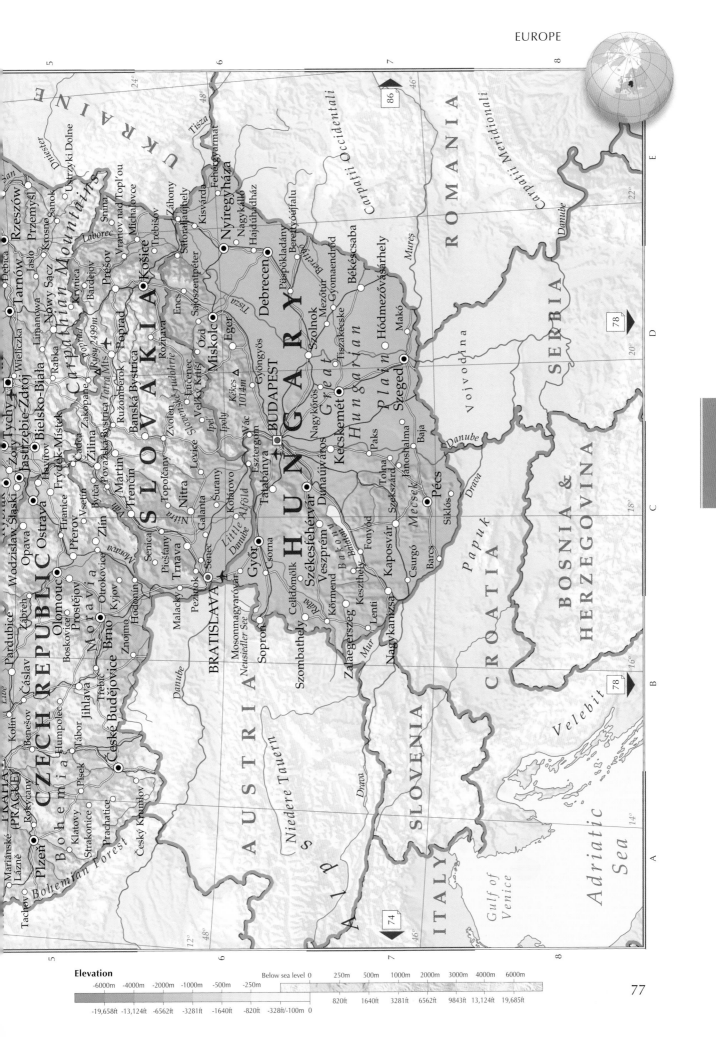

Elevation

| Below sea level 0 | 250m | 500m | 1000m | 2000m | 3000m | 4000m | 6000m |

| -6000m | -4000m | -2000m | -1000m | -500m | -250m |

| -19,658ft | -13,124ft | -6562ft | -3281ft | -1640ft | -820ft | -328ft/-100m | 0 |

| 820ft | 1640ft | 3281ft | 6562ft | 9843ft | 13,124ft | 19,685ft |

Southeast Europe

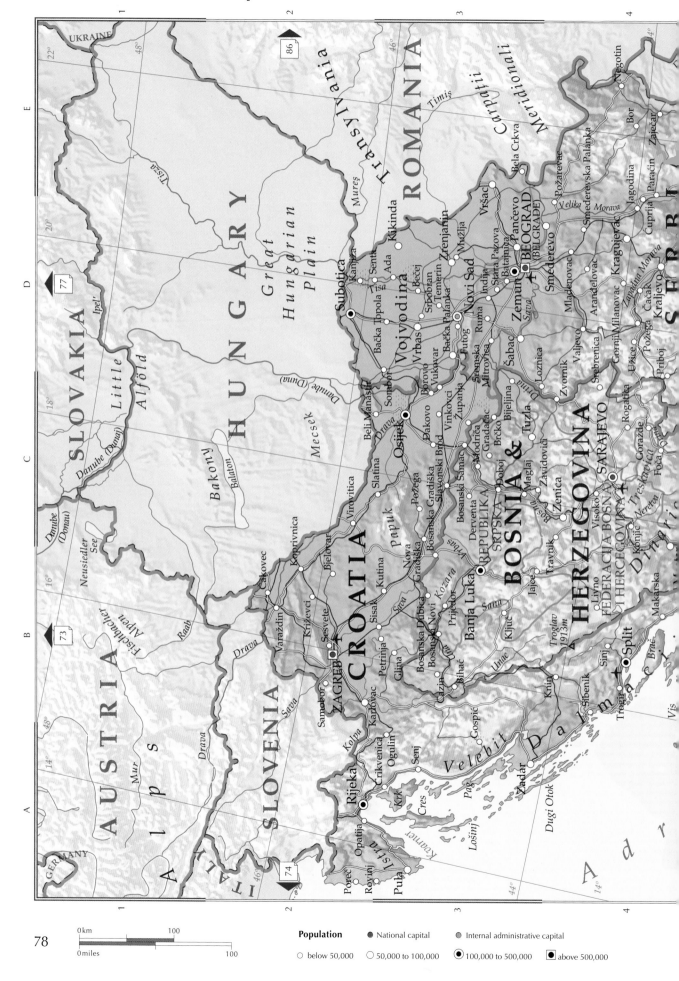

UKRAINE

SLOVAKIA

AUSTRIA

GERMANY

ITALY

SLOVENIA

HUNGARY

Great Hungarian Plain

ROMANIA

Transylvania

Carpaţii Meridionali

CROATIA

Voivodina

SERBIA

BOSNIA & HERZEGOVINA

REPUBLIKA SRPSKA

FEDERACIJA BOSNA I HERCEGOVINA

Alps

Little Alföld

Bakony

Mecsek

Danube (Dunaj)

Danube (Duna)

Tisza

Ipeľ

Raab

Neusiedler See

Balaton

Drava

Sava

Mur

Kolpa

Velebit

Istra

Papuk

Kozara

Sana

Vrbas

Bosna

Drina

Dinara

Neretva

Kornat

Adriatic

ZAGREB
BEOGRAD (BELGRADE)
SARAJEVO

Subotica
Kanjiža
Senta
Ada
Bačka Topola
Tisa
Bečej
Kikinda
Mužlja
Zrenjanin
Temerin
Srbobran
Novi Sad
Vrbas
India
Stara Pazova
Batajnica
Pančevo
Vršac
Bela Crkva
Smederevska Palanka
Požarevac
Velika
Negotin
Bor
Zaječar
Jagodina
Paraćin
Ćuprija
Zemun
Smederevo
Mladenovac
Aranđelovac
Gornji Milanovac
Valjevo
Zapadna Morava
Čačak
Kragujevac
Kraljevo
Šabac
Loznica
Zvornik
Srebrenica
Užice
Požega
Pribo
Rogatica
Goražde
Foča
Trešnjevica
Sombor
Beli Manastir
Osijek
Đakovo
Vinkovci
Borovo
Vukovar
Županja
Slavonski Brod
Šamac
Modriča
Gradačac
Brčko
Bijeljina
Tuzla
Zavidovići
Maglaj
Doboj
Derventa
Bosanski Šamac
Bosanska Dubica
Bosanski Novi
Bosanska Gradiška
Prijedor
Banja Luka
Ključ
Unac
Troglav 1913m
Livno
Sinj
Split
Trogir
Šibenik
Knin
Zadar
Dugi Otok
Pag
Krk
Cres
Lošinj
Vis
Rijeka
Opatija
Crikvenica
Senj
Ogulin
Gospić
Karlovac
Ogulin
Petrinja
Glina
Sisak
Kutina
Nova Gradiška
Slatina
Virovitica
Bjelovar
Koprivnica
Križevci
Samobor
Varaždin
Čakovec
Sesvete
Poreč
Rovinj
Pula
Zenica
Visoko
Travnik
Jajce
Bihać
Cazin
Makarska
Brač
Bosanski Brod
Vis
Komiža
Konjic
Kladanj
Trešnjevica
Muč

0 km 100
0 miles 100

78

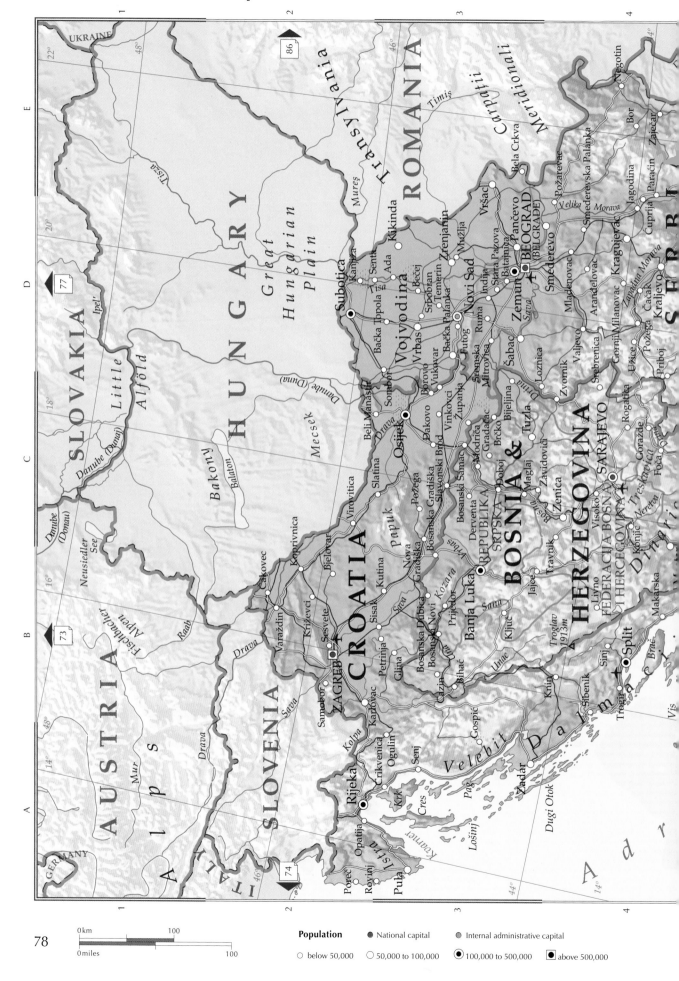

Population   ● National capital   ● Internal administrative capital

○ below 50,000   ◯ 50,000 to 100,000   ◉ 100,000 to 500,000   ◼ above 500,000

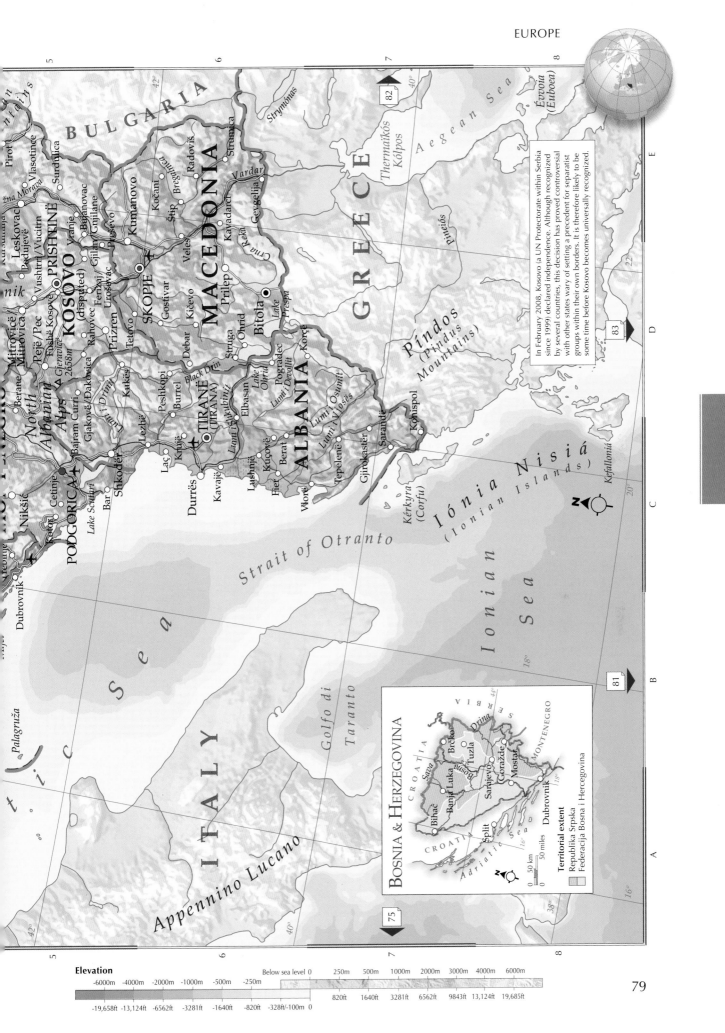

In February 2008, Kosovo (a UN Protectorate within Serbia since 1999) declared independence. Although recognized by several countries, this decision has proved controversial with other states wary of setting a precedent for separatist groups within their own borders. It is therefore likely to be some time before Kosovo becomes universally recognized.

Elevation

| Below sea level 0 | 250m | 500m | 1000m | 2000m | 3000m | 4000m | 6000m |

-6000m -4000m -2000m -1000m -500m -250m

-19,658ft -13,124ft -6562ft -3281ft -1640ft -820ft -328ft/-100m 0

820ft 1640ft 3281ft 6562ft 9843ft 13,124ft 19,685ft

The Mediterranean

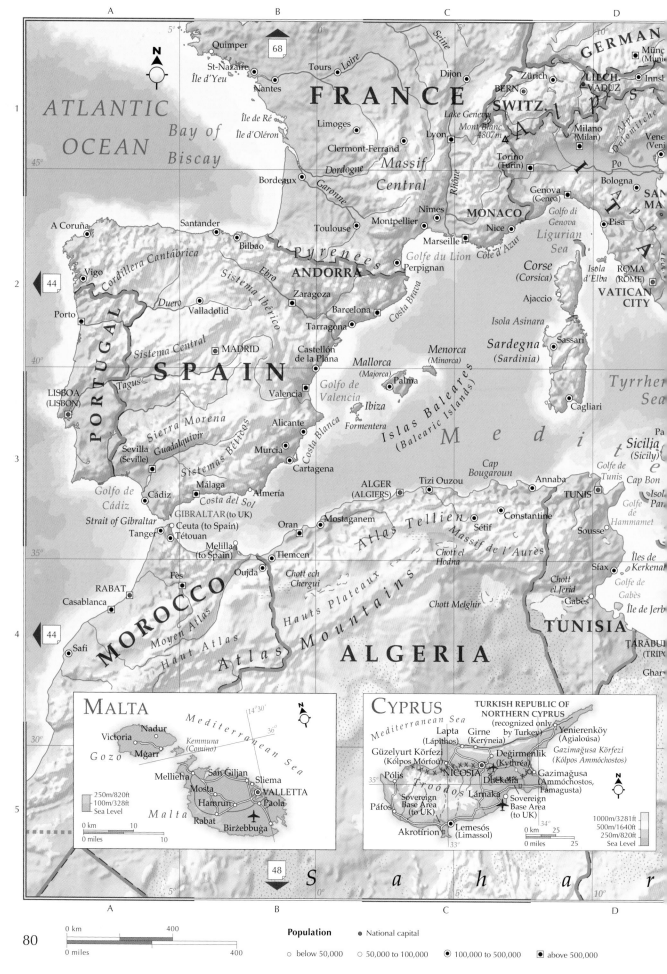

ATLANTIC OCEAN

Bay of Biscay

FRANCE

Quimper
St-Nazaire
Île d'Yeu
Nantes
Tours
Loire
Limoges
Clermont-Ferrand
Dordogne
Bordeaux
Garonne
Toulouse
Nîmes
Montpellier
Marseille
Perpignan
Golfe du Lion
Côte d'Azur

Dijon
Seine
Lyon
Mont Blanc 4807m
Rhône
Lake Geneva
BERN
SWITZ.
Zürich
LIECH.
VADUZ
Milano (Milan)
Torino (Turin)
Genova (Genoa)
Nice
MONACO
Ligurian Sea
Corse (Corsica)
Ajaccio
Isola d'Elba
ROMA (ROME)
VATICAN CITY

GERMAN
München (Muni
Innsb
Vene (Veni
Po
Bologna
SAN MA

A Coruña
Santander
Bilbao
Vigo
Porto
LISBOA (LISBON)
PORTUGAL
Cordillera Cantábrica
Duero
Valladolid
Sistema Central
MADRID
Tagus
SPAIN
Sierra Morena
Guadalquivir
Sevilla (Seville)
Sistemas Béticos
Málaga
Cádiz
Golfo de Cádiz
GIBRALTAR (to UK)
Strait of Gibraltar
Ceuta (to Spain)
Tanger
Tétouan
Melilla (to Spain)

Pyrenees
ANDORRA
Ebro
Zaragoza
Sistema Ibérico
Barcelona
Tarragona
Castellón de la Plana
Valencia
Golfo de Valencia
Alicante
Costa Blanca
Murcia
Cartagena
Almería
Costa del Sol

Costa Brava
Mallorca (Majorca)
Palma
Ibiza
Formentera
Menorca (Minorca)
Islas Baleares (Balearic Islands)
Mediterranean

Isola Asinara
Sardegna (Sardinia)
Sassari
Cagliari
Tyrrher Sea
Sicilia (Sicily)
Golfe de Tunis
Cap Bon
Isol Pan

RABAT
Casablanca
Fès
Safi
MOROCCO
Moyen Atlas
Haut Atlas
Haut Atlas

Oujda
Tlemcen
Oran
Mostaganem
ALGER (ALGIERS)
Tizi Ouzou
Sétif
Constantine
Atlas Tellien
Massif de l'Aurès
Chott el Hodna
Chott ech Chergui
Hauts Plateaux
Atlas Mountains
Chott Melghir
ALGERIA

Annaba
TUNIS
Sousse
Sfax
Gabès
Golfe de Hammamet
Chott el Jerid
Golfe de Gabès
Îles de Kerkenai
Île de Jerb
TUNISIA
ṬARĀBU (TRIPO
Ghar

MALTA

Mediterranean Sea

Victoria
Nadur
Mġarr
Gozo
Kemmuna (Comino)
Mellieħa
Mosta
San Ġiljan
Sliema
VALLETTA
Hamrun
Paola
Rabat
Birżebbuġa
Malta

250m/820ft
100m/328ft
Sea Level

0 km   10
0 miles   10

CYPRUS

TURKISH REPUBLIC OF NORTHERN CYPRUS
(recognized only by Turkey)

Mediterranean Sea
Lapta (Lápithos)
Girne (Kerýneia)
Güzelyurt Körfezi (Kólpos Mórfou)
Pólis
Değirmenlik (Kythréa)
NICOSIA
Dhekelia
Páfos
Troódos
Lárnaka
Sovereign Base Area (to UK)
Akrotírion
Lemesós (Limassol)
Sovereign Base Area (to UK)
Yenierenköy (Agialoúsa)
Gazimağusa Körfezi (Kólpos Ammóchostos)
Gazimağusa (Ammóchostos, Famagusta)

1000m/3281ft
500m/1640ft
250m/820ft
Sea Level

0 km   25
0 miles   25

Sahar

0 km   400
0 miles   400

Population   ● National capital

○ below 50,000   ○ 50,000 to 100,000   ◉ 100,000 to 500,000   ■ above 500,000

SLOVAKIA

VIEN
NNA
Danube
BUDAPEST

HUNGARY

Great Hungarian Plain

Tisza
Satu Mare

Carpathian Mountains

Bâlti

86

U K R A I N E

Kakhovs'ka Vodoskhovyshche

MOLD.
CHIŞINĂU
Dniester
Odesa

ZAGREB
BLJANA

CROATIA

Sava
Novi Sad

Târgu Mureş

R O M A N I A

Carpaţii Meridonali

Galaţi

Dnieper
Berdyans'k

Sea of Azov

Kerch
RUSS. FED.

Kryms'kyý Pivostrov (Crimea)

BOSNIA & HERZ.

BEOGRAD (BELGRADE)

SERBIA

BUCUREŞTI (BUCHAREST)

Danube
Constanţa

Sevastopol'

Novorossiysk

SARAJEVO

MON.

PODGORICA

PRISHTINË

KOSOVO
(disputed)

SKOPJE

B U L G A R I A

Balkan Mountains

SOFIYA (SOFIA)

Varna

Burgas

B l a c k S e a

(the Ukrainian territory of Crimea was annexed by Russia in 2014)

95

TIRANË (TIRANA)

MACED.

Rhodope Mountains

Edirne

İstanbul *Boğazı (Bosporus)*

Küre Dağları

Bari

ALBANIA

Pindos (Pindus) Mts

Thessaloníki (Salonica)

İstanbul

Zonguldak

Samsun

Ordu

i (Naples)

Strait of Otranto

Lecce

Golfo di Taranto

Vesuvio 1277m

Kérkyra (Corfu)

GREECE

Límnos

Marmara Denizi

Bursa

ANKARA

Kızıl Irmak

T U R K E Y

Cosenza

Ionian Sea

Lárisa

Aegean Sea

Balıkesir

Tuz Gölü

Kayseri

Catanzaro

Kefallonia

ATHÍNA (ATHENS)

Chíos

İzmir

Monte Etna 3340m

Zákynthos

Kýthira

Sámos

Kykládes (Cyclades)

Dodekánisa (Dodecanese)

Antalya

Toros Dağları

Adana

Gaziantep

Catania

Siracusa

Mirtóo Pelagos

Kritikó Pélagos (Sea of Crete)

Ródos (Rhodes)

Antalya Körfezi

İskenderun Körfezi

Euphrates

Halab (Aleppo)

LLETTA

Irákleio

Kríti (Crete)

Kárpathos

NICOSIA

CYPRUS

Lárnaka

Lemesós (Limassol)

SYRIA

LEBANON

BEYROUTH (BEIRUT)

DIMASHQ (DAMASCUS)

n S e a

Darnah

Banghāzī (Benghazi)

Mişrātah

Ţubruq

Hefa (Haifa)

ISRAEL

Tel Aviv-Yafo

JERUSALEM

Gaza

AMMAN

Dead Sea

JORDAN

Alexandria (Al Iskandarīyah)

Nile Delta

Būr Sa'īd (Port Said)

Qanāt as Suways (Suez Canal)

Libyan Plateau

CAIRO (AL QĀHIRAH)

Suez (As Suways)

Al 'Aqabah

Giza (Al Jīzah)

Elat

In 1974 Turkey occupied the northern part of Cyprus while Greek Cypriots remained in control of the south. Cyprus was effectively partitioned and a UN buffer zone currently divides the two areas. In 1983 the north of the island proclaimed itself the Turkish Republic of North Cyprus. It was only recognized by Turkey.

Munkhafaḍ al Qaṭṭārah (Qattara Depression)

Great Sand Sea

Ṣaḥārā el Sharqīya (Eastern Desert)

Sinai (Sīnā)

Khalīj as Suways (Gulf of Suez)

Nile

SAUDI ARABIA

L I B Y A

Libyan Desert

E G Y P T

50

Red Sea

81

Elevation

		Below sea level 0	250m	500m	1000m	2000m	3000m	4000m	6000m
-6000m	-4000m	-2000m	-1000m	-500m	-250m				

820ft 1640ft 3281ft 6562ft 9843ft 13,124ft 19,685ft

-19,658ft -13,124ft -6562ft -3281ft -1640ft -820ft -328ft/-100m 0

Bulgaria & Greece

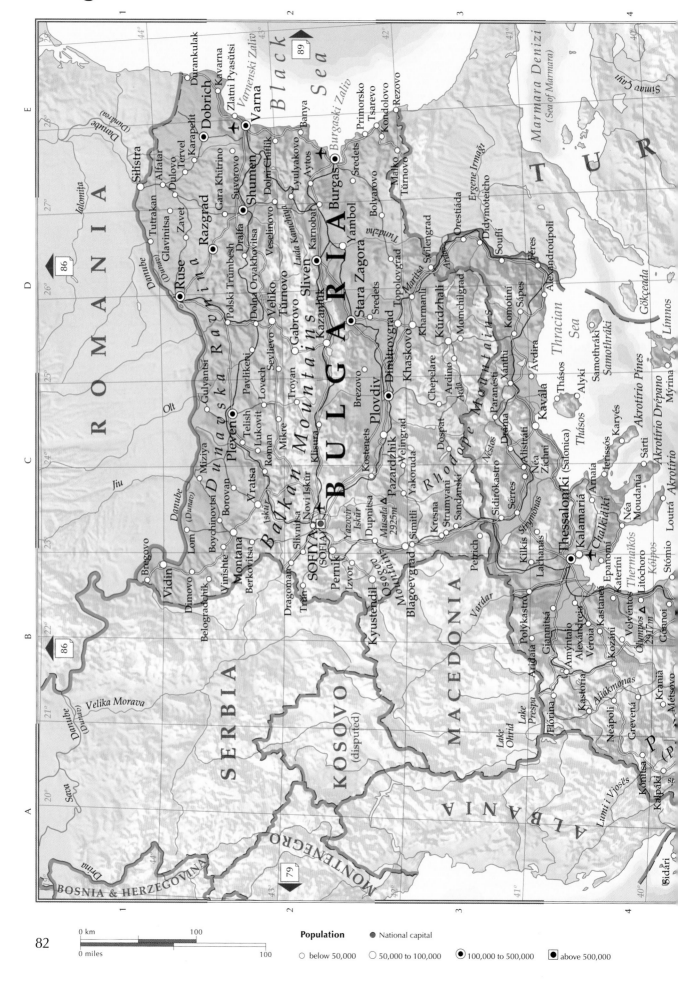

0 km   100

0 miles   100

Population

- ○ below 50,000
- ○ 50,000 to 100,000
- ◉ 100,000 to 500,000
- ■ above 500,000
- ● National capital

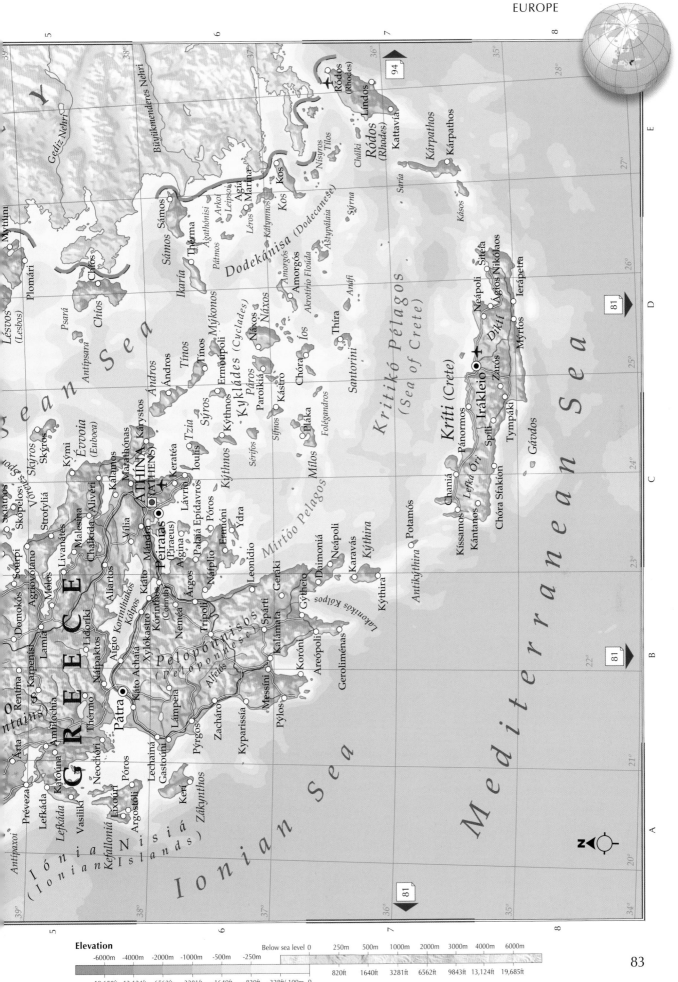

Elevation

| Below sea level 0 | 250m | 500m | 1000m | 2000m | 3000m | 4000m | 6000m |

| -6000m | -4000m | -2000m | -1000m | -500m | -250m |

| -19,658ft | -13,124ft | -6562ft | -3281ft | -1640ft | -820ft | -328ft/-100m 0 |

| 820ft | 1640ft | 3281ft | 6562ft | 9843ft | 13,124ft | 19,685ft |

The Baltic States & Belarus

Population

● National capital

○ below 50,000　◎ 50,000 to 100,000　◉ 100,000 to 500,000　■ above 500,000

0 km　100
0 miles　100

Elevation

-6000m	-4000m	-2000m	-1000m	-500m	-250m	Below sea level 0	250m	500m	1000m 2000m 3000m 4000m 6000m
-19,658ft	-13,124ft	-6562ft	-3281ft	-1640ft	-820ft	-328ft/-100m 0	820ft 1640ft 3281ft 6562ft 9843ft 13,124ft 19,685ft

Ukraine, Moldova & Romania

POLAND

Małopolska

Wyżyna Lubelska

Wisła

BELARUS

Pripet

Pripet *Pripet Marshes*

Styr

Bug

Sluch

Kovel' Sarny Olevs'k

Volodymyr-Volyns'kyy Korosten

Novovolyns'k Kivertsi Mal

Luts'k Rivne Radomy

Sokal' Dubno Novohrad-Volyns'kyy Zhytom

Zhovkva Chervonohrad Slavuta Shepetivka Berd

Yavoriv L'viv Zolochiv Kremenets' Polonne Zhytom

Horodok Zbarazh Starokostyantyniv

Sambir Khodoriv Berezhany Ternopil' Khmel'nyts'kyy

Drohobych Zhydachiv **U** **K** **R** Lypo

Boryslav Stryy Kalush Chortkiv Vinnytsya Ko

Carpathian Mountains

Tatra Mountains

SLOVAKIA

Slovenské Rudohorie

Uzhhorod Dolyna Ivano-Frankivs'k Zhmerynka Hays

Mukacheve Nadvirna Kam''yanets'-Podil's'kyy Tul

Berehove Kolomyya *Podil's'ka Vysochina*

Vynohradiv Khust Chernivtsi Mohyliv-Podil's'kyy *Dniester*

Negreşti-Oaş *Hora Hoverla 2061m* Darabani Soroca *Transnistria*

Tisza

HUNGARY

Satu Mare Baia Mare Rădăuţi Dorohoi Bălţi Rîbniţa

Carei Baia Sprie Solca Botoşani **MOLDOVA**

Marghita Borşa Suceava Fălticeni

Someş Năsăud Târgu-Neamţ Paşcani Călăraşi Orhei

Şimleu Silvaniei Zalău Bistriţa Toplita Bicaz Iaşi Ungheni Străşeni

Oradea Dej Reghin **CHIŞINĂU** Tig

Aleşd *Transylvania* Gheorgheni Piatra-Neamţ (KISHINEV) (Ber

Salonta Beiuş Cluj-Napoca Ludus Târgu Mureş Bacău Vaslui Tiraspo

Curtici Ineu Turda Cristuru Miercurea-Ciuc Hânceşti

Muntii Apuseni Abrud Aiud Mediaş Secuiesc Târgu Ocna Bârlad Comrat

Sânnicolau Mare Arad Alba Iulia Rupea **R** **O** Adjud Comrat Basara

Mureş Lipova Deva **M** **A** **N** **I** **A** Focşani Cahul Taraclia Ciadîr-L

Jimbolia Hunedoara Sibiu Făgăraş Târgu Secuiesc Tecuci Artsyz

Timiş Timişoara Cisnădie Codlea Sfântu Gheorghe Râmnicu Sărat Bolhrad *Ozero Yalpuh*

Lugoj Oţelu Roşu Haţeg Câmpulung Braşov Focşani Galaţi Ki

Bocşa *Varful Moldoveanu 2544m* Râşnov Buzău Reni Braila

Reşiţa Petroşani *Carpaţii* Sinaia Mizil Măcin Izmayil

Oraviţa Anina Târgu Jiu Câlimăneşti *Meridionali* Câmpina Tulcea

Moldova Nouă Curtea Moreni Isaccea Babadag

Orşova Motru de Argeş Câmpina Ploieşti *Lacul Raz*

Danube Drobeta-Turnu Râmnicu Vâlcea Piteşti Târgovişte Urziceni Hârşova *Lacul Sinoie*

Severin Strehaia Drăgăşani Titu Buftea Ţăndărei

Filiaşi *Wallachia* Buftea **BUCUREŞTI** Slobozia Feteşti Medgidia

SERBIA Craiova Slatina Oltenita (BUCHAREST) Călăraşi Constanţa

Calafat Bals Caracal Alexandria *Ialomiţa*

Băileşti *Jiu* Roşiori de Vede Turnu Giurgiu Techirghiol

Corabia Măgurele Zimnicea Eforie-Sud

Danube (Dunărea) *Olt* *Dunavska Ravnina* Mangalia

Velika Morava

BULGARIA

0 km 100

0 miles 100

Population ● National capital

○ below 50,000 ○ 50,000 to 100,000 ◉ 100,000 to 500,000 ● above 500,000

Elevation

| Below sea level 0 | 250m | 500m | 1000m | 2000m | 3000m | 4000m | 6000m |

-6000m -4000m -2000m -1000m -500m -250m

-19,658ft -13,124ft -6562ft -3281ft -1640ft -820ft -328ft/-100m 0

820ft 1640ft 3281ft 6562ft 9843ft 13,124ft 19,685ft

RUSSIAN FEDERATION

Black Sea

Sea of Azov

(the Ukrainian territory of Crimea was annexed by Russia in 2014)

European Russia

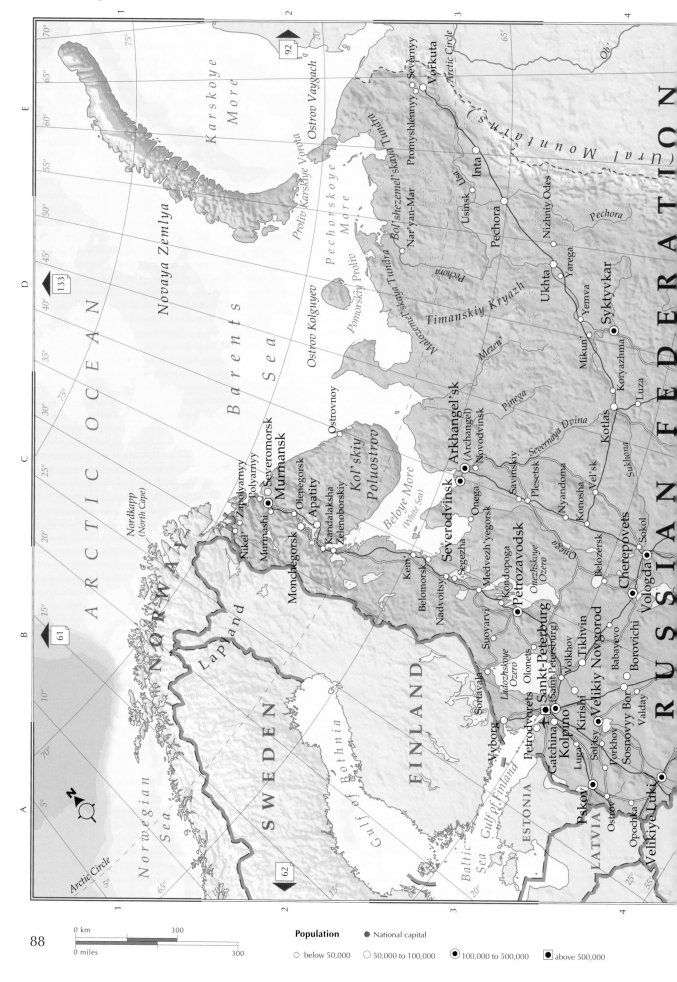

Population ● National capital

○ below 50,000   ○ 50,000 to 100,000   ◉ 100,000 to 500,000   ◾ above 500,000

0 km   300
0 miles   300

ARCTIC OCEAN

NORWAY

SWEDEN

FINLAND

ESTONIA

LATVIA

RUSSIAN FEDERATION

Novaya Zemlya

Karskoye More

Barents Sea

Pechorskoye More

Beloye More (White Sea)

Kol'skiy Poluostrov

Lapland

Ural Mountains

Norwegian Sea

Gulf of Bothnia

Baltic Sea

Gulf of Finland

Nordkapp (North Cape)

Arctic Circle

Ostrov Vaygach

Ostrov Kolguyev

Prolitv Karskiye Vorota

Pomorskiy Proliv

Proliv Karskiye Vorota

Malozemel'skaya Tundra

Bol'shezemel'skaya Tundra

Timanskiy Kryazh

Severnaya Dvina

Mezen'

Pinega

Sukhona

Onega

Ob'

Pechora

Vorkuta
Severnyy
Promyshlennyy
Inta
Usinsk
Usa
Nar'yan-Mar
Pechora
Nizhniy Odes
Yarega
Ukhta
Yemva
Syktyvkar
Mikun'
Koryazhma
Luza
Kotlas
Yel'sk

Ostrovnoy
Severomorsk
Murmansk
Nikel'
Zapolyarnyy
Polyarnyy
Murmashi
Olenegorsk
Apatity
Monchegorsk
Kandalaksha
Zelenoborskiy

Arkhangel'sk (Archangel)
Severodvinsk
Novodvinsk
Savinskiy
Plesetsk
Nyandoma
Konosha
Sokol
Vel'sk
Medvezh'yegorsk
Kem'
Belomorsk
Onega
Nadvoitsy
Segezha
Kondopoga
Petrozavodsk
Cherepovets
Vologda
Belozersk
Onezhskoye Ozero
Suoyarvi
Ladozhskoye Ozero
Sortavala
Olonets
Vyborg
Petrodvorets
Sankt-Peterburg (Saint Petersburg)
Kolpino
Gatchina
Kirishi
Volkhov
Tikhvin
Babayevo
Borovichi
Velikiy Novgorod
Sosnovyy Bor
Valday
Luga
Soltsy
Porkhov
Pskov
Ostrov
Opochka
Velikiye Luki

92
133
61
62

Uzbekistan

KAZAKHSTAN

UZBEKISTAN

Kyzyl Kum

Aral Sea

Syr Darya

Amu Darya

Kirghiz Steppe

Ustyurt Plateau

Caspian Sea

TURKMEN.

Ural'skiye Gory

Perm'
Kungur
Krasnokamsk
Chaykovskiy
Izhevsk
Neftekamsk
Glazov
Birsk
Ufa
Oktyabr'skiy
Sterlitamak
Salavat
Beloretsk
Sibay
Baymak
Orsk
Novotroitsk
Kumertau
Saraktash
Orenburg
Sol'-Iletsk

Naberezhnyye Chelny
Al'met'yevsk
Buguruslan
Buzuluk

Nizhnekamsk
Kazan'
Yoshkar-Ola
Novocheboksarsk
Cheboksary
Kanash
Sarov
Saransk
Sarapul
Tol'yatti
Dimitrovgrad
Ul'yanovsk
Chapayevsk
Balakovo

Nolinsk
Vyatka
Yaransk

Kuybyshevskoye
Vodokhranilishche

Syzran'
Vol'sk
Krasnyy Kut

Nizhniy Novgorod
Dzerzhinsk
Murom
Sarov
Penza
Kuznetsk
Saratov
Kamyshin

Vladimir
Kolomna
Ryazan'
Novomoskovsk
Michurinsk
Tambov
Borisoglebsk
Balashov
Krasnoarmeysk
Mikhaylovka
Ilovlya
Volzhskiy
Akhtubinsk

MOSCOW
Podol'sk
Serpukhov
Aleksin
Tula
Yefremov
Gryazi
Voronezh
Liski
Millerovo
Kamensk-Shakhtinskiy
Kamenka
Zimovniki
Volga

Kaluga
Shchëkino
Orël
Yelets
Lipetsk
Staryy Oskol
Gubkin
Rossosh'
Kakhtemirovka
Shebekino

Pochinok
Roslavl'
Klintsy
Bryansk
Zheleznogorsk
Kursk
Belgorod

Volgodonsk
Sal'sk
Elista
Astrakhan'

Caspian Depression

Volgograd

Kuma
Makhachkala
Kaspiysk
Derbent

Khasavyurt
Groznyy
Nal'chik
Buynaksk

Volga

Don
Donets
Donets

UKRAINE

(the Ukrainian territory of Crimea was annexed by Russia in 2014)

Dnieper
Desna
Dnieper

Sea of Azov

Novoshakhtinsk
Taganrog
Rostov-na-Donu
Staromizinskaya
Novocherkassk

Tikhoretsk
Novorossiysk
Tuapse
Sochi

Krasnodar
Maykop
Kropotkin
Stavropol'
Svetlograd
Nevinnomyssk
Cherkessk
Kislovodsk
Pyatigorsk
Prokhladnyy
Vladikavkaz
Elbrus 5642m

C A U C A S U S

GEORGIA
ARM.
AZERB.
TURKEY

Black Sea

Doğu Karadeniz Dağları
Euphrates

Elevation

Below sea level 0	250m	500m	1000m	2000m	3000m	4000m	6000m

-6000m	-4000m	-2000m	-1000m	-500m	-250m

-19,658ft	-13,124ft	-6562ft	-3281ft	-1640ft	-820ft	-328ft/-100m	0

820ft	1640ft	3281ft	6562ft	9843ft	13,124ft	19,685ft

North & West Asia

A   B   C   D

133

A R C T I

Franz Josef Land

Severnaya Z

Ostrov Komsomolets

Summer limit of pack ice

Ostrov Oktyabr'skoy Revolyutsii
Ostrov Bol'shevik

Winter limit of pack ice

Novaya Zemlya

East Novaya Zemlya Trench

Kara Sea

Poluostrov Taym

North Sibe

Kheta

Norwegian Sea

North Cape

Barents Sea

Ostrov Kolguyev

Gulf of Ob

Poluostrov Yamal

Noril'sk

Centra Siberia Platea

Kureyka

Murmansk

Kola Peninsula

Ostrov

59

Arctic Circle

White Sea

R U S S I A N F

Lower Tunguska

Archangel

Lake Onega

Northern Dvina

U r a l   M o u n t a i n s

Ob'

West Siberian Plain

S i

Stony Tunguska

Yenisey

Angara

Lake Ladoga

Saint Petersburg

Vologda

Yaroslavl'

MOSCOW

Volga

Nizhniy Novgorod

Perm'

Yekaterinburg

Irtysh

Ob'

Chulym

Tomsk

Krasnoyarsk

Kazan'

Chelyabinsk

Novosibirsk

Kaliningrad

Central Russian Upland

Ul'yanovsk

Ufa

Samara

Omsk

Novokuznetsk

Baltic Sea

KALININGRAD (to Russ. Fed.)

Voronezh

Volga

Saratov

Orenburg

Ural'sk

ASTANA

Karagandy

Istim

Semipalatinsk

Sayanskiy Khrebet

Irk

E U R O P E

Volgograd

Ural

Kirghiz Steppe

Kazakh Uplands

A

S

Altai Mountains

(the Ukrainian territory of Crimea was annexed by Russia in 2014)

Don

Rostov-na-Donu

Aral'sk

KAZAKHSTAN

Ozero Zaysan

Stavropol'

Astrakhan'

Aktau

Syr Darya

Kyzylorda

Taraz

Almaty

Danube

Black Sea

El'brus 5642m

C a u c a s u s

Caspian Sea

Ustyurt Plateau

Aral Sea

Kyzyl Kum

UZBEKISTAN

BISHKEK

Ili

Tien Shan

Jengish Chokusu/Tömür Feng 7443m

Istanbul

GEORGIA

Kure Daglari

ARMENIA

TBILISI

BAKU

Garagum

TASHKENT

KYRGYZSTAN

ANKARA

YEREVAN

AZERB.

TURKMENISTAN

Amu Darya

DUSHANBE

TAJIKISTAN

TURKEY

Lake Van

Tabriz

ASHGABAT

Gaziantep

Mosul

TEHRAN

Hindu Kush

Kunlun Mountains

Adana

Aleppo

Qom

KABUL

Jalalabad

CYPRUS

81

SYRIA

IRAQ

Isfahan

IRAN

Herat

AFGHANISTAN

Khyber Pass

BEIRUT

DAMASCUS

BAGHDAD

Iranian Plateau

LEBANON

Syrian Desert

Tigris

Euphrates

Zagros Mountains

H

i

m

a

l

a

y

a

s

ISRAEL

AMMAN

Basra

Shiraz

Zahedan

Thar Desert

JERUSALEM

JORDAN

Dead Sea -427m

KUWAIT

Bandar-e 'Abbas

Ganges

An Nafud

KUWAIT

MANAMA

Dubai

Murray Ridge

SAUDI ARABIA

BAHRAIN

DOHA

U.A.E.

MUSCAT

Tropic of Cancer

RIYADH

QATAR

ABU DHABI

Sur

Indus Fan

Nile

Jedda

Arabian Peninsula

OMAN

Ganges Fan

At Ta'if

Ar Rub' al Khali

Red Sea

Arabian Sea

Bay of Bengal

A F R I C A

SANA

YEMEN

Socotra (to Yemen)

Ta'izz

Aden

Gulf of Aden

47

A   B   C   D

0 km		800
0 miles		800

Population   ● National capital

○ below 50,000   ◎ 50,000 to 100,000   ◉ 100,000 to 500,000   ◼ above 500,000

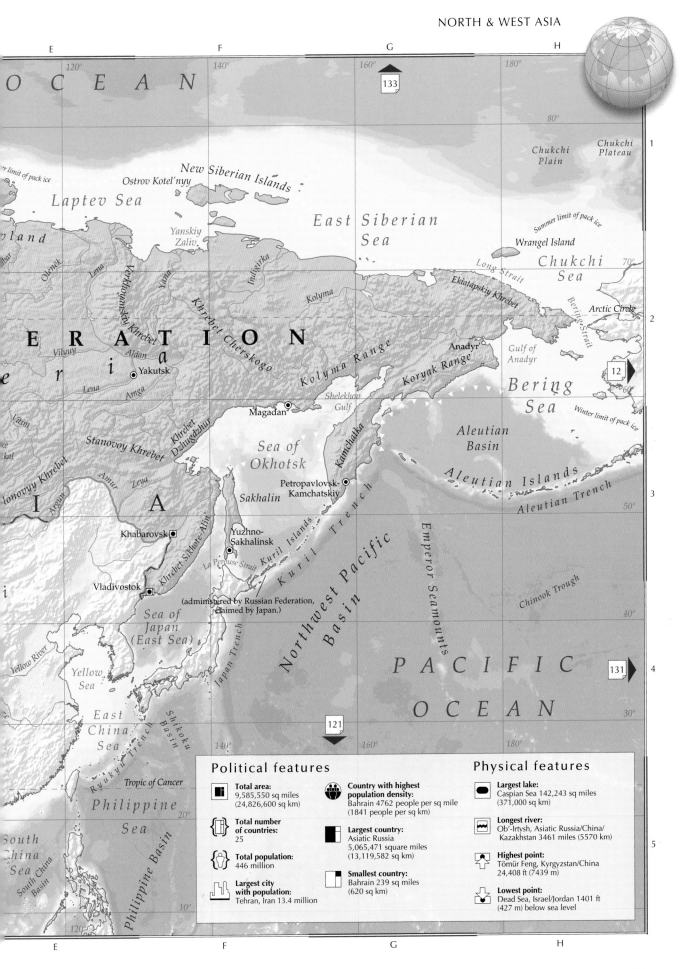

Political features

🏙 **Total area:**
9,585,550 sq miles
(24,826,600 sq km)

👥 **Country with highest population density:**
Bahrain 4762 people per sq mile (1841 people per sq km)

🔲 **Total number of countries:**
25

⬛ **Largest country:**
Asiatic Russia
5,065,471 square miles
(13,119,582 sq km)

👤 **Total population:**
446 million

◧ **Smallest country:**
Bahrain 239 sq miles
(620 sq km)

🏢 **Largest city with population:**
Tehran, Iran 13.4 million

Physical features

⬛ **Largest lake:**
Caspian Sea 142,243 sq miles
(371,000 sq km)

〰 **Longest river:**
Ob'-Irtysh, Asiatic Russia/China/
Kazakhstan 3461 miles (5570 km)

🔺 **Highest point:**
Tömür Feng, Kyrgyzstan/China
24,408 ft (7439 m)

🔻 **Lowest point:**
Dead Sea, Israel/Jordan 1401 ft
(427 m) below sea level

Russia & Kazakhstan

NETH.
NORWAY
DENMARK
SWEDEN
GERMANY
FINLAND
Baltic Sea
Gulf of Bothnia
Gulf of Finland
SVALBARD (to Norway)
Winter limit of pack ice
Summer limit of pack ice
Zemlya Franz Iosifa
ARCT
Nordkapp (North Cape)
Barents Sea
Murmansk
Kandalaksha
Kol'skiy Poluostrov
Beloye More
Novaya Zemlya
Ostrov Kolguyev
Karskoye More
Ostrov Belyy
Diks

KALININGRAD (to Russ. Fed.)
Kaliningrad
POLAND
LITH.
LAT.
EST.
Sankt-Peterburg
Pskov
Velikiy Novgorod
BELARUS
Smolensk
MOSKVA (MOSCOW)
Tver'
Ladozhskoye Ozero
Petrozavodsk
Onezhskoye Ozero
Cherepovets
Vel'sk
Severnaya Dvina
Severodvinsk
Arkhangel'sk
Nar'yan-Mar
Pechora
Poluostrov Yamal
Obskaya Guba

UKRAINE
MOLDOVA
Bryansk
Tula
Belgorod
Ryazan'
Voronezh
(the Ukrainian territory of Crimea was annexed by Russia in 2014)
Sea of Azov
Vologda
Yaroslavl'
Kineshma
Vladimir
Tambov
Nizhniy Novgorod
Kirov
Kotlas
Ukhta
Vorkuta
Salekhard
Noril
Igarka
Taz
Nadym
Nyagan'
Zapadno-Sibirskaya Ravnina

Rostov-na-Donu
Krasnodar
Sochi
Stavropol'
El'brus 5642m
Nal'chik
Vladikavkaz
Groznyy
Makhachkala
GEORGIA
ARM.
AZERBAIJAN
Caucasus
Penza
Ul'yanovsk
Saratov
Balakovo
Volgograd
Astrakhan'
Mikhaylovka
Tol'yatti
Samara
Kazan'
Izhevsk
Naberezhnyye Chelny
Ufa
Glazov
Perm'
Solikamsk
Serov
Lesnoy
Yekaterinburg
Khanty-Mansiysk
Surgut
Nizhnevartovsk
Sibirskaya
Syktyvkar
Ural'skiye Gory
RUSSIA
Chulym
Ob'
Yenisey
Irtysh

Ural'sk
Orenburg
Magnitogorsk
Aktobe (Aktyubinsk)
Atyrau
Fort-Shevchenko
Aktau
Zhanaozen
Alga
Emba
Sterlitamak
Orsk
Rudnyy
Kostanay
Kokshetau
Atbasar
Shchuchinsk
Tyumen'
Chelyabinsk
Ishim
Tobol'sk
Petropavlovsk
Omsk
Seversk
Tomsk
Krasnoy
Kemerov
Novosibirsk
Barnaul
Novokuznetsk
Ab
Tobol

Caspian Sea
Ustyurt Plateau
Aral Sea
Syr Darya
KAZAKHSTAN
ASTANA
Pavlodar
Karaganda Steppe
Temirtau
Saran'
Karagandy
Semey
Ridder
Zyryanovsk
K
Gora Belukha 4506m
Ust'-Kamenogorsk
Altai Mountain

TURKMENISTAN
UZBEKISTAN
Amu Darya
Aral'sk
Ayteke Bi
Zhosaly
Kyzylorda
Kyzyl Kum
Zhezkazgan
Kazakhskiy Melkosopochnik
Shar
Balkhash
Ozero Balkhash
Ayagoz
Ozero Zaysan
Zap

IRAN
TAJIKISTAN
AFGHANISTAN
Turkistan
Kentau
Karatau
Arys
Shymkent
Taraz
Shu
Kirghiz Range
Tekeli
Taldykorgan
Almaty (Alma-Ata)
KYRGYZSTAN
Tien Shan
CHINA

61
86
98
100

0 km 600
0 miles 600

Population
● National capital
○ below 50,000
○ 50,000 to 100,000
◉ 100,000 to 500,000
▪ above 500,000

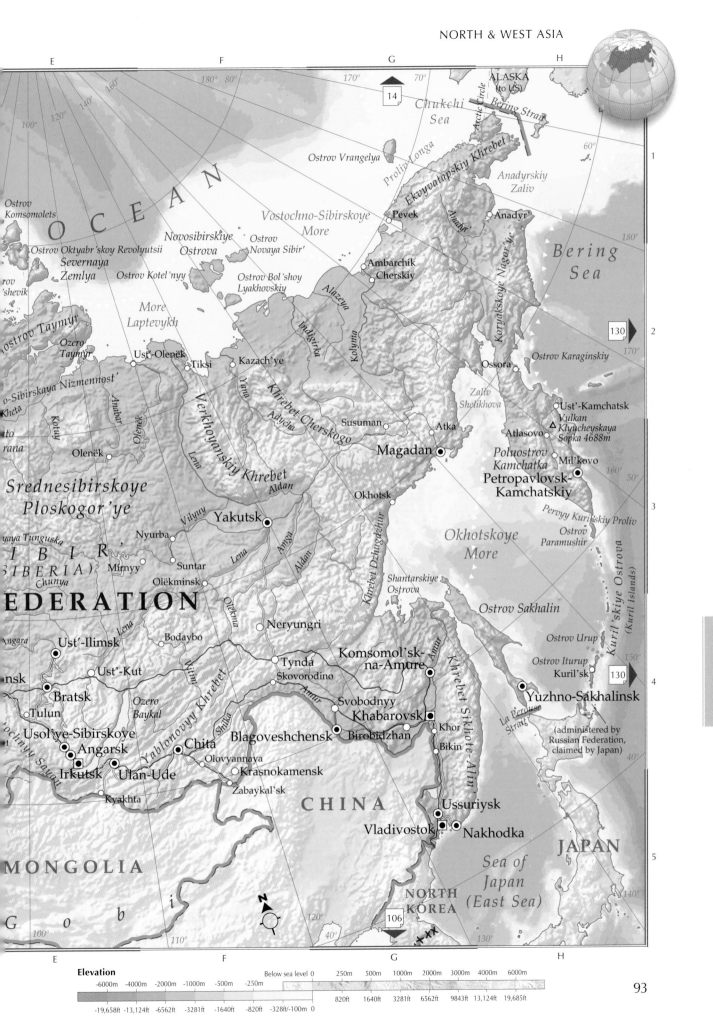

OCEAN

ALASKA
(to US)

Chukchi Sea

Bering Strait

Arctic Circle

14

Ostrov Vrangelya

Ekvyvatapskiy Khrebet

Proliv Longa

Anadyrskiy Zaliv

Ostrov Komsomolets

Vostochno-Sibirskoye More

Novosibirskiye Ostrova

Ostrov Novaya Sibir'

Pevek

Anadyr'

Anadyr'

Koryakskoye Nagor'ye

Bering Sea

180°

Ostrov Oktyabr'skoy Revolyutsii
Severnaya Zemlya

Ostrov Kotel'nyy

Ostrov Bol'shoy Lyakhovskiy

Ambarchik
Cherskiy

Alazeya

Indigirka

Kolyma

Ossora

Ostrov Karaginskiy

130

170°

'shevik

More Laptevykh

Ostrov Taymyr

Ozero Taymyr

Ust'-Olenëk

Tiksi

Kazach'ye

Yana

Khrebet Cherskogo

Adycha

Susuman

Zaliv Shelikhova

Ust'-Kamchatsk
Vulkan Klyucheyskaya Sopka 4688m

-Sibirskaya Nizmennost'

Kheta

Kotuy

Anabar

Olenëk

Atka

Atlasovo

Poluostrov Kamchatka

Mil'kovo

to rana

Olenëk

Lena

Verkhoyanskiy Khrebet

Magadan

Petropavlovsk-Kamchatskiy

160°

50°

3

Srednesibirskoye Ploskogor'ye

Nyurba

Vilyuy

Yakutsk

Aldan

Okhotsk

Okhotskoye More

Pervyy Kuril'skiy Proliv

Ostrov Paramushir

ya Tunguska

SIBIR'
(SIBERIA)

Chunya

Mirnyy

Suntar

Olëkminsk

Lena

Amga

Aldan

Shantarskiye Ostrova

Khrebet Dzhugdzhur

Ostrov Sakhalin

Kuril'skiye Ostrova (Kuril Islands)

150°

FEDERATION

Angara

Ust'-Ilimsk

Bodaybo

Neryungri

Olëkma

Vitim

Tynda

Komsomol'sk-na-Amure

Amur

Khrebet Sikhote-Alin'

Ostrov Urup

Ostrov Iturup

Kuril'sk

130

Ust'-Kut

Lena

Yablonovyy Khrebet

Skovorodino

Svobodnyy

Khabarovsk

Yuzhno-Sakhalinsk

nsk

Bratsk

Ozero Baykal

Birobidzhan

Khor

La Pérouse Strait

Tulun

Usol'ye-Sibirskoye

Shilka

Amur

Blagoveshchensk

Bikin

(administered by Russian Federation, claimed by Japan)

40°

Angarsk

Chita

Olovyannaya

Irkutsk

Ulan-Ude

Krasnokamensk

ochnyy Sayan

Kyakhta

Zabaykal'sk

CHINA

Ussuriysk

Vladivostok

Nakhodka

JAPAN

5

MONGOLIA

N

Sea of Japan (East Sea)

G *o* *b* *i*

NORTH KOREA

106

100°

110°

120°

40°

130°

140°

Elevation

| -6000m | -4000m | -2000m | -1000m | -500m | -250m | | Below sea level 0 | 250m | 500m | 1000m | 2000m | 3000m | 4000m | 6000m |

-19,658ft -13,124ft -6562ft -3281ft -1640ft -820ft -328ft/-100m 0   820ft 1640ft 3281ft 6562ft 9843ft 13,124ft 19,685ft

Turkey & The Caucasus

ROMANIA

Iacul Sinoie

Danube

UKRAINE

Kryms'kyy
Pivostriv
(Crimea)

(the Ukrainian territory of
Crimea was annexed by
Russia in 2014)

BULGARIA

Varnenski
Zaliv

Burgaski
Zaliv

B l a c k   S e a

Maritsa

Kırklareli

Edirne

İnebolu

Sinop

Cide

Gerze

Zonguldak

Bartın

Küre Dağları

Kastamonu

Bafra

Samsun

Çorlu

İstanbul Boğazı
(Bosporus)

Devrek

Karabük

Kargı

Çanik Dağları

Or

Ergene Çayı

Tekirdağ

İstanbul

İzmit

Adapazarı

Bolu

Çerkeş

Merzifon

Marmara Denizi
(Sea of Marmara)

Karabük

Gerede

Çankırı

Kızıl Irmak

Çorum

Tokat

Bandırma

Yalova

İznik Gölü

Kalecik

Alaca

Çanakkale

Bursa

Bilecik

Eskişehir

ANKARA

Bozüyük

Sorgun

Yıldızeli

Çanakkale
Boğazı
(Dardanelles)

Balıkesir

Kütahya

Kırıkkale

Polatlı

T U R K

Siv

Edremit

Ayvalık

Simav

Hirfanlı
Baraji

Şarkışla

Boğazlıyan

Lésvos

Akhisar

Gediz

Kulu

Tuz Gölü

Bünyan

Manisa

Uşak

Afyon

Cihanbeyli

Nevşehir

İncesu

Gürün

Chíos

Gediz Nehri

Aksaray

Kayseri

İzmir

Ödemiş

Akşehir

Sámos

Alaşehir

A n a t o l i a

Göksun

G ü

Aydın

Nazilli

Dinar

Kahramanm

Söke

Büyükmenderes Nehri

Denizli

Beyşehir
Gölü

Konya

Niğde

Milas

Burdur

Isparta

Ereğli

Tavas

Burdur
Gölü

Bodrum

Muğla

Suğla Gölü

Toros Dağları

Ceyhan

Gazia

Marmaris

Dalaman

Antalya

Karaman

Tarsus

Adana

Osmaniye

Fethiye

Manavgat

Mersin (İçel)

İskenderun

Kilis

Kaş

Alanya

Mut

Antakya

Kırıkhan

Finike

Antalya
Körfezi

Silifke

Dodekánisa
(Dodecánese)

Ródos
(Rhodes)

Anamur

Kárpathos

TURKISH REPUBLIC OF
NORTHERN CYPRUS
(recognized only by Turkey)

CYPRUS

Orantes

LEBANON

M e d i t e r r a n e a n

S e a

GREECE

0 km   200
0 miles   200

Population

National capital

○ below 50,000   ◎ 50,000 to 100,000   ◉ 100,000 to 500,000   ⬓ above 500,000

RUSSIAN

FEDERATION

Caspian

Sea

C a u c a s u s

Ap'khazet'i

Gagra
Gudauta
Sokhumi
Ochamchire

Mestia

Enguri

Kazbek
5047m

South
Ossetia

Greater Caucasus

Zaqatala

Xaçmaz

Quba
Siyäzän

100

Kutaisi

GEORGIA

Gori

Samtredia

Poti

Tsalka

TBILISI

Rustavi

Lesser

Kobuleti

Akhaltsikhe

C a u c a s u s

Kura

Şäki

Mingäçevir

Märäzä

Sumqayıt

Batumi
Achara

Hopa

Vanadzor

Gäncä

Yevlax

BAKI
(BAKU)

Pazar
Rize

Artvin

Gyumri

Trabzon
Of

Kars

Artik

Sevan

AZERBAIJAN

Qazimämmäd

iresun

Doğu Karadeniz Dağları

Ispir

Sarıkamış

ARMENIA

YEREVAN

Sevana Lich

Nagorno-
Karabakh

Imişli

Kura

Äli-Bayramı

nüşhane

Çoruh Nehri

Horasan

Aras

Artashat

Xankändi

Aşkale

Pasinler

Bıyükağrı Dağı
(Mount Ararat)△
5137m

Goris

Aras

Biläsuvar

thiye
*rates
Nehri)*

Erzincan

Tercan

Erzurum

Ağrı

AZERBAIJAN

Länkäran

Kemah

Doğubayazıt

Naxçıvan

Patnos

Erciş

*Daryācheh-ye
Orūmīyeh*

EY

Bingöl

Muş

Muradiye

*Reşteh-ye Kūhhā-ye Alborz
(Elburz Mountains)*

Elazığ

latya
D o ğ u

Toroslar

Tatvan

*Van
Gölü*

Van

Silvan

Bitlis

Gevaş

Siirt

Diyarbakır

Batman

Şırnak

Silverek

Mardin

*Ataturk
Baraji*

Viranşehir

Kurdistan

IRAN

98

Şanlıurfa

Ceylanpınar

Nusaybin

Tigris

ayrat
sad

Al Jazīrah

*Kūhhā-ye Zāgros
(Zagros Mountains)*

Euphrates

Jabal Bishrī

IRAQ

RIA

*Buhayrat
ath
Tharthār*

98

The Near East

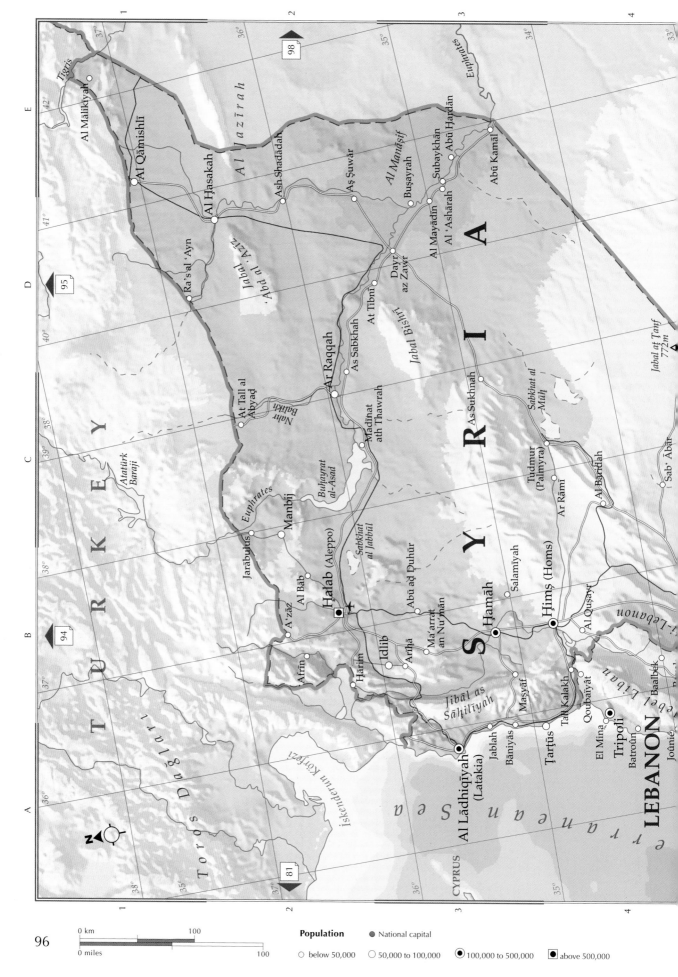

Al Māliklyah
Tigris
Al Qāmishlī
Al Ḥasakah
Al Jazīrah
Ash Shadādah
Al Maruṣif
As Ṣuwār
Busayrah
Subaykhān
Abū Ḥardān
Abū Kamal
Al Mayādīn
Al 'Asharah
Dayr az Zawr
At Tibni
Jabal Bishrī
Ra's al 'Ayn
Jabal al 'Abd al 'Azīz
At Tall al Abyaḍ
Nahr Balīkh
Ar Raqqah
As Sabkhah
Madīnat ath Thawrah
Buḥayrat al-Asad
As Sukhnah
Sabkhat al Mūh
Tudmur (Palmyra)
Jabal at Ṭanf 772m
'Sab 'Abār
Al Bardah
Ar Rāmī
Atatürk Barrajı
Euphrates
Jarābulus
Manbij
Sabkhat al Jabbūl
A'zāz
Al Bāb
Halab (Aleppo)
Abū aḍ Ḍuhūr
Ma'arrat an Nu'mān
Salamīyah
Hamāh
Ḥimṣ (Homs)
Al Quṣayr
Afrin
Ḥārim
Idlib
Arīḥā
Masyāf
Jibāl as Sāḥilīyah
Anti-Lebanon
Baalbek
Jotnié
Tall Kalakh
Qoubaiyāt
Jablah
Bāniyās
Ṭarṭūs
El Mina
Tripoli
Batroûn
Jebel Liban
Al Lādhiqīyah (Latakia)
LEBANON
İskenderun Körfezi
CYPRUS
Mediterranean Sea
Toros Dağları
TURKEY
SYRIA
Euphrates

0 km   100
0 miles   100

Population   National capital

○ below 50,000   ○ 50,000 to 100,000   ◉ 100,000 to 500,000   ■ above 500,000

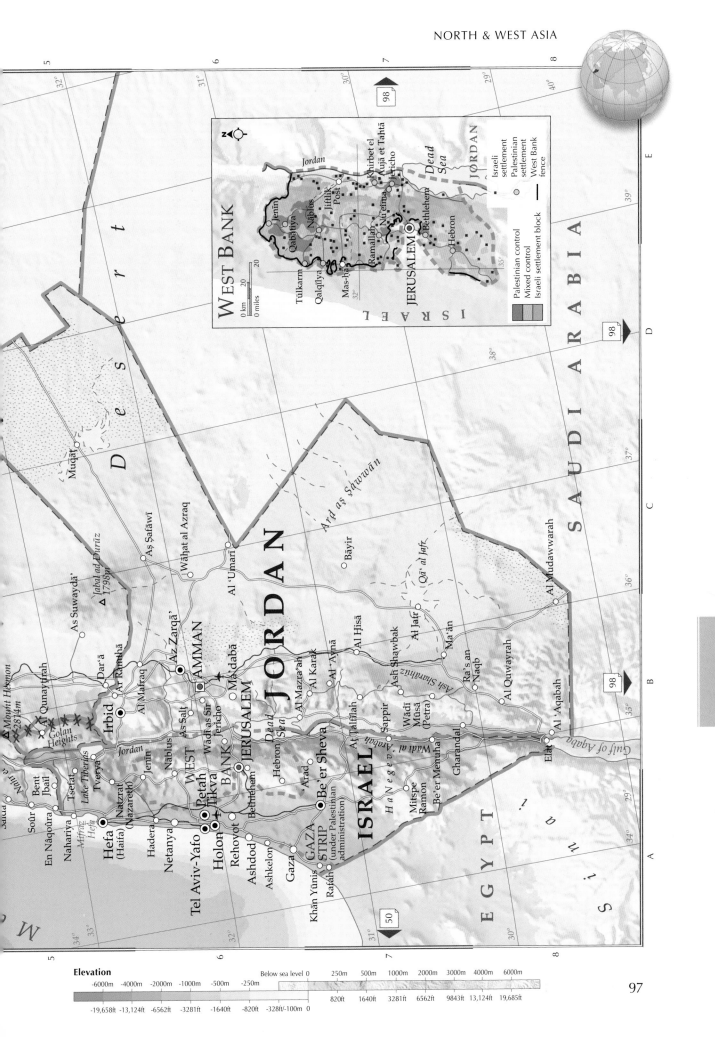

WEST BANK

Tulkarm
Qalqilya
Mas-ha
JERUSALEM

Jenin
Qabātiya
Nāblus
Jiftlik
Post
Khirbet el
'Aujā et Tahtā
Jericho
Ramallah
Ni'eima
Bethlehem
Hebron

Jordan
Dead Sea

JORDAN
ISRAEL

Israeli settlement
Palestinian settlement
West Bank fence

Palestinian control
Mixed control
Israeli settlement block

0 km   20
0 miles   20

SAUDI ARABIA

JORDAN

ISRAEL

EGYPT

Sinai

Mount Hermon
2814m
Golan Heights
Al Qunaytirah
As Suwaydā'
Jabal ad Durūz
1798m
As Safāwī
Wāhat al Azraq
Al 'Umarī
Ard as Shawān
Bāyir
Qa' al Jafr
Al Jafr
Ma'ān
Ra's an Naqb
Al Mudawwarah
Muqat
Des ert
Dar'ā
Ar Ramthā
Al Mafraq
Az Zarqā'
AMMAN
Mādabā
Al Mazra'ah
Al Karak
Al 'Aynā
Al Ḥisā
Ash Shawbak
Ash Sharāh
Sappir
Wādī Mūsā (Petra)
Al Quwayrah
Al 'Aqabah
Elat
Gulf of Aqaba
Irbid
As Salt
Wadi as Sir
Jericho
JERUSALEM
Dead Sea
Hebron
Arad
Be'er Sheva
At Tafilah
Wādī al 'Arabah
Gharandal
Be'er Menuha
Mitspe Ramon
Ha Negev
JORDAN
WEST BANK
Nāblus
Jenin
Jordan
Nahariya
En Nāqoūra
Bent Jbail
Sōur
Tsefat
Lake Tiberias
Tverya
Ḥefa
Mitraz Ḥefa
Hefa (Haifa)
Natzrat (Nazareth)
Hadera
Netanya
Petah Tikva
Holon
Tel Aviv-Yafo
Rehovot
Ashdod
Ashkelon
Gaza
Khān Yūnis
Rafah
GAZA STRIP
(under Palestinian administration)
Bethlehem
ISRAEL

Elevation

-6000m	-4000m	-2000m	-1000m	-500m	-250m	Below sea level 0	250m	500m	1000m	2000m	3000m	4000m	6000m

-19,658ft -13,124ft -6562ft -3281ft -1640ft -820ft -328ft/-100m 0   820ft 1640ft 3281ft 6562ft 9843ft 13,124ft 19,685ft

The Middle East

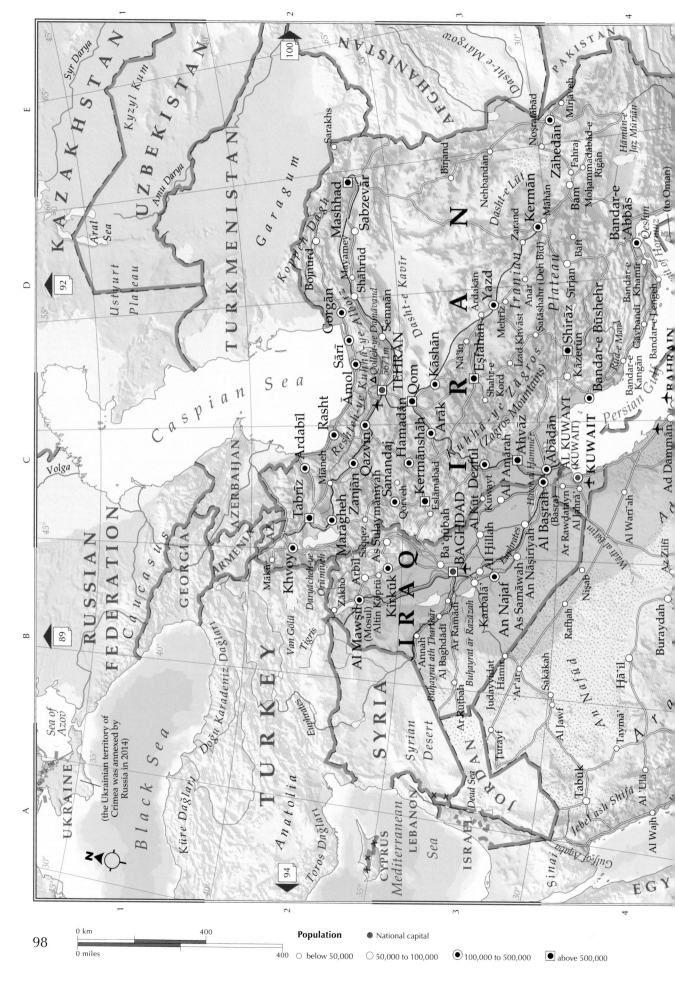

0 km   400

0 miles   400

Population   ● National capital

○ below 50,000   ○ 50,000 to 100,000   ◉ 100,000 to 500,000   ● above 500,000

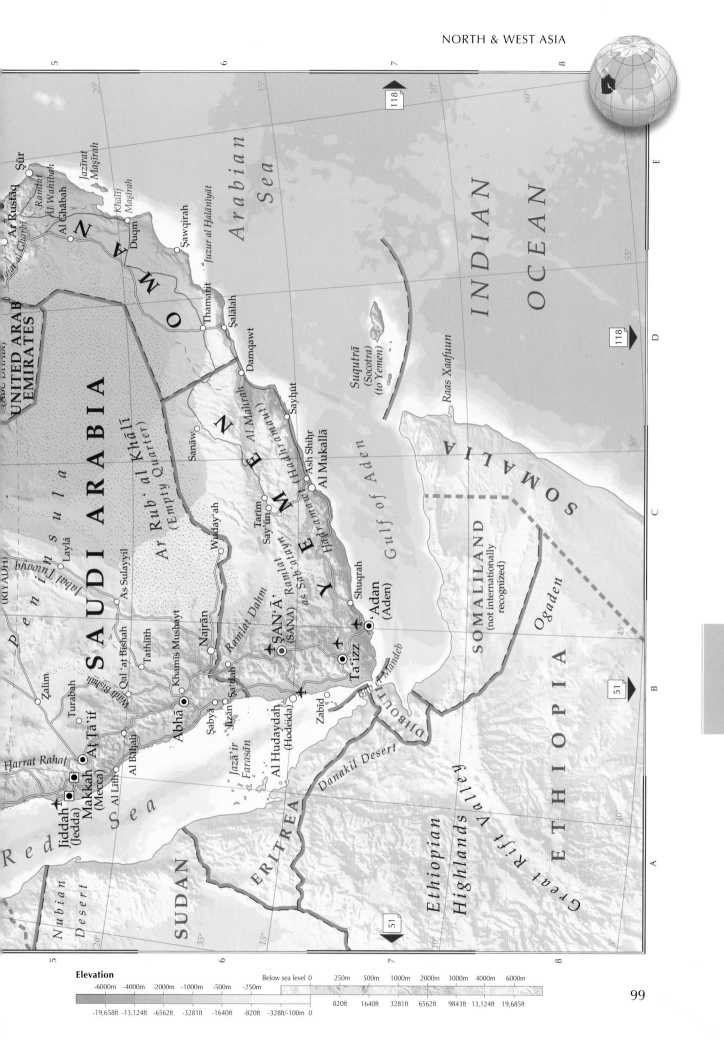

5   6   7   8

INDIAN OCEAN

Arabian Sea

O M A N

Şūr
Ar Rustāq
Ramlat
Al Wahībah
Al Ghābah
Jazīrat Maşīrah
Khalīj Maşīrah
Duqm
Şawqirah
Juzur al Halānīyāt

Thamarīt
Şalālah
Damqawt

UNITED ARAB EMIRATES

SAUDI ARABIA

Peninsula
(RIYADH)
Jabal Tuwayq

Layla
Ar Rub' al Khālī
(Empty Quarter)
As Sulayyil

Wuday'ah

Sanāw

Y E M E N

Al Mahrah
(Hadhramaut)
Sayhūt

Ash Shihr
Al Mukallā

Tarīm
Say'ūn
Ramlat
as Sab'atayn
(Hadhramaut)

Suqutrā
(Socotra)
(to Yemen)

Raas Xaafuun

Gulf of Aden

S O M A L I A

SOMALILAND
(not internationally
recognized)

Ogaden

Zalim
Turabah
Wadi Bīshah
Qal'at Bīshah
Tathlīth
Najrān
Khamīs Mushayt
Ramlat Dahm
Sa'dah

SAN'Ā'
(SANA)

Ta'izz

Shuqrah
Adan
(Aden)

Ethiopian Highlands

E T H I O P I A

At Ţā'if
Harrat Rahat
Makkah (Mecca)
Al Lith
Jiddah (Jedda)

Abhā
Şabyā
Jīzān
Al Baḩah
Jazā'ir Farasān
Al Hudaydah
(Hodeïda)
Zabīd

DJIBOUTI
Bāb el Mandeb

Danakil Desert

Great Rift Valley

Red Sea

Nubian Desert

S U D A N

E R I T R E A

5   6   7   8

Elevation

Below sea level 0   250m   500m   1000m   2000m   3000m   4000m   6000m

-6000m -4000m -2000m -1000m -500m -250m

820ft 1640ft 3281ft 6562ft 9843ft 13,124ft 19,685ft

-19,658ft -13,124ft -6562ft -3281ft -1640ft -820ft -328ft/-100m 0

Central Asia

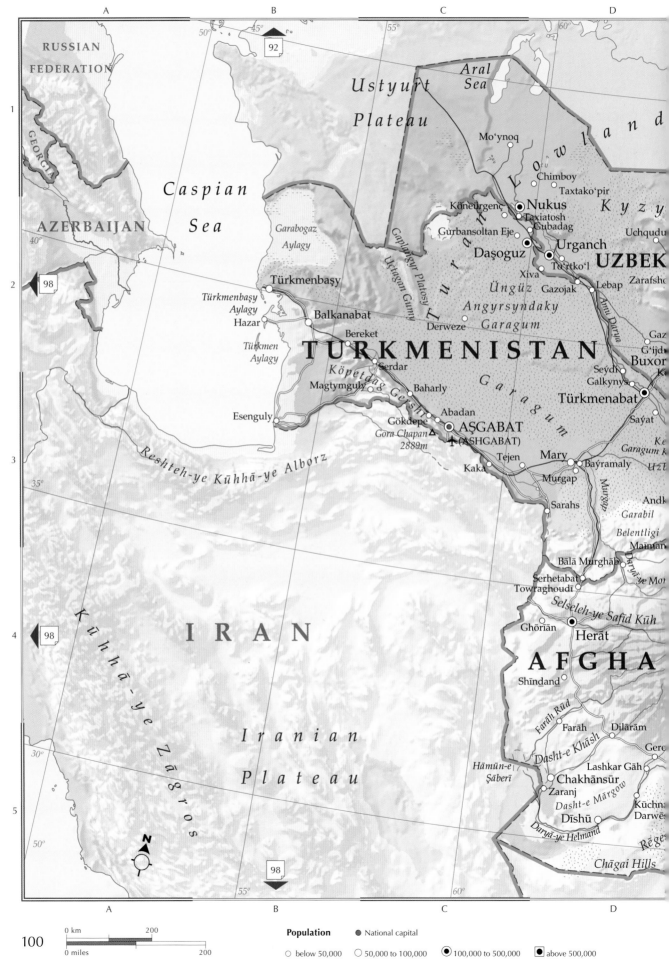

RUSSIAN
FEDERATION

GEORGIA

AZERBAIJAN

*Caspian
Sea*

*Ustyurt
Plateau*

*Aral
Sea*

T u r a n   L o w l a n d

Mo'ynoq

Chimboy

Taxtako'pir

*Garabogaz
Aylagy*

Köneürgenç

Taxiatosh

●Nukus
Çubadag

Uchqudu

Gurbansoltan Eje

Urganch

●UZBEK

Türkmenbaşy

Daşoguz

To'rtko'l

*Türkmenbaşy
Aylagy*

Xiva

Üngüz

Gazojak

Lebap

Zarafsho

Hazar

Balkanabat

Derweze

*Angyrsyndaky
Garagum*

Gaz

Bereket

*Türkmen
Aylagy*

T U R K M E N I S T A N

G'ijd

Buxor

Köpetdag Gerşi

Serdar

G a r a g u m

Seýdi

Magtymguly

Baharly

Galkynyş

Türkmenabat

Esenguly

Abadan

Gökdepe

△*Gora Chapan
2889m*

◉AŞGABAT
(ASHGABAT)

Saýat

Ke

Mary

Bayramaly

Garagum k

Tejen

Murgap

Uz

Kaka

Reshteh-ye Kūhhā-ye Alborz

Sarahs

Murgap

Andk

Garabil

Belentligi

Maïman

Bālā Murghāb

Daryā-ye Mo

Serhetabat

Towraghoudī

I R A N

Selseleh-ye Safid Kūh

Ghōriān

◉Herāt

A F G H A

Kūhhā-ye Zāgros

Shīndand

*Iranian
Plateau*

Farāh Rūd

Farāh

Dilārām

Dasht-e Khāsh

Gere

*Hāmūn-e
Şāberī*

Lashkar Gāh

Chakhānsūr

Dasht-e Mārgow

Kūchn

Zaranj

Darwēs

Dīshū

Daryā-ye Helmand

Rēge

Chāgai Hills

0 km   200
0 miles   200

Population   ● National capital

○ below 50,000   ◎ 50,000 to 100,000   ◉ 100,000 to 500,000   ■ above 500,000

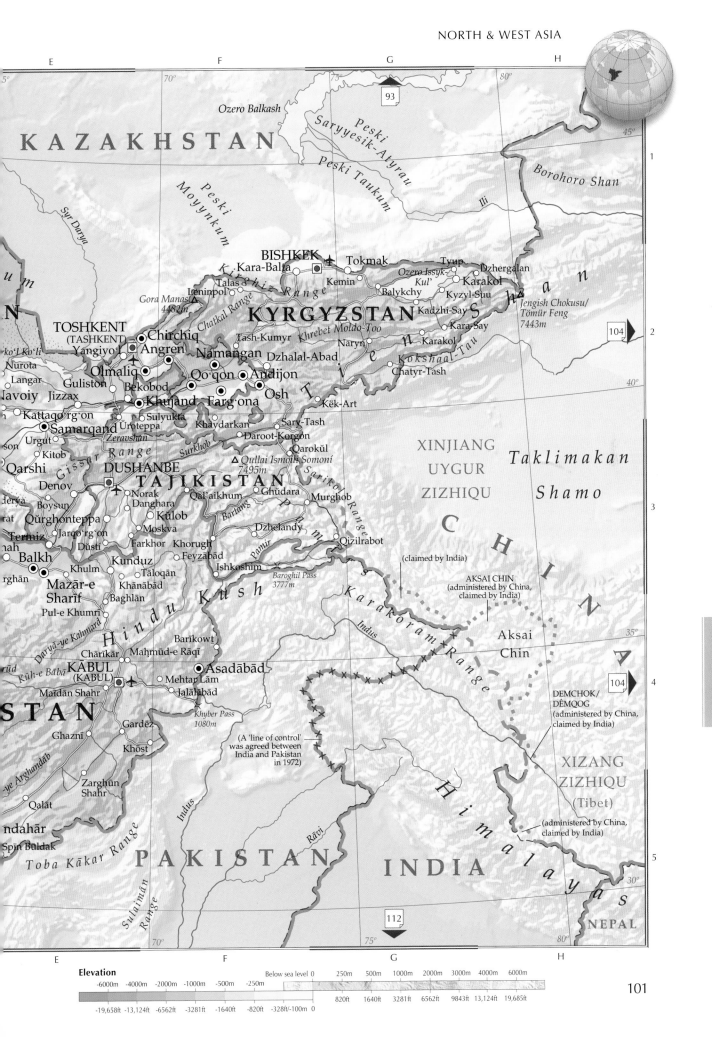

KAZAKHSTAN

Ozero Balkash

Peski Saryyesik-Atyrau

Peski Taukum

Peski Moyynkum

Ili

Borohoro Shan

93

104

BISHKEK
Kara-Balta Tokmak
Kirghiz Range Kemin
Leninpol Talas Balykchy *Ozero Issyk-Kul'* Tyup Dzhergalan
Gora Manas Karakol
4482m *Chatkal Range* Kyzyl-Suu
KYRGYZSTAN Kadzhi-Say Kara-Say *Jengish Chokusu/*
TOSHKENT Tash-Kumyr *Khrebet Moldo-Too* *Tömür Feng*
(TASHKENT) Chirchiq Naryn Karakol 7443m
Yangiyo'l Angren Namangan Dzhalal-Abad Chatyr-Tash *Kokshaal-Tau*
'ko'l Ko'li Olmaliq Qo'qon Andijon
Nurota Bekobod Osh Kёk-Art
Langar Guliston
Navoiy Jizzax Khujand Farg'ona
Kattaqo'rg'on Sulyukta Khaydarkan Sary-Tash
Samarqand Uroteppa Daroot-Korgon
Urgut *Zeravshan* *Surkhob* Qarokül
Kitob *Gissar Range* *Sarikol Range*
Qarshi Qullai Ismoili Somoni XINJIANG *Taklimakan*
son DUSHANBE 7495m UYGUR *Shamo*
Denov Norak Qal'aikhum Ghúdara ZIZHIQU
derya Danghara Murghob
Boysun *Bartang*
Qúrghonteppa Kúlob Dzhelandy *C*
Termiz Jarqo'rg'on Moskva *Pamirs* Qizilrabot (claimed by India) *H*
nah Dústi Farkhor Khorugh AKSAI CHIN *I*
Balkh Khunduz Feyzabad *Pamir* (administered by China, *N*
rghān Khulm Tāloqān Ishkoshim claimed by India) *A*
Mazār-e Khānābād *Baroghil Pass* Aksai
Sharīf Baghlān 3777m Chin
Pul-e Khumrī *Hindu Kush* *Karakoram Range*
Daryā-ye Kahmard 35°
Kūh-e Bābā Barīkowt DEMCHOK/
Chārikār Mahmūd-e Rāqī *Indus* DÊMQOG
KABUL 104 (administered by China,
(KABUL) Asadābād claimed by India)
Maīdān Shahr Mehtar Lām
STAN Jalālābād XIZANG
Gardēz *Khyber Pass* ZIZHIQU
Ghaznī 1080m (Tibet)
Khōst (A 'line of control' (administered by China,
was agreed between claimed by India)
India and Pakistan
ye Arghandāb in 1972)
Zarghūn *Himalayas*
Shahr *Indus*
Qalāt
ndahār *Rāvi*
Spīn Buldak PAKISTAN INDIA
Toba Kākar Range 112
Sulaimān Range NEPAL

30°

E   F   G   H

Elevation

| -6000m | -4000m | -2000m | -1000m | -500m | -250m | | Below sea level 0 | 250m | 500m | 1000m | 2000m | 3000m | 4000m | 6000m |

-19,658ft -13,124ft -6562ft -3281ft -1640ft -820ft -328ft/-100m 0
820ft 1640ft 3281ft 6562ft 9843ft 13,124ft 19,685ft

South & East Asia

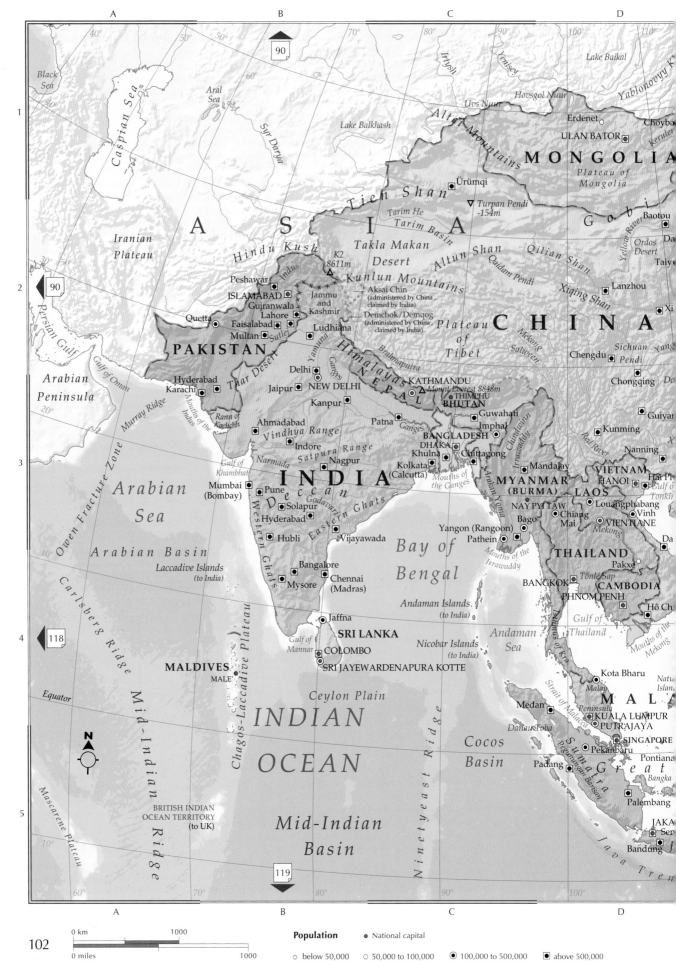

Black Sea
Caspian Sea
Aral Sea
Syr Darya
Lake Balkhash
Irtysh
Uvs Nuur
Hovsgol Nuur
Yenisey
Lake Baikal
Yablonovyy K
Altai Mountains
Erdenet
Choyba
ULAN BATOR
Keruler
MONGOLIA
Plateau of Mongolia
Gobi
Tien Shan
Ürümqi
Baotou
Ordos Desert
Da
Turpan Pendi -154m
Tarim He
Tarim Basin
Takla Makan Desert
Altun Shan
Qaidam Pendi
Qilian Shan
Xiqing Shan
Lanzhou
Taiy
Iranian Plateau
Hindu Kush
K2 8611m
Kunlun Mountains
Aksai Chin (administered by China, claimed by India)
CHINA
Xi
Peshawar
Indus
ISLAMABAD
Jammu and Kashmir
Demchok/Demqog (administered by China, claimed by India)
Plateau of Tibet
Chengdu
Sichuan Yang
Pendi
Quetta
Gujranwala
Lahore
Faisalabad
Ludhiana
Mekong
Salween
Chongqing
Do
Multan
Sutlej
Himalayas
Brahmaputra
PAKISTAN
Thar Desert
Yamuna
Ganges
Delhi
NEW DELHI
NEPAL
KATHMANDU
Mount Everest 8848m
THIMPHU
BHUTAN
Guwahati
Guiyar
Persian Gulf
Gulf of Oman
Arabian Peninsula
Hyderabad
Karachi
Jaipur
Kanpur
Patna
Ganges
Imphal
Kunming
Red River
Nanning
Murray Ridge
Rann of Kachchh
Mouths of the Indus
Ahmadabad
Vindhya Range
Indore
Narmada
Satpura Range
Nagpur
Kolkata (Calcutta)
DHAKA
BANGLADESH
Khulna
Chittagong
Mandalay
Chindwin
Irrawaddy
VIETNAM
HANOI
Hai Ph
Gulf of Tonkin
Owen Fracture Zone
Gulf of Khambhat
Arabian Sea
Mumbai (Bombay)
Pune
Deccan
Godavari
Solapur
Hyderabad
INDIA
Mouths of the Ganges
Western Ghats
Eastern Ghats
MYANMAR (BURMA)
NAY PYI TAW
Arakan Yoma
LAOS
Louangphabang
Vinh
Da
Arabian Basin
Laccadive Islands (to India)
Hubli
Vijayawada
Bay of Bengal
Yangon (Rangoon)
Pathein
Bago
Chiang Mai
VIENTIANE
Mekong
Carlsberg Ridge
Bangalore
Chennai (Madras)
Mysore
Mouths of the Irrawaddy
THAILAND
Pakxe
Andaman Islands (to India)
Tônlé Sap
BANGKOK
CAMBODIA
PHNOM PENH
Hô Ch
Jaffna
Gulf of Mannar
SRI LANKA
COLOMBO
Andaman Sea
Gulf of Thailand
Mouths of the Mekong
118
Nicobar Islands (to India)
Kota Bharu
Nat
Islan
MALDIVES
MALE
SRI JAYEWARDENAPURA KOTTE
Strait of Malacca
Malay Peninsula
MALA
Equator
Ceylon Plain
Medan
Danau Toba
KUALA LUMPUR
PUTRAJAYA
INDIAN
N
Chagos-Laccadive Plateau
Cocos Basin
SINGAPORE
Pegunungan Barisan
Pekanbaru
Pontiana
OCEAN
Ninetyeast Ridge
Padang
Sumatra
Gre a t
Bangka
Mid-Indian Ridge
BRITISH INDIAN OCEAN TERRITORY (to UK)
Palembang
Mascarene Plateau
Mid-Indian Basin
119
JAKA
Ser
Bandung
J
Java Tren

0 km   1000
0 miles   1000

Population   ● National capital

○ below 50,000   ◎ 50,000 to 100,000   ◉ 100,000 to 500,000   ■ above 500,000

E   F   G   H

130°   140°   50°   150°   160°   170°   40°   180°   30°

Qiqihar
Manchuria
Plain Harbin
Changchun
Liao He
Shenyang
Great Khingan Range
amur

Sakhalin

Kuril Islands
Kuril Trench
Japan Trench

Sapporo
Hokkaido

Lake Khanka

NORTH
KOREA
Dandong PYONGYANG
EIJING
Tianjin Dalian
SEOUL SOUTH
KOREA
SEJONG
CITY
Qingdao
azhuang
Jinan
Bo Hai
Plain
na Plain
anjing
Yellow
Sea
zhou
East China
Sea
Shanghai
Hangzhou
Nanchang
ngsha
Fuzhou
han
Yun Shan
Gaoxiong
TAIBEI
(TAIPEI)
TAIWAN
Hong Kong
cao
angzhou
ntou

JAPAN
Sea of
Japan
(East Sea)
Sendai
Nagoya TOKYO
Kyoto Yokohama
Fuji-san
3776m
Osaka
Hiroshima
Kitakyushu
Shikoku
Kyushu
Honshu
Yalu
Korea Strait
Ryukyu Islands
Taiwan Strait
Ryukyu Trench

Northwest
Pacific
Basin

Shatskiy Rise

Japan Trench

Shikoku Trench
Shikoku Basin

Kyushu Palau Ridge

PACIFIC
OCEAN

Philippine Sea

Philippine Basin

West
Mariana
Basin

Mariana Trench

Micronesia

Equator

Luzon Strait

CEL
DS
ted)
Baguio
Luzon
MANILA
PHILIPPINES
Mindoro
South China
na
sin Sea
SPRATLY ISLANDS
(disputed)
Palawan
Bacolod Cebu
Negros
Samar
Panay
Sulu
Sea
Zamboanga
BANDAR
SERI BEGAWAN
NEI
A
A

Mindanao
Davao

Celebes
Sea

Manado

Halmahera

Yap Trench

Eauripik Rise

Ontong
Java
Rise

Melanesia

Bismarck Archipelago

Solomon
Islands

Jayapura

Borneo
alikpapan
unda Islands
DONESIA
Banjarmasin
Celebes
Makassar
Flores
Sea
rabaya
Sea
Bali
Makassar Strait

Moluccas
Seram
Ambon
Buru

Banda Sea

Lesser Sunda Islands
Flores
Timor
Sumba
Timor Trough
DILI EAST TIMOR

New Guinea
Pegunungan Maoke

Solomon
Sea

Arafura
Sea

Coral
Sea

120°   130°   140°   150°   160°

AUSTRALIA

Timor
Sea

E   F   G   H

Political features

Total area:
7,936,200 sq miles
(20,554,700 sq km)

Total number of countries:
24

Total population:
3775 million

Largest city with population:
Tokyo, Japan 39.4 million

Country with highest
population density:
Singapore 22,881 people per sq mile
(8852 people per sq km)

Largest country:
China 3,705,386 sq miles
(9,596,960 sq km)

Smallest country:
Maldives 116 sq miles
(300 sq km)

Physical features

Largest lake:
Tônlé Sap, Cambodia
1000 sq miles (2850 sq km)

Longest river:
Chang Jiang (Yangtze), China
3965 miles (6380 km)

Highest point:
Mount Everest, China/Nepal
29,029 ft (8848 m)

Lowest point:
Turpan Pendi (Turfan Basin), China
505 ft (154 m) below sea level

Western China & Mongolia

A
B
C
D

70°
75°
80°
55°
85°
90°
95°
100°

92

R U S S I A N   F E

Yenisey

1

50°

Kulunda Steppe

Zapadnyy Sayan

Hövsgöl Nuur

K A Z A K H S T A N

Kazakhskiy

Melkosopochnik

Ozero Zaysan

Uvs Nuur

Ulaangom

Mö

Altai

Ölgiy

Hyargas Nuur

45°

Ozero Balkhash

Altay

Har Us Nuur

Hangayn Nuruu

Tsetserle

2

92

70°

Ulungur Hu

Hovd

Har Nuur

M O N

Altay

Bayanhongor

Karamay

Gurbantünggüt Shamo

△ Aj Bogd Uul 3802m

Borohoro Shan

Kuytun

Fukang

Jimsar

Yining

Shihezi

Ürümqi

Qitai

Turpan

Hami

Atas Bogd 2695m △

G

Ozero Issyk-Kul'

KYRGYZSTAN

Tien Shan

△ Jengish Chokusu/Tomür Feng 7443m

Korla

Bosten Hu

Turpan Pendi

Xingxingxia

Dalain He

Kuruktag

3

TAJIKISTAN

Kashi

Tarim He

Tarim Basin

XINJIANG UYGUR

Lop Nur

GANSU

AFGH.

Yengisar

Shache

ZIZHIQU

Qilian Shan

Karakoram Range

Yecheng

Pishan

Taklimakan

Ruoqiang

Altun Shan

Danghe Nanshan

(claimed by India)

Moyu

Shamo

Qaidam Pendi

Qinghai

K2 △ 8611m

Hotan

Qira

Burhan Budai Shan

Dulan

PAKISTAN

Kashmir

Kunlun Shan

Golmud

Anyêmaqen

4

112

JAMMU AND KASHMIR

AKSAI CHIN

AKSAICHIN (administered by China, claimed by India)

Qingzang Gaoyuan

C

H

QINGHAI

Bayan Har Sh

Indus

Rutog

(Plateau of Tibet)

Tongtian He

DEMCHOK/DÊMQOG (administered by China, claimed by India)

Gar Xincun

XIZANG

Gozhê

Siling Co

Amdo

Yushu

Mekong

Zanda

Tangra Yumco

ZIZHIQU

Gyaring Co

Nam Co

Nagqu

Salween

Qamdo

30°

Brahmaputra

(Tibet)

Ngangzê Co

Damxung

Nyainqêntanglha Shan

Tanggula Shan

Jinsha Jiang

Yamuna

75°

Ganges

Lhazê

Xigazê

Maizhokunggar

ARUNACHAL PRADESH (claimed by China)

Hengduan Shan

5

NEPAL

Gonggar

Lhasa

Gyangzê

H i m a l a y a s

△ Mount Everest 8848m

25°

I N D I A

80°

113

BHUTAN

I N D I A

85°

MYANMAR (BURMA)

95°

A
B
C
D

0 km ────── 400

0 miles ────── 400

Population   ● National capital   ● Internal administrative capital

○ below 50,000   ○ 50,000 to 100,000   ◉ 100,000 to 500,000   ■ above 500,000

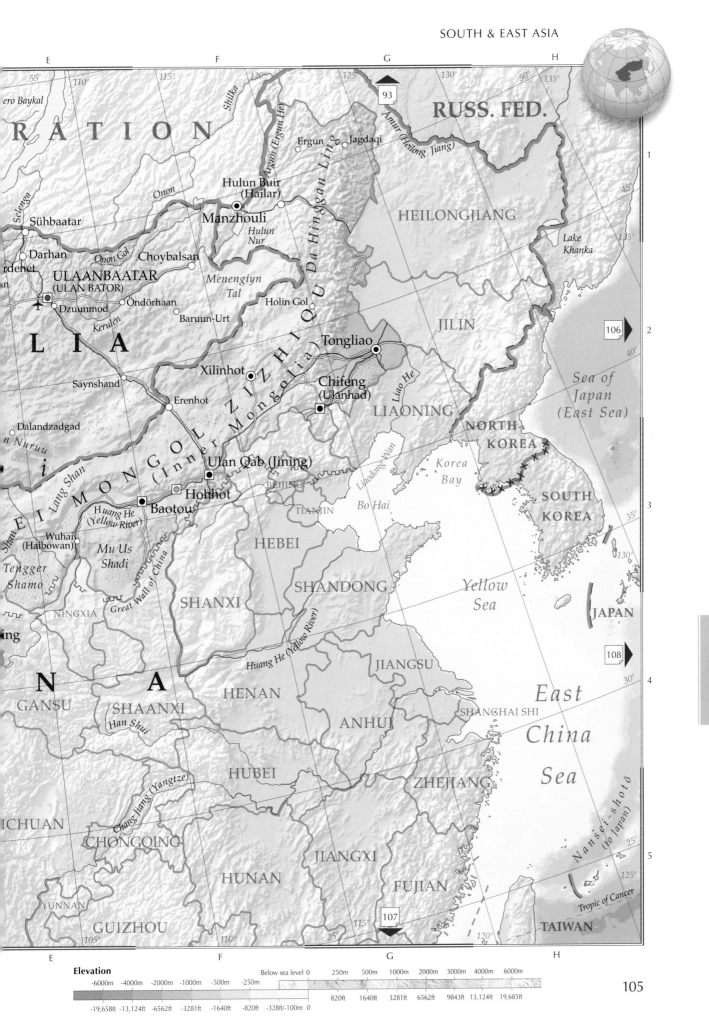

RUSS. FED.

93

RATION

HEILONGJIANG

ero Baykal

Shilka

Argun (Ergun He)

Amur (Heilong Jiang)

Ergun

Jagdaqi

Lake Khanka

Hulun Buir (Hailar)

Onon

Manzhouli

Hulun Nur

JILIN

Sühbaatar

106

Darhan

Onon Gol

Choybalsan

Da Hinggan Ling

rdenet

ULAANBAATAR (ULAN BATOR)

Menengiyn Tal

Holin Gol

Sea of Japan (East Sea)

Dzuunmod

Öndörhaan

Tongliao

LIA

Kerulen

Baruun-Urt

Xilinhot

Chifeng (Ulanhad)

Liao He

Saynshand

Erenhot

LIAONING

NORTH KOREA

Dalandzadgad

(Inner Mongolia)

MONGOL ZIZHIQU

Lihodong Wan

Korea Bay

SOUTH KOREA

n Nuruu

Ulan Qab (Jining)

BEIJING

Bo Hai

i

Lang Shan

Hohhot

TIANJIN

130°

EI

Baotou

Huang He (Yellow River)

Wuhai (Haibowan)

Mu Us Shadi

HEBEI

Yellow Sea

JAPAN

Tengger Shamo

SHANDONG

NINGXIA

Great Wall of China

108

ng

SHANXI

Huang He (Yellow River)

N

A

JIANGSU

GANSU

HENAN

East

SHAANXI

Han Shui

ANHUI

SHANGHAI SHI

China

HUBEI

ICHUAN

Chang Jiang (Yangtze)

ZHEJIANG

Sea

CHONGQING

Nansei-shotō (to Japan)

JIANGXI

HUNAN

FUJIAN

107

Tropic of Cancer

GUIZHOU

YUNNAN

TAIWAN

Elevation

Below sea level 0   250m   500m   1000m   2000m   3000m   4000m   6000m

-6000m -4000m -2000m -1000m -500m -250m

-19,658ft -13,124ft -6,562ft -3,281ft -1,640ft -820ft -328ft/-100m 0

820ft   1640ft   3281ft   6562ft   9843ft 13,124ft 19,685ft

Eastern China & Korea

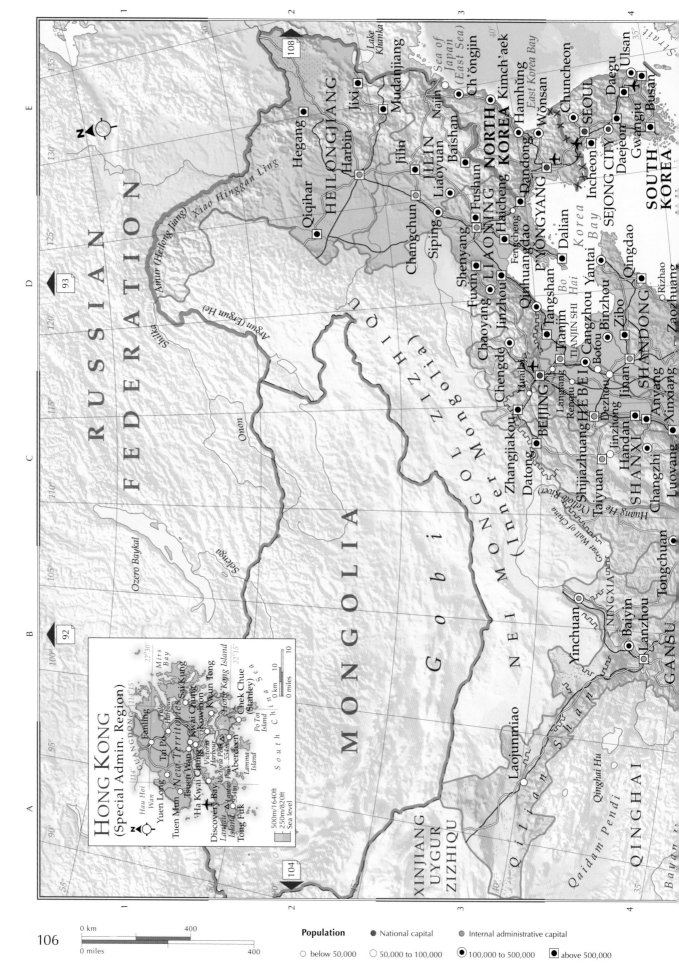

RUSSIAN FEDERATION

MONGOLIA

HEILONGJIANG

Qiqihar
Hegang
Harbin
Jixi
Mudanjiang

JILIN
Changchun
Jilin
Siping
Baishan
Liaoyuan

Lake Khanka

Sea of Japan
(East Sea)

Najin
Ch'ŏngjin
Kimch'aek
Hamhŭng
Wŏnsan

NORTH KOREA

PYONGYANG

Dandong

Fengcheng
Haicheng

LIAONING
Shenyang
Fushun
Fuxin
Chaoyang
Jinzhou

Chuncheon
SEOUL
Incheon
SEJONG CITY
Daejeon
Daegu
Gwangju
Busan
Ulsan

SOUTH KOREA

East Korea Bay

Korea Strait

Qinhuangdao
Chengde
Tangshan
Tianjin
TIANJIN SHI
Cangzhou
Yantai
Binzhou
Zibo
Jinan
Qingdao
Rizhao
Zaozhuang

Dalian

Bo Hai

SHANDONG

Zhangjiakou
Datong
BEIJING
Langfang
Renqiu
HEBEI
Dezhou
Jinzhong
Shijiazhuang
Taiyuan
Handan
Anyang
Xinxiang
SHANXI
Changzhi
Luoyang
Tongchuan

NEI MONGOL (Inner Mongolia)

Gobi

Great Wall of China

Huang He (Yellow River)

Yinchuan
NINGXIA
Baiyin
Lanzhou
GANSU

Qilian Shan

Laojunmiao

Qinghai Hu
Qaidam Pendi
QINGHAI

XINJIANG UYGUR ZIZHIQU

Amur (Heilong Jiang)
Xiao Hinggan Ling
Argun (Ergun He)
Shilka
Ozero Baykal
Onon
Selenga
Ozero Baykal

South China Sea

HONG KONG (Special Admin. Region)

GUANGDONG

Mirs Bay
Fanling
Sai Kung
Tai Po
Tsuen Wan
Kwai Chung
Ha Kwai Chung
Kowloon
Kwun Tong
Tuen Mun
Yuen Long
Hau Hoi Wan
Victoria Harbour
Discovery Bay
Lantau Island
Aberdeen
Tong Fuk
Lamma Island
Po Toi Island
Hong Kong Island
Chek Chue (Stanley)
Lantau Peak 934m
Victoria Peak 554m

500m/1640ft
250m/820ft
Sea level

0 km   10
0 miles 10

Population

○ below 50,000
○ 50,000 to 100,000
◉ 100,000 to 500,000
■ above 500,000

● National capital
● Internal administrative capital

0 km   400
0 miles 400

108
93
92
104

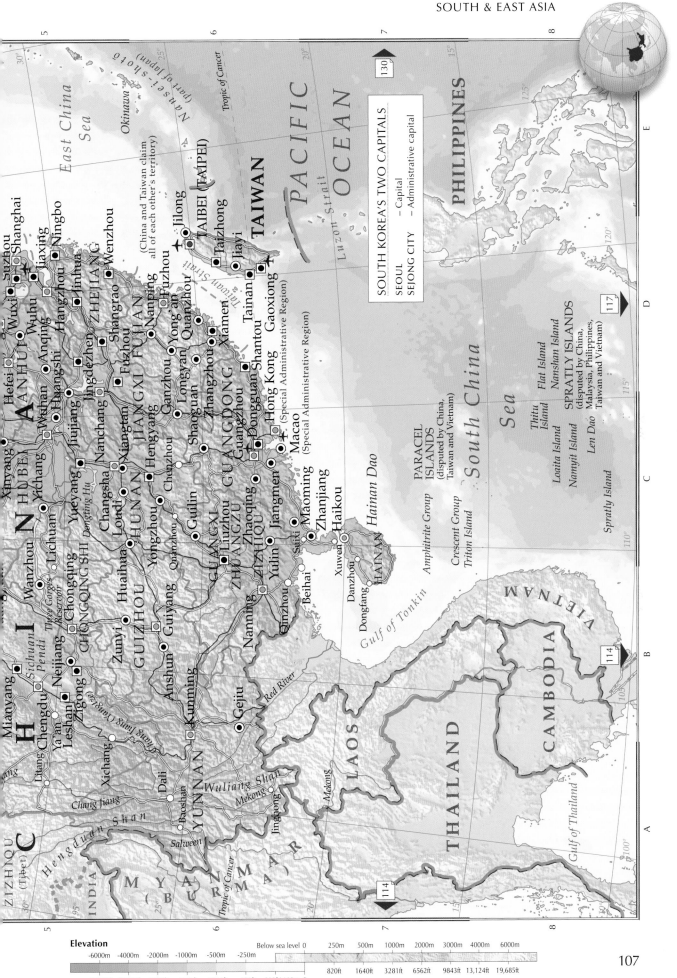

Elevation

| Below sea level 0 | 250m | 500m | 1000m | 2000m | 3000m | 4000m | 6000m |

-6000m -4000m -2000m -1000m -500m -250m

-19,658ft -13,124ft -6562ft -3281ft -1640ft -820ft -328ft/-100m 0

820ft 1640ft 3281ft 6562ft 9843ft 13,124ft 19,685ft

Japan

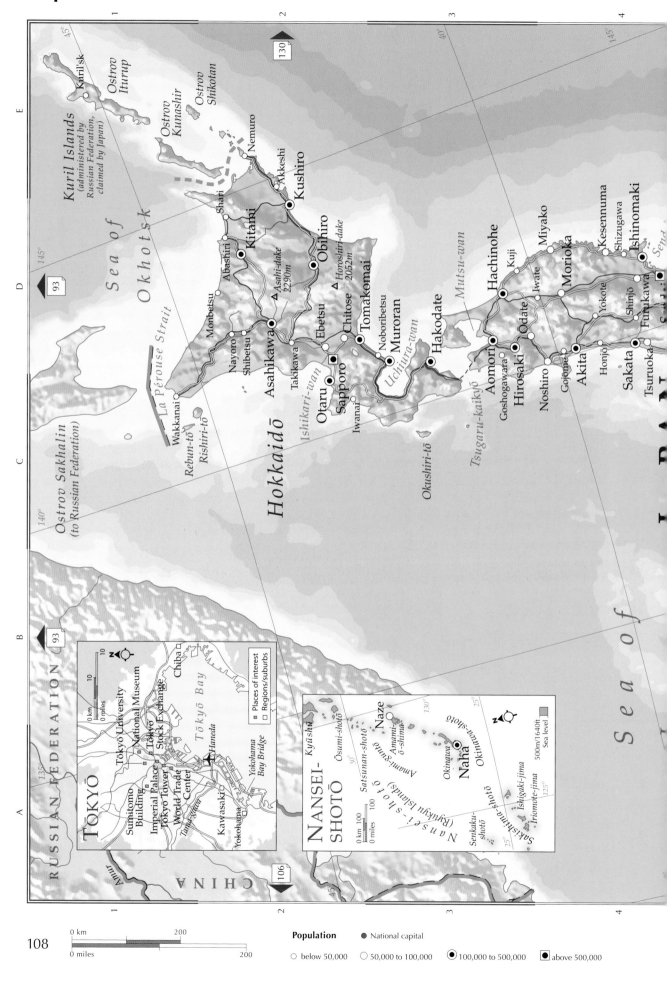

RUSSIAN FEDERATION

CHINA

Amur

Ostrov Sakhalin
(to Russian Federation)

Kuril Islands
(administered by
Russian Federation,
claimed by Japan)

Kuril'sk
Ostrov
Iturup

Ostrov
Shikotan

Ostrov
Kunashir

Nemuro

Akkeshi

Kushiro

Sea
of
Okhotsk

La Pérouse Strait

Shari

Abashiri

Kitami

Monbetsu

Nayoro

Shibetsu

Wakkanai

Rebun-tō
Rishiri-tō

Takikawa

Asahi-dake
2290m

Asahikawa

Obihiro

Horoshiri-dake
2052m

Ebetsu

Chitose

Tomakomai

Otaru

Sapporo

Iwanai

Ishikari-wan

Hokkaidō

Noboribetsu

Muroran

Uchiura-wan

Hakodate

Okushiri-tō

Tsugaru-kaikyō

Mutsu-wan

Aomori

Goshogawara

Hirosaki

Noshiro

Gojōme

Akita

Honjō

Hachinohe

Kuji

Iwate

Ōdate

Yokote

Shinjō

Furukawa

Miyako

Morioka

Shizugawa

Kesennuma

Ishinomaki

Sakata

Tsuruoka

Sea of

TŌKYŌ

Chiba

Tōkyō Bay

Tōkyō University
National Museum
Tōkyō
Stock Exchange

Sumitomo
Building

Imperial Palace
Tōkyō Tower
World Trade
Center

Kawasaki

Yokohama

Haneda

Yokohama
Bay Bridge

■ Places of interest
□ Regions/suburbs

NANSEI-SHOTŌ
(Ryūkyū Islands)

Kyūshū

Ōsumi-shotō

Satsunan-shotō

Naze

Amami-
ō-shima

Amami-guntō

Okinawa

Naha

Okinawa

Nansei-shotō

Ishigaki-jima

Iriomote-jima

Sakishima-shotō

Senkaku-shotō

500m/1640ft
Sea level

Population   ● National capital

○ below 50,000   ○ 50,000 to 100,000   ◉ 100,000 to 500,000   ■ above 500,000

0 km   200
0 miles   200

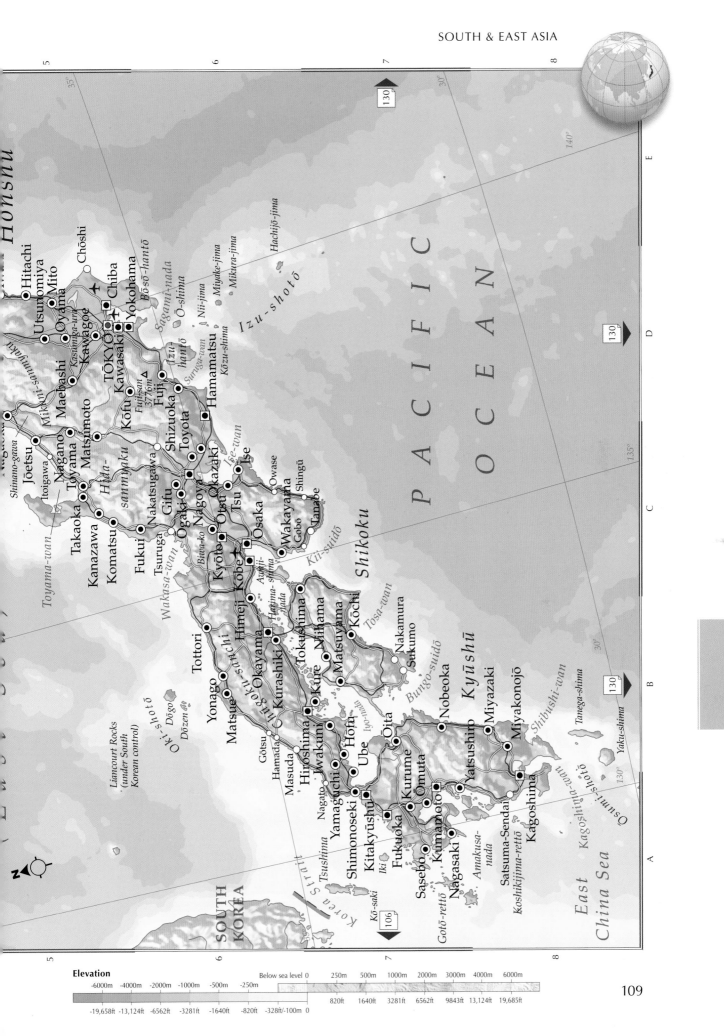

HONSHU

Hitachi
Utsunomiya
Mito
Choshi
Oyama
Chiba
Kashiwa-umi
Kawagoe
Yokohama
TŌKYŌ
Kawasaki
Bōsō-hantō
Sagami-nada
Ōshima
Izu-shotō
Hachijō-jima
Miyake-jima
Mikura-jima
Nii-jima
Izu-hantō
Kōzu-shima
Suruga-wan
Maebashi
Mikuni-sammyaku
Matsumoto
Kōfu
Fujisan 3776m
Fuji
Shizuoka
Hamamatsu
Toyota
Nagano
Toyama
Hida-sammyaku
Gifu
Nagoya
Okazaki
Tsu
Ise
Owase
Shingū
Nakatsugawa
Ōgaki
Ōtsu
Ise-wan
Jōetsu
Shinano-gawa
Itoigawa
Takaoka
Kanazawa
Komatsu
Fukui
Tsuruga
Wakasa-wan
Biwa-ko
Kyōto
Kōbe
Himeji
Awaji-shima
Harima-nada
Wakayama
Gobō
Tanabe
Kii-suidō

PACIFIC OCEAN

Osaka

Shikoku

Tottori
Yonago
Matsue
Chūgoku-sanchi
Okayama
Kurashiki
Kure
Iwakuni
Tokushima
Niihama
Matsuyama
Kōchi
Nakamura
Sukumo
Tosa-wan
Ōki-shotō
Dōgo
Dōzen
Liancourt Rocks (under South Korean control)
Gōtsu
Hamada
Masuda
Hiroshima
Hōfu
Ube
Ōita
Bungo-suidō
Iyo-nada
Kyūshū
Nobeoka
Miyazaki
Miyakonojō
Shibushi-wan
Tanega-shima
Yatsushiro
Satsuma-Sendai
Kagoshima
Koshikijima-rettō
Kagoshima-wan
Ōsumi-shotō
Yaku-shima
Ōsumi-shotō

SOUTH KOREA
Tsushima
Korea Strait
Kō-saki
Iki
Nagato
Shimonoseki
Yamaguchi
Kitakyūshū
Fukuoka
Sasebo
Nagasaki
Kurume
Ōmuta
Kumamoto
Amakusa-nada
Gotō-rettō

East China Sea

Elevation

| Below sea level | 0 | 250m | 500m | 1000m | 2000m | 3000m | 4000m | 6000m |

-6000m -4000m -2000m -1000m -500m -250m

-19,658ft -13,124ft -6562ft -3281ft -1640ft -820ft -328ft/-100m 0   820ft 1640ft 3281ft 6562ft 9843ft 13,124ft 19,685ft

South India & Sri Lanka

Kalyān
Mumbai (Bombay)
Pune
Ahmadnagar
Nānded
Jagdalp
Bārāmati
Nizāmābād
Karīmnagar
Vizianāgaram
I N D I A
Solāpur
Sāngli
Secunderābād
Visākhapa
Kolhāpur
Gulbarga
Hyderābād
Rājahm
Kāki
Belgaum
Karnātaka
Rāichūr
Krishna
Vijayawāda
Gadag
Kurnool
Andhra
Machilīpatr
Panaji
Nandyāl
Pradesh
Chīrāla
Hubli
Ongole
Tungabhadra Reservoir
Tādpatri
Kāvali
Dāvangere
Anantapur
Nellore
Shimoga
Cuddapah
Bhadrāvati
Udupi
Tumkūr
Chennai (Mad ras)
Mangalore
Bangalore
Vellore
Kānchīpuram
Kāsaragod
Mandya
Krishnagiri
Tiruppattūr
Kannur (Cannanore)
Mysore
Salem
Pondicherry
Kozhikode (Calicut)
Erode
Neyveli
Coimbatore
Tamil Nādu
Thrissur (Trichūr)
Tiruchchirāppalli
Ernākulam
Dindigul
Madurai
Kochi (Cochin)
Alappuzha (Alleppey)
Rājapālaiyam
Kollam (Quilon)
Thiruvananthapuram (Trivandrum)
Tuticorin
Nāgercoil

Deccan
Western Ghats
Eastern Ghats
Godāvari
Telangana
Coromandel Coast

Arabian Sea

Amīndīvi Islands

Lakshadweep (Laccadive Islands) (to India)
Kavaratti Island
Kalpeni Island
Nine Degree Channel
Minicoy Island
Eight Degree Channel

MALDIVES
Ihavandhippolhu Atoll
Faadhippolhu Atoll
Horsburgh Atoll
Ari Atoll
Male' Atoll
✈ MALE'
Felidhu Atoll
Mulakatholhu
Kolhumadulu
Hadhdhunmathi Atoll
North Huvadhu Atoll
South Huvadhu Atoll
Gan 118
Addu Atoll

SRI LAN
Jaffna
Mannar
Vavuniya
Trincomalee
Anuradhapura
Batticalo
Palk Strait
Gulf of Mannar
Puttalam
Matale
Negombo
Kandy
COLOMBO
SRI JAYEWARDENAPURA KOTTE
Kalutara
Ratnapura
Galle
Matara

I N D I A N

112
99
51

70° *75°* *80°*

15° *10°* *5°* *Equator*

SRI LANKA'S TWO CAPITALS	
COLOMBO	– Capital
SRI JAYEWARDENAPURA KOTTE	– Administrative capital

Population ● National capital
○ below 50,000   ○ 50,000 to 100,000   ◉ 100,000 to 500,000   ◼ above 500,000

0 km   300
0 miles   300

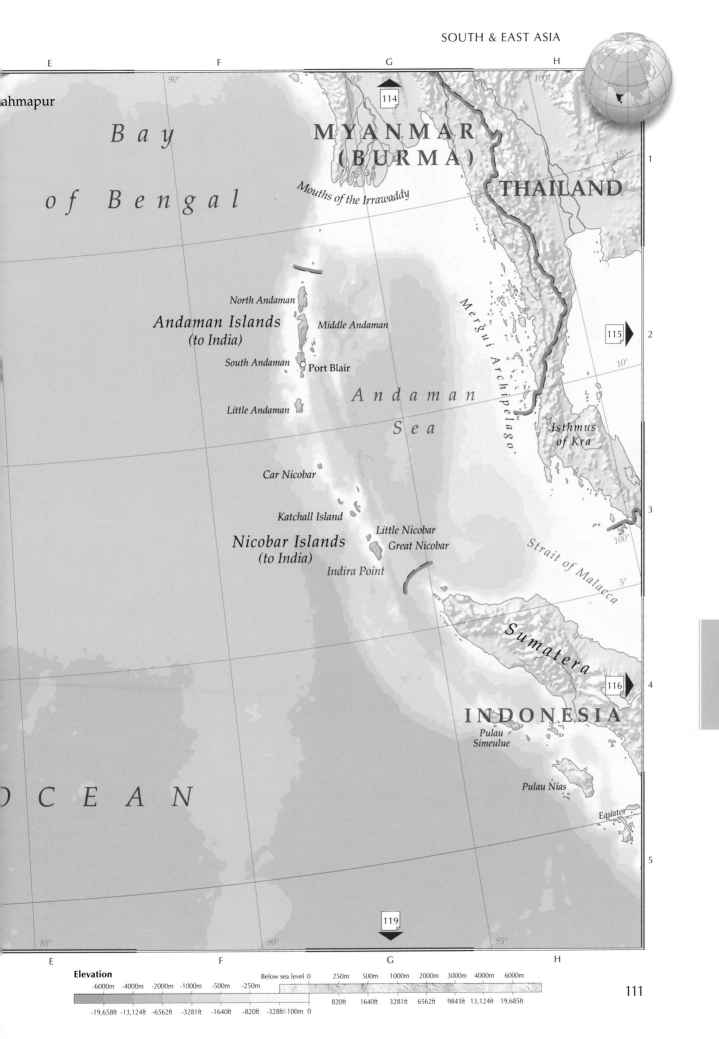

Bay

of Bengal

ahmapur

MYANMAR
(BURMA)

114

Mouths of the Irrawaddy

THAILAND

115

North Andaman

Andaman Islands
(to India)

Middle Andaman

Mergui Archipelago

South Andaman
○ Port Blair

A n d a m a n

Little Andaman

S e a

*Isthmus
of Kra*

Car Nicobar

Katchall Island

Little Nicobar
Great Nicobar

Nicobar Islands
(to India)

Strait of Malacca

Indira Point

116

Sumatera

INDONESIA

*Pulau
Simeulue*

OCEAN

Pulau Nias

Equator

119

Elevation

Below sea level 0 250m 500m 1000m 2000m 3000m 4000m 6000m

-6000m -4000m -2000m -1000m -500m -250m

820ft 1640ft 3281ft 6562ft 9843ft 13,124ft 19,685ft

-19,658ft -13,124ft -6562ft -3281ft -1640ft -820ft -328ft/-100m 0

Northern India, Pakistan & Bangladesh

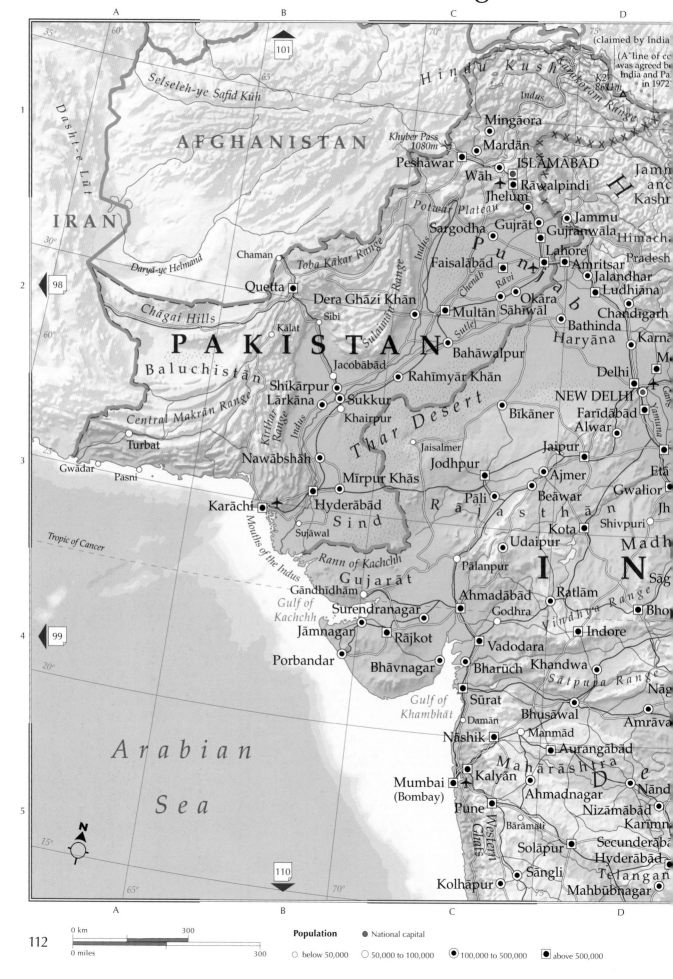

A B C D

35° 60° 65° 70° 75°

(claimed by India

(A"line of co
was agreed be
India and Pa
in 1972

Dasht-e Lūt

Selseleh-ye Safid Kūh

AFGHANISTAN

Hindu Kush

Karakoram Range

K2
8611m

1

Indus

Mingaora

Khyber Pass
1080m

Mardān

IRAN

Peshāwar

Wāh

Rāwalpindi

ISLĀMĀBĀD

Jammu
and
Kashir

Jhelum

Potwar Plateau

Jammu

Chaman

Toba Kākar Range

Indus

Sargodha

Gujrāt

Gujrānwāla

Daryā-ye Helmand

Himach

2

Quetta

Sulaimān Range

Faisalābād

Lahore

Amritsar

Pradesh

Jalandhar

Chenāb

Rāvi

Ludhiāna

Chāgai Hills

Dera Ghāzi Khan

Sibi

Multān

Sāhiwāl

Okāra

Chandīgarh

Sutlej

Bathinda

Karna

P A K I S T A N

Kālat

Haryāna

Bahāwalpur

Baluchistān

Jacobābād

Rahīmyār Khān

Delhi

M

Central Makrān Range

Shikārpur

NEW DELHI

Farīdābād

Larkāna

Sukkur

Bīkaner

Alwar

3

Turbat

Khairpur

Jaisalmer

Kirthar Range

Jaipur

Etā

Gwādar

Pasni

Nawābshāh

Jodhpur

Ajmer

Gwalior

Indus

Thar Desert

Pāli

Beāwar

Jh

Mīrpur Khās

R

a

Shivpuri

Karāchi

Hyderābād

Kota

j a s t h ā n

Madh

Tropic of Cancer

Sind

Udaipur

I

N

25°

Sujāwal

Pālanpur

Rann of Kachchh

Sāg

Mouths of the Indus

Gāndhīdhām

Gujarāt

Ahmadābād

Ratlām

20°

*Gulf of
Kachchh*

Surendranagar

Godhra

Bhoṕ

Vindhya Range

4

Jāmnagar

Rājkot

Indore

Vadodara

Porbandar

Bhāvnagar

Bharūch

Khandwa

Nāg

Sātpura Range

Sūrat

*Gulf of
Khambhāt*

Damān

Bhusāwal

Amrāva

Nāshik

Manmād

A r a b i a n

Aurangābād

M a h ā r ā s h t r a

D

S e a

Mumbai
(Bombay)

Kalyān

Ahmadnagar

Nānd

Nizāmābād

Pune

Karimn

15°

Bārāmatī

Secunderāba

Western Ghats

Hyderābād

Solāpur

Telangan

Sāngli

Mahbūbnagar

Kolhāpur

65° 70° 75°

A B C D

0 km 300

0 miles 300

Population ● National capital

○ below 50,000 ◎ 50,000 to 100,000 ⦿ 100,000 to 500,000 ■ above 500,000

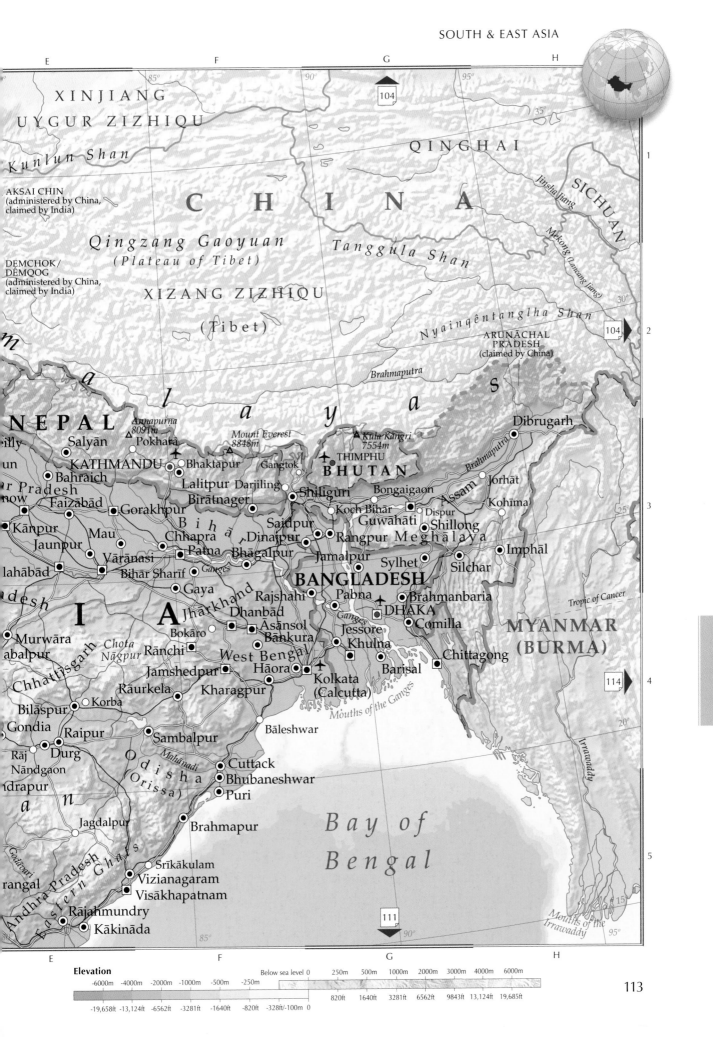

XINJIANG
UYGUR ZIZHIQU

Kunlun Shan

QINGHAI

SICHUAN

AKSAI CHIN
(administered by China,
claimed by India)

C H I N A

Qingzang Gaoyuan
(Plateau of Tibet)

Jinsha Jiang

Mekong (Lancang Jiang)

DEMCHOK/
DÊMQOG
(administered by China,
claimed by India)

XIZANG ZIZHIQU

(Tibet)

Tanggula Shan

Nyainqêntanglha Shan

ARUNĀCHAL
PRADESH
(claimed by China)

Brahmaputra

H *i* *m* *a* *l* *a* *y* *a* *s*

Dibrugarh

NEPAL

Annapurna
8091m

△ Mount Everest
8848m

△ Kula Kangri
7554m

Salyān
Pokharā

Bhaktapur
KATHMANDU
Lalitpur
Darjiling
Gangtok

THIMPHU
BHUTAN

Brahmaputra

Jorhāt

Bahraich

Biratnager
Shiliguri

Bongaigaon

Assam

Kohīma

r Pradesh
now

Faizābād
Gorakhpur

B i h a r

Saidpur

Koch Bihār

Guwāhāti

Dispur
Shillong

Kānpur
Mau
Jaunpur

Chhapra
Dinajpur
Rangpur

Meghālaya

Imphāl

Vārānasi
Patna
Bhāgalpur

Jamalpur

Sylhet

Silchar

ahābād

Bihār Sharīf

Ganges

BANGLADESH

desh

Gaya

Rajshahi
Pabna

Brahmanbaria

I
A

Jharkhand
Dhanbād
Bokāro
Āsānsol

Ganges

DHAKA

Comilla

MYANMAR
(BURMA)

Murwāra
abalpur

Chota
Nāgpur

Rānchi
Bankura

Jessore

Khulna

Chhattisgarh

Jamshedpur

West Bengal
Hāora

Barisal

Chittagong

Bilāspur
Korba
Rāurkela
Kharagpur

Kolkata
(Calcutta)

Gondia
Raipur

Rāj
Nāndgaon

Durg

Sambalpur

Bāleshwar

Mouths of the Ganges

drapur

O *d* *i* *s* *h* *a*
(Orissa)

Mahānadi

Cuttack

Bhubaneshwar

Puri

Bay of
Bengal

a
n

Jagdalpur

Brahmapur

Eastern Ghats

Godāvari

rangal

Srīkākulam
Vizianagaram
Visākhapatnam

Andhra Pradesh

Mouths of the
Irrawaddy

Irrawaddy

Rājahmundry
Kākināda

Tropic of Cancer

Elevation

| Below sea level 0 | 250m | 500m | 1000m | 2000m | 3000m | 4000m | 6000m |

-6000m -4000m -2000m -1000m -500m -250m

-19,658ft -13,124ft -6562ft -3281ft -1640ft -820ft -328ft/-100m 0

820ft 1640ft 3281ft 6562ft 9843ft 13,124ft 19,685ft

Mainland Southeast Asia

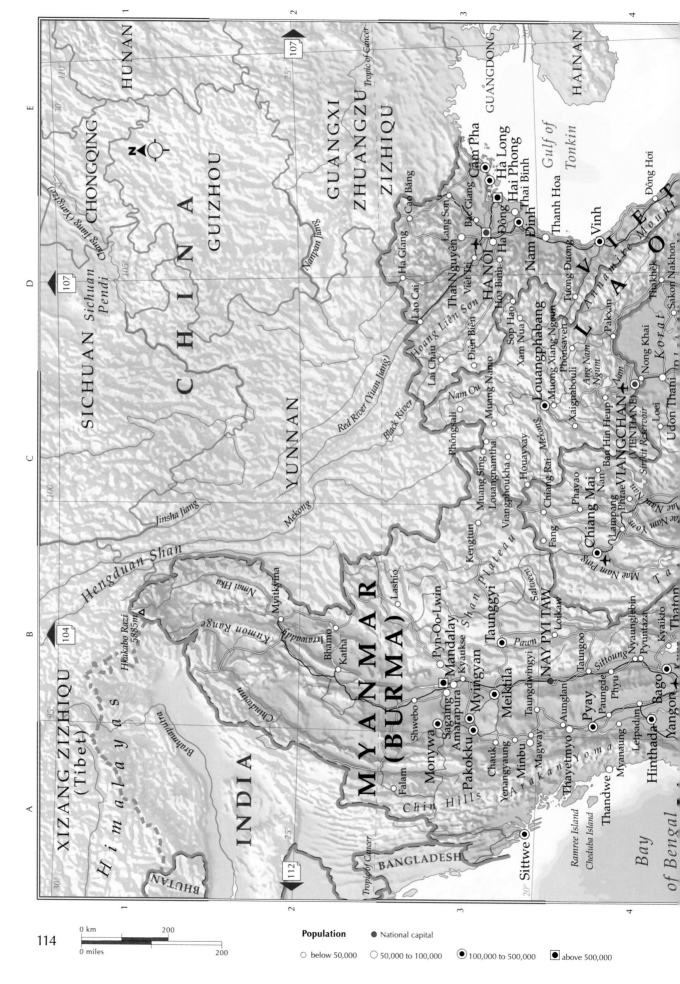

Population

- National capital
- ○ below 50,000
- ○ 50,000 to 100,000
- ◉ 100,000 to 500,000
- ■ above 500,000

0 km   200

0 miles   200

Elevation

| Below sea level | 0 | 250m | 500m | 1000m | 2000m | 3000m | 4000m | 6000m |

-6000m -4000m -2000m -1000m -500m -250m

-19,658ft -13,124ft -6562ft -3281ft -1640ft -820ft -328ft/-100m 0

820ft 1640ft 3281ft 6562ft 9843ft 13,124ft 19,685ft

Maritime Southeast Asia

SINGAPORE

0 km   10
0 miles   10

MALAYSIA

Johore Strait

Causeway
Lim Chu Kang
Bukit Panjang Hougang New Town
Choa Chu Kang
Pulau Ubin
Pulau Tekong
Changi
Bukit Timah 176m
Queenstown City Bedok New Town
Jurong Industrial Estate
Telok Blangah
Sentosa
Selat Pandan
Pulau Sudong
Pulau Pawai
Strait of Singapore

Urban areas
Open areas
Nature reserves

MYANMAR (BURMA)

115

LAOS
THAILAND
VIETNAM

Mekong

Gulf of Tonkin

Hainan Dao (to China)

PARACEL ISLANDS
(disputed by China, Taiwan and Vietnam)

South China Sea

CAMBODIA

SPRATLY ISLANDS
(disputed by China, Malaysia, Philippines, Taiwan and Vietnam)

111

Andaman Sea

Nicobar Islands (to India)

Gulf of Thailand

Mouths of the Mekong

Isthmus of Kra

Banda Aceh Sigli
Langsa
Meulaboh
Strait of Malacca
George Town Kota Bharu
Pulau Pinang Butterworth
Taiping Kuala Terengganu
Ipoh Dungun
Cukai
Kuantan
Medan
Tebingtinggi
Klang
Pematangsiantar KUALA LUMPUR
Pulau Simeulue PUTRAJAYA
Danau Toba Melaka
Kepulauan Banyak Muar
Sibolga Batu Pahat Johor Bahru
Pulau Nias SINGAPORE

Kepulauan Natuna

BANDAR SERI BEGAWAN
Kota Kinabalu
BRUNEI
Miri

Gunung Kina
Balabac

M A L A Y S I A

Bintulu
Selat Serasan
Sibu *Batang Rajang*
Sarawak
Sungai Kay
Kuching Sri Aman
Pegunungan Muller
Sungai Mahar

Equator

Pekanbaru
Kepulauan Lingga
Solok Rengat
Singkawang Sidas
Pontianak *Sungai Kapuas*
B o r n e o

Padang
Batang Hari Kualatungkal
Jambi
Selat Karimata
K a l i m a n t a n
Samarinda
Balikpapan

111

Sungaipenuh
Kepulauan Mentawai
Pulau Siberut
Pangkalpinang
Bangka
Palembang
I N
Sampit
Sungai Barito
D
Amunta Kandar

Bengkulu Lahat
Pulau Belitung
Banjarmasin
Pulau Laut

Sumatera (Sumatra)

Kotabumi

Java Sea
Mak

I N D I A N

Bandar Lampung
JAKARTA Cirebon Tegal
Serang Pekalongan
Selat Sunda Bogor Semarang
Sukabumi Kudus *Pulau Madura*
Bandung Surabaya
Tasikmalaya Probolinggo
Jember Mata
Cilacap Malang
Jawa (Java) Magelang Kediri
Yogyakarta Madiun *Bali* Denpa
Surakarta *Pulau Lombok*

O C E A N

MALAYSIA'S TWO CAPITALS

KUALA LUMPUR   – Capital
PUTRAJAYA   – Administrative capital

119

0 km   200
0 miles   200

Population ● National capital

○ below 50,000   ○ 50,000 to 100,000   ◉ 100,000 to 500,000   ◼ above 500,000

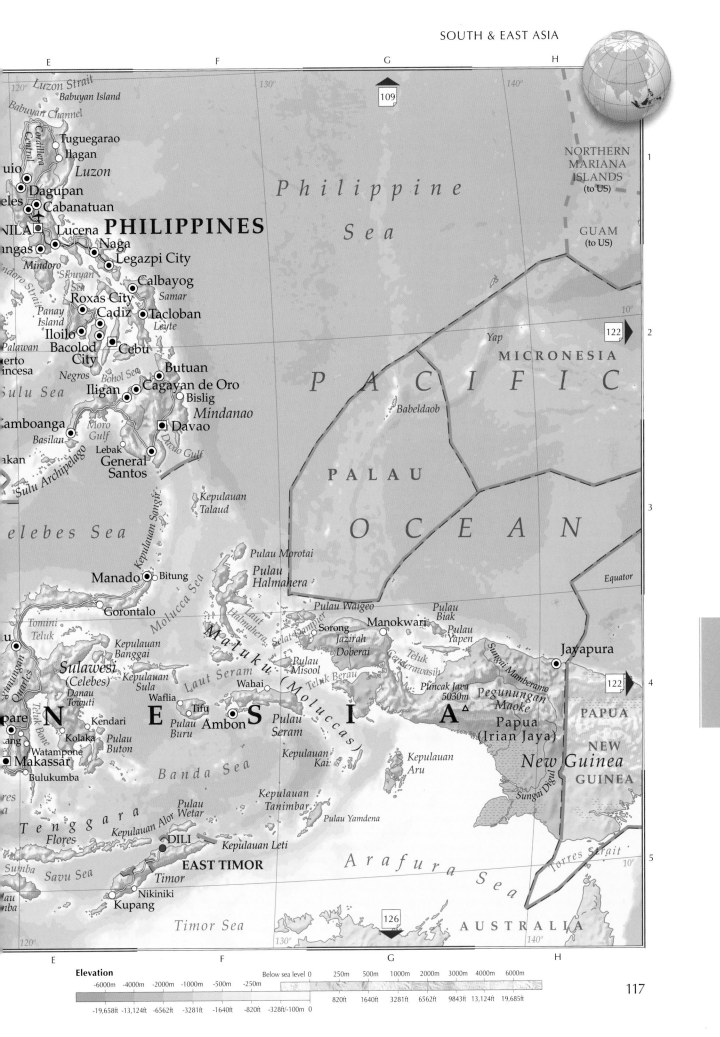

Luzon Strait
Babuyan Island
Babuyan Channel
Cordillera Central
Tuguegarao
Ilagan
Luzon
uio
Dagupan
eles
Cabanatuan
NILA
Lucena
angas
Naga
PHILIPPINES
Mindoro
Sibuyan Sea
Legazpi City
Calbayog
Samar
Roxas City
Cadiz
Panay Island
Tacloban
Leyte
Iloilo
Bacolod City
Cebu
Palawan
erto incesa
Negros
Bohol Sea
Butuan
Sulu Sea
Iligan
Cagayan de Oro
Bislig
amboanga
Moro Gulf
Mindanao
Basilan
Lebak
Davao
Davao Gulf
akan
Sulu Archipelago
General Santos

Philippine Sea

NORTHERN MARIANA ISLANDS (to US)

GUAM (to US)

Yap

MICRONESIA

P A C I F I C

Babeldaob

PALAU

O C E A N

Equator

Kepulauan Talaud

Kepulauan Sangir'

Pulau Morotai
Pulau Halmahera

elebes Sea

Manado
Bitung

Gorontalo
Laut Halmahera'
Molucca Sea

Pulau Waigeo
Selat Dampier
Sorong
Jazirah Doberai
Manokwari
Pulau Biak
Pulau Yapen
Teluk Cenderawasih

Sungai Mamberamo

Jayapura

Tomini Teluk

Sulawesi (Celebes)

Kepulauan Banggai

Kepulauan Sula

Matuku
Laut Seram
(Moluccas)
Teluk Berau

Pulau Misool

Puncak Jaya 5030m

Pegunungan Maoke

PAPUA

u

Danau Towuti

N
Kendari
E
S
Wahai
Waflia
Tifu
Pulau Buru
Ambon
Pulau Seram
Pulau
I
A
Papua (Irian Jaya)

New Guinea

NEW GUINEA

pare

Kolaka

Pulau Buton

Teluk Bone
Watampone
Makassar
Bulukumba

Banda Sea

Kepulauan Kai

Kepulauan Aru

Sungai Digul

Kepulauan Tanimbar

T e n g g a r a
Flores
Pulau Wetar
Kepulauan Alor
Pulau Yamdena

Kepulauan Leti

Sumba
Savu Sea
DILI
EAST TIMOR
Timor
Nikiniki
Kupang

A r a f u r a   S e a

Torres Strait

A U S T R A L I A

Timor Sea

109

122

122

126

117

Elevation

Below sea level 0   250m 500m 1000m 2000m 3000m 4000m 6000m

-6000m -4000m -2000m -1000m -500m -250m

-19,658ft -13,124ft -6562ft -3281ft -1640ft -820ft -328ft/-100m 0

820ft 1640ft 3281ft 6562ft 9843ft 13,124ft 19,685ft

The Indian Ocean

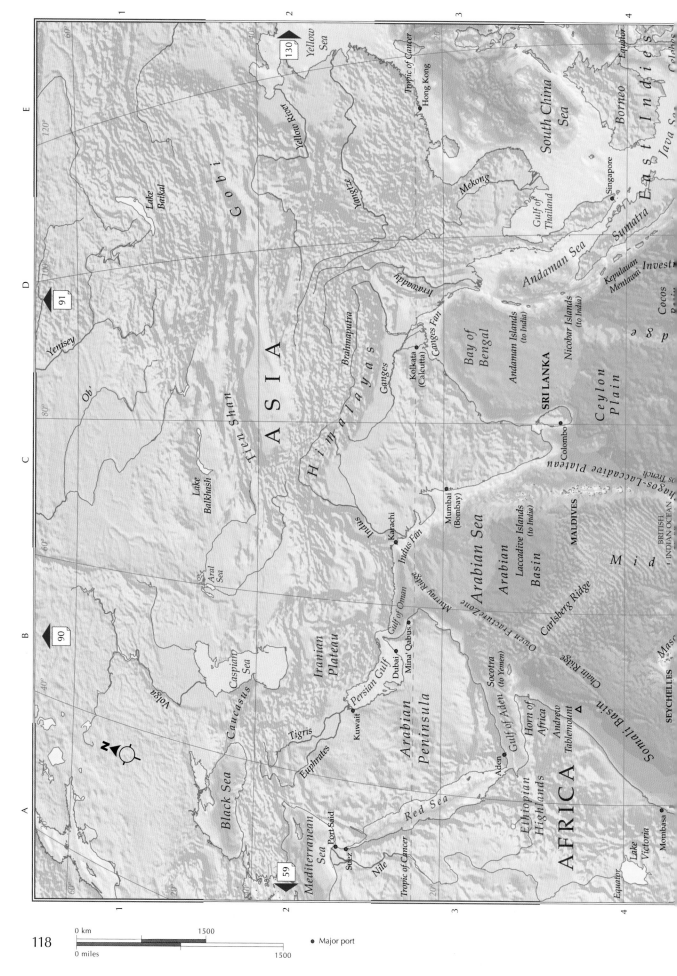

0 km		1500
0 miles		1500

● Major port

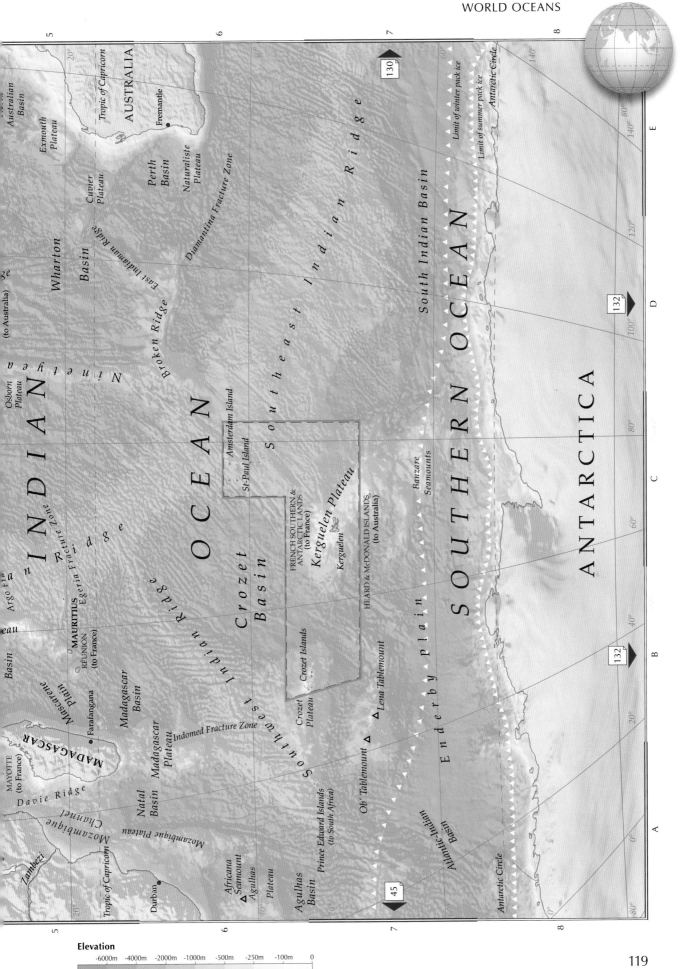

Elevation

-6000m	-4000m	-2000m	-1000m	-500m	-250m	-100m	0
-19,658ft	-13,124ft	-6562ft	-3281ft	-1640ft	-820ft	-328ft/-100m	0

Australasia & Oceania

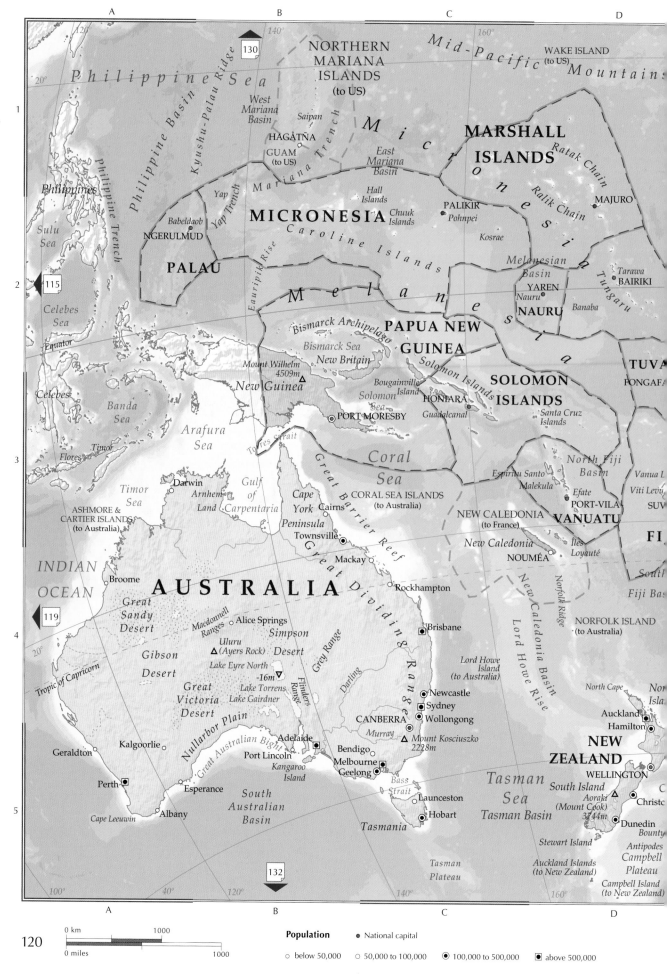

Philippine Sea

Mid-Pacific Mountains

WAKE ISLAND (to US)

NORTHERN MARIANA ISLANDS (to US)

130

MARSHALL ISLANDS

Ratak Chain

West Mariana Basin

Saipan

Micronesia

Ralik Chain

MAJURO

Philippine Basin

Kyushu-Palau Ridge

HAGÅTÑA

GUAM (to US)

East Mariana Basin

PALIKIR

Pohnpei

Philippine Trench

Philippines

Yap

Yap Trench

Mariana Trench

Hall Islands

Chuuk Islands

Kosrae

Babeldaob

MICRONESIA

Caroline Islands

Melanesian Basin

Tarawa

BAIRIKI

Tungaru

NGERULMUD

Eauripik Rise

Sulu Sea

PALAU

Melan

e

s

YAREN

Nauru

Banaba

i

NAURU

a

TUVA

FONGAFA

Celebes Sea

Equator

Bismarck Archipelago

Bismarck Sea

New Britain

PAPUA NEW GUINEA

Solomon Islands

SOLOMON ISLANDS

115

Celebes

Mount Wilhelm 4509m

New Guinea

Bougainville Island

Solomon Sea

HONIARA

North Fiji Basin

Banda Sea

Timor

Flores

Arafura Sea

PORT MORESBY

Guadalcanal

Santa Cruz Islands

Espíritu Santo

Malekula

Efate

Vanua

Viti Lev

SUV

Timor Sea

Darwin

Torres Strait

Coral Sea

CORAL SEA ISLANDS (to Australia)

NEW CALEDONIA (to France)

PORT-VILA

VANUATU

FI

ASHMORE & CARTIER ISLANDS (to Australia)

Arnhem Land

Gulf of Carpentaria

Cape York

Cairns

New Caledonia

Îles Loyauté

South

Fiji Bas

INDIAN OCEAN

Peninsula

Townsville

Great Barrier Reef

NOUMÉA

Broome

AUSTRALIA

Mackay

Rockhampton

New Caledonia Basin

Lord Howe Rise

Norfolk Ridge

NORFOLK ISLAND (to Australia)

119

Great Sandy Desert

Macdonnell Ranges

Alice Springs

Simpson Desert

Grey Range

Brisbane

Lord Howe Island (to Australia)

North Cape

Nor

Isla

Gibson Desert

Uluru (Ayers Rock)

Great Dividing Range

Auckland

Tropic of Capricorn

Lake Eyre North

-16m

Lake Torrens

Flinders Range

Darling

Newcastle

Sydney

Hamilton

Great Victoria Desert

Lake Gairdner

Murray

CANBERRA

Wollongong

NEW ZEALAND

Geraldton

Kalgoorlie

Nullarbor Plain

Adelaide

Bendigo

Mount Kosciuszko 2228m

WELLINGTON

Perth

Esperance

Great Australian Bight

Port Lincoln

Kangaroo Island

Melbourne

Geelong

Bass Strait

Tasman Sea

South Island

Aoraki (Mount Cook) 3744m

Christc

Cape Leeuwin

Albany

South Australian Basin

Launceston

Hobart

Tasmania

Tasman Basin

Dunedin

Bounty

Stewart Island

Antipodes

Campbell Plateau

132

Tasman Plateau

Auckland Islands (to New Zealand)

Campbell Island (to New Zealand)

0 km   1000

0 miles   1000

Population   • National capital

○ below 50,000   ○ 50,000 to 100,000   ◉ 100,000 to 500,000   ▣ above 500,000

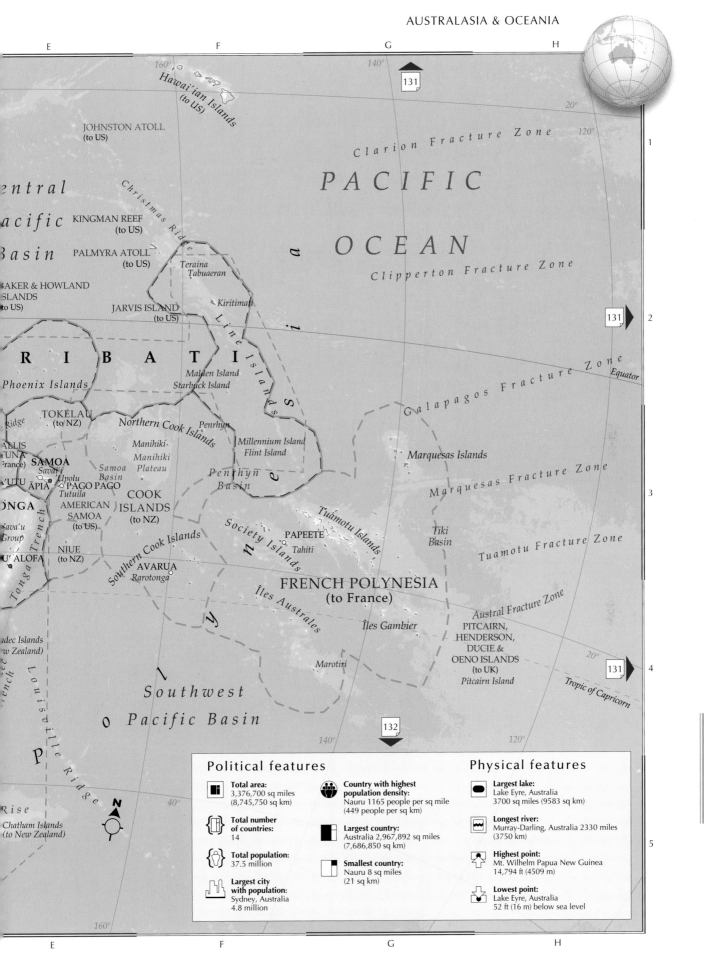

PACIFIC

OCEAN

Clarion Fracture Zone

Clipperton Fracture Zone

Galapagos Fracture Zone

Equator

Marquesas Fracture Zone

Tuamotu Fracture Zone

Austral Fracture Zone

Tropic of Capricorn

Christmas Ridge

Hawai'ian Islands
(to US)

JOHNSTON ATOLL
(to US)

KINGMAN REEF
(to US)

PALMYRA ATOLL
(to US)

Teraina
Tabuaeran

Kiritimati

JARVIS ISLAND
(to US)

Line Islands

Malden Island
Starbuck Island

Millennium Island
Flint Island

Marquesas Islands

Tiki
Basin

Tuamotu Islands

entral

acific

Basin

BAKER & HOWLAND
ISLANDS
(to US)

KIRIBATI

Phoenix Islands

TOKELAU
(to NZ)

Northern Cook Islands

Penrhyn

Penrhyn
Basin

Manihiki
Manihiki
Plateau

Ridge

ALLIS
UNA
(France)

SAMOA
Savai'i
Upolu
APIA
Tutuila
PAGO PAGO

Samoa
Basin

COOK
ISLANDS
(to NZ)

AMERICAN
SAMOA
(to US)

Society Islands

PAPEETE
Tahiti

FRENCH POLYNESIA
(to France)

Îles Australes

Îles Gambier

PITCAIRN,
HENDERSON,
DUCIE &
OENO ISLANDS
(to UK)
Pitcairn Island

Marotiri

TONGA

Vava'u
Group

NUKU'ALOFA

NIUE
(to NZ)

Southern Cook Islands

AVARUA
Rarotonga

Tonga Trench

Southwest

Pacific Basin

Louisville Ridge

dec Islands
w Zealand)

Rise

Chatham Islands
(to New Zealand)

N

Political features

Total area:
3,376,700 sq miles
(8,745,750 sq km)

**Total number
of countries:**
14

Total population:
37.5 million

**Largest city
with population:**
Sydney, Australia
4.8 million

**Country with highest
population density:**
Nauru 1165 people per sq mile
(449 people per sq km)

Largest country:
Australia 2,967,892 sq miles
(7,686,850 sq km)

Smallest country:
Nauru 8 sq miles
(21 sq km)

Physical features

Largest lake:
Lake Eyre, Australia
3700 sq miles (9583 sq km)

Longest river:
Murray-Darling, Australia 2330 miles
(3750 km)

Highest point:
Mt. Wilhelm Papua New Guinea
14,794 ft (4509 m)

Lowest point:
Lake Eyre, Australia
52 ft (16 m) below sea level

The Southwest Pacific

A B C D

1
2
3
4
5

130

117

124

127

Saipan
Tinian
Rota
NORTHERN
MARIANA
ISLANDS
(to US)
GUAM
(to US)
HAGÁTÑA

MARSHALL
ISLANDS

Micronesia

MICRONESIA

Yap

Babeldaob
NGERULMUD
PALAU

Chuuk
Islands
PALIKIR
Pohnpei

Caroline Islands

Kosrae

Enewetak
Atoll
Bikini Atoll
Rongelap
Atoll
Ailuk Ato
Ujelang Atoll
Kwajalein
Atoll
Namu Atoll
Ailinglaplap Atoll
Jaluit Atoll
Ebon Atoll
Wotje At
Maloe
Atoll
Maj
Ato
Mili At
Ma

Ratak Chain
Ratik Chain

Equator

Admiralty
Islands
St. Matthias Group

BAIRIK

Tar

Abemo
Non

YAREN
NAURU
Banaba

Bismarck Archipelago
Bismarck Sea
New Guinea
New Ireland
PAPUA NEW GUINEA
Madang
Mount Wilhelm
4509m
Central Range
Lae
New
Britain
Bougainville
Island

Melanesia

INDONESIA

Solomon Sea
Owen Stanley Range
Gulf of
Papua
PORT MORESBY
Torres Strait

Choiseul
Santa Isabel
New Georgia
Islands
Malaita
SOLOMON
ISLANDS
HONIARA
Guadalcanal

Solomon Islands

Arafura Sea

D'Entrecasteaux
Islands
Louisiade
Archipelago

San Cristobal
Rennell
Santa Cruz
Islands

Arnhem
Land
Groote
Eylandt
Gulf of
Carpentaria
Cape
York
Peninsula
Barkly Tableland

Great Barrier Reef

Coral Sea

CORAL SEA ISLANDS
(to Australia)

Banks Islands
Espiritu Santo
Malekula
Maéwo
Pentecost
Ambrym
Epi
Efate
PORT-VILA
VANUATU
Erromango
Tanna
Aneityum

NEW
CALEDONIA
(to France)
Ouvéa
Lifou
Maré
New
Caledonia
NOUMÉA
Îles Loyauté

NORTHERN
TERRITORY
Tropic of Capricorn
Macdonnell
Ranges
QUEENSLAND
Great Dividing Range
AUSTRALIA

140°
150°
160°
170°
10°
20°

Population

National capital

0 km   750
0 miles   750

below 50,000   50,000 to 100,000   100,000 to 500,000   above 500,000

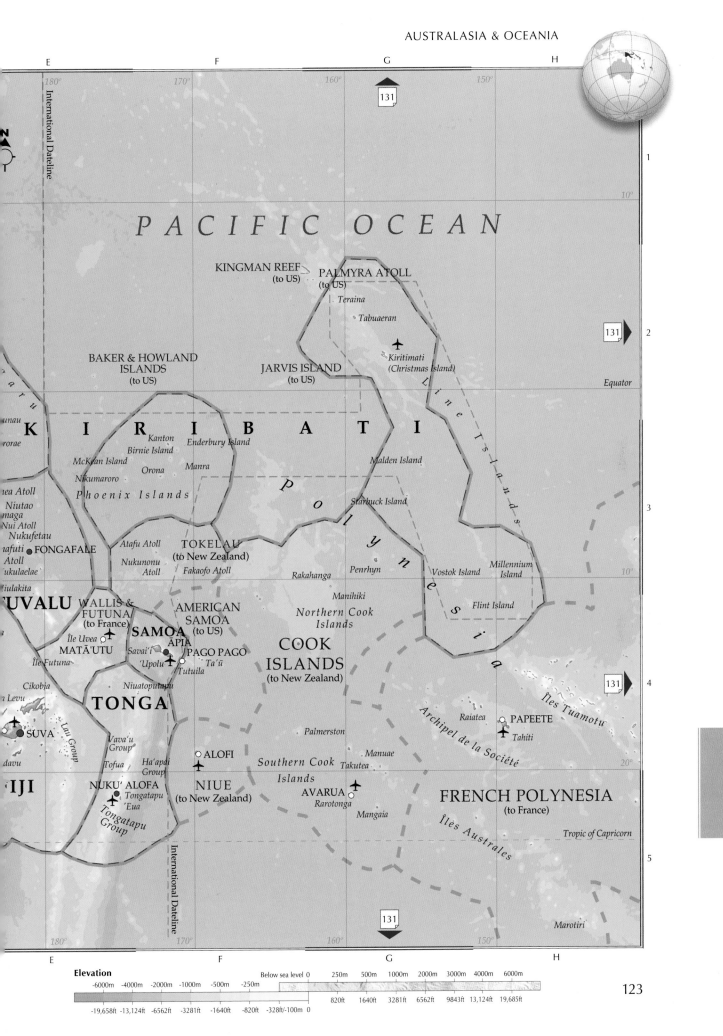

PACIFIC OCEAN

KINGMAN REEF
(to US)
PALMYRA ATOLL
(to US)

Teraina

Tabuaeran

BAKER & HOWLAND
ISLANDS
(to US)

JARVIS ISLAND
(to US)

*Kiritimati
(Christmas Island)*

Equator

K I R I B A T I

Kanton

Birnie Island

Enderbury Island

McKean Island

Orona

Manra

Malden Island

Nikumaroro

Phoenix Islands

P
o
l
y
n
e
s
i
a

Starbuck Island

Niutao

...maga

Nui Atoll

Nukufetau

...*ea Atoll*

...*afuti* ● FONGAFALE

Atoll

...ukulaelae

Atafu Atoll

TOKELAU
(to New Zealand)

*Nukunonu
Atoll*

Fakaofo Atoll

Rakahanga

Penrhyn

Vostok Island

*Millennium
Island*

Flint Island

...*iulakita*

WALLIS &
FUTUNA
(to France)

AMERICAN
SAMOA
(to US)

Manihiki

Northern Cook
Islands

TUVALU

Île Uvea ○

MATĀ'UTU

SAMOA

ĀPIA

PAGO PAGO

Savai'i

'Upolu

Ta'ū

Tutuila

COOK
ISLANDS
(to New Zealand)

Île Futuna

Cikobia

Niuatoputapu

...a *Levu*

TONGA

● SUVA

Lau Group

*Vava'u
Group*

Palmerston

Manuae

Raiatea

○ PAPEETE

Takutea

Tahiti

Archipel de la Société

Îles Tuamotu

...*davu*

Tofua

*Ha'apai
Group*

○ ALOFI

Southern Cook
Islands

...IJI

NUKU' ALOFA

Tongatapu

'Eua

NIUE
(to New Zealand)

AVARUA ○

Rarotonga

FRENCH POLYNESIA
(to France)

*Tongatapu
Group*

Mangaia

Îles Australes

Tropic of Capricorn

Marotiri

International Dateline

180°

170°

160°

150°

10°

Equator

10°

20°

131

131

131

131

Elevation

| -6000m | -4000m | -2000m | -1000m | -500m | -250m | Below sea level 0 | 250m | 500m | 1000m | 2000m | 3000m | 4000m | 6000m |

-19,658ft -13,124ft -6562ft -3281ft -1640ft -820ft -328ft/-100m 0 820ft 1640ft 3281ft 6562ft 9843ft 13,124ft 19,685ft

Western Australia

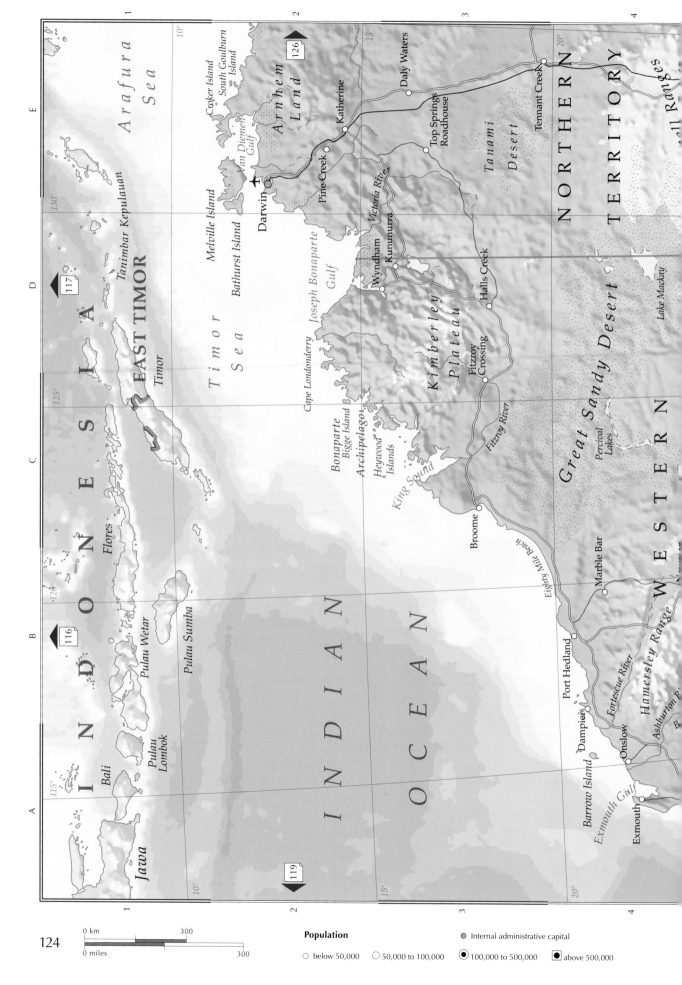

INDONESIA

Jawa

Bali

Pulau Lombok

Pulau Wetar

Pulau Sumba

Flores

Tanimbar Kepulauan

Timor

EAST TIMOR

116

117

119

126

Arafura Sea

Croker Island

South Goulburn Island

Melville Island

Bathurst Island

Van Diemen Gulf

Arnhem Land

Darwin

Katherine

Pine Creek

Timor Sea

Cape Londonderry

Joseph Bonaparte Gulf

Wyndham

Kununurra

Daly Waters

Top Springs Roadhouse

Tennant Creek

Victoria River

Tanami Desert

NORTHERN

TERRITORY

Bonaparte Archipelago

Bigge Island

Heywood Islands

King Sound

Kimberley Plateau

Halls Creek

Fitzroy Crossing

Fitzroy River

Great Sandy Desert

Percival Lakes

Lake Mackay

Broome

Eighty Mile Beach

Marble Bar

Port Hedland

Fortescue River

Dampier

Onslow

Hamersley Range

Ashburton R.

Barrow Island

Exmouth Gulf

Exmouth

WESTERN

INDIAN

OCEAN

124

| 0 km | | 300 |
| 0 miles | | 300 |

Population

○ below 50,000   ○ 50,000 to 100,000   ◉ 100,000 to 500,000   ◼ above 500,000

● Internal administrative capital

Elevation

-6000m	-4000m	-2000m	-1000m	-500m	-250m	Below sea level 0	250m	500m	1000m 2000m 3000m 4000m 6000m

| -19,658ft | -13,124ft | -6562ft | -3281ft | -1640ft | -820ft | -328ft/-100m 0 | 820ft | 1640ft | 3281ft 6562ft 9843ft 13,124ft 19,685ft |

Eastern Australia

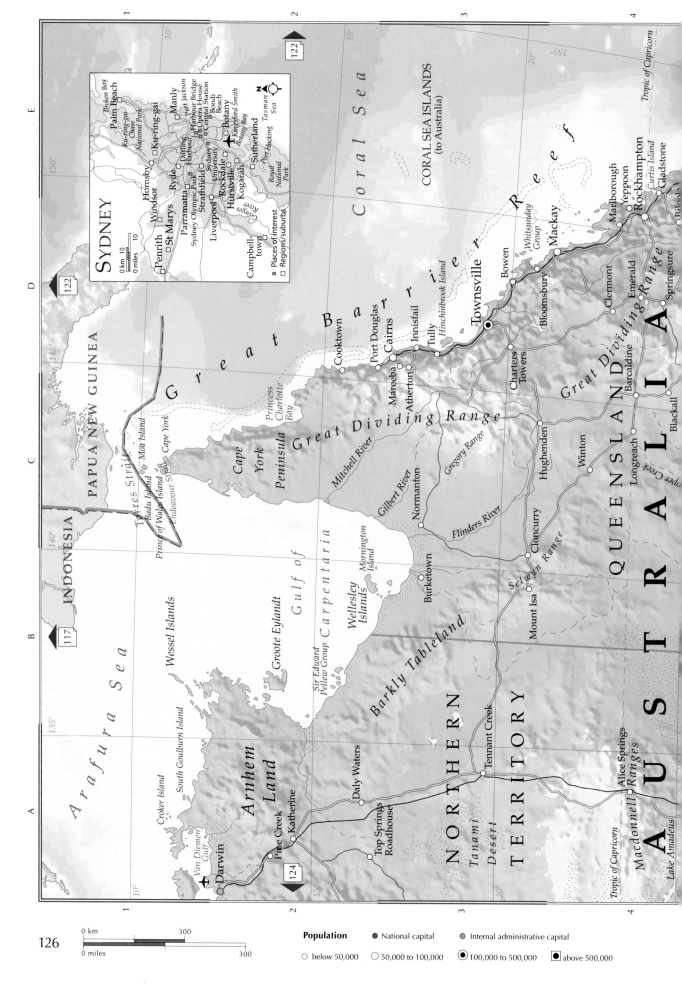

SYDNEY

Broken Bay
Palm Beach
Ku-ring-gai Chase National Park
Manly
Ku-ring-gai
Port Jackson
Harbour Bridge
Opera House
Central Station
Darling Harbour
Hornsby
Windsor
St Marys
Ryde
Parramatta
Sydney Olympic Park
Strathfield
University
Sydney
Liverpool
Penrith
Rockdale
Hurstville
Kogarah
Campbell town
Bondi Beach
Botany
Botany Bay
Sutherland
Port Hacking
Royal National Park
Georges River

0 km   10
0 miles   10

■ Places of interest
□ Regions/suburbs

INDONESIA

PAPUA NEW GUINEA

Arafura Sea

Croker Island
South Goulburn Island
Wessel Islands
Groote Eylandt
Van Diemen Gulf
Darwin
Pine Creek
Katherine
Daly Waters
Top Springs Roadhouse
Tanami Desert

Arnhem Land

NORTHERN TERRITORY

Tennant Creek
Alice Springs
Macdonnell Ranges
Lake Amadeus
Tropic of Capricorn

AUSTRALIA

Torres Strait
Badu Island
Moa Island
Prince of Wales Island
Endeavour Strait
Cape York
Cape York Peninsula
Princess Charlotte Bay

Gulf of Carpentaria
Sir Edward Pellew Group
Wellesley Islands
Mornington Island
Burketown
Barkly Tableland
Mount Isa
Selwyn Range
Cloncurry

Great Dividing Range
Mitchell River
Gilbert River
Normanton
Flinders River
Gregory Range
Hughenden
Winton
Longreach
Barcaldine
Blackall
Cooper Creek

QUEENSLAND

Cooktown
Port Douglas
Cairns
Mareeba
Atherton
Innisfail
Tully
Hinchinbrook Island
Townsville
Charters Towers
Bowen
Bloomsbury
Whitsunday Group
Mackay
Clermont
Emerald
Springsure
Marlborough
Yeppoon
Rockhampton
Curtis Island
Gladstone
Biloela

Coral Sea

Great Barrier Reef

CORAL SEA ISLANDS
(to Australia)

Great Dividing Range

Tasman Sea

0 km   300
0 miles   300

Population   ● National capital   ● Internal administrative capital
○ below 50,000   ○ 50,000 to 100,000   ◉ 100,000 to 500,000   ◼ above 500,000

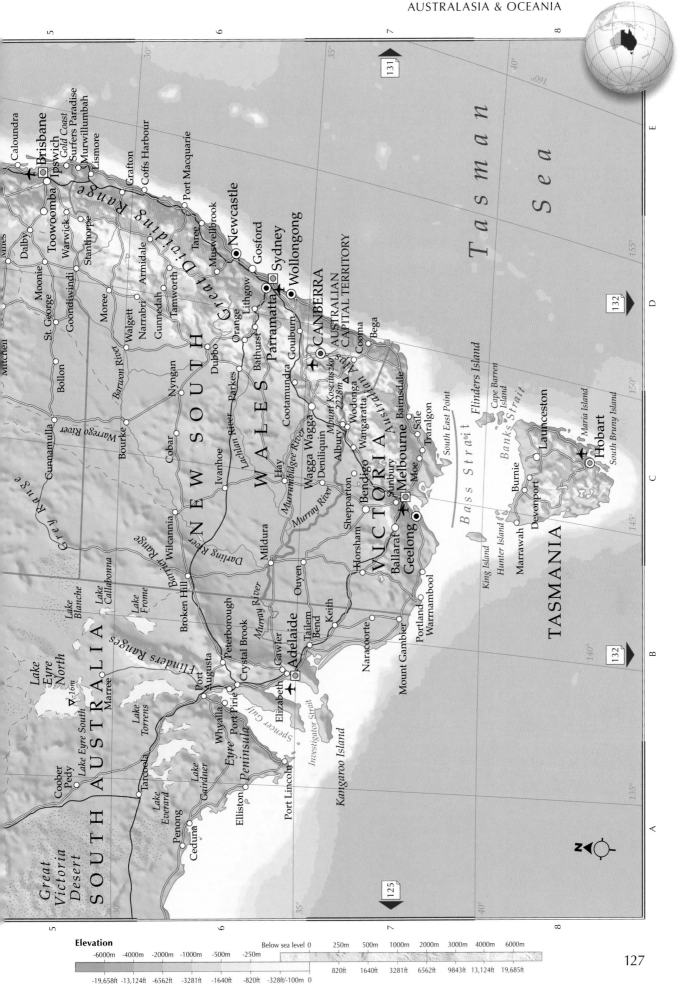

Elevation

| Below sea level 0 | 250m | 500m | 1000m | 2000m | 3000m | 4000m | 6000m |

| -6000m | -4000m | -2000m | -1000m | -500m | -250m |

| -19,658ft | -13,124ft | -6562ft | -3281ft | -1640ft | -820ft | -328ft/-100m 0 |

| 820ft | 1640ft | 3281ft | 6562ft | 9843ft | 13,124ft | 19,685ft |

New Zealand

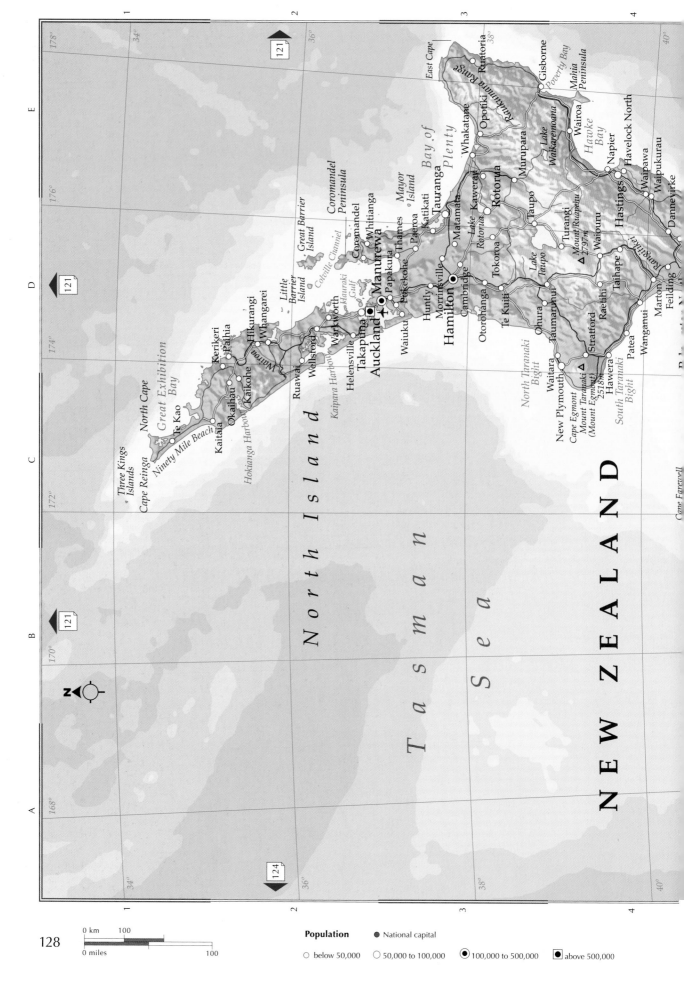

0 km   100

0 miles   100

Population   ● National capital

○ below 50,000   ○ 50,000 to 100,000   ◉ 100,000 to 500,000   ◼ above 500,000

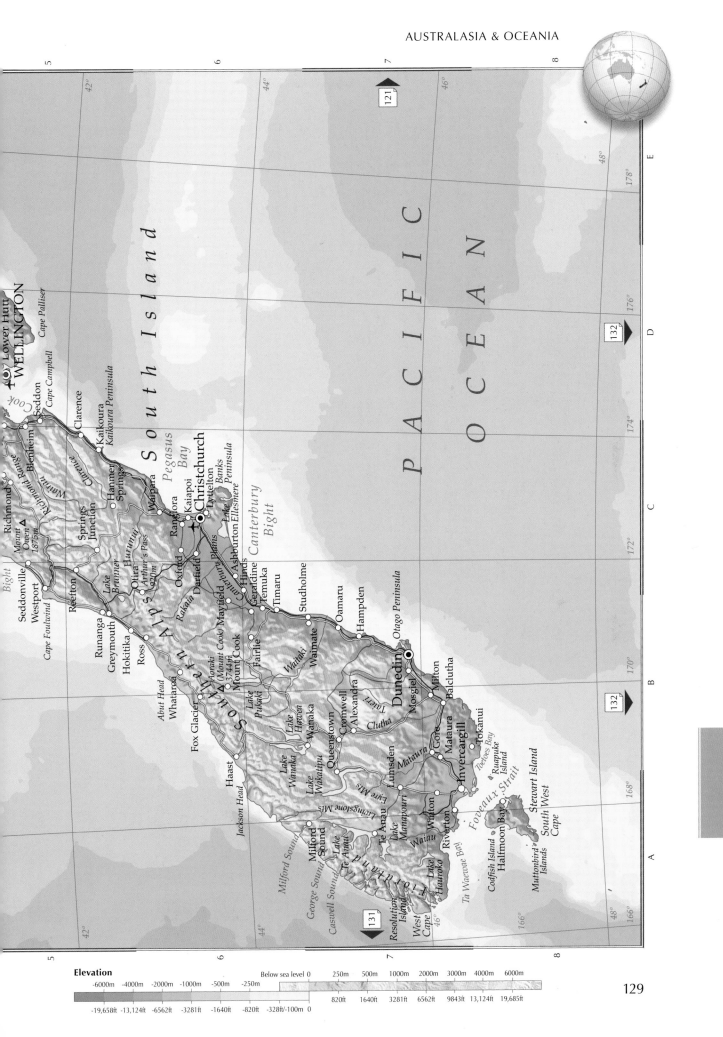

Elevation

							Below sea level 0	250m	500m	1000m	2000m	3000m	4000m	6000m
-6000m	-4000m	-2000m	-1000m	-500m	-250m									
-19,658ft	-13,124ft	-6562ft	-3281ft	-1640ft	-820ft	-328ft/-100m 0		820ft	1640ft	3281ft	6562ft	9843ft	13,124ft	19,685ft

The Pacific Ocean

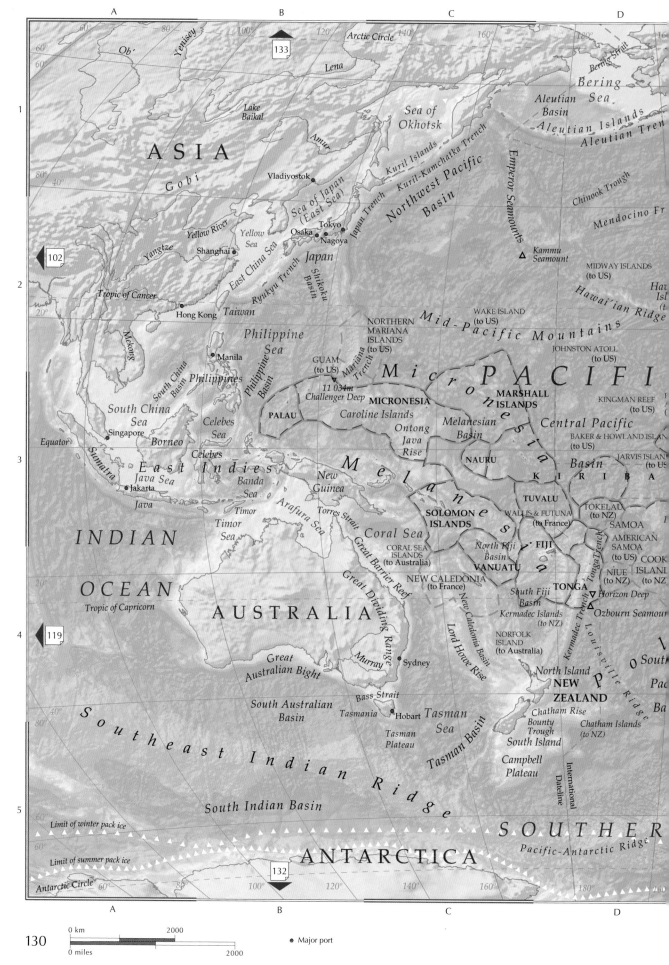

NORTH
AMERICA

ATLANTIC

OCEAN

44 ▶

Rocky Mountains

Arctic Circle

Hudson
Bay

Labrador
Sea

Vancouver

Great Lakes

Cascadia
Basin

Appalachian Mountains

Colorado

San Francisco

Mississippi

Long Beach

ay Fracture Zone

Gulf of California

Gulf of
Mexico

Tropic of Cancer

kai Fracture Zone

Greater Antilles

Lesser Antilles

Clarion Fracture Zone

Middle America Trench

Caribbean Sea

OCEAN

CLIPPERTON ISLAND
(to France)

Clipperton Fracture Zone

Guatemala
Basin

Panama City

N

Cocos Ridge

Galapagos Fracture Zone

Gallego Rise

Galápagos Islands
(to Ecuador)

Amazon

Equator

SOUTH
AMERICA

Marquesas
Islands

Peru Basin

Peru-Chile Trench

Marquesas
Fracture Zone

Bauer
Basin

Galapagos
Rise

Callao

Tiki
Basin

Mendaña Fracture Zone

FRENCH
POLYNESIA
(to France)

Austral
Fracture Zone

Nazca Ridge

Îles Gambier

PITCAIRN,
HENDERSON,
DUCIE &
OENO ISLANDS
(to UK)

Sala y Gomez
(to Chile)

Sala y Gomez Ridge

Easter Fracture Zone

Tropic of Capricorn

East Pacific Rise

Easter Island
(to Chile)

Isla San Félix
(to Chile)

Isla San Ambrosio
(to Chile)

Chile Basin

Paraná

45 ▶

Australes

Islas Juan Fernández
(to Chile)

Valparaiso

Andes

Agassiz Fracture Zone

Challenger Fracture Zone

ATLANTIC

Chile Rise

OCEAN

Eltanin Fracture Zone

Mornington
Abyssal
Plain

Cape Horn

OCEAN

Southeast Pacific Basin

Bellingshausen Plain

Drake Passage

PETER I ØY
(to Norway)

Amundsen Plain

Antarctic Circle

133

132

Elevation

| -6000m | -4000m | -2000m | -1000m | -500m | -250m | -100m | 0 |

| -19,658ft | -13,124ft | -6562ft | -3281ft | -1640ft | -820ft | -328ft/-100m | 0 |

131

Antarctica

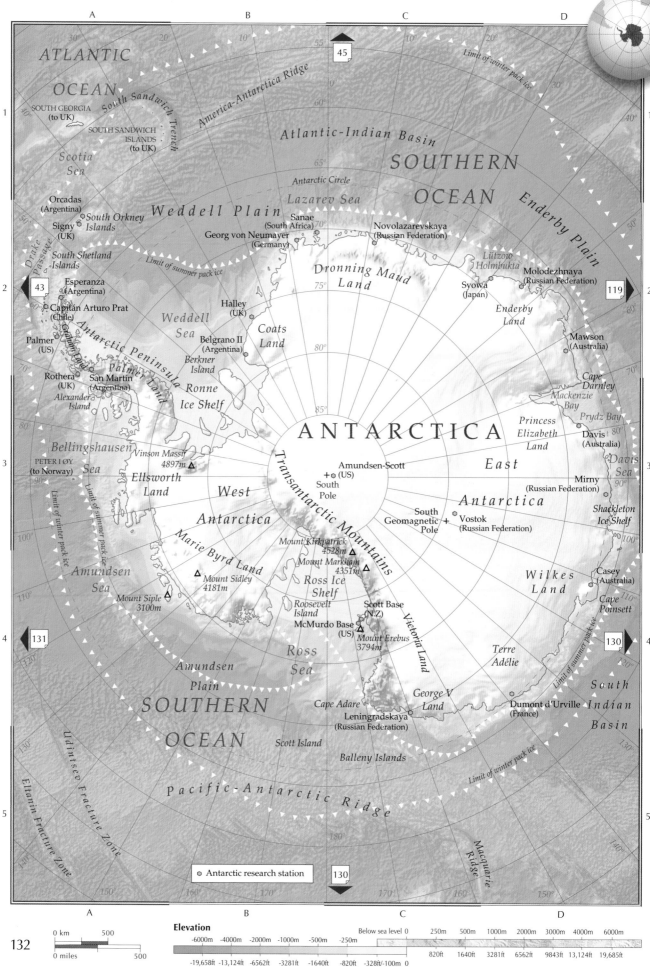

ATLANTIC OCEAN

SOUTH GEORGIA (to UK)

South Sandwich Trench

SOUTH SANDWICH ISLANDS (to UK)

Scotia Sea

America-Antarctica Ridge

Atlantic-Indian Basin

SOUTHERN OCEAN

Enderby Plain

Antarctic Circle

Lazarev Sea

Weddell Plain

Orcadas (Argentina)

South Orkney Islands

Signy (UK)

Drake Passage

South Shetland Islands

Sanae (South Africa)

Georg von Neumayer (Germany)

Novolazarevskaya (Russian Federation)

Lützow Holmbukta

Molodezhnaya (Russian Federation)

Dronning Maud Land

Syowa (Japan)

Enderby Land

Esperanza (Argentina)

Capitán Arturo Prat (Chile)

Halley (UK)

Weddell Sea

Belgrano II (Argentina)

Coats Land

Mawson (Australia)

Cape Darnley

Palmer (US)

Antarctic Peninsula

Graham Land

Palmer Land

Berkner Island

Ronne Ice Shelf

Mackenzie Bay

Prydz Bay

Rothera (UK)

San Martín (Argentina)

Alexander Island

ANTARCTICA

Princess Elizabeth Land

Davis (Australia)

Bellingshausen Sea

Vinson Massif 4897m △

East

Davis Sea

PETER I ØY (to Norway)

Ellsworth Land

West Antarctica

Amundsen-Scott (US)

South Pole

Antarctica

Mirny (Russian Federation)

Transantarctic Mountains

South Geomagnetic Pole

Vostok (Russian Federation)

Shackleton Ice Shelf

Amundsen Sea

Marie Byrd Land

Mount Kirkpatrick 4528m △

Mount Markham 4351m △

Casey (Australia)

△ Mount Sidley 4181m

Ross Ice Shelf

Wilkes Land

Cape Poinsett

Mount Siple 3100m △

Roosevelt Island

Scott Base (N.Z.)

McMurdo Base (US) △

Mount Erebus 3794m

Victoria Land

Terre Adélie

Amundsen Plain

Ross Sea

SOUTHERN OCEAN

Cape Adare

George V Land

Dumont d'Urville (France)

South Indian Basin

Leningradskaya (Russian Federation)

Scott Island

Balleny Islands

Pacific-Antarctic Ridge

Macquarie Ridge

Udintsev Fracture Zone

Eltanin Fracture Zone

Limit of winter pack ice

Limit of summer pack ice

○ Antarctic research station

Limit of winter pack ice

132

0 km 500
0 miles 500

Elevation

							Below sea level 0	250m	500m	1000m	2000m	3000m	4000m	6000m
-6000m	-4000m	-2000m	-1000m	-500m	-250m									
-19,658ft	-13,124ft	-6562ft	-3281ft	-1640ft	-820ft	-328ft/-100m	0		820ft	1640ft	3281ft	6562ft	9843ft	13,124ft	19,685ft

Arctic Ocean

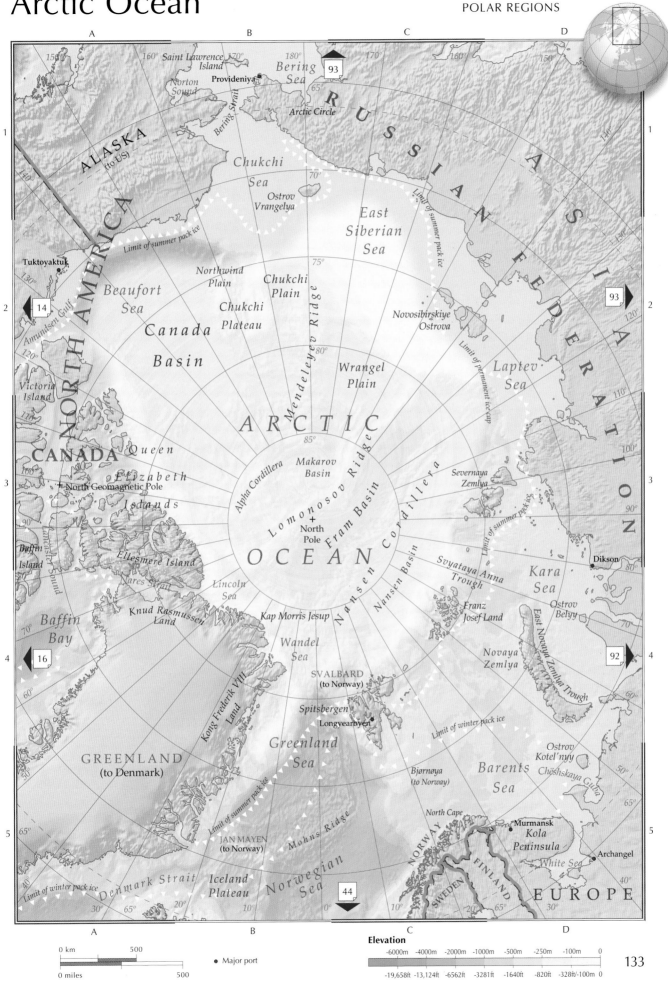

ALASKA
(to US)

NORTH AMERICA

CANADA

Saint Lawrence
Island

Norton
Sound

Providenlya

Bering
Sea

Arctic Circle

RUSSIAN FEDERATION

Chukchi
Sea

Ostrov
Vrangelya

East
Siberian
Sea

Tuktoyaktuk

Limit of summer pack ice

Beaufort
Sea

Northwind
Plain

Chukchi
Plain

Mendeleyev Ridge

Limit of summer pack ice

Amundsen Gulf

Canada
Basin

Chukchi
Plateau

Novosibirskiye
Ostrova

Laptev
Sea

Limit of permanent ice cap

Victoria
Island

Wrangel
Plain

Queen

Elizabeth

North Geomagnetic Pole

Islands

ARCTIC

Alpha Cordillera

Makarov
Basin

Lomonosov Ridge

Severnaya
Zemlya

Nansen Cordillera

Baffin
Island

Ellesmere Island

Nares Strait

Lincoln
Sea

North
Pole

Fram Basin

Dikson

Kara
Sea

Ostrov
Belyy

Baffin
Bay

Knud Rasmussen
Land

Kap Morris Jesup

OCEAN

Nansen Basin

Svyataya Anna
Trough

Franz
Josef Land

East Novaya Zemlya Trough

Wandel
Sea

SVALBARD
(to Norway)

Novaya
Zemlya

GREENLAND
(to Denmark)

Kong Frederik VIII Land

Spitsbergen

Longyearbyen

Greenland
Sea

Bjørnøya
(to Norway)

Limit of winter pack ice

Ostrov
Kotel'nyy

Barents
Sea

Chëshskaya Guba

Limit of summer pack ice

JAN MAYEN
(to Norway)

Mohns Ridge

North Cape

NORWAY

Murmansk

Kola
Peninsula

Archangel

White Sea

Denmark Strait

Limit of winter pack ice

Iceland
Plateau

Norwegian
Sea

FINLAND

SWEDEN

EUROPE

Elevation

-6000m	-4000m	-2000m	-1000m	-500m	-250m	-100m	0
-19,658ft	-13,124ft	-6562ft	-3281ft	-1640ft	-820ft	-328ft/-100m	0

0 km   500

0 miles   500

● Major port

133

Country Profiles

This Factfile is intended as a guide to a world that is continually changing as political fashions and personalities come and go. Nevertheless, all the material in these factfiles has been researched from the most up-to-date and authoritative sources to give an incisive portrait of the geographical, political, and social characteristics that make each country so unique.

There are currently 196 independent countries in the world - more than at any previous time - and over 50 dependencies. Antarctica is the only land area on Earth that is not officially part of, and does not belong to, any single country.

Country profile key

Formation Date of formation denotes the date of political origin or independence of a state, i.e. its emergence as a recognizable entity in the modern political world / date current borders were established

Population Total population / population density – based on total *land* area

Languages An asterisk (*) denotes the official language(s)

Calorie consumption Average number of kilocalories consumed daily per person

---

AFGHANISTAN
Central Asia

Page 100 D4

Landlocked in Central Asia, Afghanistan has suffered decades of conflict. The Islamist taliban, ousted by a US-led offensive in 2001, continue to resist subsequent elected governments.

Official name Islamic Republic of Afghanistan
Formation 1919 / 1919
Capital Kabul
Population 30.6 million / 122 people per sq mile (47 people per sq km)
Total area 250,000 sq miles (647,500 sq km)
Languages Pashtu*, Tajik, Dari*, Farsi, Uzbek, Turkmen
Religions Sunni Muslim 80%, Shi'a Muslim 19%, Other 1%
Ethnic mix Pashtun 38%, Tajik 25%, Hazara 19%, Uzbek and Turkmen 15%, Other 3%
Government Nonparty system
Currency Afghani = 100 puls
Literacy rate 32%
Calorie consumption 2107 kilocalories

ALBANIA
Southeast Europe

Page 79 C6

Lying at the southeastern end of the Adriatic Sea, Albania – or the "land of the eagles" – underwent upheavals after 1991 to emerge from its communist-period isolation.

Official name Republic of Albania
Formation 1912 / 1921
Capital Tirana
Population 3.2 million / 302 people per sq mile (117 people per sq km)
Total area 11,100 sq miles (28,748 sq km)
Languages Albanian*, Greek
Religions Sunni Muslim 70%, Albanian Orthodox 20%, Roman Catholic 10%
Ethnic mix Albanian 98%, Greek 1%, Other 1%
Government Parliamentary system
Currency Lek = 100 qindarka (qintars)
Literacy rate 97%
Calorie consumption 3023 kilocalories

ALGERIA
North Africa

Page 48 C3

Lying mostly in the Sahara, this former French colony was riven by civil war after Islamists were denied electoral victory in 1992. Fighting has subsided but Islamic extremists remain a threat.

Official name People's Democratic Republic of Algeria
Formation 1962 / 1962
Capital Algiers
Population 39.2 million / 43 people per sq mile (16 people per sq km)
Total area 919,590 sq miles (2,381,740 sq km)
Languages Arabic*, Tamazight (Kabyle, Shawia, Tamashek), French
Religions Sunni Muslim 99%, Christian & Jewish 1%
Ethnic mix Arab 75%, Berber 24%, European & Jewish 1%
Government Presidential system
Currency Algerian dinar = 100 centimes
Literacy rate 73%
Calorie consumption 3220 kilocalories

ANDORRA
Southwest Europe

Page 69 B6

A tiny landlocked principality, Andorra lies between France and Spain, high in the eastern Pyrenees. Its economy, based on tourism, also features low tax and duty-free shopping.

Official name Principality of Andorra
Formation 1278 / 1278
Capital Andorra la Vella
Population 85,293 / 474 people per sq mile (183 people per sq km)
Total area 181 sq miles (468 sq km)
Languages Spanish, Catalan*, French, Portuguese
Religions Roman Catholic 94%, Other 6%
Ethnic mix Spanish 46%, Andorran 28%, Other 18%, French 8%
Government Parliamentary system
Currency Euro = 100 cents
Literacy rate 99%
Calorie consumption Not available

ANGOLA
Southern Africa

Page 56 B2

An oil- and diamond-rich former Portuguese colony in Southwest Africa, Angola is badly scarred by decades of civil war, though fighting formally ended in 2002.

Official name Republic of Angola
Formation 1975 / 1975
Capital Luanda
Population 21.5 million / 45 people per sq mile (17 people per sq km)
Total area 481,351 sq miles (1,246,700 sq km)
Languages Portuguese*, Umbundu, Kimbundu, Kikongo
Religions Roman Catholic 68%, Protestant 20%, Indigenous beliefs 12%
Ethnic mix Ovimbundu 37%, Kimbundu 25%, Other 25%, Bakongo 13%
Government Presidential system
Currency Readjusted kwanza = 100 lwei
Literacy rate 71%
Calorie consumption 2400 kilocalories

ANTIGUA & BARBUDA
West Indies

Page 33 H3

Lying on the Atlantic edge of the Leeward Islands, Antigua was in turn a Spanish, French, and British colony. Tourism is the economic mainstay, with the best beaches on Barbuda.

Official name Antigua and Barbuda
Formation 1981 / 1981
Capital St. John's
Population 90,156 / 530 people per sq mile (205 people per sq km)
Total area 170 sq miles (442 sq km)
Languages English*, English patois
Religions Anglican 45%, Other Protestant 42%, Roman Catholic 10%, Other 2%, Rastafarian 1%
Ethnic mix Black African 95%, Other 5%
Government Parliamentary system
Currency East Caribbean dollar = 100 cents
Literacy rate 99%
Calorie consumption 2396 kilocalories

ARGENTINA
South America

Page 43 B5

From semiarid lowlands, through fertile grasslands, to the glacial southern tip of South America, Argentina has enjoyed democratic rule since 1983 but struggled with high foreign debts.

Official name Republic of Argentina
Formation 1816 / 1816
Capital Buenos Aires
Population 41.4 million / 39 people per sq mile (15 people per sq km)
Total area 1,068,296 sq miles (2,766,890 sq km)
Languages Spanish*, Italian, Amerindian languages
Religions Roman Catholic 70%, Other 18%, Protestant 9%, Muslim 2%, Jewish 1%
Ethnic mix Indo-European 97%, Mestizo 2%, Amerindian 1%
Government Presidential system
Currency Argentine peso = 100 centavos
Literacy rate 98%
Calorie consumption 3155 kilocalories

ARMENIA
Southwest Asia

Page 95 F3

The smallest of the ex-Soviet republics, landlocked Armenia lies in the Lesser Caucasus mountains. It was the first country to adopt Christianity as the state religion, in the 4th century AD.

Official name Republic of Armenia
Formation 1991 / 1991
Capital Yerevan
Population 3 million / 261 people per sq mile (101 people per sq km)
Total area 11,506 sq miles (29,800 sq km)
Languages Armenian*, Azeri, Russian
Religions Armenian Apostolic Church (Orthodox) 88%, Other 6%, Armenian Catholic Church 6%
Ethnic mix Armenian 98%, Other 1%, Yezidi 1%
Government Parliamentary system
Currency Dram = 100 luma
Literacy rate 99%
Calorie consumption 2809 kilocalories

AUSTRALIA
Australasia & Oceania

Page 125 B5

An island continent between the Indian and Pacific oceans, Australia was settled by Europeans from 1788, but recent immigrants are mostly Asian. Minerals underpin the economy.

Official name Commonwealth of Australia
Formation 1901 / 1901
Capital Canberra
Population 23.3 million / 8 people per sq mile (3 people per sq km)
Total area 2,967,893 sq miles (7,686,850 sq km)
Languages English*, Italian, Cantonese, Greek, Arabic, Vietnamese, Aboriginal languages
Religions Roman Catholic 26%, Nonreligious 19%, Anglican 19%, Other 23%, Other Christian 13%
Ethnic mix European 90%, Asian 7%, Aboriginal 2%, Other 1%
Government Parliamentary system
Currency Australian dollar = 100 cents
Literacy rate 99%
Calorie consumption 3265 kilocalories

AUSTRIA
Central Europe

Page 73 D7

Nestled in Central Europe, Austria was created after the Austro-Hungarian Empire was defeated in World War I. Absorbed into Hitler's Germany in 1938, it re-emerged in 1955.

Official name Republic of Austria
Formation 1918 / 1919
Capital Vienna
Population 8.5 million / 266 people per sq mile (103 people per sq km)
Total area 32,378 sq miles (83,858 sq km)
Languages German*, Croatian, Slovenian, Hungarian (Magyar)
Religions Roman Catholic 78%, Nonreligious 9%, Other 8%, Protestant 5%
Ethnic mix Austrian 93%, Croat, Slovene, and Hungarian 6%, Other 1%
Government Parliamentary system
Currency Euro = 100 cents
Literacy rate 99%
Calorie consumption 3784 kilocalories

AZERBAIJAN
Southwest Asia

Page 95 G2

On the west coast of the Caspian Sea, oil-rich Azerbaijan regained its independence from the USSR in 1991. A territorial dispute with Armenia remains unresolved.

Official name Republic of Azerbaijan
Formation 1991 / 1991
Capital Baku
Population 9.4 million / 281 people per sq mile (109 people per sq km)
Total area 33,436 sq miles (86,600 sq km)
Languages Azeri*, Russian
Religions Shi'a Muslim 68%, Sunni Muslim 26%, Russian Orthodox 3%, Armenian Apostolic Church (Orthodox) 2%, Other 1%
Ethnic mix Azeri 91%, Other 3%, Armenian 2%, Russian 2%, Lazs 2%
Government Presidential system
Currency New manat = 100 gopik
Literacy rate 99%
Calorie consumption 2952 kilocalories

BAHAMAS, THE
West Indies

Page 32 C1

Located in the western Atlantic, off the Florida coast, the Bahamas comprise some 700 islands and 2400 cays; only 30 are inhabited. Financial services and shipping support the economy.

Official name Commonwealth of the Bahamas
Formation 1973 / 1973
Capital Nassau
Population 400,000 / 103 people per sq mile (40 people per sq km)
Total area 5382 sq miles (13,940 sq km)
Languages English*, English Creole, French Creole
Religions Baptist 32%, Anglican 20%, Roman Catholic 19%, Other 17%, Methodist 6%, Church of God 6%
Ethnic mix Black African 85%, European 12%, Asian and Hispanic 3%
Government Parliamentary system
Currency Bahamian dollar = 100 cents
Literacy rate 96%
Calorie consumption 2575 kilocalories

BAHRAIN
Southwest Asia

Page 98 C4

Only three of Bahrain's 33 islands lying between the Qatar peninsula and Saudi Arabian are inhabited. The first Gulf emirate to export oil, reserves are expected to last another 10 to 15 years.

Official name Kingdom of Bahrain
Formation 1971 / 1971
Capital Manama
Population 1.3 million / 4762 people per sq mile (1841 people per sq km)
Total area 239 sq miles (620 sq km)
Languages Arabic*
Religions Muslim (mainly Shi'a) 99%, Other 1%
Ethnic mix Bahraini 63%, Asian 19%, Other Arab 10%, Iranian 8%
Government Monarchical / parliamentary system
Currency Bahraini dinar = 1000 fils
Literacy rate 95%
Calorie consumption Not available

BANGLADESH
South Asia

Page 113 G3

Low-lying Bangladesh on the Bay of Bengal suffers annual monsoon flooding. It seceded from Pakistan in 1971. Political instability and corruption are ongoing problems.

Official name People's Republic of Bangladesh
Formation 1971 / 1971
Capital Dhaka
Population 157 million / 3029 people per sq mile (1169 people per sq km)
Total area 55,598 sq miles (144,000 sq km)
Languages Bengali*, Urdu, Chakma, Marma (Magh), Garo, Khasi, Santhali, Tripuri, Mro
Religions Muslim (mainly Sunni) 88%, Hindu 11%, Other 1%
Ethnic mix Bengali 98%, Other 2%
Government Parliamentary system
Currency Taka = 100 poisha
Literacy rate 59%
Calorie consumption 2430 kilocalories

BARBADOS
West Indies

Page 33 H4

The most easterly of the Windward Islands, Barbados was under British rule from the 1620s. A sugar exporter in the 18th century, it now relies on tourism and financial services.

Official name Barbados
Formation 1966 / 1966
Capital Bridgetown
Population 300,000 / 1807 people per sq mile (698 people per sq km)
Total area 166 sq miles (430 sq km)
Languages Bajan (Barbadian English), English*
Religions Anglican 40%, Other 24%, Nonreligious 17%, Pentecostal 8%, Methodist 7%, Roman Catholic 4%
Ethnic mix Black African 92%, Other 3%, White 3%, Mixed race 2%
Government Parliamentary system
Currency Barbados dollar = 100 cents
Literacy rate 99%
Calorie consumption 3047 kilocalories

BELARUS
Eastern Europe

Page 85 B6

Landlocked in eastern Europe, forested Belarus, which means "White Russia," was reluctant to become independent of the USSR in 1991, and has been slow to reform its economy since.

Official name Republic of Belarus
Formation 1991 / 1991
Capital Minsk
Population 9.4 million / 117 people per sq mile (45 people per sq km)
Total area 80,154 sq miles (207,600 sq km)
Languages Belarussian*, Russian*
Religions Orthodox Christian 80%, Roman Catholic 14%, Other 4%, Protestant 2%
Ethnic mix Belarussian 81%, Russian 11%, Polish 4%, Ukrainian 2%, Other 2%
Government Presidential system
Currency Belarussian rouble = 100 kopeks
Literacy rate 99%
Calorie consumption 3253 kilocalories

BELGIUM
Northwest Europe

Page 65 B6

Located in Northwest Europe, Belgium has forests and canals in the south and canals in the flat north. Its history and politics are marked by the division between its Flemish and Walloon communities.

Official name Kingdom of Belgium
Formation 1830 / 1919
Capital Brussels
Population 11.1 million / 876 people per sq mile (338 people per sq km)
Total area 11,780 sq miles (30,510 sq km)
Languages Dutch*, French*, German*
Religions Roman Catholic 88%, Other 10%, Muslim 2%
Ethnic mix Fleming 58%, Walloon 33%, Other 6%, Italian 2%, Moroccan 1%
Government Parliamentary system
Currency Euro = 100 cents
Literacy rate 99%
Calorie consumption 3793 kilocalories

BELIZE
Central America

Page 30 B1

The last Central American country to gain independence, this former British colony lies on the eastern shore of the Yucatan Peninsula. Offshore is the world's second-largest barrier reef.

Official name Belize
Formation 1981 / 1981
Capital Belmopan
Population 300,000 / 34 people per sq mile (13 people per sq km)
Total area 8867 sq miles (22,966 sq km)
Languages English Creole, Spanish, English*, Mayan, Garifuna (Carib)
Religions Roman Catholic 62%, Other 16%, Anglican 12%, Methodist 6%, Mennonite 4%
Ethnic mix Mestizo 49%, Creole 25%, Maya 11%, Garifuna 6%, Other 6%, Asian Indian 3%
Government Parliamentary system
Currency Belizean dollar = 100 cents
Literacy rate 75%
Calorie consumption 2757 kilocalories

BENIN
West Africa

Page 53 F4

Stretching north from the West African coast, this ex-French colony suffered military rule after independence but in recent decades has been a leading example of African democratization.

Official name Republic of Benin
Formation 1960 / 1960
Capital Porto-Novo
Population 10.3 million / 241 people per sq mile (93 people per sq km)
Total area 43,483 sq miles (112,620 sq km)
Languages Fon, Bariba, Yoruba, Adja, Houeda, Somba, French*
Religions Indigenous beliefs and Voodoo 50%, Christian 30%, Muslim 20%
Ethnic mix Fon 41%, Other 21%, Adja 16%, Yoruba 12%, Bariba 10%
Government Presidential system
Currency CFA franc = 100 centimes
Literacy rate 29%
Calorie consumption 2594 kilocalories

BHUTAN
South Asia

Page 113 G3

This landlocked Buddhist kingdom, perched in the eastern Himalayas between India and China, is carefully protecting its cultural identity from modernization and the outside world.

Official name Kingdom of Bhutan
Formation 1656 / 1865
Capital Thimphu
Population 800,000 / 44 people per sq mile (17 people per sq km)
Total area 18,147 sq miles (47,000 sq km)
Languages Dzongkha*, Nepali, Assamese
Religions Mahayana Buddhist 75%, Hindu 25%
Ethnic mix Drukpa 50%, Nepalese 35%, Other 15%
Government Monarchical / parliamentary system
Currency Ngultrum = 100 chetrum
Literacy rate 53%
Calorie consumption Not available

BOLIVIA
South America

Page 39 F3

Bolivia lies landlocked high in central South America. Mineral riches once made it the region's wealthiest state, but wars, coups, and poor governance have reduced it to the poorest.

Official name Plurinational State of Bolivia
Formation 1825 / 1938
Capital La Paz (administrative); Sucre (judicial)
Population 10.7 million / 26 people per sq mile (10 people per sq km)
Total area 424,162 sq miles (1,098,580 km)
Languages Aymara*, Quechua*, Spanish*
Religions Roman Catholic 93%, Other 7%
Ethnic mix Quechua 37%, Aymara 32%, Mixed race 13%, European 10%, Other 8%
Government Presidential system
Currency Boliviano = 100 centavos
Literacy rate 94%
Calorie consumption 2254 kilocalories

BOSNIA & HERZEGOVINA
Southeast Europe

Page 78 B3

In the mountainous western Balkans this state, born out of the bitter conflicts of Yugoslavia's collapse, has two key concerns: balancing ethnic rivalries, and integrating with Europe.

Official name Bosnia and Herzegovina
Formation 1992 / 1992
Capital Sarajevo
Population 3.8 million / 192 people per sq mile (74 people per sq km)
Total area 19,741 sq miles (51,129 sq km)
Languages Bosnian*, Serbian*, Croatian*
Religions Muslim (mainly Sunni) 40%, Orthodox Christian 31%, Roman Catholic 15%, Other 14%
Ethnic mix Bosniak 48%, Serb 34%, Croat 16%, Other 2%
Government Parliamentary system
Currency Marka = 100 pfeninga
Literacy rate 98%
Calorie consumption 3130 kilocalories

BOTSWANA
Southern Africa

Page 56 C3

Botswana, once the British protectorate of Bechuanaland, lies landlocked in Southern Africa. Diamonds provide it with a relatively prosperous economy, but the rate of HIV infection is high.

Official name Republic of Botswana
Formation 1966 / 1966
Capital Gaborone
Population 2 million / 9 people per sq mile (4 people per sq km)
Total area 231,803 sq miles (600,370 sq km)
Languages Setswana, English*, Shona, San, Khoikhoi, isiNdebele
Religions Christian (mainly Protestant) 70%, Nonreligious 20%, Traditional beliefs 6%, Other (including Muslim) 4%
Ethnic mix Tswana 79%, Kalanga 11%, Other 10%
Government Presidential system
Currency Pula = 100 thebe
Literacy rate 87%
Calorie consumption 2285 kilocalories

BRAZIL
South America

Page 40 C2

Brazil covers more than half of South America and is the site of the world's largest rain forest. It has immense natural resources and produces a third of the world's coffee.

Official name Federative Republic of Brazil
Formation 1822 / 1828
Capital Brasília
Population 200 million / 61 people per sq mile (24 people per sq km)
Total area 3,286,470 sq miles (8,511,965 sq km)
Languages Portuguese*, German, Italian, Spanish, Polish, Japanese, Amerindian languages
Religions Roman Catholic 74%, Protestant 15%, Atheist 7%, Other 3%, Afro-American Spiritist 1%
Ethnic mix White 54%, Mixed race 38%, Black 6%, Other 2%
Government Presidential system
Currency Real = 100 centavos
Literacy rate 91%
Calorie consumption 3287 kilocalories

BRUNEI
Southeast Asia

Page 116 D3

On the northwest coast of the island of Borneo, Brunei is surrounded and divided in two by the Malaysian state of Sarawak. Oil and gas revenues have brought a high standard of living.

Official name Brunei Darussalam
Formation 1984 / 1984
Capital Bandar Seri Begawan
Population 400,000 / 197 people per sq mile (76 people per sq km)
Total area 2228 sq miles (5770 sq km)
Languages Malay*, English, Chinese
Religions Muslim (mainly Sunni) 66%, Buddhist 14%, Christian 10%, Other 10%
Ethnic mix Malay 67%, Chinese 16%, Other 11%, Indigenous 6%
Government Monarchy
Currency Brunei dollar = 100 cents
Literacy rate 95%
Calorie consumption 2949 kilocalories

BULGARIA
Southeast Europe

Page 82 C2

Bulgaria is located on the western shore of the Black Sea. After the fall of its communist regime in 1990, economic and political reform were slow, but EU membership was achieved in 2007.

Official name Republic of Bulgaria
Formation 1908 / 1947
Capital Sofia
Population 7.2 million / 169 people per sq mile (65 people per sq km)
Total area 42,822 sq miles (110,910 sq km)
Languages Bulgarian*, Turkish, Romani
Religions Bulgarian Orthodox 83%, Muslim 12%, Other 4%, Roman Catholic 1%
Ethnic mix Bulgarian 84%, Turkish 9%, Roma 5%, Other 2%
Government Parliamentary system
Currency Lev = 100 stotinki
Literacy rate 98%
Calorie consumption 2877 kilocalories

BURKINA FASO
West Africa

Page 53 E4

Known as Upper Volta until 1984, Burkina Faso is landlocked in the semiarid Sahel of West Africa. It has been under military rule for most of its post-independence history.

Official name Burkina Faso
Formation 1960 / 1960
Capital Ouagadougou
Population 16.9 million / 160 people per sq mile (62 people per sq km)
Total area 105,869 sq miles (274,200 sq km)
Languages Mossi, Fulani, French*, Tuareg, Dyula, Songhai
Religions Muslim 55%, Christian 25%, Traditional beliefs 20%
Ethnic mix Mossi 48%, Other 21%, Peul 10%, Lobi 7%, Bobo 7%, Mandé 7%
Government Transitional regime
Currency CFA franc = 100 centimes
Literacy rate 29%
Calorie consumption 2655 kilocalories

BURUNDI
Central Africa

Page 51 B7

Small, landlocked Burundi lies just south of the Equator, on the Nile–Congo watershed. A decade of brutal conflict between Hutu and Tutsi from 1993 led to power-sharing in governance.

Official name Republic of Burundi
Formation 1962 / 1962
Capital Bujumbura
Population 10.2 million / 1030 people per sq mile (398 people per sq km)
Total area 10,745 sq miles (27,830 sq km)
Languages Kirundi*, French*, Kiswahili
Religions Roman Catholic 62%, Traditional beliefs 23%, Muslim 10%, Protestant 5%
Ethnic mix Hutu 85%, Tutsi 14%, Twa 1%
Government Presidential system
Currency Burundian franc = 100 centimes
Literacy rate 87%
Calorie consumption 1604 kilocalories

CAMBODIA
Southeast Asia

Page 115 D5

This ancient Southeast Asian kingdom suffered the brutal totalitarian Khmer Rouge regime in the 1970s and then a decade of Vietnamese puppet rule. Free elections were only held in 1993.

Official name Kingdom of Cambodia
Formation 1953 / 1953
Capital Phnom Penh
Population 15.1 million / 222 people per sq mile (86 people per sq km)
Total area 69,900 sq miles (181,040 sq km)
Languages Khmer*, French, Chinese, Vietnamese, Cham
Religions Buddhist 93%, Muslim 6%, Christian 1%
Ethnic mix Khmer 90%, Vietnamese 5%, Other 4%, Chinese 1%
Government Parliamentary system
Currency Riel = 100 sen
Literacy rate 74%
Calorie consumption 2411 kilocalories

CAMEROON
Central Africa

Page 54 A4

A former trading hub on the central West African coast, Cameroon was effectively a one-party state for 30 years. Elections since 1992 have brought no change in leadership.

Official name Republic of Cameroon
Formation 1960 / 1961
Capital Yaoundé
Population 22.3 million / 124 people per sq mile (48 people per sq km)
Total area 183,567 sq miles (475,400 sq km)
Languages Bamileke, Fang, Fulani, French*, English*
Religions Roman Catholic 35%, Traditional beliefs 25%, Muslim 22%, Protestant 18%
Ethnic mix Cameroon highlanders 31%, Other 21%, Equatorial Bantu 19%, Kirdi 11%, Fulani 10%, Northwestern Bantu 8%
Government Presidential system
Currency CFA franc = 100 centimes
Literacy rate 71%
Calorie consumption 2586 kilocalories

CANADA
North America

Page 15 E4

The world's second-largest country spans six time zones, extends north from its US border into the Arctic, and is rich in natural resources. Separatism is strong in French-speaking Québec.

Official name Canada
Formation 1867 / 1949
Capital Ottawa
Population 35.2 million / 10 people per sq mile (4 people per sq km)
Total area 3,855,171 sq miles (9,984,670 sq km)
Languages English*, French*, Chinese, Italian, German, Ukrainian, Portuguese, Inuktitut, Cree
Religions Roman Catholic 44%, Protestant 29%, Other and nonreligious 27%
Ethnic mix British, French, and other European 87%, Asian 9%, Amerindian, Métis, and Inuit 4%
Government Parliamentary system
Currency Canadian dollar = 100 cents
Literacy rate 99%
Calorie consumption 3419 kilocalories

CAPE VERDE
Atlantic Ocean

Page 52 A2

The mostly volcanic islands that make up Cape Verde lie off Africa's west coast. A Portuguese colony until 1975, it has been a stable democracy since its first multiparty elections in 1991.

Official name Republic of Cape Verde
Formation 1975 / 1975
Capital Praia
Population 500,000 / 321 people per sq mile (124 people per sq km)
Total area 1557 sq miles (4033 sq km)
Languages Portuguese Creole, Portuguese*
Religions Roman Catholic 97%, Other 2%, Protestant (Church of the Nazarene) 1%
Ethnic mix Mestiço 71%, African 28%, European 1%
Government Presidential / parliamentary system
Currency Escudo = 100 centavos
Literacy rate 85%
Calorie consumption 2716 kilocalories

CENTRAL AFRICAN REPUBLIC
Central Africa

Page 54 C4

A landlocked plateau dividing the Chad and Congo river basins, the CAR has been plagued by rebellions since military rule ended in 1993. The arid north is sparsely populated.

Official name Central African Republic
Formation 1960 / 1960
Capital Bangui
Population 4.6 million / 19 people per sq mile (7 people per sq km)
Total area 240,534 sq miles (622,984 sq km)
Languages Sango, Banda, Gbaya, French*
Religions Traditional beliefs 35%, Roman Catholic 25%, Protestant 25%, Muslim 15%
Ethnic mix Baya 33%, Banda 27%, Other 17%, Mandjia 13%, Sara 10%
Government Transitional regime
Currency CFA franc = 100 centimes
Literacy rate 37%
Calorie consumption 2154 kilocalories

CHAD
Central Africa

Page 54 C3

Landlocked in north Central Africa, Chad has been torn by intermittent periods of civil war since it gained independence from France in 1960. It became a net oil exporter in 2003.

Official name Republic of Chad
Formation 1960 / 1960
Capital N'Djaména
Population 12.8 million / 26 people per sq mile (10 people per sq km)
Total area 495,752 sq miles (1,284,000 sq km)
Languages French*, Sara, Arabic*, Maba
Religions Muslim 51%, Christian 35%, Animist 7%, Traditional beliefs 7%
Ethnic mix Other 30%, Sara 28%, Mayo-Kebbi 12%, Arab 12%, Ouaddai 9%, Kanem-Bornou 9%
Government Presidential system
Currency CFA franc = 100 centimes
Literacy rate 37%
Calorie consumption 2061 kilocalories

CHILE
South America

Page 42 B3

Extending in a ribbon down the Pacific coast of South America, Chile restored democracy in 1989 after a referendum rejected its military dictator. It is the world's largest copper producer.

Official name Republic of Chile
Formation 1818 / 1883
Capital Santiago
Population 17.6 million / 61 people per sq mile (24 people per sq km)
Total area 292,258 sq miles (756,950 sq km)
Languages Spanish*, Amerindian languages
Religions Roman Catholic 89%, Other and nonreligious 11%
Ethnic mix Mestizo and European 90%, Other Amerindian 9%, Mapuche 1%
Government Presidential system
Currency Chilean peso = 100 centavos
Literacy rate 99%
Calorie consumption 2989 kilocalories

CHINA
East Asia

Page 104 C4

This vast East Asian country, home to a fifth of the global population, became a communist state in 1949. It has now emerged as one of the world's major political and economic powers.

Official name People's Republic of China
Formation 960 / 1999
Capital Beijing
Population 1.39 billion / 385 people per sq mile (149 people per sq km)
Total area 3,705,386 sq miles (9,596,960 sq km)
Languages Mandarin*, Wu, Cantonese, Hsiang, Min, Hakka, Kan
Religions Nonreligious 59%, Traditional beliefs 20%, Other 13%, Buddhist 6%, Muslim 2%
Ethnic mix Han 92%, Other 4%, Zhuang 1%, Hui 1%, Manchu 1%, Miao 1%
Government One-party state
Currency Renminbi (known as yuan) = 10 jiao = 100 fen
Literacy rate 95%
Calorie consumption 3074 kilocalories

COLOMBIA
South America

Page 36 B3

Lying in northwest South America, Colombia has suffered civil war since 1964, with over three million internal refugees. It is noted for coffee, gold, emeralds, and narcotics trafficking.

Official name Republic of Colombia
Formation 1819 / 1903
Capital Bogotá
Population 48.3 million / 120 people per sq mile (47 people per sq km)
Total area 439,733 sq miles (1,138,910 sq km)
Languages Spanish*, Wayuu, Páez, and other Amerindian languages
Religions Roman Catholic 95%, Other 5%
Ethnic mix Mestizo 58%, White 20%, European–African 14%, African 4%, African–Amerindian 3%, Amerindian 1%
Government Presidential system
Currency Colombian peso = 100 centavos
Literacy rate 94%
Calorie consumption 2593 kilocalories

COMOROS
Indian Ocean

Page 57 F2

The Comoros islands lie between Mozambique and Madagascar. There have been many coups and secession attempts by the smaller islands since independence from France in 1975.

Official name Union of the Comoros
Formation 1975 / 1975
Capital Moroni
Population 700,000 / 813 people per sq mile (314 people per sq km)
Total area 838 sq miles (2170 sq km)
Languages Arabic*, Comoran*, French*
Religions Muslim (mainly Sunni) 98%, Other 1%, Roman Catholic 1%
Ethnic mix Comoran 97%, Other 3%
Government Presidential system
Currency Comoros franc = 100 centimes
Literacy rate 76%
Calorie consumption 2139 kilocalories

CONGO
Central Africa

Page 55 B5

Astride the Equator in Central Africa, this former French colony emerged from 26 years of Marxist-Leninist rule in 1990, though the Marxist-era dictator seized power again in 1997.

Official name Republic of the Congo
Formation 1960 / 1960
Capital Brazzaville
Population 4.4 million / 33 people per sq mile (13 people per sq km)
Total area 132,046 sq miles (342,000 sq km)
Languages Kongo, Teke, Lingala, French*
Religions Traditional beliefs 50%, Roman Catholic 35%, Protestant 13%, Muslim 2%
Ethnic mix Bakongo 51%, Teke 17%, Other 16%, Mbochi 11%, Mbédé 5%
Government Presidential system
Currency CFA franc = 100 centimes
Literacy rate 79%
Calorie consumption 2195 kilocalories

CONGO, DEM. REP.
Central Africa

Page 55 C6

Straddling the Equator in east Central Africa, mineral-rich Dem. Rep. Congo is Africa's second-largest country. The former Belgian colony has endured years of corrupt rule and conflict.

Official name Democratic Republic of the Congo
Formation 1960 / 1960
Capital Kinshasa
Population 67.5 million / 77 people per sq mile (30 people per sq km)
Total area 905,563 sq miles (2,345,410 sq km)
Languages Kiswahili, Tshiluba, Kikongo, Lingala, French*
Religions Roman Catholic 50%, Protestant 20%, Traditional beliefs and other 20%, Muslim 10%
Ethnic mix Other 55%, Mongo, Luba, Kongo, and Mangbetu-Azande 45%
Government Presidential system
Currency Congolese franc = 100 centimes
Literacy rate 61%
Calorie consumption 1585 kilocalories

COSTA RICA
Central America

Page 31 E4

Costa Rica is the most stable country in Central America. It abolished its army in 1948 and its neutrality in foreign affairs is long-standing, but it has very strong ties with the US.

Official name Republic of Costa Rica
Formation 1838 / 1838
Capital San José
Population 4.9 million / 249 people per sq mile (96 people per sq km)
Total area 19,730 sq miles (51,100 sq km)
Languages Spanish*, English Creole, Bribri, Cabecar
Religions Roman Catholic 71%, Evangelical 14%, Nonreligious 11%, Other 4%
Ethnic mix Mestizo and European 94%, Black 3%, Chinese 1%, Other 1%, Amerindian 1%
Government Presidential system
Currency Costa Rican colón = 100 céntimos
Literacy rate 97%
Calorie consumption 2898 kilocalories

CÔTE D'IVOIRE (IVORY COAST)
West Africa

Page 52 D4

One of the larger countries on the West African coast, this ex-French colony is the world's biggest cocoa producer. Coups and recent conflicts have destroyed its reputation for stability.

Official name Republic of Côte d'Ivoire
Formation 1960 / 1960
Capital Yamoussoukro
Population 20.3 million / 165 people per sq mile (64 people per sq km)
Total area 124,502 sq miles (322,460 sq km)
Languages Akan, French*, Krou, Voltaique
Religions Muslim 38%, Roman Catholic 25%, Traditional beliefs 25%, Protestant 6%, Other 6%
Ethnic mix Akan 42%, Voltaique 18%, Mandé du Nord 17%, Krou 11%, Mandé du Sud 10%, Other 2%
Government Presidential system
Currency CFA franc = 100 centimes
Literacy rate 41%
Calorie consumption 2781 kilocalories

CROATIA
Southeast Europe

Page 78 B2

Post-independence fighting afflicted this former Yugoslav republic until 1995. It is now capitalizing on its location on the eastern Adriatic coast and joined the EU in 2013.

Official name Republic of Croatia
Formation 1991 / 1991
Capital Zagreb
Population 4.3 million / 197 people per sq mile (76 people per sq km)
Total area 21,831 sq miles (56,542 sq km)
Languages Croatian*
Religions Roman Catholic 88%, Other 7%, Orthodox Christian 4%, Muslim 1%
Ethnic mix Croat 90%, Serb 5%, Other 5%
Government Parliamentary system
Currency Kuna = 100 lipa
Literacy rate 99%
Calorie consumption 3052 kilocalories

CUBA
West Indies

Page 32 C2

Cuba is the largest island in the Caribbean and the only communist country in the Americas. It was led by Fidel Castro for almost 40 years until he stepped down in 2008.

Official name Republic of Cuba
Formation 1902 / 1902
Capital Havana
Population 11.3 million / 264 people per sq mile (102 people per sq km)
Total area 42,803 sq miles (110,860 sq km)
Languages Spanish*
Religions Nonreligious 49%, Roman Catholic 40%, Atheist 6%, Other 4%, Protestant 1%
Ethnic mix Mulatto (mixed race) 51%, White 37%, Black 11%, Chinese 1%
Government One-party state
Currency Cuban peso = 100 centavos
Literacy rate 99%
Calorie consumption 3277 kilocalories

CYPRUS

Page 80 C5

Cyprus lies south of Turkey in the eastern Mediterranean. Since 1974, it has been partitioned between the Turkish-occupied north and the Greek south (which joined the EU in 2004).

Official name Republic of Cyprus
Formation 1960 / 1960
Capital Nicosia
Population 1.1 million / 308 people per sq mile (119 people per sq km)
Total area 3571 sq miles (9250 sq km)
Languages Greek*, Turkish*
Religions Orthodox Christian 78%, Muslim 18%, Other 4%
Ethnic mix Greek 81%, Turkish 11%, Other 8%
Government Presidential system
Currency Euro = 100 cents (In TRNC, new Turkish lira = 100 kurus)
Literacy rate 99%
Calorie consumption 2661 kilocalories

CZECH REPUBLIC
Central Europe

Page 77 A5

Landlocked in Central Europe, and formerly part of communist Czechoslovakia, it peacefully dissolved its federal union with Slovakia in 1993, and joined the EU in 2004.

Official name Czech Republic
Formation 1993 / 1993
Capital Prague
Population 10.7 million / 351 people per sq mile (136 people per sq km)
Total area 30,450 sq miles (78,866 sq km)
Languages Czech*, Slovak, Hungarian (Magyar)
Religions Roman Catholic 39%, Atheist 38%, Other 18%, Protestant 3%, Hussite 2%
Ethnic mix Czech 90%, Moravian 4%, Other 4%, Slovak 2%
Government Parliamentary system
Currency Czech koruna = 100 haleru
Literacy rate 99%
Calorie consumption 3292 kilocalories

DENMARK
Northern Europe

Page 63 A7

Denmark occupies the low-lying Jutland peninsula and over 400 islands. In the 1930s it set up one of the first welfare systems. Greenland and the Faroe Islands are self-governing territories.

Official name Kingdom of Denmark
Formation 950 / 1944
Capital Copenhagen
Population 5.6 million / 342 people per sq mile (132 people per sq km)
Total area 16,639 sq miles (43,094 sq km)
Languages Danish*
Religions Evangelical Lutheran 95%, Roman Catholic 3%, Muslim 2%
Ethnic mix Danish 96%, Other (including Scandinavian and Turkish) 3%, Faroese and Inuit 1%
Government Parliamentary system
Currency Danish krone = 100 øre
Literacy rate 99%
Calorie consumption 3363 kilocalories

DJIBOUTI
East Africa

Page 50 D4

Once known as French Somaliland, this city state with a desert hinterland lies on the coast of the Horn of Africa. Its economy relies on its Red Sea port, a vital trade link for landlocked Ethiopia.

Official name Republic of Djibouti
Formation 1977 / 1977
Capital Djibouti
Population 900,000 / 101 people per sq mile (39 people per sq km)
Total area 8494 sq miles (22,000 sq km)
Languages Somali, Afar, French*, Arabic*
Religions Muslim (mainly Sunni) 94%, Christian 6%
Ethnic mix Issa 60%, Afar 35%, Other 5%
Government Presidential system
Currency Djibouti franc = 100 centimes
Literacy rate 70%
Calorie consumption 2526 kilocalories

DOMINICA
West Indies

Page 33 H4

This Caribbean island, known for its lush flora and fauna, resisted European colonization until the 18th century, when it came first under French and then British rule.

Official name Commonwealth of Dominica
Formation 1978 / 1978
Capital Roseau
Population 73,286 / 253 people per sq mile (98 people per sq km)
Total area 291 sq miles (754 sq km)
Languages French Creole, English*
Religions Roman Catholic 77%, Protestant 15%, Other 8%
Ethnic mix Black 87%, Mixed race 9%, Carib 3%, Other 1%
Government Parliamentary system
Currency East Caribbean dollar = 100 cents
Literacy rate 88%
Calorie consumption 3047 kilocalories

DOMINICAN REPUBLIC
West Indies

Page 33 E2

Occupying the eastern two-thirds of the island of Hispaniola, the Dominican Republic is the Caribbean's top tourist destination and largest economy. Ties with the US are strong.

Official name Dominican Republic
Formation 1865 / 1865
Capital Santo Domingo
Population 10.4 million / 557 people per sq mile (215 people per sq km)
Total area 18,679 sq miles (48,380 sq km)
Languages Spanish*, French Creole
Religions Roman Catholic 95%, Other and nonreligious 5%
Ethnic mix Mixed race 73%, European 16%, Black African 11%
Government Presidential system
Currency Dominican Republic peso = 100 centavos
Literacy rate 91%
Calorie consumption 2597 kilocalories

EAST TIMOR
Southeast Asia

Page 116 F5

This former Portuguese colony on the island of Timor in the East Indies was invaded by Indonesia in 1975. In 1999 it voted for independence, achieved in 2002 after a turbulent transition.

Official name Democratic Republic of Timor-Leste
Formation 2002 / 2002
Capital Dili
Population 1.1 million / 195 people per sq mile (75 people per sq km)
Total area 5756 sq miles (14,874 sq km)
Languages Tetum* (Portuguese/Austronesian), Bahasa Indonesia, Portuguese*
Religions Roman Catholic 95%, Other 5%
Ethnic mix Papuan groups approx. 85%, Indonesian groups approx. 13%, Chinese 2%
Government Parliamentary system
Currency US dollar = 100 cents
Literacy rate 58%
Calorie consumption 2083 kilocalories

ECUADOR
South America

Page 38 A2

Once part of the Inca heartland on the northwest coast of South America, Ecuador is the world's leading banana producer. Its territory includes the wildlife-rich Galapagos Islands.

Official name Republic of Ecuador
Formation 1830 / 1942
Capital Quito
Population 15.7 million / 147 people per sq mile (57 people per sq km)
Total area 109,483 sq miles (283,560 sq km)
Languages Spanish*, Quechua, other Amerindian languages
Religions Roman Catholic 95%, Protestant, Jewish, and other 5%
Ethnic mix Mestizo 77%, White 11%, Amerindian 7%, Black African 5%
Government Presidential system
Currency US dollar = 100 cents
Literacy rate 93%
Calorie consumption 2477 kilocalories

EGYPT
North Africa

Page 50 B2

Egypt lies in Africa's northeast corner; the fertile Nile valley divides desert lands. Nearly 50 years of de facto military rule was interrupted in 2011 by the "Arab Spring" popular uprising.

Official name Arab Republic of Egypt
Formation 1936 / 1982
Capital Cairo
Population 82.1 million / 214 people per sq mile (82 people per sq km)
Total area 386,660 sq miles (1,001,450 sq km)
Languages Arabic*, French, English, Berber
Religions Muslim (mainly Sunni) 90%, Coptic Christian and other 9%, Other Christian 1%
Ethnic mix Egyptian 99%, Nubian, Armenian, Greek, and Berber 1%
Government Transitional regime
Currency Egyptian pound = 100 piastres
Literacy rate 74%
Calorie consumption 3557 kilocalories

EL SALVADOR
Central America

Page 30 B3

El Salvador is Central America's smallest country. Since a 12-year war between the US-backed army and left-wing guerrillas ended in 1992, crime and gang violence have been key issues.

Official name Republic of El Salvador
Formation 1841 / 1841
Capital San Salvador
Population 6.3 million / 788 people per sq mile (304 people per sq km)
Total area 8124 sq miles (21,040 sq km)
Languages Spanish*
Religions Roman Catholic 80%, Evangelical 18%, Other 2%
Ethnic mix Mestizo 90%, White 9%, Amerindian 1%
Government Presidential system
Currency Salvadorean colón = 100 centavos; US dollar = 100 cents
Literacy rate 86%
Calorie consumption 2513 kilocalories

EQUATORIAL GUINEA
Central Africa

Page 55 A5

Equatorial Guinea comprises the Rio Muni mainland in west Central Africa and five islands. Free elections were first held in 1988, but the former ruling party still dominates.

Official name Republic of Equatorial Guinea
Formation 1968 / 1968
Capital Malabo
Population 800,000 / 74 people per sq mile (29 people per sq km)
Total area 10,830 sq miles (28,051 sq km)
Languages Spanish*, Fang, Bubi, French*
Religions Roman Catholic 90%, Other 10%
Ethnic mix Fang 85%, Other 11%, Bubi 4%
Government Presidential system
Currency CFA franc = 100 centimes
Literacy rate 94%
Calorie consumption Not available

ERITREA
East Africa

Page 50 C4

Lying on the shores of the Red Sea, this former Italian colony was annexed by Ethiopia in 1952. It successfully seceded in 1993, following a 30-year war for independence.

Official name State of Eritrea
Formation 1993 / 2002
Capital Asmara
Population 6.3 million / 139 people per sq mile (54 people per sq km)
Total area 46,842 sq miles (121,320 sq km)
Languages Tigrinya*, English*, Tigre, Afar, Arabic*, Saho, Bilen, Kunama, Nara, Hadareb
Religions Christian 50%, Muslim 48%, Other 2%
Ethnic mix Tigray 50%, Tigre 31%, Other 9%, Saho 5%, Afar 5%
Government Presidential / parliamentary system
Currency Nakfa = 100 cents
Literacy rate 70%
Calorie consumption 1640 kilocalories

ESTONIA
Northeast Europe

Page 84 D2

The smallest, richest, most developed Baltic state has emphasized advanced IT and integration with Europe since renouncing the Soviet model. Joined the EU in 2004.

Official name Republic of Estonia
Formation 1991 / 1991
Capital Tallinn
Population 1.3 million / 75 people per sq mile (29 people per sq km)
Total area 17,462 sq miles (45,226 sq km)
Languages Estonian*, Russian
Religions Evangelical Lutheran 56%, Orthodox Christian 25%, Other 19%
Ethnic mix Estonian 69%, Russian 25%, Other 4%, Ukrainian 2%
Government Parliamentary system
Currency Euro = 100 cents
Literacy rate 99%
Calorie consumption 3214 kilocalories

ETHIOPIA
East Africa

Page 51 C5

Ethiopia, the only African country to escape colonization, was a Marxist regime in 1974–1991. Now landlocked in the Horn of Africa, it has suffered economic, civil, and natural crises.

Official name Federal Democratic Republic of Ethiopia
Formation 1896 / 2002
Capital Addis Ababa
Population 94.1 million / 220 people per sq mile (85 people per sq km)
Total area 435,184 sq miles (1,127,127 sq km)
Languages Amharic*, Tigrinya, Galla, Sidamo, Somali, English, Arabic
Religions Orthodox Christian 40%, Muslim 40%, Traditional beliefs 15%, Other 5%
Ethnic mix Oromo 40%, Amhara 25%, Other 35%
Government Parliamentary system
Currency Birr = 100 cents
Literacy rate 39%
Calorie consumption 2105 kilocalories

FIJI
Australasia & Oceania

Page 123 E5

Fiji is a volcanic archipelago of 882 islands in the southern Pacific Ocean. Tensions between ethnic Fijians and Indo-Fijians provoked coups in 1987 and 2000. Sugar is the main export.

Official name Republic of the Fiji Islands
Formation 1970 / 1970
Capital Suva
Population 900,000 / 128 people per sq mile (49 people per sq km)
Total area 7054 sq miles (18,270 sq km)
Languages Fijian, English*, Hindi, Urdu, Tamil, Telugu
Religions Hindu 38%, Methodist 37%, Roman Catholic 9%, Muslim 8%, Other 8%
Ethnic mix Melanesian 51%, Indian 44%, Other 5%
Government Parliamentary system
Currency Fiji dollar = 100 cents
Literacy rate 94%
Calorie consumption 2930 kilocalories

FINLAND
Northern Europe

Page 62 D4

A low-lying country of forests and lakes, Finland joins Scandinavia to Russia. Its language is related to only two others in Europe. Finnish women were the first in Europe to get the vote, in 1906.

Official name Republic of Finland
Formation 1917 / 1947
Capital Helsinki
Population 5.4 million / 46 people per sq mile (18 people per sq km)
Total area 130,127 sq miles (337,030 sq km)
Languages Finnish*, Swedish*, Sámi
Religions Evangelical Lutheran 83%, Other 15%, Roman Catholic 1%, Orthodox Christian 1%
Ethnic mix Finnish 93%, Other (including Sámi) 7%
Government Parliamentary system
Currency Euro = 100 cents
Literacy rate 99%
Calorie consumption 3285 kilocalories

FRANCE
Western Europe

Page 68 B4

Straddling Western Europe from the English Channel to the Mediterranean Sea, France was Europe's first modern republic. It is now one of the world's leading industrial powers.

Official name French Republic
Formation 987 / 1919
Capital Paris
Population 64.3 million / 303 people per sq mile (117 people per sq km)
Total area 211,208 sq miles (547,030 sq km)
Languages French*, Provençal, German, Breton, Catalan, Basque
Religions Roman Catholic 88%, Muslim 8%, Protestant 2%, Jewish 1%, Buddhist 1%
Ethnic mix French 90%, North African (mainly Algerian) 6%, German (Alsace) 2%, Other 2%
Government Presidential / parliamentary system
Currency Euro = 100 cents
Literacy rate 99%
Calorie consumption 3524 kilocalories

GABON
Central Africa

Page 55 A5

A former French colony straddling the Equator on Central Africa's west coast, it returned to multiparty politics in 1990, after 22 years of one-party rule. The economy relies on oil revenue.

Official name Gabonese Republic
Formation 1960 / 1960
Capital Libreville
Population 1.7 million / 17 people per sq mile (7 people per sq km)
Total area 103,346 sq miles (267,667 sq km)
Languages Fang, French*, Punu, Sira, Nzebi, Mpongwe
Religions Christian (mainly Roman Catholic) 55%, Traditional beliefs 40%, Other 4%, Muslim 1%
Ethnic mix Fang 26%, Shira-punu 24%, Other 16%, Foreign residents 15%, Nzabi-duma 11%, Mbédé-Teke 8%
Government Presidential system
Currency CFA franc = 100 centimes
Literacy rate 82%
Calorie consumption 2781 kilocalories

GAMBIA
West Africa

Page 52 B3

A narrow state along the Gambia River on Africa's west coast and surrounded by Senegal, Gambia was renowned for its stability until a coup in 1994; the coup leader remains in power.

Official name Republic of the Gambia
Formation 1965 / 1965
Capital Banjul
Population 1.8 million / 466 people per sq mile (180 people per sq km)
Total area 4363 sq miles (11,300 sq km)
Languages Mandinka, Fulani, Wolof, Jola, Soninke, English*
Religions Sunni Muslim 90%, Christian 8%, Traditional beliefs 2%
Ethnic mix Mandinka 42%, Fulani 18%, Wolof 16%, Jola 10%, Serahuli 9%, Other 5%
Government Presidential system
Currency Dalasi = 100 butut
Literacy rate 52%
Calorie consumption 2849 kilocalories

GEORGIA
Southwest Asia

Page 95 F2

Located in the Caucasus on the Black Sea's eastern shore, Georgia is noted for its wine. Conflict broke out after the breakup of the USSR; the northern provinces have de facto autonomy.

Official name Georgia
Formation 1991 / 1991
Capital Tbilisi
Population 4.3 million / 160 people per sq mile (62 people per sq km)
Total area 26,911 sq miles (69,700 sq km)
Languages Georgian*, Russian, Azeri, Armenian, Mingrelian, Ossetian, Abkhazian
Religions Georgian Orthodox 74%, Russian Orthodox 10%, Muslim 10%, Other 6%
Ethnic mix Georgian 84%, Armenian 6%, Azeri 6%, Russian 2%, Other 1%, Ossetian 1%
Government Presidential system
Currency Lari = 100 tetri
Literacy rate 99%
Calorie consumption 2731 kilocalories

GHANA
West Africa

Page 53 E5

Once known as the Gold Coast, Ghana was the first colony in West Africa to gain independence. In recent decades multiparty democracy has been consolidated despite economic issues.

Official name Republic of Ghana
Formation 1957 / 1957
Capital Accra
Population 25.9 million / 292 people per sq mile (113 people per sq km)
Total area 92,100 sq miles (238,540 sq km)
Languages Twi, Fanti, Ewe, Ga, Adangbe, Gurma, Dagomba (Dagbani), English*
Religions Christian 69%, Muslim 16%, Traditional beliefs 9%, Other 6%
Ethnic mix Akan 49%, Mole-Dagbani 17%, Ewe 13%, Other 9%, Ga and Ga-Adangbe 8%, Guan 4%
Government Presidential system
Currency Cedi = 100 pesewas
Literacy rate 72%
Calorie consumption 3003 kilocalories

GREECE
Southeast Europe

Page 83 A5

The southernmost Balkan nation has a mountainous mainland and over 2000 islands, engendering its seafaring tradition. High state debt has led to recent unpopular austerity measures.

Official name Hellenic Republic
Formation 1829 / 1947
Capital Athens
Population 11.1 million / 220 people per sq mile (85 people per sq km)
Total area 50,942 sq miles (131,940 sq km)
Languages Greek*, Turkish, Macedonian, Albanian
Religions Orthodox Christian 98%, Other 1%, Muslim 1%
Ethnic mix Greek 98%, Other 2%
Government Parliamentary system
Currency Euro = 100 cents
Literacy rate 97%
Calorie consumption 3433 kilocalories

GERMANY
Northern Europe

Page 72 B4

Germany is Europe's major economic power and a leading influence in the EU. Divided after World War II, its democratic west and communist east were re-unified in 1990.

Official name Federal Republic of Germany
Formation 1871 / 1990
Capital Berlin
Population 82.7 million / 613 people per sq mile (237 people per sq km)
Total area 137,846 sq miles (357,021 sq km)
Languages German*, Turkish
Religions Protestant 34%, Roman Catholic 33%, Other 30%, Muslim 3%
Ethnic mix German 92%, Other 3%, Other European 3%, Turkish 2%
Government Parliamentary system
Currency Euro = 100 cents
Literacy rate 99%
Calorie consumption 3539 kilocalories

GRENADA
West Indies

Page 33 G5

The most southerly Windward gained worldwide notoriety in 1983, when the US invaded to sever its growing links with Cuba. It is the world's second-biggest nutmeg producer.

Official name Grenada
Formation 1974 / 1974
Capital St. George's
Population 109,590 / 837 people per sq mile (322 people per sq km)
Total area 131 sq miles (340 sq km)
Languages English*, English Creole
Religions Roman Catholic 68%, Anglican 17%, Other 15%
Ethnic mix Black African 82%, Mulatto (mixed race) 13%, East Indian 3%, Other 2%
Government Parliamentary system
Currency East Caribbean dollar = 100 cents
Literacy rate 96%
Calorie consumption 2453 kilocalories

GUATEMALA
Central America

Page 30 A2

Once the heart of the Mayan civilization, the largest and most populous state on the Central American isthmus is consolidating its fledgling democracy after years of civil war and army rule.

Official name Republic of Guatemala
Formation 1838 / 1838
Capital Guatemala City
Population 15.5 million / 370 people per sq mile (143 people per sq km)
Total area 42,042 sq miles (108,890 sq km)
Languages Quiché, Mam, Cakchiquel, Kekchí, Spanish*
Religions Roman Catholic 65%, Protestant 33%, Other and nonreligious 2%
Ethnic mix Amerindian 60%, Mestizo 30%, Other 10%
Government Presidential system
Currency Quetzal = 100 centavos
Literacy rate 78%
Calorie consumption 2502 kilocalories

GUINEA
West Africa

Page 52 C4

A former French colony on Africa's west coast, Guinea chose a Marxist path, then came under army rule. The 2010 polls brought fresh hope, before the Ebola epidemic struck in 2014.

Official name Republic of Guinea
Formation 1958 / 1958
Capital Conakry
Population 11.7 million / 123 people per sq mile (48 people per sq km)
Total area 94,925 sq miles (245,857 sq km)
Languages Pulaar, Malinké, Soussou, French*
Religions Muslim 85%, Christian 8%, Traditional beliefs 7%
Ethnic mix Peul 40%, Malinké 30%, Soussou 20%, Other 10%
Government Presidential system
Currency Guinea franc = 100 centimes
Literacy rate 25%
Calorie consumption 2553 kilocalories

GUINEA-BISSAU
West Africa

Page 52 B4

Known as Portuguese Guinea in colonial times, Guinea-Bissau is situated on Africa's west coast. One of the world's poorest countries, it has now become a transit point for cocaine trafficking.

Official name Republic of Guinea-Bissau
Formation 1974 / 1974
Capital Bissau
Population 1.7 million / 157 people per sq mile (60 people per sq km)
Total area 13,946 sq miles (36,120 sq km)
Languages Portuguese Creole, Balante, Fulani, Malinké, Portuguese*
Religions Traditional beliefs 50%, Muslim 40%, Christian 10%
Ethnic mix Balante 30%, Fulani 20%, Other 16%, Mandyako 14%, Mandinka 13%, Papel 7%
Government Presidential system
Currency CFA franc = 100 centimes
Literacy rate 57%
Calorie consumption 2304 kilocalories

GUYANA
South America

Page 37 F3

A land of rain forest, mountains, coastal plains, and savanna, Guyana is South America's only English-speaking state. It became a republic in 1970, four years after independence from Britain.

Official name Cooperative Republic of Guyana
Formation 1966 / 1966
Capital Georgetown
Population 800,000 / 11 people per sq mile (4 people per sq km)
Total area 83,000 sq miles (214,970 sq km)
Languages English Creole, Hindi, Tamil, Amerindian languages, English*
Religions Christian 57%, Hindu 28%, Muslim 10%, Other 5%
Ethnic mix East Indian 43%, Black African 30%, Mixed race 17%, Amerindian 9%, Other 1%
Government Presidential system
Currency Guyanese dollar = 100 cents
Literacy rate 85%
Calorie consumption 2648 kilocalories

HAITI
West Indies

Page 32 D3

The western third of the Caribbean island of Hispaniola, Haiti became the world's first black republic in 1804. Natural disasters and periodic anarchy perpetuate its endemic poverty.

Official name Republic of Haiti
Formation 1804 / 1884
Capital Port-au-Prince
Population 10.3 million / 968 people per sq mile (374 people per sq km)
Total area 10,714 sq miles (27,750 sq km)
Languages French Creole*, French*
Religions Roman Catholic 55%, Protestant 28%, Other (including Voodoo) 16%, Nonreligious 1%
Ethnic mix Black African 95%, Mulatto (mixed race) and European 5%
Government Presidential system
Currency Gourde = 100 centimes
Literacy rate 49%
Calorie consumption 2105 kilocalories

HONDURAS
Central America

Page 30 C2

Straddling the Central American isthmus, Honduras returned to civilian rule in 1984, after a succession of military regimes. Crime is high and it has the world's worst murder rate.

Official name Republic of Honduras
Formation 1838 / 1838
Capital Tegucigalpa
Population 8.1 million / 187 people per sq mile (72 people per sq km)
Total area 43,278 sq miles (112,090 sq km)
Languages Spanish*, Garifuna (Carib), English Creole
Religions Roman Catholic 97%, Protestant 3%
Ethnic mix Mestizo 90%, Black African 5%, Amerindian 4%, White 1%
Government Presidential system
Currency Lempira = 100 centavos
Literacy rate 85%
Calorie consumption 2651 kilocalories

HUNGARY
Central Europe

Page 77 C6

Hungary is bordered by seven states in Central Europe. After the fall of communism in 1989, it introduced political and economic reforms and joined the EU in 2004.

Official name Republic of Hungary
Formation 1918 / 1947
Capital Budapest
Population 10 million / 280 people per sq mile (108 people per sq km)
Total area 35,919 sq miles (93,030 sq km)
Languages Hungarian* (Magyar)
Religions Roman Catholic 52%, Calvinist 16%, Other 15%, Nonreligious 14%, Lutheran 3%
Ethnic mix Magyar 90%, Roma 4%, German 3%, Serb 2%, Other 1%
Government Parliamentary system
Currency Forint = 100 fillér
Literacy rate 99%
Calorie consumption 2968 kilocalories

ICELAND
Northwest Europe

Page 61 E4

This northerly island outpost of Europe, sitting on the mid-Atlantic ridge, has stunning, sparsely inhabited volcanic terrain. Its economy crashed heavily in the 2008 global credit crunch.

Official name Republic of Iceland
Formation 1944 / 1944
Capital Reykjavík
Population 300,000 / 8 people per sq mile (3 people per sq km)
Total area 39,768 sq miles (103,000 sq km)
Languages Icelandic*
Religions Evangelical Lutheran 84%, Other (mostly Christian) 10%, Nonreligious 3%, Roman Catholic 3%
Ethnic mix Icelandic 94%, Other 5%, Danish 1%
Government Parliamentary system
Currency Icelandic króna = 100 aurar
Literacy rate 99%
Calorie consumption 3339 kilocalories

INDIA
South Asia

Page 112 D4

The Indian subcontinent, divided from the rest of Asia by the Himalayas, was once the jewel of the British empire. India is the world's largest democracy and second most populous country.

Official name Republic of India
Formation 1947 / 1947
Capital New Delhi
Population 1.25 billion / 1091 people per sq mile (421 people per sq km)
Total area 1,269,338 sq miles (3,287,590 sq km)
Languages Hindi*, English*, Urdu, Bengali, Marathi, Telugu, Tamil, Bihari, Gujarati, Kanarese
Religions Hindu 81%, Muslim 13%, Sikh 2%, Christian 2%, Buddhist 1%, Other 1%
Ethnic mix Indo-Aryan 72%, Dravidian 25%, Mongoloid and other 3%
Government Parliamentary system
Currency Indian rupee = 100 paise
Literacy rate 63%
Calorie consumption 2459 kilocalories

INDONESIA
Southeast Asia

Page 116 C4

The world's largest archipelago spans over 3100 miles (5000 km), from the Indian to the Pacific Ocean. Formerly the Dutch East Indies, it produces palm oil, rubber, spices, and natural gas.

Official name Republic of Indonesia
Formation 1949 / 1999
Capital Jakarta
Population 250 million / 360 people per sq mile (139 people per sq km)
Total area 741,096 sq miles (1,919,440 sq km)
Languages Javanese, Sundanese, Madurese, Bahasa Indonesia*, Dutch
Religions Sunni Muslim 86%, Christian 9%, Hindu 2%, Other 2%, Buddhist 1%
Ethnic mix Javanese 41%, Other 29%, Sundanese 15%, Coastal Malays 12%, Madurese 3%
Government Presidential system
Currency Rupiah = 100 sen
Literacy rate 93%
Calorie consumption 2713 kilocalories

IRAN
Southwest Asia

Page 98 C3

After the 1979 Islamist revolution led by Ayatollah Khomeini deposed the shah, this Middle Eastern country became the world's largest theocracy. It has large oil and natural gas reserves.

Official name Islamic Republic of Iran
Formation 1502 / 1990
Capital Tehran
Population 77.4 million / 123 people per sq mile (47 people per sq km)
Total area 636,293 sq miles (1,648,000 sq km)
Languages Farsi*, Azeri, Luri, Gilaki, Mazanderani, Kurdish, Turkmen, Arabic, Baluchi
Religions Shi'a Muslim 89%, Sunni Muslim 9%, Other 2%
Ethnic mix Persian 51%, Azari 24%, Other 10%, Lur and Bakhtiari 8%, Kurdish 7%
Government Islamic theocracy
Currency Iranian rial = 100 dinars
Literacy rate 84%
Calorie consumption 3058 kilocalories

IRAQ
Southwest Asia

Page 98 B3

Oil-rich Iraq is situated in the central Middle East. A US-led invasion in 2003 toppled Saddam Hussein's regime, but sectarian violence since then has caused political and social turmoil.

Official name Republic of Iraq
Formation 1932 / 1990
Capital Baghdad
Population 33.8 million / 200 people per sq mile (77 people per sq km)
Total area 168,753 sq miles (437,072 sq km)
Languages Arabic*, Kurdish*, Turkic languages, Armenian, Assyrian
Religions Shi'a Muslim 60%, Sunni Muslim 35%, Other (including Christian) 5%
Ethnic mix Arab 80%, Kurdish 15%, Turkmen 3%, Other 2%
Government Parliamentary system
Currency New Iraqi dinar = 1000 fils
Literacy rate 79%
Calorie consumption 2489 kilocalories

IRELAND
Northwest Europe

Page 67 A6

British rule ended in 1922 for 80% of the island of Ireland, which became the Irish Republic in 1949. The economy is now recovering after suffering heavily in the 2008 global financial crisis.

Official name Ireland
Formation 1922 / 1922
Capital Dublin
Population 4.6 million / 173 people per sq mile (67 people per sq km)
Total area 27,135 sq miles (70,280 sq km)
Languages English*, Irish*
Religions Roman Catholic 87%, Other and nonreligious 10%, Anglican 3%
Ethnic mix Irish 99%, Other 1%
Government Parliamentary system
Currency Euro = 100 cents
Literacy rate 99%
Calorie consumption 3591 kilocalories

ISRAEL
Southwest Asia

Page 97 A7

In 1948 this Jewish state was carved out of Palestine on the east coast of the Mediterranean. It has gained land from its Arab neighbors, and the status of the Palestinians remains unresolved.

Official name State of Israel
Formation 1948 / 1994
Capital Jerusalem (not internationally recognized)
Population 7.7 million / 981 people per sq mile (379 people per sq km)
Total area 8019 sq miles (20,770 sq km)
Languages Hebrew*, Arabic*, Yiddish, German, Russian, Polish, Romanian, Persian
Religions Jewish 76%, Muslim (mainly Sunni) 16%, Druze 2%, Christian 2%
Ethnic mix Jewish 76%, Arab 20%, Other 4%
Government Parliamentary system
Currency Shekel = 100 agorot
Literacy rate 98%
Calorie consumption 3619 kilocalories

ITALY
Southern Europe

Page 74 B3

A boot-shaped peninsula jutting into the Mediterranean, Italy is a world leader in product design, fashion, and textiles. Divisions exist between the industrial north and poorer south.

Official name Italian Republic
Formation 1861 / 1947
Capital Rome
Population 61 million / 537 people per sq mile (207 people per sq km)
Total area 116,305 sq miles (301,230 sq km)
Languages Italian*, German, French, Rhaeto-Romanic, Sardinian
Religions Roman Catholic 85%, Other and nonreligious 13%, Muslim 2%
Ethnic mix Italian 94%, Other 4%, Sardinian 2%
Government Parliamentary system
Currency Euro = 100 cents
Literacy rate 99%
Calorie consumption 3539 kilocalories

JAMAICA
West Indies

Page 32 C3

Colonized by Spain and then Britain, Jamaica was the first Caribbean island to gain independence in the postwar era. Jamaican popular music culture developed reggae, ska, and dancehall.

Official name Jamaica
Formation 1962 / 1962
Capital Kingston
Population 2.8 million / 670 people per sq mile (259 people per sq km)
Total area 4243 sq miles (10,990 sq km)
Languages English Creole, English*
Religions Other and nonreligious 45%, Other Protestant 20%, Church of God 18%, Baptist 10%, Anglican 7%
Ethnic mix Black African 91%, Mulatto (mixed race) 7%, European and Chinese 1%, East Indian 1%
Government Parliamentary system
Currency Jamaican dollar = 100 cents
Literacy rate 88%
Calorie consumption 2789 kilocalories

JAPAN
East Asia

Page 108 C4

Japan has four main islands and over 3000 smaller ones. Rebuilding after defeat in World War II, by 1990 it was the world's second-biggest economy. It retains its emperor as head of state.

Official name Japan
Formation 1590 / 1972
Capital Tokyo
Population 127 million / 874 people per sq mile (338 people per sq km)
Total area 145,882 sq miles (377,835 sq km)
Languages Japanese*, Korean, Chinese
Religions Shinto and Buddhist 76%, Buddhist 16%, Other (including Christian) 8%
Ethnic mix Japanese 99%, Other (mainly Korean) 1%
Government Parliamentary system
Currency Yen = 100 sen
Literacy rate 99%
Calorie consumption 2719 kilocalories

JORDAN
Southwest Asia

Page 97 B6

This Middle Eastern kingdom stretches from the east bank of the Jordan River into largely uninhabited desert. Calls for greater democratization have engendered some reforms.

Official name Hashemite Kingdom of Jordan
Formation 1946 / 1967
Capital Amman
Population 7.3 million / 213 people per sq mile (82 people per sq km)
Total area 35,637 sq miles (92,300 sq km)
Languages Arabic*
Religions Sunni Muslim 92%, Christian 6%, Other 2%
Ethnic mix Arab 98%, Circassian 1%, Armenian 1%
Government Monarchy
Currency Jordanian dinar = 1000 fils
Literacy rate 98%
Calorie consumption 3149 kilocalories

KAZAKHSTAN
Central Asia

Page 92 B4

Second-largest of the former Soviet republics, mineral-rich Kazakhstan is Central Asia's major economic power. The former communist leader remains in charge, facing little opposition.

Official name Republic of Kazakhstan
Formation 1991 / 1991
Capital Astana
Population 16.4 million / 16 people per sq mile (6 people per sq km)
Total area 1,049,150 sq miles (2,717,300 sq km)
Languages Kazakh*, Russian, Ukrainian, German, Uzbek, Tatar, Uighur
Religions Muslim (mainly Sunni) 47%, Orthodox Christian 44%, Other 7%, Protestant 2%
Ethnic mix Kazakh 57%, Russian 27%, Other 8%, Ukrainian 3%, Uzbek 3%, German 2%
Government Presidential system
Currency Tenge = 100 tiyn
Literacy rate 99%
Calorie consumption 3107 kilocalories

KENYA
East Africa

Page 51 C6

Straddling the Equator on Africa's east coast, Kenya has known both stable periods and internal strife since independence in 1963. Corruption is now a key political issue.

Official name Republic of Kenya
Formation 1963 / 1963
Capital Nairobi
Population 44.4 million / 203 people per sq mile (78 people per sq km)
Total area 224,961 sq miles (582,650 sq km)
Languages Kiswahili*, English*, Kikuyu, Luo, Kalenjin, Kamba
Religions Christian 80%, Muslim 10%, Traditional beliefs 9%, Other 1%
Ethnic mix Other 28%, Kikuyu 22%, Luo 14%, Luhya 14%, Kamba 11%, Kalenjin 11%
Government Presidential system
Currency Kenya shilling = 100 cents
Literacy rate 72%
Calorie consumption 2189 kilocalories

KIRIBATI
Australasia & Oceania

Page 123 F3

Part of the British colony of the Gilbert and Ellice Islands until independence in 1979, Kiribati comprises 33 islands in the mid-Pacific Ocean. Phosphate deposits on Banaba ran out in 1980.

Official name Republic of Kiribati
Formation 1979 / 1979
Capital Bairiki (Tarawa Atoll)
Population 103,248 / 377 people per sq mile (145 people per sq km)
Total area 277 sq miles (717 sq km)
Languages English*, Kiribati
Religions Roman Catholic 55%, Kiribati Protestant Church 36%, Other 9%
Ethnic mix Micronesian 99%, Other 1%
Government Presidential system
Currency Australian dollar = 100 cents
Literacy rate 99%
Calorie consumption 3022 kilocalories

KOSOVO (not fully recognized)
Southeast Europe

Page 79 D5

NATO intervention in 1999 ended ethnic cleansing by the Serbs of Kosovo's majority Albanian population, and nine years later the region unilaterally declared independence from Serbia.

Official name Republic of Kosovo
Formation 2008 / 2008
Capital Priština
Population 1.8 million / 427 people per sq mile (165 people per sq km)
Total area 4212 sq miles (10,908 sq km)
Languages Albanian*, Serbian*, Bosniak, Gorani, Roma, Turkish
Religions Muslim 92%, Orthodox Christian 4%, Roman Catholic 4%
Ethnic mix Albanian 92%, Serb 4%, Bosniak and Gorani 2%, Roma 1%, Turkish 1%
Government Parliamentary system
Currency Euro = 100 cents
Literacy rate 92%
Calorie consumption Not available

KUWAIT
Southwest Asia

Page 98 C4

Kuwait, on the Persian Gulf, was a British protectorate from 1914 to 1961. Oil-rich since the 1950s, it was annexed briefly in 1990 by Iraq but US-led intervention restored the ruling amir.

Official name State of Kuwait
Formation 1961 / 1961
Capital Kuwait City
Population 3.4 million / 494 people per sq mile (191 people per sq km)
Total area 6880 sq miles (17,820 sq km)
Languages Arabic*, English
Religions Sunni Muslim 45%, Shi'a Muslim 40%, Christian, Hindu, and other 15%
Ethnic mix Kuwaiti 45%, Other Arab 35%, South Asian 9%, Other 7%, Iranian 4%
Government Monarchy
Currency Kuwaiti dinar = 1000 fils
Literacy rate 96%
Calorie consumption 3471 kilocalories

KYRGYZSTAN
Central Asia

Page 101 F2

This mountainous, landlocked state in Central Asia is the most rural of the ex-Soviet republics. Popular protests ousted the long-term president in 2005 and his successor in 2010.

Official name Kyrgyz Republic
Formation 1991 / 1991
Capital Bishkek
Population 5.5 million / 72 people per sq mile (28 people per sq km)
Total area 76,641 sq miles (198,500 sq km)
Languages Kyrgyz*, Russian*, Uzbek, Tatar, Ukrainian
Religions Muslim (mainly Sunni) 70%, Orthodox Christian 30%
Ethnic mix Kyrgyz 69%, Uzbek 14%, Russian 9%, Other 6%, Uighur 1%, Dungan 1%
Government Presidential system
Currency Som = 100 tyiyn
Literacy rate 99%
Calorie consumption 2828 kilocalories

LAOS
Southeast Asia

Page 114 D4

Landlocked Laos suffered a long civil war after French rule ended, and was badly bombed by US forces engaged in Vietnam. It has been under communist rule since 1975.

Official name Lao People's Democratic Republic
Formation 1953 / 1953
Capital Vientiane
Population 6.8 million / 76 people per sq mile (29 people per sq km)
Total area 91,428 sq miles (236,800 sq km)
Languages Lao*, Mon-Khmer, Yao, Vietnamese, Chinese, French
Religions Buddhist 65%, Other (including animist) 34%, Christian 1%
Ethnic mix Lao Loum 66%, Lao Theung 30%, Other 2%, Lao Soung 2%
Government One-party state
Currency Kip = 100 at
Literacy rate 73%
Calorie consumption 2356 kilocalories

LATVIA
Northeast Europe

Page 84 C3

Situated on the low-lying eastern shores of the Baltic Sea, Latvia, like its Baltic neighbors, regained its independence at the collapse of the USSR in 1991. It retains a large Russian population.

Official name Republic of Latvia
Formation 1991 / 1991
Capital Riga
Population 2.1 million / 84 people per sq mile (33 people per sq km)
Total area 24,938 sq miles (64,589 sq km)
Languages Latvian*, Russian
Religions Other 43%, Lutheran 24%, Roman Catholic 18%, Orthodox Christian 15%
Ethnic mix Latvian 62%, Russian 27%, Other 4%, Belarussian 3%, Ukrainian 2%, Polish 2%
Government Parliamentary system
Currency Euro = 100 cents
Literacy rate 99%
Calorie consumption 3293 kilocalories

LEBANON
Southwest Asia

Page 96 A4

Lebanon is dwarfed by its two powerful neighbors, Syria and Israel. Muslims and Christians fought a 14-year civil war until agreeing to share power in 1989, however, instability continues.

Official name Republic of Lebanon
Formation 1941 / 1941
Capital Beirut
Population 4.8 million / 1215 people per sq mile (469 people per sq km)
Total area 4015 sq miles (10,400 sq km)
Languages Arabic*, French, Armenian, Assyrian
Religions Muslim 60%, Christian 39%, Other 1%
Ethnic mix Arab 95%, Armenian 4%, Other 1%
Government Parliamentary system
Currency Lebanese pound = 100 piastres
Literacy rate 90%
Calorie consumption 3181 kilocalories

LESOTHO
Southern Africa

Page 56 D5

Lesotho lies within South Africa, on whom it is economically dependent. Elections in 1993 ended military rule, but South Africa has had to intervene in politics since. AIDS is a problem.

Official name Kingdom of Lesotho
Formation 1966 / 1966
Capital Maseru
Population 2.1 million / 179 people per sq mile (69 people per sq km)
Total area 11,720 sq miles (30,355 sq km)
Languages English*, Sesotho*, isiZulu
Religions Christian 90%, Traditional beliefs 10%
Ethnic mix Sotho 99%, European and Asian 1%
Government Parliamentary system
Currency Loti = 100 lisente; South African rand = 100 cents
Literacy rate 76%
Calorie consumption 2595 kilocalories

LIBERIA
West Africa

Page 52 C5

Facing the Atlantic Ocean, Liberia is Africa's oldest republic, founded in 1847 by freed US slaves. Recovery from the 1990s' civil war has been set back by the 2014 Ebola epidemic.

Official name Republic of Liberia
Formation 1847 / 1847
Capital Monrovia
Population 4.3 million / 116 people per sq mile (45 people per sq km)
Total area 43,000 sq miles (111,370 sq km)
Languages Kpelle, Vai, Bassa, Kru, Grebo, Kissi, Gola, Loma, English*
Religions Traditional beliefs 40%, Christian 40%, Muslim 20%
Ethnic mix Indigenous tribes (12 groups) 49%, Kpellé 20%, Bassa 16%, Gio 8%, Krou 7%
Government Presidential system
Currency Liberian dollar = 100 cents
Literacy rate 43%
Calorie consumption 2251 kilocalories

LIBYA
North Africa

Page 49 F3

On the Mediterranean coast, Libya was ruled from 1969 by the idiosyncratic Col. Gaddafi. The 2011 "Arab Spring" turned to civil war, toppling his regime, but leaving Libya in anarchy.

Official name State of Libya
Formation 1951 / 1951
Capital Tripoli
Population 6.2 million / 9 people per sq mile (4 people per sq km)
Total area 679,358 sq miles (1,759,540 sq km)
Languages Arabic*, Tuareg
Religions Muslim (mainly Sunni) 97%, Other 3%
Ethnic mix Arab and Berber 97%, Other 3%
Government Transitional regime
Currency Libyan dinar = 1000 dirhams
Literacy rate 90%
Calorie consumption 3211 kilocalories

LIECHTENSTEIN
Central Europe

Page 73 B7

Tucked in the Alps between Switzerland and Austria, Liechtenstein became an independent principality of the Holy Roman Empire in 1719. Switzerland handles its foreign affairs and defense.

Official name Principality of Liechtenstein
Formation 1719 / 1719
Capital Vaduz
Population 37,000 / 597 people per sq mile (231 people per sq km)
Total area 62 sq miles (160 sq km)
Languages German*, Alemannish dialect, Italian
Religions Roman Catholic 79%, Other 13%, Protestant 8%
Ethnic mix Liechtensteiner 66%, Other 12%, Swiss 10%, Austrian 6%, German 3%, Italian 3%
Government Parliamentary system
Currency Swiss franc = 100 rappen/centimes
Literacy rate 99%
Calorie consumption Not available

LITHUANIA
Northeast Europe

Page 84 B4

A flat land of lakes, moors, and bogs, Lithuania is the largest of the three Baltic states. It has historical ties to Poland and was the first former Soviet republic to declare independence.

Official name Republic of Lithuania
Formation 1991 / 1991
Capital Vilnius
Population 3 million / 119 people per sq mile (46 people per sq km)
Total area 25,174 sq miles (65,200 sq km)
Languages Lithuanian*, Russian
Religions Roman Catholic 77%, Other 17%, Russian Orthodox 4%, Protestant 1%, Old Believers 1%
Ethnic mix Lithuanian 85%, Polish 7%, Russian 6%, Belarussian 1%, Other 1%
Government Parliamentary system
Currency Litas = 100 centu
Literacy rate 99%
Calorie consumption 3463 kilocalories

LUXEMBOURG
Northwest Europe

Page 65 D8

Part of the forested Ardennes plateau in Northwest Europe, Luxembourg is Europe's last independent duchy and one of its richest states. It is a banking center and hosts EU institutions.

Official name Grand Duchy of Luxembourg
Formation 1867 / 1867
Capital Luxembourg-Ville
Population 500,000 / 501 people per sq mile (193 people per sq km)
Total area 998 sq miles (2586 sq km)
Languages Luxembourgish*, German*, French*
Religions Roman Catholic 97%, Protestant, Orthodox Christian, and Jewish 3%
Ethnic mix Luxembourger 62%, Foreign residents 38%
Government Parliamentary system
Currency Euro = 100 cents
Literacy rate 99%
Calorie consumption 3568 kilocalories

MACEDONIA
Southeast Europe

Page 79 D6

This ex-Yugoslav state is landlocked in the southern Balkans. Its EU candidacy is held back over Greek fears that its name implies a claim to its own northern province of Macedonia.

Official name Republic of Macedonia
Formation 1991 / 1991
Capital Skopje
Population 2.1 million / 212 people per sq mile (82 people per sq km)
Total area 9781 sq miles (25,333 sq km)
Languages Macedonian*, Albanian*, Turkish, Romani, Serbian
Religions Orthodox Christian 65%, Muslim 29%, Roman Catholic 4%, Other 2%
Ethnic mix Macedonian 64%, Albanian 25%, Turkish 4%, Roma 3%, Other 2%, Serb 2%
Government Presidential / parliamentary system
Currency Macedonian denar = 100 deni
Literacy rate 98%
Calorie consumption 2923 kilocalories

MADAGASCAR
Indian Ocean

Page 57 F4

Off Africa's southeast coast, this former French colony is the world's fourth-largest island. Free elections in 1993 ended 18 years of socialism, but power struggles have blighted politics since.

Official name Republic of Madagascar
Formation 1960 / 1960
Capital Antananarivo
Population 22.9 million / 102 people per sq mile (39 people per sq km)
Total area 226,656 sq miles (587,040 sq km)
Languages Malagasy*, French*, English*
Religions Traditional beliefs 52%, Christian (mainly Roman Catholic) 41%, Muslim 7%
Ethnic mix Other Malay 46%, Merina 26%, Betsimisaraka 15%, Betsileo 12%, Other 1%
Government Presidential / parliamentary system
Currency Ariary = 5 iraimbilanja
Literacy rate 64%
Calorie consumption 2092 kilocalories

MALAWI
Southern Africa

Page 57 E1

This landlocked former British colony lies along the Great Rift Valley and Lake Nyasa, Africa's third-largest lake. Multiparty elections in 1994 ended three decades of single-party rule.

Official name Republic of Malawi
Formation 1964 / 1964
Capital Lilongwe
Population 16.4 million / 451 people per sq mile (174 people per sq km)
Total area 45,745 sq miles (118,480 sq km)
Languages Chewa, Lomwe, Yao, Ngoni, English*
Religions Protestant 55%, Roman Catholic 20%, Muslim 20%, Traditional beliefs 5%
Ethnic mix Bantu 99%, Other 1%
Government Presidential system
Currency Malawi kwacha = 100 tambala
Literacy rate 61%
Calorie consumption 2334 kilocalories

MALAYSIA
Southeast Asia

Page 116 B3

Three separate territories, Peninsular Malaysia, and Sarawak and Sabah on Borneo, make up Malaysia. Relations between indigenous Malays and the Chinese minority dominate politics.

Official name Federation of Malaysia
Formation 1963 / 1965
Capital Kuala Lumpur; Putrajaya (administrative)
Population 29.7 million / 234 people per sq mile (90 people per sq km)
Total area 127,316 sq miles (329,750 sq km)
Languages Bahasa Malaysia*, Malay, Chinese, Tamil, English
Religions Muslim (mainly Sunni) 61%, Buddhist 19%, Christian 9%, Hindu 6%, Other 5%
Ethnic mix Malay 53%, Chinese 26%, Indigenous tribes 12%, Indian 8%, Other 1%
Government Parliamentary system
Currency Ringgit = 100 sen
Literacy rate 93%
Calorie consumption 2855 kilocalories

MALDIVES
Indian Ocean

Page 110 A4

Of this group of over 1000 small low-lying coral islands in the Indian Ocean, only 200 are inhabited. A few families dominate politics and have reversed the electoral upsets of 2008 and 2009.

Official name Republic of Maldives
Formation 1965 / 1965
Capital Male'
Population 300,000 / 2586 people per sq mile (1000 people per sq km)
Total area 116 sq miles (300 sq km)
Languages Dhivehi* (Maldivian), Sinhala, Tamil, Arabic
Religions Sunni Muslim 100%
Ethnic mix Arab–Sinhalese–Malay 100%
Government Presidential system
Currency Rufiyaa = 100 laari
Literacy rate 98%
Calorie consumption 2722 kilocalories

MALI
West Africa

Page 53 E2

Mali's power as a trans-Saharan trading empire peaked 700 years ago. Modern Mali, a one-party state until 1992, called in former colonial power France to suppress Islamist rebels in 2013.

Official name Republic of Mali
Formation 1960 / 1960
Capital Bamako
Population 15.3 million / 32 people per sq mile (13 people per sq km)
Total area 478,764 sq miles (1,240,000 sq km)
Languages Bambara, Fulani, Senufo, Soninke, French*
Religions Muslim (mainly Sunni) 90%, Traditional beliefs 6%, Christian 4%
Ethnic mix Bambara 52%, Other 14%, Fulani 11%, Saracolé 7%, Soninka 7%, Tuareg 5%, Mianka 4%
Government Presidential system
Currency CFA franc = 100 centimes
Literacy rate 34%
Calorie consumption 2833 kilocalories

MALTA
Southern Europe

Page 80 A5

The Maltese archipelago lies off Sicily. Only Malta, Kemmuna, and Gozo are inhabited. Its mid-Mediterranean location has made it a gateway for illegal migration from Africa to Europe.

Official name Republic of Malta
Formation 1964 / 1964
Capital Valletta
Population 400,000 / 3226 people per sq mile (1250 people per sq km)
Total area 122 sq miles (316 sq km)
Languages Maltese*, English*
Religions Roman Catholic 98%, Other and nonreligious 2%
Ethnic mix Maltese 96%, Other 4%
Government Parliamentary system
Currency Euro = 100 cents
Literacy rate 92%
Calorie consumption 3389 kilocalories

MARSHALL ISLANDS
Australasia & Oceania

Page 122 D1

This group of 34 atolls was under US rule as part of the UN Trust Territory of the Pacific Islands until 1986. The economy depends on US aid and rent for the US missile base on Kwajalein.

Official name Republic of the Marshall Islands
Formation 1986 / 1986
Capital Majuro
Population 69,747 / 996 people per sq mile (385 people per sq km)
Total area 70 sq miles (181 sq km)
Languages Marshallese*, English*, Japanese, German
Religions Protestant 90%, Roman Catholic 8%, Other 2%
Ethnic mix Micronesian 90%, Other 10%
Government Presidential system
Currency US dollar = 100 cents
Literacy rate 91%
Calorie consumption Not available

MAURITANIA
West Africa

Page 52 C2

Two-thirds of this former French colony is desert. The Maures oppress the black minority. Multiparty elections from 1991 returned the military leader to power until a coup in 2005.

Official name Islamic Republic of Mauritania
Formation 1960 / 1960
Capital Nouakchott
Population 3.9 million / 10 people per sq mile (4 people per sq km)
Total area 397,953 sq miles (1,030,700 sq km)
Languages Arabic*, Hassaniyah Arabic, Wolof, French
Religions Sunni Muslim 100%
Ethnic mix Maure 81%, Wolof 7%, Tukolor 5%, Other 4%, Soninka 3%
Government Presidential system
Currency Ouguiya = 5 khoums
Literacy rate 46%
Calorie consumption 2791 kilocalories

MAURITIUS
Indian Ocean

Page 57 H3

East of Madagascar in the Indian Ocean, Mauritius became a republic 24 years after independence from Britain. Its diversified economy includes tourism, financial services, and outsourcing.

Official name Republic of Mauritius
Formation 1968 / 1968
Capital Port Louis
Population 1.2 million / 1671 people per sq mile (645 people per sq km)
Total area 718 sq miles (1860 sq km)
Languages French Creole, Hindi, Urdu, Tamil, Chinese, English*, French
Religions Hindu 48%, Roman Catholic 24%, Muslim 17%, Protestant 9%, Other 2%
Ethnic mix Indo-Mauritian 68%, Creole 27%, Sino-Mauritian 3%, Franco-Mauritian 2%
Government Parliamentary system
Currency Mauritian rupee = 100 cents
Literacy rate 89%
Calorie consumption 3055 kilocalories

MEXICO
North America

Page 28 D3

Located between the US and the Central American states, Mexico was a Spanish colony for 300 years. Sprawling Mexico City is built on the site of the Aztec capital, Tenochtitlán.

Official name United Mexican States
Formation 1836 / 1848
Capital Mexico City
Population 122 million / 166 people per sq mile (64 people per sq km)
Total area 761,602 sq miles (1,972,550 sq km)
Languages Spanish*, Nahuatl, Mayan, Zapotec, Mixtec, Otomi, Totonac, Tzotzil, Tzeltal
Religions Roman Catholic 77%, Other 14%, Protestant 6%, Nonreligious 3%
Ethnic mix Mestizo 60%, Amerindian 30%, European 9%, Other 1%
Government Presidential system
Currency Mexican peso = 100 centavos
Literacy rate 94%
Calorie consumption 3024 kilocalories

MICRONESIA
Australasia & Oceania

Page 122 B1

The Federated States of Micronesia, situated in the western Pacific, comprises 607 islands and atolls grouped into four main island states. The economy relies on US aid.

Official name Federated States of Micronesia
Formation 1986 / 1986
Capital Palikir (Pohnpei Island)
Population 106,104 / 392 people per sq mile (151 people per sq km)
Total area 271 sq miles (702 sq km)
Languages Trukese, Pohnpeian, Kosraean, Yapese, English*
Religions Roman Catholic 50%, Protestant 47%, Other 3%
Ethnic mix Chuukese 49%, Pohnpeian 24%, Other 14%, Kosraean 6%, Yapese 5%, Asian 2%
Government Nonparty system
Currency US dollar = 100 cents
Literacy rate 81%
Calorie consumption Not available

MOLDOVA
Southeast Europe

Page 86 D3

The smallest and most densely populated of the ex-Soviet republics, Moldova has strong linguistic and cultural ties with Romania to the west. It exports tobacco, wine, and fruit.

Official name Republic of Moldova
Formation 1991 / 1991
Capital Chisinau
Population 3.5 million / 269 people per sq mile (104 people per sq km)
Total area 13,067 sq miles (33,843 sq km)
Languages Moldovan*, Ukrainian, Russian
Religions Orthodox Christian 93%, Other 6%, Baptist 1%
Ethnic mix Moldovan 84%, Ukrainian 7%, Gagauz 5%, Russian 2%, Bulgarian 1%, Other 1%
Government Parliamentary system
Currency Moldovan leu = 100 bani
Literacy rate 99%
Calorie consumption 2837 kilocalories

MONACO
Southern Europe

Page 69 E6

The destiny of this tiny enclave on France's Côte d'Azur was changed in 1863 when its prince opened a casino. A jet-set image and thriving service sector define its modern identity.

Official name Principality of Monaco
Formation 1861 / 1861
Capital Monaco-Ville
Population 36,136 / 48,181 people per sq mile (18,531 people per sq km)
Total area 0.75 sq miles (1.95 sq km)
Languages French*, Italian, Monégasque, English
Religions Roman Catholic 89%, Protestant 6%, Other 5%
Ethnic mix French 47%, Other 21%, Italian 16%, Monégasque 16%
Government Monarchical / parliamentary system
Currency Euro = 100 cents
Literacy rate 99%
Calorie consumption Not available

MONGOLIA
East Asia

Page 104 D2

Vast Mongolia is sparsely populated and mostly desert. Under the sway of its giant neighbors, Russia and China, it was communist from independence from China in 1924 until 1990.

Official name Mongolia
Formation 1924 / 1924
Capital Ulan Bator
Population 2.8 million / 5 people per sq mile (2 people per sq km)
Total area 604,247 sq miles (1,565,000 sq km)
Languages Khalkha Mongolian*, Kazakh, Chinese, Russian
Religions Tibetan Buddhist 50%, Nonreligious 40%, Shamanist and Christian 6%, Muslim 4%
Ethnic mix Khalkh 95%, Kazakh 4%, Other 1%
Government Presidential / parliamentary system
Currency Tugrik (tögrög) = 100 möngö
Literacy rate 98%
Calorie consumption 2463 kilocalories

MONTENEGRO
Southeast Europe

Page 79 C5

Part of the former Yugoslavia, the tiny republic of Montenegro broke away from Serbia in 2006. Its attractive coast and mountains are a big tourist draw. It hopes to join the EU soon.

Official name Montenegro
Formation 2006 / 2006
Capital Podgorica
Population 600,000 / 113 people per sq mile (43 people per sq km)
Total area 5332 sq miles (13,812 sq km)
Languages Montenegrin*, Serbian, Albanian, Bosniak, Croatian
Religions Orthodox Christian 74%, Muslim 18%, Other 4%, Roman Catholic 4%
Ethnic mix Montenegrin 43%, Serb 32%, Other 12%, Bosniak 8%, Albanian 5%
Government Parliamentary system
Currency Euro = 100 cents
Literacy rate 98%
Calorie consumption 3568 kilocalories

MOROCCO
North Africa

Page 48 C2

A former French colony in northwest Africa, Morocco has occupied the disputed territory of Western Sahara since 1975. The king has handed more power to parliament since 2011.

Official name Kingdom of Morocco
Formation 1956 / 1969
Capital Rabat
Population 33 million / 192 people per sq mile (74 people per sq km)
Total area 172,316 sq miles (446,300 sq km)
Languages Arabic*, Tamazight (Berber), French, Spanish
Religions Muslim (mainly Sunni) 99%, Other (mostly Christian) 1%
Ethnic mix Arab 70%, Berber 29%, European 1%
Government Monarchical / parliamentary system
Currency Moroccan dirham = 100 centimes
Literacy rate 67%
Calorie consumption 3334 kilocalories

MOZAMBIQUE
Southern Africa

Page 57 E3

Mozambique, on the southeast African coast, frequently suffers both floods and droughts. It was torn by civil war from 1977 to 1992 as the Marxist state fought South African-backed rebels.

Official name Republic of Mozambique
Formation 1975 / 1975
Capital Maputo
Population 25.8 million / 85 people per sq mile (33 people per sq km)
Total area 309,494 sq miles (801,590 sq km)
Languages Makua, Xitsonga, Sena, Lomwe, Portuguese*
Religions Traditional beliefs 56%, Christian 30%, Muslim 14%
Ethnic mix Makua Lomwe 47%, Tsonga 23%, Malawi 12%, Shona 11%, Yao 4%, Other 3%
Government Presidential system
Currency New metical = 100 centavos
Literacy rate 51%
Calorie consumption 2267 kilocalories

MYANMAR (BURMA)
Southeast Asia

Page 114 A3

Myanmar, on the eastern shores of the Bay of Bengal and the Andaman Sea, has suffered years of ethnic conflict and repressive military rule since independence from Britain in 1948.

Official name Republic of the Union of Myanmar
Formation 1948 / 1948
Capital Nay Pyi Taw
Population 53.3 million / 210 people per sq mile (81 people per sq km)
Total area 261,969 sq miles (678,500 sq km)
Languages Burmese* (Myanmar), Shan, Karen, Rakhine, Chin, Yangbye, Kachin, Mon
Religions Buddhist 89%, Christian 4%, Muslim 4%, Other 2%, Animist 1%
Ethnic mix Burman (Bamah) 68%, Other 12%, Shan 9%, Karen 7%, Rakhine 4%
Government Presidential system
Currency Kyat = 100 pyas
Literacy rate 93%
Calorie consumption 2528 kilocalories

NAMIBIA
Southern Africa

Page 56 B3

On Africa's southwest coast, this mineral-rich ex-German colony was governed by South Africa from 1915 to 1990. The white minority controls the economy, a legacy of apartheid.

Official name Republic of Namibia
Formation 1990 / 1994
Capital Windhoek
Population 2.3 million / 7 people per sq mile (3 people per sq km)
Total area 318,694 sq miles (825,418 sq km)
Languages Ovambo, Kavango, English*, Bergdama, German, Afrikaans
Religions Christian 90%, Traditional beliefs 10%
Ethnic mix Ovambo 50%, Other tribes 22%, Kavango 9%, Herero 7%, Damara 7%, Other 5%
Government Presidential system
Currency Namibian dollar = 100 cents; South African rand = 100 cents
Literacy rate 76%
Calorie consumption 2086 kilocalories

NAURU
Australasia & Oceania

Page 122 D3

The world's smallest republic, 2480 miles (4000 km) northeast of Australia, grew rich from its phosphate deposits, but these have almost run out and poor investment has caused financial crisis.

Official name Republic of Nauru
Formation 1968 / 1968
Capital None
Population 9434 / 1165 people per sq mile (449 people per sq km)
Total area 8.1 sq miles (21 sq km)
Languages Nauruan*, Kiribati, Chinese, Tuvaluan, English
Religions Nauruan Congregational Church 60%, Roman Catholic 35%, Other 5%
Ethnic mix Nauruan 93%, Chinese 5%, Other Pacific islanders 1%, European 1%
Government Nonparty system
Currency Australian dollar = 100 cents
Literacy rate 95%
Calorie consumption Not available

NEPAL
South Asia

Page 113 E3

Nestled in the Himalayas, Nepal had an absolute monarch until 1990. Unstable coalitions typify politics. Abolition of the monarchy was a condition for ending the Maoist rebellion in 2008.

Official name Federal Democratic Republic of Nepal
Formation 1769 / 1769
Capital Kathmandu
Population 27.8 million / 526 people per sq mile (203 people per sq km)
Total area 54,363 sq miles (140,800 sq km)
Languages Nepali*, Maithili, Bhojpuri
Religions Hindu 81%, Buddhist 11%, Muslim 4%, Other (including Christian) 4%
Ethnic mix Other 52%, Chhetri 16%, Hill Brahman 13%, Magar 7%, Tharu 7%, Tamang 5%
Government Transitional regime
Currency Nepalese rupee = 100 paisa
Literacy rate 57%
Calorie consumption 2580 kilocalories

NETHERLANDS
Northwest Europe

Page 64 C3

Astride the delta of four major rivers in northwest Europe, the Netherlands was ruled by Spain until 1648. It has a long trading tradition, and Rotterdam remains the world's largest port.

Official name Kingdom of the Netherlands
Formation 1648 / 1839
Capital Amsterdam; The Hague (administrative)
Population 16.8 million / 1283 people per sq mile (495 people per sq km)
Total area 16,033 sq miles (41,526 sq km)
Languages Dutch*, Frisian
Religions Roman Catholic 36%, Other 34%, Protestant 27%, Muslim 3%
Ethnic mix Dutch 82%, Other 12%, Surinamese 2%, Turkish 2%, Moroccan 2%
Government Parliamentary system
Currency Euro = 100 cents
Literacy rate 99%
Calorie consumption 3147 kilocalories

NEW ZEALAND
Australasia & Oceania

Page 128 A4

This former British colony, on the Pacific Rim, has a volcanic, more populous North Island and a mountainous South Island. It was the first country to give women the vote, in 1893.

Official name New Zealand
Formation 1947 / 1947
Capital Wellington
Population 4.5 million / 43 people per sq mile (17 people per sq km)
Total area 103,737 sq miles (268,680 sq km)
Languages English*, Maori*
Religions Anglican 24%, Other 22%, Presbyterian 18%, Nonreligious 16%, Roman Catholic 15%, Methodist 5%
Ethnic mix European 75%, Maori 15%, Other 7%, Samoan 3%
Government Parliamentary system
Currency New Zealand dollar = 100 cents
Literacy rate 99%
Calorie consumption 3170 kilocalories

NICARAGUA
Central America

Page 30 D3

Nicaragua lies at the heart of Central America. Left-wing Sandinistas threw out a brutal dictator in 1978, then faced conflict with US-backed Contras. Polls have since swung back and forth.

Official name Republic of Nicaragua
Formation 1838 / 1838
Capital Managua
Population 6.1 million / 133 people per sq mile (51 people per sq km)
Total area 49,998 sq miles (129,494 sq km)
Languages Spanish*, English Creole, Miskito
Religions Roman Catholic 80%, Protestant Evangelical 17%, Other 3%
Ethnic mix Mestizo 69%, White 17%, Black 9%, Amerindian 5%
Government Presidential system
Currency Córdoba oro = 100 centavos
Literacy rate 78%
Calorie consumption 2564 kilocalories

NIGER
West Africa

Page 53 G3

Landlocked Niger is linked to the sea by the River Niger. This ex-French colony has suffered coups, military rule, civil unrest, and severe droughts. It is one of the poorest countries in the world.

Official name Republic of Niger
Formation 1960 / 1960
Capital Niamey
Population 17.8 million / 36 people per sq mile (14 people per sq km)
Total area 489,188 sq miles (1,267,000 sq km)
Languages Hausa, Djerma, Fulani, Tuareg, Teda, French*
Religions Muslim 99%, Other (including Christian) 1%
Ethnic mix Hausa 53%, Djerma and Songhai 21%, Tuareg 11%, Fulani 7%, Kanuri 6%, Other 2%
Government Presidential system
Currency CFA franc = 100 centimes
Literacy rate 16%
Calorie consumption 2546 kilocalories

NIGERIA
West Africa

Page 53 G4

Nigeria has Africa's largest population, whose religious and ethnic rivalries have brought down both civilian and military regimes in the past. Islamic extremists are one current challenge.

Official name Federal Republic of Nigeria
Formation 1960 / 1961
Capital Abuja
Population 174 million / 494 people per sq mile (191 people per sq km)
Total area 356,667 sq miles (923,768 sq km)
Languages Hausa, English*, Yoruba, Ibo
Religions Muslim 50%, Christian 40%, Traditional beliefs 10%
Ethnic mix Other 29%, Hausa 21%, Yoruba 21%, Ibo 18%, Fulani 11%
Government Presidential system
Currency Naira = 100 kobo
Literacy rate 51%
Calorie consumption 2724 kilocalories

NORTH KOREA
East Asia

Page 106 E3

The maverick communist state in Korea's northern half has been isolated from the outside world since 1948. Its shattered state-run economy leaves people short of food and power.

Official name Democratic People's Republic of Korea
Formation 1948 / 1953
Capital Pyongyang
Population 24.9 million / 536 people per sq mile (207 people per sq km)
Total area 46,540 sq miles (120,540 sq km)
Languages Korean*
Religions Atheist 100%
Ethnic mix Korean 100%
Government One-party state
Currency North Korean won = 100 chon
Literacy rate 99%
Calorie consumption 2103 kilocalories

NORWAY
Northern Europe

Page 63 A5

Lying on the rugged western coast of Scandinavia, most people live in southern, coastal areas. Oil and gas wealth has brought one of the world's best standards of living.

Official name Kingdom of Norway
Formation 1905 / 1905
Capital Oslo
Population 5 million / 42 people per sq mile (16 people per sq km)
Total area 125,181 sq miles (324,220 sq km)
Languages Norwegian* (Bokmål "book language" and Nynorsk "new Norsk"), Sámi
Religions Evangelical Lutheran 88%, Other and nonreligious 8%, Muslim 2%, Roman Catholic 1%, Pentecostal 1%
Ethnic mix Norwegian 93%, Other 6%, Sámi 1%
Government Parliamentary system
Currency Norwegian krone = 100 øre
Literacy rate 99%
Calorie consumption 3484 kilocalories

OMAN
Southwest Asia

Page 99 D6

Situated on the eastern corner of the Arabian Peninsula, Oman is the least developed of the Gulf states, despite modest oil exports. The current sultan has been in power since 1970.

Official name Sultanate of Oman
Formation 1951 / 1951
Capital Muscat
Population 3.6 million / 44 people per sq mile (17 people per sq km)
Total area 82,031 sq miles (212,460 sq km)
Languages Arabic*, Baluchi, Farsi, Hindi, Punjabi
Religions Ibadi Muslim 75%, Other Muslim and Hindu 25%
Ethnic mix Arab 88%, Baluchi 4%, Indian and Pakistani 3%, Persian 3%, African 2%
Government Monarchy
Currency Omani rial = 1000 baisa
Literacy rate 87%
Calorie consumption Not available

PAKISTAN
South Asia

Page 112 B2

Once part of British India, Pakistan was created in 1947 as a Muslim state. Today, this nuclear-armed country is struggling to deal with complex domestic and international tensions.

Official name Islamic Republic of Pakistan
Formation 1947 / 1971
Capital Islamabad
Population 182 million / 612 people per sq mile (236 people per sq km)
Total area 310,401 sq miles (803,940 sq km)
Languages Punjabi, Sindhi, Pashtu, Urdu*, Baluchi, Brahui
Religions Sunni Muslim 77%, Shi'a Muslim 20%, Hindu 2%, Christian 1%
Ethnic mix Punjabi 56%, Pathan (Pashtun) 15%, Sindhi 14%, Mohajir 7%, Other 4%, Baluchi 4%
Government Parliamentary system
Currency Pakistani rupee = 100 paisa
Literacy rate 55%
Calorie consumption 2428 kilocalories

PALAU
Australasia & Oceania

Page 122 A2

This archipelago of over 200 islands, only ten of which are inhabited, lies in the western Pacific Ocean. Until 1994 it was under US administration. The economy relies on US aid and tourism.

Official name Republic of Palau
Formation 1994 / 1994
Capital Ngerulmud
Population 21,108 / 108 people per sq mile (42 people per sq km)
Total area 177 sq miles (458 sq km)
Languages Palauan*, English*, Japanese, Angaur, Tobi, Sonsorolese
Religions Christian 66%, Modekngei 34%
Ethnic mix Palauan 74%, Filipino 16%, Other 6%, Chinese and other Asian 4%
Government Nonparty system
Currency US dollar = 100 cents
Literacy rate 99%
Calorie consumption Not available

PANAMA
Central America

Page 31 F5

The US invaded Central America's southernmost country in 1989 to oust its dictator. The Panama Canal is a vital shortcut for shipping between the Atlantic and Pacific oceans.

Official name Republic of Panama
Formation 1903 / 1903
Capital Panama City
Population 3.9 million / 133 people per sq mile (51 people per sq km)
Total area 30,193 sq miles (78,200 sq km)
Languages English Creole, Spanish*, Amerindian languages, Chibchan languages
Religions Roman Catholic 84%, Protestant 15%, Other 1%
Ethnic mix Mestizo 70%, Black 14%, White 10%, Amerindian 6%
Government Presidential system
Currency Balboa = 100 centésimos; US dollar
Literacy rate 94%
Calorie consumption 2644 kilocalories

PAPUA NEW GUINEA
Australasia & Oceania

Page 122 B3

The world's most linguistically diverse country, mineral-rich PNG occupies the east of the island of New Guinea and several other island groups. It was administered by Australia before 1975.

Official name Independent State of Papua New Guinea
Formation 1975 / 1975
Capital Port Moresby
Population 7.3 million / 42 people per sq mile (16 people per sq km)
Total area 178,703 sq miles (462,840 sq km)
Languages Pidgin English, Papuan, English*, Motu, 800 (est.) native languages
Religions Protestant 60%, Roman Catholic 37%, Other 3%
Ethnic mix Melanesian and mixed race 100%
Government Parliamentary system
Currency Kina = 100 toea
Literacy rate 63%
Calorie consumption 2193 kilocalories

PARAGUAY
South America

Page 42 D2

South America's longest dictatorship held power in landlocked Paraguay from 1954 to 1989. Now under democratic rule, the country's economy is still largely agricultural.

Official name Republic of Paraguay
Formation 1811 / 1938
Capital Asunción
Population 6.8 million / 44 people per sq mile (17 people per sq km)
Total area 157,046 sq miles (406,750 sq km)
Languages Guaraní*, Spanish*, German
Religions Roman Catholic 90%, Protestant (including Mennonite) 10%
Ethnic mix Mestizo 91%, Other 7%, Amerindian 2%
Government Presidential system
Currency Guaraní = 100 céntimos
Literacy rate 94%
Calorie consumption 2698 kilocalories

PERU
South America

Page 38 C3

On the Pacific coast of South America, Peru was once the heart of the Inca empire, before the Spanish conquest in the 16th century. It elected its first Amerindian president in 2001.

Official name Republic of Peru
Formation 1824 / 1941
Capital Lima
Population 30.4 million / 62 people per sq mile (24 people per sq km)
Total area 496,223 sq miles (1,285,200 sq km)
Languages Spanish*, Quechua*, Aymara
Religions Roman Catholic 81%, Other 19%
Ethnic mix Amerindian 45%, Mestizo 37%, White 15%, Other 3%
Government Presidential system
Currency New sol = 100 céntimos
Literacy rate 94%
Calorie consumption 2624 kilocalories

PHILIPPINES
Southeast Asia

Page 117 E1

This 7107-island archipelago between the South China Sea and the Pacific is subject to earthquakes and volcanic activity. After 21 years of dictatorship, democracy was restored in 1986.

Official name Republic of the Philippines
Formation 1946 / 1946
Capital Manila
Population 98.4 million / 855 people per sq mile (330 people per sq km)
Total area 115,830 sq miles (300,000 sq km)
Languages Filipino*, English*, Tagalog, Cebuano, Ilocano, Hiligaynon, many other local languages
Religions Roman Catholic 81%, Protestant 9%, Muslim 5%, Other (including Buddhist) 5%
Ethnic mix Other 34%, Tagalog 28%, Cebuano 13%, Ilocano 9%, Hiligaynon 8%, Bisaya 8%
Government Presidential system
Currency Philippine peso = 100 centavos
Literacy rate 95%
Calorie consumption 2608 kilocalories

POLAND
Northern Europe

Page 76 B3

Poland's low-lying plains extend from the Baltic Sea into the heart of Europe. It has undergone massive political and economic change since the fall of communism. It joined the EU in 2004.

Official name Republic of Poland
Formation 1918 / 1945
Capital Warsaw
Population 38.2 million / 325 people per sq mile (125 people per sq km)
Total area 120,728 sq miles (312,685 sq km)
Languages Polish*
Religions Roman Catholic 93%, Other and nonreligious 5%, Orthodox Christian 2%
Ethnic mix Polish 98%, Other 2%
Government Parliamentary system
Currency Zloty = 100 groszy
Literacy rate 99%
Calorie consumption 3485 kilocalories

PORTUGAL
Southwest Europe

Page 70 B3

Portugal, on the Iberian Peninsula, is the westernmost country in mainland Europe. Isolated under 44 years of dictatorship until 1974, it modernized fast after joining the EU in 1986.

Official name Republic of Portugal
Formation 1139 / 1640
Capital Lisbon
Population 10.6 million / 299 people per sq mile (115 people per sq km)
Total area 35,672 sq miles (92,391 sq km)
Languages Portuguese*
Religions Roman Catholic 92%, Protestant 4%, Nonreligious 3%, Other 1%
Ethnic mix Portuguese 98%, African and other 2%
Government Parliamentary system
Currency Euro = 100 cents
Literacy rate 94%
Calorie consumption 3456 kilocalories

QATAR
Southwest Asia

Page 98 C4

Projecting north from the Arabian Peninsula into the Persian Gulf, Qatar is mostly flat, semiarid desert. Massive reserves of oil and gas have made it one of the world's wealthiest states.

Official name State of Qatar
Formation 1971 / 1971
Capital Doha
Population 2.2 million / 518 people per sq mile (200 people per sq km)
Total area 4416 sq miles (11,437 sq km)
Languages Arabic*
Religions Muslim (mainly Sunni) 95%, Other 5%
Ethnic mix Qatari 20%, Other Arab 20%, Indian 20%, Nepalese 13%, Filipino 10%, Other 10%, Pakistani 7%
Government Monarchy
Currency Qatar riyal = 100 dirhams
Literacy rate 97%
Calorie consumption Not available

ROMANIA
Southeast Europe

Page 86 B4

Romania lies on the western shores of the Black Sea. Its communist regime was overthrown in 1989 and, despite a slow transition to a free-market economy, it joined the EU in 2007.

Official name Romania
Formation 1878 / 1947
Capital Bucharest
Population 21.7 million / 244 people per sq mile (94 people per sq km)
Total area 91,699 sq miles (237,500 sq km)
Languages Romanian*, Hungarian (Magyar), Romani, German
Religions Romanian Orthodox 87%, Protestant 5%, Roman Catholic 5%, Other 3%
Ethnic mix Romanian 89%, Magyar 7%, Roma 3%, Other 1%
Government Presidential system
Currency New Romanian leu = 100 bani
Literacy rate 99%
Calorie consumption 3363 kilocalories

RUSSIAN FEDERATION
Europe / Asia

Page 92 D4

The world's largest country, with vast mineral and energy reserves, Russia dominated the former USSR and is still a major power. It has over 150 ethnic groups, many with their own territory.

Official name Russian Federation
Formation 1480 / 1991
Capital Moscow
Population 143 million / 22 people per sq mile (8 people per sq km)
Total area 6,592,735 sq miles (17,075,200 sq km)
Languages Russian*, Tatar, Ukrainian, Chavash, various other national languages
Religions Orthodox Christian 75%, Muslim 14%, Other 11%
Ethnic mix Russian 80%, Other 12%, Tatar 4%, Ukrainian 2%, Chavash 1%, Bashkir 1%
Government Presidential / parliamentary system
Currency Russian rouble = 100 kopeks
Literacy rate 99%
Calorie consumption 3358 kilocalories

RWANDA
Central Africa

Page 51 B6

Rwanda lies just south of the Equator in Central Africa. Ethnic violence flared into genocide in 1994, when almost a million died. The main victims, the Tutsi, dominate government now.

Official name Republic of Rwanda
Formation 1962 / 1962
Capital Kigali
Population 11.8 million / 1225 people per sq mile (473 people per sq km)
Total area 10,169 sq miles (26,338 sq km)
Languages Kinyarwanda*, French*, Kiswahili, English*
Religions Christian 94%, Muslim 5%, Traditional beliefs 1%
Ethnic mix Hutu 85%, Tutsi 14%, Other (including Twa) 1%
Government Presidential system
Currency Rwanda franc = 100 centimes
Literacy rate 66%
Calorie consumption 2148 kilocalories

ST KITTS AND NEVIS
West Indies

Page 33 G3

Saint Kitts and Nevis are part of the Caribbean Leeward Islands. A former British colony, the country is a popular tourist destination. Less-developed Nevis is famed for its hot springs.

Official name Federation of Saint Christopher and Nevis
Formation 1983 / 1983
Capital Basseterre
Population 51,134 / 368 people per sq mile (142 people per sq km)
Total area 101 sq miles (261 sq km)
Languages English*, English Creole
Religions Anglican 33%, Methodist 29%, Other 22%, Moravian 9%, Roman Catholic 7%
Ethnic mix Black 95%, Mixed race 3%, White 1%, Other and Amerindian 1%
Government Parliamentary system
Currency East Caribbean dollar = 100 cents
Literacy rate 98%
Calorie consumption 2507 kilocalories

ST LUCIA
West Indies

Page 33 G4

One of the most beautiful Caribbean Windward Islands, Saint Lucia retains both French and British influences from its colonial history. Tourism and fruit production dominate the economy.

Official name Saint Lucia
Formation 1979 / 1979
Capital Castries
Population 162,781 / 690 people per sq mile (267 people per sq km)
Total area 239 sq miles (620 sq km)
Languages English*, French Creole
Religions Roman Catholic 90%, Other 10%
Ethnic mix Black 83%, Mulatto (mixed race) 13%, Asian 3%, Other 1%
Government Parliamentary system
Currency East Caribbean dollar = 100 cents
Literacy rate 95%
Calorie consumption 2629 kilocalories

ST VINCENT & THE GRENADINES
West Indies

Page 33 G4

Formerly ruled by Britain, these volcanic islands form part of the Caribbean Windward Islands. The economy relies on tourism and bananas, and it is the world's largest arrowroot producer.

Official name Saint Vincent and the Grenadines
Formation 1979 / 1979
Capital Kingstown
Population 103,220 / 788 people per sq mile (304 people per sq km)
Total area 150 sq miles (389 sq km)
Languages English*, English Creole
Religions Anglican 47%, Methodist 28%, Roman Catholic 13%, Other 12%
Ethnic mix Black 66%, Mulatto (mixed race) 19%, Other 12%, Carib 2%, Asian 1%
Government Parliamentary system
Currency East Caribbean dollar = 100 cents
Literacy rate 88%
Calorie consumption 2960 kilocalories

SAMOA
Australasia & Oceania

Page 123 F4

Samoa, in the southern Pacific, was ruled by New Zealand before 1962. Four of the nine islands are inhabited. The traditional Samoan way of life is communal and conservative.

Official name Independent State of Samoa
Formation 1962 / 1962
Capital Apia
Population 200,000 / 183 people per sq mile (71 people per sq km)
Total area 1104 sq miles (2860 sq km)
Languages Samoan*, English*
Religions Christian 99%, Other 1%
Ethnic mix Polynesian 91%, Euronesian 7%, Other 2%
Government Parliamentary system
Currency Tala = 100 sene
Literacy rate 99%
Calorie consumption 2872 kilocalories

SAN MARINO
Southern Europe

Page 74 C3

Perched on the slopes of Monte Titano in the Italian Appennino, San Marino has been a city-state since the 4th century AD, and was recognized as independent by the pope in 1631.

Official name Republic of San Marino
Formation 1631 / 1631
Capital San Marino
Population 32,448 / 1352 people per sq mile (532 people per sq km)
Total area 23.6 sq miles (61 sq km)
Languages Italian*
Religions Roman Catholic 93%, Other and nonreligious 7%
Ethnic mix Sammarinese 88%, Italian 10%, Other 2%
Government Parliamentary system
Currency Euro = 100 cents
Literacy rate 99%
Calorie consumption Not available

SAO TOME & PRINCIPE
West Africa

Page 55 A5

This ex-Portuguese colony off Africa's west coast has two main islands and smaller islets. Multiparty democracy, adopted in 1990, ended 15 years of Marxism. Cocoa is the main export.

Official name Democratic Republic of São Tomé and Príncipe
Formation 1975 / 1975
Capital São Tomé
Population 200,000 / 539 people per sq mile (208 people per sq km)
Total area 386 sq miles (1001 sq km)
Languages Portuguese Creole, Portuguese*
Religions Roman Catholic 84%, Other 16%
Ethnic mix Black 90%, Portuguese and Creole 10%
Government Presidential system
Currency Dobra = 100 céntimos
Literacy rate 70%
Calorie consumption 2676 kilocalories

SAUDI ARABIA
Southwest Asia

Page 99 B5

The desert kingdom of Saudi Arabia, rich in oil and gas, covers an area the size of Western Europe. It includes Islam's holiest cities, Medina and Mecca. Women's rights are restricted.

Official name Kingdom of Saudi Arabia
Formation 1932 / 1932
Capital Riyadh
Population 28.8 million / 35 people per sq mile (14 people per sq km)
Total area 756,981 sq miles (1,960,582 sq km)
Languages Arabic*
Religions Sunni Muslim 85%, Shi'a Muslim 15%
Ethnic mix Arab 72%, Foreign residents (mostly south and southeast Asian) 20%, Afro-Asian 8%
Government Monarchy
Currency Saudi riyal = 100 halalat
Literacy rate 94%
Calorie consumption 3122 kilocalories

SENEGAL
West Africa

Page 52 B3

This ex-French colony was ruled by one party for 40 years after independence, despite the adoption of multipartyism in 1981. Its capital, Dakar, stands on the westernmost cape of Africa.

Official name Republic of Senegal
Formation 1960 / 1960
Capital Dakar
Population 14.1 million / 190 people per sq mile (73 people per sq km)
Total area 75,749 sq miles (196,190 sq km)
Languages Wolof, Pulaar, Serer, Diola, Mandinka, Malinké, Soninké, French*
Religions Sunni Muslim 95%, Christian (mainly Roman Catholic) 4%, Traditional beliefs 1%
Ethnic mix Wolof 43%, Serer 15%, Peul 14%, Other 14%, Toucouleur 9%, Diola 5%
Government Presidential system
Currency CFA franc = 100 centimes
Literacy rate 52%
Calorie consumption 2426 kilocalories

SERBIA
Southeast Europe

Page 78 D4

The former Yugoslavia began breaking up in 1991, and Serbia has found itself the sole successor republic. It refuses to acknowledge the 2008 secession of Albanian-dominated Kosovo.

Official name Republic of Serbia
Formation 2006 / 2008
Capital Belgrade
Population 9.5 million / 318 people per sq mile (123 people per sq km)
Total area 29,905 sq miles (77,453 sq km)
Languages Serbian*, Hungarian (Magyar)
Religions Orthodox Christian 85%, Roman Catholic 6%, Other 6%, Muslim 3%
Ethnic mix Serb 83%, Other 10%, Magyar 4%, Bosniak 2%, Roma 1%
Government Parliamentary system
Currency Serbian dinar = 100 para
Literacy rate 98%
Calorie consumption 2724 kilocalories

SEYCHELLES
Indian Ocean

Page 57 G1

This ex-British colony spans 115 islands in the Indian Ocean. Multiparty polls in 1993 ended 14 years of one-party rule. Unique flora includes the world's largest seed, the coco-de-mer.

Official name Republic of Seychelles
Formation 1976 / 1976
Capital Victoria
Population 90,846 / 874 people per sq mile (336 people per sq km)
Total area 176 sq miles (455 sq km)
Languages French Creole*, English*, French*
Religions Roman Catholic 82%, Anglican 6%, Other (including Muslim) 6%, Other Christian 3%, Hindu 2%, Seventh-day Adventist 1%
Ethnic mix Creole 89%, Indian 5%, Other 4%, Chinese 2%
Government Presidential system
Currency Seychelles rupee = 100 cents
Literacy rate 92%
Calorie consumption 2426 kilocalories

SIERRA LEONE
West Africa

Page 52 C4

Founded in 1787 as a British colony for freed slaves, Sierra Leone gained independence in 1961. Recovery from civil war in the 1990s was set back in 2014 by West Africa's Ebola epidemic.

Official name Republic of Sierra Leone
Formation 1961 / 1961
Capital Freetown
Population 6.1 million / 221 people per sq mile (85 people per sq km)
Total area 27,698 sq miles (71,740 sq km)
Languages Mende, Temne, Krio, English*
Religions Muslim 60%, Christian 30%, Traditional beliefs 10%
Ethnic mix Mende 35%, Temne 32%, Other 21%, Limba 8%, Kuranko 4%
Government Presidential system
Currency Leone = 100 cents
Literacy rate 44%
Calorie consumption 2333 kilocalories

SINGAPORE
Southeast Asia

Page 116 A1

A city state linked to the southern tip of the Malay Peninsula by a causeway, Singapore is one of Asia's major commercial centers. Politics has been dominated for decades by one party.

Official name Republic of Singapore
Formation 1965 / 1965
Capital Singapore
Population 5.4 million / 22,881 people per sq mile (8852 people per sq km)
Total area 250 sq miles (648 sq km)
Languages Mandarin*, Malay*, Tamil*, English*
Religions Buddhist 55%, Taoist 22%, Muslim 16%, Hindu, Christian, and Sikh 7%
Ethnic mix Chinese 74%, Malay 14%, Indian 9%, Other 3%
Government Parliamentary system
Currency Singapore dollar = 100 cents
Literacy rate 96%
Calorie consumption Not available

SLOVAKIA
Central Europe

Page 77 C6

After 900 years of Hungarian control, Slovakia was the less-developed half of communist Czechoslovakia in the 20th century. It became independent in 1993 and joined the EU in 2004.

Official name Slovak Republic
Formation 1993 / 1993
Capital Bratislava
Population 5.5 million / 290 people per sq mile (112 people per sq km)
Total area 18,859 sq miles (48,845 sq km)
Languages Slovak*, Hungarian (Magyar), Czech
Religions Roman Catholic 69%, Other 13%, Nonreligious 13%, Greek Catholic (Uniate) 4%, Orthodox Christian 1%
Ethnic mix Slovak 86%, Magyar 10%, Roma 2%, Other 1%, Czech 1%
Government Parliamentary system
Currency Euro = 100 cents
Literacy rate 99%
Calorie consumption 2902 kilocalories

SLOVENIA
Central Europe

Page 73 D8

The northernmost of the ex-Yugoslav republics was the first to break away, with little violence, in 1991. It always had the closest links with Western Europe, and joined the EU in 2004.

Official name Republic of Slovenia
Formation 1991 / 1991
Capital Ljubljana
Population 2.1 million / 269 people per sq mile (104 people per sq km)
Total area 7820 sq miles (20,253 sq km)
Languages Slovenian*
Religions Roman Catholic 58%, Other 28%, Atheist 10%, Muslim 2%, Orthodox Christian 2%
Ethnic mix Slovene 83%, Other 12%, Serb 2%, Croat 2%, Bosniak 1%
Government Parliamentary system
Currency Euro = 100 cents
Literacy rate 99%
Calorie consumption 3173 kilocalories

SOLOMON ISLANDS
Australasia & Oceania

Page 122 C3

This archipelago of around 1000 islands scattered in the southwest Pacific was formerly ruled by Britain. Most people live on six main islands. Ethnic conflict from 1998 led to devolved governance.

Official name Solomon Islands
Formation 1978 / 1978
Capital Honiara
Population 600,000 / 56 people per sq mile (21 people per sq km)
Total area 10,985 sq miles (28,450 sq km)
Languages English*, Pidgin English, Melanesian Pidgin, around 120 native languages
Religions Church of Melanesia (Anglican) 34%, Roman Catholic 19%, South Seas Evangelical Church 17%, Methodist 11%, Other 19%
Ethnic mix Melanesian 93%, Polynesian 4%, Other 3%
Government Parliamentary system
Currency Solomon Islands dollar = 100 cents
Literacy rate 77%
Calorie consumption 2473 kilocalories

SOMALIA
East Africa

Page 51 E5

Italian and British Somaliland were united to create this semiarid state on the Horn of Africa. Anarchy since 1991 has caused mass hunger, a refugee crisis, and ineffective central authority.

Official name Federal Republic of Somalia
Formation 1960 / 1960
Capital Mogadishu
Population 10.5 million / 43 people per sq mile (17 people per sq km)
Total area 246,199 sq miles (637,657 sq km)
Languages Somali*, Arabic*, English, Italian
Religions Sunni Muslim 99%, Christian 1%
Ethnic mix Somali 85%, Other 15%
Government Nonparty system
Currency Somali shilin = 100 senti
Literacy rate 24%
Calorie consumption 1696 kilocalories

SOUTH AFRICA
Southern Africa

Page 56 C4

Mineral-rich South Africa was settled by the Dutch and the British. Multiracial polls in 1994 ended decades of white minority rule and apartheid. AIDS, poverty, and crime are problems.

Official name Republic of South Africa
Formation 1934 / 1994
Capital Pretoria; Cape Town; Bloemfontein
Population 52.8 million / 112 people per sq mile (43 people per sq km)
Total area 471,008 sq miles (1,219,912 sq km)
Languages English*, isiZulu*, isiXhosa*, Afrikaans*, Sepedi*, Setswana*, Sesotho*, Xitsonga*, siSwati*, Tshivenda*, isiNdebele*
Religions Christian 68%, Traditional beliefs and animist 29%, Muslim 2%, Hindu 1%
Ethnic mix Black 80%, White 9%, Colored 9%, Asian 2%
Government Parliamentary system
Currency Rand = 100 cents
Literacy rate 94%
Calorie consumption 3007 kilocalories

SOUTH KOREA
East Asia

Page 106 E4

Allied with the US, the southern half of the Korean peninsula was separated from the communist North in 1948. It is the world's leading shipbuilder and a major force in high-tech industries.

Official name Republic of Korea
Formation 1948 / 1953
Capital Seoul; Sejong City (administrative)
Population 49.3 million / 1293 people per sq mile (499 people per sq km)
Total area 38,023 sq miles (98,480 sq km)
Languages Korean*
Religions Mahayana Buddhist 47%, Protestant 38%, Roman Catholic 11%, Confucianist 3%, Other 1%
Ethnic mix Korean 100%
Government Presidential system
Currency South Korean won = 100 chon
Literacy rate 99%
Calorie consumption 3329 kilocalories

SOUTH SUDAN
East Africa

Page 51 B5

Landlocked and little developed, this mostly Christian region seceded from the mainly Muslim north of Sudan in 2011 after years of civil war. Oil production is the economic mainstay.

Official name Republic of South Sudan
Formation 2011 / 2011
Capital Juba
Population 11.3 million / 45 people per sq mile (18 people per sq km)
Total area 248,777 sq miles (644,329 sq km)
Languages Arabic, Dinka, Nuer, Zande, Bari, Shilluk, Lotuko, English*
Religions Over 50% Christian/traditional beliefs
Ethnic mix Dinka 40%, Nuer 15%, Shilluk/Anwak 10%, Azande 10%, Arab 10%, Bari 10%, Other 5%
Government Transitional regime
Currency South Sudan Pound = 100 piastres
Literacy rate 37%
Calorie consumption Not available

SPAIN
Southwest Europe

Page 70 D2

At the gateway to the Mediterranean, Spain became a world power once united in 1492. A vigorous regionalism now exists, with separatist movements in the Basque Country and Catalonia.

Official name Kingdom of Spain
Formation 1492 / 1713
Capital Madrid
Population 46.9 million / 243 people per sq mile (94 people per sq km)
Total area 194,896 sq miles (504,782 sq km)
Languages Spanish*, Catalan*, Galician*, Basque*
Religions Roman Catholic 96%, Other 4%
Ethnic mix Castilian Spanish 72%, Catalan 17%, Galician 6%, Basque 2%, Other 2%, Roma 1%
Government Parliamentary system
Currency Euro = 100 cents
Literacy rate 98%
Calorie consumption 3183 kilocalories

SRI LANKA
South Asia

Page 110 D3

A former British colony, the island republic of Sri Lanka is separated from India by the narrow Palk Strait. A brutal 26-year civil war between the Sinhalese and Tamils ended in 2009.

Official name Democratic Socialist Republic of Sri Lanka
Formation 1948 / 1948
Capital Colombo; Sri Jayewardenapura Kotte
Population 21.3 million / 852 people per sq mile (329 people per sq km)
Total area 25,332 sq miles (65,610 sq km)
Languages Sinhala*, Tamil*, Sinhala-Tamil, English
Religions Buddhist 69%, Hindu 15%, Muslim 8%, Christian 8%
Ethnic mix Sinhalese 74%, Tamil 18%, Moor 7%, Other 1%
Government Presidential / parliamentary system
Currency Sri Lanka rupee = 100 cents
Literacy rate 91%
Calorie consumption 2488 kilocalories

SUDAN
East Africa

Page 50 B4

On the west coast of the Red Sea, Sudan has been ruled by a military Islamic regime since a coup in 1989. In 2011, it lost its southern third (and most of its oil reserves) after years of civil war.

Official name Republic of the Sudan
Formation 1956 / 2011
Capital Khartoum
Population 38 million / 53 people per sq mile (20 people per sq km)
Total area 718,722 sq miles (1,861,481 sq km)
Languages Arabic*, Nubian, Beja, Fur
Religions Almost 100% Muslim (mainly Sunni)
Ethnic mix Arab 60%, Other 18%, Nubian 10%, Beja 8%, Fur 3%, Zaghawa 1%
Government Presidential system
Currency New Sudanese pound = 100 piastres
Literacy rate 73%
Calorie consumption 2346 kilocalories

SURINAME
South America

Page 37 G3

This former Dutch colony on the north coast of South America has some of the world's richest bauxite reserves. Democracy was restored in 1991, after almost 11 years of military rule.

Official name Republic of Suriname
Formation 1975 / 1975
Capital Paramaribo
Population 500,000 / 8 people per sq mile (3 people per sq km)
Total area 63,039 sq miles (163,270 sq km)
Languages Sranan (creole), Dutch*, Javanese, Sarnami Hindi, Saramaccan, Chinese, Carib
Religions Hindu 27%, Protestant 25%, Roman Catholic 23%, Muslim 20%, Traditional beliefs 5%
Ethnic mix East Indian 27%, Creole 18%, Black 15%, Javanese 15%, Mixed race 13%, Other 12%
Government Presidential / parliamentary system
Currency Surinamese dollar = 100 cents
Literacy rate 95%
Calorie consumption 2727 kilocalories

SWAZILAND
Southern Africa

Page 56 D4

This tiny kingdom, ruled by Britain until 1968, depends economically on its neighbor South Africa. Its absolute monarch has banned political parties. It has the world's highest rate of HIV.

Official name Kingdom of Swaziland
Formation 1968 / 1968
Capital Mbabane
Population 1.2 million / 181 people per sq mile (70 people per sq km)
Total area 6704 sq miles (17,363 sq km)
Languages English*, siSwati*, isiZulu, Xitsonga
Religions Traditional beliefs 40%, Other 30%, Roman Catholic 20%, Muslim 10%
Ethnic mix Swazi 97%, Other 3%
Government Monarchy
Currency Lilangeni = 100 cents
Literacy rate 83%
Calorie consumption 2275 kilocalories

SWEDEN
Northern Europe

Page 62 B4

Densely forested Sweden is the largest and most populous Scandinavian country and stretches into the Arctic Circle. Its strong industrial base helps to fund its extensive welfare system.

Official name Kingdom of Sweden
Formation 1523 / 1921
Capital Stockholm
Population 9.6 million / 60 people per sq mile (23 people per sq km)
Total area 173,731 sq miles (449,964 sq km)
Languages Swedish*, Finnish, Sámi
Religions Evangelical Lutheran 75%, Other 13%, Muslim 5%, Other Protestant 5%, Roman Catholic 2%
Ethnic mix Swedish 86%, Foreign-born or first-generation immigrant 12%, Finnish & Sámi 2%
Government Parliamentary system
Currency Swedish krona = 100 öre
Literacy rate 99%
Calorie consumption 3160 kilocalories

SWITZERLAND
Central Europe

Page 73 A7

One of the world's richest countries, with a long tradition of neutrality, this mountainous nation lies at the center of Europe geographically, but outside it politically, having not joined the EU.

Official name Swiss Confederation
Formation 1291 / 1857
Capital Bern
Population 8.1 million / 528 people per sq mile (204 people per sq km)
Total area 15,942 sq miles (41,290 sq km)
Languages German*, Swiss-German, French*, Italian*, Romansch*
Religions Roman Catholic 42%, Protestant 35%, Other and nonreligious 19%, Muslim 4%
Ethnic mix German 64%, French 20%, Other 9.5%, Italian 6%, Romansch 0.5%
Government Parliamentary system
Currency Swiss franc = 100 rappen/centimes
Literacy rate 99%
Calorie consumption 3487 kilocalories

SYRIA
Southwest Asia

Page 96 B3

Syria's borders were drawn in 1941 at the end of French rule. Suppression of pro-democracy protests in 2011 erupted into civil war; in 2014 Islamist militias took control of the Euphrates Valley.

Official name Syrian Arab Republic
Formation 1941 / 1967
Capital Damascus
Population 21.9 million / 308 people per sq mile (119 people per sq km)
Total area 71,498 sq miles (184,180 sq km)
Languages Arabic*, French, Kurdish, Armenian, Circassian, Turkic languages, Assyrian, Aramaic
Religions Sunni Muslim 74%, Alawi 12%, Christian 10%, Druze 3%, Other 1%
Ethnic mix Arab 90%, Kurdish 9%, Armenian, Turkmen, and Circassian 1%
Government Presidential system
Currency Syrian pound = 100 piastres
Literacy rate 85%
Calorie consumption 3106 kilocalories

TAIWAN
East Asia

Page 107 D6

China's nationalist government fled to Taiwan in 1949 when ousted by the communists. China regards the island, 80 miles (130 km) southeast of the mainland, as a renegade province.

Official name Republic of China (ROC)
Formation 1949 / 1949
Capital Taipei
Population 23.3 million / 1871 people per sq mile (722 people per sq km)
Total area 13,892 sq miles (35,980 sq km)
Languages Amoy Chinese, Mandarin Chinese*, Hakka Chinese
Religions Buddhist, Confucianist, and Taoist 93%, Christian 5%, Other 2%
Ethnic mix Han (pre-20th-century migration) 84%, Han (20th-century migration) 14%, Aboriginal 2%
Government Presidential system
Currency Taiwan dollar = 100 cents
Literacy rate 98%
Calorie consumption 2959 kilocalories

TAJIKISTAN
Central Asia

Page 101 F3

The poorest of the ex-Soviet republics lies landlocked on the western slopes of the Pamirs. Tajiks are of Persian (Iranian) origin rather than Turkic like their Central Asian neighbors.

Official name Republic of Tajikistan
Formation 1991 / 1991
Capital Dushanbe
Population 8.2 million / 148 people per sq mile (57 people per sq km)
Total area 55,251 sq miles (143,100 sq km)
Languages Tajik*, Uzbek, Russian
Religions Sunni Muslim 95%, Shi'a Muslim 3%, Other 2%
Ethnic mix Tajik 80%, Uzbek 15%, Other 3%, Russian 1%, Kyrgyz 1%
Government Presidential system
Currency Somoni = 100 diram
Literacy rate 99%
Calorie consumption 2101 kilocalories

TANZANIA
East Africa

Page 51 B7

This East African state was formed in 1964 by the union of Tanganyika and Zanzibar. A third of its area is game reserve or national park, including Africa's highest peak, Mt. Kilimanjaro.

Official name United Republic of Tanzania
Formation 1964 / 1964
Capital Dodoma
Population 49.3 million / 144 people per sq mile (56 people per sq km)
Total area 364,898 sq miles (945,087 sq km)
Languages Kiswahili*, Sukuma, Chagga, Nyamwezi, Hehe, Makonde, Yao, Sandawe, English*
Religions Christian 63%, Muslim 35%, Other 2%
Ethnic mix Native African (over 120 tribes) 99%, European, Asian, and Arab 1%
Government Presidential system
Currency Tanzanian shilling = 100 cents
Literacy rate 68%
Calorie consumption 2167 kilocalories

THAILAND
Southeast Asia

Page 115 C5

Thailand lies at the heart of the Indochinese Peninsula. Formerly Siam, it has been an independent kingdom for most of its history. The military has frequently intervened in politics.

Official name Kingdom of Thailand
Formation 1238 / 1907
Capital Bangkok
Population 67 million / 340 people per sq mile (131 people per sq km)
Total area 198,455 sq miles (514,000 sq km)
Languages Thai*, Chinese, Malay, Khmer, Mon, Karen, Miao
Religions Buddhist 95%, Muslim 4%, Other (including Christian) 1%
Ethnic mix Thai 83%, Chinese 12%, Malay 3%, Khmer and Other 2%
Government Transitional regime
Currency Baht = 100 satang
Literacy rate 96%
Calorie consumption 2757 kilocalories

TOGO
West Africa

Page 53 F4

Togo lies sandwiched between Ghana and Benin in West Africa. Its long-term military leader, and then his son and successor, have won every election held there since 1993.

Official name Republic of Togo
Formation 1960 / 1960
Capital Lomé
Population 6.8 million / 324 people per sq mile (125 people per sq km)
Total area 21,924 sq miles (56,785 sq km)
Languages Ewe, Kabye, Gurma, French*
Religions Christian 47%, Traditional beliefs 33%, Muslim 14%, Other 6%
Ethnic mix Ewe 46%, Other African 41%, Kabye 12%, European 1%
Government Presidential system
Currency CFA franc = 100 centimes
Literacy rate 60%
Calorie consumption 2366 kilocalories

TONGA
Australasia & Oceania

Page 123 E4

Northeast of New Zealand, Tonga is a 170-island archipelago, 45 of which are inhabited. Politics is effectively controlled by the king, though limited democratic reforms are taking place.

Official name Kingdom of Tonga
Formation 1970 / 1970
Capital Nuku'alofa
Population 106,322 / 382 people per sq mile (148 people per sq km)
Total area 289 sq miles (748 sq km)
Languages English*, Tongan*
Religions Free Wesleyan 41%, Other 17%, Roman Catholic 16%, Church of Jesus Christ of Latter-day Saints 14%, Free Church of Tonga 12%
Ethnic mix Tongan 98%, Other 2%
Government Monarchy
Currency Pa'anga (Tongan dollar) = 100 seniti
Literacy rate 99%
Calorie consumption Not available

TRINIDAD AND TOBAGO
West Indies

Page 33 H5

This former British colony is the most southerly of the Windward Islands, just 9 miles (15 km) off the coast of Venezuela. Politics is mainly polarized by race. Oil and gas are exported.

Official name Republic of Trinidad and Tobago
Formation 1962 / 1962
Capital Port-of-Spain
Population 1.3 million / 656 people per sq mile (253 people per sq km)
Total area 1980 sq miles (5128 sq km)
Languages English Creole, English*, Hindi, French, Spanish
Religions Other 30%, Roman Catholic 26%, Hindu 23%, Anglican 8%, Baptist 7%, Muslim 6%
Ethnic mix East Indian 40%, Black 38%, Mixed race 20%, White and Chinese 1%, Other 1%
Government Parliamentary system
Currency Trinidad and Tobago dollar = 100 cents
Literacy rate 99%
Calorie consumption 2889 kilocalories

TUNISIA
North Africa

Page 49 E2

North Africa's smallest country, one of the more liberal yet stable Arab states, had only two post-independence rulers until the "Arab Spring" of 2011. Moderate Islamists then won power.

Official name Republic of Tunisia
Formation 1956 / 1956
Capital Tunis
Population 11 million / 183 people per sq mile (71 people per sq km)
Total area 63,169 sq miles (163,610 sq km)
Languages Arabic*, French
Religions Muslim (mainly Sunni) 98%, Christian 1%, Jewish 1%
Ethnic mix Arab and Berber 98%, European 1%, Jewish 1%
Government Transitional regime
Currency Tunisian dinar = 1000 millimes
Literacy rate 80%
Calorie consumption 3362 kilocalories

TURKEY
Asia / Europe

Page 94 B3

With land in Europe and Asia, Turkey guards the entrance to the Black Sea. The secular/Islamic divide is key to its politics. It is the only Muslim member of NATO, and hopes to join the EU.

Official name Republic of Turkey
Formation 1923 / 1939
Capital Ankara
Population 74.9 million / 252 people per sq mile (97 people per sq km)
Total area 301,382 sq miles (780,580 sq km)
Languages Turkish*, Kurdish, Arabic, Circassian, Armenian, Greek, Georgian, Ladino
Religions Muslim (mainly Sunni) 99%, Other 1%
Ethnic mix Turkish 70%, Kurdish 20%, Other 8%, Arab 2%
Government Parliamentary system
Currency Turkish lira = 100 kurus
Literacy rate 95%
Calorie consumption 3680 kilocalories

TURKMENISTAN
Central Asia

Page 100 B2

Stretching from the Caspian Sea into Central Asia's deserts, this ex-Soviet state exploits vast gas reserves. The pre-independence president built a personality cult and ruled until 2007.

Official name Turkmenistan
Formation 1991 / 1991
Capital Ashgabat
Population 5.2 million / 28 people per sq mile (11 people per sq km)
Total area 188,455 sq miles (488,100 sq km)
Languages Turkmen*, Uzbek, Russian, Kazakh, Tatar
Religions Sunni Muslim 89%, Orthodox Christian 9%, Other 2%
Ethnic mix Turkmen 85%, Other 6%, Uzbek 5%, Russian 4%
Government Presidential system
Currency New manat = 100 tenge
Literacy rate 99%
Calorie consumption 2883 kilocalories

TUVALU
Australasia & Oceania

Page 123 E3

Known as the Ellice Islands under British rule, Tuvalu is a chain of nine atolls in the Central Pacific. It has the world's smallest GNI, but made substantial earnings leasing its ".tv" internet suffix.

Official name Tuvalu
Formation 1978 / 1978
Capital Fongafale (Funafuti Atoll)
Population 10,698 / 1070 people per sq mile (411 people per sq km)
Total area 10 sq miles (26 sq km)
Languages Tuvaluan, Kiribati, English*
Religions Church of Tuvalu 97%, Other 1%, Baha'i 1%, Seventh-day Adventist 1%
Ethnic mix Polynesian 96%, Micronesian 4%
Government Nonparty system
Currency Australian dollar = 100 cents; Tuvaluan dollar = 100 cents
Literacy rate 98%
Calorie consumption Not available

UGANDA
East Africa

Page 51 B6

Landlocked Uganda faced ethnic strife under 1970s' dictator Idi Amin. From 1986, reconciliation was aided by two decades of "no-party" democracy, but insurgency continued in the north.

Official name Republic of Uganda
Formation 1962 / 1962
Capital Kampala
Population 37.6 million / 488 people per sq mile (188 people per sq km)
Total area 91,135 sq miles (236,040 sq km)
Languages Luganda, Nkole, Chiga, Lango, Acholi, Teso, Lugbara, English*
Religions Christian 85%, Muslim (mainly Sunni) 12%, Other 3%
Ethnic mix Other 50%, Baganda 17%, Banyakole 10%, Basoga 9%, Bakiga 7%, Iteso 7%
Government Presidential system
Currency Uganda shilling = 100 cents
Literacy rate 73%
Calorie consumption 2279 kilocalories

UKRAINE
Eastern Europe

Page 86 C2

Bordered by seven states, fertile Ukraine was the "breadbasket" of the USSR. Its political divide between pro-Russian sentiment and assertive nationalism exploded into civil war in 2014.

Official name Ukraine
Formation 1991 / 1991
Capital Kiev
Population 45.2 million / 194 people per sq mile (75 people per sq km)
Total area 223,089 sq miles (603,700 sq km)
Languages Ukrainian*, Russian, Tatar
Religions Christian (mainly Orthodox) 95%, Other 5%
Ethnic mix Ukrainian 78%, Russian 17%, Other 5%
Government Presidential system
Currency Hryvna = 100 kopiykas
Literacy rate 99%
Calorie consumption 3142 kilocalories

UNITED ARAB EMIRATES
Southwest Asia

Page 99 D5

Bordering the Persian Gulf on the north of the Arabian Peninsula, the United Arab Emirates is a federation of seven states. Wealth once relied on pearls, but oil and gas are now exported.

Official name United Arab Emirates
Formation 1971 / 1972
Capital Abu Dhabi
Population 9.3 million / 288 people per sq mile (111 people per sq km)
Total area 32,000 sq miles (82,880 sq km)
Languages Arabic*, Farsi, Indian and Pakistani languages, English
Religions Muslim (mainly Sunni) 96%, Christian, Hindu, and other 4%
Ethnic mix Asian 60%, Emirian 25%, Other Arab 12%, European 3%
Government Monarchy
Currency UAE dirham = 100 fils
Literacy rate 90%
Calorie consumption 3215 kilocalories

UNITED KINGDOM
Northwest Europe

Page 67 C5

Lying across the English Channel from France, the UK comprises England, Wales, Scotland, and Northern Ireland. Its prominent role in world affairs is a legacy of its once-vast empire.

Official name United Kingdom of Great Britain and Northern Ireland
Formation 1707 / 1922
Capital London
Population 63.1 million / 676 people per sq mile (261 people per sq km)
Total area 94,525 sq miles (244,820 sq km)
Languages English*, Welsh* (in Wales), Gaelic
Religions Anglican 45%, Other & nonreligious 39%, Roman Catholic 9%, Presbyterian 4%, Muslim 3%
Ethnic mix English 80%, Scottish 9%, West Indian, Asian, & other 5%, Welsh 3%, Northern Irish 3%
Government Parliamentary system
Currency Pound sterling = 100 pence
Literacy rate 99%
Calorie consumption 3414 kilocalories

UNITED STATES
North America

Page 13 B5

Stretching across the most temperate part of North America, and with many natural resources, the USA is the sole truly global superpower and has the world's largest economy.

Official name United States of America
Formation 1776 / 1959
Capital Washington D.C.
Population 320 million / 90 people per sq mile (35 people per sq km)
Total area 3,717,792 sq miles (9,626,091 sq km)
Languages English*, Spanish, Chinese, French, Polish, German, Tagalog, Vietnamese, Italian, Korean, Russian
Religions Protestant 52%, Roman Catholic 25%, Other & nonreligious 20%, Jewish 2%, Muslim 1%
Ethnic mix White 60%, Hispanic 17%, Black American/African 14%, Asian 6%, Other 3%
Government Presidential system
Currency US dollar = 100 cents
Literacy rate 99%
Calorie consumption 3639 kilocalories

URUGUAY
South America

Page 42 D4

Uruguay, in southeastern South America, has much rich low-lying pasture land and is a major wool exporter. Military rule from 1973 to 1985 has given way to democracy.

Official name Eastern Republic of Uruguay
Formation 1828 / 1828
Capital Montevideo
Population 3.4 million / 50 people per sq mile (19 people per sq km)
Total area 68,039 sq miles (176,220 sq km)
Languages Spanish*
Religions Roman Catholic 66%, Other and nonreligious 30%, Jewish 2%, Protestant 2%
Ethnic mix White 90%, Mestizo 6%, Black 4%
Government Presidential system
Currency Uruguayan peso = 100 centésimos
Literacy rate 98%
Calorie consumption 2939 kilocalories

UZBEKISTAN
Central Asia

Page 100 D2

The most populous of the Central Asian republics lies on the ancient Silk Road between Asia and Europe. Today, its main exports are cotton and gold. Its pre-independence ruler retains power.

Official name Republic of Uzbekistan
Formation 1991 / 1991
Capital Tashkent
Population 28.9 million / 167 people per sq mile (65 people per sq km)
Total area 172,741 sq miles (447,400 sq km)
Languages Uzbek*, Russian, Tajik, Kazakh
Religions Sunni Muslim 88%, Orthodox Christian 9%, Other 3%
Ethnic mix Uzbek 80%, Other 6%, Russian 6%, Tajik 5%, Kazakh 3%
Government Presidential system
Currency Som = 100 tiyin
Literacy rate 99%
Calorie consumption 2675 kilocalories

VATICAN CITY
Southern Europe

Page 75 A8

The Vatican City, seat of the Roman Catholic Church, is a walled enclave in Rome. It is the world's smallest country. Its head, the pope, is elected for life by a college of cardinals.

Official name State of the Vatican City
Formation 1929 / 1929
Capital Vatican City
Population 839 / 4935 people per sq mile (1907 people per sq km)
Total area 0.17 sq miles (0.44 sq km)
Languages Italian*, Latin*
Religions Roman Catholic 100%
Ethnic mix Most resident lay persons are Italian
Government Papal state
Currency Euro = 100 cents
Literacy rate 99%
Calorie consumption Not available

VIETNAM
Southeast Asia

Page 114 D4

The eastern strip of the Indochinese Peninsula, Vietnam was partitioned in 1954, and only reunited after the communist north's victory in the devastating 1962–75 Vietnam War.

Official name Socialist Republic of Vietnam
Formation 1976 / 1976
Capital Hanoi
Population 91.7 million / 730 people per sq mile (282 people per sq km)
Total area 127,243 sq miles (329,560 sq km)
Languages Vietnamese*, Chinese, Thai, Khmer, Muong, Nung, Miao, Yao, Jarai
Religions Other 74%, Buddhist 14%, Roman Catholic 7%, Cao Dai 3%, Protestant 2%
Ethnic mix Vietnamese 86%, Other 8%, Thai 2%, Muong 2%, Tay 2%
Government One-party state
Currency Dông = 10 hao = 100 xu
Literacy rate 94%
Calorie consumption 2703 kilocalories

ZAMBIA
Southern Africa

Page 56 C2

Landlocked in southern Africa, copper-rich Zambia (once known as Northern Rhodesia) has seen its politics dogged by corruption both before and after the end of single-party rule in 1991.

Official name Republic of Zambia
Formation 1964 / 1964
Capital Lusaka
Population 14.5 million / 51 people per sq mile (20 people per sq km)
Total area 290,584 sq miles (752,614 sq km)
Languages Bemba, Tonga, Nyanja, Lozi, Lala-Bisa, Nsenga, English*
Religions Christian 63%, Traditional beliefs 36%, Muslim and Hindu 1%
Ethnic mix Bemba 34%, Other African 26%, Tonga 16%, Nyanja 14%, Lozi 9%, European 1%
Government Presidential system
Currency New Zambian kwacha = 100 ngwee
Literacy rate 61%
Calorie consumption 1937 kilocalories

VANUATU
Australasia & Oceania

Page 122 D4

This South Pacific archipelago of 82 islands and islets boasts the world's highest per capita density of languages. Until independence, it was under joint Anglo-French rule.

Official name Republic of Vanuatu
Formation 1980 / 1980
Capital Port Vila
Population 300,000 / 64 people per sq mile (25 people per sq km)
Total area 4710 sq miles (12,200 sq km)
Languages Bislama* (Melanesian pidgin), English*, French*, other indigenous languages
Religions Presbyterian 37%, Other 19%, Roman Catholic 15%, Anglican 15%, Traditional beliefs 8%, Seventh-day Adventist 6%
Ethnic mix ni-Vanuatu 94%, European 4%, Other 2%
Government Parliamentary system
Currency Vatu = 100 centimes
Literacy rate 83%
Calorie consumption 2820 kilocalories

VENEZUELA
South America

Page 36 D2

Located on the Caribbean coast of South America, Venezuela has the continent's most urbanized society, and some of the largest known oil deposits outside the Middle East.

Official name Bolivarian Republic of Venezuela
Formation 1830 / 1830
Capital Caracas
Population 30.4 million / 89 people per sq mile (34 people per sq km)
Total area 352,143 sq miles (912,050 sq km)
Languages Spanish*, Amerindian languages
Religions Roman Catholic 96%, Protestant 2%, Other 2%
Ethnic mix Mestizo 69%, White 20%, Black 9%, Amerindian 2%
Government Presidential system
Currency Bolívar fuerte = 100 céntimos
Literacy rate 96%
Calorie consumption 2880 kilocalories

YEMEN
Southwest Asia

Page 99 C7

The Arab world's only Marxist regime and a military-run republic united in 1990 to form Yemen, stretching across southern Arabia. Islamist militants and tribal insurgency threaten its stability.

Official name Republic of Yemen
Formation 1990 / 1990
Capital Sana
Population 24.4 million / 112 people per sq mile (43 people per sq km)
Total area 203,849 sq miles (527,970 sq km)
Languages Arabic*
Religions Sunni Muslim 55%, Shi'a Muslim 42%, Christian, Hindu, and Jewish 3%
Ethnic mix Arab 99%, Afro-Arab, Indian, Somali, and European 1%
Government Transitional regime
Currency Yemeni rial = 100 fils
Literacy rate 66%
Calorie consumption 2185 kilocalories

ZIMBABWE
Southern Africa

Page 56 D3

Full independence from Britain in 1980 ended 15 years of troubled white-minority rule. Poor governance, violent land redistribution, and severe drought have destroyed the economy.

Official name Republic of Zimbabwe
Formation 1980 / 1980
Capital Harare
Population 14.1 million / 94 people per sq mile (36 people per sq km)
Total area 150,803 sq miles (390,580 sq km)
Languages Shona, isiNdebele, English*
Religions Syncretic 50%, Christian 25%, Traditional beliefs 24%, Other (including Muslim) 1%
Ethnic mix Shona 71%, Ndebele 16%, Other African 11%, White 1%, Asian 1%
Government Presidential system
Currency Zimbabwe dollar suspended in 2009; nine other currencies are legal tender
Literacy rate 84%
Calorie consumption 2210 kilocalories

Overseas Territories and Dependencies

Despite the rapid process of decolonization since the end of the Second World War, around 10 million people in more than 50 territories around the world continue to live under the protection of a parent state.

AUSTRALIA

ASHMORE & CARTIER ISLANDS
Indian Ocean
Claimed 1931
Capital not applicable
Area 2 sq miles (5 sq km)
Population None

CHRISTMAS ISLAND
Indian Ocean
Claimed 1958
Capital The Settlement
Area 52 sq miles (135 sq km)
Population 1530

COCOS ISLANDS
Indian Ocean
Claimed 1955
Capital West Island
Area 5.5 sq miles (14 sq km)
Population 596

CORAL SEA ISLANDS
Southwest Pacific
Claimed 1969
Capital None
Area Less than 1.2 sq miles (3 sq km)
Population below 10 (scientists)

HEARD & McDONALD ISLANDS
Indian Ocean
Claimed 1947
Capital not applicable
Area 161 sq miles (417 sq km)
Population None

NORFOLK ISLAND
Southwest Pacific
Claimed 1774
Capital Kingston
Area 13.3 sq miles (34 sq km)
Population 2210

DENMARK

FAROE ISLANDS
North Atlantic
Claimed 1380
Capital Tórshavn
Area 540 sq miles (1399 sq km)
Population 49,469

GREENLAND
North Atlantic
Claimed 1380
Capital Nuuk
Area 840,000 sq miles (2,175,516 sq km)
Population 56,483

FRANCE

CLIPPERTON ISLAND
East Pacific
Claimed 1935
Capital not applicable
Area 2.7 sq miles (7 sq km)
Population None

FRENCH GUIANA
South America
Claimed 1817
Capital Cayenne
Area 35,135 sq miles (90,996 sq km)
Population 250,109

FRENCH POLYNESIA
South Pacific
Claimed 1843
Capital Papeete
Area 1,608 sq miles (4165 sq km)
Population 276,831

GUADELOUPE
West Indies
Claimed 1635
Capital Basse-Terre
Area 629 sq miles (1628 sq km)
Population 405,739

MARTINIQUE
West Indies
Claimed 1635
Capital Fort-de-France
Area 425 sq miles (1100 sq km)
Population 386,486

MAYOTTE
Indian Ocean
Claimed 1843
Capital Mamoudzou
Area 144 sq miles (374 sq km)
Population 212,645

NEW CALEDONIA
Southwest Pacific
Claimed 1853
Capital Nouméa
Area 7,374 sq miles (19,103 sq km)
Population 262,000

RÉUNION
Indian Ocean
Claimed 1638
Capital Saint-Denis
Area 970 sq miles (2512 sq km)
Population 840,974

ST. PIERRE & MIQUELON
North America
Claimed 1604
Capital Saint-Pierre
Area 93 sq miles (242 sq km)
Population 5716

WALLIS & FUTUNA
South Pacific
Claimed 1842
Capital Matá'Utu
Area 106 sq miles (274 sq km)
Population 15,561

NETHERLANDS

ARUBA
West Indies
Claimed 1643
Capital Oranjestad
Area 75 sq miles (194 sq km)
Population 102,911

BONAIRE
West Indies
Claimed 1816
Capital Kralendijk
Area 113 sq miles (294 sq km)
Population 18,413

CURAÇAO
West Indies
Claimed 1815
Capital Willemstad
Area 171 sq miles (444 sq km)
Population 153,500

SABA
West Indies
Claimed 1816
Capital The Bottom
Area 5 sq miles (13 sq km)
Population 1846

SINT-EUSTATIUS
West Indies
Claimed 1784
Capital Oranjestad
Area 8 sq miles (21 sq km)
Population 4020

SINT-MAARTEN
West Indies
Claimed 1648
Capital Phillipsburg
Area 13 sq miles (34 sq km)
Population 39,689

NEW ZEALAND

COOK ISLANDS
South Pacific
Claimed 1901
Capital Avarua
Area 91 sq miles (235 sq km)
Population 13,700

NIUE
South Pacific
Claimed 1901
Capital Alofi
Area 102 sq miles (264 sq km)
Population 1190

TOKELAU
South Pacific
Claimed 1926
Capital not applicable
Area 4 sq miles (10 sq km)
Population 1337

NORWAY

BOUVET ISLAND
South Atlantic
Claimed 1928
Capital not applicable
Area 22 sq miles (58 sq km)
Population None

JAN MAYEN
North Atlantic
Claimed 1929
Capital not applicable
Area 147 sq miles (381 sq km)
Population 18 (scientists)

PETER I ISLAND
Antarctica
Claimed 1931
Capital not applicable
Area 69 sq miles (180 sq km)
Population None

SVALBARD
Arctic Ocean
Claimed 1920
Capital Longyearbyen
Area 24,289 sq miles (62,906 sq km)
Population 1872

UNITED KINGDOM

ANGUILLA
West Indies
Claimed 1650
Capital The Valley
Area 37 sq miles (96 sq km)
Population 16,086

ASCENSION ISLAND
South Atlantic
Claimed 1673
Capital Georgetown
Area 34 sq miles (88 sq km)
Population 880

BERMUDA
North Atlantic
Claimed 1612
Capital Hamilton
Area 20 sq miles (53 sq km)
Population 65,024

BRITISH INDIAN OCEAN TERRITORY
Indian Ocean
Claimed 1814
Capital Diego Garcia
Area 23 sq miles (60 sq km)
Population 4000

BRITISH VIRGIN ISLANDS
West Indies
Claimed 1672
Capital Road Town
Area 59 sq miles (153 sq km)
Population 32,680

CAYMAN ISLANDS
West Indies
Claimed 1670
Capital George Town
Area 100 sq miles (259 sq km)
Population 58,435

FALKLAND ISLANDS
South Atlantic
Claimed 1832
Capital Stanley
Area 4699 sq miles (12,173 sq km)
Population 2840

GIBRALTAR
Southwest Europe
Claimed 1713
Capital Gibraltar
Area 2.5 sq miles (6.5 sq km)
Population 29,185

GUERNSEY
Northwest Europe
Claimed 1066
Capital St Peter Port
Area 25 sq miles (65 sq km)
Population 65,849

ISLE OF MAN
Northwest Europe
Claimed 1765
Capital Douglas
Area 221 sq miles (572 sq km)
Population 85,888

JERSEY
Northwest Europe
Claimed 1066
Capital St. Helier
Area 45 sq miles (116 sq km)
Population 96,513

MONTSERRAT
West Indies
Claimed 1632
Capital Brades *(de facto)*; Plymouth *(de jure)*
Area 40 sq miles (102 sq km)
Population 5215

PITCAIRN GROUP OF ISLANDS
South Pacific
Claimed 1887
Capital Adamstown
Area 18 sq miles (47 sq km)
Population 48

ST. HELENA
South Atlantic
Claimed 1673
Capital Jamestown
Area 47 sq miles (122 sq km)
Population 3800

SOUTH GEORGIA &
THE SOUTH SANDWICH ISLANDS
South Atlantic
Capital not applicable
Claimed 1775
Area 1387 sq miles (3592 sq km)
Population None

TRISTAN DA CUNHA
South Atlantic
Claimed 1612
Capital Edinburgh
Area 38 sq miles (98 sq km)
Population 264

TURKS & CAICOS ISLANDS
West Indies
Claimed 1766
Capital Cockburn Town
Area 166 sq miles (430 sq km)
Population 33,098

UNITED STATES OF AMERICA

AMERICAN SAMOA
South Pacific
Claimed 1900
Capital Pago Pago
Area 75 sq miles (195 sq km)
Population 55,165

BAKER & HOWLAND ISLANDS
Central Pacific
Claimed 1856
Capital not applicable
Area 0.54 sq miles (1.4 sq km)
Population None

GUAM
West Pacific
Claimed 1898
Capital Hagåtña
Area 212 sq miles (549 sq km)
Population 165,124

JARVIS ISLAND
Central Pacific
Claimed 1856
Capital not applicable
Area 1.7 sq miles (4.5 sq km)
Population None

NORTHERN MARIANA ISLANDS
West Pacific
Claimed 1947
Capital Saipan
Area 177 sq miles (457 sq km)
Population 53,855

PALMYRA ATOLL
Central Pacific
Claimed 1898
Capital not applicable
Area 5 sq miles (12 sq km)
Population None

PUERTO RICO
West Indies
Claimed 1898
Capital San Juan
Area 3515 sq miles (9104 sq km)
Population 3.62 million

VIRGIN ISLANDS
West Indies
Claimed 1917
Capital Charlotte Amalie
Area 137 sq miles (355 sq km)
Population 104,737

WAKE ISLAND
Central Pacific
Claimed 1898
Capital not applicable
Area 2.5 sq miles (6.5 sq km)
Population 150 (US air base)

Geographical comparisons

Largest countries

Russian Federation	6,592,735 sq miles	(17,075,200 sq km)
Canada	3,855,171 sq miles	(9,984,670 sq km)
USA	3,717,792 sq miles	(9,626,091 sq km)
China	3,705,386 sq miles	(9,596,960 sq km)
Brazil	3,286,470 sq miles	(8,511,965 sq km)
Australia	2,967,893 sq miles	(7,686,850 sq km)
India	1,269,338 sq miles	(3,287,590 sq km)
Argentina	1,068,296 sq miles	(2,766,890 sq km)
Kazakhstan	1,049,150 sq miles	(2,717,300 sq km)
Algeria	919,590 sq miles	(2,381,740 sq km)

Smallest countries

Vatican City	0.17 sq miles	(0.44 sq km)
Monaco	0.75 sq miles	(1.95 sq km)
Nauru	8 sq miles	(21 sq km)
Tuvalu	10 sq miles	(26 sq km)
San Marino	24 sq miles	(61 sq km)
Liechtenstein	62 sq miles	(160 sq km)
Marshall Islands	70 sq miles	(181 sq km)
St. Kitts & Nevis	101 sq miles	(261 sq km)
Maldives	116 sq miles	(300 sq km)
Malta	122 sq miles	(316 sq km)

Largest islands

Greenland	840,000 sq miles (2,175,600 sq km)
New Guinea	312,000 sq miles (808,000 sq km)
Borneo	292,222 sq miles (757,050 sq km)
Madagascar	226,656 sq miles (587,040 sq km)
Sumatra	202,300 sq miles (524,000 sq km)
Baffin Island	183,800 sq miles (476,000 sq km)
Honshu	88,800 sq miles (230,000 sq km)
Britain	88,700 sq miles (229,800 sq km)
Victoria Island	81,900 sq miles (212,000 sq km)
Ellesmere Island	75,700 sq miles (196,000 sq km)

Richest countries

(GNI per capita, in US$)

Monaco	186,950
Liechtenstein	136,770
Norway	102,610
Switzerland	86,600
Qatar	85,550
Luxembourg	71,810
Australia	65,520
Denmark	61,160
Sweden	59,240
Singapore	54,040

Poorest countries

(GNI per capita, in US$)

Malawi	270
Burundi	280
Somalia	288
Central African Republic	320
Congo, Democratic Republic	400
Niger	410
Liberia	410
Madagascar	440
Guinea	460
Ethiopia	470

Most populous countries

China	1.386 billion
India	1.252 billion
USA	320 million
Indonesia	250 million
Brazil	200 million
Pakistan	182 million

Most populous countries *continued*

Nigeria	174 million
Bangladesh	157 million
Russian Federation	143 million
Japan	127 million

Least populous countries

Vatican City	839
Nauru	9434
Tuvalu	10,698
Palau	21,108
San Marino	32,448
Monaco	36,136
Liechtenstein	37,000
St. Kitts & Nevis	51,134
Marshall Islands	69,747
Dominica	73,286

Most densely populated countries

Monaco	48,181 people per sq mile (18,531 per sq km)
Singapore	22,881 people per sq mile (8852 per sq km)
Vatican City	4935 people per sq mile (1907 per sq km)
Bahrain	4762 people per sq mile (1841 per sq km)
Malta	3226 people per sq mile (1250 per sq km)
Bangladesh	3029 people per sq mile (1169 per sq km)
Maldives	2586 people per sq mile (1000 per sq km)
Taiwan	1871 people per sq mile (722 per sq km)
Barbados	1807 people per sq mile (698 per sq km)
Mauritius	1671 people per sq mile (645 per sq km)

Most sparsely populated countries

Mongolia	5 people per sq mile (2 per sq km)
Namibia	7 people per sq mile (3 per sq km)
Iceland	8 people per sq mile (3 per sq km)
Suriname	8 people per sq mile (3 per sq km)
Australia	8 people per sq mile (3 per sq km)
Botswana	9 people per sq mile (4 per sq km)
Libya	9 people per sq mile (4 per sq km)
Mauritania	10 people per sq mile (4 per sq km)
Canada	10 people per sq mile (4 per sq km)
Guyana	11 people per sq mile (4 per sq km)

Most widely spoken languages

1. Chinese (Mandarin)
2. Spanish
3. English
4. Hindi
5. Arabic
6. Portuguese
7. Bengali
8. Russian
9. Japanese
10. Javanese

Largest conurbations

Tokyo (Japan)	39,400,000
Guangzhou (China)	32,600,000
Shanghai (China)	29,600,000
Jakarta (Indonesia)	27,000,000
Delhi (India)	25,300,000
Seoul (South Korea)	24,200,000
Karachi (Pakistan)	23,200,000
Mumbai (India)	22,600,000
Manila (Philippines)	22,500,000
Mexico City (Mexico)	22,200,000
New York (USA)	21,800,000
São Paulo (Brazil)	21,700,000
Beijing (China)	19,900,000
Osaka (Japan)	17,800,000
Los Angeles (USA)	17,300,000
Dhaka (Bangladesh)	16,700,000
Moscow (Russian Federation)	16,700,000
Cairo (Egypt)	16,400,000

Largest conurbations *continued*

Kolkata (India)	15,800,000
Buenos Aires (Argentina)	15,700,000
Bangkok (Thailand)	14,900,000
Istanbul (Turkey)	14,000,000
London (UK)	14,000,000
Lagos (Nigeria)	13,500,000
Tehran (Iran)	13,400,000

Longest rivers

Nile (Northeast Africa)	4160 miles	(6695 km)
Amazon (South America)	4049 miles	(6516 km)
Yangtze (China)	3915 miles	(6299 km)
Mississippi/Missouri (USA)	3710 miles	(5969 km)
Ob'-Irtysh (Russian Federation)	3461 miles	(5570 km)
Yellow River (China)	3395 miles	(5464 km)
Congo (Central Africa)	2900 miles	(4667 km)
Mekong (Southeast Asia)	2749 miles	(4425 km)
Lena (Russian Federation)	2734 miles	(4400 km)
Mackenzie (Canada)	2640 miles	(4250 km)
Yenisey (Russian Federation)	2541 miles	(4090 km)

Highest mountains
(Height above sea level)

Everest	29,035 ft	(8850 m)
K2	28,253 ft	(8611 m)
Kanchenjunga I	28,210 ft	(8598 m)
Makalu I	27,767 ft	(8463 m)
Cho Oyu	26,907 ft	(8201 m)
Dhaulagiri I	26,796 ft	(8167 m)
Manaslu I	26,783 ft	(8163 m)
Nanga Parbat I	26,661 ft	(8126 m)
Annapurna I	26,547 ft	(8091 m)
Gasherbrum I	26,471 ft	(8068 m)

Largest bodies of inland water
(Area & depth)

Caspian Sea	143,243 sq miles (371,000 sq km)	3215 ft (980 m)
Lake Superior	32,151 sq miles (83,270 sq km)	1289 ft (393 m)
Lake Victoria	26,560 sq miles (68,880 sq km)	328 ft (100 m)
Lake Huron	23,436 sq miles (60,700 sq km)	751 ft (229 m)
Lake Michigan	22,402 sq miles (58,020 sq km)	922 ft (281 m)
Lake Tanganyika	12,703 sq miles (32,900 sq km)	4700 ft (1435 m)
Great Bear Lake	12,274 sq miles (31,790 sq km)	1047 ft (319 m)
Lake Baikal	11,776 sq miles (30,500 sq km)	5712 ft (1741 m)
Great Slave Lake	10,981 sq miles (28,440 sq km)	459 ft (140 m)
Lake Erie	9915 sq miles (25,680 sq km)	197 ft (60 m)

Deepest ocean features

Challenger Deep, Mariana Trench (Pacific)	36,201 ft	(11,034 m)
Vityaz III Depth, Tonga Trench (Pacific)	35,704 ft	(10,882 m)
Vityaz Depth, Kurile-Kamchatka Trench (Pacific)	34,588 ft	(10,542 m)
Cape Johnson Deep, Philippine Trench (Pacific)	34,441 ft	(10,497 m)
Kermadec Trench (Pacific)	32,964 ft	(10,047 m)
Ramapo Deep, Japan Trench (Pacific)	32,758 ft	(9984 m)
Milwaukee Deep, Puerto Rico Trench (Atlantic)	30,185 ft	(9200 m)
Argo Deep, Torres Trench (Pacific)	30,070 ft	(9165 m)
Meteor Depth, South Sandwich Trench (Atlantic)	30,000 ft	(9144 m)
Planet Deep, New Britain Trench (Pacific)	29,988 ft	(9140 m)

Greatest waterfalls
(Mean flow of water)

Boyoma (Congo, Dem. Rep.)	600,400 cu. ft/sec	(17,000 cu.m/sec)
Khône (Laos/Cambodia)	410,000 cu. ft/sec	(11,600 cu.m/sec)
Niagara (USA/Canada)	195,000 cu. ft/sec	(5500 cu.m/sec)
Grande (Uruguay)	160,000 cu. ft/sec	(4500 cu.m/sec)
Paulo Afonso (Brazil)	100,000 cu. ft/sec	(2800 cu.m/sec)
Urubupunga (Brazil)	97,000 cu. ft/sec	(2750 cu.m/sec)
Iguaçu (Argentina/Brazil)	62,000 cu. ft/sec	(1700 cu.m/sec)
Maribondo (Brazil)	53,000 cu. ft/sec	(1500 cu.m/sec)
Victoria (Zimbabwe)	39,000 cu. ft/sec	(1100 cu.m/sec)

Greatest waterfalls *continued*

Kabalega (Uganda)	42,000 cu. ft/sec	(1200 cu.m/sec)
Churchill (Canada)	35,000 cu. ft/sec	(1000 cu.m/sec)
Cauvery (India)	33,000 cu. ft/sec	(900 cu.m/sec)

Highest waterfalls

Angel (Venezuela)	3212 ft	(979 m)
Tugela (South Africa)	3110 ft	(948 m)
Utigard (Norway)	2625 ft	(800 m)
Mongefossen (Norway)	2539 ft	(774 m)
Mtarazi (Zimbabwe)	2500 ft	(762 m)
Yosemite (USA)	2425 ft	(739 m)
Ostre Mardola Foss (Norway)	2156 ft	(657 m)
Tyssestrengane (Norway)	2119 ft	(646 m)
*Cuquenan (Venezuela)	2001 ft	(610 m)
Sutherland (New Zealand)	1903 ft	(580 m)
*Kjellfossen (Norway)	1841 ft	(561 m)

* indicates that the total height is a single leap

Largest deserts

Sahara	3,450,000 sq miles	(9,065,000 sq km)
Gobi	500,000 sq miles	(1,295,000 sq km)
Ar Rub al Khali	289,600 sq miles	(750,000 sq km)
Great Victorian	249,800 sq miles	(647,000 sq km)
Sonoran	120,000 sq miles	(311,000 sq km)
Kalahari	120,000 sq miles	(310,800 sq km)
Garagum	115,800 sq miles	(300,000 sq km)
Takla Makan	100,400 sq miles	(260,000 sq km)
Namib	52,100 sq miles	(135,000 sq km)
Thar	33,670 sq miles	(130,000 sq km)

NB – Most of Antarctica is a polar desert, with only 2 inches (50 mm) of precipitation annually

Hottest inhabited places

Djibouti (Djibouti)	86.0°F	(30.0°C)
Timbouctou (Mali)	84.7°F	(29.3°C)
Tirunelveli (India)	84.7°F	(29.3°C)
Tuticorin (India)	84.7°F	(29.3°C)
Nellore (India)	84.5°F	(29.2°C)
Santa Marta (Colombia)	84.5°F	(29.2°C)
Aden (Yemen)	84.0°F	(29.0°C)
Madurai (India)	84.0°F	(29.0°C)
Niamey (Niger)	84.0°F	(29.0°C)

Driest inhabited places

Aswân (Egypt)	0.02 in	(0.5 mm)
Luxor (Egypt)	0.03 in	(0.7 mm)
Arica (Chile)	0.04 in	(1.1 mm)
Ica (Peru)	0.10 in	(2.3 mm)
Antofagasta (Chile)	0.20 in	(4.9 mm)
El Minya (Egypt)	0.20 in	(5.1 mm)
Asyût (Egypt)	0.20 in	(5.2 mm)
Callao (Peru)	0.50 in	(12.0 mm)
Trujillo (Peru)	0.55 in	(14.0 mm)
El Faiyûm (Egypt)	0.80 in	(19.0 mm)

Wettest inhabited places

Buenaventura (Colombia)	265 in	(6743 mm)
Monrovia (Liberia)	202 in	(5131 mm)
Pago Pago (American Samoa)	196 in	(4990 mm)
Moulmein (Myanmar)	191 in	(4852 mm)
Lae (Papua New Guinea)	183 in	(4645 mm)
Baguio (Luzon I., Philippines)	180 in	(4573 mm)
Sylhet (Bangladesh)	176 in	(4457 mm)
Padang (Sumatra, Indonesia)	166 in	(4225 mm)
Bogor (Java, Indonesia)	166 in	(4225 mm)
Conakry (Guinea)	171 in	(4341 mm)

Aa – Almere

A

Aa *see* Gauja
Aachen *72 A4 Dut.* Aken, *Fr.* Aix-la-Chapelle; *anc.* Aquae Grani, Aquisgranum. Nordrhein-Westfalen, W Germany
Aaiún *see* Laâyoune
Aalborg *63 B7 var.* Ålborg, Ålborg-Nørresundby; *anc.* Alburgum. Nordjylland, N Denmark
Aalen *73 B6* Baden-Württemberg, S Germany
Aalsmeer *64 C3* Noord-Holland, C Netherlands
Aalst *65 B6* Oost-Vlaanderen, C Belgium
Aalten *64 E4* Gelderland, E Netherlands
Aalter *65 B5* Oost-Vlaanderen, NW Belgium
Aanaarjävri *see* Inarijärvi
Äänekoski *63 D5* Länsi-Suomi, W Finland
Aar *see* Aare
Aare *73 A7 var.* Aar. *river* W Switzerland
Aarhus *see* Århus
Aarlen *see* Arlon
Aat *see* Ath
Aba *55 E5* Orientale, NE Dem. Rep. Congo
Aba *53 G5* Abia, S Nigeria
Abā as Su'ūd *see* Najrān
Abaco Island *see* Great Abaco, N Bahamas
Ābādān *98 C4* Khūzestān, SW Iran
Abadan *100 C3 prev.* Bezmein, Büzmeýin, *Rus.* Byuzmeyin. Ahal Welaýaty, C Turkmenistan
Abai *see* Blue Nile
Abakan *92 D4* Respublika Khakasiya, S Russian Federation
Abancay *38 D4* Apurímac, SE Peru
Abariringa *see* Kanton
Abashiri *108 D2 var.* Abasiri. Hokkaidō, NE Japan
Abasiri *see* Abashiri
Ābay Wenz *see* Blue Nile
Abbaia *see* Ābaya Hāyk'
Abbatis Villa *see* Abbeville
Abbazia *see* Opatija
Abbeville *68 C2 anc.* Abbatis Villa. Somme, N France
'Abd al 'Azīz, Jabal *96 D2 mountain range* NE Syria
Abéché *54 C3 var.* Abécher, Abeshr. Ouaddaï, SE Chad
Abécher *see* Abéché
Abela *see* Ávila
Abellinum *see* Avellino
Abemama *122 D2 var.* Apamama; *prev.* Roger Simpson Island. *atoll* Tungaru, W Kiribati
Abengourou *53 E5* E Côte d'Ivoire
Aberbrothock *see* Arbroath
Abercorn *see* Mbala
Aberdeen *66 D3 anc.* Devana. NE Scotland, United Kingdom
Aberdeen *23 E2* South Dakota, N USA
Aberdeen *24 B2* Washington, NW USA
Abergwaun *see* Fishguard
Abertawe *see* Swansea
Aberystwyth *67 C6* W Wales, United Kingdom
Abeshr *see* Abéché
Abhā *99 B6* 'Asīr, SW Saudi Arabia
Abidavichy *85 D7 Rus.* Obidovichi. Mahilyowskaya Voblasts', E Belarus
Abidjan *53 E5* S Côte d'Ivoire
Abilene *27 F3* Texas, SW USA
Abingdon *see* Pinta, Isla
Abkhazia *see* Ap'khazet'i
Åbo *see* Turku
Aboisso *53 E5* SE Côte d'Ivoire
Abo, Massif d' *54 B1 mountain range* NW Chad
Abomey *53 F5* S Benin
Abou-Déïa *54 C3* Salamat, SE Chad
Aboudouhour *see* Abū aḑ Ḑuhūr
Abou Kémal *see* Abū Kamāl
Abrantes *70 B3 var.* Abrántes. Santarém, C Portugal
Abrashlare *see* Brezovo
Abrolhos Bank *34 E4 undersea bank* W Atlantic Ocean
Abrova *85 B6 Rus.* Obrovo. Brestskaya Voblasts', SW Belarus
Abrud *86 B4 Ger.* Gross-Schlatten, *Hung.* Abrudbánya. Alba, SW Romania
Abrudbánya *see* Abrud
Abruzzese, Appennino *74 C4 mountain range* C Italy
Absaroka Range *22 B2 mountain range* Montana/Wyoming, NW USA
Abū aḑ Ḑuhūr *96 B3 Fr.* Aboudouhour. Idlib, NW Syria
Abu Dhabi *see* Abū Ẓabī
Abu Hamed *50 C3* River Nile, N Sudan
Abū Ḩardān *96 E3 var.* Hajîne. Dayr az Zawr, E Syria
Abuja *53 G4 country capital* (Nigeria) Federal Capital District, C Nigeria
Abū Kamāl *96 E3 Fr.* Abou Kémal. Dayr az Zawr, E Syria
Abula *see* Ávila
Abunã, Rio *40 C2 var.* Río Abuná. *river* Bolivia/Brazil
Abut Head *129 B6 headland* South Island, New Zealand
Abuye Meda *50 D4 mountain* C Ethiopia
Abū Ẓabī *99 C5 var.* Abū Ẓabī, *Eng.* Abu Dhabi. *country capital* (United Arab Emirates) Abū Ẓaby, C United Arab Emirates
Abū Ẓabī *see* Abū Ẓabī
Abyaḑ, Al Baḩr al *see* White Nile
Abyla *see* Ávila
Abyssinia *see* Ethiopia
Acalayong *55 A5* SW Equatorial Guinea
Acaponeta *28 D4* Nayarit, C Mexico
Acapulco *29 E5 var.* Acapulco de Juárez. Guerrero, S Mexico
Acapulco de Juárez *see* Acapulco
Acarai Mountains *37 F4 Sp.* Serra Acaraí. *mountain range* Brazil/Guyana
Acaraí, Serra *see* Acarai Mountains
Acarigua *36 D2* Portuguesa, N Venezuela
Accra *53 E5 country capital* (Ghana) SE Ghana
Achacachi *39 E4* La Paz, W Bolivia
Achara *95 F2 var.* Ajaria. *autonomous republic* SW Georgia
Acklins Island *32 C2 island* SE Bahamas
Aconcagua, Cerro *42 B4 mountain* W Argentina
Açores/Açores, Arquipélago dos/Açores, Ilhas dos *see* Azores
A Coruña *70 B1 Cast.* La Coruña, *Eng.* Corunna; *anc.* Caronium. Galicia, NW Spain
Acre *40 C2 off.* Estado do Acre. *state* W Brazil
Açu *see* Assu
Acunum Acusio *see* Montélimar
Ada *78 D3* Vojvodina, N Serbia
Ada *27 G2* Oklahoma, C USA
Ada Bazar *see* Adapazarı
Adalia *see* Antalya
Adalia, Gulf of *see* Antalya Körfezi
Adama *see* Nazrēt
'Adan *99 B7 Eng.* Aden. SW Yemen
Adana *94 D4 var.* Seyhan. Adana, S Turkey
Adâncata *see* Horlivka
Adapazarı *94 B2 prev.* Ada Bazar. Sakarya, NW Turkey
Adare, Cape *132 B4 cape* Antarctica
Ad Dahna *98 C4 desert* E Saudi Arabia
Ad Dakhla *48 A4 var.* Dakhla. SW Western Sahara
Ad Dalanj *see* Dilling
Ad Damar *see* Ed Damer
Ad Damazīn *see* Ed Damazin
Ad Dāmir *see* Ed Damer
Ad Dammān *98 C4 var.* Dammām. Ash Sharqīyah, NE Saudi Arabia
Ad Dāmūr *see* Damoûr
Ad Dawḩah *98 C4 Eng.* Doha. *country capital* (Qatar) C Qatar
Aḑ Ḑiffah *see* Libyan Plateau
Addis Ababa *see* Ādīs Ābeba
Addoo Atoll *see* Addu Atoll
Addu Atoll *110 A5 var.* Addoo Atoll, Seenu Atoll. *atoll* S Maldives
Adelaide *127 B6 state capital* South Australia
Adelsberg *see* Postojna
Aden *see* 'Adan
Aden, Gulf of *99 C7 gulf* SW Arabian Sea
Adige *74 C2 Ger.* Etsch. *river* N Italy
Adirondack Mountains *19 F2 mountain range* New York, NE USA
Ādīs Ābeba *51 C5 Eng.* Addis Ababa. *country capital* (Ethiopia) Ādīs Ābeba, C Ethiopia
Adıyaman *95 E4* Adıyaman, SE Turkey
Adjud *86 C4* Vrancea, E Romania
Admiralty Islands *122 B3 island group* N Papua New Guinea
Adra *71 E5* Andalucía, S Spain
Adrar *48 D3* C Algeria
Adrian *18 C3* Michigan, N USA
Adrianople/Adrianopolis *see* Edirne
Adriatico, Mare *see* Adriatic Sea
Adriatic Sea *81 E2 Alb.* Deti Adriatik, *It.* Mare Adriatico, *SCr.* Jadransko More, *Slvn.* Jadransko Morje. *sea* N Mediterranean Sea
Adriatik, Deti *see* Adriatic Sea
Adycha *93 F2 river* NE Russian Federation
Aegean Sea *83 C5 Gk.* Aigaíon Pelagos, Aigaío Pélagos, *Turk.* Ege Denizi. *sea* NE Mediterranean Sea
Aegviidu *84 D2 Ger.* Charlottenhof. Harjumaa, NW Estonia
Aegyptus *see* Egypt
Aelana *see* Al 'Aqabah
Aelok *see* Ailuk Atoll
Aelōnlaplap *see* Ailinglaplap Atoll
Aemona *see* Ljubljana
Aeolian Islands *75 C6 var.* Isole Lipari, *Eng.* Aeolian Islands, Lipari Islands. *island group* S Italy
Aeolian Islands *see* Eolie, Isole
Æsernia *see* Isernia
Afar Depression *see* Danakil Desert
Afars et des Issas, Territoire Français des *see* Djibouti
Afghānestān, Dowlat-e Eslāmī-ye *see* Afghanistan
Afghanistan *100 C4 off.* Islamic Republic of Afghanistan, *Per.* Dowlat-e Eslāmī-ye Afghānestān; *prev.* Republic of Afghanistan. *country* C Asia
Afmadow *51 D6* Jubbada Hoose, S Somalia
Africa *46 continent*
Africa, Horn of *46 E4 physical region* Ethiopia/Somalia
Africana Seamount *119 A6 seamount* SW Indian Ocean
'Afrīn *96 B2* Ḩalab, N Syria
Afyon *94 B3 prev.* Afyonkarahisar. Afyon, W Turkey
Agadès *see* Agadez
Agadez *53 G3 prev.* Agadès. Agadez, C Niger
Agadir *48 B3* SW Morocco
Agana/Agaña *see* Hagåtña
Āgaro *51 C5* Oromīya, C Ethiopia
Agassiz Fracture Zone *121 G5 fracture zone* S Pacific Ocean
Agatha *see* Agde
Agathónisi *83 D6 island* Dodekánisa, Greece, Aegean Sea
Agde *69 C6 anc.* Agatha. Hérault, S France
Agedabia *see* Ajdābiyā
Agen *69 B5 anc.* Aginnum. Lot-et-Garonne, SW France
Agendicum *see* Sens
Aghri Dagh *see* Büyükağrı Dağı
Agiá *82 B4 var.* Ayiá. Thessalía, C Greece
Agialoúsa *see* Yenierenköy
Agía Marína *83 E6* Léros, Dodekánisa, Greece, Aegean Sea
Aginnum *see* Agen
Ágios Efstrátios *82 D4 var.* Áyios Evstrátios, Hagios Evstrátios. *island* E Greece
Ágios Nikólaos *83 D8 var.* Áyios Nikólaos. Kríti, Greece, E Mediterranean Sea
Āgra *112 D3* Uttar Pradesh, N India
Agra and Oudh, United Provinces of *see* Uttar Pradesh
Agram *see* Zagreb
Ağrı *95 F3 var.* Karaköse; *prev.* Karakılısse. Ağrı, NE Turkey
Agri Dagi *see* Büyükağrı Dağı
Agrigento *75 C7 Gk.* Akragas; *prev.* Girgenti. Sicilia, Italy, C Mediterranean Sea
Agriovótano *83 C5* Évvoia, C Greece
Agropoli *75 D5* Campania, S Italy
Aguachica *36 B2* Cesar, N Colombia
Aguadulce *31 F5* Coclé, S Panama
Agua Prieta *28 B1* Sonora, NW Mexico
Aguascalientes *28 D4* Aguascalientes, C Mexico
Aguaytía *38 C3* Ucayali, C Peru
Aguilas *71 E4* Murcia, SE Spain
Aguililla *28 D4* Michoacán, SW Mexico
Agulhas Basin *47 D8 undersea basin* SW Indian Ocean
Agulhas, Cape *56 C5 headland* SW South Africa
Agulhas Plateau *45 D6 undersea plateau* SW Indian Ocean
Ahaggar *53 F2 high plateau region* SE Algeria
Ahlen *72 B4* Nordrhein-Westfalen, W Germany
Ahmadābād *112 C4 var.* Ahmedabad. Gujarāt, W India
Ahmadnagar *112 C5 var.* Ahmednagar. Mahārāshtra, W India
Ahmedabad *see* Ahmadābād
Ahmednagar *see* Ahmadnagar
Ahuachapán *30 B3* Ahuachapán, W El Salvador
Ahvāz *98 C3 var.* Ahwāz; *prev.* Nāsiri. Khūzestān, SW Iran
Ahvenanmaa *see* Åland
Ahwāz *see* Ahvāz
Aigaíon Pelagos/Aigaío Pélagos *see* Aegean Sea
Aígina *83 C6 var.* Aíyina, Egina. Aígina, C Greece
Aígio *83 B5 var.* Egio; *prev.* Aíyion. Dytikí Ellás, S Greece
Aiken *21 E2* South Carolina, SE USA
Ailinglaplap Atoll *122 D2 var.* Aelōnlaplap. *atoll* Ralik Chain, S Marshall Islands
Ailuk Atoll *122 D1 var.* Aelok. *atoll* Ratak Chain, NE Marshall Islands
Ainaži *84 D3 Est.* Heinaste, *Ger.* Hainasch. Limbaži, N Latvia
'Aïn Ben Tili *52 D1* Tiris Zemmour, N Mauritania
Aintab *see* Gaziantep
Aïoun el Atrous/Aïoun el Atroûss *see* 'Ayoûn el 'Atroûs
Aiquile *39 F4* Cochabamba, C Bolivia
Aïr *see* Aïr, Massif de l'
Air du Azbine *see* Aïr, Massif de l'
Aïr, Massif de l' *53 G2 var.* Aïr, Air du Azbine, Asben. *mountain range* NC Niger
Aiud *86 B4 Ger.* Strassburg, *Hung.* Nagyenyed; *prev.* Engeten. Alba, SW Romania
Aix *see* Aix-en-Provence
Aix-en-Provence *69 D6 var.* Aix; *anc.* Aquae Sextiae. Bouches-du-Rhône, SE France
Aix-la-Chapelle *see* Aachen
Aíyina *see* Aígina
Aíyion *see* Aígio
Aizkraukle *84 C4* Aizkraukle, S Latvia
Ajaccio *69 E7* Corse, France, C Mediterranean Sea
Ajaria *see* Achara
Ajastan *see* Armenia
Aj Bogd Uul *104 D2 mountain* SW Mongolia
Ajdābiyā *49 G2 var.* Agedabia, Ajdābiyah. NE Libya
Ajdābiyah *see* Ajdābiyā
Ajjinena *see* El Geneina
Ajmer *112 D3 var.* Ajmere. Rājasthān, N India
Ajo *26 A3* Arizona, SW USA
Akaba *see* Al 'Aqabah
Akamagaseki *see* Shimonoseki
Akasha *50 B3* Northern, N Sudan
Akchâr *52 C2 desert* W Mauritania
Aken *see* Aachen
Akermanceaster *see* Bath
Akhaltsikhe *95 F2 prev.* Akhalts'ikhe. SW Georgia
Akhalts'ikhe *see* Akhaltsikhe
Akhisar *94 A3* Manisa, W Turkey
Akhmīm *50 B2 var.* Akhmîm; *anc.* Panopolis. C Egypt
Akhtubinsk *89 C7* Astrakhanskaya Oblast', SW Russian Federation
Akhtyrka *see* Okhtyrka
Akimiski Island *16 C3 island* Nunavut, C Canada
Akinovka *87 F4* Zaporiz'ka Oblast', S Ukraine
Akita *108 D4* Akita, Honshū, C Japan
Akjoujt *52 C2 prev.* Fort-Repoux. Inchiri, W Mauritania
Akkeshi *108 E2* Hokkaidō, NE Japan
Aklavik *14 D3* Northwest Territories, NW Canada
Akmola *see* Astana
Akmolinsk *see* Astana
Aknavásár *see* Târgu Ocna
Akpatok Island *17 E1 island* Nunavut, E Canada
Akragas *see* Agrigento
Akron *18 D4* Ohio, N USA
Akrotiri *see* Akrotírion
Akrotírion *80 C5 var.* Akrotiri. *UK air base* S Cyprus
Aksai Chin *102 B2 Chin.* Aksayqin. *disputed region* China/India
Aksaray *94 C4* Aksaray, C Turkey
Aksayqin *see* Aksai Chin
Akşehir *94 B4* Konya, W Turkey
Aktash *see* Oqtosh
Aktau *92 A4 Kaz.* Aqtaū; *prev.* Shevchenko. Mangistau, W Kazakhstan
Aktjubinsk/Aktyubinsk *see* Aktobe
Aktobe *92 B4 Kaz.* Aqtöbe; *prev.* Aktjubinsk, Aktyubinsk. Aktyubinsk, NW Kazakhstan
Aktsyabrski *85 C7 Rus.* Oktyabr'skiy; *prev.* Karpilovka. Homyel'skaya Voblasts', SE Belarus
Aktyubinsk *see* Aktobe
Akula *55 C5* Equateur, NW Dem. Rep. Congo
Akureyri *61 E4* Nordhurland Eystra, N Iceland
Akyab *see* Sittwe
Alabama *20 C2 off.* State of Alabama, *also known as* Camellia State, Heart of Dixie, The Cotton State, Yellowhammer State. *state* S USA
Alabama River *20 C3 river* Alabama, S USA
Alaca *94 C3* Çorum, N Turkey
Alacant *see* Alicante
Alagoas *41 G2 off.* Estado de Alagoas. *region* E Brazil
Alagoas *41 G2 off.* Estado de Alagoas. *state* E Brazil
Alais *see* Alès
Alajuela *31 E4* Alajuela, C Costa Rica
Alakanuk *14 C2* Alaska, USA
Al 'Alamayn *50 B1 var.* El 'Alamein. N Egypt
Al 'Amārah *98 C3 var.* Amara. Maysān, E Iraq
Alamo *25 D6* Nevada, W USA
Alamogordo *26 D3* New Mexico, SW USA
Alamosa *22 C5* Colorado, C USA
Åland *63 C6 var.* Aland Islands, *Fin.* Ahvenanmaa. *island group* SW Finland
Aland Islands *see* Åland
Aland Sea *see* Ålands Hav
Ålands Hav *63 C6 var.* Aland Sea. *strait* Baltic Sea/Gulf of Bothnia
Alanya *94 C4* Antalya, S Turkey
Alappuzha *110 C3 var.* Alleppey. Kerala, SW India
Al 'Aqabah *97 B8 var.* Akaba, Aqaba, 'Aqaba; *anc.* Aelana, Elath. Al 'Aqabah, SW Jordan
Al 'Arabīyah as Su'ūdīyah *see* Saudi Arabia
Alasca, Golfo de *see* Alaska, Gulf of
Alaşehir *94 A4* Manisa, W Turkey
Al 'Ashārah *96 E3 var.* Ashara. Dayr az Zawr, E Syria
Alaska *14 C3 off.* State of Alaska, *also known as* Land of the Midnight Sun, The Last Frontier, Seward's Folly; *prev.* Russian America. *state* NW USA
Alaska, Gulf of *14 C4 var.* Golfo de Alasca. *gulf* Canada/USA
Alaska Peninsula *14 C3 peninsula* Alaska, USA
Alaska Range *12 B2 mountain range* Alaska, USA
Al-Asnam *see* Chlef
Alattio *see* Alta
Al Awaynāt *see* Al 'Uwaynāt
Alaykel'/Alay-Kuu *see* Këk-Art
Al 'Aynā *97 B7* Al Karak, W Jordan
Alazeya *93 G2 river* NE Russian Federation
Al Bāb *96 B2* Ḩalab, N Syria
Albacete *71 E3* Castilla-La Mancha, C Spain
Al Baghdādī *98 B3 var.* Khān al Baghdādī. Al Anbār, SW Iraq
Al Bāha *see* Al Bāḩah
Al Bāḩah *99 B5 var.* Al Bāha. Al BāḨah, SW Saudi Arabia
Al Baḩrayn *see* Bahrain
Alba Iulia *86 B4 Ger.* Weissenburg, *Hung.* Gyulafehérvár; *prev.* Bălgrad, Karlsburg, Károly-Fehérvár. Alba, W Romania
Albania *79 C7 off.* Republic of Albania, *Alb.* Republika e Shqipqërisë, Shqipëria; *prev.* People's Socialist Republic of Albania. *country* SE Europe
Albania *see* Aubagne
Albany *125 B7* Western Australia
Albany *20 D3* Georgia, SE USA
Albany *19 F3 state capital* New York, NE USA
Albany *24 B3* Oregon, NW USA
Albany *16 C3 river* Ontario, S Canada
Alba Regia *see* Székesfehérvár
Al Bāridah *96 C4 var.* Bāridah. Ḩimş, C Syria
Al Başrah *98 C3 Eng.* Basra, *hist.* Busra, Bussora. Al Başrah, SE Iraq
Al Batrūn *see* Batroûn
Al Bayḑā' *49 G2 var.* Beida. NE Libya
Albemarle Island *see* Isabela, Isla
Albemarle Sound *21 G1 inlet* W Atlantic Ocean
Albergaria-a-Velha *70 B2* Aveiro, N Portugal
Albert *68 C3* Somme, N France
Alberta *15 E4 province* SW Canada
Albert Edward Nyanza *see* Edward, Lake
Albert, Lake *51 B6 var.* Albert Nyanza, Lac Mobutu Sese Seko. *lake* Uganda/Dem. Rep. Congo
Albert Lea *23 F3* Minnesota, N USA
Albert Nyanza *see* Albert, Lake
Albertville *see* Kalemie
Albi *69 C6 anc.* Albiga. Tarn, S France
Albiga *see* Albi
Ålborg *see* Aalborg
Ålborg-Nørresundby *see* Aalborg
Alborz, Reshteh-ye Kūhhā-ye *98 C2 Eng.* Elburz Mountains. *mountain range* N Iran
Albuquerque *26 D2* New Mexico, SW USA
Al Burayqah *see* Marsá al Burayqah
Alburgum *see* Aalborg
Albury *127 C7* New South Wales, SE Australia
Alcácer do Sal *70 B4* Setúbal, W Portugal
Alcalá de Henares *71 E3 Ar.* Alkal'a; *anc.* Complutum. Madrid, C Spain
Alcamo *75 C7* Sicilia, Italy, C Mediterranean Sea
Alcañiz *71 F2* Aragón, NE Spain
Alcántara, Embalse de *70 C3 reservoir* W Spain
Alcaudete *70 D4* Andalucía, S Spain
Alcázar *see* Ksar-el-Kebir
Alcazarquivir *see* Ksar-el-Kebir
Alcoi *see* Alcoy
Alcoy *71 F4 Cat.* Alcoi. País Valenciano, E Spain
Aldabra Group *57 G2 island group* SW Seychelles
Aldan *93 F3 river* NE Russian Federation
al Dar al Baida *see* Rabat
Alderney *68 A2 island* Channel Islands
Aleg *52 C3* Brakna, SW Mauritania
Aleksandriya *see* Oleksandriya
Aleksandropol' *see* Gyumri
Aleksandrovka *see* Oleksandrivka
Aleksandrovsk *see* Zaporizhzhya
Aleksin *89 B5* Tul'skaya Oblast', W Russian Federation
Aleksinac *78 E4* Serbia, SE Serbia
Alençon *68 B3* Orne, N France
Alenquer *41 E2* Pará, NE Brazil
Alep/Aleppo *see* Ḩalab
Alert *15 F1* Ellesmere Island, Nunavut, N Canada
Alès *69 C6 prev.* Alais. Gard, S France
Aleşd *86 B3 Hung.* Élesd. Bihor, SW Romania
Alessandria *74 B2 Fr.* Alexandrie. Piemonte, N Italy
Ålesund *63 A5* Møre og Romsdal, S Norway
Aleutian Basin *91 G3 undersea basin* Bering Sea
Aleutian Islands *14 A3 island group* Alaska, USA
Aleutian Range *12 A2 mountain range* Alaska, USA
Aleutian Trench *91 H3 trench* S Bering Sea
Alexander Archipelago *14 D4 island group* Alaska, USA
Alexander City *20 D2* Alabama, S USA
Alexander Island *132 A3 island* Antarctica
Alexander Range *see* Kirghiz Range
Alexandra *129 B7* Otago, South Island, New Zealand
Alexándreia *82 B4 var.* Alexándria. Kentrikí Makedonía, N Greece
Alexandretta *see* İskenderun
Alexandretta, Gulf of *see* İskenderun Körfezi
Alexandria *50 B1 Ar.* Al Iskandarīyah. N Egypt
Alexandria *86 C5* Teleorman, S Romania
Alexandria *20 B3* Louisiana, S USA
Alexandria *23 F2* Minnesota, N USA
Alexándria *see* Alexándreia
Alexandrie *see* Alessandria
Alexandroúpoli *82 D3 var.* Alexandroúpolis, *Turk.* Dedeagaç, Dedeagach. Anatolikí Makedonía kai Thráki, NE Greece
Alexandroúpolis *see* Alexandroúpoli
Al Fāshir *see* El Fasher
Alfatar *82 E1* Silistra, NE Bulgaria
Alfeiós *83 B6 prev.* Alfiós; *anc.* Alpheius, Alpheus. *river* S Greece
Alfiós *see* Alfeiós
Alföld *see* Great Hungarian Plain
Al-Furāt *see* Euphrates
Alga *92 B4 Kaz.* Algha. Aktyubinsk, NW Kazakhstan
Algarve *70 B4 cultural region* S Portugal
Algeciras *70 C5* Andalucía, SW Spain
Algemesí *71 F3* País Valenciano, E Spain
Al-Genain *see* El Geneina
Alger *49 E1 var.* Algiers, El Djazaïr, Al Jazair. *country capital* (Algeria) N Algeria
Algeria *48 C3 off.* Democratic and Popular Republic of Algeria. *country* N Africa
Algeria, Democratic and Popular Republic of *see* Algeria
Algerian Basin *58 C5 var.* Balearic Plain. *undersea basin* W Mediterranean Sea
Algha *see* Alga
Al Ghābah *99 E5 var.* Ghaba. C Oman
Alghero *75 A5* Sardegna, Italy, C Mediterranean Sea
Al Ghurdaqah *50 C2 var.* Hurghada, Ghurdaqah. E Egypt
Algiers *see* Alger
Al Golea *see* El Goléa
Algona *23 F3* Iowa, C USA
Al Hajar al Gharbi *99 D5 mountain range* N Oman
Al Hamad *see* Syrian Desert
Al Ḩasakah *96 D2 var.* Al Hasijah, El Haseke, *Fr.* Hassetché. Al Ḩasakah, NE Syria
Al Hasijah *see* Al Ḩasakah
Al Ḩillah *98 B3 var.* Hilla. Bābil, C Iraq
Al Ḩisā *97 B7* Aţ Ţafīlah, W Jordan
Al Ḩudaydah *99 B6 Eng.* Hodeida. W Yemen
Al Hufūf *98 C4 var.* Hofuf. Ash Sharqīyah, NE Saudi Arabia
Aliákmon *see* Aliákmonas
Aliákmonas *82 B4 prev.* Aliákmon; *anc.* Haliacmon. *river* N Greece
Alíartos *83 C5* Stereá Ellás, C Greece
Alicante *71 F4 Cat.* Alacant, *Lat.* Lucentum. País Valenciano, SE Spain
Alice *27 G5* Texas, SW USA
Alice Springs *126 A4* Northern Territory, C Australia
Alifu Atoll *see* Ari Atoll
Aligandí *31 G4* Kuna Yala, NE Panama
Aliki *see* Alykí
Alima *55 B6 river* C Congo
Al Imārāt al 'Arabīyahal Muttaḩidah *see* United Arab Emirates
Alindao *54 C4* Basse-Kotto, S Central African Republic
Aliquippa *18 D4* Pennsylvania, NE USA
Al Iskandarīyah *see* Alexandria
Al Ismā'īlīya *50 B1 var.* Ismailia, Ismâ'ilîya. N Egypt
Alistráti *82 C3* Kentrikí Makedonía, NE Greece
Alivéri *83 C5 var.* Alivérion. Évvoia, C Greece
Alivérion *see* Alivéri
Al Jabal al Akhḑar *49 G2 mountain range* NE Libya
Al Jafr *97 B7* Ma'ān, S Jordan
Al Jaghbūb *49 H3* NE Libya
Al Jahrā' *98 C4 var.* Al Jahrah, Jahra. C Kuwait
Al Jahrah *see* Al Jahrā'
Al Jamāhīrīyah al 'Arabīyah al Lībīyah ash Sha'bīyah al Ishtirākīy *see* Libya
Al Jawf *98 B4 off.* Jauf. Al Jawf, NW Saudi Arabia
Al Jawlān *see* Golan Heights
Al Jazair *see* Alger
Al Jazirah *96 E2 physical region* Iraq/Syria
Al Jīzah *see* Gîza
Al Junaynah *see* El Geneina
Alkal'a *see* Alcalá de Henares
Al Karak *97 B7 var.* El Kerak, Karak, Kerak; *anc.* Kir Moab, Kir of Moab. Al Karak, W Jordan
Al-Kasr al-Kebir *see* Ksar-el-Kebir
Al Khalīl *see* Hebron
Al Khārijah *50 B2 var.* El Khârga. C Egypt
Al Khums *49 F2 var.* Homs, Khoms, Khums. NW Libya
Alkmaar *64 C2* Noord-Holland, NW Netherlands
Al Kufrah *49 H4* SE Libya
Al Kūt *98 C3 var.* Kūt al 'Amārah, Kut al Imara. Wāsiţ, E Iraq
Al-Kuwait *see* Al Kuwayt
Al Kuwayt *98 C4 var.* Al-Kuwait, *Eng.* Kuwait, Kuwait City; *prev.* Qurein. *country capital* (Kuwait) E Kuwait
Al Lādhiqīyah *96 A3 Eng.* Latakia, *Fr.* Lattaquié; *anc.* Laodicea, Laodicea ad Mare. Al Lādhiqīyah, W Syria
Allahābād *113 E3* Uttar Pradesh, N India
Allanmyo *see* Aunglan
Allegheny Plateau *19 E3 mountain range* New York/Pennsylvania, NE USA
Allenstein *see* Olsztyn
Allentown *19 F4* Pennsylvania, NE USA
Alleppey *see* Alappuzha
Alliance *22 D3* Nebraska, C USA
Al Līth *99 B5* Makkah, SW Saudi Arabia
Al Lubnān *see* Lebanon
Alma-Ata *see* Almaty
Almada *70 B4* Setúbal, W Portugal
Al Madīnah *99 A5 Eng.* Medina. Al Madīnah, W Saudi Arabia
Al Mafraq *97 B6 var.* Mafraq. Al Mafraq, N Jordan
Al Mahdīyah *see* Mahdia
Al Mahrah *99 C6 mountain range* E Yemen
Al Majma'ah *98 B4* Ar Riyāḑ, C Saudi Arabia
Al Mālikīyah *96 E1 var.* Malkiye. Al Ḩasakah, N Syria
Almalyk *see* Olmaliq
Al Mamlakah *see* Morocco
Al Mamlaka al Urdunīya al Hashemīyah *see* Jordan
Al Manāmah *98 C4 Eng.* Manama. *country capital* (Bahrain) N Bahrain
Al Manāşif *96 E3 mountain range* E Syria
Almansa *71 F4* Castilla-La Mancha, C Spain
Al-Mariyya *see* Almería
Al Marj *49 G2 var.* Barka, *It.* Barce. NE Libya
Almaty *92 C5 var.* Alma-Ata. Almaty, SE Kazakhstan
Al Mawşil *98 B2 Eng.* Mosul. Nīnawá, N Iraq
Al Mayādīn *96 D3 var.* Mayadin, *Fr.* Meyadine. Dayr az Zawr, E Syria
Al Mazra' *see* Al Mazra'ah
Al Mazra'ah *97 B6 var.* Al Mazra', Mazra'a. Al Karak, W Jordan
Almelo *64 E3* Overijssel, E Netherlands
Almendra, Embalse de *70 C2 reservoir* Castilla-León, NW Spain
Almendralejo *70 C4* Extremadura, W Spain
Almere *64 C3 var.* Almere-stad. Flevoland, C Netherlands

Almere-stad – Arīhā

Almere-stad see Almere
Almería 71 E5 Ar. Al-Mariyya; anc. Unci, Lat. Portus Magnus. Andalucía, S Spain
Al'met'yevsk 89 D5 Respublika Tatarstan, W Russian Federation
Al Mīnā' see El Mina
Al Minyā 50 B2 var. El Minya, Minya. C Egypt
Almirante 31 E4 Bocas del Toro, NW Panama
Al Mudawwarah 97 B8 Ma'ān, SW Jordan
Al Mukallā 99 C6 var. Mukalla. SE Yemen
Al Obayyid see El Obeid
Alofi 123 F4 dependent territory capital (Niue) W Niue
Aloha State see Hawaii
Aloja 84 D3 Limbaži, N Latvia
Alónnisos 83 C5 island Vóreies Sporádes, Greece, Aegean Sea
Álora 70 D5 Andalucía, S Spain
Alor, Kepulauan 117 E5 island group E Indonesia
Al Oued see El Oued
Alpen see Alps
Alpena 18 D2 Michigan, N USA
Alpes see Alps
Alpha Cordillera 133 B3 var. Alpha Ridge. seamount range Arctic Ocean
Alpha Ridge see Alpha Cordillera
Alpheius see Alfeiós
Alphen see Alphen aan den Rijn
Alphen aan den Rijn 64 C3 var. Alphen. Zuid-Holland, C Netherlands
Alpheus see Alfeiós
Alpi see Alps
Alpine 27 E4 Texas, SW USA
Alps 80 C1 Fr. Alpes, Ger. Alpen, It. Alpi. mountain range C Europe
Al Qaḍārif see Gedaref
Al Qāhirah see Cairo
Al Qāmishlī 96 E1 var. Kamishli, Qamishly. Al Ḥasakah, NE Syria
Al Qaşrayn see Kasserine
Al Qayrawān see Kairouan
Al-Qsar al-Kbir see Ksar-el-Kebir
Al Qubayyāt see Qoubaïyât
Al Quds/Al Quds ash Sharīf see Jerusalem
Alqueva, Barragem do 70 C4 reservoir Portugal/ Spain
Al Qunayţirah 97 B5 var. El Kuneitra, El Quneitra, Kuneitra, Qunaytra. Al Qunayţirah, SW Syria
Al Quşayr 96 B4 var. El Quseir, Quşayr, Fr. Kousseir. Ḥimş, W Syria
Al Quwayrah 97 B8 var. El Quweira. Al 'Aqabah, SW Jordan
Alsace 68 E3 Ger. Elsass; anc. Alsatia. cultural region NE France
Alsatia see Alsace
Alsdorf 72 A4 Nordrhein-Westfalen, W Germany
Alt see Olt
Alta 62 D2 Fin. Alattio. Finnmark, N Norway
Altai see Altai Mountains
Altai Mountains 104 C2 var. Altai, Chin. Altay Shan, Rus. Altay. mountain range Asia/Europe
Altamaha River 21 E3 river Georgia, SE USA
Altamira 41 E2 Pará, NE Brazil
Altamura 75 E5 anc. Lupatia. Puglia, SE Italy
Altar, Desierto de 28 A1 var. Sonoran Desert. desert Mexico/USA
Altar, Desierto de see Sonoran Desert
Altay 104 C2 Xinjiang Uygur Zizhiqu, NW China
Altay 104 D2 prev. Yösönbulag. Govĭ-Altay, W Mongolia
Altay Altai Mountains, Asia/Europe
Altay Shan see Altai Mountains
Altbetsche see Bečej
Altenburg see Bucureşti, Romania
Altin Köprü 98 B3 var. Altun Kupri. At Ta'mīn, N Iraq
Altiplano 39 F4 physical region W South America
Altkanischa see Kanjiža
Alton 18 B5 Illinois, N USA
Alton 18 B4 Missouri, C USA
Altoona 19 E4 Pennsylvania, NE USA
Alto Paraná see Paraná
Altpasua see Stara Pazova
Alt-Schwanenburg see Gulbene
Altsohl see Zvolen
Altun Kupri see Altin Köprü
Altun Shan 104 C3 var. Altyn Tagh. mountain range NW China
Altus 27 F2 Oklahoma, C USA
Altyn Tagh see Altun Shan
Al Ubayyiḍ see El Obeid
Alūksne 84 D3 Ger. Marienburg. Alūksne, NE Latvia
Al 'Ulā 98 A4 Al Madīnah, NW Saudi Arabia
Al 'Umarī 97 C6 'Ammān, E Jordan
Alupka 87 F5 Respublika Krym, S Ukraine
Al Uqşur see Luxor
Al Urdunn see Jordan
Alushta 87 F5 Respublika Krym, S Ukraine
Al 'Uwaynāt 49 F4 var. Al Awaynāt. SW Libya
Alva 27 F1 Oklahoma, C USA
Alvarado 29 F4 Veracruz-Llave, E Mexico
Alvin 27 H4 Texas, SW USA
Al Wajh 98 A4 Tabūk, NW Saudi Arabia
Alwar 112 D3 Rājasthān, N India
Al Warī'ah 98 C4 Ash Sharqīyah, N Saudi Arabia
Al Yaman see Yemen
Alykí 82 C4 var. Aliki. Thásos, N Greece
Alytus 85 B5 Pol. Olita. Alytus, S Lithuania
Alzette 65 D8 river S Luxembourg
Amadeus, Lake 125 D5 seasonal lake Northern Territory, C Australia
Amadi 51 B5 W Equatoria, S South Sudan
Amadjuak Lake 15 G3 lake Baffin Island, Nunavut, N Canada
Amakusa-nada 109 A7 gulf SW Japan
Åmål 63 B6 Västra Götaland, S Sweden
Amami-gunto 108 A3 island group SW Japan
Amami-o-shim 108 A3 island S Japan
Amantea 75 D6 Calabria, SW Italy
Amapá 41 E1 off. Estado de Amapá; prev. Território de Amapá. region NE Brazil
Amapá 41 E1 off. Estado de Amapá; prev. Território de Amapá. state NE Brazil
Amapá, Estado de see Amapá
Amapá, Território de see Amapá
Amara see Al 'Amārah
Amarapura 114 B3 Mandalay, C Myanmar (Burma)
Amarillo 27 E2 Texas, SW USA
Amay 65 C6 Liège, E Belgium
Amazon 41 E1 Sp. Amazonas. river Brazil/Peru
Amazonas see Amazon
Amazon Basin 40 D2 basin N South America
Amazon, Mouths of the 41 F1 delta NE Brazil
Ambam 55 B5 Sud, S Cameroon
Ambanja 57 G2 Antsirañana, N Madagascar
Ambarchik 93 G2 Respublika Sakha (Yakutiya), NE Russian Federation
Ambato 38 B1 Tungurahua, C Ecuador
Ambérieu-en-Bugey 69 D5 Ain, E France
Ambianum see Amiens
Amboasary 57 F4 Toliara, S Madagascar
Amboina see Ambon
Ambon 117 F4 prev. Amboina, Amboyna. Pulau Ambon, E Indonesia
Ambositra 57 G3 Fianarantsoa, SE Madagascar
Amboyna see Ambon
Ambracia see Árta
Ambre, Cap d' see Bobaomby, Tanjona
Ambrim see Ambrym
Ambriz 56 A1 Bengo, NW Angola
Ambrym 122 D4 var. Ambrim. island C Vanuatu
Amchitka Island 14 A2 island Aleutian Islands, Alaska, USA
Amdo 104 C5 Xizang Zizhiqu, W China
Ameland 64 D1 Fris. It Amelân. island Waddeneilanden, N Netherlands
Amelân, It see Ameland
America see United States of America
America-Antarctica Ridge 45 C7 undersea ridge S Atlantic Ocean
America in Miniature see Maryland
American Falls Reservoir 24 E4 reservoir Idaho, NW USA
American Samoa 123 E4 US unincorporated territory W Polynesia
Amersfoort 64 D3 Utrecht, C Netherlands
Ames 23 F3 Iowa, C USA
Amfilochía 83 A5 var. Amfilokhía. Dytikí Ellás, C Greece
Amfilokhía see Amfilochía
Amga 93 F3 river NE Russian Federation
Amherst 17 F4 Nova Scotia, SE Canada
Amherst see Kyaikkami
Amida see Diyarbakır
Amiens 68 C3 anc. Ambianum, Samarobriva. Somme, N France
Amíndaion/Amindeo see Amýntaio
Amindivi Islands 110 A2 island group Lakshadweep, India, N Indian Ocean
Amirante Islands 57 G1 var. Amirantes Group. island group C Seychelles
Amirantes Group see Amirante Islands
Amistad, Presa de la see Amistad Reservoir
Amistad Reservoir 27 F4 var. Presa de la Amistad. reservoir Mexico/USA
Amisus see Samsun
Ammaia see Portalegre
'Ammān 97 B6 anc. Philadelphia, Bibl. Rabbah Ammon, Rabbath Ammon. country capital (Jordan) 'Ammān, NW Jordan
Ammassalik 60 D4 var. Angmagssalik. Tunu, S Greenland
Ammóchostos see Gazimağusa
Ammóchostos, Kólpos see Gazimağusa Körfezi
Amnok-kang see Yalu
Amoea see Portalegre
Amoentai see Amuntai
Āmol 98 D2 var. Amul. Māzandarān, N Iran
Amorgós 83 D6 Amorgós, Kykládes, Greece, Aegean Sea
Amorgós 83 D6 island Kykládes, Greece, Aegean Sea
Amos 16 D4 Québec, SE Canada
Amourj 52 D3 Hodh ech Chargui, SE Mauritania
Amoy see Xiamen
Ampato, Nevado 39 E4 mountain S Peru
Amposta 71 F2 Cataluña, NE Spain
Amraoti see Amrāvati
Amrāvati 112 D4 prev. Amraoti. Mahārāshtra, C India
Amritsar 112 D2 Punjab, N India
Amstelveen 64 C3 Noord-Holland, C Netherlands
Amsterdam 64 C3 country capital (Netherlands) Noord-Holland, C Netherlands
Amsterdam Island 119 C6 island NE French Southern and Antarctic Lands
Am Timan 54 C3 Salamat, SE Chad
Amu Darya 100 D2 Rus. Amudar'ya, Taj. Dar"yoi Amu, Turkm. Amyderya, Uzb. Amudaryo; anc. Oxus. river C Asia
Amu-Dar'ya see Amyderýa
Amudar'ya/Amudaryo/Amu, Dar"yoi see Amu Darya
Amul see Āmol
Amund Ringnes Island 15 F2 Island Nunavut, N Canada
Amundsen Basin see Fram Basin
Amundsen Plain 132 A4 abyssal plain S Pacific Ocean
Amundsen-Scott 132 B3 US research station Antarctica
Amundsen Sea 132 A4 sea S Pacific Ocean
Amuntai 116 D4 prev. Amoentai. Borneo, C Indonesia
Amur 93 G4 Chin. Heilong Jiang. river China/ Russian Federation
Amvrosiyevka see Amvrosiyivka
Amvrosiyivka 87 H3 Rus. Amvrosiyevka. Donets'ka Oblast', SE Ukraine
Amyderýa 101 E3 Rus. Amu-Dar'ya. Lebap Welaýaty, NE Turkmenistan
Amyderya see Amu Darya
Amýntaio 82 B4 var. Amindeo; prev. Amíndaion. Dytikí Makedonía, N Greece
Anabar 93 E2 river NE Russian Federation
An Abhainn Mhór see Blackwater
Anaco 37 E2 Anzoátegui, NE Venezuela
Anaconda 22 B2 Montana, NW USA
Anacortes 24 B1 Washington, NW USA
Anadolu Dağları see Doğu Karadeniz Dağları
Anadyr' 93 H1 Chukotskiy Avtonomnyy Okrug, NE Russian Federation
Anadyr' 93 G1 river NE Russian Federation
Anadyr, Gulf of see Anadyrskiy Zaliv
Anadyrskiy Zaliv 93 H1 Eng. Gulf of Anadyr. gulf NE Russian Federation
Anáfi 83 D7 anc. Anaphe. island Kykládes, Greece, Aegean Sea
'Ānah see 'Annah
Anaheim 24 E2 California, W USA
Anaiza see 'Unayzah
Analalava 57 G2 Mahajanga, NW Madagascar
Anamur 94 C5 İçel, S Turkey
Anantapur 110 C2 Andhra Pradesh, S India
Anaphe see Anáfi
Anápolis 41 F3 Goiás, C Brazil
Anār 98 D3 Kermān, C Iran
Anatolia 94 C4 plateau C Turkey
Anatom see Aneityum
Añatuya 42 C3 Santiago del Estero, N Argentina
An Bhearú see Barrow
Anchorage 14 C3 Alaska, USA
Ancona 74 C3 Marche, C Italy
Ancud 43 B6 prev. San Carlos de Ancud. Los Lagos, S Chile
Ancyra see Ankara
Åndalsnes 63 A5 Møre og Romsdal, S Norway
Andalucía 70 D4 cultural region S Spain
Andalusia 20 C3 Alabama, S USA
Andaman Islands 102 B4 island group India, NE Indian Ocean
Andaman Sea 102 C4 sea NE Indian Ocean
Andenne 65 C6 Namur, SE Belgium
Anderlues 65 B7 Hainaut, S Belgium
Anderson 18 C4 Indiana, N USA
Andes 42 B3 mountain range W South America
Andhra Pradesh 113 E5 cultural region E India
Andijon 101 F2 Rus. Andizhan. Andijon Viloyati, E Uzbekistan
Andikíthira see Antikýthira
Andipaxi see Antípaxoi
Andípsara see Antípsara
Ándissa see Ántissa
Andizhan see Andijon
Andkhvoy 100 D3 Fāryāb, N Afghanistan
Andorra 69 A7 off. Principality of Andorra, Cat. Valls d'Andorra, Fr. Vallée d'Andorre. country SW Europe
Andorra see Andorra la Vella
Andorra la Vella 69 A8 var. Andorra, Fr. Andorre la Vielle, Sp. Andorra la Vieja. country capital (Andorra) C Andorra
Andorra la Vieja see Andorra la Vella
Andorra, Principality of see Andorra
Andorra, Valls d'/Andorra, Vallée d' see Andorra
Andorre la Vielle see Andorra la Vella
Andover 67 D7 S England, United Kingdom
Andøya 62 C2 island C Norway
Andreanof Islands 14 A3 island group Aleutian Islands, Alaska, USA
Andrews 27 E3 Texas, SW USA
Andrew Tablemount 118 A4 var. Gora Andryu. seamount W Indian Ocean
Andria 75 D5 Puglia, SE Italy
An Droichead Nua see Newbridge
Andropov see Rybinsk
Ándros 83 D6 Ándros, Kykládes, Greece, Aegean Sea
Ándros 83 C6 island Kykládes, Greece, Aegean Sea
Andros Island 32 B2 island NW Bahamas
Andros Town 32 C1 Andros Island, NW Bahamas
Andryu, Gora see Andrew Tablemount
Aneityum 122 D5 var. Anatom; prev. Kéamu. island S Vanuatu
Änewetak see Enewetak Atoll
Angara 93 E4 river C Russian Federation
Angarsk 93 E4 Irkutskaya Oblast', S Russian Federation
Ånge 63 C5 Västernorrland, C Sweden
Ángel de la Guarda, Isla 28 B2 island NW Mexico
Angeles 117 E1 off. Angeles City. Luzon, N Philippines
Angeles City see Angeles
Angel Falls 37 E3 Eng. Angel Falls. waterfall E Venezuela
Angel Falls see Ángel, Salto
Angerburg see Węgorzewo
Ångermanälven 62 C4 river N Sweden
Angermünde 72 D3 Brandenburg, NE Germany
Angers 68 B4 anc. Juliomagus. Maine-et-Loire, NW France
Anglesey 67 C5 island NW Wales, United Kingdom
Anglet 69 A6 Pyrénées-Atlantiques, SW France
Angleton 27 H4 Texas, SW USA
Anglia see England
Anglo-Egyptian Sudan see Sudan
Angmagssalik see Ammassalik
Ang Nam Ngum 114 C4 lake C Laos
Angola 56 B2 off. Republic of Angola; prev. People's Republic of Angola, Portuguese West Africa. country SW Africa
Angola Basin 47 B5 undersea basin E Atlantic Ocean
Angola, People's Republic of see Angola
Angola, Republic of see Angola
Angora see Ankara
Angostura see Ciudad Bolívar
Angostura, Presa de la 29 G5 reservoir SE Mexico
Angoulême 69 B5 anc. Iculisma. Charente, W France
Angoumois 69 B5 cultural region W France
Angra Pequena see Lüderitz
Angren 101 F2 Toshkent Viloyati, E Uzbekistan
Anguilla 33 G3 UK dependent territory E West Indies
Anguilla Cays 32 B2 islets SW Bahamas
Anhui 106 C5 var. Anhui Sheng, Anhwei, Wan. province E China
AnhuiSheng/Anhwei Wan see Anhui
Anicium see le Puy
Anina 86 A4 Ger. Steierdorf, Hung. Stájerlakanina; prev. Ştaierdorf-Anina, Steierdorf-Anina, Steyerlak-Anina. Caraş-Severin, SW Romania
Anjou 68 B4 cultural region NW France
Anjouan 57 F2 var. Ndzouani, Nzwani. island SE Comoros
Ankara 94 C3 prev. Angora; anc. Ancyra. country capital (Turkey) Ankara, C Turkey
Ankeny 23 F3 Iowa, C USA
Anklam 72 D2 Mecklenburg-Vorpommern, NE Germany
An Mhuir Cheilteach see Celtic Sea
Annaba 49 E1 prev. Bône. NE Algeria
An Nafud 98 B4 desert NW Saudi Arabia
'Annah 98 B3 var. 'Ānah. Al Anbār, NW Iraq
An Najaf 98 B3 var. Najaf. An Najaf, S Iraq
Annamite Mountains 114 D4 Fr. Annamitique, Chaîne. mountain range C Laos
Annamitique, Chaîne see Annamite Mountains
Annapolis 19 F4 state capital Maryland, NE USA
Annapurna 113 E3 mountain C Nepal
An Nāqūrah see En Nâqoûra
Ann Arbor 18 C3 Michigan, N USA
An Nāşirīyah 98 C3 var. Nasiriya. Dhī Qār, SE Iraq
Anneciacum see Annecy
Annecy 69 D5 anc. Anneciacum. Haute-Savoie, E France
An Nīl al Abyaḍ see White Nile
An Nīl al Azraq see Blue Nile
Anniston 20 D2 Alabama, S USA
Annotto Bay 32 B4 C Jamaica
An Ómaigh see Omagh
Anqing 106 D5 Anhui, E China
Anse La Raye 33 F1 NW Saint Lucia
Anshun 106 B6 Guizhou, S China
Ansongo 53 E3 Gao, E Mali
An Srath Bán see Strabane
Antakya 94 D4 anc. Antioch, Antiochia. Hatay, S Turkey
Antalaha 57 G2 Antsirañana, NE Madagascar
Antalya 94 B4 prev. Adalia; anc. Attaleia, Bibl. Attalia. Antalya, SW Turkey
Antalya, Gulf of 94 B4 var. Gulf of Adalia, Eng. Gulf of Antalya. gulf SW Turkey
Antalya, Gulf of see Antalya Körfezi
Antananarivo 57 G3 prev. Tananarive. country capital (Madagascar) Antananarivo, C Madagascar
Antarctica 132 B3 continent
Antarctic Peninsula 132 A2 peninsula Antarctica
Antep see Gaziantep
Antequera 70 D5 anc. Anticaria, Antiquaria. Andalucía, S Spain
Antequera see Oaxaca
Antibes 69 D6 anc. Antipolis. Alpes-Maritimes, SE France
Anticaria see Antequera
Anticosti, Île d' 17 F3 Eng. Anticosti Island. island Québec, E Canada
Anticosti Island see Anticosti, Île d'
Antigua 33 G3 island S Antigua and Barbuda, Leeward Islands
Antigua and Barbuda 33 G3 country E West Indies
Antikythira 83 B7 var. Andikíthira. island S Greece
Anti-Lebanon 96 B4 var. Jebel esh Sharqi, Ar. Al Jabal ash Sharqī, Fr. Anti-Liban. mountain range Lebanon/Syria
Anti-Liban see Anti-Lebanon
Antioch see Antakya
Antiochia see Antakya
Antípaxoi 83 A5 var. Andipaxi. island Iónia Nisiá, Greece, C Mediterranean Sea
Antipodes Islands 120 D5 island group S New Zealand
Antipolis see Antibes
Antípsara 83 D5 var. Andípsara. island E Greece
Antiquaria see Antequera
Ántissa 83 D5 var. Ándissa. Lésvos, E Greece
An tIúr see Newry
Antivari see Bar
Antofagasta 42 B2 Antofagasta, N Chile
Antony 68 E2 Hauts-de-Seine, N France
An tSionainn see Shannon
Antsirañana 57 G2 province N Madagascar
Antsohihy 57 G2 Mahajanga, NW Madagascar
An-tung see Dandong
Antwerpen 65 C5 Eng. Antwerp, Fr. Anvers. Antwerpen, N Belgium
Anuradhapura 110 D3 North Central Province, C Sri Lanka
Anvers see Antwerpen
Anyang 106 C4 Henan, C China
A'nyêmaqên Shan 104 D4 mountain range C China
Anykščiai 84 C4 Utena, E Lithuania
Anzio 75 C5 Lazio, C Italy
Ao Krung Thep 115 C5 var. Krung Thep Mahanakhon, Eng. Bangkok. country capital (Thailand) Bangkok, C Thailand
Aomori 108 D3 Aomori, Honshū, C Japan
Aóos see Vjosës, Lumi i
Aoraki 129 B6 prev. Aorangi, Mount Cook. mountain South Island, New Zealand
Aorangi see Aoraki
Aosta 74 A1 anc. Augusta Praetoria. Valle d'Aosta, NW Italy
Aoukâr 52 D3 var. Aouker. plateau C Mauritania
Aouk, Bahr 54 C4 river Central African Republic/ Chad
Aouker see Aoukâr
Aozou 54 C1 Borkou-Ennedi-Tibesti, N Chad
Apalachee Bay 20 D3 bay Florida, SE USA
Apalachicola River 20 D3 river Florida, SE USA
Apamama see Abemama
Apaporis, Río 36 C4 river Brazil/Colombia
Apatity 88 C2 Murmanskaya Oblast', NW Russian Federation
Ape 84 D3 Alūksne, NE Latvia
Apeldoorn 64 D3 Gelderland, E Netherlands
Apennines 74 E2 Eng. Apennines. mountain range Italy/San Marino
Apennines see Appennino
Āpia 123 F4 country capital (Samoa) Upolu, SE Samoa
Ap'khazet'i 95 E1 var. Abkhazia. autonomous republic NW Georgia
Apoera 37 G3 Sipaliwini, NW Suriname
Apostle Islands 18 B1 island group Wisconsin, N USA
Appalachian Mountains 13 D5 mountain range E USA
Appingedam 64 E1 Groningen, NE Netherlands
Appleton 18 B2 Wisconsin, N USA
Apulia see Puglia
Apure, Río 36 C2 river W Venezuela
Apurímac, Río 38 D3 river S Peru
Apuseni, Munţii 86 A4 mountain range W Romania
Aqaba/'Aqaba see Al 'Aqabah
Aqaba, Gulf of 98 A4 var. Gulf of Elat, Ar. Khalīj al 'Aqabah; anc. Sinus Aelaniticus. gulf NE Red Sea
'Aqabah, Khalīj al see Aqaba, Gulf of
Āqchah 101 E3 var. Āqcheh. Jowzjān, N Afghanistan
Āqcheh see Āqchah
Aqmola see Astana
Aqtöbe see Aktobe
Aquae Augustae see Dax
Aquae Calidae see Bath
Aquae Flaviae see Chaves
Aquae Grani see Aachen
Aquae Sextiae see Aix-en-Provence
Aquae Solis see Bath
Aquae Tarbelicae see Dax
Aquidauana 41 E4 Mato Grosso do Sul, S Brazil
Aquila/Aquila degli Abruzzi see L'Aquila
Aquisgranum see Aachen
Aquitaine 69 B6 cultural region SW France
'Arabah, Wadi al 97 B7 Heb. Ha'Arava. dry watercourse Israel/Jordan
Arabian Basin 102 A4 undersea basin N Arabian Sea
Arabian Desert see Sahara el Sharqîya
Arabian Peninsula 99 B5 peninsula SW Asia
Arabian Sea 102 A3 sea NW Indian Ocean
Arabicus, Sinus see Red Sea
'Arabī, Khalīj al see Persian Gulf
'Arabīyah as Su'ūdīyah, Al Mamlakah al see Saudi Arabia
'Arabīyah Jumhūrīyah, Mişr al see Egypt
Arab Republic of Egypt see Egypt
Aracaju 41 G3 state capital Sergipe, E Brazil
Araçuai 41 F3 Minas Gerais, SE Brazil
Arad 97 B7 Southern, S Israel
Arad 86 A4 Arad, W Romania
Arafura Sea 120 A3 Ind. Laut Arafuru. sea W Pacific Ocean
Arafuru, Laut see Arafura Sea
Aragón 71 E2 autonomous community E Spain
Araguaia, Río 41 E3 var. Araguaya. river C Brazil
Araguari 41 F3 Minas Gerais, SE Brazil
Araguaya see Araguaia, Río
Ara Jovis see Aranjuez
Arāk 98 C3 prev. Sultānābād. Markazī, W Iran
Arakan Yoma 114 A3 mountain range W Myanmar (Burma)
Araks/Arak's see Aras
Aral see Aralsk, Kazakhstan
Aral Sea 100 C1 Kaz. Aral Tengizi, Rus. Aral'skoye More, Uzb. Orol Dengizi. inland sea Kazakhstan/ Uzbekistan
Aral'sk 92 B4 Kaz. Aral. Kzylorda, SW Kazakhstan
Aral'skoye More/Aral Tengizi see Aral Sea
Aranda de Duero 70 D2 Castilla-León, N Spain
Aranđelovac 78 D4 prev. Arandjelovac. Serbia, C Serbia
Arandjelovac see Aranđelovac
Aranjuez 70 D3 anc. Ara Jovis. Madrid, C Spain
Araouane 53 E2 Tombouctou, N Mali
'Ar'ar 98 B3 Al Ḥudūd ash Shamālīyah, NW Saudi Arabia
Ararat, Mount see Büyükağrı Dağı
Aras 95 G3 Arm. Arak's, Az. Araz Nehri, Per. Rūd-e Aras, Rus. Araks; prev. Araxes. river SW Asia
Aras, Rūd-e see Aras
Arauca 36 C2 Arauca, NE Colombia
Arauca, Río 36 C2 river Colombia/Venezuela
Arausio see Orange
Araxes see Aras
Araz Nehri see Aras
Arbela see Arbīl
Arbīl 98 B2 var. Erbil, Irbīl, Kurd. Hawlēr; anc. Arbela. Arbīl, N Iraq
Arbroath 66 D3 anc. Aberbrothock. E Scotland, United Kingdom
Arbuzinka see Arbuzynka
Arbuzynka 87 E3 Rus. Arbuzinka. Mykolayivs'ka Oblast', S Ukraine
Arcachon 69 B5 Gironde, SW France
Arcae Remorum see Châlons-en-Champagne
Arcata 24 A4 California, W USA
Archangel see Arkhangel'sk
Archangel Bay see Chëshskaya Guba
Archidona 70 D5 Andalucía, S Spain
Arco 74 C2 Trentino-Alto Adige, N Italy
Arctic Mid Oceanic Ridge see Nansen Cordillera
Arctic Ocean 133 B3 ocean
Arda 82 C3 var. Ardhas, Gk. Ardas. river Bulgaria/ Greece
Ardabīl 98 C2 var. Ardebil. Ardabīl, NW Iran
Ardakān 98 D3 Yazd, C Iran
Ardas 82 D3 var. Ardhas, Bul. Arda. river Bulgaria/Greece
Arḍ aş Şawwān 97 C7 var. Ardh es Suwwān. plain S Jordan
Ardeal see Transylvania
Ardebil see Ardabīl
Ardèche 69 C5 cultural region E France
Ardennes 65 C8 physical region Belgium/France
Ardhas see Arda/Ardas
Ardh es Suwwān see Arḍ aş Şawwān
Ardino 82 D3 Kŭrdzhali, S Bulgaria
Ard Mhacha see Armagh
Ardmore 27 G2 Oklahoma, C USA
Arel see Arlon
Arelas/Arelate see Arles
Arendal 63 A6 Aust-Agder, S Norway
Arensburg see Kuressaare
Arenys de Mar 71 G2 Cataluña, NE Spain
Areópoli 83 B7 prev. Areópolis. Pelopónnisos, S Greece
Areópolis see Areópoli
Arequipa 39 E4 Arequipa, SE Peru
Arezzo 74 C3 anc. Arretium. Toscana, C Italy
Argalastí 83 C5 Thessalía, C Greece
Argenteuil 68 D1 Val-d'Oise, N France
Argentina 43 B5 off. Argentine Republic. country S South America
Argentina Basin see Argentine Basin
Argentine Basin 35 C7 var. Argentina Basin. undersea basin SW Atlantic Ocean
Argentine Republic see Argentina
Argentine Rise see Falkland Plateau
Argentoratum see Strasbourg
Arghandab, Darya-ye 101 E5 river SE Afghanistan
Argirocastro see Gjirokastër
Argo 50 B3 Northern, N Sudan
Argo Fracture Zone 119 C5 tectonic Feature C Indian Ocean
Árgos 83 B6 Pelopónnisos, S Greece
Argostóli 83 A5 var. Argostólion. Kefallonía, Iónia Nisiá, Greece, C Mediterranean Sea
Argostólion see Argostóli
Argun 103 E1 Chin. Ergun He, Rus. Argun'. river China/Russian Federation
Argyrokastron see Gjirokastër
Århus 63 B7 var. Aarhus. Århus, C Denmark
Aria see Herāt
Ari Atoll 110 A4 var. Alifu Atoll. atoll C Maldives
Arica 42 B1 hist. San Marcos de Arica. Tarapacá, N Chile
Aridaía 82 B3 var. Aridea, Aridhaía. Dytikí Makedonía, N Greece
Aridea see Aridaía
Aridhaía see Aridaía
Arīhā 96 B3 Eng. Jericho. Al Karak, W Jordan

Ariminum – Balaton, Lake

Ariminum *see* Rimini
Arinsal *69 A7* NW Andorra Europe
Arizona *26 A2* *off.* State of Arizona, *also known as* Copper State, Grand Canyon State. *state* SW USA
Arkansas *20 A1* *off.* State of Arkansas, *also known as* The Land of Opportunity. *state* S USA
Arkansas City *23 F5* Kansas, C USA
Arkansas River *27 G1* *river* C USA
Arkhangel'sk *92 B2* *Eng.* Archangel. Arkhangel'skaya Oblast', NW Russian Federation
Arkoí *83 E6* *island* Dodekánisa, Greece, Aegean Sea
Arles *69 D6* *var.* Arles-sur-Rhône; *anc.* Arelas, Arelate. Bouches-du-Rhône, SE France
Arles-sur-Rhône *see* Arles
Arlington *27 G2* Texas, SW USA
Arlington *19 E4* Virginia, NE USA
Arlon *65 D8* *Dut.* Aarlen, *Ger.* Arel, *Lat.* Orolaunum. Luxembourg, SE Belgium
Armagh *67 B5* *Ir.* Ard Mhacha. S Northern Ireland, United Kingdom
Armagnac *69 B6* *cultural region* S France
Armenia *36 B3* Quindío, W Colombia
Armenia *95 F3* *off.* Republic of Armenia, *var.* Ajastan, *Arm.* Hayastani Hanrapetut'yun; *prev.* Armenian Soviet Socialist Republic. *country* SW Asia
Armenian Soviet Socialist Republic *see* Armenia
Armenia, Republic of *see* Armenia
Armidale *127 D6* New South Wales, SE Australia
Armstrong *16 B3* Ontario, S Canada
Armyans'k *87 F4* *Rus.* Armyansk. Respublika Krym, S Ukraine
Arnaía *82 C4* *Cont.* Arnea. Kentrikí Makedonía, N Greece
Arnaud *60 A3* *river* Québec, E Canada
Arnea *see* Arnaía
Arnedo *71 E2* La Rioja, N Spain
Arnhem *64 D4* Gelderland, SE Netherlands
Arnhem Land *126 A2* *physical region* Northern Territory, N Australia
Arno *74 B3* *river* C Italy
Arnold *23 G4* Missouri, C USA
Arnswalde *see* Choszczno
Aroe Islands *see* Aru, Kepulauan
Arorae *123 E3* *atoll* Tungaru, W Kiribati
Arrabona *see* Győr
Ar Rahad *see* Er Rahad
Ar Ramādī *98 B3* *var.* Ramadi, Rumadiya. Al Anbār, SW Iraq
Ar Rāmī *96 C4* Ḥimş, C Syria
Ar Ramthā *97 B5* *var.* Ramtha. Irbid, N Jordan
Arran, Isle of *66 C4* *island* SW Scotland, United Kingdom
Ar Raqqah *96 C2* *var.* Rakka; *anc.* Nicephorium. Ar Raqqah, N Syria
Arras *68 C2* *anc.* Nemetocenna. Pas-de-Calais, N France
Ar Rawḍatayn *98 C4* *var.* Raudhatain. N Kuwait
Arretium *see* Arezzo
Arriaca *see* Guadalajara
Arriaga *29 G5* Chiapas, SE Mexico
Ar Riyāḍ *99 C5* *Eng.* Riyadh. *country capital* (Saudi Arabia) Ar Riyāḍ, C Saudi Arabia
Ar Rub 'al Khali *99 C6* *Eng.* Empty Quarter, Great Sandy Desert. *desert* SW Asia
Ar Rustāq *99 E5* *var.* Rostak, Rustaq. N Oman
Ar Ruṭbah *98 B3* *var.* Rutba. Al Anbār, SW Iraq
Árta *83 A5* *anc.* Ambracia. Ípeiros, W Greece
Artashat *95 F3* S Armenia
Artemisa *32 B2* La Habana, W Cuba
Artesia *26 D3* New Mexico, SW USA
Arthur's Pass *129 C6* *pass* South Island, New Zealand
Artigas *42 D3* *prev.* San Eugenio, San Eugenio del Cuareim. Artigas, N Uruguay
Art'ik *95 F2* W Armenia
Artois *68 C2* *cultural region* N France
Artsiz *see* Artsyz
Artsyz *86 D4* *Rus.* Artsiz. Odes'ka Oblast', SW Ukraine
Artvin *95 F2* Artvin, NE Turkey
Arua *51 B6* NW Uganda
Aruba *36 C1* *var.* Oruba. *Dutch autonomous region* S West Indies
Aru Islands *see* Aru, Kepulauan
Aru, Kepulauan *117 G4* *Eng.* Aru Islands; *prev.* Aroe Islands. *island group* E Indonesia
Arunāchal Pradesh *113 G3* *prev.* North East Frontier Agency, North East Frontier Agency of Assam. *cultural region* NE India
Arusha *51 C7* Arusha, N Tanzania
Arviat *15 G4* *prev.* Eskimo Point. Nunavut, C Canada
Arvidsjaur *62 C4* Norrbotten, N Sweden
Arys' *92 B5* *Kaz.* Arys. Yuzhnyy Kazakhstan, S Kazakhstan
Arys *see* Arys'
Asadābād *101 F4* *var.* Asaḏābād; *prev.* Chaghasarāy. Konar, E Afghanistan
Asaḏābād *see* Asadābād
Asad, Buḩayrat al *96 C2* *Eng.* Lake Assad. *lake* N Syria
Asahi-dake *108 D2* *mountain* Hokkaidō, N Japan
Asahikawa *108 D2* Hokkaidō, N Japan
Asamankese *53 E5* SE Ghana
Āsānsol *113 F4* West Bengal, NE India
Asben *see* Aïr, Massif de l'
Ascension Fracture Zone *47 A5* *tectonic Feature* C Atlantic Ocean
Ascension Island *45 C5* *UK dependent territory* C Atlantic Ocean
Ascoli Piceno *74 C4* *anc.* Asculum Picenum. Marche, C Italy
Asculum Picenum *see* Ascoli Piceno
'Aseb *50 D4* *var.* Assab, *Amh.* Āseb. SE Eritrea
Assen *64 E2* Drenthe, NE Netherlands
Aşgabat *100 C3* *prev.* Ashgabat, Ashkhabad, Poltoratsk. *country capital* (Turkmenistan) Ahal Welaýaty, C Turkmenistan
Ashara *see* Al 'Ashārah
Ashburton *129 C6* Canterbury, South Island, New Zealand
Ashburton River *124 A4* *river* Western Australia
Ashdod *97 A6* *anc.* Azotos, *Lat.* Azotus. Central, W Israel
Asheville *21 E1* North Carolina, SE USA
Ashgabat *see* Aşgabat
Ashkelon *97 A6* *prev.* Ashqelon. Southern, C Israel
Ashkhabad *see* Aşgabat
Ashland *24 B4* Oregon, NW USA
Ashland *18 B1* Wisconsin, N USA
Ashmore and Cartier Islands *120 A3* *Australian external territory* E Indian Ocean
Ashmyany *85 C5* *Rus.* Oshmyany. Hrodzyenskaya Voblasts', W Belarus
Ashqelon *see* Ashkelon
Ash Shadādah *96 D2* *var.* Ash Shaddādah, Jisr ash Shadadi, Shaddādī, Shedadi, Tell Shedadi. Al Ḥasakah, NE Syria
Ash Shaddādah *see* Ash Shadādah
Ash Sharah *97 B7* *var.* Esh Sharā. *mountain range* W Jordan
Ash Shāriqah *98 D4* *Eng.* Sharjah. Ash Shāriqah, NE United Arab Emirates
Ash Shawbak *97 B7* Ma'ān, W Jordan
Ash Shiḩr *99 C6* SE Yemen
Asia *90* *continent*
Asinara *74 A4* *island* W Italy
Asipovichy *85 D6* *Rus.* Osipovichi. Mahilyowskaya Voblasts', C Belarus
Aşkale *95 E3* Erzurum, NE Turkey
Askersund *63 C6* Örebro, C Sweden
Asmara *see* Asmera
Asmera *50 C4* *var.* Asmara. *country capital* (Eritrea) C Eritrea
Aspadana *see* Eşfahān
Asphaltites, Lacus *see* Dead Sea
Aspinwall *see* Colón
Assab *see* 'Aseb
As Sabkhah *96 D2* *var.* Sabkha. Ar Raqqah, NE Syria
Assad, Lake *see* Asad, Buḩayrat al
Aş Şafāwī *97 C6* Al Mafraq, N Jordan
Aş Şaḩrā' ash Sharqīyah *see* Sahara el Sharqîya
As Salamīyah *see* Salamīyah
'Assal, Lac *46 E4* *lake* C Djibouti
As Salṭ *97 B6* *var.* Salt. Al Balqā', NW Jordan
Assamaka *see* Assamakka
Assamakka *53 F2* *var.* Assamaka. Agadez, NW Niger
As Samāwah *98 B3* *var.* Samawa. Al Muthanná, S Iraq
Assenede *65 B5* Oost-Vlaanderen, NW Belgium
Assiout *see* Asyūţ
Assiut *see* Asyūţ
Assling *see* Jesenice
Assouan *see* Aswān
Assu *41 G2* *var.* Açu. Rio Grande do Norte, E Brazil
Assuan *see* Aswān
As Sukhnah *96 C3* *var.* Sukhne, *Fr.* Soukhné. Ḥimş, C Syria
As Sulaymānīyah *98 C3* *var.* Sulaimaniya, *Kurd.* Slēmānī. As Sulaymānīyah, NE Iraq
As Sulayyil *99 B5* Ar Riyāḍ, S Saudi Arabia
Aş Şuwār *96 D2* *var.* Şuwār. Dayr az Zawr, E Syria
As Suwaydā' *97 B5* *var.* El Suweida, Es Suweida, Suweida, *Fr.* Soueida. As Suwaydā', SW Syria
As Suways *see* Suez
Asta Colonia *see* Asti
Astacus *see* İzmit
Astana *92 C4* *prev.* Akmola, Akmolinsk, Tselinograd, Aqmola. *country capital* (Kazakhstan) Akmola, N Kazakhstan
Asta Pompeia *see* Asti
Astarabad *see* Gorgān
Asterābād *see* Gorgān
Asti *74 A2* *anc.* Asta Colonia, Asta Pompeia, Hasta Colonia, Hasta Pompeia. Piemonte, NW Italy
Astigi *see* Ecija
Astipálaia *see* Astypálaia
Astorga *70 C1* *anc.* Asturica Augusta. Castilla-León, N Spain
Astrabad *see* Gorgān
Astrakhan' *89 C7* Astrakhanskaya Oblast', SW Russian Federation
Asturias *70 C1* *autonomous community* NW Spain
Asturias *see* Oviedo
Asturica Augusta *see* Astorga
Astypálaia *83 D7* *var.* Astipálaia, *It.* Stampalia. *island* Kykládes, Greece, Aegean Sea
Asunción *42 D2* *country capital* (Paraguay) Central, S Paraguay
Aswān *50 B2* *var.* Assouan, Assuan, Aswân; *anc.* Syene. SE Egypt
Aswân *see* Aswān
Asyūţ *50 B2* *var.* Assiout, Assiut, Asyût, Siut; *anc.* Lycopolis. C Egypt
Asyût *see* Asyūţ
Atacama Desert *42 B2* *Eng.* Atacama Desert. *desert* N Chile
Atacama Desert *see* Atacama, Desierto de
Atafu Atoll *123 E3* *island* NW Tokelau
Atamyrat *100 D3* *prev.* Kerki. Lebap Welaýaty, E Turkmenistan
Aţâr *52 C2* Adrar, W Mauritania
Atas Bogd *104 D3* *mountain* SW Mongolia
Atascadero *25 B7* California, W USA
Atatürk Baraji *95 E4* *reservoir* S Turkey
Atbara *50 C3* *var.* 'Aṭbārah. River Nile, NE Sudan
'Aṭbārah/'Aṭbarah, Nahr *see* Atbara
Atbasar *92 C4* Akmola, N Kazakhstan
Atchison *23 F4* Kansas, C USA
Aternum *see* Pescara
Ath *65 B6* *var.* Aat. Hainaut, SW Belgium
Athabasca *15 E5* Alberta, SW Canada
Athabasca *15 E5* *var.* Athabaska. *river* Alberta, SW Canada
Athabasca, Lake *15 F4* *lake* Alberta/Saskatchewan, SW Canada
Athabaska *see* Athabasca
Athenae *see* Athína
Athens *21 E2* Georgia, SE USA
Athens *18 D4* Ohio, N USA
Athens *27 G3* Texas, SW USA
Athens *see* Athína
Atherton *126 D3* Queensland, NE Australia
Athína *83 C6* *Eng.* Athens, *prev.* Athínai; *anc.* Athenae. *country capital* (Greece) Attikí, C Greece
Athínai *see* Athína
Athlone *67 B5* *Ir.* Baile Átha Luain. C Ireland
Ath Thawrah *see* Madīnat ath Thawrah
Ati *54 C3* Batha, C Chad
Atikokan *16 B4* Ontario, S Canada
Atka *93 G3* Magadanskaya Oblast', E Russian Federation
Atka *14 A3* Atka Island, Alaska, USA
Atlanta *20 D2* *state capital* Georgia, SE USA
Atlanta *27 H2* Texas, SW USA
Atlantic City *19 F4* New Jersey, NE USA
Atlantic-Indian Basin *45 D7* *undersea basin* SW Indian Ocean
Atlantic-Indian Ridge *47 B8* *undersea ridge* SW Indian Ocean
Atlantic Ocean *44 B4* *ocean*
Atlas Mountains *48 C2* *mountain range* NW Africa
Atlasovo *93 H3* Kamchatskaya Oblast', E Russian Federation
Atlas Saharien *48 D2* *var.* Saharan Atlas. *mountain range* Algeria/Morocco
Atlas, Tell *see* Atlas Tellien
Atlas Tellien *80 C3* *Eng.* Tell Atlas. *mountain range* N Algeria
Atlin *14 D4* British Columbia, W Canada
Aţ Ţafīlah *97 B7* *var.* Et Tafila, Tafila. Aţ Ţafīlah, W Jordan
Aţ Ţā'if *99 B5* Makkah, W Saudi Arabia
Attaleia/Attalia *see* Antalya
At Tall al Abyaḍ *96 C2* *var.* Tall al Abyaḍ, Tell Abyad, *Fr.* Tell Abiad. Ar Raqqah, N Syria
Aţ Ţanf *96 D4* Ḥimş, S Syria
Attapu *115 E5* *var.* Samakhixai, Attopeu. Attapu, S Laos
Attawapiskat *16 C3* Ontario, C Canada
Attawapiskat *16 C3* *river* Ontario, S Canada
At Tibnī *96 D2* *var.* Tibnī. Dayr az Zawr, NE Syria
Attopeu *see* Attapu
Attu Island *14 A2* *island* Aleutian Islands, Alaska, USA
Atyrau *92 B4* *prev.* Gur'yev. Atyrau, W Kazakhstan
Aubagne *69 D6* *anc.* Albania. Bouches-du-Rhône, SE France
Aubange *65 D8* Luxembourg, SE Belgium
Aubervilliers *68 E1* Seine-St-Denis, Île-de-France, N France Europe
Auburn *24 B2* Washington, NW USA
Auch *69 B6* *Lat.* Augusta Auscorum, Elimberrum. Gers, S France
Auckland *128 D2* Auckland, North Island, New Zealand
Auckland Islands *120 C5* *island group* S New Zealand
Audern *see* Audru
Audincourt *68 E4* Doubs, E France
Audru *84 D2* *Ger.* Audern. Pärnumaa, SW Estonia
Augathella *127 D5* Queensland, E Australia
Augsbourg *see* Augsburg
Augsburg *73 C6* *Fr.* Augsbourg; *anc.* Augusta Vindelicorum. Bayern, S Germany
Augusta *125 A7* Western Australia
Augusta *21 E2* Georgia, SE USA
Augusta *19 G2* *state capital* Maine, NE USA
Augusta *see* London
Augusta Auscorum *see* Auch
Augusta Emerita *see* Mérida
Augusta Praetoria *see* Aosta
Augusta Trajana *see* Stara Zagora
Augusta Treverorum *see* Trier
Augusta Vangionum *see* Worms
Augusta Vindelicorum *see* Augsburg
Augustobona Tricassium *see* Troyes
Augustodurum *see* Bayeux
Augustoritum Lemovicensium *see* Limoges
Augustów *76 E2* *Rus.* Avgustov. Podlaskie, NE Poland
Aulie Ata/Auliye-Ata *see* Taraz
Aunglan *114 B4* *var.* Allanmyo, Myaydo. Magway, C Myanmar (Burma)
Auob *56 B4* *var.* Oup. *river* Namibia/South Africa
Aurangābād *112 D5* Mahārāshtra, C India
Auray *68 A3* Morbihan, NW France
Aurelia Aquensis *see* Baden-Baden
Aurelianum *see* Orléans
Aurès, Massif de l' *80 C4* *mountain range* NE Algeria
Aurillac *69 C5* Cantal, C France
Aurium *see* Ourense
Aurora *37 F2* NW Guyana
Aurora *22 D4* Colorado, C USA
Aurora *18 B3* Illinois, N USA
Aurora *23 G5* Missouri, C USA
Aurora *see* Maéwo, Vanuatu
Aus *56 B4* Karas, SW Namibia
Ausa *see* Vic
Aussig *see* Ústí nad Labem
Austin *23 G3* Minnesota, N USA
Austin *27 G3* *state capital* Texas, SW USA
Australes, Archipel des *see* Australes, Îles
Australes et Antarctiques Françaises, Terres *see* French Southern and Antarctic Lands
Australes, Îles *121 F4* *var.* Archipel des Australes, Îles Tubuai, Tubuai Islands, *Eng.* Austral Islands. *island group* SW French Polynesia
Austral Fracture Zone *121 H4* *tectonic feature* S Pacific Ocean
Australia *120 A4* *off.* Commonwealth of Australia. *country*
Australia, Commonwealth of *see* Australia
Australian Alps *127 C7* *mountain range* SE Australia
Australian Capital Territory *127 D7* *prev.* Federal Capital Territory. *territory* SE Australia
Australie, Bassin Nord de l' *see* North Australian Basin
Austral Islands *see* Australes, Îles
Austrava *see* Ostrov
Austria *73 D7* *off.* Republic of Austria, *Ger.* Österreich. *country* C Europe
Austria, Republic of *see* Austria
Autesiodorum *see* Auxerre
Autissiodorum *see* Auxerre
Autricum *see* Chartres
Auvergne *69 C5* *cultural region* C France
Auxerre *68 C4* *anc.* Autesiodorum, Autissiodorum. Yonne, C France
Avaricum *see* Bourges
Avarua *123 G5* *dependent territory capital* (Cook Islands) Rarotonga, S Cook Islands
Avasfelsőfalu *see* Negreşti-Oaş
Ávdira *82 C3* Anatolikí Makedonía kai Thráki, NE Greece
Aveiro *70 B2* *anc.* Talabriga. Aveiro, W Portugal
Avela *see* Ávila
Avellino *75 D5* *anc.* Abellinum. Campania, S Italy
Avenio *see* Avignon
Avesta *63 C6* Dalarna, C Sweden
Aveyron *69 C6* *river* S France
Avezzano *74 C4* Abruzzo, C Italy
Avgustov *see* Augustów
Aviemore *66 C3* N Scotland, United Kingdom
Avignon *69 D6* *anc.* Avenio. Vaucluse, SE France
Ávila *70 D3* *var.* Avila; *anc.* Abela, Abula, Abyla, Avela. Castilla-León, C Spain
Avilés *70 C1* Asturias, NW Spain
Avranches *68 B3* Manche, N France
Avveel *see* Ivalo, Finland
Avvil *see* Ivalo
Awaji-shima *109 C6* *island* SW Japan
Āwash *51 D5* Āfar, NE Ethiopia
Awbārī *49 F3* SW Libya
Ax *see* Dax
Axel *65 B5* Zeeland, SW Netherlands
Axel Heiberg Island *15 E1* *var.* Axel Heiburg. *island* Nunavut, N Canada
Axel Heiburg *see* Axel Heiberg Island
Axiós *see* Vardar
Ayacucho *38 D4* Ayacucho, S Peru
Ayagoz *92 C5* *var.* Ayaguz, *Kaz.* Ayakoz. *river* E Kazakhstan
Ayamonte *70 C4* Andalucía, S Spain
Ayaviri *39 E4* Puno, S Peru
Aydarko'l Ko'li *101 E2* *Rus.* Ozero Aydarkul'. *lake* C Uzbekistan
Aydarkul', Ozero *see* Aydarko'l Ko'li
Aydın *94 A4* *var.* Aïdin; *anc.* Tralles Aydin. Aydın, SW Turkey
Ayers Rock *see* Uluru
Ayeyarwady *see* Irrawaddy
Ayiá *see* Agiá
Áyios Evstrátios *see* Ágios Efstrátios
Áyios Nikólaos *see* Ágios Nikólaos
Ayorou *53 E3* Tillabéri, W Niger
'Ayoûn el 'Atroûs *52 D3* *var.* Aïoun el Atrous, Aïoun el Atroûss. Hodh el Gharbi, SE Mauritania
Ayr *66 C4* W Scotland, United Kingdom
Ayteke Bi *92 B4* *Kaz.* Zhangaqazaly; *prev.* Novokazalinsk. Kzylorda, SW Kazakhstan
Aytos *82 E2* Burgas, E Bulgaria
Ayutthaya *115 C5* *var.* Phra Nakhon Si Ayutthaya. Phra Nakhon Si Ayutthaya, C Thailand
Ayvalık *94 A3* Balıkesir, W Turkey
Azahar, Costa del *71 F3* *coastal region* E Spain
Azaouâd *53 E3* *desert* C Mali
Azärbaycan/Azärbaycan Respublikasī *see* Azerbaijan
A'zāz *96 B2* Ḥalab, NW Syria
Azerbaijan *95 G2* *off.* Azerbaijani Republic, *Az.* Azärbaycan, Azärbaycan Respublikası; *prev.* Azerbaijan SSR. *country* SE Asia
Azerbaijani Republic *see* Azerbaijan
Azerbaijan SSR *see* Azerbaijan
Azimabad *see* Patna
Azizie *see* Telish
Azogues *38 B2* Cañar, S Ecuador
Azores *70 A4* *var.* Açores, Ilhas dos Açores, *Port.* Arquipélago dos Açores. *island group* Portugal, NE Atlantic Ocean
Azores-Biscay Rise *58 A3* *undersea rise* E Atlantic Ocean
Azotos/Azotus *see* Ashdod
Azoum, Bahr *54 C3* *seasonal river* SE Chad
Azov, Sea of *81 H1* *Rus.* Azovskoye More, *Ukr.* Azovs'ke More. *sea* NE Black Sea
Azovs'ke More/Azovskoye More *see* Azov, Sea of
Azraq, Wāḩat al *97 C6* *oasis* N Jordan
Aztec *26 C1* New Mexico, SW USA
Azuaga *70 C4* Extremadura, W Spain
Azuero, Península de *31 F5* *peninsula* S Panama
Azul *43 D5* Buenos Aires, E Argentina
Azur, Côte d' *69 E6* *coastal region* SE France
'Azza *see* Gaza
Az Zagāzīq *50 B1* *var.* Az Zaqāzīq *var.* Zagazig. N Egypt
Az Zaqāzīq *see* Az Zagāzīq
Az Zarqā' *97 B6* NW Jordan
Az Zāwiyah *49 F2* *var.* Zawia. NW Libya
Az Zilfī *98 B4* Ar Riyāḍ, N Saudi Arabia

B

Baalbek *96 B4* *var.* Ba'labakk; *anc.* Heliopolis. E Lebanon
Baardheere *51 D6* *var.* Bardere, *It.* Bardera. Gedo, SW Somalia
Baarle-Hertog *65 C5* Antwerpen, N Belgium
Baarn *64 C3* Utrecht, C Netherlands
Babadag *86 D5* Tulcea, SE Romania
Babahoyo *38 B2* *prev.* Bodegas. Los Ríos, C Ecuador
Bābā, Kūh-e *101 E4* *mountain range* C Afghanistan
Babayevo *88 B4* Vologodskaya Oblast', NW Russian Federation
Babeldaob *122 A1* *var.* Babeldaop, Babelthuap. *island* N Palau
Babeldaop *see* Babeldaob
Bab el Mandeb *99 B7* *strait* Gulf of Aden/Red Sea
Babelthuap *see* Babeldaob
Babian Jiang *see* Black River
Babruysk *85 D7* *Rus.* Bobruysk. Mahilyowskaya Voblasts', E Belarus
Babuyan Channel *117 E1* *channel* N Philippines
Babuyan Islands *117 E1* *island group* N Philippines
Bacabal *41 F2* Maranhão, E Brazil
Bacău *86 C4* *Hung.* Bákó. Bacău, NE Romania
Băc Bô, Vinh *see* Tonkin, Gulf of
Băc Giang *114 D3* Ha Băc, N Vietnam
Bacheykava *85 D5* *Rus.* Bocheykovo. Vitsyebskaya Voblasts', N Belarus
Back *15 F3* *river* Nunavut, N Canada
Bačka Palanka *78 D3* *prev.* Palanka. Serbia, NW Serbia
Bačka Topola *78 D3* *Hung.* Topolya; *prev. Hung.* Bácstopolya. Vojvodina, N Serbia
Bac Liêu *115 D6* *var.* Vinh Loi. Minh Hai, S Vietnam
Bacolod *103 E4* *off.* Bacolod City. Negros, C Philippines
Bacolod City *see* Bacolod
Bácsszenttamás *see* Srbobran
Bácstopolya *see* Bačka Topola
Bactra *see* Balkh
Badajoz *70 C4* *anc.* Pax Augusta. Extremadura, W Spain
Baden-Baden *73 B6* *anc.* Aurelia Aquensis. Baden-Württemberg, SW Germany
Bad Freienwalde *72 D3* Brandenburg, NE Germany
Badger State *see* Wisconsin
Bad Hersfeld *72 B4* Hessen, C Germany
Bad Homburg *see* Bad Homburg vor der Höhe
Bad Homburg vor der Höhe *73 B5* *var.* Bad Homburg. Hessen, W Germany
Bad Ischl *73 D7* Oberösterreich, N Austria
Bad Krozingen *73 A6* Baden-Württemberg, SW Germany
Badlands *22 D2* *physical region* North Dakota/South Dakota, N USA
Badu Island *126 C1* *island* Queensland, NE Australia
Bad Vöslau *73 E6* Niederösterreich, NE Austria
Baeterrae/Baeterrae Septimanorum *see* Béziers
Baetic Cordillera/Baetic Mountains *see* Béticos, Sistemas
Bafatá *52 C4* C Guinea-Bissau
Baffin Bay *15 G2* *bay* Canada/Greenland
Baffin Island *15 G2* *island* Nunavut, NE Canada
Bafing *52 C3* *river* W Africa
Bafoussam *54 A4* Ouest, W Cameroon
Bafra *94 D2* Samsun, N Turkey
Bāft *98 D4* Kermān, S Iran
Bagaces *30 D4* Guanacaste, NW Costa Rica
Bagdad *see* Baghdad
Bagé *41 E5* Rio Grande do Sul, S Brazil
Baghdad *98 B3* *var.* Bagdad, *country capital* (Iraq) Baghdad, C Iraq
Baghlān *101 E3* Baghlān, NE Afghanistan
Bago *114 B4* *var.* Pegu. Bago, SW Myanmar (Burma)
Bagoé *52 D4* *river* Côte d'Ivoire/Mali
Bagrationovsk *84 A4* *Ger.* Preussisch Eylau. Kaliningradskaya Oblast', W Russian Federation
Bagrax Hu *see* Bosten Hu
Baguio *117 E1* *off.* Baguio City. Luzon, N Philippines
Baguio City *see* Baguio
Bagzane, Monts *53 F3* *mountain* N Niger
Bahama Islands *see* Bahamas
Bahamas, The *32 C2* *off.* Commonwealth of the Bahamas. *country* N West Indies
Bahamas *13 D6* *var.* Bahama Islands. *island group* N West Indies
Bahamas, Commonwealth of the *see* Bahamas, The
Baharly *100 C3* *var.* Bäherden, *Rus.* Bakharden; *prev.* Bakherden. Ahal Welaýaty, C Turkmenistan
Bahāwalpur *112 C2* Punjab, E Pakistan
Bäherden *see* Baharly
Bahia *41 F3* *off.* Estado da Bahia. *region* E Brazil
Bahia *41 F3* *off.* Estado da Bahia. *state* E Brazil
Bahía Blanca *43 C5* Buenos Aires, E Argentina
Bahia, Estado da *see* Bahia
Bahir Dar *50 C4* *var.* Bahr Dar, Bahrdar Giyorgis. Āmara, N Ethiopia
Bahraich *113 E3* Uttar Pradesh, N India
Bahrain *98 C4* *off.* State of Bahrain, Dawlat al Bahrayn, *Ar.* Al Baḩrayn, *prev.* Bahrein; *anc.* Tylos, Tyros. *country* SW Asia
Bahrain, State of *see* Bahrain
Bahrayn, Dawlat al *see* Bahrain
Bahr Dar/Bahrdar Giyorgis *see* Bahir Dar
Bahrein *see* Bahrain
Bahr el, Azraq *see* Blue Nile
Bahr Tabariya, Sea of *see* Tiberias, Lake
Bahushewsk *85 E6* *Rus.* Bogushëvsk. Vitsyebskaya Voblasts', NE Belarus
Baia Mare *86 B3* *Ger.* Frauenbach, *Hung.* Nagybánya; *prev.* Neustadt. Maramureş, NW Romania
Baia Sprie *86 B3* *Ger.* Mittelstadt, *Hung.* Felsőbánya. Maramureş, NW Romania
Baïbokoum *54 B4* Logone-Oriental, SW Chad
Baidoa *see* Baydhabo
Baie-Comeau *17 E3* Québec, SE Canada
Baikal, Lake *93 E4* *Eng.* Lake Baikal. *lake* S Russian Federation
Baikal, Lake *see* Baykal, Ozero
Baile Átha Cliath *see* Dublin
Baile Átha Luain *see* Athlone
Bailén *70 D4* Andalucía, S Spain
Baile na Mainistreach *see* Newtownabbey
Băileşti *86 B5* Dolj, SW Romania
Ba Illi *54 B3* Chari-Baguirmi, SW Chad
Bainbridge *20 D3* Georgia, SE USA
Bā'ir *see* Bāyir
Baireuth *see* Bayreuth
Bairiki *122 D2* *country capital* (Kiribati) Tarawa, NW Kiribati
Bairnsdale *127 C7* Victoria, SE Australia
Baishan *107 E3* *prev.* Hunjiang. Jilin, NE China
Baiyin *106 B4* Gansu, C China
Baja *77 C7* Bács-Kiskun, S Hungary
Baja California *26 A4* *Eng.* Lower California. *peninsula* NW Mexico
Baja California Norte *28 B2* *state* NW Mexico
Bajo Boquete *see* Boquete
Bajram Curri *79 D5* Kukës, N Albania
Bakala *54 C4* Ouaka, C Central African Republic
Bakan *see* Shimonoseki
Baker *24 C3* Oregon, NW USA
Baker and Howland Islands *123 E2* *US unincorporated territory* W Polynesia
Baker Lake *15 F3* Nunavut, N Canada
Bakersfield *25 C7* California, W USA
Bakharden *see* Baharly
Bakhchisaray *see* Bakhchysaray
Bakhchysaray *87 F5* *Rus.* Bakhchisaray. Respublika Krym, S Ukraine
Bakherden *see* Baharly
Bakhmach *87 F1* Chernihivs'ka Oblast', N Ukraine
Bākhtarān *see* Kermānshāh
Bakı *95 H2* *Eng.* Baku. *country capital* (Azerbaijan) E Azerbaijan
Bákó *see* Bacău
Bakony *77 C7* *Eng.* Bakony Mountains, *Ger.* Bakonywald. *mountain range* W Hungary
Bakony Mountains/Bakonywald *see* Bakony
Baku *see* Bakı
Bakwanga *see* Mbuji-Mayi
Balabac Island *107 C8* *island* W Philippines
Balabac, Selat *see* Balabac Strait
Balabac Strait *116 D2* *var.* Selat Balabac. *strait* Malaysia/Philippines
Ba'labakk *see* Baalbek
Balaguer *71 F2* Cataluña, NE Spain
Balakovo *89 C6* Saratovskaya Oblast', W Russian Federation
Bālā Morghāb *see* Bālā Murghāb
Bālā Murghāb *100 D4* *prev.* Bālā Morghāb. Laghmān, NW Afghanistan
Balashov *89 B6* Saratovskaya Oblast', W Russian Federation
Balasore *see* Bāleshwar
Balaton, Lake *77 C7* *var.* Lake Balaton, *Ger.* Plattensee. *lake* W Hungary
Balaton, Lake *see* Balaton

Balbina, Represa – Berezhany

Balbina, Represa 40 D1 reservoir NW Brazil
Balboa 31 G4 Panamá, C Panama
Balcarce 43 D5 Buenos Aires, E Argentina
Balclutha 129 B7 Otago, South Island, New Zealand
Baldy Mountain 22 C1 mountain Montana, NW USA
Bâle see Basel
Balearic Plain see Algerian Basin
Baleares Major see Mallorca
Balearic Islands 71 G3 Eng. Balearic Islands. island group Spain, W Mediterranean Sea
Balearic Islands see Baleares, Islas
Balearis Minor see Menorca
Baleine, Rivière à la 17 E2 river Québec, E Canada
Balen 65 C5 Antwerpen, N Belgium
Bāleshwar 113 F4 prev. Balasore. Odisha, E India
Bălgrad see Alba Iulia
Bali 116 D5 island C Indonesia
Balıkesir 94 A3 Balıkesir, W Turkey
Balīkh, Nahr 96 C2 river N Syria
Balikpapan 116 D4 Borneo, C Indonesia
Balkanabat 100 B2 Rus. Nebitdag. Balkan Welaýaty, W Turkmenistan
Balkan Mountains 82 C2 Bul./SCr. Stara Planina. mountain range Bulgaria/Serbia
Balkh 101 E3 anc. Bactra. Balkh, N Afghanistan
Balkhash 92 C5 Kaz. Balqash. Karagandy, SE Kazakhstan
Balkhash, Lake see Balkhash, Ozero
Balkhash, Ozero 92 C5 Eng. Lake Balkhash, Kaz. Balqash. lake SE Kazakhstan
Balladonia 125 C6 Western Australia
Ballarat 127 C7 Victoria, SE Australia
Balleny Islands 132 B5 island group Antarctica
Ballinger 27 F3 Texas, SW USA
Balochistān see Baluchistān
Balqash see Balkhash/Balkhash, Ozero
Balş 86 B5 Olt, S Romania
Balsas 41 F2 Maranhão, E Brazil
Balsas, Río 29 E5 var. Río Mexcala. river S Mexico
Bal'shavik 85 D7 Rus. Bol'shevik. Homyel'skaya Voblasts', SE Belarus
Balta 86 D3 Odes'ka Oblast', SW Ukraine
Bălţi 86 D3 Rus. Bel'tsy. N Moldova
Baltic Port see Paldiski
Baltic Sea 63 C7 Ger. Ostee, Rus. Baltiskoye More. sea N Europe
Baltimore 19 F4 Maryland, NE USA
Baltischport/Baltiski see Paldiski
Baltiskoye More see Baltic Sea
Baltkrievija see Belarus
Baluchistān 112 B3 var. Balochistān, Beluchistan. province SW Pakistan
Balvi 84 D4 Balvi, NE Latvia
Balykchy 101 G2 Kir. Ysyk-Köl; prev. Issyk-Kul', Rybach'ye. Issyk-Kul'skaya Oblast', NE Kyrgyzstan
Balzers 72 E2 S Liechtenstein
Bam 98 E4 Kermān, SE Iran
Bamako 52 D4 country capital (Mali) Capital District, SW Mali
Bambari 54 C4 Ouaka, C Central African Republic
Bamberg 73 C5 Bayern, SE Germany
Bamenda 54 A4 Nord-Ouest, W Cameroon
Banaba 122 D2 var. Ocean Island. island Tungaru, W Kiribati
Banaras see Vārānasi
Bandaaceh 116 A3 var. Banda Atjeh; prev. Koetaradja, Kutaradja, Kutaraja. Sumatera, W Indonesia
Banda Atjeh see Bandaaceh
Banda, Laut see Banda Sea
Bandama 52 D5 var. Bandama Fleuve. river S Côte d'Ivoire
Bandama Fleuve see Bandama
Bandar 'Abbās see Bandar-e 'Abbās
Bandarbeyla 51 E5 var. Bender Beila, Bender Beyla. Bari, NE Somalia
Bandar-e 'Abbās 98 D4 var. Bandar 'Abbās; prev. Gombroon. Hormozgān, S Iran
Bandar-e Būshehr 98 C4 var. Būshehr, Eng. Bushire. Būshehr, S Iran
Bandar-e Kangān 98 D4 var. Kangān. Būshehr, S Iran
Bandar-e Khamīr 98 D4 Hormozgān, S Iran
Bandar-e Langeh see Bandar-e Lengeh
Bandar-e Lengeh 98 D4 var. Bandar-e Langeh, Lingeh. Hormozgān, S Iran
Bandar Kassim see Boosaaso
Bandar Lampung 116 C4 var. Bandarlampung, Tanjungkarang-Telukbetung; prev. Tandjoengkarang, Tanjungkarang, Teloekbetoeng, Telukbetung. Sumatera, W Indonesia
Bandarlampung see Bandar Lampung
Bandar Maharani see Muar
Bandar Masulipatnam see Machilīpatnam
Bandar Penggaram see Batu Pahat
Bandar Seri Begawan 116 D3 prev. Brunei Town. country capital (Brunei) N Brunei
Banda Sea 117 F5 var. Laut Banda. sea E Indonesia
Bandiagara 53 E3 Mopti, C Mali
Bandırma 94 A3 var. Penderma. Balıkesir, NW Turkey
Bandjarmasin see Banjarmasin
Bandoeng see Bandung
Bandundu 55 C6 prev. Banningville. Bandundu, W Dem. Rep. Congo
Bandung 116 C5 prev. Bandoeng. Jawa, C Indonesia
Bangalore 110 C2 var. Bengalooru. state capital Karnātaka, S India
Bangassou 54 D4 Mbomou, SE Central African Republic
Banggai, Kepulauan 117 E4 island group C Indonesia
Banghāzī 49 G2 Eng. Bengazi, Benghazi, It. Bengasi. NE Libya
Bangka, Pulau 116 C4 island W Indonesia
Bangkok see Ao Krung Thep
Bangkok, Bight of see Krung Thep, Ao
Bangladesh 113 G3 off. People's Republic of Bangladesh; prev. East Pakistan. country S Asia
Bangladesh, People's Republic of see Bangladesh
Bangor 67 C6 NW Wales, United Kingdom
Bangor 67 B5 Ir. Beannchar. E Northern Ireland, United Kingdom
Bangor 19 G2 Maine, NE USA
Bang Pla Soi see Chon Buri
Bangui 55 B5 country capital (Central African Republic) Ombella-Mpoko, SW Central African Republic
Bangweulu, Lake 51 B8 var. Lake Bengweulu. lake N Zambia
Ban Hat Yai see Hat Yai
Ban Hin Heup 114 C4 Viangchan, C Laos
Ban Houayxay/Ban Houei Sai see Houayxay
Ban Hua Hin 115 C6 var. Hua Hin. Prachuap Khiri Khan, SW Thailand
Bani 52 D3 river S Mali
Banias see Bāniyās
Banijska Palanka see Glina
Banī Suwayf 50 B2 var. Beni Suef. N Egypt
Bāniyās 96 B3 var. Banias, Baniyas, Paneas. Tarṭūs, W Syria
Banjak, Kepulauan see Banyak, Kepulauan
Banja Luka 78 B3 Republika Srpska, NW Bosnia and Herzegovina
Banjarmasin 116 D4 prev. Bandjarmasin. Borneo, C Indonesia
Banjul 52 B3 prev. Bathurst. country capital (Gambia) W Gambia
Banks, Îles see Banks Islands
Banks Island 15 E2 island Northwest Territories, NW Canada
Banks Islands 122 D4 Fr. Îles Banks. island group N Vanuatu
Banks Lake 24 B1 reservoir Washington, NW USA
Banks Peninsula 129 C6 peninsula South Island, New Zealand
Banks Strait 127 C8 strait SW Tasman Sea
Bānkura 113 F4 West Bengal, NE India
Ban Mak Khaeng see Udon Thani
Banmo see Bhamo
Banningville see Bandundu
Bañolas see Banyoles
Ban Pak Phanang see Pak Phanang
Ban Sichon see Sichon
Banská Bystrica 77 C6 Ger. Neusohl, Hung. Besztercebánya. Banskobystricky Kraj, C Slovakia
Bantry Bay 67 A7 Ir. Bá Bheanntraí. bay SW Ireland
Banya 82 E2 Burgas, E Bulgaria
Banyak, Kepulauan 116 A3 prev. Kepulauan Banjak. island group NW Indonesia
Banyo 54 B4 Adamaoua, NW Cameroon
Banyoles 71 G2 var. Bañolas. Cataluña, NE Spain
Banzare Seamounts 119 C7 seamount range S Indian Ocean
Banzart see Bizerte
Baoji 106 B4 var. Pao-chi, Paoki. Shaanxi, C China
Baoro 54 B4 Nana-Mambéré, W Central African Republic
Baoshan 106 A6 var. Pao-shan. Yunnan, SW China
Baotou 105 F3 var. Pao-t'ou, Paotow. Nei Mongol Zizhiqu, N China
Ba'qūbah 98 B3 var. Qubba. Diyālá, C Iraq
Baquerizo Moreno see Puerto Baquerizo Moreno
Bar 79 C5 It. Antivari. S Montenegro
Baraawe 51 D6 It. Brava. Shabeellaha Hoose, S Somalia
Bārāmati 112 C5 Mahārāshtra, W India
Baranavichy 85 B6 Pol. Baranowicze, Rus. Baranovichi. Brestskaya Voblasts', SW Belarus
Baranovichi/Baranowicze see Baranavichy
Barbados 33 G1 country SE West Indies
Barbastro 71 F2 Aragón, NE Spain
Barbate de Franco 70 C5 Andalucía, S Spain
Barbuda 33 G3 island N Antigua and Barbuda
Barcaldine 126 C4 Queensland, E Australia
Barcarozsnyó see Râşnov
Barcău see Berettyó
Barce see Al Marj
Barcelona 71 G2 anc. Barcino, Barcinona. Cataluña, E Spain
Barcelona 37 E2 Anzoátegui, NE Venezuela
Barcino/Barcinona see Barcelona
Barcoo see Cooper Creek
Barcs 77 C7 Somogy, SW Hungary
Bardaï 54 C1 Borkou-Ennedi-Tibesti, N Chad
Bardejov 77 D5 Ger. Bartfeld, Hung. Bártfa. Presovský Kraj, E Slovakia
Bardera/Bardere see Baardheere
Barduli see Barletta
Bareilly 113 E3 var. Bareli. Uttar Pradesh, N India
Bareli see Bareilly
Barendrecht 64 C4 Zuid-Holland, SW Netherlands
Barentin 68 C3 Seine-Maritime, N France
Barentsburg 61 G2 Spitsbergen, W Svalbard
Barentsevo More/Barents Havet see Barents Sea
Barentsøya 61 G2 island E Svalbard
Barents Sea 88 C2 Nor. Barents Havet, Rus. Barentsevo More. sea Arctic Ocean
Bar Harbor 19 H2 Mount Desert Island, Maine, NE USA
Bari 75 E5 var. Bari delle Puglie; anc. Barium. Puglia, SE Italy
Bāridah see Al Bāridah
Bari delle Puglie see Bari
Barikot see Barīkowṯ
Barīkowṯ 101 F4 var. Barikot. Konar, NE Afghanistan
Barillas 30 A2 var. Santa Cruz Barillas. Huehuetenango, NW Guatemala
Barinas 36 C2 Barinas, W Venezuela
Barisal 113 G4 Barisal, S Bangladesh
Barisan, Pegunungan 116 B4 mountain range Sumatera, W Indonesia
Barito, Sungai 116 D4 river Borneo, C Indonesia
Barium see Bari
Barka see Al Marj
Barkly Tableland 126 B3 plateau Northern Territory/Queensland, N Australia
Bârlad 86 D4 prev. Bîrlad. Vaslui, E Romania
Barlavento, Ilhas de 52 A2 var. Windward Islands. island group N Cape Verde
Bar-le-Duc 68 D3 var. Bar-sur-Ornain. Meuse, NE France
Barlee, Lake 125 B6 lake Western Australia
Barlee Range 124 A4 mountain range Western Australia
Barletta 75 D5 anc. Barduli. Puglia, SE Italy
Barlinek 76 B3 Ger. Berlinchen. Zachodnio-pomorskie, NW Poland
Barmen-Elberfeld see Wuppertal
Barmouth 67 C6 NW Wales, United Kingdom
Barnaul 92 D4 Altayskiy Kray, C Russian Federation
Barnet 67 A7 United Kingdom
Barnstaple 67 C7 SW England, United Kingdom
Baroda see Vadodara
Baroghil Pass 101 F3 var. Kowtal-e Barowghīl. pass Afghanistan/Pakistan
Baron'ki 85 E7 Rus. Boron'ki. Mahilyowskaya Voblasts', E Belarus
Barowghīl, Kowtal-e see Baroghil Pass
Barquisimeto 36 C2 Lara, NW Venezuela
Barra 66 B3 island NW Scotland, United Kingdom
Barra de Río Grande 31 E3 Región Autónoma Atlántico Sur, E Nicaragua
Barranca 38 C3 Lima, W Peru
Barrancabermeja 36 B2 Santander, N Colombia
Barranquilla 36 B1 Atlántico, N Colombia
Barreiro 70 B4 Setúbal, W Portugal
Barrier Range 127 C6 hill range New South Wales, SE Australia
Barrow 14 D2 Alaska, USA
Barrow 67 B6 Ir. An Bhearú. river SE Ireland
Barrow-in-Furness 67 C5 NW England, United Kingdom
Barrow Island 124 A4 island Western Australia
Barstow 25 C7 California, W USA
Bar-sur-Ornain see Bar-le-Duc
Bartang 101 F3 river SE Tajikistan
Bartenstein see Bartoszyce
Bártfa/Bartfeld see Bardejov
Bartica 37 F3 N Guyana
Bartın 94 C2 Bartın, NW Turkey
Bartlesville 27 G1 Oklahoma, C USA
Bartoszyce 76 D2 Ger. Bartenstein. Warmińsko-mazurskie, NE Poland
Baruun-Urt 105 F2 Sühbaatar, E Mongolia
Barú, Volcán 31 E5 var. Volcán de Chiriquí. volcano W Panama
Barwon River 127 D5 river New South Wales, SE Australia
Barysaw 85 D6 Rus. Borisov. Minskaya Voblasts', NE Belarus
Basarabeasca 86 D4 Rus. Bessarabka. SE Moldova
Basel 73 A7 Eng. Basle, Fr. Bâle. Basel-Stadt, NW Switzerland
Basilan 117 E3 island Sulu Archipelago, SW Philippines
Basle see Basel
Basra see Al Başrah
Bassano del Grappa 74 C2 Veneto, NE Italy
Bassein see Pathein
Basseterre 33 G4 country capital (Saint Kitts and Nevis) Saint Kitts, Saint Kitts and Nevis
Basse-Terre 33 G3 dependent territory capital (Guadeloupe) Basse Terre, SW Guadeloupe
Basse Terre 33 G4 island W Guadeloupe
Bassikounou 52 D3 Hodh ech Chargui, SE Mauritania
Bass, Îlots de see Marotiri
Bass Strait 127 C7 strait SE Australia
Bassum 72 B3 Niedersachsen, NW Germany
Bastia 69 E7 Corse, France, C Mediterranean Sea
Bastogne 65 D7 Luxembourg, SE Belgium
Bastrop 20 B2 Louisiana, S USA
Bastyn' 85 B7 Rus. Bostyn'. Brestskaya Voblasts', SW Belarus
Basuo see Dongfang
Basutoland see Lesotho
Bata 55 A5 NW Equatorial Guinea
Batae Coritanorum see Leicester
Batajnica 78 D3 Vojvodina, N Serbia
Batangas 117 E2 off. Batangas City. Luzon, N Philippines
Batangas City see Batangas
Batavia see Jakarta
Bătdâmbâng 115 C5 prev. Battambang. Bătdâmbâng, NW Cambodia
Batéké, Plateaux 55 B6 plateau S Congo
Bath 67 D7 hist. Akermanceaster; anc. Aquae Calidae, Aquae Solis. SW England, United Kingdom
Bathinda 112 D2 Punjab, NW India
Bathsheba 33 G1 E Barbados
Bathurst 127 D6 New South Wales, SE Australia
Bathurst 17 F4 New Brunswick, SE Canada
Bathurst see Banjul
Bathurst Island 124 D2 island Northern Territory, N Australia
Bathurst Island 15 F2 island Parry Islands, Nunavut, N Canada
Batin, Wadi al 98 C4 dry watercourse SW Asia
Batman 95 E4 var. İluh. Batman, SE Turkey
Batna 49 E2 NE Algeria
Baton Rouge 20 B3 state capital Louisiana, S USA
Batroûn 96 A4 var. Al Batrūn. N Lebanon
Battambang see Bătdâmbâng
Batticaloa 110 D3 Eastern Province, E Sri Lanka
Battipaglia 75 D5 Campania, S Italy
Battle Born State see Nevada
Batumi 95 F2 prev. Bat'umi. W Georgia
Bat'umi see Batumi
Batu Pahat 116 B3 prev. Bandar Penggaram. Johor, Peninsular Malaysia
Bauchi 53 G4 Bauchi, NE Nigeria
Bauer Basin 131 F3 undersea basin E Pacific Ocean
Bauska 84 C3 Ger. Bauske. Bauska, S Latvia
Bauske see Bauska
Bautzen 72 D4 Lus. Budyšin. Sachsen, E Germany
Bauzanum see Bolzano
Bavaria see Bayern
Bavarian Alps 73 C7 Ger. Bayrische Alpen. mountain range Austria/Germany
Bavière see Bayern
Bavispe, Río 28 C2 river NW Mexico
Bawīţī 50 B2 var. Bawîţi. N Egypt
Bawîţi see Bawīţī
Bawku 53 E4 N Ghana
Bayamo 32 C3 Granma, E Cuba
Bayan Har Shan 104 D4 var. Bayan Khar. mountain range C China
Bayanhongor 104 D2 Bayanhongor, C Mongolia
Bayan Khar see Bayan Har Shan
Bayano, Lago 31 G4 lake E Panama
Bay City 18 C3 Michigan, N USA
Bay City 27 G4 Texas, SW USA
Baydhabo 51 D6 var. Baydhowa, Isha Baydhabo, It. Baidoa. Bay, SW Somalia
Baydhowa see Baydhabo
Bayern 73 C6 Eng. Bavaria, Fr. Bavière. state SE Germany
Bayeux 68 B3 anc. Augustodurum. Calvados, N France
Bāyir 97 C7 var. Bā'ir. Ma'ān, S Jordan
Bay Islands 30 C1 Eng. Bay Islands. island group N Honduras
Bay Islands see Bahía, Islas de la
Baymak 89 D6 Respublika Bashkortostan, W Russian Federation
Bayonne 69 A6 anc. Lapurdum. Pyrénées-Atlantiques, SW France
Bayou State see Mississippi
Bayram-Ali see Baýramaly
Baýramaly 100 D3 var. Bayramaly; prev. Bayram-Ali. Mary Welaýaty, S Turkmenistan
Bayreuth 73 C5 var. Baireuth. Bayern, SE Germany
Bayrische Alpen see Bavarian Alps
Bayrūt see Beyrouth
Bay State see Massachusetts
Baysun see Boysun
Bayt Laḩm see Bethlehem
Baytown 27 H4 Texas, SW USA
Baza 71 E4 Andalucía, S Spain
Bazargic see Dobrich
Bazin see Pezinok
Beagle Channel 43 C8 channel Argentina/Chile
Béal Feirste see Belfast
Beannchar see Bangor, Northern Ireland, UK
Bear Island see Bjørnøya
Bear Lake 24 E4 lake Idaho/Utah, NW USA
Beas de Segura 71 E4 Andalucía, S Spain
Beata, Isla 33 E3 island SW Dominican Republic
Beatrice 23 F4 Nebraska, C USA
Beaufort Sea 14 D2 sea Arctic Ocean
Beaufort-Wes see Beaufort West
Beaufort West 56 C5 Afr. Beaufort-Wes. Western Cape, SW South Africa
Beaumont 27 H3 Texas, SW USA
Beaune 68 D4 Côte d'Or, C France
Beauvais 68 C3 anc. Bellovacum, Caesaromagus. Oise, N France
Beaver Island 18 C2 island Michigan, N USA
Beaver Lake 27 H1 reservoir Arkansas, C USA
Beaver River 27 F1 river Oklahoma, C USA
Beaver State see Oregon
Beāwar 112 C3 Rājasthān, N India
Bečej 78 D3 Ger. Altbetsche, Hung. Óbecse, Rácz-Becse; prev. Magyar-Becse, Stari Bečej. Vojvodina, N Serbia
Béchar 48 D2 prev. Colomb-Béchar. W Algeria
Beckley 18 D5 West Virginia, NE USA
Bécs see Wien
Bedford 67 D6 E England, United Kingdom
Bedum 64 E1 Groningen, NE Netherlands
Beehive State see Utah
Be'er Menuha 97 B7 prev. Be'ér Menuḫa. Southern, S Israel
Be'ér Menuḫa see Be'er Menuha
Beernem 65 A5 West-Vlaanderen, NW Belgium
Beersheba see Be'er Sheva
Be'er Sheva 97 A7 var. Beersheba, Ar. Bir es Saba; prev. Be'ér Sheva'. Southern, S Israel
Be'ér Sheva' see Be'er Sheva
Beesel 65 D5 Limburg, SE Netherlands
Beeville 27 G4 Texas, SW USA
Bega 127 D7 New South Wales, SE Australia
Begoml' see Byahoml'
Begovat see Bekobod
Behagle see Laï
Behar see Bihār
Beibu Wan see Tonkin, Gulf of
Beida see Al Bayḑā'
Beihai 106 B6 Guangxi Zhuangzu Zizhiqu, S China
Beijing 106 C3 var. Pei-ching, Peking; prev. Pei-p'ing. country capital (China) Beijing Shi, E China
Beilen 64 E2 Drenthe, NE Netherlands
Beira 57 E3 Sofala, C Mozambique
Beirut see Beyrouth
Beit Lekhem see Bethlehem
Beiuş 86 B3 Hung. Belényes. Bihor, NW Romania
Beja 70 B4 anc. Pax Julia. Beja, SE Portugal
Béjar 70 C3 Castilla-León, N Spain
Bejraburi see Phetchaburi
Bekabad see Bekobod
Békás see Bicaz
Bek-Budi see Qarshi
Békéscsaba 77 D7 Rom. Bichiş-Ciaba. Békés, SE Hungary
Bekobod 101 E2 Rus. Bekabad; prev. Begovat. Toshkent Viloyati, E Uzbekistan
Bela Crkva 78 E3 Ger. Weisskirchen, Hung. Fehértemplom. Vojvodina, W Serbia
Belarus 85 B6 off. Republic of Belarus, var. Belorussia, Latv. Baltkrievija; prev. Belorussian SSR, Rus. Belorusskaya SSR. country E Europe
Belarus, Republic of see Belarus
Belau see Palau
Belaya Tserkov' see Bila Tserkva
Bełchatów 76 C4 var. Belchatow. Łódzski, C Poland
Belchatow see Bełchatów
Belcher, Îles see Belcher Islands
Belcher Islands 16 C2 Fr. Îles Belcher. island group Nunavut, SE Canada
Beledweyne 51 D5 var. Belet Huen, It. Belet Uen. Hiiraan, C Somalia
Belém 41 F1 var. Pará. state capital Pará, N Brazil
Belén 30 D4 Rivas, SW Nicaragua
Belen 26 D2 New Mexico, SW USA
Belényes see Beiuş
Belet Huen/Belet Uen see Beledweyne
Belfast 67 B5 Ir. Béal Feirste. national capital E Northern Ireland, United Kingdom
Belfield 22 D2 North Dakota, N USA
Belfort 68 E4 Territoire-de-Belfort, E France
Belgard see Białogard
Belgaum 110 B1 Karnātaka, W India
Belgian Congo see Congo (Democratic Republic of)
België/Belgique see Belgium
Belgium 65 B6 off. Kingdom of Belgium, Dut. België, Fr. Belgique. country NW Europe
Belgium, Kingdom of see Belgium
Belgorod 89 A6 Belgorodskaya Oblast', W Russian Federation
Belgrano II 132 A2 Argentinian research station Antarctica
Belice see Belize/Belize City
Beligrad see Berat
Beli Manastir 78 C3 Hung. Pélmonostor; prev. Monostor. Osijek-Baranja, NE Croatia
Bélinga 55 B5 Ogooué-Ivindo, NE Gabon
Belitung, Pulau 116 C4 island W Indonesia
Belize 30 B1 Sp. Belice; prev. British Honduras, Colony of Belize. country Central America
Belize 30 B1 river Belize/Guatemala
Belize see Belize City
Belize City 30 C1 var. Belize, Sp. Belice. Belize, NE Belize
Belize, Colony of see Belize
Beljak see Villach
Belkofski 14 B3 Alaska, USA
Belle Île 68 A4 island NW France
Belle Isle, Strait of 17 G3 strait Newfoundland and Labrador, E Canada
Bellenz see Bellinzona
Belleville 18 B4 Illinois, N USA
Bellevue 23 F4 Iowa, C USA
Bellevue 24 B2 Washington, NW USA
Bellingham 24 B1 Washington, NW USA
Belling Hausen Mulde see Southeast Pacific Basin
Bellingshausen Abyssal Plain see Bellingshausen Plain
Bellingshausen Plain 131 F5 var. Bellingshausen Abyssal Plain. abyssal plain SE Pacific Ocean
Bellingshausen Sea 132 A3 sea Antarctica
Bellinzona 73 B8 Ger. Bellenz. Ticino, S Switzerland
Bello 36 B2 Antioquia, W Colombia
Bello Horizonte see Belo Horizonte
Bellovacum see Beauvais
Bellville 56 B5 Western Cape, SW South Africa
Belmopan 30 C1 country capital (Belize) Cayo, C Belize
Belogradchik 82 B1 Vidin, NW Bulgaria
Belo Horizonte 41 F4 prev. Bello Horizonte. state capital Minas Gerais, SE Brazil
Belomorsk 88 B3 Respublika Kareliya, NW Russian Federation
Beloretsk 89 D6 Respublika Bashkortostan, W Russian Federation
Belorussia/Belorussian SSR see Belarus
Belorusskaya Gryada see Byelaruskaya Hrada
Belorusskaya SSR see Belarus
Beloshchel'ye see Nar'yan-Mar
Belostok see Białystok
Belovár see Bjelovar
Beloye More 88 C3 Eng. White Sea. sea NW Russian Federation
Belozërsk 88 B4 Vologodskaya Oblast', NW Russian Federation
Belton 27 G3 Texas, SW USA
Bel'tsy see Bălţi
Beluchistan see Baluchistān
Belukha, Gora 92 D5 mountain Kazakhstan/Russian Federation
Belynichi see Byalynichy
Belyy, Ostrov 92 D2 island N Russian Federation
Bemaraha 57 F3 var. Plateau du Bemaraha. mountain range W Madagascar
Bemaraha, Plateau du see Bemaraha
Bemidji 23 F1 Minnesota, N USA
Bemmel 64 D4 Gelderland, SE Netherlands
Benaco see Garda, Lago di
Benares see Vārānasi
Benavente 70 D2 Castilla-León, N Spain
Bend 24 B3 Oregon, NW USA
Bender see Tighina
Bender Beila/Bender Beyla see Bandarbeyla
Bender Cassim/Bender Qaasim see Boosaaso
Bendern 72 E1 NW Liechtenstein Europe
Bendery see Tighina
Bendigo 127 C7 Victoria, SE Australia
Beneschau see Benešov
Beneški Zaliv see Venice, Gulf of
Benešov 77 B5 Ger. Beneschau. Středočeský Kraj, W Czech Republic
Benevento 75 D5 anc. Beneventum, Malventum. Campania, S Italy
Beneventum see Benevento
Bengal, Bay of 102 C4 bay N Indian Ocean
Bengalooru see Bangalore
Bengasi see Banghāzī
Bengazi see Banghāzī
Bengbu 106 D5 var. Peng-pu. Anhui, E China
Benghazi see Banghāzī
Bengkulu 116 B4 prev. Bengkoeloe, Benkoelen, Benkulen. Sumatera, W Indonesia
Benguela 56 A2 var. Benguella. Benguela, W Angola
Benguella see Benguela
Bengweulu, Lake see Bangweulu, Lake
Ben Hope 66 B2 mountain N Scotland, United Kingdom
Beni 55 E5 Nord-Kivu, NE Dem. Rep. Congo
Benidorm 71 F4 País Valenciano, SE Spain
Beni-Mellal 48 C2 C Morocco
Benin 53 F4 off. Republic of Benin; prev. Dahomey. country W Africa
Benin, Bight of 53 F5 gulf W Africa
Benin City 53 F5 Edo, SW Nigeria
Benin, Republic of see Benin
Beni, Río 39 E3 river N Bolivia
Beni Suef see Banī Suwayf
Ben Nevis 66 C3 mountain N Scotland, United Kingdom
Bénoué see Benue
Benson 26 B3 Arizona, SW USA
Bent Jbaïl 97 A5 var. Bint Jubayl. S Lebanon
Benton 20 B1 Arkansas, C USA
Benue 54 B4 Fr. Bénoué. river Cameroon/Nigeria
Beograd 78 D3 Eng. Belgrade. Serbia, N Serbia
Berane 79 D5 prev. Ivangrad. E Montenegro
Berat 79 C6 var. Berati, SCr. Beligrad. Berat, C Albania
Berătău see Berettyó
Berati see Berat
Berau, Teluk 117 G4 var. MacCluer Gulf. bay Papua, E Indonesia
Berbera 50 D4 Sahil, NW Somalia
Berbérati 55 B5 Mambéré-Kadéï, SW Central African Republic
Berck-Plage 68 C2 Pas-de-Calais, N France
Berdichev see Berdychiv
Berdyans'k 87 G4 Rus. Berdyansk; prev. Osipenko. Zaporiz'ka Oblast', SE Ukraine
Berdychiv 86 D2 Rus. Berdichev. Zhytomyrs'ka Oblast', N Ukraine
Beregovo/Beregszász see Berehove
Berehove 86 B3 Cz. Berehovo, Hung. Beregszász, Rus. Beregovo. Zakarpats'ka Oblast', W Ukraine
Berehovo see Berehove
Bereket 100 B2 prev. Rus. Gazandzhyk, Kazandzhik, Turkm. Gazanjyk. Balkan Welaýaty, W Turkmenistan
Beretău see Berettyó
Berettyó 77 D7 Rom. Barcău; prev. Berătău, Beretău. river Hungary/Romania
Berettyóújfalu 77 D6 Hajdú-Bihar, E Hungary
Berezhany 86 C2 Pol. Brzeżany. Ternopil's'ka Oblast', W Ukraine

Bragança – Cape Verde

Bragança *70 C2 Eng.* Braganza; *anc.* Julio Briga. Bragança, NE Portugal
Braganza *see* Bragança
Brahestad *see* Raahe
Brahmanbaria *113 G4* Chittagong, E Bangladesh
Brahmapur *113 F5* Odisha, E India
Brahmaputra *113 H3 var.* Padma, Tsangpo, *Ben.* Jamuna, *Chin.* Yarlung Zangbo Jiang, *Ind.* Bramaputra, Dihang, Siang. *river* S Asia
Brăila *86 D4* Brăila, E Romania
Braine-le-Comte *65 B6* Hainaut, SW Belgium
Brainerd *23 F2* Minnesota, N USA
Brak *see* Birāk
Bramaputra *see* Brahmaputra
Brampton *16 D5* Ontario, S Canada
Branco, Rio *34 C3 river* N Brazil
Brandberg *56 A3 mountain* NW Namibia
Brandenburg *72 C3 var.* Brandenburg an der Havel. Brandenburg, NE Germany
Brandenburg an der Havel *see* Brandenburg
Brandon *15 F5* Manitoba, S Canada
Braniewo *76 D2 Ger.* Braunsberg. Warmińsko-mazurskie, N Poland
Brasil *see* Brazil
Brasília *41 F3 country capital* (Brazil) Distrito Federal, C Brazil
Brasil, República Federativa do *see* Brazil
Braşov *86 C4 Ger.* Kronstadt, *Hung.* Brassó; *prev.* Oraşul Stalin. Braşov, C Romania
Brassó *see* Braşov
Bratislava *77 C6 Ger.* Pressburg, *Hung.* Pozsony. *country capital* (Slovakia) Bratislavský Kraj, W Slovakia
Bratsk *93 E4* Irkutskaya Oblast', C Russian Federation
Brattia *see* Brač
Braunsberg *see* Braniewo
Braunschweig *72 C4 Eng./Fr.* Brunswick. Niedersachsen, N Germany
Brava *see* Baraawe
Brava, Costa *71 H2 coastal region* NE Spain
Bravo del Norte, Río/Bravo, Río *see* Grande, Rio
Bravo, Río *28 C1 river* Mexico/USA North America
Brawley *25 D8* California, W USA
Brazil *40 C2 off.* Federative Republic of Brazil, *Port.* República Federativa do Brasil, *Sp.* Brasil; *prev.* United States of Brazil. *country* South America
Brazil Basin *45 C5 var.* Brazilian Basin, Brazil'skaya Kotlovina. *undersea basin* W Atlantic Ocean
Brazil, Federative Republic of *see* Brazil
Brazilian Basin *see* Brazil Basin
Brazilian Highlands *see* Central, Planalto
Brazil'skaya Kotlovina *see* Brazil Basin
Brazil, United States of *see* Brazil
Brazos River *27 G3 river* Texas, SW USA
Brazza *see* Brač
Brazzaville *55 B6 country capital* (Congo) Capital District, S Congo
Brčko *78 C3* Republika Srpska, NE Bosnia and Herzegovina
Brecht *65 C5* Antwerpen, N Belgium
Brecon Beacons *67 C6 mountain range* S Wales, United Kingdom
Breda *64 C4* Noord-Brabant, S Netherlands
Bree *65 D5* Limburg, NE Belgium
Bregalnica *79 E6 river* E FYR Macedonia
Bregenz *35 B7 anc.* Brigantium. Vorarlberg, W Austria
Bregovo *82 B1* Vidin, NW Bulgaria
Bremen *72 B3 Fr.* Brême. Bremen, NW Germany
Bremerhaven *72 B3* Bremen, NW Germany
Bremerton *24 B2* Washington, NW USA
Brenham *27 G3* Texas, SW USA
Brenner, Col du/Brennero, Passo del *see* Brenner Pass
Brenner Pass *74 C1 var.* Brenner Sattel, *Fr.* Col du Brenner, *Ger.* Brennerpass, *It.* Passo del Brennero. *pass* Austria/Italy
Brennerpass *see* Brenner Pass
Brenner Sattel *see* Brenner Pass
Brescia *74 B2 anc.* Brixia. Lombardia, N Italy
Breslau *see* Wrocław
Bressanone *74 C1 Ger.* Brixen. Trentino-Alto Adige, N Italy
Brest *85 A6 Pol.* Brześć nad Bugiem, *Rus.* Brest-Litovsk; *prev.* Brześć Litewski. Brestskaya Voblasts', SW Belarus
Brest *68 A3* Finistère, NW France
Brest-Litovsk *see* Brest
Bretagne *68 A3 Eng.* Brittany, *Lat.* Britannia Minor. *cultural region* NW France
Brewster, Kap *see* Kangikajik
Brewton *20 C3* Alabama, S USA
Brezhnev *see* Naberezhnyye Chelny
Brezovo *82 D2 prev.* Abrashlare. Plovdiv, C Bulgaria
Bria *54 D4* Haute-Kotto, C Central African Republic
Briançon *69 D5 anc.* Brigantio. Hautes-Alpes, SE France
Bricgstow *see* Bristol
Bridgeport *19 F3* Connecticut, NE USA
Bridgetown *33 G2 country capital* (Barbados) SW Barbados
Bridlington *67 D5* E England, United Kingdom
Bridport *67 D7* S England, United Kingdom
Brieg *see* Brzeg
Brig *73 A7 Fr.* Brigue, *It.* Briga. Valais, SW Switzerland
Briga *see* Brig
Brigantio *see* Briançon
Brigantium *see* Bregenz
Brigham City *22 B3* Utah, W USA
Brighton *67 E7* SE England, United Kingdom
Brighton *22 D4* Colorado, C USA
Brigue *see* Brig
Brindisi *75 E5 anc.* Brundisium, Brundusium. Puglia, SE Italy
Briovera *see* St-Lô
Brisbane *127 E5 state capital* Queensland, E Australia
Bristol *67 D7 anc.* Bricgstow. SW England, United Kingdom
Bristol *19 F3* Connecticut, NE USA
Bristol *18 D5* Tennessee, S USA
Bristol Bay *14 B3 bay* Alaska, USA
Bristol Channel *67 C7 inlet* England/Wales, United Kingdom
Britain *58 C3 var.* Great Britain. *island* United Kingdom
Britannia Minor *see* Bretagne
British Columbia *14 D4 Fr.* Colombie-Britannique. *province* SW Canada
British Guiana *see* Guyana
British Honduras *see* Belize
British Indian Ocean Territory *119 B5 UK dependent territory* C Indian Ocean
British Isles *67 island group* NW Europe
British North Borneo *see* Sabah
British Solomon Islands Protectorate *see* Solomon Islands
British Virgin Islands *33 F3 var.* Virgin Islands. *UK dependent territory* E West Indies
Brittany *see* Bretagne
Briva Curretia *see* Brive-la-Gaillarde
Briva Isarae *see* Pontoise
Brive *see* Brive-la-Gaillarde
Brive-la-Gaillarde *69 C5 prev.* Brive; *anc.* Briva Curretia. Corrèze, C France
Brixen *see* Bressanone
Brixia *see* Brescia
Brno *77 B5 Ger.* Brünn. Jihomoravský Kraj, SE Czech Republic
Broceni *84 B3* Saldus, SW Latvia
Brod/Bród *see* Slavonski Brod
Brodeur Peninsula *15 F2 peninsula* Baffin Island, Nunavut, NE Canada
Brod na Savi *see* Slavonski Brod
Brodnica *76 C3 Ger.* Buddenbrock. Kujawski-pomorskie, C Poland
Broek-in-Waterland *64 C3* Noord-Holland, C Netherlands
Broken Arrow *27 G1* Oklahoma, C USA
Broken Bay *126 E1 bay* New South Wales, SE Australia
Broken Hill *127 B6* New South Wales, SE Australia
Broken Ridge *119 D6 undersea plateau* S Indian Ocean
Bromberg *see* Bydgoszcz
Bromley *67 B8* United Kingdom
Brookhaven *20 B3* Mississippi, S USA
Brookings *23 F3* South Dakota, N USA
Brooks Range *14 D2 mountain range* Alaska, USA
Brookton *125 B6* Western Australia
Broome *124 B3* Western Australia
Broomfield *22 D4* Colorado, C USA
Broucsella *see* Brussel/Bruxelles
Brovary *87 E2* Kyyivs'ka Oblast', N Ukraine
Brownfield *27 E2* Texas, SW USA
Brownsville *27 G5* Texas, SW USA
Brownwood *27 F3* Texas, SW USA
Brozha *85 D7* Mahilyowskaya Voblasts', E Belarus
Bruges *see* Brugge
Brugge *65 A5 Fr.* Bruges. West-Vlaanderen, NW Belgium
Brummen *64 D3* Gelderland, E Netherlands
Brundisium/Brundusium *see* Brindisi
Brunei *116 D3 off.* Brunei Darussalam, *Mal.* Negara Brunei Darussalam. *country* SE Asia
Brunei Darussalam *see* Brunei
Brunei Town *see* Bandar Seri Begawan
Brünn *see* Brno
Brunner, Lake *129 C5 lake* South Island, New Zealand
Brunswick *21 E3* Georgia, SE USA
Brunswick *see* Braunschweig
Brusa *see* Bursa
Brus Laguna *30 D2* Gracias a Dios, E Honduras
Brussa *see* Bursa
Brussel *65 C6 var.* Brussels, *Fr.* Bruxelles, *Ger.* Brüssel; *anc.* Broucsella. *country capital* (Belgium) Brussels, C Belgium
Brüssel/Brussels *see* Brussel/Bruxelles
Brüx *see* Most
Bruxelles *see* Brussel
Bryan *27 G3* Texas, SW USA
Bryansk *89 A5* Bryanskaya Oblast', W Russian Federation
Brzeg *76 C4 Ger.* Brieg; *anc.* Civitas Altae Ripae. Opolskie, S Poland
Brześć Litewski/Brześć nad Bugiem *see* Brest
Brzeżany *see* Berezhany
Bucaramanga *36 B2* Santander, N Colombia
Buchanan *52 C5 prev.* Grand Bassa. SW Liberia
Buchanan, Lake *27 F3 reservoir* Texas, SW USA
Bucharest *see* Bucureşti
Buckeye State *see* Ohio
Bu Craa *see* Bou Craa
Bucureşti *86 C5 Eng.* Bucharest, *Ger.* Bukarest, *prev.* Altenburg; *anc.* Cetatea Damboviţei. *country capital* (Romania) Bucureşti, S Romania
Buda-Kashalyova *85 D7 Rus.* Buda-Koshelëvo. Homyel'skaya Voblasts', SE Belarus
Buda-Koshelëvo *see* Buda-Kashalyova
Budapest *77 C6 off.* Budapest Főváros, *SCr.* Budimpešta. *country capital* (Hungary) Pest, N Hungary
Budapest Főváros *see* Budapest
Budaun *112 D3* Uttar Pradesh, N India
Buddenbrock *see* Brodnica
Budimpešta *see* Budapest
Budweis *see* České Budějovice
Budyšin *see* Bautzen
Buena Park *24 E2* California, W USA North America
Buenaventura *36 A3* Valle del Cauca, W Colombia
Buena Vista *39 G4* Santa Cruz, C Bolivia
Buena Vista *71 H5* S Gibraltar Europe
Buenos Aires *42 D4 hist.* Santa Maria del Buen Aire. *country capital* (Argentina) Buenos Aires, E Argentina
Buenos Aires *31 E5* Puntarenas, SE Costa Rica
Buenos Aires, Lago *43 B6 var.* Lago General Carrera. *lake* Argentina/Chile
Buffalo *19 E3* New York, NE USA
Buffalo Narrows *15 F4* Saskatchewan, C Canada
Buff Bay *32 B5* E Jamaica
Buftea *86 C5* Ilfov, S Romania
Bug *59 E3 Bel.* Zakhodni Buh, *Eng.* Western Bug, *Rus.* Zapadnyy Bug, *Ukr.* Zakhidnyy Buh. *river* E Europe
Buga *36 B3* Valle del Cauca, W Colombia
Bughotu *see* Santa Isabel
Buguruslan *89 D6* Orenburgskaya Oblast', W Russian Federation
Buitenzorg *see* Bogor
Bujalance *70 D4* Andalucía, S Spain
Bujanovac *79 E5* SE Serbia
Bujnurd *see* Bojnūrd
Bujumbura *51 B7 prev.* Usumbura. *country capital* (Burundi) W Burundi
Bukarest *see* Bucureşti
Bukavu *55 E6 prev.* Costermansville. Sud-Kivu, E Dem. Rep. Congo
Bukhara *see* Buxoro
Bukoba *51 B6* Kagera, NW Tanzania
Bülach *73 B7* Zürich, NW Switzerland
Bulawayo *56 D3* Matabeleland North, SW Zimbabwe
Bulgan *105 E2* Bulgan, N Mongolia
Bulgaria *82 C2 off.* Republic of Bulgaria, *Bul.* Bŭlgariya; *prev.* People's Republic of Bulgaria. *country* SE Europe
Bulgaria, People's Republic of *see* Bulgaria
Bulgaria, Republic of *see* Bulgaria
Bŭlgariya *see* Bulgaria
Bullion State *see* Missouri
Bull Shoals Lake *20 B1 reservoir* Arkansas/Missouri, C USA
Bulukumba *117 E4 prev.* Boeloekoemba. Sulawesi, C Indonesia
Bumba *55 D5* Equateur, N Dem. Rep. Congo
Bunbury *125 A7* Western Australia
Bundaberg *126 E4* Queensland, E Australia
Bungo-suido *109 B7 strait* SW Japan
Bunia *55 E5* Orientale, NE Dem. Rep. Congo
Bünyan *94 D3* Kayseri, C Turkey
Buraida *see* Buraydah
Buraydah *98 B4 var.* Buraida. Al Qaşīm, N Saudi Arabia
Burdigala *see* Bordeaux
Burdur *94 B4 var.* Buldur. Burdur, SW Turkey
Burdur Gölü *94 B4 salt lake* SW Turkey
Burë *50 C4* Āmara, N Ethiopia
Burgas *82 E2 var.* Bourgas. Burgas, E Bulgaria
Burgaski Zaliv *82 E2 gulf* E Bulgaria
Burgos *70 D2* Castilla-León, N Spain
Burgundy *see* Bourgogne
Burhan Budai Shan *104 D4 mountain range* C China
Buriram *115 D5 var.* Buri Ram, Puriramya. Buri Ram, E Thailand
Buri Ram *see* Buriram
Burjassot *71 F3* País Valenciano, E Spain
Burkburnett *27 F2* Texas, SW USA
Burketown *126 B3* Queensland, NE Australia
Burkina *see* Burkina Faso
Burkina Faso *53 E4 off.* Burkina Faso; *var.* Burkina; *prev.* Upper Volta. *country* W Africa
Burley *24 D4* Idaho, NW USA
Burlington *23 G4* Iowa, C USA
Burlington *19 F2* Vermont, NE USA
Burma *see* Myanmar
Burnie *127 C8* Tasmania, SE Australia
Burns *24 C3* Oregon, NW USA
Burnside *15 F3 river* Nunavut, NW Canada
Burnsville *23 F2* Minnesota, N USA
Burrel *79 D6 var.* Burreli. Dibër, C Albania
Burreli *see* Burrel
Burriana *see* Borriana
Bursa *94 B3 var.* Brussa, *prev.* Brusa; *anc.* Prusa. Bursa, NW Turkey
Būr Sa'īd *50 B1 var.* Port Said. N Egypt
Burtnieks *84 C3 var.* Burtnieks Ezers. *lake* N Latvia
Burtnieks Ezers *see* Burtnieks
Burundi *51 B7 off.* Republic of Burundi; *prev.* Kingdom of Burundi, Urundi. *country* C Africa
Burundi, Kingdom of *see* Burundi
Burundi, Republic of *see* Burundi
Buru, Pulau *117 F4 prev.* Boeroe. *island* E Indonesia
Busan *107 E4 off.* Busan Gwang-yeoksi, *prev.* Pusan, *Jap.* Fusan. SE South Korea
Busan Gwang-yeoksi *see* Busan
Buşayrah *96 D3* Dayr az Zawr, E Syria
Büshehr/Bushire *see* Bandar-e Büshehr
Busra *see* Al Başrah, Iraq
Busselton *125 A7* Western Australia
Bussora *see* Al Başrah
Buta *55 D5* Orientale, N Dem. Rep. Congo
Butembo *55 E5* Nord-Kivu, NE Dem. Rep. Congo
Butler *19 E4* Pennsylvania, NE USA
Buton, Pulau *117 E4 var.* Pulau Butung; *prev.* Boetoeng. *island* C Indonesia
Bütow *see* Bytów
Butte *22 B2* Montana, NW USA
Butterworth *116 B3* Pinang, Peninsular Malaysia
Button Islands *17 E1 island group* Nunavut, NE Canada
Butuan *117 F2 off.* Butuan City. Mindanao, S Philippines
Butuan City *see* Butuan
Butung, Pulau *see* Buton, Pulau
Butuntum *see* Bitonto
Buulobarde *51 D5 var.* Buulo Berde. Hiiraan, C Somalia
Buulo Berde *see* Buulobarde
Buur Gaabo *51 D6* Jubbada Hoose, S Somalia
Buxoro *100 D2 var.* Bokhara, *Rus.* Bukhara. Buxoro Viloyati, C Uzbekistan
Buynaksk *89 B8* Respublika Dagestan, SW Russian Federation
Büyükağrı Dağı *95 F3 var.* Aghri Dagh, Agri Dagi, Koh I Noh, Masis, *Eng.* Great Ararat, Mount Ararat. *mountain* E Turkey
Büyükmenderes Nehri *94 A4 river* SW Turkey
Buzău *86 C4* Buzău, SE Romania
Būzmeyin *see* Abadan
Buzuluk *89 D6* Orenburgskaya Oblast', W Russian Federation
Byahoml' *85 D5 Rus.* Begoml'. Vitsyebskaya Voblasts', N Belarus
Byalynichy *85 D6 Rus.* Belynichi. Mahilyowskaya Voblasts', E Belarus
Byan Tumen *see* Choybalsan
Byarezina *85 D6 prev.* Byerezino, *Rus.* Berezina. *river* C Belarus
Bydgoszcz *76 C3 Ger.* Bromberg. Kujawski-pomorskie, C Poland
Byelaruskaya Hrada *85 B6 Rus.* Belorusskaya Gryada. *ridge* N Belarus
Byerezino *see* Byarezina
Byron Island *see* Nikunau
Bystrovka *see* Kemin
Bytča *77 C5* Žilinský Kraj, N Slovakia
Bytom *77 C5 Ger.* Beuthen. Śląskie, S Poland
Bytów *76 C2 Ger.* Bütow. Pomorskie, N Poland
Byuzmeyin *see* Abadan
Byval'ki *85 D8* Homyel'skaya Voblasts', SE Belarus
Byzantium *see* İstanbul

C

Caála *56 B2 var.* Kaala, Robert Williams, *Port.* Vila Robert Williams. Huambo, C Angola
Caazapá *42 D3* Caazapá, S Paraguay
Caballo Reservoir *26 C3 reservoir* New Mexico, SW USA
Cabanaquinta *70 D1 var.* Cabañaquinta. Asturias, N Spain
Cabañaquinta *see* Cabanaquinta
Cabanatuan *117 E1 off.* Cabanatuan City. Luzon, N Philippines
Cabanatuan City *see* Cabanatuan
Cabillonum *see* Chalon-sur-Saône
Cabimas *36 C1* Zulia, NW Venezuela
Cabinda *56 A1 var.* Kabinda. Cabinda, NW Angola
Cabinda *56 A1 var.* Kabinda. *province* NW Angola
Cahora Bassa, Albufeira de *56 D2 var.* Lake Cabora Bassa. *reservoir* NW Mozambique
Cabora Bassa, Lake *see* Cahora Bassa, Albufeira de
Caborca *28 B1* Sonora, NW Mexico
Cabot Strait *17 G4 strait* E Canada
Cabo Verde, Ilhas do *see* Cape Verde
Cabras, Ilha das *54 E2 island* S Sao Tome and Principe, Africa, E Atlantic Ocean
Cabrera, Illa de *71 G3 island* E Spain
Cáceres *70 C3 Ar.* Qazris. Extremadura, W Spain
Cachimbo, Serra do *41 E2 mountain range* C Brazil
Caconda *56 B2* Huíla, C Angola
Čadca *77 C5 Hung.* Csaca. Žilinský Kraj, N Slovakia
Cadillac *18 C2* Michigan, N USA
Cadiz *117 E2 off.* Cadiz City. Negros, C Philippines
Cádiz *70 C5 anc.* Gades, Gadier, Gadir, Gadire. Andalucía, SW Spain
Cadiz City *see* Cadiz
Cádiz, Golfo de *70 B5 Eng.* Gulf of Cadiz. *gulf* Portugal/Spain
Cadiz, Gulf of *see* Cádiz, Golfo de
Cadurcum *see* Cahors
Caen *68 B3* Calvados, N France
Caene/Caenepolis *see* Qinā
Caerdydd *see* Cardiff
Caer Glou *see* Gloucester
Caer Gybi *see* Holyhead
Caerleon *see* Chester
Caer Luel *see* Carlisle
Caesaraugusta *see* Zaragoza
Caesarea Mazaca *see* Kayseri
Caesarobriga *see* Talavera de la Reina
Caesarodunum *see* Tours
Caesaromagus *see* Beauvais
Caesena *see* Cesena
Cafayate *42 C2* Salta, N Argentina
Cagayan de Oro *117 E2 off.* Cagayan de Oro City. Mindanao, S Philippines
Cagayan de Oro City *see* Cagayan de Oro
Cagliari *75 A6 anc.* Caralis. Sardegna, Italy, C Mediterranean Sea
Caguas *33 F3* E Puerto Rico
Cahors *69 C5 anc.* Cadurcum. Lot, S France
Cahul *86 D4 Rus.* Kagul. S Moldova
Caicos Passage *32 D2 strait* Bahamas/Turks and Caicos Islands
Caiffa *see* Hefa
Cailungo *74 E1* N San Marino
Caiphas *see* Hefa
Cairns *126 D3* Queensland, NE Australia
Cairo *50 B2 var.* El Qâhira, *Ar.* Al Qāhirah. *country capital* (Egypt) N Egypt
Caisleán an Bharraigh *see* Castlebar
Cajamarca *38 B3 prev.* Caxamarca. Cajamarca, NW Peru
Čakovec *78 B2 Ger.* Csakathurn, *Hung.* Csáktornya; *prev. Ger.* Tschakathurn. Medimurje, N Croatia
Calabar *53 G5* Cross River, S Nigeria
Calabozo *36 D2* Guárico, C Venezuela
Calafat *86 B5* Dolj, SW Romania
Calafate *see* El Calafate
Calahorra *71 E2* La Rioja, N Spain
Calais *68 C2* Pas-de-Calais, N France
Calais *19 H2* Maine, NE USA
Calais, Pas de *see* Dover, Strait of
Calama *42 B2* Antofagasta, N Chile
Cālāras *see* Călăraşi
Călăraşi *86 D3 var.* Cālāras, *Rus.* Kalarash. C Moldova
Călăraşi *86 C5* Călăraşi, SE Romania
Calatayud *71 E2* Aragón, NE Spain
Calbayog *117 E2 off.* Calbayog City. Samar, C Philippines
Calbayog City *see* Calbayog
Calcutta *see* Kolkata
Caldas da Rainha *70 B3* Leiria, W Portugal
Caldera *42 B3* Atacama, N Chile
Caldwell *24 C3* Idaho, NW USA
Caledonia *30 C1* Corozal, N Belize
Caleta Olivia *43 B6* Santa Cruz, SE Argentina
Calgary *15 E5* Alberta, SW Canada
Cali *36 B3* Valle del Cauca, W Colombia
Calicut *see* Kozhikode
California *25 B7 off.* State of California, *also known as* El Dorado, The Golden State. *state* W USA
California, Golfo de *28 B2 Eng.* Gulf of California; *prev.* Sea of Cortez. *gulf* W Mexico
California, Gulf of *see* California, Golfo de
Călimăneşti *86 B4* Vâlcea, SW Romania
Calisia *see* Kalisz
Callabonna, Lake *127 B5 lake* South Australia
Callao *38 C4* Callao, W Peru
Callatis *see* Mangalia
Callosa de Segura *71 F4* País Valenciano, E Spain
Calmar *see* Kalmar
Caloundra *127 E5* Queensland, E Australia
Caltanissetta *75 C7* Sicilia, Italy, C Mediterranean Sea
Caluula *50 E4* Bari, NE Somalia
Camabatela *56 B1* Cuanza Norte, NW Angola
Camacupa *56 B2 var.* General Machado, *Port.* Vila General Machado. Bié, C Angola
Camagüey *32 C2 prev.* Puerto Príncipe. Camagüey, C Cuba
Camagüey, Archipiélago de *32 C2 island group* C Cuba
Camana *39 E4 var.* Camaná. Arequipa, SW Peru
Camargue *69 D6 physical region* SE France
Ca Mau *115 D6 var.* Quan Long. Minh Hai, S Vietnam
Cambay, Gulf of *see* Khambhāt, Gulf of
Camberia *see* Chambéry
Cambodia *115 D5 off.* Kingdom of Cambodia, *var.* Democratic Kampuchea, Roat Kampuchea, *Cam.* Kampuchea; *prev.* People's Democratic Republic of Kampuchea. *country* SE Asia
Cambodia, Kingdom of *see* Cambodia
Cambrai *68 C2 Flem.* Kambryk, *prev.* Cambray; *anc.* Cameracum. Nord, N France
Cambray *see* Cambrai
Cambrian Mountains *67 C6 mountain range* C Wales, United Kingdom
Cambridge *32 A4* W Jamaica
Cambridge *128 D3* Waikato, North Island, New Zealand
Cambridge *67 E6 Lat.* Cantabrigia. E England, United Kingdom
Cambridge *19 F4* Maryland, NE USA
Cambridge *18 D4* Ohio, NE USA
Cambridge Bay *15 F3 var.* Ikaluktutiak. Victoria Island, Nunavut, NW Canada
Camden *20 B2* Arkansas, C USA
Camellia State *see* Alabama
Cameracum *see* Cambrai
Cameroon *54 A4 off.* Republic of Cameroon, *Fr.* Cameroun. *country* W Africa
Cameroon, Republic of *see* Cameroon
Cameroun *see* Cameroon
Camocim *41 F2* Ceará, E Brazil
Camopi *37 H3* E French Guiana
Campamento *30 C2* Olancho, C Honduras
Campania *75 D5 Eng.* Champagne. *region* S Italy
Campbell, Cape *129 D5 headland* South Island, New Zealand
Campbell Island *120 D5 island* S New Zealand
Campbell Plateau *120 D5 undersea plateau* SW Pacific Ocean
Campbell River *14 D5* Vancouver Island, British Columbia, SW Canada
Campeche *29 G4* Campeche, SE Mexico
Campeche, Bahía de *29 F4 Eng.* Bay of Campeche. *bay* E Mexico
Campeche, Bay of *see* Campeche, Bahía de
Câm Pha *114 E3* Quang Ninh, N Vietnam
Câmpina *86 C4 prev.* Cîmpina. Prahova, SE Romania
Campina Grande *41 G2* Paraíba, E Brazil
Campinas *41 F4* São Paulo, S Brazil
Campobasso *75 D5* Molise, C Italy
Campo Criptana *see* Campo de Criptana
Campo de Criptana *71 E3 var.* Campo Criptana. Castilla-La Mancha, C Spain
Campo dos Goytacazes *41 F4 var.* Campos. Rio de Janeiro, SE Brazil
Campo Grande *41 E4 state capital* Mato Grosso do Sul, SW Brazil
Campos *see* Campo dos Goytacazes
Câmpulung *86 B4 prev.* Câmpulung-Muşcel, Cîmpulung. Argeş, S Romania
Câmpulung-Muşcel *see* Câmpulung
Campus Stellae *see* Santiago de Compostela
Cam Ranh *115 E6* Khanh Hoa, S Vietnam
Canada *12 B4 country* N North America
Canada Basin *12 C2 undersea basin* Arctic Ocean
Canadian River *27 E2 river* SW USA
Çanakkale *94 A3 var.* Dardanelli; *prev.* Chanak, Kale Sultanie. Çanakkale, W Turkey
Cananea *28 B1* Sonora, NW Mexico
Canarreos, Archipiélago de los *32 B2 island group* W Cuba
Canary Islands *48 A2 Eng.* Canary Islands. *island group* Spain, NE Atlantic Ocean
Canary Islands *see* Canarias, Islas
Cañas *30 D4* Guanacaste, NW Costa Rica
Canaveral, Cape *21 E4 headland* Florida, SE USA
Canavieiras *41 G3* Bahia, E Brazil
Canberra *120 C4 country capital* (Australia) Australian Capital Territory, SE Australia
Cancún *29 H3* Quintana Roo, SE Mexico
Candia *see* Irákleio
Canea *see* Chaniá
Cangzhou *106 D4* Hebei, E China
Caniapiscau *17 E2 river* Québec, E Canada
Caniapiscau, Réservoir de *16 D3 reservoir* Québec, C Canada
Canik Dağları *94 D2 mountain range* N Turkey
Canillo *69 A7* Canillo, C Andorra Europe
Çankırı *94 C3 var.* Chankiri; *anc.* Gangra, Germanicopolis. Çankın, N Turkey
Cannanore *see* Kannur
Cannes *69 D6* Alpes-Maritimes, SE France
Canoas *41 E5* Rio Grande do Sul, S Brazil
Canon City *22 C5* Colorado, C USA
Cantabria *70 D1 autonomous community* N Spain
Cantábrica, Cordillera *70 C1 mountain range* N Spain
Cantabrigia *see* Cambridge
Cantaura *37 E2* Anzoátegui, NE Venezuela
Canterbury *67 E7 hist.* Cantwaraburh; *anc.* Durovernum, *Lat.* Cantuaria. SE England, United Kingdom
Canterbury Bight *129 C6 bight* South Island, New Zealand
Canterbury Plains *129 C6 plain* South Island, New Zealand
Cân Thơ *115 E6* Cân Thơ, S Vietnam
Canton *20 B2* Mississippi, S USA
Canton *18 D4* Ohio, N USA
Canton *see* Guangzhou
Canton Island *see* Kanton
Cantuaria/Cantwaraburh *see* Canterbury
Canyon *27 E2* Texas, SW USA
Cao Bằng *114 D3 var.* Caobang. Cao Bằng, N Vietnam
Caobang *see* Cao Bằng
Cap-Breton, Île du *see* Cape Breton Island
Cape Barren Island *127 C8 island* Furneaux Group, Tasmania, SE Australia
Cape Basin *47 B7 undersea basin* S Atlantic Ocean
Cape Breton Island *17 G4 Fr.* Île du Cap-Breton. *island* Nova Scotia, SE Canada
Cape Charles *19 F5* Virginia, NE USA
Cape Coast *53 E5 prev.* Cape Coast Castle. S Ghana
Cape Coast Castle *see* Cape Coast
Cape Girardeau *23 H5* Missouri, C USA
Capelle aan den IJssel *64 C4* Zuid-Holland, SW Netherlands
Cape Palmas *see* Harper
Cape Saint Jacques *see* Vung Tau
Cape Town *56 B5 var.* Ekapa, *Afr.* Kaapstad, Kapstad. *country capital* (South Africa-legislative capital) Western Cape, SW South Africa
Cape Verde *52 A2 off.* Republic of Cape Verde, *Port.* Cabo Verde, Ilhas do Cabo Verde. *country* E Atlantic Ocean

China – Cowan, Lake

China 102 C2 off. People's Republic of China, Chin. Chung-hua Jen-min Kung-ho-kuo, Zhonghua Renmin Gongheguo; prev. Chinese Empire. country E Asia
Chi-nan/Chinan see Jinan
Chinandega 30 C3 Chinandega, NW Nicaragua
China, People's Republic of see China
China, Republic of see Taiwan
Chincha Alta 38 D4 Ica, SW Peru
Chin-chiang see Quanzhou
Chin-chou/Chinchow see Jinzhou
Chindwin see Chindwinn
Chindwinn 114 B2 var. Chindwin. river N Myanmar (Burma)
Chinese Empire see China
Ch'ing Hai see Qinghai Hu, China
Chinghai see Qinghai
Chingola 56 D2 Copperbelt, C Zambia
Ching-Tao/Ch'ing-tao see Qingdao
Chinguețți 52 C2 var. Chinguetti. Adrar, C Mauritania
Chin Hills 114 A3 mountain range W Myanmar (Burma)
Chinhsien see Jinzhou
Chinnereth see Tiberias, Lake
Chinook Trough 91 H4 trough N Pacific Ocean
Chioggia 74 C2 anc. Fossa Claudia. Veneto, NE Italy
Chíos 83 D5 var. Hios, Khíos, It. Scio, Turk. Sakiz-Adasi. Chíos, E Greece
Chíos 83 D5 var. Khíos. island E Greece
Chipata 56 D2 prev. Fort Jameson. Eastern, E Zambia
Chiquián 38 C3 Ancash, W Peru
Chiquimula 30 B2 Chiquimula, SE Guatemala
Chīrāla 110 D1 Andhra Pradesh, E India
Chirchik see Chirchiq
Chirchiq 101 E2 Rus. Chirchik. Toshkent Viloyati, E Uzbekistan
Chiriqui Gulf 31 E5 Eng. Chiriqui Gulf. gulf SW Panama
Chiriqui Gulf see Chiriquí, Golfo de
Chiriquí, Laguna de 31 E5 lagoon NW Panama
Chiriquí, Volcán de see Barú, Volcán
Chirripó, Cerro see Chirripó Grande, Cerro
Chirripó Grande, Cerro 30 D4 var. Cerro Chirripó. mountain SE Costa Rica
Chisec 30 B2 Alta Verapaz, C Guatemala
Chisholm 23 F1 Minnesota, N USA
Chisimaio/Chisimayu see Kismaayo
Chişinău 86 D4 Rus. Kishinev. country capital (Moldova) C Moldova
Chita 93 F4 Chitinskaya Oblast', S Russian Federation
Chitangwiza see Chitungwiza
Chitato 56 C1 Lunda Norte, NE Angola
Chitina 14 D3 Alaska, USA
Chitose 108 D2 var. Titose. Hokkaidō, NE Japan
Chitré 31 F5 Herrera, S Panama
Chittagong 113 G4 Ben. Chāttagām. Chittagong, SE Bangladesh
Chitungwiza 56 D3 prev. Chitangwiza. Mashonaland East, NE Zimbabwe
Chkalov see Orenburg
Chlef 48 D2 var. Ech Cheliff, Ech Chleff; prev. Al-Asnam, El Asnam, Orléansville. NW Algeria
Chocolate Mountains 25 D8 mountain range California, W USA
Chodorów see Khodoriv
Chodzież 76 C3 Wielkopolskie, C Poland
Choele Choel 43 C5 Río Negro, C Argentina
Choiseul 122 C3 var. Lauru. island NW Solomon Islands
Chojnice 76 C2 Ger. Konitz. Pomorskie, N Poland
Ch'ok'ē 50 C4 var. Choke Mountains. mountain range NW Ethiopia
Choke Mountains see Ch'ok'ē
Cholet 68 B4 Maine-et-Loire, NW France
Choluteca 30 C3 Choluteca, S Honduras
Choluteca, Río 30 C3 river SW Honduras
Choma 56 D2 Southern, S Zambia
Chomutov 76 A4 Ger. Komotau. Ústecký Kraj, NW Czech Republic
Chona 91 E2 river C Russian Federation
Chon Buri 115 C5 prev. Bang Pla Soi. Chon Buri, S Thailand
Chone 38 A1 Manabí, W Ecuador
Ch'ŏngjin 107 E3 NE North Korea
Chongqing 107 B5 var. Ch'ung-ching, Ch'ung-ch'ing, Chungking, Pahsien, Tchongking, Yuzhou. Chongqing Shi, C China
Chongqing Shi 107 B5 province C China
Chonnacht see Connaught
Chonos, Archipiélago de los 43 A6 island group S Chile
Chóra 83 D7 Kykládes, Greece, Aegean Sea
Chóra Sfakíon 83 C8 var. Sfákia. Kríti, Greece, E Mediterranean Sea
Chorne More see Black Sea
Chornomors'ke 87 E4 Rus. Chernomorskoye. Respublika Krym, S Ukraine
Chorokh/Chorokhi see Çoruh Nehri
Chortkiv 86 C2 Rus. Chortkov. Ternopil's'ka Oblast', W Ukraine
Chortkov see Chortkiv
Chorzów 77 C5 Ger. Königshütte; prev. Królewska Huta. Śląskie, S Poland
Chośebuz see Cottbus
Chōsen-kaikyō see Korea Strait
Chōshi 109 D5 var. Tyôsi. Chiba, Honshū, S Japan
Chosŏn-minjujuŭi-inmin-kanghwaguk see North Korea
Choszczno 76 B3 Ger. Arnswalde. Zachodnio-pomorskie, NW Poland
Chota Nagpur 113 E4 plateau N India
Choûm 52 C2 Adrar, C Mauritania
Choybalsan 105 F2 prev. Byan Tumen. Dornod, E Mongolia
Christchurch 129 C6 Canterbury, South Island, New Zealand
Christiana 32 B5 C Jamaica
Christiania see Oslo
Christiansand see Kristiansand
Christianshåb see Qasigiannguit
Christiansund see Kristiansund
Christmas Island 119 D5 Australian external territory E Indian Ocean
Christmas Island see Kiritimati
Christmas Ridge 121 E1 undersea ridge C Pacific Ocean
Chuan see Sichuan
Ch'uan-chou see Quanzhou
Chubek see Moskva
Chubut, Río 43 B6 river SE Argentina
Ch'u-chiang see Shaoguan
Chudskoye Ozero see Peipus, Lake
Chugoku-sanchi 109 B6 mountain range Honshū, SW Japan
Chuí see Chuy
Chukai see Cukai
Chukchi Plain 133 B2 abyssal plain Arctic Ocean
Chukchi Plateau 12 C2 undersea plateau Arctic Ocean
Chukchi Sea 12 B2 Rus. Chukotskoye More. sea Arctic Ocean
Chukotskoye More see Chukchi Sea
Chula Vista 25 C8 California, W USA
Chulucanas 38 B2 Piura, NW Peru
Chulym 92 D4 river C Russian Federation
Chumphon 115 C6 var. Jumporn. Chumphon, SW Thailand
Chuncheon 107 E4 prev. Ch'unch'ŏn, Jap. Shunsen. N South Korea
Ch'unch'ŏn see Chuncheon
Ch'ung-ch'ing/Ch'ung-ching see Chongqing
Chung-hua Jen-min Kung-ho-kuo see China
Chungking see Chongqing
Chunya 93 E3 river C Russian Federation
Chuquicamata 42 B2 Antofagasta, N Chile
Chuquisaca see Sucre
Chur 73 B7 Fr. Coire, It. Coira, Rmsch. Cuera, Quera; anc. Curia Rhaetorum. Graubünden, E Switzerland
Churchill 15 G4 Manitoba, C Canada
Churchill 16 B2 river Manitoba/Saskatchewan, C Canada
Churchill 17 F2 river Newfoundland and Labrador, E Canada
Chuska Mountains 26 C1 mountain range Arizona/New Mexico, SW USA
Chusovoy 89 D5 Permskaya Oblast', NW Russian Federation
Chust see Khust
Chuuk Islands 122 B2 var. Hogoley Islands; prev. Truk Islands. island group Caroline Islands, C Micronesia
Chuy 42 E4 var. Chuí. Rocha, E Uruguay
Chyhyryn 87 E2 Rus. Chigirin. Cherkas'ka Oblast', N Ukraine
Ciadîr-Lunga 86 D4 var. Ceadâr-Lunga, Rus. Chadyr-Lunga. S Moldova
Cide 94 C2 Kastamonu, N Turkey
Ciechanów 76 D3 prev. Zichenau. Mazowieckie, C Poland
Ciego de Ávila 32 C2 Ciego de Ávila, C Cuba
Ciénaga 36 B1 Magdalena, N Colombia
Cienfuegos 32 B2 Cienfuegos, C Cuba
Cieza 71 E4 Murcia, SE Spain
Cihanbeyli 94 C3 Konya, C Turkey
Cikobia 123 E4 prev. Thikombia. island N Fiji
Cilacap 116 C5 prev. Tjilatjap. Jawa, C Indonesia
Cill Airne see Killarney
Cill Chainnigh see Kilkenny
Cilli see Celje
Cill Mhantáin see Wicklow
Cimpina see Câmpina
Cîmpulung see Câmpulung
Cina Selatan, Laut see South China Sea
Cincinnati 18 C4 Ohio, N USA
Ciney 65 C7 Namur, SE Belgium
Cinto, Monte 69 E7 mountain Corse, France, C Mediterranean Sea
Cintra see Sintra
Cipolletti 43 B5 Río Negro, C Argentina
Cirebon 116 C4 prev. Tjirebon. Jawa, S Indonesia
Cirkvenica see Crikvenica
Cirò Marina 75 E6 Calabria, S Italy
Cirquenizza see Crikvenica
Cisnădie 86 B4 Ger. Heltau, Hung. Nagydisznód. Sibiu, SW Romania
Citharista see la Ciotat
Citlaltépetl see Orizaba, Volcán Pico de
Citrus Heights 25 B5 California, W USA
Ciudad Acuña see Villa Acuña
Ciudad Bolívar 37 E2 prev. Angostura. Bolívar, E Venezuela
Ciudad Camargo 28 D2 Chihuahua, N Mexico
Ciudad Cortés see Cortés
Ciudad Darío 30 D3 var. Dario. Matagalpa, W Nicaragua
Ciudad de Dolores Hidalgo see Dolores Hidalgo
Ciudad de Guatemala 30 B2 Eng. Guatemala City; prev. Santiago de los Caballeros. country capital (Guatemala) Guatemala, C Guatemala
Ciudad del Carmen see Carmen
Ciudad del Este 42 E2 prev. Ciudad Presidente Stroessner, Presidente Stroessner, Puerto Presidente Stroessner. Alto Paraná, SE Paraguay
Ciudad Delicias see Delicias
Ciudad de México see México
Ciudad de Panama see Panamá
Ciudad Guayana 37 E2 prev. San Tomé de Guayana, Santo Tomé de Guayana. Bolívar, NE Venezuela
Ciudad Guzmán 28 D4 Jalisco, SW Mexico
Ciudad Hidalgo 29 G5 Chiapas, SE Mexico
Ciudad Juárez 28 C1 Chihuahua, N Mexico
Ciudad Lerdo 28 D3 Durango, C Mexico
Ciudad Madero 29 E3 var. Villa Cecilia. Tamaulipas, C Mexico
Ciudad Mante 29 E3 Tamaulipas, C Mexico
Ciudad Miguel Alemán 29 E2 Tamaulipas, C Mexico
Ciudad Obregón 28 B2 Sonora, NW Mexico
Ciudad Ojeda 36 C1 Zulia, NW Venezuela
Ciudad Porfirio Díaz see Piedras Negras
Ciudad Presidente Stroessner see Ciudad del Este
Ciudad Quesada see Quesada
Ciudad Real 70 D3 Castilla-La Mancha, C Spain
Ciudad-Rodrigo 70 C3 Castilla-León, N Spain
Ciudad Trujillo see Santo Domingo
Ciudad Valles 29 E3 San Luis Potosí, C Mexico
Ciudad Victoria 29 E3 Tamaulipas, C Mexico
Ciutadella 71 H3 var. Ciutadella de Menorca. Menorca, Spain, W Mediterranean Sea
Ciutadella Ciutadella de Menorca see Ciutadella
Civitanova Marche 74 D3 Marche, C Italy
Civitas Altae Ripae see Brzeg
Civitas Carnutum see Chartres
Civitas Eburovicum see Évreux
Civitavecchia 74 C4 anc. Centum Cellae, Trajani Portus. Lazio, C Italy
Claremore 27 G1 Oklahoma, C USA
Clarence 129 C5 Canterbury, South Island, New Zealand
Clarence 129 C5 river South Island, New Zealand
Clarence Town 32 D2 Long Island, C Bahamas
Clarinda 23 F4 Iowa, C USA
Clarion Fracture Zone 131 E2 tectonic feature NE Pacific Ocean
Clarión, Isla 28 A5 island W Mexico
Clark Fork 22 A1 river Idaho/Montana, NW USA
Clark Hill Lake 21 E2 var. J.Storm Thurmond Reservoir. reservoir Georgia/South Carolina, SE USA
Clarksburg 18 D4 West Virginia, NE USA
Clarksdale 20 B2 Mississippi, S USA
Clarksville 20 C1 Tennessee, S USA
Clausentum see Southampton
Clayton 27 E1 New Mexico, SW USA
Clearwater 21 E4 Florida, SE USA
Clearwater Mountains 24 D2 mountain range Idaho, NW USA
Cleburne 27 G3 Texas, SW USA
Clermont 126 D4 Queensland, E Australia
Clermont-Ferrand 69 C5 Puy-de-Dôme, C France
Cleveland 18 D3 Ohio, N USA
Cleveland 20 D1 Tennessee, S USA
Clifton 26 C2 Arizona, SW USA
Clinton 20 B2 Mississippi, S USA
Clinton 27 F1 Oklahoma, C USA
Clipperton Fracture Zone 131 E3 tectonic feature E Pacific Ocean
Clipperton Island 13 A7 French dependency of French Polynesia E Pacific Ocean
Cloncurry 126 B3 Queensland, C Australia
Clonmel 67 B6 Ir. Cluain Meala. S Ireland
Cloppenburg 72 B3 Niedersachsen, NW Germany
Cloquet 23 G2 Minnesota, N USA
Cloud Peak 22 C3 mountain Wyoming, C USA
Clovis 27 E2 New Mexico, SW USA
Cluain Meala see Clonmel
Cluj see Cluj-Napoca
Cluj-Napoca 86 B3 Ger. Klausenburg, Hung. Kolozsvár; prev. Cluj. Cluj, NW Romania
Clutha 129 B7 river South Island, New Zealand
Clyde 66 C4 river W Scotland, United Kingdom
Coari 40 D2 Amazonas, N Brazil
Coast Mountains 14 D4 Fr. Chaîne Côtière. mountain range Canada/USA
Coast Ranges 24 A4 mountain range W USA
Coats Island 15 G3 island Nunavut, NE Canada
Coats Land 132 B2 physical region Antarctica
Coatzacoalcos 29 G4 var. Quetzalcoalco; prev. Puerto México. Veracruz-Llave, E Mexico
Cobán 30 B2 Alta Verapaz, C Guatemala
Cobar 127 C6 New South Wales, SE Australia
Cobija 39 E3 Pando, NW Bolivia
Coblence/Coblenz see Koblenz
Coburg 73 C5 Bayern, SE Germany
Coca see Puerto Francisco de Orellana
Cocanada see Kākināda
Cochabamba 39 F4 hist. Oropeza. Cochabamba, C Bolivia
Cochin see Kochi
Cochinos, Bahía de 32 B2 Eng. Bay of Pigs. bay SE Cuba
Cochrane 16 C4 Ontario, S Canada
Cochrane 43 B7 Aisén, S Chile
Cocibolca see Nicaragua, Lago de
Cockade State see Maryland
Cockburn Town 33 E2 San Salvador, E Bahamas
Cockpit Country, The 32 A4 physical region W Jamaica
Cocobeach 55 A5 Estuaire, NW Gabon
Coconino Plateau 26 B1 plain Arizona, SW USA
Coco, Río 31 E2 var. Río Wanki, Segoviao Wangkí. river Honduras/Nicaragua
Cocos Basin 102 C5 undersea basin E Indian Ocean
Cocos Island Ridge see Cocos Ridge
Cocos Islands 119 D5 island group E Indian Ocean
Cocos Ridge 13 C8 var. Cocos Island Ridge. undersea ridge E Pacific Ocean
Cod, Cape 19 G3 headland Massachusetts, NE USA
Codfish Island 129 A8 island SW New Zealand
Codlea 86 C4 Ger. Zeiden, Hung. Feketehalom. Braşov, C Romania
Cody 22 C2 Wyoming, C USA
Coeur d'Alene 24 C2 Idaho, NW USA
Coevorden 64 E2 Drenthe, NE Netherlands
Coffs Harbour 127 E6 New South Wales, SE Australia
Cognac 69 B5 anc. Compniacum. Charente, W France
Cohalm see Rupea
Coiba, Isla de 31 E5 island SW Panama
Coihaique 43 B6 var. Coyhaique. Aisén, S Chile
Coimbatore 110 C3 Tamil Nādu, S India
Coimbra 70 B3 anc. Conimbria, Conimbriga. Coimbra, W Portugal
Coín 70 D5 Andalucía, S Spain
Coira/Coire see Chur
Coirib, Loch see Corrib, Lough
Colby 23 E4 Kansas, C USA
Colchester 67 E6 Connecticut, NE USA
Coleman 27 F3 Texas, SW USA
Coleraine 66 B4 Ir. Cúil Raithin. N Northern Ireland, United Kingdom
Colesberg 56 C5 Northern Cape, C South Africa
Colima 28 D4 Colima, S Mexico
Coll 66 B3 island W Scotland, United Kingdom
College Station 27 G3 Texas, SW USA
Collie 125 A7 Western Australia
Collipo see Leiria
Colmar 68 E4 Ger. Kolmar. Haut-Rhin, NE France
Cöln see Köln
Cologne see Köln
Colomb-Béchar see Béchar
Colombia 36 B3 off. Republic of Colombia. country N South America
Colombian Basin 34 A1 undersea basin SW Caribbean Sea
Colombia, Republic of see Colombia
Colombie-Britannique see British Columbia
Colombo 110 C4 country capital (Sri Lanka) Western Province, W Sri Lanka
Colón 31 G4 prev. Aspinwall. Colón, C Panama
Colón, Archipiélago de see Galápagos Islands
Colón Ridge 13 B8 undersea ridge E Pacific Ocean
Colorado 22 C4 off. State of Colorado, also known as Centennial State, Silver State. state C USA
Colorado City 27 F3 Texas, SW USA
Colorado Plateau 26 B1 plateau W USA
Colorado, Río 43 C5 river E Argentina
Colorado, Río see Colorado River
Colorado River 13 B5 var. Río Colorado. river Mexico/USA
Colorado River 27 G4 river Texas, SW USA
Colorado Springs 22 D5 Colorado, C USA
Columbia 19 E4 Maryland, NE USA
Columbia 23 G4 Missouri, C USA
Columbia 21 E2 state capital South Carolina, SE USA
Columbia 20 C1 Tennessee, S USA
Columbia River 24 B3 river Canada/USA
Columbia Plateau 24 C3 plateau Idaho/Oregon, NW USA
Columbus 20 D2 Georgia, SE USA
Columbus 18 C4 Indiana, N USA
Columbus 20 C2 Mississippi, S USA
Columbus 23 F4 Nebraska, C USA
Columbus 18 D4 state capital Ohio, N USA
Colville Channel 128 D2 channel North Island, New Zealand
Colville River 14 D2 river Alaska, USA
Comacchio 74 C3 var. Commachio; anc. Comactium. Emilia-Romagna, N Italy
Comactium see Comacchio
Comalcalco 29 G4 Tabasco, SE Mexico
Coma Pedrosa, Pic de 69 A7 mountain NW Andorra
Comarapa 39 F4 Santa Cruz, C Bolivia
Comayagua 30 C2 Comayagua, W Honduras
Comer See see Como, Lago di
Comilla 113 G4 Ben. Kumillā. Chittagong, E Bangladesh
Comino see Kemmuna
Comitán 29 G5 var. Comitán de Domínguez. Chiapas, SE Mexico
Comitán de Domínguez see Comitán
Commachio see Comacchio
Commissioner's Point 20 A5 headland W Bermuda
Communism Peak 101 F3 prev. Qullai Kommunizm. mountain E Tajikistan
Como 74 B2 anc. Comum. Lombardia, N Italy
Comodoro Rivadavia 43 B6 Chubut, SE Argentina
Como, Lake 74 B2 var. Lario, Eng. Lake Como, Ger. Comer See. lake N Italy
Como, Lake see Como, Lago di
Comores, République Fédérale Islamique des see Comoros
Comoros 57 F2 off. Federal Islamic Republic of the Comoros, Fr. République Fédérale Islamique des Comores. country W Indian Ocean
Comoros, Federal Islamic Republic of the see Comoros
Compiègne 68 C3 Oise, N France
Complutum see Alcalá de Henares
Compniacum see Cognac
Compostella see Santiago de Compostela
Comrat 86 D4 Rus. Komrat. S Moldova
Comum see Como
Conakry 52 C4 country capital (Guinea) SW Guinea
Conca see Cuenca
Concarneau 68 A3 Finistère, NW France
Concepción 39 G3 Santa Cruz, E Bolivia
Concepción 43 B5 Bío Bío, C Chile
Concepción 42 D2 var. Villa Concepción. Concepción, C Paraguay
Concepción see La Concepción
Concepción de la Vega see La Vega
Conchos, Río 26 D4 river NW Mexico
Conchos, Río 28 D2 river C Mexico
Concord 19 G3 state capital New Hampshire, NE USA
Concordia 42 D4 Entre Ríos, E Argentina
Concordia 23 E4 Kansas, C USA
Côn Đao see Côn Đao Son
Côn Đao Son 115 E7 var. Côn Đao, Con Son. island S Vietnam
Condate see Rennes, Ille-et-Vilaine, France
Condate see St-Claude, Jura, France
Condega 30 D3 Estelí, NW Nicaragua
Condivincum see Nantes
Confluentes see Koblenz
Công Hoa Xa Hôi Chu Nghia Viêt Nam see Vietnam
Congo 55 D5 off. Republic of the Congo, Fr. Moyen-Congo; prev. Middle Congo. country C Africa
Congo 55 C6 off. Democratic Republic of Congo; prev. Zaire, Belgian Congo, Congo (Kinshasa). country C Africa
Congo 55 C6 var. Kongo, Fr. Zaire. river C Africa
Congo Basin 55 C6 drainage basin W Dem. Rep. Congo
Congo/Congo (Kinshasa) see Congo (Democratic Republic of)
Coni see Cuneo
Conimbria/Conimbriga see Coimbra
Conjeeveram see Kānchīpuram
Connacht see Connaught
Connaught 67 A5 var. Connacht, Ir. Chonnacht, Cúige. province W Ireland
Connecticut 19 F3 off. State of Connecticut, also known as Blue Law State, Constitution State, Land of Steady Habits, Nutmeg State. state NE USA
Connecticut 19 G3 river Canada/USA
Conroe 27 G3 Texas, SW USA
Consentia see Cosenza
Consolación del Sur 32 A2 Pinar del Río, W Cuba
Con Son see Côn Đao Son
Constance see Konstanz
Constance, Lake 73 B7 Ger. Bodensee. lake C Europe
Constanța 86 D5 var. Küstendje, Eng. Constanza, Ger. Konstanza, Turk. Küstence. Constanța, SE Romania
Constantia see Coutances
Constantia see Konstanz
Constantine 49 E2 var. Qacentina, Ar. Qoussantîna. NE Algeria
Constantinople see İstanbul
Constantiola see Oltenița
Constanz see Konstanz
Constanza see Constanța
Constitution State see Connecticut
Coo see Kos
Coober Pedy 127 A5 South Australia
Cookeville 20 D1 Tennessee, S USA
Cook Islands 123 F4 territory in free association with New Zealand S Pacific Ocean
Cook, Mount see Aoraki
Cook Strait 129 D5 var. Raukawa. strait New Zealand
Cooktown 126 D2 Queensland, NE Australia
Coolgardie 125 B6 Western Australia
Cooma 127 D7 New South Wales, SE Australia
Coomassie see Kumasi
Coon Rapids 23 F2 Minnesota, N USA
Cooper Creek 126 C4 var. Barcoo, Cooper's Creek. seasonal river Queensland/ South Australia, Australia
Cooper's Creek see Cooper Creek
Coos Bay 24 A3 Oregon, NW USA
Cootamundra 127 D6 New South Wales, SE Australia
Copacabana 39 E4 La Paz, W Bolivia
Copenhagen see København
Copiapó 42 B3 Atacama, N Chile
Copperas Cove 27 G3 Texas, SW USA
Coppermine see Kugluktuk
Copper State see Arizona
Coquilhatville see Mbandaka
Coquimbo 42 B3 Coquimbo, N Chile
Corabia 86 B5 Olt, S Romania
Coral Harbour 15 G3 Southampton Island, Nunavut, NE Canada
Coral Sea 120 B3 sea SW Pacific Ocean
Coral Sea Islands 122 B4 Australian external territory SW Pacific Ocean
Corantijn Rivier see Courantyne River
Corcovado, Golfo 43 B6 gulf S Chile
Corcyra Nigra see Korčula
Cordele 20 D3 Georgia, SE USA
Córdoba 42 C3 Córdoba, C Argentina
Córdoba 29 F4 Veracruz-Llave, E Mexico
Córdoba 70 D4 var. Cordoba, Eng. Cordova; anc. Corduba. Andalucía, SW Spain
Cordova 14 C3 Alaska, USA
Cordova/Cordoba see Córdoba
Corduba see Córdoba
Corentyne River see Courantyne River
Corfu see Kérkyra
Coria 70 C3 Extremadura, W Spain
Corinth 20 C1 Mississippi, S USA
Corinth see Kórinthos
Corinth, Gulf of/Corinthiacus Sinus see Korinthiakós Kólpos
Corinthus see Kórinthos
Corinto 30 C3 Chinandega, NW Nicaragua
Cork 67 A6 Ir. Corcaigh. S Ireland
Çorlu 94 A2 Tekirdağ, NW Turkey
Corner Brook 17 G3 Newfoundland, Newfoundland and Labrador, E Canada
Cornhusker State see Nebraska
Corn Islands 31 E3 var. Corn Islands. island group SE Nicaragua
Corn Islands see Maíz, Islas del
Cornwallis Island 15 F2 Island Nunavut, N Canada
Coro 36 C1 prev. Santa Ana de Coro. Falcón, NW Venezuela
Corocoro 39 F4 La Paz, W Bolivia
Coromandel 128 D2 Waikato, North Island, New Zealand
Coromandel Coast 110 D2 coast E India
Coromandel Peninsula 128 D2 peninsula North Island, New Zealand
Coronado, Bahía de 30 D5 bay S Costa Rica
Coronel Dorrego 43 C5 Buenos Aires, E Argentina
Coronel Oviedo 42 D2 Caaguazú, SE Paraguay
Corozal 30 C1 Corozal, N Belize
Corpus Christi 27 G4 Texas, SW USA
Corrales 26 D2 New Mexico, SW USA
Corrib, Lough 67 A5 Ir. Loch Coirib. lake W Ireland
Corrientes 42 D3 Corrientes, NE Argentina
Corriza see Korçë
Corsica 69 E7 Eng. Corsica. island France, C Mediterranean Sea
Corsica see Corse
Corsicana 27 G3 Texas, SW USA
Cortegana 70 C4 Andalucía, S Spain
Cortés 31 E5 var. Ciudad Cortés. Puntarenas, SE Costa Rica
Cortez, Sea of see California, Golfo de
Cortina d'Ampezzo 74 C1 Veneto, NE Italy
Coruche 70 B3 Santarém, C Portugal
Çoruh Nehri 95 E3 Geor. Chorokh, Rus. Chorokhi. river Georgia/Turkey
Çorum 94 D3 var. Chorum. Çorum, N Turkey
Corunna see A Coruña
Corvallis 24 B3 Oregon, NW USA
Corvo 70 A5 var. Ilha do Corvo. island Azores, Portugal, NE Atlantic Ocean
Corvo, Ilha do see Corvo
Cos see Kos
Cosenza 75 D6 anc. Consentia. Calabria, SW Italy
Cosne-Cours-sur-Loire 68 C4 Nièvre, Bourgogne, C France Europe
Costa Mesa 24 D2 California, W USA North America
Costa Rica 31 E4 off. Republic of Costa Rica. country Central America
Costa Rica, Republic of see Costa Rica
Costermansville see Bukavu
Cotagaita 39 F5 Potosí, S Bolivia
Côte d'Ivoire 52 D4 off. République de la Côte d'Ivoire; Eng. Ivory Coast. country W Africa
Côte d'Ivoire, République de la see Côte d'Ivoire
Côte d'Or 68 D4 cultural region C France
Côte Française des Somalis see Djibouti
Côtière, Chaîne see Coast Mountains
Cotonou 53 F5 var. Kotonu. S Benin
Cotrone see Crotone
Cotswold Hills 67 D6 var. Cotswolds. hill range S England, United Kingdom
Cotswolds see Cotswold Hills
Cottbus 72 D4 Lus. Chośebuz; prev. Kottbus. Brandenburg, E Germany
Cotton State, The see Alabama
Cotyora see Ordu
Couentrey see Coventry
Council Bluffs 23 F4 Iowa, C USA
Courantyne River 37 G4 var. Corantijn Rivier, Corentyne River. river Guyana/Suriname
Courland Lagoon 84 A4 Ger. Kurisches Haff, Rus. Kurskiy Zaliv. lagoon Lithuania/Russian Federation
Courtrai see Kortrijk
Coutances 68 B3 anc. Constantia. Manche, N France
Couvin 65 C7 Namur, S Belgium
Coventry 67 D6 anc. Couentrey. C England, United Kingdom
Covilhã 70 C3 Castelo Branco, E Portugal
Cowan, Lake 125 B6 lake Western Australia

D

Djajapura – Ellen, Mount

Djajapura see Jayapura
Djakarta see Jakarta
Djakovo see Đakovo
Djambala 55 B6 Plateaux, C Congo
Djambi see Jambi
Djambi see Hari, Batang
Djanet 49 E4 prev. Fort Charlet. SE Algeria
Djéblé see Jablah
Djelfa 48 D2 var. El Djelfa. N Algeria
Djéma 54 D4 Haut-Mbomou, E Central African Republic
Djember see Jember
Djérablous see Jarābulus
Djerba 49 F2 var. Djerba, Jazīrat Jarbah. island E Tunisia
Djerba see Jerba, Île de
Djérem 54 B4 river C Cameroon
Djevdjelija see Gevgelija
Djibouti 50 D4 var. Jibuti. country capital (Djibouti) E Djibouti
Djibouti 50 D4 off. Republic of Djibouti, var. Jibuti; prev. French Somaliland, French Territory of the Afars and Issas, Fr. Côte Française des Somalis, Territoire Français des Afars et des Issas. country E Africa
Djibouti, Republic of see Djibouti
Djokjakarta see Yogyakarta
Djourab, Erg du 54 C2 desert N Chad
Djúpivogur 61 E5 Austurland, SE Iceland
Dmitriyevsk see Makiyivka
Dnepr see Dnieper
Dneprodzerzhinsk see Romaniv
Dneprodzerzhinskoye Vodokhranilishche see Dniprodzerzhyns'ke Vodoskhovyshche
Dnepropetrovsk see Dnipropetrovs'k
Dneprorudnoye see Dniprorudne
Dnestr see Dniester
Dnieper 59 F4 Bel. Dnyapro, Rus. Dnepr, Ukr. Dnipro. river E Europe
Dnieper Lowland 87 E2 Bel. Prydnyaprowskaya Nizina, Ukr. Prydniprovs'ka Nyzovyna. lowlands Belarus/Ukraine
Dniester 59 E4 Rom. Nistru, Rus. Dnestr, Ukr. Dnister; anc. Tyras. river Moldova/Ukraine
Dnipro see Dnieper
Dniprodzerzhyns'k see Romaniv
Dniprodzerzhyns'ke Vodoskhovyshche 87 F3 Rus. Dneprodzerzhinskoye Vodokhranilishche. reservoir C Ukraine
Dnipropetrovs'k 87 F3 Rus. Dnepropetrovsk; prev. Yekaterinoslav. Dnipropetrovs'ka Oblast', E Ukraine
Dniprorudne 87 F3 Rus. Dneprorudnoye. Zaporiz'ka Oblast', SE Ukraine
Dnister see Dniester
Dnyapro see Dnieper
Doba 54 C4 Logone-Oriental, S Chad
Döbeln 72 D4 Sachsen, E Germany
Doberai Peninsula 117 G4 Dut. Vogelkop. peninsula Papua, E Indonesia
Doboj 78 C3 Republiks Srpska, N Bosnia and Herzegovina
Dobre Miasto 76 D2 Ger. Guttstadt. Warmińsko-mazurskie, NE Poland
Dobrich 82 E1 Rom. Bazargic; prev. Tolbukhin. Dobrich, NE Bulgaria
Dobrush 85 D7 Homyel'skaya Voblasts', SE Belarus
Dobryn' see Dabryn'
Dodecanese see Dodekánisa
Dodekánisa 83 D6 var. Nóties Sporádes, Eng. Dodecanese; prev. Dhodhekánisos, Dodekanisos. island group SE Greece
Dodekanisos see Dodekánisa
Dodge City 23 E5 Kansas, C USA
Dodoma 47 D5 country capital (Tanzania) Dodoma, C Tanzania
Dogana 74 E1 NE San Marino Europe
Dogo 109 B6 island Oki-shotō, SW Japan
Dogondoutchi 53 F3 Dosso, SW Niger
Dogrular see Pravda
Doğubayazıt 95 F3 Ağrı, E Turkey
Doğu Karadeniz Dağları 95 E3 var. Anadolu Dağları. mountain range NE Turkey
Doha see Ad Dawḩah
Doire see Londonderry
Dokdo see Liancourt Rocks
Dokkum 64 D1 Friesland, N Netherlands
Dokuchayevs'k 87 G3 var. Dokuchayevsk. Donets'ka Oblast', SE Ukraine
Dokuchayevsk see Dokuchayevs'k
Doldrums Fracture Zone 44 C4 fracture zone W Atlantic Ocean
Dôle 68 D4 Jura, E France
Dolina see Dolyna
Dolinskaya see Dolyns'ka
Dolisie 55 B6 prev. Loubomo. Niari, S Congo
Dolna Oryakhovitsa 82 D2 prev. Polikrayshte. Veliko Tŭrnovo, N Bulgaria
Dolni Chiflik 82 E2 prev. Rudnik. Varna, E Bulgaria
Dolomites 74 C1 var. Dolomiti, Eng. Dolomites. mountain range NE Italy
Dolomites/Dolomiti see Dolomitiche, Alpi
Dolores 42 D4 Buenos Aires, E Argentina
Dolores 30 B1 Petén, N Guatemala
Dolores 42 D4 Soriano, SW Uruguay
Dolores Hidalgo 29 E4 var. Ciudad de Dolores Hidalgo. Guanajuato, C Mexico
Dolyna 86 B2 Rus. Dolina. Ivano-Frankivs'ka Oblast', W Ukraine
Dolyns'ka 87 F3 Rus. Dolinskaya. Kirovohrads'ka Oblast', S Ukraine
Domachëvo/Domaczewo see Damachava
Dombås 63 B5 Oppland, S Norway
Domel Island see Letsôk-aw Kyun
Domesnes, Cape see Kolkasrags
Domeyko 42 B3 Atacama, N Chile
Dominica 33 H4 off. Commonwealth of Dominica. country E West Indies
Dominica Channel see Martinique Passage
Dominica, Commonwealth of see Dominica
Dominican Republic 33 E2 country C West Indies
Domokós 83 B5 var. Dhomokós. Stereá Ellás, C Greece
Don 89 B6 var. Duna, Tanais. river SW Russian Federation
Donau see Danube
Donauwörth 73 C6 Bayern, S Germany
Don Benito 70 C3 Extremadura, W Spain
Doncaster 67 D5 anc. Danum. N England, United Kingdom
Dondo 56 B1 Cuanza Norte, NW Angola
Donegal 67 B5 Ir. Dún na nGall. Donegal, NW Ireland
Donegal Bay 67 A5 Ir. Bá Dhún na nGall. bay NW Ireland
Donets 87 G2 river Russian Federation/Ukraine
Donets'k 87 G3 Rus. Donetsk; prev. Stalino. Donets'ka Oblast', E Ukraine
Dongfang 106 B7 var. Basuo. Hainan, S China
Dongguan 106 C6 Guangdong, S China
Đông Hà 114 E4 Quang Tri, C Vietnam
Dong Hai see East China Sea
Đông Hơi 114 D4 Quang Binh, C Vietnam
Dongliao see Liaoyuan
Dongola 50 B3 var. Donqola, Dunqulah. Northern, N Sudan
Dongou 55 C5 Likouala, NE Congo
Dong Rak, Phanom see Dângrêk, Chuŏr Phnum
Dongting Hu 106 C5 var. Tung-t'ing Hu. lake S China
Donostia 71 E1 País Vasco, N Spain see also San Sebastián
Donqola see Dongola
Doolow 51 D5 Sumalē, E Ethiopia
Doornik see Tournai
Door Peninsula 18 C2 peninsula Wisconsin, N USA
Dooxo Nugaaleed 51 E5 var. Nogal Valley. valley E Somalia
Dordogne 69 B5 cultural region SW France
Dordogne 69 B5 river W France
Dordrecht 64 C4 var. Dordt, Dort. Zuid-Holland, SW Netherlands
Dordt see Dordrecht
Dorohoi 86 C3 Botoşani, NE Romania
Dorotea 62 C4 Västerbotten, N Sweden
Dorpat see Tartu
Dorre Island 125 A5 island Western Australia
Dort see Dordrecht
Dortmund 72 A4 Nordrhein-Westfalen, W Germany
Dos Hermanas 70 C4 Andalucía, S Spain
Dospad Dagh see Rhodope Mountains
Dospat 82 C3 Smolyan, S Bulgaria
Dothan 20 D3 Alabama, S USA
Dotnuva 84 B4 Kaunas, C Lithuania
Douai 68 C2 prev. Douay; anc. Duacum. Nord, N France
Douala 55 A5 var. Duala. Littoral, W Cameroon
Douay see Douai
Douglas 67 C5 dependent territory capital (Isle of Man) E Isle of Man
Douglas 26 C3 Arizona, SW USA
Douglas 22 D3 Wyoming, C USA
Douma see Dūmā
Douro see Duero
Douvres see Dover
Dover 67 E7 Fr. Douvres, Lat. Dubris Portus. SE England, United Kingdom
Dover 19 F4 state capital Delaware, NE USA
Dover, Strait of 68 C2 var. Straits of Dover, Fr. Pas de Calais. strait England, United Kingdom/France
Dover, Straits of see Dover, Strait of
Dovrefjell 63 B5 plateau S Norway
Downpatrick 67 B5 Ir. Dún Pádraig. SE Northern Ireland, United Kingdom
Dozen 109 B6 island Oki-shotō, SW Japan
Drâa, Hammada du see Dra, Hamada du
Drač/Drac see Durrës
Drachten 64 D2 Friesland, N Netherlands
Drăgăşani 86 B5 Vâlcea, SW Romania
Dragoman 82 B2 Sofiya, W Bulgaria
Dra, Hamada du 48 C3 var. Hammada du Drâa, Haut Plateau du Dra. plateau W Algeria
Dra, Haut Plateau du see Dra, Hamada du
Drahichyn 85 B6 Pol. Drohiczyn Poleski, Rus. Drogichin. Brestskaya Voblasts', SW Belarus
Drakensberg 56 D5 mountain range Lesotho/South Africa
Drake Passage 35 B8 passage Atlantic Ocean/Pacific Ocean
Dralfa 82 D2 Tŭrgovishte, N Bulgaria
Dráma 82 C3 var. Dhráma. Anatolikí Makedonía kai Thráki, NE Greece
Dramburg see Drawsko Pomorskie
Drammen 63 B6 Buskerud, S Norway
Drau see Drava
Drava 78 C3 var. Drau, Eng. Drave, Hung. Dráva. river C Europe
Dráva/Drave see Drau/Drava
Drawsko Pomorskie 76 B3 Ger. Dramburg. Zachodnio-pomorskie, NW Poland
Drépano, Akrotírio 82 C4 var. Akrotírio Dhrepanon. headland N Greece
Drepanum see Trapani
Dresden 72 D4 Sachsen, E Germany
Drin see Drinit, Lumi i
Drina 78 C3 river Bosnia and Herzegovina/Serbia
Drinit, Lumi i 79 D5 var. Drin. river NW Albania
Drinit të Zi, Lumi i see Black Drin
Drissa see Drysa
Drobeta-Turnu Severin 86 B5 prev. Turnu Severin. Mehedinţi, SW Romania
Drogheda 67 B5 Ir. Droichead Átha. NE Ireland
Drogichin see Drahichyn
Drogobych see Drohobych
Drohiczyn Poleski see Drahichyn
Drohobych 86 B2 Pol. Drohobycz, Rus. Drogobych. L'vivs'ka Oblast', NW Ukraine
Drohobycz see Drohobych
Droichead Átha see Drogheda
Drôme 69 D5 cultural region E France
Dronning Maud Land 132 B2 physical region Antarctica
Drontheim see Trondheim
Drug see Durg
Druk-yul see Bhutan
Drummondville 17 E4 Québec, SE Canada
Druskienniki see Druskininkai
Druskininkai 85 B5 Pol. Druskienniki. Alytus, S Lithuania
Dryden 16 B3 Ontario, C Canada
Drysa 85 D5 Rus. Drissa. river N Belarus
Duacum see Douai
Duala see Douala
Dubai see Dubayy
Dubăsari 86 D3 Rus. Dubossary. NE Moldova
Dubawnt 15 F4 river Nunavut, NW Canada
Dubayy 98 D4 Eng. Dubai. Dubayy, NE United Arab Emirates
Dubbo 127 D6 New South Wales, SE Australia
Dublin 67 B5 Ir. Baile Átha Cliath; anc. Eblana. country capital (Ireland) Dublin, E Ireland
Dublin 21 E2 Georgia, SE USA
Dubno 86 C2 Rivnens'ka Oblast', NW Ukraine
Dubossary see Dubăsari
Dubris Portus see Dover
Dubrovnik 79 B5 It. Ragusa. Dubrovnik-Neretva, SE Croatia
Dubuque 23 G3 Iowa, C USA
Dudelange 65 D8 var. Forge du Sud, Ger. Dudelingen. Luxembourg, S Luxembourg
Dudelingen see Dudelange
Duero 70 D2 Port. Douro. river Portugal/Spain
Duesseldorf see Düsseldorf
Duffel 65 C5 Antwerpen, C Belgium
Dugi Otok 78 A4 var. Isola Grossa, It. Isola Lunga. island W Croatia
Duinekerke see Dunkerque
Duisburg 72 A4 prev. Duisburg-Hamborn. Nordrhein-Westfalen, W Germany
Duisburg-Hamborn see Duisburg
Duiven 64 D4 Gelderland, E Netherlands
Duk Faiwil 51 B5 Jonglei, C South Sudan
Dulan 104 D4 var. Qagan Us. Qinghai, C China
Dulce, Golfo 31 E5 gulf S Costa Rica
Dulce, Golfo see Izabal, Lago de
Dülmen 72 A4 Nordrhein-Westfalen, W Germany
Dulovo 82 E1 Silistra, NE Bulgaria
Duluth 23 G2 Minnesota, N USA
Dūmā 97 B5 Fr. Douma. Dimashq, SW Syria
Dumas 27 E1 Texas, SW USA
Dumfries 66 C4 S Scotland, United Kingdom
Dumont d'Urville 132 C4 French research station Antarctica
Dumyāt 50 B1 var. Dumyât, Eng. Damietta. N Egypt
Duna see Danube, C Europe
Düna see Western Dvina
Duna see Don, Russian Federation
Dünaburg see Daugavpils
Dunaj see Wien, Austria
Dunaj see Danube, C Europe
Dunapentele see Dunaújváros
Dunărea see Danube
Dunaújváros 77 C7 prev. Dunapentele, Sztálinváros. Fejér, C Hungary
Dunav see Danube
Dunavska Ravnina 82 C2 Eng. Danubian Plain. lowlands N Bulgaria
Duncan 27 G2 Oklahoma, C USA
Dundalk 67 B5 Ir. Dún Dealgan. Louth, NE Ireland
Dún Dealgan see Dundalk
Dundee 56 D4 KwaZulu/Natal, E South Africa
Dundee 66 C4 E Scotland, United Kingdom
Dunedin 129 B7 Otago, South Island, New Zealand
Dunfermline 66 C4 C Scotland, United Kingdom
Dungu 55 E5 Orientale, NE Dem. Rep. Congo
Dungun 116 B3 var. Kuala Dungun. Terengganu, Peninsular Malaysia
Dunholme see Durham
Dunkerque 68 C2 Eng. Dunkirk, Flem. Duinekerke; prev. Dunquerque. Nord, N France
Dunkirk see Dunkerque
Dún Laoghaire 67 B6 Eng. Dunleary; prev. Kingstown. E Ireland
Dunleary see Dún Laoghaire
Dún Pádraig see Downpatrick
Dunquerque see Dunkerque
Dunqulah see Dongola
Dupnitsa 82 C2 prev. Marek, Stanke Dimitrov. Kyustendil, W Bulgaria
Duqm 99 E5 var. Daqm. E Oman
Durance 69 D6 river SE France
Durango 28 D3 var. Victoria de Durango. Durango, W Mexico
Durango 22 C5 Colorado, C USA
Durankulak 82 E1 Rom. Răcari; prev. Blatnitsa, Duranulac. Dobrich, NE Bulgaria
Durant 27 G2 Oklahoma, C USA
Duranulac see Durankulak
Durazzo see Durrës
Durban 56 D4 var. Port Natal. KwaZulu/Natal, E South Africa
Durbe 84 B3 Ger. Durben. Liepāja, W Latvia
Durben see Durbe
Durg 113 E4 prev. Drug. Chhattīsgarh, C India
Durham 67 D5 hist. Dunholme. N England, United Kingdom
Durham 21 F1 North Carolina, SE USA
Durocortorum see Reims
Durostorum see Silistra
Durovernum see Canterbury
Durrës 79 C6 var. Durrësi, Dursi, It. Durazzo, SCr. Drač, Turk. Draç. Durrës, W Albania
Durrësi see Durrës
Dursi see Durrës
Durūz, Jabal ad 97 C5 mountain SW Syria
D'Urville Island 128 C4 island C New Zealand
Dusa Mareb/Dusa Marreb see Dhuusa Marreeb
Dushanbe 101 E3 var. Dyushambe; prev. Stalinabad, Taj. Stalinobod. country capital (Tajikistan) W Tajikistan
Düsseldorf 72 A4 var. Duesseldorf. Nordrhein-Westfalen, W Germany
Dŭstí 101 E3 Rus. Dusti. SW Tajikistan
Dutch East Indies see Indonesia
Dutch Guiana see Suriname
Dutch Harbor 14 B3 Unalaska Island, Alaska, USA
Dutch New Guinea see Papua
Duzdab see Zāhedān
Dvina Bay see Chëshskaya Guba
Dvinsk see Daugavpils
Dyanev see Galkynyş
Dyersburg 20 C1 Tennessee, S USA
Dyushambe see Dushanbe
Dza Chu see Mekong
Dzaudzhikau see Vladikavkaz
Dzerzhinsk 89 C5 Nizhegorodskaya Oblast', W Russian Federation
Dzerzhinskiy see Nar'yan-Mar
Dzhalal-Abad 101 F2 Kir. Jalal-Abad. Dzhalal-Abadskaya Oblast', W Kyrgyzstan
Dzhambul see Taraz
Dzhankoy 87 F4 Respublika Krym, S Ukraine
Dzharkurgan see Jarqo'rg'on
Dzhelandy 101 F3 SE Tajikistan
Dzhergalan 101 G2 Kir. Jyrgalan. Issyk-Kul'skaya Oblast', NE Kyrgyzstan
Dzhezkazgan see Zhezkazgan
Dzhizak see Jizzax
Dzhugdzhur, Khrebet 93 G3 mountain range E Russian Federation
Dzhusaly see Zhosaly
Działdowo 76 D3 Warmińsko-Mazurskie, C Poland
Dzuunmod 105 E2 Töv, C Mongolia
Dzüün Soyonï Nuruu see Vostochnyy Sayan
Dzvina see Western Dvina

E

Eagle Pass 27 F4 Texas, SW USA
East Açores Fracture Zone see East Azores Fracture Zone
East Antarctica 132 C3 var. Greater Antarctica. physical region Antarctica
East Australian Basin see Tasman Basin
East Azores Fracture Zone 44 C3 var. East Açores Fracture Zone. tectonic feature E Atlantic Ocean
Eastbourne 67 E7 SE England, United Kingdom
East Cape 128 E3 headland North Island, New Zealand
East China Sea 103 E2 Chin. Dong Hai. sea W Pacific Ocean
Easter Fracture Zone 131 G4 tectonic feature E Pacific Ocean
Easter Island see Pascua, Isla de
Eastern Desert 46 D3 var. Aş Şahrā' ash Sharqīyah, Eng. Arabian Desert, Eastern Desert. desert E Egypt
Eastern Desert see Sahara el Sharqîya
Eastern Ghats 102 B3 mountain range SE India
Eastern Sayans see Vostochnyy Sayan
Eastern Sierra Madre see Madre Oriental, Sierra
East Falkland 43 D8 var. Isla Soledad. island E Falkland Islands
East Frisian Islands 72 A3 Eng. East Frisian Islands. island group NW Germany
East Frisian Islands see Ostfriesische Inseln
East Grand Forks 23 E1 Minnesota, N USA
East Indiaman Ridge 119 D6 undersea ridge E Indian Ocean
East Indies 130 A3 island group SE Asia
East Kilbride 66 C4 S Scotland, United Kingdom
East Korea Bay 107 E3 bay E North Korea
Eastleigh 67 D7 S England, United Kingdom
East London 56 D5 Afr. Oos-Londen; prev. Emonti, Port Rex. Eastern Cape, S South Africa
Eastmain 16 D3 river Québec, C Canada
East Mariana Basin 120 B1 undersea basin W Pacific Ocean
East Novaya Zemlya Trough 90 C1 var. Novaya Zemlya Trough. trough W Kara Sea
East Pacific Rise 131 F4 undersea rise E Pacific Ocean
East Pakistan see Bangladesh
East Saint Louis 18 B4 Illinois, N USA
East Scotia Basin 45 C7 undersea basin SE Scotia Sea
East Sea 108 A4 var. Sea of Japan, Rus. Yapanskoye More. Sea NW Pacific Ocean
East Siberian Sea see Vostochno-Sibirskoye More
East Timor 117 F5 var. Loro Sae; prev. Portuguese Timor, Timor Timur. country S Indonesia
Eau Claire 18 A2 Wisconsin, N USA
Eau Claire, Lac à L' see St. Clair, Lake
Eauripik Rise 120 B2 undersea rise W Pacific Ocean
Ebensee 73 D6 Oberösterreich, N Austria
Eberswalde-Finow 72 D3 Brandenburg, E Germany
Ebetsu 108 D2 var. Ebetu. Hokkaidō, NE Japan
Ebetu see Ebetsu
Eblana see Dublin
Ebolowa 55 A5 Sud, S Cameroon
Ebon Atoll 122 D2 var. Epoon. atoll Ralik Chain, S Marshall Islands
Ebora see Évora
Eboracum see York
Ebro 71 E2 river NE Spain
Eburacum see York
Ebusus see Eivissa
Ecbatana see Hamadān
Ech Cheliff/Ech Chleff see Chlef
Echo Bay 15 E3 Northwest Territories, NW Canada
Echt 65 D5 Limburg, SE Netherlands
Ecija 70 D4 anc. Astigi. Andalucía, SW Spain
Eckengraf see Viesīte
Ecuador 38 B1 off. Republic of Ecuador. country NW South America
Ecuador, Republic of see Ecuador
Ed Da'ein 50 A4 Southern Darfur, W Sudan
Ed Damazin 50 C4 var. Ad Damazīn. Blue Nile, E Sudan
Ed Damer 50 C3 var. Ad Dāmir, Ad Damar. River Nile, NE Sudan
Ed Debba 50 B3 Northern, N Sudan
Ede 64 D4 Gelderland, C Netherlands
Ede 53 F5 Osun, SW Nigeria
Edéa 55 A5 Littoral, SW Cameroon
Edessa see Şanlıurfa
Edfu see Idfū
Edgeøya 61 G2 island S Svalbard
Edgware 67 A7 Harrow, SE England, United Kingdom
Edinburg 27 G5 Texas, SW USA
Edinburgh 66 C4 national capital S Scotland, United Kingdom
Edingen see Enghien
Edirne 94 A2 Eng. Adrianople; anc. Adrianopolis, Hadrianopolis. Edirne, NW Turkey
Edmonds 24 B2 Washington, NW USA
Edmonton 15 E5 province capital Alberta, SW Canada
Edmundston 17 E4 New Brunswick, SE Canada
Edna 27 G4 Texas, SW USA
Edolo 74 B1 Lombardia, N Italy
Edremit 94 A3 Balıkesir, NW Turkey
Edward, Lake 55 E5 var. Albert Edward Nyanza, Edward Nyanza, Lac Idi Amin, Lake Rutanzige. lake Uganda/Dem. Rep. Congo
Edward Nyanza see Edward, Lake
Edwards Plateau 27 F3 plain Texas, SW USA
Edzo 53 E4 prev. Rae-Edzo. Northwest Territories, NW Canada
Eeklo 65 B5 var. Eekloo. Oost-Vlaanderen, NW Belgium
Eekloo see Eeklo
Eems see Ems
Eersel 65 C5 Noord-Brabant, S Netherlands
Eesti Vabariik see Estonia
Efate 122 D4 var. Efate, Fr. Vaté; prev. Sandwich Island. island C Vanuatu
Efate see Efate
Effingham 18 B4 Illinois, N USA
Eforie-Sud 86 D5 Constanţa, E Romania
Egadi Is. 75 B7 island group S Italy
Ege Denizi see Aegean Sea
Eger 77 D6 Ger. Erlau. Heves, NE Hungary
Eger see Cheb, Czech Republic
Egeria Fracture Zone 119 C5 tectonic feature W Indian Ocean
Éghezèe 65 C6 Namur, C Belgium
Egina see Aígina
Egio see Aígio
Egmont see Taranaki, Mount
Egmont, Cape 128 C4 headland North Island, New Zealand
Egoli see Johannesburg
Egypt 50 B2 off. Arab Republic of Egypt, Ar. Jumhūrīyah Mişr al 'Arabīyah, prev. United Arab Republic; anc. Aegyptus. country NE Africa
Eibar 71 E1 País Vasco, N Spain
Eibergen 64 E3 Gelderland, E Netherlands
Eidfjord 63 A5 Hordaland, S Norway
Eier-Berg see Suur Munamägi
Eifel 73 A5 plateau W Germany
Eiger 73 B7 mountain C Switzerland
Eigg 66 B3 island W Scotland, United Kingdom
Eight Degree Channel 110 B3 channel India/Maldives
Eighty Mile Beach 124 B4 beach Western Australia
Eijsden 65 D6 Limburg, SE Netherlands
Eilat see Elat
Eindhoven 65 D5 Noord-Brabant, S Netherlands
Eipel see Ipel'
Éire see Ireland
Éireann, Muir see Irish Sea
Eisenhüttenstadt 72 D4 Brandenburg, E Germany
Eisenmarkt see Hunedoara
Eisenstadt 73 E6 Burgenland, E Austria
Eisleben 72 C4 Sachsen-Anhalt, C Germany
Eivissa 71 G3 var. Iviza, Cast. Ibiza; anc. Ebusus. Ibiza, Spain, W Mediterranean Sea
Ejea de los Caballeros 71 E2 Aragón, NE Spain
Ejin Qi see Dalain Hob
Ekapa see Cape Town
Ekaterinodar see Krasnodar
Ekvyvatapskiy Khrebet 93 G1 mountain range NE Russian Federation
El 'Alamein see Al 'Alamayn
El Asnam see Chlef
Elat 97 B8 var. Eilat, Elath. Southern, S Israel
Elat, Gulf of see Aqaba, Gulf of
Elath see Elat, Israel
Elath see Al 'Aqabah, Jordan
El'Atrun 50 B3 Northern Darfur, NW Sudan
Elâzığ 95 E3 var. Elâzig, Elâziz. Elâzığ, E Turkey
Elba 74 B4 island Archipelago Toscano, C Italy
Elbasan 79 D6 var. Elbasani. Elbasan, C Albania
Elbasani see Elbasan
Elbe 58 D3 Cz. Labe. river Czech Republic/Germany
Elbert, Mount 22 C4 mountain Colorado, C USA
Elbing see Elbląg
Elbląg 76 C2 var. Elblag, Ger. Elbing. Warmińsko-Mazurskie, NE Poland
El Boulaida/El Boulaïda see Blida
El'brus 89 A8 var. Gora El'brus. mountain SW Russian Federation
El'brus, Gora see El'brus
El Burgo de Osma 71 E2 Castilla-León, C Spain
Elburz Mountains see Alborz, Reshteh-ye Kūhhā-ye
El Cajon 25 C8 California, W USA
El Calafate 43 B7 var. Calafate. Santa Cruz, S Argentina
El Callao 37 E2 Bolívar, E Venezuela
El Campo 27 G4 Texas, SW USA
El Carmen de Bolívar 36 B2 Bolívar, NW Colombia
El Cayo see San Ignacio
El Centro 25 D8 California, W USA
Elche 71 F4 Cat. Elx; anc. Ilici, Lat. Illicis. País Valenciano, E Spain
Elda 71 F4 País Valenciano, E Spain
El Djazaïr see Alger
El Djelfa see Djelfa
Eldorado 42 E3 Misiones, NE Argentina
El Dorado 28 C3 Sinaloa, C Mexico
El Dorado 20 B2 Arkansas, C USA
El Dorado 23 F5 Kansas, C USA
El Dorado 37 F2 Bolívar, E Venezuela
El Dorado see California
Eldoret 51 C6 Rift Valley, W Kenya
Elektrostal' 89 B5 Moskovskaya Oblast', W Russian Federation
Elemi Triangle 51 B5 disputed region Kenya/Sudan
Elephant Butte Reservoir 26 C2 reservoir New Mexico, SW USA
Élesd see Aleşd
Eleuthera Island 32 C1 island N Bahamas
El Fasher 50 A4 var. Al Fāshir. Northern Darfur, W Sudan
El Ferrol/El Ferrol del Caudillo see Ferrol
El Gedaref see Gedaref
El Geneina 50 A4 var. Ajjinena, Al-Genain, Al Junaynah. Western Darfur, W Sudan
Elgin 66 C3 NE Scotland, United Kingdom
Elgin 18 B3 Illinois, N USA
El Gîza see Gîza
El Goléa 48 D3 var. Al Golea. C Algeria
El Ḩank 52 D1 cliff N Mauritania
El Haseke see Al Ḩasakah
Elimberrum see Auch
Eliocroca see Lorca
Élisabethville see Lubumbashi
Elista 89 B7 Respublika Kalmykiya, SW Russian Federation
Elizabeth 127 B6 South Australia
Elizabeth City 21 G1 North Carolina, SE USA
Elizabethtown 18 C5 Kentucky, S USA
El-Jadida 48 C2 prev. Mazagan. W Morocco
Ełk 76 E2 Ger. Lyck. Warmińsko-mazurskie, NE Poland
Elk City 27 F1 Oklahoma, C USA
El Khalil see Hebron
El Khârga see Al Khārijah
Elkhart 18 C3 Indiana, N USA
El Khartûm see Khartoum
Elk River 23 F2 Minnesota, N USA
El Kuneitra see Al Qunayţirah
Ellás see Greece
Ellef Ringnes Island 15 E1 island Nunavut, N Canada
Ellen, Mount 22 B5 mountain Utah, W USA

Ellensburg - Forchheim

Ellensburg 24 B2 Washington, NW USA
Ellesmere Island 15 F1 *island* Queen Elizabeth Islands, Nunavut, N Canada
Ellesmere, Lake 129 C6 *lake* South Island, New Zealand
Ellice Islands *see* Tuvalu
Elliston 127 A6 South Australia
Ellsworth Land 132 A3 *physical region* Antarctica
El Mahbas 48 B3 *var.* Mahbés. SW Western Sahara
El Mina 96 B4 *var.* Al Mīnā'. N Lebanon
El Minya *see* Al Minyā
Elmira 19 E3 New York, NE USA
El Mreyyé 52 D2 *desert* E Mauritania
Elmshorn 72 B3 Schleswig-Holstein, N Germany
El Muglad 50 B4 Western Kordofan, C Sudan
El Obeid 50 B4 *var.* Al Obayyid, Al Ubayyiḍ. Northern Kordofan, C Sudan
El Ouâdi *see* El Oued
El Oued 49 E2 *var.* Al Oued, El Ouâdi, El Wad. NE Algeria
Eloy 26 B2 Arizona, SW USA
El Paso 26 D3 Texas, SW USA
El Porvenir 31 G4 Kuna Yala, N Panama
El Progreso 30 C2 Yoro, NW Honduras
El Puerto de Santa María 70 C5 Andalucía, S Spain
El Qâhira *see* Cairo
El Quneitra *see* Al Qunayṭirah
El Quseir *see* Al Quṣayr
El Quweira *see* Al Quwayrah
El Rama 31 E3 Región Autónoma Atlántico Sur, SE Nicaragua
El Real 31 H5 *var.* El Real de Santa María. Darién, SE Panama
El Real de Santa María *see* El Real
El Reno 27 F1 Oklahoma, C USA
El Salvador 30 B3 *off.* Republica de El Salvador. *country* Central America
El Salvador, Republica de *see* El Salvador
Elsass *see* Alsace
El Sáuz 28 C2 Chihuahua, N Mexico
El Serrat 69 A7 N Andorra Europe
Elst 64 D4 Gelderland, E Netherlands
El Sueco 28 C2 Chihuahua, N Mexico
El Suweida *see* As Suwaydā'
El Suweis *see* Suez
Eltanin Fracture Zone 131 E5 *tectonic feature* SE Pacific Ocean
El Tigre 37 E2 Anzoátegui, NE Venezuela
Elvas 70 C4 Portalegre, C Portugal
El Vendrell 71 G2 Cataluña, NE Spain
El Vigía 36 C2 Mérida, NW Venezuela
El Wad *see* El Oued
Elwell, Lake 22 B1 *reservoir* Montana, NW USA
Elx *see* Elche
Ely 25 D5 Nevada, W USA
El Yopal *see* Yopal
Emajõgi 84 D3 *Ger.* Embach. *river* SE Estonia
Emāmrūd *see* Shāhrūd
Emāmshahr *see* Shāhrūd
Emba 92 B4 *Kaz.* Embi. Aktyubinsk, W Kazakhstan
Embach *see* Emajõgi
Embi *see* Emba
Emden 72 A3 Niedersachsen, NW Germany
Emerald 126 D4 Queensland, E Australia
Emerald Isle *see* Montserrat
Emesa *see* Ḥimṣ
Emmaste 84 C2 Hiiumaa, W Estonia
Emmeloord 64 D2 Flevoland, N Netherlands
Emmen 64 E2 Drenthe, NE Netherlands
Emmendingen 73 A6 Baden-Württemberg, SW Germany
Emona *see* Ljubljana
Emonti *see* East London
Emory Peak 27 E4 *mountain* Texas, SW USA
Empalme 28 B2 Sonora, NW Mexico
Emperor Seamounts 91 G3 *seamount range* NW Pacific Ocean
Empire State of the South *see* Georgia
Emporia 23 F5 Kansas, C USA
Empty Quarter *see* Ar Rub 'al Khālī
Ems 72 A3 *Dut.* Eems. *river* NW Germany
Enareträsk *see* Inarijärvi
Encamp 69 A8 Encamp, C Andorra Europe
Encarnación 42 D3 Itapúa, S Paraguay
Encinitas 25 C8 California, W USA
Encs 77 D6 Borsod-Abaúj-Zemplén, NE Hungary
Endeavour Strait 126 C1 *strait* Queensland, NE Australia
Enderbury Island 123 F3 *atoll* Phoenix Islands, C Kiribati
Enderby Land 132 C2 *physical region* Antarctica
Enderby Plain 132 D2 *abyssal plain* S Indian Ocean
Endersdorf *see* Jędrzejów
Enewetak Atoll 122 C1 *var.* Ānewetak, Eniwetok. *atoll* Ralik Chain, W Marshall Islands
Enfield 67 A7 United Kingdom
Engeten *see* Aiud
Enghien 65 B6 *Dut.* Edingen. Hainaut, SW Belgium
England 67 D5 *Lat.* Anglia. *cultural region* England, United Kingdom
Englewood 22 D4 Colorado, C USA
English Channel 67 D8 *var.* The Channel, *Fr.* la Manche. *channel* NW Europe
Engure 84 C3 Tukums, W Latvia
Engures Ezers 84 B3 *lake* NW Latvia
Enguri 95 F1 *Rus.* Inguri. *river* NW Georgia
Enid 27 F1 Oklahoma, C USA
Enikale Strait *see* Kerch Strait
Eniwetok *see* Enewetak Atoll
En Nâqoûra 97 A5 *var.* An Nāqūrah. SW Lebanon
En Nazira *see* Natzrat
Ennedi 54 D2 *plateau* E Chad
Ennis 67 A6 *Ir.* Inis. Clare, W Ireland
Ennis 27 G3 Texas, SW USA
Enniskillen 67 B5 *var.* Inniskilling, *Ir.* Inis Ceithleann. SW Northern Ireland, United Kingdom
Enns 73 D6 *river* C Austria
Enschede 64 E3 Overijssel, E Netherlands
Ensenada 28 A1 Baja California Norte, NW Mexico
Entebbe 51 B6 S Uganda
Entroncamento 70 B3 Santarém, C Portugal
Enugu 53 G5 Enugu, S Nigeria
Epanomí 82 B4 Kentrikí Makedonía, N Greece
Epéna 55 B5 Likouala, NE Congo
Eperies/Eperjes *see* Prešov
Epi 122 D4 *var.* Épi. *island* C Vanuatu
Épi *see* Epi
Épinal 68 D4 Vosges, NE France
Epiphania *see* Ḥamāh
Epoon *see* Ebon Atoll
Epsom 67 A8 United Kingdom
Equality State *see* Wyoming
Equatorial Guinea 55 A5 *off.* Equatorial Guinea, Republic of. *country* C Africa
Equatorial Guinea, Republic of *see* Equatorial Guinea
Erautini *see* Johannesburg
Erbil *see* Arbīl
Erciş 95 F3 Van, E Turkey
Erdély *see* Transylvania
Erdélyi-Havasok *see* Carpaţii Meridionalii
Erdenet 105 E2 Orhon, N Mongolia
Erdi 54 C2 *plateau* NE Chad
Erdi Ma 54 D2 *desert* NE Chad
Erebus, Mount 132 B4 *volcano* Ross Island, Antarctica
Ereğli 94 C4 Konya, S Turkey
Erenhot 105 F2 *var.* Erlian. Nei Mongol Zizhiqu, NE China
Erfurt 72 C4 Thüringen, C Germany
Ergene Çayı *see* Ergene Irmağī
Ergene Irmağī 94 A2 *var.* Ergene Çayı. *river* NW Turkey
Ergun 105 F1 *var.* Labudalin; *prev.* Ergun Youqi. Nei Mongol Zizhiqu, N China
Ergun He *see* Argun
Ergun Youqi *see* Ergun
Erie 18 D3 Pennsylvania, NE USA
Érié, Lac *see* Erie, Lake
Erie, Lake 18 D3 *Fr.* Lac Érié. *lake* Canada/USA
Eritrea 50 C4 *off.* State of Eritrea, Ērtra. *country* E Africa
Eritrea, State of *see* Eritrea
Erivan *see* Yerevan
Erlangen 73 C5 Bayern, S Germany
Erlau *see* Eger
Erlian *see* Erenhot
Ermelo 64 D3 Gelderland, C Netherlands
Ermióni 83 C6 Pelopónnisos, S Greece
Ermoúpoli 83 D6 *var.* Hermoupolis; *prev.* Ermoúpolis. Sýyros, Kykládes, Greece, Aegean Sea
Ermoúpolis *see* Ermoúpoli
Ernākulam 110 C3 Kerala, SW India
Erode 110 C2 Tamil Nādu, SE India
Erquelinnes 65 B7 Hainaut, S Belgium
Er-Rachidia 48 C2 *var.* Ksar al Soule. E Morocco
Er Rahad 50 B4 *var.* Ar Rahad. Northern Kordofan, C Sudan
Erromango 122 D4 *island* S Vanuatu
Ertis *see* Irtysh, C Asia
Ērtra *see* Eritrea
Erzerum *see* Erzurum
Erzgebirge 73 C5 *Cz.* Krušné Hory, *Eng.* Ore Mountains. *mountain range* Czech Republic/Germany
Erzincan 95 E3 *var.* Erzinjan. Erzincan, E Turkey
Erzinjan *see* Erzincan
Erzurum 95 E3 *prev.* Erzerum. Erzurum, NE Turkey
Esbjerg 63 A7 Ribe, W Denmark
Esbo *see* Espoo
Escaldes 69 A8 Escaldes Engordany, C Andorra Europe
Escanaba 18 C2 Michigan, N USA
Escaut *see* Scheldt
Esch-sur-Alzette 65 D8 Luxembourg, S Luxembourg
Esclaves, Grand Lac des *see* Great Slave Lake
Escondido 25 C8 California, W USA
Escuinapa 28 D3 *var.* Escuinapa de Hidalgo. Sinaloa, C Mexico
Escuinapa de Hidalgo *see* Escuinapa
Escuintla 30 B2 Escuintla, S Guatemala
Escuintla 29 G5 Chiapas, SE Mexico
Esenguly 100 B3 *Rus.* Gasan-Kuli. Balkan Welaýaty, W Turkmenistan
Eşfahān 98 C3 *Eng.* Isfahan; *anc.* Aspadana. Eşfahān, C Iran
Esh Sharā *see* Ash Sharāh
Esil *see* Ishim, Kazakhstan/Russian Federation
Eskimo Point *see* Arviat
Eskişehir 94 B3 *var.* Eskishehr. Eskişehir, W Turkey
Eskishehr *see* Eskişehir
Eslāmābād 98 C3 *var.* Eslāmābād-e Gharb; *prev.* Harunabad, Shāhābād. Kermānshāhān, W Iran
Eslāmābād-e Gharb *see* Eslāmābād
Esmeraldas 38 A1 Esmeraldas, N Ecuador
Esna *see* Isnā
España *see* Spain
Espanola 26 D1 New Mexico, SW USA
Esperance 125 B7 Western Australia
Esperanza 28 B2 Sonora, NW Mexico
Esperanza 132 A2 *Argentinian research station* Antarctica
Espinal 36 B3 Tolima, C Colombia
Espinhaço, Serra do 34 D4 *mountain range* SE Brazil
Espírito Santo 41 F4 *off.* Estado do Espírito Santo. *region* E Brazil
Espírito Santo 41 F4 *off.* Estado do Espírito Santo. *state* E Brazil
Espírito Santo, Estado do *see* Espírito Santo
Espiritu Santo 122 C4 *var.* Santo. *island* W Vanuatu
Espoo 63 D6 *Swe.* Esbo. Etelä-Suomi, S Finland
Esquel 43 B6 Chubut, SW Argentina
Essaouira 48 B2 *prev.* Mogador. W Morocco
Esseg *see* Osijek
Es Semara *see* Smara
Essen 65 C5 Antwerpen, N Belgium
Essen 72 A4 *var.* Essen an der Ruhr. Nordrhein-Westfalen, W Germany
Essen an der Ruhr *see* Essen
Essequibo River 37 F3 *river* C Guyana
Es Suweida *see* As Suwaydā'
Estacado, Llano 27 E2 *plain* New Mexico/Texas, SW USA
Estados, Isla de los 43 C8 *prev. Eng.* Staten Island. *island* S Argentina
Estância 41 G3 Sergipe, E Brazil
Estelí 30 D3 Estelí, NW Nicaragua
Estella 71 E1 *Bas.* Lizarra. Navarra, N Spain
Estepona 70 D5 Andalucía, S Spain
Estevan 15 F5 Saskatchewan, S Canada
Estland *see* Estonia
Estonia 84 D2 *off.* Republic of Estonia, *Est.* Eesti Vabariik, *Ger.* Estland, *Latv.* Igaunija; *prev.* Estonian SSR, *Rus.* Estonskaya SSR. *country* NE Europe
Estonian SSR *see* Estonia
Estonia, Republic of *see* Estonia
Estonskaya SSR *see* Estonia
Estrela, Serra da 70 C3 *mountain range* C Portugal
Estremadura *see* Extremadura
Estremoz 70 C4 Évora, S Portugal
Eszék *see* Osijek
Esztergom 77 C6 *Ger.* Gran; *anc.* Strigonium. Komárom-Esztergom, N Hungary
Étalle 65 D8 Luxembourg, SE Belgium
Etāwah 112 D3 Uttar Pradesh, N India
Ethiopia 51 C5 *off.* Federal Democratic Republic of Ethiopia; *prev.* Abyssinia, People's Democratic Republic of Ethiopia. *country* E Africa
Ethiopia, Federal Democratic Republic of *see* Ethiopia
Ethiopian Highlands 51 C5 *var.* Ethiopian Plateau. *plateau* N Ethiopia
Ethiopian Plateau *see* Ethiopian Highlands
Ethiopia, People's Democratic Republic of *see* Ethiopia
Etna, Monte 75 C7 *Eng.* Mount Etna. *volcano* Sicilia, Italy, C Mediterranean Sea
Etna, Mount *see* Etna, Monte
Etosha Pan 56 B3 *salt lake* N Namibia
Etoumbi 55 B5 Cuvette Ouest, NW Congo
Etsch *see* Adige
Et Tafila *see* Aţ Ţafīlah
Ettelbrück 65 D8 Diekirch, C Luxembourg
'Eua 123 E5 *prev.* Middleburg Island. *island* Tongatapu Group, SE Tonga
Euboea 83 C5 *Lat.* Euboea. *island* C Greece
Euboea *see* Évvoia
Eucla 125 D6 Western Australia
Euclid 18 D3 Ohio, N USA
Eufaula Lake 27 G1 *var.* Eufaula Reservoir. *reservoir* Oklahoma, C USA
Eufaula Reservoir *see* Eufaula Lake
Eugene 24 B3 Oregon, NW USA
Eumolpias *see* Plovdiv
Eupen 65 D6 Liège, E Belgium
Euphrates 90 B4 *Ar.* Al-Furāt, *Turk.* Firat Nehri. *river* SW Asia
Eureka 25 A5 California, W USA
Eureka 22 A1 Montana, NW USA
Europa Point 71 H5 *headland* S Gibraltar
Europe 58 *continent*
Eutin 72 C2 Schleswig-Holstein, N Germany
Euxine Sea *see* Black Sea
Evansdale 23 G3 Iowa, C USA
Evanston 18 B3 Illinois, N USA
Evanston 22 B4 Wyoming, C USA
Evansville 18 B5 Indiana, N USA
Eveleth 23 G1 Minnesota, N USA
Everard, Lake 127 A6 *salt lake* South Australia
Everest, Mount 104 B5 *Chin.* Qomolangma Feng, *Nep.* Sagarmāthā. *mountain* China/Nepal
Everett 24 B2 Washington, NW USA
Everglades, The 21 F5 *wetland* Florida, SE USA
Evje 63 A6 Aust-Agder, S Norway
Evmolpia *see* Plovdiv
Évora 70 B4 *anc.* Ebora, *Lat.* Liberalitas Julia. Évora, C Portugal
Évreux 68 C3 *anc.* Civitas Eburovicum. Eure, N France
Évros *see* Maritsa
Évry 68 E2 Essonne, N France
Ewarton 32 B5 C Jamaica
Excelsior Springs 23 F4 Missouri, C USA
Exe 67 C7 *river* SW England, United Kingdom
Exeter 67 C7 *anc.* Isca Damnoniorum. SW England, United Kingdom
Exmoor 67 C7 *moorland* SW England, United Kingdom
Exmouth 124 A4 Western Australia
Exmouth 67 C7 SW England, United Kingdom
Exmouth Gulf 124 A4 *gulf* Western Australia
Exmouth Plateau 119 E5 *undersea plateau* E Indian Ocean
Extremadura 70 C3 *var.* Estremadura. *autonomous community* W Spain
Exuma Cays 32 C1 *islets* C Bahamas
Exuma Sound 32 C1 *sound* C Bahamas
Eyre Mountains 129 A7 *mountain range* South Island, New Zealand
Eyre North, Lake 127 A5 *salt lake* South Australia
Eyre Peninsula 127 A6 *peninsula* South Australia
Eyre South, Lake 127 A5 *salt lake* South Australia
Ezo *see* Hokkaidō

F

Faadhippolhu Atoll 110 B4 *var.* Fadiffolu, Lhaviyani Atoll. *atoll* N Maldives
Fabens 26 D3 Texas, SW USA
Fada 54 C2 Borkou-Ennedi-Tibesti, E Chad
Fada-Ngourma 53 E4 E Burkina
Fadiffolu *see* Faadhippolhu Atoll
Faenza 74 C3 *anc.* Faventia. Emilia-Romagna, N Italy
Faeroe Islands *see* Faero Islands
Færøerne *see* Faroe Islands
Faetano 74 E2 E San Marino
Făgăraş 86 C4 *Ger.* Fogarasch, *Hung.* Fogaras. Braşov, C Romania
Fagibina, Lake *see* Faguibine, Lac
Fagne 65 C7 *hill range* S Belgium
Faguibine, Lac 53 E3 *var.* Lake Fagibina. *lake* NW Mali
Fahlun *see* Falun
Fahraj 98 E4 Kermān, SE Iran
Faial 70 A5 *var.* Ilha do Faial. *island* Azores, Portugal, NE Atlantic Ocean
Faial, Ilha do *see* Faial
Faifo *see* Hôi An
Fairbanks 14 D3 Alaska, USA
Fairfield 25 B6 California, W USA
Fair Isle 66 D2 *island* NE Scotland, United Kingdom
Fairlie 129 B6 Canterbury, South Island, New Zealand
Fairmont 23 F3 Minnesota, N USA
Faisalābād 112 C2 *prev.* Lyallpur. Punjab, NE Pakistan
Faizābād 113 E3 Uttar Pradesh, N India
Faizabad/Faizābād *see* Feyẕābād
Fakaofo Atoll 123 F3 *island* SE Tokelau
Falam 114 A3 Chin State, W Myanmar (Burma)
Falconara Marittima 74 C3 Marche, C Italy
Falkenau an der Eger *see* Sokolov
Falkland Islands 43 D7 *var.* Falklands, Islas Malvinas. *UK dependent territory* SW Atlantic Ocean
Falkland Plateau 35 D7 *var.* Argentine Rise. *undersea feature* SW Atlantic Ocean
Falklands *see* Falkland Islands
Falknov nad Ohří *see* Sokolov
Fallbrook 25 C8 California, W USA
Falmouth 32 A4 W Jamaica
Falmouth 67 C7 SW England, United Kingdom
Falster 63 B8 *island* SE Denmark
Fălticeni 86 C3 *Hung.* Falticsén. Suceava, NE Romania
Falticsén *see* Fălticeni
Falun 63 C6 *var.* Fahlun. Kopparberg, C Sweden
Famagusta *see* Gazimağusa
Famagusta Bay *see* Gazimağusa Körfezi
Famenne 65 C7 *physical region* SE Belgium
Fang 114 C3 Chiang Mai, NW Thailand
Fanning Island *see* Tabuaeran
Fanø 74 C3 *island* W Denmark
Farafangana 57 G4 Fianarantsoa, SE Madagascar
Farāh 100 D4 *var.* Farah, Fararud. Farāh, W Afghanistan
Farah Rud 100 D4 *river* W Afghanistan
Faranah 52 C4 Haute-Guinée, S Guinea
Fararud *see* Farāh
Farasan, Jaza'ir 99 A6 *island group* SW Saudi Arabia
Farewell, Cape 128 C4 *headland* South Island, New Zealand
Farewell, Cape *see* Nunap Isua
Fargo 23 F2 North Dakota, N USA
Farg'ona 101 F2 *Rus.* Fergana; *prev.* Novyy Margilan. Farg'ona Viloyati, E Uzbekistan
Faribault 23 F2 Minnesota, N USA
Farīdābād 112 D3 Haryāna, N India
Farkhor 101 E3 *Rus.* Parkhar. SW Tajikistan
Farmington 23 G5 Missouri, C USA
Farmington 26 C1 New Mexico, SW USA
Faro 70 B5 Faro, S Portugal
Faroe-Iceland Ridge 58 C1 *undersea ridge* NW Norwegian Sea
Faroe Islands 61 E5 *var.* Faero Islands, *Dan.* Færøerne, *Faer.* Føroyar. *Danish external territory* N Atlantic Ocean
Faroe-Shetland Trough 58 C2 *trough* NE Atlantic Ocean
Farquhar Group 57 G2 *island group* S Seychelles
Fars, Khalīj-e *see* Persian Gulf
Farvel, Kap *see* Nunap Isua
Fastiv 87 E2 *Rus.* Fastov. Kyyivs'ka Oblast', NW Ukraine
Fastov *see* Fastiv
Fauske 62 C3 Nordland, C Norway
Faventia *see* Faenza
Faxa Bay *see* Faxaflói
Faxaflói 60 D5 *Eng.* Faxa Bay. *bay* W Iceland
Faya 54 C2 *prev.* Faya-Largeau, Largeau. Borkou-Ennedi-Tibesti, N Chad
Faya-Largeau *see* Faya
Fayetteville 20 A1 Arkansas, C USA
Fayetteville 21 F1 North Carolina, SE USA
Fdérick *see* Fdérik
Fdérik 52 C2 *var.* Fdérick, *Fr.* Fort Gouraud. Tiris Zemmour, NW Mauritania
Fear, Cape 21 F2 *headland* Bald Head Island, North Carolina, SE USA
Fécamp 68 B3 Seine-Maritime, N France
Fédala *see* Mohammedia
Federal Capital Territory *see* Australian Capital Territory
Fehérgyarmat 77 E6 Szabolcs-Szatmár-Bereg, E Hungary
Fehértemplom *see* Bela Crkva
Fehmarn 72 C2 *island* N Germany
Fehmarn Belt 72 C2 *Dan.* Femern Bælt, *Ger.* Fehmarnbelt. *strait* Denmark /Germany
Fehmarnbelt Fehmarn Belt/Femer Bælt
Feijó 40 C2 Acre, W Brazil
Feilding 128 D4 Manawatu-Wanganui, North Island, New Zealand
Feira *see* Feira de Santana
Feira de Santana 41 G3 *var.* Feira. Bahia, E Brazil
Feketehalom *see* Codlea
Felanitx 71 G3 Mallorca, Spain, W Mediterranean Sea
Felicitas Julia *see* Lisboa
Felidhu Atoll 110 B4 *atoll* C Maldives
Felipe Carrillo Puerto 29 H4 Quintana Roo, SE Mexico
Felixstowe 67 E6 E England, United Kingdom
Fellin *see* Viljandi
Felsőbánya *see* Baia Sprie
Felsőmuzslya *see* Mužlja
Femunden 63 B5 *lake* S Norway
Fénérive *see* Fenoarivo Atsinanana
Fengcheng 106 D3 *var.* Feng-cheng, Fenghwangcheng. Liaoning, NE China
Feng-cheng *see* Fengcheng
Fenghwangcheng *see* Fengcheng
Fengtien *see* Shenyang, China
Fengtien *see* Liaoning, China
Fenoarivo Atsinanana 57 G3 *Fr.* Fénérive. Toamasina, E Madagascar
Fens, The 67 E6 *wetland* E England, United Kingdom
Feodosiya 87 F5 *var.* Kefe, *It.* Kaffa; *anc.* Theodosia. Respublika Krym, S Ukraine
Ferdinand *see* Montana, Bulgaria
Ferdinandsberg *see* Oţelu Roşu
Féres 82 D3 Anatolikí Makedonía kai Thráki, NE Greece
Fergana *see* Farg'ona
Fergus Falls 23 F2 Minnesota, N USA
Ferizaj 79 D5 *Serb.* Uroševac. C Kosovo
Ferkessédougou 52 D4 N Côte d'Ivoire
Fermo 74 C4 *anc.* Firmum Picenum. Marche, C Italy
Fernandina, Isla 38 A5 *var.* Narborough Island. *island* Galápagos Islands, Ecuador, E Pacific Ocean
Fernando de Noronha 41 H2 *island* E Brazil
Fernando Po/Fernando Póo *see* Bioco, Isla de
Ferrara 74 C2 *anc.* Forum Alieni. Emilia-Romagna, N Italy
Ferreñafe 38 B3 Lambayeque, W Peru
Ferro *see* Hierro
Ferrol 70 B1 *var.* El Ferrol; *prev.* El Ferrol del Caudillo. Galicia, NW Spain
Fertő *see* Neusiedler See
Ferwerd *see* Ferwert
Ferwert 64 D1 *Dutch.* Ferwerd. Friesland, N Netherlands
Fès 48 C2 *Eng.* Fez. N Morocco
Feteşti 86 D5 Ialomiţa, SE Romania
Fethiye 94 B4 Muğla, SW Turkey
Fetlar 66 D1 *island* NE Scotland, United Kingdom
Feuilles, Rivière aux 16 D2 *river* Québec, E Canada
Feyẕābād 101 F3 *var.* Faizabad, Faizābād, Feyzābād, Fyzabad. Badakhshān, NE Afghanistan
Feyzābād *see* Feyẕābād
Fez *see* Fès
Fianarantsoa 57 F3 Fianarantsoa, C Madagascar
Fianga 54 B4 Mayo-Kébbi, SW Chad
Fier 79 C6 *var.* Fieri. Fier, SW Albania
Fieri *see* Fier
Figeac 69 C5 Lot, S France
Figig *see* Figuig
Figueira da Foz 70 B3 Coimbra, W Portugal
Figueres 71 G2 Cataluña, E Spain
Figuig 48 D2 *var.* Figig. E Morocco
Fiji 123 E5 *off.* Sovereign Democratic Republic of Fiji, *Fij.* Viti. *country* SW Pacific Ocean
Fiji, Sovereign Democratic Republic of *see* Fiji
Filadelfia 30 D4 Guanacaste, W Costa Rica
Filiaşi 86 B5 Dolj, SW Romania
Filipstad 63 B6 Värmland, C Sweden
Finale Ligure 74 A3 Liguria, NW Italy
Finchley 67 A7 United Kingdom
Findlay 18 C4 Ohio, N USA
Finike 94 B4 Antalya, SW Turkey
Finland 62 D4 *off.* Republic of Finland, *Fin.* Suomen Tasavalta, Suomi. *country* N Europe
Finland, Gulf of 63 D6 *Est.* Soome Laht, *Fin.* Suomenlahti, *Ger.* Finnischer Meerbusen, *Rus.* Finskiy Zaliv, *Swe.* Finska Viken. *gulf* E Baltic Sea
Finland, Republic of *see* Finland
Finnischer Meerbusen *see* Finland, Gulf of
Finnmarksvidda 62 D2 *physical region* N Norway
Finska Viken/Finskiy Zaliv *see* Finland, Gulf of
Finsterwalde 72 D4 Brandenburg, E Germany
Fiordland 129 A7 *physical region* South Island, New Zealand
Fiorina 74 E1 NE San Marino
Firat Nehri *see* Euphrates
Firenze 74 C3 *Eng.* Florence; *anc.* Florentia. Toscana, C Italy
Firmum Picenum *see* Fermo
First State *see* Delaware
Fischbacher Alpen 73 E7 *mountain range* E Austria
Fischhausen *see* Primorsk
Fish 56 B4 *var.* Vis. *river* S Namibia
Fishguard 67 C6 *Wel.* Abergwaun. SW Wales, United Kingdom
Fisterra, Cabo 70 B1 *headland* NW Spain
Fitzroy Crossing 124 C3 Western Australia
Fitzroy River 124 C3 *river* Western Australia
Fiume *see* Rijeka
Flagstaff 26 B2 Arizona, SW USA
Flanders 65 A6 *Dut.* Vlaanderen, *Fr.* Flandre. *cultural region* Belgium/France
Flandre *see* Flanders
Flathead Lake 22 B1 *lake* Montana, NW USA
Flat Island 106 C8 *island* NE Spratly Islands
Flatts Village 20 B5 *var.* The Flatts Village. C Bermuda
Flensburg 72 B2 Schleswig-Holstein, N Germany
Flessingue *see* Vlissingen
Flickertail State *see* North Dakota
Flinders Island 127 C8 *island* Furneaux Group, Tasmania, SE Australia
Flinders Ranges 127 B6 *mountain range* South Australia
Flinders River 126 C3 *river* Queensland, NE Australia
Flin Flon 15 F5 Manitoba, C Canada
Flint 18 C3 Michigan, N USA
Flint Island 123 G4 *island* Line Islands, E Kiribati
Floreana, Isla *see* Santa María, Isla
Florence 20 C1 Alabama, S USA
Florence 21 F2 South Carolina, SE USA
Florence *see* Firenze
Florencia 36 B4 Caquetá, S Colombia
Florentia *see* Firenze
Flores 30 B1 Petén, N Guatemala
Flores 117 E5 *island* Nusa Tenggara, C Indonesia
Flores 70 A5 *island* Azores, Portugal, NE Atlantic Ocean
Flores, Laut *see* Flores Sea
Flores Sea 116 D5 *Ind.* Laut Flores. *sea* C Indonesia
Floriano 41 F2 Piauí, E Brazil
Florianópolis 41 F5 *prev.* Destêrro. *state capital* Santa Catarina, S Brazil
Florida 42 D4 Florida, S Uruguay
Florida 21 E4 *off.* State of Florida, *also known as* Peninsular State, Sunshine State. *state* SE USA
Florida Bay 21 E5 *bay* Florida, SE USA
Florida Keys 21 E5 *island group* Florida, SE USA
Florida, Straits of 32 B1 *strait* Atlantic Ocean/Gulf of Mexico
Flórina 82 B4 *var.* Phlórina. Dytikí Makedonía, N Greece
Florissant 23 G4 Missouri, C USA
Floúda, Akrotírio 83 D7 *headland* Astypálaia, Kykládes, Greece, Aegean Sea
Flushing *see* Vlissingen
Flylân *see* Vlieland
Foča 78 C4 *var.* Srbinje. SE Bosnia and Herzegovina
Focşani 86 C4 Vrancea, E Romania
Fogaras/Fogarasch *see* Făgăraş
Foggia 75 D5 Puglia, SE Italy
Fogo 52 A3 *island* Ilhas de Sotavento, SW Cape Verde
Foix 69 B6 Ariège, S France
Folégandros 83 C7 *island* Kykládes, Greece, Aegean Sea
Foleyet 16 C4 Ontario, S Canada
Foligno 74 C4 Umbria, C Italy
Folkestone 67 E7 SE England, United Kingdom
Fond du Lac 18 B2 Wisconsin, N USA
Fongafale 123 E3 *var.* Funafuti. *country capital* (Tuvalu) Funafuti Atoll, SE Tuvalu
Fonseca, Golfo de *see* Fonseca, Gulf of
Fonseca, Gulf of 30 C3 *Sp.* Golfo de Fonseca. *gulf* C Central America
Fontainebleau 68 C3 Seine-et-Marne, N France
Fontenay-le-Comte 68 B4 Vendée, NW France
Fontvieille 69 B8 SW Monaco Europe
Fonyód 77 C7 Somogy, W Hungary
Foochow *see* Fuzhou
Foochow *see* Fuzhou
Forchheim 73 C5 Bayern, SE Germany

Forel, Mont – Girin

Forel, Mont 60 D4 mountain SE Greenland
Forfar 66 C3 E Scotland, United Kingdom
Forge du Sud see Dudelange
Forlì 74 C3 anc. Forum Livii. Emilia-Romagna, N Italy
Formentera 71 G4 anc. Ophiusa, Lat. Frumentum. island Islas Baleares, Spain, W Mediterranean Sea
Formosa 42 D2 Formosa, NE Argentina
Formosa/Formo'sa see Taiwan
Formosa, Serra 41 E3 mountain range C Brazil
Formosa Strait see Taiwan Strait
Føroyar see Faroe Islands
Forrest City 20 B1 Arkansas, C USA
Fort Albany 16 C3 Ontario, C Canada
Fortaleza 39 F2 Pando, N Bolivia
Fortaleza 41 G2 prev. Ceará. state capital Ceará, NE Brazil
Fort-Archambault see Sarh
Fort-Bayard see Zhanjiang
Fort-Cappolani see Tidjikja
Fort Charlet see Djanet
Fort-Chimo see Kuujjuaq
Fort Collins 22 D4 Colorado, C USA
Fort-Crampel see Kaga Bandoro
Fort Davis 27 E3 Texas, SW USA
Fort-de-France 33 H4 prev. Fort-Royal. dependent territory capital (Martinique) W Martinique
Fort Dodge 23 F3 Iowa, C USA
Fortescue River 124 A4 river Western Australia
Fort-Foureau see Kousséri
Fort Frances 16 B4 Ontario, S Canada
Fort Good Hope 15 E3 var. Rádeyilikóé. Northwest Territories, NW Canada
Fort Gouraud see Fdérik
Forth 66 C4 river C Scotland, United Kingdom
Forth, Firth of 66 C4 estuary E Scotland, United Kingdom
Fortín General Eugenio Garay see General Eugenio A. Garay
Fort Jameson see Chipata
Fort-Lamy see Ndjamena
Fort Lauderdale 21 F5 Florida, SE USA
Fort Liard 15 E4 var. Liard. Northwest Territories, W Canada
Fort Madison 23 G4 Iowa, C USA
Fort McMurray 15 E4 Alberta, C Canada
Fort McPherson 14 D3 var. McPherson. Northwest Territories, NW Canada
Fort Morgan 22 D4 Colorado, C USA
Fort Myers 21 E5 Florida, SE USA
Fort Nelson 15 E4 British Columbia, W Canada
Fort Peck Lake 22 C1 reservoir Montana, NW USA
Fort Pierce 21 F4 Florida, SE USA
Fort Providence 15 E4 var. Providence. Northwest Territories, W Canada
Fort-Repoux see Akjoujt
Fort Rosebery see Mansa
Fort Rousset see Owando
Fort-Royal see Fort-de-France
Fort St. John 15 E4 British Columbia, W Canada
Fort Scott 23 F5 Kansas, C USA
Fort Severn 16 C2 Ontario, C Canada
Fort-Shevchenko 92 A4 Mangistau, W Kazakhstan
Fort-Sibut see Sibut
Fort Simpson 15 E4 var. Simpson. Northwest Territories, W Canada
Fort Smith 15 E4 Northwest Territories, W Canada
Fort Smith 20 B1 Arkansas, C USA
Fort Stockton 27 E3 Texas, SW USA
Fort-Trinquet see Bîr Mogreïn
Fort Vermilion 15 E4 Alberta, W Canada
Fort Victoria see Masvingo
Fort Walton Beach 20 C3 Florida, SE USA
Fort Wayne 18 C4 Indiana, N USA
Fort William 66 C3 N Scotland, United Kingdom
Fort Worth 27 G2 Texas, SW USA
Fort Yukon 14 D3 Alaska, USA
Forum Alieni see Ferrara
Forum Livii see Forlì
Fossa Claudia see Chioggia
Fougamou 55 A6 Ngounié, C Gabon
Fougères 68 B3 Ille-et-Vilaine, NW France
Fou-hsin see Fuxin
Foulwind, Cape 129 B5 headland South Island, New Zealand
Foumban 54 A4 Ouest, NW Cameroon
Fou-shan see Fushun
Foveaux Strait 129 A8 strait S New Zealand
Foxe Basin 15 G3 sea Nunavut, N Canada
Fox Glacier 129 B6 West Coast, South Island, New Zealand
Fraga 71 F2 Aragón, NE Spain
Fram Basin 133 C3 var. Amundsen Basin. undersea basin Arctic Ocean
France 68 B4 off. French Republic, It./Sp. Francia; prev. Gaul, Gaule, Lat. Gallia. country W Europe
Franceville 55 B6 var. Massoukou, Masuku. Haut-Ogooué, E Gabon
Francfort see Frankfurt am Main
Franche-Comté 68 D4 cultural region E France
Francia see France
Francis Case, Lake 23 E3 reservoir South Dakota, N USA
Francisco Escárcega 29 G4 Campeche, SE Mexico
Francistown 56 D3 North East, NE Botswana
Franconian Jura see Fränkische Alb
Frankenalb see Fränkische Alb
Frankenstein/Frankenstein in Schlesien see Ząbkowice Śląskie
Frankfort 18 C5 state capital Kentucky, S USA
Frankfort on the Main see Frankfurt am Main
Frankfurt see Frankfurt am Main, Germany
Frankfurt see Słubice, Poland
Frankfurt am Main 73 B5 var. Frankfurt, Fr. Francfort; prev. Eng. Frankfort on the Main. Hessen, SW Germany
Frankfurt an der Oder 72 D3 Brandenburg, E Germany
Fränkische Alb 73 C6 var. Frankenalb, Eng. Franconian Jura. mountain range S Germany
Franklin 20 C1 Tennessee, S USA
Franklin D. Roosevelt Lake 24 C1 reservoir Washington, NW USA
Franz Josef Land 92 D1 Eng. Franz Josef Land. island group N Russian Federation
Franz Josef Land see Frantsa-Iosifa, Zemlya
Fraserburgh 66 D3 NE Scotland, United Kingdom
Fraser Island 126 E4 var. Great Sandy Island. island Queensland, E Australia
Frauenbach see Baia Mare
Frauenburg see Saldus, Latvia
Fredericksburg 19 E5 Virginia, NE USA
Fredericton 17 F4 province capital New Brunswick, SE Canada
Frederikshåb see Paamiut
Fredrikshald see Halden
Fredrikstad 63 B6 Østfold, S Norway
Freeport 32 C1 Grand Bahama Island, N Bahamas
Freeport 27 H4 Texas, SW USA
Free State see Maryland
Freetown 52 C4 country capital (Sierra Leone) W Sierra Leone
Freiburg see Freiburg im Breisgau, Germany
Freiburg im Breisgau 73 A6 var. Freiburg, Fr. Fribourg-en-Brisgau. Baden-Württemberg, SW Germany
Freiburg in Schlesien see Świebodzice
Fremantle 125 A6 Western Australia
Fremont 23 F4 Nebraska, C USA
French Guiana 37 H3 var. Guiana, Guyane. French overseas department N South America
French Guinea see Guinea
French Polynesia 121 F4 French overseas territory S Pacific Ocean
French Republic see France
French Somaliland see Djibouti
French Southern and Antarctic Lands 119 B7 Fr. Terres Australes et Antarctiques Françaises. French overseas territory S Indian Ocean
French Sudan see Mali
French Territory of the Afars and Issas see Djibouti
French Togoland see Togo
Fresnillo 28 D3 var. Fresnillo de González Echeverría. Zacatecas, C Mexico
Fresnillo de González Echeverría see Fresnillo
Fresno 25 C6 California, W USA
Frías 42 C3 Catamarca, N Argentina
Fribourg-en-Brisgau see Freiburg im Breisgau
Friedek-Mistek see Frýdek-Místek
Friedrichshafen 73 B7 Baden-Württemberg, S Germany
Friendly Islands see Tonga
Frisches Haff see Vistula Lagoon
Frobisher Bay 60 B3 inlet Baffin Island, Nunavut, NE Canada
Frobisher Bay see Iqaluit
Frohavet 62 B4 sound C Norway
Frome, Lake 127 B6 salt lake South Australia
Frontera 29 G4 Tabasco, SE Mexico
Frontignan 69 C6 Hérault, S France
Frostviken see Kvarnbergsvattnet
Frøya 62 A4 island W Norway
Frumentum see Formentera
Frunze see Bishkek
Frýdek-Místek 77 C5 Ger. Friedek-Mistek. Moravskoslezský Kraj, E Czech Republic
Fu-chien see Fujian
Fu-chou see Fuzhou
Fuengirola 70 D5 Andalucía, S Spain
Fuerte Olimpo 42 D2 var. Olimpo. Alto Paraguay, NE Paraguay
Fuerte, Río 26 C5 river C Mexico
Fuerteventura 48 B3 island Islas Canarias, Spain, NE Atlantic Ocean
Fuhkien see Fujian
Fu-hsin see Fuxin
Fuji 109 D6 var. Huzi. Shizuoka, Honshū, S Japan
Fujian 106 D6 var. Fu-chien, Fuhkien, Fukien, Min, Fujian Sheng. province SE China
Fujian Sheng see Fujian
Fuji, Mount/Fujiyama see Fuji-san
Fuji-san 109 C6 var. Fujiyama, Eng. Mount Fuji. mountain Honshū, SE Japan
Fukang 104 C2 Xinjiang Uygur Zizhiqu, W China
Fukien see Fujian
Fukui 109 C6 var. Hukui. Fukui, Honshū, SW Japan
Fukuoka 109 A7 var. Hukuoka, hist. Najima. Fukuoka, Kyūshū, SW Japan
Fukushima 108 D4 var. Hukusima. Fukushima, Honshū, C Japan
Fulda 73 B5 Hessen, C Germany
Funafuti see Fongafale
Funafuti Atoll 123 E3 atoll C Tuvalu
Funchal 48 A2 Madeira, Portugal, NE Atlantic Ocean
Fundy, Bay of 17 F5 bay Canada/USA
Fünen see Fyn
Fünfkirchen see Pécs
Furnes see Veurne
Fürth 73 C5 Bayern, S Germany
Furukawa 108 D4 var. Hurukawa, Ōsaki. Miyagi, Honshū, C Japan
Fusan see Busan
Fushë Kosovë 79 D5 Serb. Kosovo Polje. C Kosovo
Fushun 106 D3 var. Fou-shan, Fu-shun. Liaoning, NE China
Fu-shun see Fushun
Fusin see Fuxin
Füssen 73 C7 Bayern, S Germany
Futog 78 D3 Vojvodina, NW Serbia
Futuna, Île 123 E4 island S Wallis and Futuna
Fuxin 106 D3 var. Fou-hsin, Fu-hsin, Fusin. Liaoning, NE China
Fuzhou 106 D6 var. Foochow, Fu-chou. province capital Fujian, SE China
Fyn 63 B8 Ger. Fünen. island C Denmark
FYR Macedonia/FYROM see Macedonia, FYR
Fyzabad see Feyzābād

G

Gaafu Alifu Atoll see North Huvadhu Atoll
Gaalkacyo 51 E5 var. Galka'yo, It. Galcaio. Mudug, C Somalia
Gabela 56 B2 Cuanza Sul, W Angola
Gaberones see Gaborone
Gabès 49 E2 var. Qābis. E Tunisia
Gabès, Golfe de 49 F2 Ar. Khalīj Qābis. gulf E Tunisia
Gabon 55 B6 off. Gabonese Republic. country C Africa
Gabonese Republic see Gabon
Gaborone 56 C4 prev. Gaberones. country capital (Botswana) South East, SE Botswana
Gabrovo 82 D2 Gabrovo, N Bulgaria
Gadag 110 C1 Karnātaka, W India
Gades/Gadier/Gadir/Gadire see Cádiz
Gadsden 20 D2 Alabama, S USA
Gaeta 75 C5 Lazio, C Italy
Gaeta, Golfo di 75 C5 var. Gulf of Gaeta. gulf C Italy
Gaeta, Gulf of see Gaeta, Golfo di
Gäfle see Gävle
Gafsa 49 E2 var. Qafṣah. W Tunisia
Gagnoa 52 D5 C Côte d'Ivoire
Gagra 95 E1 NW Georgia
Gaillac 69 C6 var. Gaillac-sur-Tarn. Tarn, S France
Gaillac-sur-Tarn see Gaillac
Gaillimh see Galway
Gaillimhe, Cuan na see Galway Bay
Gainesville 21 E3 Florida, SE USA
Gainesville 20 D2 Georgia, SE USA
Gainesville 27 G2 Texas, SW USA
Lake Gairdner 127 A6 salt lake South Australia
Gaizina Kalns see Gaiziņkalns
Gaiziņkalns 84 C3 var. Gaizina Kalns. mountain E Latvia
Galán, Cerro 42 B3 mountain NW Argentina
Galanta 77 C6 Hung. Galánta. Trnavský Kraj, W Slovakia
Galapagos Fracture Zone 131 E3 tectonic feature E Pacific Ocean
Galápagos Islands 131 F3 var. Islas de los Galápagos, Sp. Archipiélago de Colón, Eng. Galapagos Islands, Tortoise Islands. island group Ecuador, E Pacific Ocean
Galapagos Islands see Galápagos Islands
Galápagos, Islas de los see Galápagos Islands
Galapagos Rise 131 F3 undersea rise E Pacific Ocean
Galashiels 66 C4 SE Scotland, United Kingdom
Galaṭi 86 D4 Ger. Galatz. Galaṭi, E Romania
Galatz see Galaṭi
Galcaio see Gaalkacyo
Galesburg 18 B3 Illinois, N USA
Galicia 70 B1 anc. Gallaecia. autonomous community NW Spain
Galicia Bank 58 B4 undersea bank E Atlantic Ocean
Galilee, Sea of see Tiberias, Lake
Galka'yo see Gaalkacyo
Galkynyş 100 D3 prev. Rus. Deynau, Dyanev, Turkm. Dänew. Lebap Welaýaty, NE Turkmenistan
Gallaecia see Galicia
Galle 110 D4 prev. Point de Galle. Southern Province, SW Sri Lanka
Gallego Rise 131 F3 undersea rise E Pacific Ocean
Gallegos see Río Gallegos
Gallia see France
Gallipoli 75 E6 Puglia, SE Italy
Gällivare 62 C3 Lapp. Váhtjer. Norrbotten, N Sweden
Gallup 26 C1 New Mexico, SW USA
Galtat-Zemmour 48 B3 C Western Sahara
Galveston 27 H4 Texas, SW USA
Galway 67 A5 Ir. Gaillimh. W Ireland
Galway Bay 67 A6 Ir. Cuan na Gaillimhe. bay W Ireland
Gámas see Kaamanen
Gambell 14 C2 Saint Lawrence Island, Alaska, USA
Gambia 52 B3 off. Republic of The Gambia, The Gambia. country W Africa
Gambia 52 C3 Fr. Gambie. river W Africa
Gambia, Republic of The see Gambia
Gambia, The see Gambia
Gambie see Gambia
Gambier, Îles 121 G4 island group E French Polynesia
Gamboma 55 B6 Plateaux, E Congo
Gamlakarleby see Kokkola
Gan 110 B5 Addu Atoll, C Maldives
Gan see Gansu, China
Gan see Jiangxi, China
Ganaane see Juba
Gäncä 95 G2 Rus. Gyandzha; prev. Kirovabad, Yelisavetpol. W Azerbaijan
Gand see Gent
Gandajika 55 D7 Kasai-Oriental, S Dem. Rep. Congo
Gander 17 G3 Newfoundland and Labrador, SE Canada
Gāndhīdhām 112 C4 Gujarāt, W India
Gandia 71 F3 prev. Gandía. País Valenciano, E Spain
Gandía see Gandia
Ganges 113 F3 Ben. Padma. river Bangladesh/India
Ganges Cone see Ganges Fan
Ganges Fan 118 D3 var. Ganges Cone. undersea fan N Bay of Bengal
Ganges, Mouths of the 113 G4 delta Bangladesh/India
Gangra see Çankırı
Gangtok 113 F3 state capital Sikkim, N India
Gansos, Lago dos see Goose Lake
Gansu 106 B4 var. Gan, Gansu Sheng, Kansu. province N China
Gansu Sheng see Gansu
Gantsevichi see Hantsavichy
Ganzhou 106 D6 Jiangxi, S China
Gao 53 E3 Gao, E Mali
Gaocheng see Litang
Gaoual 52 C4 N Guinea
Gaoxiong 106 D6 var. Kaohsiung, Jap. Takao, Takow. S Taiwan
Gap 69 D5 anc. Vapincum. Hautes-Alpes, SE France
Gaplaňgyr Platosy 100 C2 Rus. Plato Kaplangky. ridge Turkmenistan/Uzbekistan
Gar see Gar Xincun
Garabil Belentligi 100 D3 Rus. Vozvyshennost' Karabil'. mountain range S Turkmenistan
Garabogaz Aylagy 100 B2 Rus. Zaliv Kara-Bogaz-Gol. bay NW Turkmenistan
Garachiné 31 G5 Darién, SE Panama
Garagum 100 C3 var. Garagumy, Qara Qum, Eng. Black Sand Desert, Kara Kum; prev. Peski Karakumy. desert C Turkmenistan
Garagum Canal 100 D3 var. Kara Kum Canal, Rus. Karagumskiy Kanal, Karakumskiy Kanal. canal C Turkmenistan
Garagumy see Garagum
Gara Khitrino 82 D2 Shumen, NE Bulgaria
Gárassavon see Kaaresuvanto
Garda, Lago di 74 C2 var. Benaco, Eng. Lake Garda, Ger. Gardasee. lake NE Italy
Garda, Lake see Garda, Lago di
Gardasee see Garda, Lago di
Garden City 23 E5 Kansas, C USA
Garden State, The see New Jersey
Gardēz 101 E4 prev. Gardīz. E Afghanistan
Gardīz see Gardēz
Gardner Island see Nikumaroro
Garegegasnjárga see Karigasniemi
Gargždai 84 B3 Klaipėda, W Lithuania
Garissa 51 D6 Coast, E Kenya
Garland 27 G2 Texas, SW USA
Garoe see Garoowe
Garonne 69 B5 anc. Garumna. river S France
Garoowe 51 E5 var. Garoe. Nugaal, N Somalia
Garoua 54 B4 var. Garua. Nord, N Cameroon
Garrygala see Magtymguly
Garry Lake 15 F3 lake Nunavut, N Canada
Garsen 51 D6 Coast, S Kenya
Garua see Garoua
Garumna see Garonne
Garwolin 76 D4 Mazowieckie, E Poland
Gar Xincun 104 A4 prev. Gar. Xizang Zizhiqu, W China
Gary 18 B3 Indiana, N USA
Garzón 36 B4 Huila, S Colombia
Gasan-Kuli see Esenguly
Gascogne 69 B6 Eng. Gascony. cultural region S France
Gascony see Gascogne
Gascoyne River 125 A5 river Western Australia
Gaspé 17 F3 Québec, SE Canada
Gaspé, Péninsule de 17 E4 var. Péninsule de la Gaspésie. peninsula Québec, SE Canada
Gaspésie, Péninsule de la see Gaspé, Péninsule de
Gastonia 21 E1 North Carolina, SE USA
Gastoúni 83 B6 Dytikí Elláda, S Greece
Gatchina 88 B4 Leningradskaya Oblast', NW Russian Federation
Gatineau 16 D4 Québec, SE Canada
Gatooma see Kadoma
Gatún, Lake 31 F4 reservoir C Panama
Gauhāti see Guwāhāti
Gauja 84 D3 Ger. Aa. river Estonia/Latvia
Gaul/Gaule see France
Gauteng see Johannesburg, South Africa
Gāvbandī 98 D4 Hormozgān, S Iran
Gávdos 83 C8 island SE Greece
Gavere 65 B6 Oost-Vlaanderen, NW Belgium
Gävle 63 C6 var. Gäfle; prev. Gefle. Gävleborg, C Sweden
Gawler 127 B6 South Australia
Gaya 113 F3 Bihār, N India
Gaya see Kyjov
Gayndah 127 E5 Queensland, E Australia
Gaysin see Haysyn
Gaza 97 A6 Ar. Ghazzah, Heb. 'Azza. NE Gaza Strip
Gaz-Achak see Gazojak
Gazandzhyk/Gazanjyk see Bereket
Gaza Strip 97 A7 Ar. Qita Ghazzah. disputed region SW Asia
Gaziantep 94 D4 var. Gazi Antep; prev. Aintab, Antep. Gaziantep, S Turkey
Gazi Antep see Gaziantep
Gazimağusa 80 D5 var. Famagusta, Gk. Ammóchostos. E Cyprus
Gazimağusa Körfezi 80 C5 var. Famagusta Bay, Gk. Kólpos Ammóchostos. bay E Cyprus
Gazli 100 D2 Buxoro Viloyati, C Uzbekistan
Gazojak 100 D2 Rus. Gaz-Achak. Lebap Welaýaty, NE Turkmenistan
Gbanga 52 D5 var. Gbarnga. N Liberia
Gbarnga see Gbanga
Gdańsk 76 C2 Fr. Dantzig, Ger. Danzig. Pomorskie, N Poland
Gdan'skaya Bukhta/Gdańsk, Gulf of see Danzig, Gulf of
Gdańska, Zakota see Danzig, Gulf of
Gdingen see Gdynia
Gdynia 76 C2 Ger. Gdingen. Pomorskie, N Poland
Gedaref 50 C4 var. Al Qaḍārif, El Gedaref. Gedaref, E Sudan
Gediz 94 B3 Kütahya, W Turkey
Gediz Nehri 94 A3 river W Turkey
Geel 65 C5 var. Gheel. Antwerpen, N Belgium
Geelong 127 C7 Victoria, SE Australia
Ge'e'mu see Golmud
Gefle see Gävle
Geilo 63 A5 Buskerud, S Norway
Gejiu 106 B6 var. Kochiu. Yunnan, S China
Gëkdepe see Gökdepe
Gela 75 C7 prev. Terranova di Sicilia. Sicilia, Italy, C Mediterranean Sea
Geldermalsen 64 C4 Gelderland, C Netherlands
Geleen 65 D6 Limburg, SE Netherlands
Gelib see Jilib
Gellinsor 51 E5 Mudug, C Somalia
Gembloux 65 C6 Namur, Belgium
Gemena 55 C5 Equateur, NW Dem. Rep. Congo
Gem of the Mountains see Idaho
Gemona del Friuli 74 D2 Friuli-Venezia Giulia, NE Italy
Gem State see Idaho
Genalē Wenz see Juba
Genck see Genk
General Alvear 42 B4 Mendoza, W Argentina
General Carrera, Lago see Buenos Aires, Lago
General Eugenio A. Garay 42 C1 var. Fortín General Eugenio Garay; prev. Yrendagüé. Nueva Asunción, NW Paraguay
General José F.Uriburu see Zárate
General Machado see Camacupa
General Santos 117 F3 off. General Santos City. Mindanao, S Philippines
General Santos City see General Santos
Gênes see Genova
Geneva see Genève
Geneva, Lake 73 A7 Fr. Lac de Genève, Lac Léman, le Léman, Ger. Genfer See. lake France/Switzerland
Genève 73 A7 Eng. Geneva, Ger. Genf, It. Ginevra. Genève, SW Switzerland
Genève, Lac de see Geneva, Lake
Genf see Genève
Genfer See see Geneva, Lake
Genichesk see Heniches'k
Genk 65 D6 var. Genck. Limburg, NE Belgium
Gennep 64 D4 Limburg, SE Netherlands
Genoa see Genova
Genoa, Gulf of 74 A3 Eng. Gulf of Genoa. gulf NW Italy
Genoa, Gulf of see Genova, Golfo di
Genova 80 D1 Eng. Genoa; anc. Genua, Fr. Gênes. Liguria, NW Italy
Genovesa, Isla 38 B5 var. Tower Island. island Galápagos Islands, Ecuador, E Pacific Ocean
Gent 65 B5 Eng. Ghent, Fr. Gand. Oost-Vlaanderen, NW Belgium
Genua see Genova
Geok-Tepe see Gökdepe
George 56 C5 Western Cape, S South Africa
George 60 A4 river Newfoundland and Labrador/Québec, E Canada
George, Lake 21 E3 lake Florida, SE USA
Georgenburg see Jurbarkas
Georges Bank 13 D5 undersea bank W Atlantic Ocean
George Sound 129 A7 sound South Island, New Zealand
Georges River 126 D2 river New South Wales, E Australia
Georgetown 37 F2 country capital (Guyana) N Guyana
George Town 32 C2 Great Exuma Island, C Bahamas
George Town 32 B3 var. Georgetown. dependent territory capital (Cayman Islands) Grand Cayman, SW Cayman Islands
George Town 116 B3 var. Penang, Pinang. Pinang, Peninsular Malaysia
Georgetown 21 F2 South Carolina, SE USA
Georgetown see George Town
George V Land 132 C4 physical region Antarctica
Georgia 95 F2 off. Republic of Georgia, Geor. Sak'art'velo, Rus. Gruzinskaya SSR, Gruziya. country SW Asia
Georgia 20 D2 off. State of Georgia, also known as Empire State of the South, Peach State. state SE USA
Georgian Bay 18 D2 lake bay Ontario, S Canada
Georgia, Republic of see Georgia
Georgia, Strait of 24 A1 strait British Columbia, W Canada
Georgi Dimitrov see Kostenets
Georgiu-Dezh see Liski
Georg von Neumayer 132 A2 German research station Antarctica
Gera 72 C4 Thüringen, E Germany
Geráki 83 B6 Pelopónnisos, S Greece
Geraldine 129 B6 Canterbury, South Island, New Zealand
Geraldton 125 A6 Western Australia
Geral, Serra 35 D5 mountain range S Brazil
Gerede 94 C2 Bolu, N Turkey
Gereshk 100 D5 Helmand, SW Afghanistan
Gering 22 D3 Nebraska, C USA
German East Africa see Tanzania
Germanicopolis see Çankırı
Germanicum, Mare/German Ocean see North Sea
German Southwest Africa see Namibia
Germany 72 B4 off. Federal Republic of Germany, Bundesrepublik Deutschland, Ger. Deutschland. country N Europe
Germany, Federal Republic of see Germany
Geroliménas 83 B7 Pelopónnisos, S Greece
Gerona see Girona
Gerpinnes 65 C7 Hainaut, S Belgium
Gerunda see Girona
Gerze 94 D2 Sinop, N Turkey
Gesoriacum see Boulogne-sur-Mer
Gessoriacum see Boulogne-sur-Mer
Getafe 70 D3 Madrid, C Spain
Gevaş 95 F3 Van, SE Turkey
Gevgeli see Gevgelija
Gevgelija 79 E6 var. Đevđelija, Djevdjelija, Turk. Gevgeli. SE Macedonia
Ghaba see Al Ghābah
Ghana 53 E5 off. Republic of Ghana. country W Africa
Ghanzi 56 C3 var. Khanzi. Ghanzi, W Botswana
Gharandal 97 B7 Al 'Aqabah, SW Jordan
Gharbt, Jabal al see Liban, Jebel
Ghardaïa 48 D2 N Algeria
Gharvān see Gharyān
Gharyān 49 F2 var. Gharvān. NW Libya
Ghawdex see Gozo
Ghaznī 101 E4 var. Ghazni. Ghaznī, E Afghanistan
Ghazzah see Gaza
Gheel see Geel
Ghent see Gent
Gheorgheni 86 C4 prev. Gheorghieni, Sîn-Miclăuş, Ger. Niklasmarkt, Hung. Gyergyószentmiklós. Harghita, C Romania
Gheorghieni see Gheorgheni
Ghōriān 100 D4 prev. Ghūrīān. Herāt, W Afghanistan
Ghūdara 101 F3 var. Gudara, Rus. Kudara. SE Tajikistan
Ghurdaqah see Al Ghurdaqah
Ghūrīān see Ghōriān
Giamame see Jamaame
Giannitsá 82 B4 var. Yiannitsá. Kentrikí Makedonía, N Greece
Gibraltar 71 G4 UK dependent territory SW Europe
Gibraltar, Bay of 71 G5 bay Gibraltar/Spain Europe Mediterranean Sea Atlantic Ocean
Gibraltar, Détroit de/Gibraltar, Estrecho de see Gibraltar, Strait of
Gibraltar, Strait of 70 C5 Fr. Détroit de Gibraltar, Sp. Estrecho de Gibraltar. strait Atlantic Ocean/Mediterranean Sea
Gibson Desert 125 B5 desert Western Australia
Giedraičiai 85 C5 Utena, E Lithuania
Giessen 73 B5 Hessen, W Germany
Gifu 109 C6 var. Gihu. Gifu, Honshū, SW Japan
Giganta, Sierra de la 28 B3 mountain range NW Mexico
Gihu see Gifu
G'ijduvon 100 D2 Rus. Gizhduvon. Buxoro Viloyati, C Uzbekistan
Gijón 70 D1 var. Xixón. Asturias, NW Spain
Gila River 26 A2 river Arizona, SW USA
Gilbert Islands see Tungaru
Gilbert River 126 C3 river Queensland, NE Australia
Gilf Kebir Plateau see Haḍabat al Jilf al Kabīr
Gillette 22 D3 Wyoming, C USA
Gilolo see Halmahera, Pulau
Gilroy 25 B6 California, W USA
Gimie, Mount 33 F1 mountain C Saint Lucia
Gimma see Jima
Ginevra see Genève
Gingin 125 A6 Western Australia
Giohar see Jawhar
Gipeswic see Ipswich
Girardot 36 B3 Cundinamarca, C Colombia
Giresun 95 E2 var. Kerasunt; anc. Cerasus, Pharnacia. Giresun, NE Turkey
Girgenti see Agrigento
Girin see Jilin

Girne – Gýtheio

Girne 80 C5 Gk. Kerýneia, Kyrenia. N Cyprus
Giron see Kiruna
Girona 71 G2 var. Gerona; anc. Gerunda. Cataluña, NE Spain
Gisborne 128 E3 Gisborne, North Island, New Zealand
Gissar Range 101 E3 Rus. Gissarskiy Khrebet. mountain range Tajikistan/Uzbekistan
Gissarskiy Khrebet see Gissar Range
Githio see Gýtheio
Giulianova 74 D4 Abruzzi, C Italy
Giumri see Gyumri
Giurgiu 86 C5 Giurgiu, S Romania
Gîza 50 B1 var. Al Jīzah, El Gîza, Gizeh. N Egypt
Gizhduvon see G'ijduvon
Giżycko 76 D2 Ger. Lötzen. Warmińsko-Mazurskie, NE Poland
Gjakovë 79 D5 Serb. Đakovica. W Kosovo
Gjilan 79 D5 Serb. Gnjilane. E Kosovo
Gjinokastër see Gjirokastër
Gjirokastër 79 C7 var. Gjirokastra; prev. Gjinokastër, Gk. Argyrokastron, It. Argirocastro. Gjirokastër, S Albania
Gjirokastra see Gjirokastër
Gjoa Haven 15 F3 var. Uqsuqtuuq. King William Island, Nunavut, NW Canada
Gjøvik 63 B5 Oppland, S Norway
Glace Bay 17 G4 Cape Breton Island, Nova Scotia, SE Canada
Gladstone 126 E4 Queensland, E Australia
Gláma 63 B5 var. Glommen. river S Norway
Glasgow 66 C4 S Scotland, United Kingdom
Glavinitsa 82 D1 prev. Pravda, Dogrular. Silistra, NE Bulgaria
Glavn'a Morava see Velika Morava
Glazov 89 D5 Udmurtskaya Respublika, NW Russian Federation
Gleiwitz see Gliwice
Glendale 26 B2 Arizona, SW USA
Glendive 22 D2 Montana, NW USA
Glens Falls 19 F3 New York, NE USA
Glevum see Gloucester
Glina 78 B3 var. Banijska Palanka. Sisak-Moslavina, NE Croatia
Glittertind 63 A5 mountain S Norway
Gliwice 77 C5 Ger. Gleiwitz. Śląskie, S Poland
Globe 26 B2 Arizona, SW USA
Globino see Hlobyne
Glogau see Głogów
Głogów 76 B4 Ger. Glogau, Glogow. Dolnośląskie, SW Poland
Glogow see Głogów
Glomma see Gláma
Glommen see Gláma
Gloucester 67 D6 hist. Caer Glou, Lat. Glevum. C England, United Kingdom
Głowno 76 D4 Łódź, C Poland
Glubokoye see Hlybokaye
Glukhov see Hlukhiv
Gnesen see Gniezno
Gniezno 76 C3 Ger. Gnesen. Weilkopolskie, C Poland
Gnjilane see Gjilan
Gobabis 56 B3 Omaheke, E Namibia
Gobi 104 D3 desert China/Mongolia
Gobō 109 C6 Wakayama, Honshū, SW Japan
Godāvari 102 B3 var. Godavari. river C India
Godavari see Godāvari
Godhavn see Qeqertarsuaq
Godhra 112 C4 Gujarāt, W India
Göding see Hodonín
Godoy Cruz 42 B4 Mendoza, W Argentina
Godthaab/Godthåb see Nuuk
Godwin Austen, Mount see K2
Goede Hoop, Kaap de see Good Hope, Cape of
Goeie Hoop, Kaap die see Good Hope, Cape of
Goeree 64 B4 island SW Netherlands
Goes 65 B5 Zeeland, SW Netherlands
Goettingen see Göttingen
Gogebic Range 18 B1 hill range Michigan/Wisconsin, N USA
Goiânia 41 E3 prev. Goyania. state capital Goiás, C Brazil
Goiás 41 E3 off. Estado de Goiás; prev. Goiaz, Goyaz. region C Brazil
Goiás 41 E3 off. Estado de Goiás; prev. Goiaz, Goyaz. state C Brazil
Goiás, Estado de see Goiás
Goiaz see Goiás
Goidhoo Atoll see Horsburgh Atoll
Gojōme 108 D4 Akita, Honshū, NW Japan
Gökçeada 82 D4 var. Imroz Adasī, Gk. Imbros. island NW Turkey
Gökdepe 100 C3 Rus. Gëkdepe, Geok-Tepe. Ahal Welaýaty, C Turkmenistan
Göksun 94 D4 Kahramanmaraş, C Turkey
Gol 63 A5 Buskerud, S Norway
Golan Heights 97 B5 Ar. Al Jawlān, Heb. HaGolan. mountain range SW Syria
Golaya Pristan see Hola Prystan'
Gołdap 76 E2 Ger. Goldap. Warmińsko-Mazurskie, NE Poland
Gold Coast 127 E5 cultural region Queensland, E Australia
Golden Bay 128 C4 bay South Island, New Zealand
Golden State, The see California
Goldingen see Kuldīga
Goldsboro 21 F1 North Carolina, SE USA
Goleniów 76 B3 Ger. Gollnow. Zachodnio-pomorskie, NW Poland
Gollnow see Goleniów
Golmo see Golmud
Golmud 104 D4 var. Ge'e'mu, Golmo, Chin. Ko-erh-mu. Qinghai, C China
Golovanevsk see Holovanivs'k
Golub-Dobrzyń 76 C3 Kujawski-pomorskie, C Poland
Goma 55 E6 Nord-Kivu, NE Dem. Rep. Congo
Gombi 53 H4 Adamawa, E Nigeria
Gombroon see Bandar-e 'Abbās
Gomel' see Homyel'
Gomera 48 A3 island Islas Canarias, Spain, NE Atlantic Ocean
Gómez Palacio 28 D3 Durango, C Mexico
Gonaïves 32 D3 var. Les Gonaïves. N Haiti
Gonâve, Île de la 32 D3 island C Haiti
Gondar see Gonder
Gonder 50 C4 var. Gondar. Āmara, NW Ethiopia
Gondia 113 E4 Mahārāshtra, C India
Gonggar 104 C5 var. Gyixong. Xizang Zizhiqu, W China
Gongola 53 G4 river E Nigeria
Gongtang see Damxung
Gonni/Gónnos see Gónnoi
Gónnoi 82 B4 var. Gonni, Gónnos; prev. Derelí. Thessalía, C Greece
Good Hope, Cape of 56 B5 Afr. Kaap de Goede Hoop, Kaap die Goeie Hoop. headland SW South Africa
Goodland 22 D4 Kansas, C USA
Goondiwindi 127 D5 Queensland, E Australia
Goor 64 E3 Overijssel, E Netherlands
Goose Green 43 D7 var. Prado del Ganso. East Falkland, Falkland Islands
Goose Lake 24 B4 var. Lago dos Gansos. lake California/Oregon, W USA
Gopher State see Minnesota
Göppingen 73 B6 Baden-Württemberg, SW Germany
Góra Kalwaria 92 D4 Mazowieckie, C Poland
Gorakhpur 113 E3 Uttar Pradesh, N India
Gorany see Harany
Goražde 78 C4 Federacija Bosna I Hercegovina, SE Bosnia and Herzegovina
Gorbovichi see Harbavichy
Goré 54 C4 Logone-Oriental, S Chad
Gorē 51 C5 Oromīya, C Ethiopia
Gore 129 B7 Southland, South Island, New Zealand
Gorgān 98 D2 var. Astarabad, Astrabad, Gurgan, prev. Asterābād; anc. Hyrcania. Golestān, N Iran
Gori 95 F2 C Georgia
Gorinchem 64 C4 var. Gorkum. Zuid-Holland, C Netherlands
Goris 95 G3 SE Armenia
Gorki see Horki
Gor'kiy see Nizhniy Novgorod
Gorkum see Gorinchem
Görlitz 72 D4 Sachsen, E Germany
Görlitz see Zgorzelec
Gorlovka see Horlivka
Gorna Dzhumaya see Blagoevgrad
Gornja Mužlja see Mužlja
Gornji Milanovac 78 C4 Serbia, C Serbia
Gorodets see Haradzyets
Gorodishche see Horodyshche
Gorodnya see Horodnya
Gorodok see Haradok
Gorodok/Gorodok Yagellonski see Horodok
Gorontalo 117 E4 Sulawesi, C Indonesia
Gorontalo, Teluk see Tomini, Gulf of
Gorssel 64 D3 Gelderland, E Netherlands
Goryn see Horyn'
Gorzów Wielkopolski 76 B3 Ger. Landsberg, Landsberg an der Warthe. Lubuskie, W Poland
Gosford 127 D6 New South Wales, SE Australia
Goshogawara 108 D3 var. Gosyogawara. Aomori, Honshū, C Japan
Gospić 78 A3 Lika-Senj, C Croatia
Gostivar 79 D6 W FYR Macedonia
Gosyogawara see Goshogawara
Göteborg 63 B7 Eng. Gothenburg. Västra Götaland, S Sweden
Gotel Mountains 53 G5 mountain range E Nigeria
Gotha 72 C4 Thüringen, C Germany
Gothenburg see Göteborg
Gotland 63 C7 island SE Sweden
Goto-retto 109 A7 island group SW Japan
Gotska Sandön 84 B1 island SE Sweden
Gōtsu 109 B6 var. Gôtu. Shimane, Honshū, SW Japan
Göttingen 72 B4 var. Goettingen. Niedersachsen, C Germany
Gottschee see Kočevje
Gottwaldov see Zlín
Gôtu see Gōtsu
Gouda 64 C4 Zuid-Holland, C Netherlands
Gough Fracture Zone 45 C6 tectonic feature S Atlantic Ocean
Gough Island 47 B8 island Tristan da Cunha, S Atlantic Ocean
Gouin, Réservoir 16 D4 reservoir Québec, SE Canada
Goulburn 127 D6 New South Wales, SE Australia
Goundam 53 E3 Tombouctou, NW Mali
Gouré 53 G3 Zinder, SE Niger
Goverla, Gora see Hoverla, Hora
Governador Valadares 41 F4 Minas Gerais, SE Brazil
Govi Altayn Nuruu 105 E3 mountain range S Mongolia
Goya 42 D3 Corrientes, NE Argentina
Goyania see Goiânia
Goyaz see Goiás
Goz Beïda 54 C3 Ouaddaï, SE Chad
Gozo 75 C8 var. Ghawdex. island N Malta
Graciosa 70 A5 var. Ilha Graciosa. island Azores, Portugal, NE Atlantic Ocean
Graciosa, Ilha see Graciosa
Gradačac 78 C3 Federacija Bosna I Hercegovina, N Bosnia and Herzegovina
Gradaús, Serra dos 41 E3 mountain range C Brazil
Gradiška see Bosanska Gradiška
Grafton 127 E5 New South Wales, SE Australia
Grafton 23 E1 North Dakota, N USA
Graham Land 132 A2 physical region Antarctica
Grajewo 76 E3 Podlaskie, NE Poland
Grampian Mountains 66 C3 mountain range C Scotland, United Kingdom
Gran see Esztergom, Hungary
Granada 30 D3 Granada, SW Nicaragua
Granada 70 D5 Andalucía, S Spain
Gran Canaria 48 A3 var. Grand Canary. island Islas Canarias, Spain, NE Atlantic Ocean
Gran Chaco 42 D2 var. Chaco. lowland plain South America
Grand Bahama Island 32 B1 island N Bahamas
Grand Banks of Newfoundland 12 E4 undersea basin NW Atlantic Ocean
Grand Bassa see Buchanan
Grand Canary see Gran Canaria
Grand Canyon 26 A1 canyon Arizona, SW USA
Grand Canyon State see Arizona
Grand Cayman 32 B3 island SW Cayman Islands
Grand Duchy of Luxembourg see Luxembourg
Grande, Bahía 43 B7 bay S Argentina
Grande-Comor see Ngazidja
Grande de Chiloé, Isla see Chiloé, Isla de
Grande Prairie 15 E4 Alberta, W Canada
Grand Erg Occidental 48 D3 desert W Algeria
Grand Erg Oriental 49 E3 desert Algeria/Tunisia
Rio Grande 29 E2 var. Río Bravo, Sp. Río Bravo del Norte, Bravo del Norte. river Mexico/USA
Grande Terre 33 G3 island E West Indies
Grand Falls 17 G3 Newfoundland, Newfoundland and Labrador, SE Canada
Grand Forks 23 E1 North Dakota, N USA
Grandichi see Hrandzichy
Grand Island 23 E4 Nebraska, C USA
Grand Junction 22 C4 Colorado, C USA
Grand Paradis see Gran Paradiso
Grand Rapids 18 C3 Michigan, N USA
Grand Rapids 23 F1 Minnesota, N USA
Grand-Saint-Bernard, Col du see Great Saint Bernard Pass
Grand-Santi 37 G3 W French Guiana
Granite State see New Hampshire
Gran Lago see Nicaragua, Lago de
Gran Malvina see West Falkland
Gran Paradiso 74 A2 Fr. Grand Paradis. mountain NW Italy
Gran San Bernardo, Passo di see Great Saint Bernard Pass
Gran Santiago see Santiago
Grants 26 C2 New Mexico, SW USA
Grants Pass 24 B4 Oregon, NW USA
Granville 68 B3 Manche, N France
Gratianopolis see Grenoble
Gratz see Graz
Graudenz see Grudziądz
Graulhet 69 C6 Tarn, S France
Grave 64 D4 Noord-Brabant, SE Netherlands
Grayling 14 C2 Alaska, USA
Graz 73 E7 prev. Gratz. Steiermark, SE Austria
Great Abaco 32 C1 var. Abaco Island. island N Bahamas
Great Alfold see Great Hungarian Plain
Great Ararat see Büyükağrı Dağı
Great Australian Bight 125 D7 bight S Australia
Great Barrier Island 128 D2 island N New Zealand
Great Barrier Reef 126 D2 reef Queensland, NE Australia
Great Basin 25 C5 basin W USA
Great Bear Lake 15 E3 Fr. Grand Lac de l'Ours. lake Northwest Territories, NW Canada
Great Belt 63 B8 var. Store Bælt, Eng. Great Belt, Storebelt. channel Baltic Sea/Kattegat
Great Belt see Storebælt
Great Bend 23 E5 Kansas, C USA
Great Britain see Britain
Great Dividing Range 126 D4 mountain range NE Australia
Greater Antilles 32 D3 island group West Indies
Greater Caucasus 95 G2 mountain range Azerbaijan/Georgia/Russian Federation Asia/Europe
Greater Sunda Islands 102 D5 var. Sunda Islands. island group Indonesia
Great Exhibition Bay 128 C1 inlet North Island, New Zealand
Great Exuma Island 32 C2 island C Bahamas
Great Falls 22 B1 Montana, NW USA
Great Grimsby see Grimsby
Great Hungarian Plain 77 C7 var. Great Alfold, Plain of Hungary, Hung. Alföld. plain SE Europe
Great Inagua 32 D2 var. Inagua Islands. island S Bahamas
Great Indian Desert see Thar Desert
Great Khingan Range see Da Hinggan Ling
Great Lake see Tônlé Sap
Great Lakes 13 C5 lakes Ontario, Canada/USA
Great Lakes State see Michigan
Great Meteor Seamount see Great Meteor Tablemount
Great Meteor Tablemount 44 B3 var. Great Meteor Seamount. seamount E Atlantic Ocean
Great Nicobar 111 G3 island Nicobar Islands, India, NE Indian Ocean
Great Plain of China 103 E2 plain E China
Great Plains 23 E3 var. High Plains. plains Canada/USA
Great Rift Valley 51 C5 var. Rift Valley. depression Asia/Africa
Great Ruaha 51 C7 river S Tanzania
Great Saint Bernard Pass 74 A1 Fr. Col du Grand-Saint-Bernard, It. Passo del Gran San Bernardo. pass Italy/Switzerland
Great Salt Desert see Kavīr, Dasht-e
Great Salt Lake 22 A3 salt lake Utah, W USA
Great Salt Lake Desert 22 A4 plain Utah, W USA
Great Sand Sea 49 H3 desert Egypt/Libya
Great Sandy Desert 124 C4 desert Western Australia
Great Sandy Desert see Ar Rub 'al Khālī
Great Sandy Island see Fraser Island
Great Slave Lake 15 E4 Fr. Grand Lac des Esclaves. lake Northwest Territories, NW Canada
Great Socialist People's Libyan Arab Jamahiriya see Libya
Great Sound 20 A5 sound Bermuda, NW Atlantic Ocean
Great Victoria Desert 125 C5 desert South Australia/Western Australia
Great Wall of China 106 C4 ancient monument N China Asia
Great Yarmouth 67 E6 var. Yarmouth. E England, United Kingdom
Grebenka see Hrebinka
Gredos, Sierra de 70 D3 mountain range W Spain
Greece 83 A5 off. Hellenic Republic, Gk. Ellás; anc. Hellas. country SE Europe
Greeley 22 D4 Colorado, C USA
Green Bay 18 B2 Wisconsin, N USA
Green Bay 18 B2 lake bay Michigan/Wisconsin, N USA
Greeneville 21 E1 Tennessee, S USA
Greenland 60 D3 Dan. Grønland, Inuit Kalaallit Nunaat. Danish external territory NE North America
Greenland Sea 61 F2 sea Arctic Ocean
Green Mountains 19 G2 mountain range Vermont, NE USA
Green Mountain State see Vermont
Greenock 66 C4 W Scotland, United Kingdom
Green River 22 B3 Wyoming, C USA
Green River 18 C5 river Kentucky, C USA
Green River 22 B4 river Utah, W USA
Greensboro 21 F1 North Carolina, SE USA
Greenville 20 B2 Mississippi, S USA
Greenville 21 F1 North Carolina, SE USA
Greenville 21 E1 South Carolina, SE USA
Greenville 27 G2 Texas, SW USA
Greenwich 67 B8 United Kingdom
Greenwood 20 B2 Mississippi, S USA
Greenwood 21 E2 South Carolina, SE USA
Gregory Range 126 C3 mountain range Queensland, E Australia
Greifenberg/Greifenberg in Pommern see Gryfice
Greifswald 72 D2 Mecklenburg-Vorpommern, NE Germany
Grenada 20 C2 Mississippi, S USA
Grenada 33 G5 country SE West Indies
Grenadines, The 33 H4 island group Grenada/St Vincent and the Grenadines
Grenoble 69 D5 anc. Cularo, Gratianopolis. Isère, E France
Gresham 24 B3 Oregon, NW USA
Grevená 82 B4 Dytikí Makedonía, N Greece
Grevenmacher 65 E8 Grevenmacher, E Luxembourg
Greymouth 129 B5 West Coast, South Island, New Zealand
Grey Range 127 C5 mountain range New South Wales/Queensland, E Australia
Greytown see San Juan del Norte
Griffin 20 D2 Georgia, SE USA
Grimari 54 C4 Ouaka, C Central African Republic
Grimsby 67 E5 prev. Great Grimsby. E England, United Kingdom
Grobin see Grobiņa
Grobiņa 84 B3 Ger. Grobin. Liepāja, W Latvia
Gródek Jagielloński see Horodok
Grodno see Hrodna
Grodzisk Wielkopolski 76 B3 Wielkopolskie, C Poland
Groesbeek 64 D4 Gelderland, SE Netherlands
Grójec 76 D4 Mazowieckie, C Poland
Groningen 64 E1 Groningen, NE Netherlands
Grønland see Greenland
Groote Eylandt 126 B2 island Northern Territory, N Australia
Grootfontein 56 B3 Otjozondjupa, N Namibia
Groot Karasberge 56 B4 mountain range S Namibia
Gros Islet 33 F1 N Saint Lucia
Grossa, Isola see Dugi Otok
Grossbetschkerek see Zrenjanin
Grosse Morava see Velika Morava
Grosser Sund see Suur Väin
Grosseto 74 B4 Toscana, C Italy
Grossglockner 73 C7 mountain W Austria
Grosskanizsa see Nagykanizsa
Gross-Karol see Carei
Grosskikinda see Kikinda
Grossmichel see Michalovce
Gross-Schlatten see Abrud
Grosswardein see Oradea
Groznyy 89 B8 Chechenskaya Respublika, SW Russian Federation
Grudovo see Sredets
Grudziądz 76 C3 Ger. Graudenz. Kujawsko-pomorskie, C Poland
Grums 63 B6 Värmland, C Sweden
Grünberg/Grünberg in Schlesien see Zielona Góra
Grüneberg see Zielona Góra
Gruzinskaya SSR/Gruziya see Georgia
Gryazi 89 B6 Lipetskaya Oblast', W Russian Federation
Gryfice 76 B2 Ger. Greifenberg, Greifenberg in Pommern. Zachodnio-pomorskie, NW Poland
Guabito 31 E4 Bocas del Toro, NW Panama
Guadalajara 28 D4 Jalisco, C Mexico
Guadalajara 71 E3 Ar. Wad Al-Hajarah; anc. Arriaca. Castilla-La Mancha, C Spain
Guadalcanal 122 C3 island C Solomon Islands
Guadalquivir 70 D4 river W Spain
Guadalupe 28 D3 Zacatecas, C Mexico
Guadalupe Peak 26 D3 mountain Texas, SW USA
Guadalupe River 27 G4 river SW USA
Guadarrama, Sierra de 71 E2 mountain range C Spain
Guadeloupe 33 H3 French overseas department E West Indies
Guadiana 70 C4 river Portugal/Spain
Guadix 71 E4 Andalucía, S Spain
Guaimaca 30 C2 Francisco Morazán, C Honduras
Guajira, Península de la 36 B1 peninsula N Colombia
Gualaco 30 D2 Olancho, C Honduras
Gualán 30 B2 Zacapa, C Guatemala
Gualdicciolo 74 D1 NW San Marino
Gualeguaychú 42 D4 Entre Ríos, E Argentina
Guam 122 B1 US unincorporated territory W Pacific Ocean
Guamúchil 28 C3 Sinaloa, C Mexico
Guanabacoa 32 B2 La Habana, W Cuba
Guanajuato 29 E4 Guanajuato, C Mexico
Guanare 36 C2 Portuguesa, N Venezuela
Guanare, Río 36 D2 river W Venezuela
Guangdong 106 C6 var. Guangdong Sheng, Kuang-tung, Kwangtung, Yue. province S China
Guangdong Sheng see Guangdong
Guangju see Gwangju
Guangxi see Guangxi Zhuangzu Zizhiqu
Guangxi Zhuangzu Zizhiqu 106 C6 var. Guangxi, Gui, Kuang-hsi, Kwangsi, Eng. Kwangsi Chuang Autonomous Region. autonomous region S China
Guangyuan 106 B5 var. Kuang-yuan, Kwangyuan. Sichuan, C China
Guangzhou 106 C6 var. Kuang-chou, Kwangchow, Eng. Canton. province capital Guangdong, S China
Guantánamo 32 D3 Guantánamo, SE Cuba
Guantánamo, Bahía de 32 D3 Eng. Guantanamo Bay. US military base SE Cuba
Guantanamo Bay see Guantánamo, Bahía de
Guaporé, Río 40 D3 var. Río Iténez. river Bolivia/Brazil
Guarda 70 C3 Guarda, N Portugal
Guarumal 31 F5 Veraguas, S Panama
Guasave 28 C3 Sinaloa, C Mexico
Guatemala 30 A2 off. Republic of Guatemala. country Central America
Guatemala Basin 13 B7 undersea basin E Pacific Ocean
Guatemala City see Ciudad de Guatemala
Guatemala, Republic of see Guatemala
Guaviare 34 B2 off. Comisaría Guaviare. province S Colombia
Guaviare, Comisaría see Guaviare
Guaviare, Río 36 D3 river E Colombia
Guayanas, Macizo de las see Guiana Highlands
Guayaquil 38 A2 var. Santiago de Guayaquil. Guayas, SW Ecuador
Guayaquil, Golfo de 38 A2 var. Gulf of Guayaquil. gulf SW Ecuador
Guayaquil, Gulf of see Guayaquil, Golfo de
Guaymas 28 B2 Sonora, NW Mexico
Gubadag 100 C2 Turkm. Tel'man; prev. Tel'mansk. Daşoguz Welaýaty, N Turkmenistan
Guben 72 D4 var. Wilhelm-Pieck-Stadt. Brandenburg, E Germany
Gudara see Ghŭdara
Gudaut'a 95 E1 NW Georgia
Guéret 68 C4 Creuse, C France
Guernsey 67 D8 British Crown Dependency Channel Islands, NW Europe
Guerrero Negro 28 A2 Baja California Sur, NW Mexico
Gui see Guangxi Zhuangzu Zizhiqu
Guiana see French Guiana
Guiana Highlands 40 D1 var. Macizo de las Guayanas. mountain range N South America
Guiba see Juba
Guidder see Guider
Guider 54 B4 var. Guidder. Nord, N Cameroon
Guidimouni 53 G3 Zinder, S Niger
Guildford 67 D7 SE England, United Kingdom
Guilin 106 C6 var. Kuei-lin, Kweilin. Guangxi Zhuangzu Zizhiqu, S China
Guimarães 70 B2 var. Guimaráes. Braga, N Portugal
Guimaráes see Guimarães
Guinea 52 C4 off. Republic of Guinea, var. Guinée; prev. French Guinea, People's Revolutionary Republic of Guinea. country W Africa
Guinea Basin 47 A5 undersea basin E Atlantic Ocean
Guinea-Bissau 52 B4 off. Republic of Guinea-Bissau, Fr. Guinée-Bissau, Port. Guiné-Bissau; prev. Portuguese Guinea. country W Africa
Guinea-Bissau, Republic of see Guinea-Bissau
Guinea, Gulf of 46 B4 Fr. Golfe de Guinée. gulf E Atlantic Ocean
Guinea, People's Revolutionary Republic of see Guinea
Guinea, Republic of see Guinea
Guiné-Bissau see Guinea-Bissau
Guinée see Guinea
Guinée-Bissau see Guinea-Bissau
Guinée, Golfe de see Guinea, Gulf of
Güiria 37 E1 Sucre, NE Venezuela
Guiyang 106 B6 var. Kuei-Yang, Kuei-yang, Kueyang, Kweiyang; prev. Kweichu. province capital Guizhou, S China
Guizhou 106 B6 Guangdong, SE China
Gujarāt 112 C4 var. Gujerat. cultural region W India
Gujerat see Gujarāt
Gujrānwāla 112 D2 Punjab, NE Pakistan
Gujrāt 112 D2 Punjab, E Pakistan
Gulbarga 110 C1 Karnātaka, C India
Gulbene 84 D3 Ger. Alt-Schwanenburg. Gulbene, NE Latvia
Gulf of Liaotung see Liaodong Wan
Gulfport 20 C3 Mississippi, S USA
Gulf, The see Persian Gulf
Gulistan see Guliston
Guliston 101 E2 Rus. Gulistan. Sirdaryo Viloyati, E Uzbekistan
Gulja see Yining
Gulkana 14 D3 Alaska, USA
Gulu 51 B6 N Uganda
Gulyantsi 82 C1 Pleven, N Bulgaria
Guma see Pishan
Gumbinnen see Gusev
Gumpolds see Humpolec
Gümülcine/Gümüljina see Komotiní
Gümüşane see Gümüşhane
Gümüşhane 95 E3 var. Gümüşane, Gumushkhane. Gümüşhane, NE Turkey
Gumushkhane see Gümüşhane
Güney Doğu Toroslar 95 E4 mountain range SE Turkey
Gunnbjørn Fjeld 60 D4 var. Gunnbjörns Bjerge. mountain C Greenland
Gunnbjörns Bjerge see Gunnbjørn Fjeld
Gunnedah 127 D6 New South Wales, SE Australia
Gunnison 22 C5 Colorado, C USA
Gurbansoltan Eje 100 C2 prev. Ýylanly, Rus. Il'yaly. Daşoguz Welaýaty, N Turkmenistan
Gurbantünggüt Shamo 104 B2 desert W China
Gurgan see Gorgān
Guri, Embalse de 37 E2 reservoir E Venezuela
Gurkfeld see Krško
Gurktaler Alpen 73 D7 mountain range S Austria
Gürün 94 D3 Sivas, C Turkey
Gur'yev/Gur'yevskaya Oblast' see Atyrau
Gusau 53 G4 Zamfara, NW Nigeria
Gusev 84 B4 Ger. Gumbinnen. Kaliningradskaya Oblast', W Russian Federation
Gustavus 14 D4 Alaska, USA
Güstrow 72 C3 Mecklenburg-Vorpommern, NE Germany
Guta/Gúta see Kolárovo
Gütersloh 72 B4 Nordrhein-Westfalen, W Germany
Gutta see Kolárovo
Guttstadt see Dobre Miasto
Guwāhāti 113 G3 prev. Gauhāti. Assam, NE India
Guyana 37 F3 off. Co-operative Republic of Guyana; prev. British Guiana. country N South America
Guyana, Co-operative Republic of see Guyana
Guyane see French Guiana
Guymon 27 E1 Oklahoma, C USA
Güzelyurt Körfezi 80 C5 Gk. Kólpos Mórfu, Morphou. W Cyprus
Gvardeysk 84 A4 Ger. Tapaiu. Kaliningradskaya Oblast', W Russian Federation
Gwādar 112 A3 var. Gwadur. Baluchistān, SW Pakistan
Gwadur see Gwādar
Gwalior 112 D3 Madhya Pradesh, C India
Gwanda 56 D3 Matabeleland South, SW Zimbabwe
Gwangju 107 E4 off. Gwangju Gwang-yeoksi, prev. Kwangju, var. Guangju, Kwangchu, Jap. Kōshū. SW South Korea
Gwangju Gwang-yeoksi see Gwangju
Gwy see Wye
Gyandzha see Gäncä
Gyangzê 104 C5 Xizang Zizhiqu, W China
Gyaring Co 104 C5 lake W China
Gyêgu see Yushu
Gyergyószentmiklós see Gheorgheni
Gyixong see Gonggar
Gympie 127 E5 Queensland, E Australia
Gyomaendrőd 77 D7 Békés, SE Hungary
Gyöngyös 77 D6 Heves, NE Hungary
Győr 77 C6 Ger. Raab, Lat. Arrabona. Győr-Moson-Sopron, NW Hungary
Gýtheio 83 B6 var. Githio; prev. Yíthion. Pelopónnisos, S Greece

Gyulafehérvár – Horsburgh Atoll

Gyulafehérvár see Alba Iulia
Gyumri 95 F2 var. Giumri, Rus. Kumayri; prev. Aleksandropol', Leninakan. W Armenia
Gyzyrlabat see Serdar

H

Haabai see Ha'apai Group
Haacht 65 C6 Vlaams Brabant, C Belgium
Haaksbergen 64 E3 Overijssel, E Netherlands
Ha'apai Group 123 F4 var. Haabai. island group C Tonga
Haapsalu 84 D2 Ger. Hapsal. Läänemaa, W Estonia
Ha'Arava see 'Arabah, Wādī al
Haarlem 64 C3 prev. Harlem. Noord-Holland, W Netherlands
Haast 129 B6 West Coast, South Island, New Zealand
Hachijo-jima 109 D6 island Izu-shotō, SE Japan
Hachinohe 108 D3 Aomori, Honshū, C Japan
Haḍabat al Jilf al Kabīr 50 A2 var. Gilf Kebir Plateau. plateau SW Egypt
Hadama see Nazrēt
Hadejia 53 G4 Jigawa, N Nigeria
Hadejia 53 G3 river N Nigeria
Hadera 97 A6 var. Khadera; prev. Ḥadera. Haifa, C Israel
Ḥadera see Hadera
Hadhdhunmathi Atoll 110 A5 atoll S Maldives
Ha Đông 114 D3 var. Hadong. Ha Tây, N Vietnam
Hadong see Ha Đông
Hadramaut see Ḥaḍramawt
Ḥaḍramawt 99 C6 Eng. Hadramaut. mountain range S Yemen
Hadrianopolis see Edirne
Haerbin/Haerhpin/Ha-erh-pin see Harbin
Hafnia see Denmark
Hafnia see København
Hafren see Severn
Hafun, Ras see Xaafuun, Raas
Hagåtña 122 B1 , var. Agaña. dependent territory capital (Guam) NW Guam
Hagerstown 19 E4 Maryland, NE USA
Ha Giang 114 D3 Ha Giang, N Vietnam
Hagios Evstrátios see Ágios Efstrátios
HaGolan see Golan Heights
Hagondange 68 D3 Moselle, NE France
Haguenau 68 E3 Bas-Rhin, NE France
Haibowan see Wuhai
Haicheng 106 D3 Liaoning, NE China
Haidarabad see Hyderābād
Haifa see Hefa
Haifa, Bay of see Mifrats Hefa
Haifong see Hai Phong
Haikou 106 C7 var. Hai-k'ou, Hoihow, Fr. Hoï-Hao. province capital Hainan, S China
Hai-k'ou see Haikou
Ḥā'il 98 B4 Ḥā'il, NW Saudi Arabia
Hailuoto 62 D4 Swe. Karlö. island W Finland
Hainan 106 B7 var. Hainan Sheng, Qiong. province S China
Hainan Dao 106 C7 island S China
Hainan Sheng see Hainan
Hainasch see Ainaži
Haines 14 D4 Alaska, USA
Hainichen 72 D4 Sachsen, E Germany
Hai Phong 114 D3 var. Haifong, Haiphong. N Vietnam
Haiphong see Hai Phong
Haiti 32 D3 off. Republic of Haiti. country C West Indies
Haiti, Republic of see Haiti
Haiya 50 C3 Red Sea, NE Sudan
Hajdúhadház 77 D6 Hajdú-Bihar, E Hungary
Hajîne see Abū Ḥardān
Hajnówka 76 E3 Ger. Hermhausen. Podlaskie, NE Poland
Hakodate 108 D3 Hokkaidō, NE Japan
Hal see Halle
Ḥalab 96 B2 Eng. Aleppo, Fr. Alep; anc. Beroea. Ḥalab, NW Syria
Hala'ib Triangle 50 C3 region Egypt/Sudan
Ḥalāniyāt, Juzur al 99 D6 var. Jazā'ir Bin Ghalfān, Eng. Kuria Muria Islands. island group S Oman
Halberstadt 72 C4 Sachsen-Anhalt, C Germany
Halden 63 B6 prev. Fredrikshald. Østfold, S Norway
Halfmoon Bay 129 A8 var. Oban. Stewart Island, Southland, New Zealand
Haliacmon see Aliákmonas
Halifax 17 F4 province capital Nova Scotia, SE Canada
Halkida see Chalkída
Halle 65 B6 Fr. Hal. Vlaams Brabant, C Belgium
Halle 72 C4 var. Halle an der Saale. Sachsen-Anhalt, C Germany
Halle an der Saale see Halle
Halle-Neustadt 72 C4 Sachsen-Anhalt, C Germany
Halley 132 B2 UK research station Antarctica
Hall Islands 120 B2 island group C Micronesia
Halls Creek 124 C3 Western Australia
Halmahera, Laut 117 F3 Eng. Halmahera Sea; sea E Indonesia
Halmahera, Pulau 117 F3 prev. Djailolo, Gilolo, Jailolo. island E Indonesia
Halmahera Sea see Halmahera, Laut
Halmstad 63 B7 Halland, S Sweden
Ha Long 114 E3 prev. Hông Gai; var. Hon Gai, Hongay. Quang Ninh, N Vietnam
Hälsingborg see Helsingborg
Hamada 109 B6 Shimane, Honshū, SW Japan
Hamadān 98 C3 anc. Ecbatana. Hamadān, W Iran
Ḥamāh 96 B3 var. Hama; anc. Epiphania, Bibl. Hamath. Ḥamāh, W Syria
Hamamatsu 109 D6 var. Hamamatu. Shizuoka, Honshū, S Japan
Hamamatu see Hamamatsu
Hamar 63 B5 prev. Storhammer. Hedmark, S Norway
Hamath see Ḥamāh
Hamburg 72 B3 Hamburg, N Germany
Hamd, Wadi al 98 A4 dry watercourse W Saudi Arabia
Hämeenlinna 63 D5 Swe. Tavastehus. Etelä-Suomi, S Finland
HaMela h, Yam see Dead Sea
Hamersley Range 124 A4 mountain range Western Australia
Hamhŭng 107 E3 C North Korea
Hami 104 C3 var. Ha-mi, Uigh. Kumul, Qomul. Xinjiang Uygur Zizhiqu, NW China
Ha-mi see Hami
Hamilton 20 A5 dependent territory capital (Bermuda) C Bermuda
Hamilton 16 D5 Ontario, S Canada
Hamilton 128 D3 Waikato, North Island, New Zealand
Hamilton 66 C4 S Scotland, United Kingdom
Hamilton 20 C2 Alabama, S USA
Hamīm, Wadi al 49 G2 river NE Libya
Hamïs Musait see Khamīs Mushayt
Hamm 72 B4 var. Hamm in Westfalen. Nordrhein-Westfalen, W Germany
Ḥammāmāt, Khalīj al see Hammamet, Golfe de
Hammamet, Golfe de 80 D3 Ar. Khalīj al Ḥammāmāt. gulf NE Tunisia
Ḥammār, Hawr al 98 C3 lake SE Iraq
Hamm in Westfalen see Hamm
Hampden 129 B7 Otago, South Island, New Zealand
Hampstead 67 A7 Maryland, USA
Ħamrun 80 B5 C Malta
Hāmūn, Daryācheh-ye see Şāberī, Hāmūn-e/ Sīstān, Daryācheh-ye
Hamwih see Southampton
Hânceşti see Hînceşti
Hancewicze see Hantsavichy
Handan 106 C4 var. Han-tan. Hebei, E China
Haneda 108 A2 (Tōkyō) Tōkyō, Honshū, S Japan
HaNegev 97 A7 Eng. Negev. desert S Israel
Hanford 25 C6 California, W USA
Hangayn Nuruu 104 D2 mountain range C Mongolia
Hang-chou/Hangchow see Hangzhou
Hangö see Hanko
Hangzhou 106 D5 var. Hang-chou, Hangchow. province capital Zhejiang, SE China
Hania see Chaniá
Hanka, Lake see Khanka, Lake
Hanko 63 D6 Swe. Hangö. Etelä-Suomi, SW Finland
Han-kou/Han-k'ou/Hankow see Wuhan
Hanmer Springs 129 C5 Canterbury, South Island, New Zealand
Hannibal 23 G4 Missouri, C USA
Hannover 72 B3 Eng. Hanover. Niedersachsen, NW Germany
Hanöbukten 63 B7 bay S Sweden
Ha Nôi 114 D3 Eng. Hanoi, Fr. Hanoï. country capital (Vietnam) N Vietnam
Hanover see Hannover
Han Shui 105 E4 river C China
Han-tan see Handan
Hantsavichy 85 B6 Pol. Hancewicze, Rus. Gantsevichi. Brestskaya Voblasts', SW Belarus
Hanyang see Wuhan
Hanzhong 106 B5 Shaanxi, C China
Hāora 113 F4 prev. Howrah. West Bengal, NE India
Haparanda 62 D4 Norrbotten, N Sweden
Hapsal see Haapsalu
Haradok 85 E5 Rus. Gorodok. Vitsyebskaya Voblasts', N Belarus
Haradzyets 85 B6 Rus. Gorodets. Brestskaya Voblasts', SW Belarus
Haramachi 108 D4 Fukushima, Honshū, E Japan
Harany 85 D5 Rus. Gorany. Vitsyebskaya Voblasts', N Belarus
Harare 56 D3 prev. Salisbury. country capital (Zimbabwe) Mashonaland East, NE Zimbabwe
Harbavichy 85 E6 Rus. Gorbovichi. Mahilyowskaya Voblasts', E Belarus
Harbel 52 C5 W Liberia
Harbin 107 E2 var. Haerbin, Ha-erh-pin, Kharbin; prev. Haerhpin, Pingkiang, Pinkiang. province capital Heilongjiang, NE China
Hardangerfjorden 63 A6 fjord S Norway
Hardangervidda 63 A6 plateau S Norway
Hardenberg 64 E3 Overijssel, E Netherlands
Harelbeke 65 A6 var. Harlebeke. West-Vlaanderen, W Belgium
Harem see Ḥārim
Haren 64 E2 Groningen, NE Netherlands
Hārer 51 D5 E Ethiopia
Hargeisa see Hargeysa
Hargeysa 51 D5 var. Hargeisa. Woqooyi Galbeed, NW Somalia
Hariana see Haryāna
Hari, Batang 116 B4 prev. Djambi. river Sumatera, W Indonesia
Ḥārim 96 B2 var. Harem. Idlib, W Syria
Harima-nada 109 B6 sea S Japan
Harirud 101 E4 var. Tedzhen, Turkm. Tejen. river Afghanistan/Iran
Harlan 23 F3 Iowa, C USA
Harlebeke see Harelbeke
Harlem see Haarlem
Harlingen 64 D2 Fris. Harns. Friesland, N Netherlands
Harlingen 27 G5 Texas, SW USA
Harlow 67 E6 E England, United Kingdom
Harney Basin 24 B4 basin Oregon, NW USA
Härnösand 63 C5 var. Hernösand. Västernorrland, C Sweden
Harns see Harlingen
Harper 52 D5 var. Cape Palmas. NE Liberia
Harricana 16 D3 river Québec, SE Canada
Harris 66 B3 physical region NW Scotland, United Kingdom
Harrisburg 19 E4 state capital Pennsylvania, NE USA
Harrisonburg 19 E4 Virginia, NE USA
Harrison, Cape 17 F2 headland Newfoundland and Labrador, E Canada
Harris Ridge see Lomonosov Ridge
Harrogate 67 D5 N England, United Kingdom
Hârşova 86 D5 prev. Hîrşova. Constanţa, SE Romania
Harstad 62 C2 Troms, N Norway
Hartford 19 G3 state capital Connecticut, NE USA
Hartlepool 67 D5 N England, United Kingdom
Harunabad see Eslāmābād
Har Us Gol 104 C2 lake Hovd, W Mongolia
Har Us Nuur 104 C2 lake NW Mongolia
Harwich 67 E6 E England, United Kingdom
Haryāna 112 D2 var. Hariana. cultural region N India
Hashemite Kingdom of Jordan see Jordan
Hasselt 65 C6 Limburg, NE Belgium
Hassetché see Al Ḥasakah
Hasta Colonia/Hasta Pompeia see Asti
Hastings 128 E4 Hawke's Bay, North Island, New Zealand
Hastings 67 E7 SE England, United Kingdom
Hastings 23 E4 Nebraska, C USA
Haţeg 86 B4 Ger. Wallenthal, Hung. Hátszeg; prev. Hatzeg, Hötzing. Hunedoara, SW Romania
Hátszeg see Haţeg
Hattem 64 D3 Gelderland, E Netherlands
Hatteras, Cape 21 G1 headland North Carolina, SE USA
Hatteras Plain 13 D6 abyssal plain W Atlantic Ocean
Hattiesburg 20 C3 Mississippi, S USA
Hatton Bank see Hatton Ridge
Hatton Ridge 58 B2 var. Hatton Bank. undersea ridge N Atlantic Ocean
Hat Yai 115 C7 var. Ban Hat Yai. Songkhla, SW Thailand
Hatzeg see Haţeg
Hatzfeld see Jimbolia
Haugesund 63 A6 Rogaland, S Norway
Haukeligrend 63 A6 Telemark, S Norway
Haukivesi 63 E5 lake SE Finland
Hauraki Gulf 128 D2 gulf North Island, N New Zealand
Hauroko, Lake 129 A7 lake South Island, New Zealand
Haut Atlas 48 C2 Eng. High Atlas. mountain range C Morocco
Hautes Fagnes 65 D6 Ger. Hohes Venn. mountain range E Belgium
Hauts Plateaux 48 D2 plateau Algeria/Morocco
Hauzenberg 73 D6 Bayern, SE Germany
Havana 13 D6 Illinois, N USA
Havana see La Habana
Havant 67 D7 S England, United Kingdom
Havelock 21 F1 North Carolina, SE USA
Havelock North 128 E4 Hawke's Bay, North Island, New Zealand
Haverfordwest 67 C6 SW Wales, United Kingdom
Havířov 77 C5 Moravskoslezský Kraj, E Czech Republic
Havre 22 C1 Montana, NW USA
Havre see le Havre
Havre-St-Pierre 17 F3 Québec, E Canada
Hawaii 25 A8 off. State of Hawaii, also known as Aloha State, Paradise of the Pacific, var. Hawai'i. state USA, C Pacific Ocean
Hawai'i 25 B8 var. Hawaii. island Hawai'ian Islands, USA, C Pacific Ocean
Hawai'ian Islands 130 D2 prev. Sandwich Islands. island group Hawaii, USA
Hawaiian Ridge 130 H4 undersea ridge N Pacific Ocean
Hawea, Lake 129 B6 lake South Island, New Zealand
Hawera 128 D4 Taranaki, North Island, New Zealand
Hawick 66 C4 SE Scotland, United Kingdom
Hawke Bay 128 E4 bay North Island, New Zealand
Hawkeye State see Iowa
Hawlēr see Arbīl
Hawthorne 25 C6 Nevada, W USA
Hay 127 C6 New South Wales, SE Australia
HaYarden see Jordan
Hayastani Hanrapetut'yun see Armenia
Hayes 16 B2 river Manitoba, C Canada
Hay River 15 E4 Northwest Territories, W Canada
Hays 23 E5 Kansas, C USA
Haysyn 86 D3 Rus. Gaysin. Vinnyts'ka Oblast', C Ukraine
Hazar 100 B2 prev. Rus. Cheleken. Balkan Welaýaty, W Turkmenistan
Heard and McDonald Islands 119 B7 Australian external territory S Indian Ocean
Hearst 16 C4 Ontario, S Canada
Heart of Dixie see Alabama
Heathrow 67 A8 (London) SE England, United Kingdom
Hebei 106 C4 var. Hebei Sheng, Hopeh, Hopei, Ji; prev. Chihli. province E China
Hebei Sheng see Hebei
Hebron 97 A6 var. Al Khalīl, El Khalil, Heb. Hevron; anc. Kiriath-Arba. S West Bank
Heemskerk 64 C3 Noord-Holland, W Netherlands
Heerde 64 D3 Gelderland, E Netherlands
Heerenveen 64 D2 Fris. It Hearrenfean. Friesland, N Netherlands
Heerhugowaard 64 C2 Noord-Holland, NW Netherlands
Heerlen 65 D6 Limburg, SE Netherlands
Heerwegen see Polkowice
Hefa 97 A5 var. Haifa, hist. Caiffa, Caiphas; anc. Sycaminum. Haifa, N Israel
Hefa, Mifraẓ see Mifrats Hefa
Hefei 106 D5 var. Hofei, hist. Luchow. province capital Anhui, E China
Hegang 107 E2 Heilongjiang, NE China
Hei see Heilongjiang
Heide 72 B2 Schleswig-Holstein, N Germany
Heidelberg 73 B5 Baden-Württemberg, SW Germany
Heidenheim see Heidenheim an der Brenz
Heidenheim an der Brenz 73 B6 var. Heidenheim. Baden-Württemberg, S Germany
Hei-ho see Nagqu
Heilbronn 73 B6 Baden-Württemberg, SW Germany
Heiligenbeil see Mamonovo
Heilongjiang 106 D2 var. Hei, Heilongjiang Sheng, Hei-lung-chiang, Heilungkiang. province NE China
Heilong Jiang see Amur
Heilongjiang Sheng see Heilongjiang
Heiloo 64 C3 Noord-Holland, NW Netherlands
Heilsberg see Lidzbark Warmiński
Hei-lung-chiang/Heilungkiang see Heilongjiang
Heimdal 63 B5 Sør-Trøndelag, S Norway
Heinaste see Ainaži
Hekimhan 94 D3 Malatya, C Turkey
Helena 22 B2 state capital Montana, NW USA
Helensville 128 D2 Auckland, North Island, New Zealand
Helgoland Bay see Helgoländer Bucht
Helgoländer Bucht 72 A2 var. Helgoland Bay, Heligoland Bight. bay NW Germany
Heligoland Bight see Helgoländer Bucht
Heliopolis see Baalbek
Hellas see Greece
Hellenic Republic see Greece
Hellevoetsluis 64 B4 Zuid-Holland, SW Netherlands
Hellín 71 E4 Castilla-La Mancha, C Spain
Darya-ye Helmand 100 D5 var. Rūd-e Hīrmand. river Afghanistan/Iran
Helmantica see Salamanca
Helmond 65 D5 Noord-Brabant, S Netherlands
Helsingborg 63 B7 prev. Hälsingborg. Skåne, S Sweden
Helsingfors see Helsinki
Helsinki 63 D6 Swe. Helsingfors. country capital (Finland) Etelä-Suomi, S Finland
Heltau see Cisnădie
Helvetia see Switzerland
Henan 106 C5 var. Henan Sheng, Honan, Yu. province C China
Henderson 18 B5 Kentucky, S USA
Henderson 25 D7 Nevada, W USA
Henderson 27 H3 Texas, SW USA
Hendū Kosh see Hindu Kush
Hengchow see Hengyang
Hengduan Shan 106 A5 mountain range SW China
Hengelo 64 E3 Overijssel, E Netherlands
Hengnan see Hengyang
Hengyang 106 C6 var. Hengnan, Heng-yang; prev. Hengchow. Hunan, S China
Heng-yang see Hengyang
Heniches'k 87 F4 Rus. Genichesk. Khersons'ka Oblast', S Ukraine
Hennebont 68 A3 Morbihan, NW France
Henrique de Carvalho see Saurimo
Henzada see Hinthada
Herakleion see Irákleio
Herāt 100 D4 var. Herat; anc. Aria. Herāt, W Afghanistan
Heredia 31 E4 Heredia, C Costa Rica
Hereford 27 E2 Texas, SW USA
Herford 72 B4 Nordrhein-Westfalen, NW Germany
Héristal see Herstal
Herk-de-Stad 65 C6 Limburg, NE Belgium
Herlen Gol/Herlen He see Kerulen
Hermannstadt see Sibiu
Hermansverk 63 A5 Sogn Og Fjordane, S Norway
Hermhausen see Hajnówka
Hermiston 24 C2 Oregon, NW USA
Hermon, Mount 97 B5 Ar. Jabal ash Shaykh. mountain S Syria
Hermosillo 28 B2 Sonora, NW Mexico
Hermoupolis see Ermoúpoli
Hernösand see Härnösand
Herrera del Duque 70 D3 Extremadura, W Spain
Herselt 65 C5 Antwerpen, C Belgium
Herstal 65 D6 Fr. Héristal. Liège, E Belgium
Herzogenbusch see 's-Hertogenbosch
Hesse see Hessen
Hessen 73 B5 Eng./Fr. Hesse. state C Germany
Hevron see Hebron
Heydebrech see Kędzierzyn-Kozle
Heydekrug see Šilutė
Heywood Islands 124 C3 island group Western Australia
Hibbing 23 F1 Minnesota, N USA
Hibernia see Ireland
Hidalgo del Parral 28 C2 var. Parral. Chihuahua, N Mexico
Hida-sanmyaku 109 C5 mountain range Honshū, S Japan
Hierosolyma see Jerusalem
Hierro 48 A3 var. Ferro. island Islas Canarias, Spain, NE Atlantic Ocean
High Atlas see Haut Atlas
High Plains see Great Plains
High Point 21 E1 North Carolina, SE USA
Hiiumaa 84 C2 Ger. Dagden, Swe. Dagö. island W Estonia
Hikurangi 128 D2 Northland, North Island, New Zealand
Hildesheim 72 B4 Niedersachsen, N Germany
Hilla see Al Ḥillah
Hillaby, Mount 33 G1 mountain N Barbados
Hill Bank 30 C1 Orange Walk, N Belize
Hillegom 64 C3 Zuid-Holland, W Netherlands
Hilo 25 B8 Hawaii, USA, C Pacific Ocean
Hilton Head Island 21 E2 South Carolina, SE USA
Hilversum 64 C3 Noord-Holland, C Netherlands
Himalaya/Himalaya Shan see Himalayas
Himalayas 113 E2 var. Himalaya, Chin. Himalaya Shan. mountain range S Asia
Himeji 109 C6 var. Himezi. Hyōgo, Honshū, SW Japan
Himezi see Himeji
Ḥimş 96 B4 var. Homs; anc. Emesa. Ḥimş, C Syria
Hînceşti 86 D4 var. Hânceşti; prev. Kotovsk. C Moldova
Hinchinbrook Island 126 D3 island Queensland, NE Australia
Hinds 129 C6 Canterbury, South Island, New Zealand
Hindu Kush 101 F4 Per. Hendū Kosh. mountain range Afghanistan/Pakistan
Hinesville 21 E3 Georgia, SE USA
Hinnøya 62 C3 Lapp. Iinnasuolu. island C Norway
Hinson Bay 20 A5 bay W Bermuda W Atlantic Ocean
Hinthada 114 B4 var. Henzada. Ayeyarwady, SW Myanmar (Burma)
Hios see Chíos
Hirfanlı Barajı 94 C3 reservoir C Turkey
Hirmand, Rūd-e see Helmand, Daryā-ye
Hirosaki 108 D3 Aomori, Honshū, C Japan
Hiroshima 109 B6 var. Hirosima. Hiroshima, Honshū, SW Japan
Hirschberg/Hirschberg im Riesengebirge/ Hirschberg in Schlesien see Jelenia Góra
Hirson 68 D3 Aisne, N France
Hîrşova see Hârşova
Hispalis see Sevilla
Hispana/Hispania see Spain
Hispaniola 34 B1 island Dominion Republic/Haiti
Hitachi 109 D5 var. Hitati. Ibaraki, Honshū, S Japan
Hitati see Hitachi
Hitra 62 A4 prev. Hitteren. island S Norway
Hitteren see Hitra
Hjälmaren 63 C6 Eng. Lake Hjalmar. lake C Sweden
Hjalmar, Lake see Hjälmaren
Hjørring 63 B7 Nordjylland, N Denmark
Hkakabo Razi 114 B1 mountain Myanmar (Burma)/China
Hlobyne 87 F2 Rus. Globino. Poltavs'ka Oblast', NE Ukraine
Hlukhiv 87 F1 Rus. Glukhov. Sums'ka Oblast', NE Ukraine
Hlybokaye 85 D5 Rus. Glubokoye. Vitsyebskaya Voblasts', N Belarus
Hoa Binh 114 D3 Hoa Binh, N Vietnam
Hoang Lien Son 114 D3 mountain range N Vietnam
Hobart 127 C8 prev. Hobarton, Hobart Town. state capital Tasmania, SE Australia
Hobarton/Hobart Town see Hobart
Hobbs 27 E3 New Mexico, SW USA
Hobro 63 A7 Nordjylland, N Denmark
Hô Chi Minh 115 E6 var. Ho Chi Minh City; prev. Saigon. S Vietnam
Ho Chi Minh City see Hô Chi Minh
Hodeida see Al Ḥudaydah
Hódmezővásárhely 77 D7 Csongrád, SE Hungary
Hodna, Chott El 80 C4 var. Chott el-Hodna, Ar. Shatt al-Hodna. salt lake N Algeria
Hodna, Chott el-/Hodna, Shatt al- see Hodna, Chott El
Hodonín 77 C5 Ger. Göding. Jihomoravský Kraj, SE Czech Republic
Hoei see Huy
Hoey see Huy
Hof 73 C5 Bayern, SE Germany
Hofei see Hefei
Hōfu 109 B7 Yamaguchi, Honshū, SW Japan
Hofuf see Al Hufūf
Hogoley Islands see Chuuk Islands
Hohensalza see Inowrocław
Hohenstadt see Zábřeh
Hohes Venn see Hautes Fagnes
Hohe Tauern 73 C7 mountain range W Austria
Hohhot 105 F3 var. Huhehot, Huhohaote, Mong. Kukukhoto; prev. Kweisui, Kwesui. Nei Mongol Zizhiqu, N China
Hôi An 115 E5 prev. Faifo. Quang Nam-Đa Năng, C Vietnam
Hoï-Hao/Hoihow see Haikou
Hokianga Harbour 128 C2 inlet SE Tasman Sea
Hokitika 129 B5 West Coast, South Island, New Zealand
Hokkaido 108 C2 prev. Ezo, Yeso, Yezo. island NE Japan
Hola Prystan' 87 E4 Rus. Golaya Pristan. Khersons'ka Oblast', S Ukraine
Holbrook 26 B2 Arizona, SW USA
Holetown 33 G1 prev. Jamestown. W Barbados
Holguín 32 C2 Holguín, SE Cuba
Hollabrunn 73 E6 Niederösterreich, NE Austria
Holland see Netherlands
Hollandia see Jayapura
Holly Springs 20 C1 Mississippi, S USA
Holman 15 E3 Victoria Island, Northwest Territories, N Canada
Holmsund 62 D4 Västerbotten, N Sweden
Holon 97 A6 var. Kholon; prev. Ḥolon. Tel Aviv, C Israel
Ḥolon see Holon
Holovanivs'k 87 E3 Rus. Golovanevsk. Kirovohrads'ka Oblast', C Ukraine
Holstebro 63 A7 Ringkøbing, W Denmark
Holsteinborg/Holsteinsborg/Holstenborg/ Holstensborg see Sisimiut
Holyhead 67 C5 Wel. Caer Gybi. NW Wales, United Kingdom
Hombori 53 E3 Mopti, S Mali
Homs see Al Khums, Libya
Homs see Ḥimş
Homyel' 85 D7 Rus. Gomel'. Homyel'skaya Voblasts', SE Belarus
Honan see Luoyang, China
Honan see Henan, China
Hondo 27 F4 Texas, SW USA
Hondo see Honshū
Honduras 30 C2 off. Republic of Honduras. country Central America
Honduras, Golfo de see Honduras, Gulf of
Honduras, Gulf of 30 C2 Sp. Golfo de Honduras. gulf W Caribbean Sea
Honduras, Republic of see Honduras
Hønefoss 63 B6 Buskerud, S Norway
Honey Lake 25 B5 lake California, W USA
Hon Gai see Ha Long
Hongay see Ha Long
Hông Gai see Ha Long
Hông Hà, Sông see Red River
Hong Kong 106 A1 Hong Kong, S China
Hong Kong Island 106 B2 island S China Asia
Honiara 122 C3 country capital (Solomon Islands) Guadalcanal, C Solomon Islands
Honjō 108 D4 var. Honzyô, Yurihonjō. Akita, Honshū, C Japan
Honolulu 25 A8 state capital O'ahu, Hawaii, USA, C Pacific Ocean
Honshu 109 E5 var. Hondo, Honsyû. island SW Japan
Honsyû see Honshū
Honte see Westerschelde
Honzyô see Honjō
Hoogeveen 64 E2 Drenthe, NE Netherlands
Hoogezand-Sappemeer 64 E2 Groningen, NE Netherlands
Hoorn 64 C2 Noord-Holland, NW Netherlands
Hoosier State see Indiana
Hopa 95 E2 Artvin, NE Turkey
Hope 14 C3 British Columbia, SW Canada
Hopedale 17 F2 Newfoundland and Labrador, NE Canada
Hopeh/Hopei see Hebei
Hopkinsville 18 B5 Kentucky, S USA
Horasan 95 F3 Erzurum, NE Turkey
Horizon Deep 130 D4 trench W Pacific Ocean
Horki 85 E6 Rus. Gorki. Mahilyowskaya Voblasts', E Belarus
Horlivka 87 G3 Rom. Adâncata, Rus. Gorlovka. Donets'ka Oblast', E Ukraine
Hormoz, Tangeh-ye see Hormuz, Strait of
Hormuz, Strait of 98 D4 var. Strait of Ormuz, Per. Tangeh-ye Hormoz. strait Iran/Oman
Cape Horn 43 C8 Eng. Cape Horn. headland S Chile
Horn, Cape see Hornos, Cabo de
Hornsby 126 E1 New South Wales, SE Australia
Horodnya 87 E1 Rus. Gorodnya. Chernihivs'ka Oblast', NE Ukraine
Horodok 86 B2 Pol. Gródek Jagielloński, Rus. Gorodok, Gorodok Yagellonski. L'vivs'ka Oblast', NW Ukraine
Horodyshche 87 E2 Rus. Gorodishche. Cherkas'ka Oblast', C Ukraine
Horoshiri-dake 108 D2 var. Horosiri Dake. mountain Hokkaidō, N Japan
Horosiri Dake see Horoshiri-dake
Horsburgh Atoll 110 A4 var. Goidhoo Atoll. atoll N Maldives

Horseshoe Bay – Jabbūl, Sabkhat al

Horseshoe Bay *20 A5 bay* W Bermuda W Atlantic Ocean
Horseshoe Seamounts *58 A4 seamount range* E Atlantic Ocean
Horsham *127 B7* Victoria, SE Australia
Horst *65 D5* Limburg, SE Netherlands
Horten *63 B6* Vestfold, S Norway
Horyn' *85 B7 Rus.* Goryn. *river* NW Ukraine
Hosingen *65 D7* Diekirch, NE Luxembourg
Hospitalet *see* L'Hospitalet de Llobregat
Hotan *104 B4 var.* Khotan, *Chin.* Ho-t'ien. Xinjiang Uygur Zizhiqu, NW China
Ho-t'ien *see* Hotan
Hoting *62 C4* Jämtland, C Sweden
Hot Springs *20 B1* Arkansas, C USA
Hötzing *see* Haţeg
Houayxay *114 C3 var.* Ban Houayxay. Bokèo, N Laos
Houghton *18 B1* Michigan, N USA
Houilles *69 B5* Yvelines, Île-de-France, N France Europe
Houlton *19 H1* Maine, NE USA
Houma *20 B3* Louisiana, S USA
Houston *27 H4* Texas, SW USA
Hovd *104 C2 var.* Khovd, Kobdo; *prev.* Jirgalanta. Hovd, W Mongolia
Hove *67 E7* SE England, United Kingdom
Hoverla, Hora *86 C3 Rus.* Gora Goverla. *mountain* W Ukraine
Hovsgol, Lake *see* Hövsgöl Nuur
Hövsgöl Nuur *104 D1 var.* Lake Hovsgol. *lake* N Mongolia
Howa, Ouadi *see* Howar, Wādi
Howar, Wadi *50 A3 var.* Ouadi Howa. *river* Chad/Sudan
Howrah *see* Hāora
Hoy *66 C2 island* N Scotland, United Kingdom
Hoyerswerda *72 D4 Lus.* Wojerecy. Sachsen, E Germany
Hpa-an *114 B4 var.* Pa-an. Kayin State, S Myanmar (Burma)
Hpyu *see* Phyu
Hradec Králové *77 B5 Ger.* Königgrätz. Královéhradecký Kraj, N Czech Republic
Hrandzichy *85 B5 Rus.* Grandichi. Hrodzyenskaya Voblasts', W Belarus
Hranice *77 C5 Ger.* Mährisch-Weisskirchen. Olomoucký Kraj, E Czech Republic
Hrebinka *87 E2 Rus.* Grebenka. Poltavs'ka Oblast', NE Ukraine
Hrodna *85 B5 Pol.* Grodno. Hrodzyenskaya Voblasts', W Belarus
Hrvatska *see* Croatia
Hsia-men *see* Xiamen
Hsiang-t'an *see* Xiangtan
Hsi Chiang *see* Xi Jiang
Hsing-K'ai Hu *see* Khanka, Lake
Hsi-ning/Hsining *see* Xining
Hsinking *see* Changchun
Hsin-yang *see* Xinyang
Hsu-chou *see* Xuzhou
Htawei *see* Dawei
Huacho *38 C4* Lima, W Peru
Hua Hin *see* Ban Hua Hin
Huaihua *106 C5* Hunan, S China
Huailai *106 C3 var.* Shacheng. Hebei, E China
Huainan *106 D5 var.* Huai-nan, Hwainan. Anhui, E China
Huai-nan *see* Huainan
Huajuapan *29 F5 var.* Huajuapan de León. Oaxaca, SE Mexico
Huajuapan de León *see* Huajuapan
Hualapai Peak *26 A2 mountain* Arizona, SW USA
Huallaga, Río *38 C3 river* N Peru
Huambo *56 B2 Port.* Nova Lisboa. Huambo, C Angola
Huancavelica *38 D4* Huancavelica, SW Peru
Huancayo *38 D3* Junín, C Peru
Huang Hai *see* Yellow Sea
Huang He *106 C4 var.* Yellow River. *river* C China
Huangshi *106 C5 var.* Huang-shih, Hwangshih. Hubei, C China
Huang-shih *see* Huangshi
Huanta *38 D4* Ayacucho, C Peru
Huánuco *38 C3* Huánuco, C Peru
Huanuni *39 F4* Oruro, W Bolivia
Huaral *38 C4* Lima, W Peru
Huarás *see* Huaraz
Huaraz *38 C3 var.* Huarás. Ancash, W Peru
Huarmey *38 C3* Ancash, W Peru
Huatabampo *28 C2* Sonora, NW Mexico
Hubli *102 B3* Karnātaka, SW India
Huddersfield *67 D5* N England, United Kingdom
Hudiksvall *63 C5* Gävleborg, C Sweden
Hudson Bay *15 G4 bay* NE Canada
Hudson, Détroit d' *see* Hudson Strait
Hudson Strait *15 H3 Fr.* Détroit d'Hudson. *strait* Northwest Territories/Québec, NE Canada
Hudur *see* Xuddur
Huê *114 E4* Th,a Thiên-Huê, C Vietnam
Huehuetenango *30 A2* Huehuetenango, W Guatemala
Huelva *70 C4 anc.* Onuba. Andalucía, SW Spain
Huesca *71 F2 anc.* Osca. Aragón, NE Spain
Huéscar *71 E4* Andalucía, S Spain
Hughenden *126 C3* Queensland, NE Australia
Hugo *27 G2* Oklahoma, C USA
Huhehot/Huhohaote *see* Hohhot
Huíla Plateau *56 B2 plateau* S Angola
Huixtla *29 G5* Chiapas, SE Mexico
Hulingol *105 G2 prev.* Huolin Gol. Nei Mongol Zizhiqu, N China
Hull *16 D4* Québec, SE Canada
Hull *see* Kingston upon Hull
Hull Island *see* Orona
Hulst *65 B5* Zeeland, SW Netherlands
Hulun *see* Hulun Buir
Hulun Buir *105 F1 var.* Hailar; *prev.* Hulun. Nei Mongol Zizhiqu, N China
Hu-lun Ch'ih *see* Hulun Nur
Hulun Nur *105 F1 var.* Hu-lun Ch'ih; *prev.* Dalai Nor. *lake* NE China
Humaitá *40 D2* Amazonas, N Brazil
Humboldt River *25 C5 river* Nevada, W USA
Humphreys Peak *26 B1 mountain* Arizona, SW USA
Humpolec *77 B5 Ger.* Gumpolds, Humpoletz. Vysočina, C Czech Republic
Humpoletz *see* Humpolec
Hunan *106 C6 var.* Hunan Sheng, Xiang. *province* S China
Hunan Sheng *see* Hunan
Hunedoara *86 B4 Ger.* Eisenmarkt, *Hung.* Vajdahunyad. Hunedoara, SW Romania
Hünfeld *73 B5* Hessen, C Germany
Hungarian People's Republic *see* Hungary
Hungary *77 C6 off.* Republic of Hungary, *Ger.* Ungarn, *Hung.* Magyarország, *Rom.* Ungaria, *SCr.* Madarska, *Ukr.* Uhorshchyna; *prev.* Hungarian People's Republic. *country* C Europe
Hungary, Plain of *see* Great Hungarian Plain
Hungary, Republic of *see* Hungary
Hunjiang *see* Baishan
Hunter Island *127 B8 island* Tasmania, SE Australia
Huntington *18 D4* West Virginia, NE USA
Huntington Beach *25 B8* California, W USA
Huntly *128 D3* Waikato, North Island, New Zealand
Huntsville *20 D1* Alabama, S USA
Huntsville *27 G3* Texas, SW USA
Huolin Gol *see* Hulingol
Hurghada *see* Al Ghurdaqah
Huron *23 E3* South Dakota, N USA
Huron, Lake *18 D2 lake* Canada/USA
Hurukawa *see* Furukawa
Hurunui *129 C5 river* South Island, New Zealand
Húsavík *61 E4* Nordhurland Eystra, NE Iceland
Husté *see* Khust
Husum *72 B2* Schleswig-Holstein, N Germany
Huszt *see* Khust
Hutchinson *23 E5* Kansas, C USA
Hutchinson Island *21 F4 island* Florida, SE USA
Huy *65 C6 Dut.* Hoei, Hoey. Liège, E Belgium
Huzi *see* Fuji
Hvannadalshnúkur *61 E5 volcano* S Iceland
Hvar *78 B4 It.* Lesina; *anc.* Pharus. *island* S Croatia
Hwainan *see* Huainan
Hwange *56 D3 prev.* Wankie. Matabeleland North, W Zimbabwe
Hwang-Hae *see* Yellow Sea
Hwangshih *see* Huangshi
Hyargas Nuur *104 C2 lake* NW Mongolia
Hyderābād *112 D5 var.* Haidarabad. *state capital* Telangana, C India
Hyderābād *112 B3 var.* Haidarabad. Sind, SE Pakistan
Hyères *69 D6* Var, SE France
Hyères, Îles d' *69 D6 island group* S France
Hypanis *see* Kuban'
Hyrcania *see* Gorgān
Hyvinge *see* Hyvinkää
Hyvinkää *63 D5 Swe.* Hyvinge. Etelä-Suomi, S Finland

I

Iader *see* Zadar
Ialomiţa *86 C5 river* SE Romania
Iaşi *86 D3 Ger.* Jassy. Iaşi, NE Romania
Ibadan *53 F5* Oyo, SW Nigeria
Ibagué *36 B3* Tolima, C Colombia
Ibar *78 D4 Alb.* Ibër. *river* C Serbia
Ibarra *38 B1 var.* San Miguel de Ibarra. Imbabura, N Ecuador
Ibër *see* Ibar
Iberia *see* Spain
Iberian Mountains *see* Ibérico, Sistema
Iberian Peninsula *58 B4 physical region* Portugal/Spain
Iberian Plain *58 B4 abyssal plain* E Atlantic Ocean
Ibérica, Cordillera *see* Ibérico, Sistema
Ibérico, Sistema *71 E2 var.* Cordillera Ibérica, *Eng.* Iberian Mountains. *mountain range* NE Spain
Ibiza *see* Eivissa
Ibo *see* Sassandra
Ica *38 D4* Ica, SW Peru
Icaria *see* Ikaría
Içá, Rio *see* Putumayo, Río
İçel *see* Mersin
Iceland *61 E4 off.* Republic of Iceland, *Dan.* Island, *Icel.* Ísland. *country* N Atlantic Ocean
Iceland Basin *58 B1 undersea basin* N Atlantic Ocean
Icelandic Plateau *see* Iceland Plateau
Iceland Plateau *133 B6 var.* Icelandic Plateau. *undersea plateau* S Greenland Sea
Iceland, Republic of *see* Iceland
Iconium *see* Konya
Iculisma *see* Angoulême
Idabel *27 H2* Oklahoma, C USA
Idaho *24 D3 off.* State of Idaho, *also known as* Gem of the Mountains, Gem State. *state* NW USA
Idaho Falls *24 E3* Idaho, NW USA
Idensalmi *see* Iisalmi
Idfū *50 B2 var.* Edfu. SE Egypt
Idi Amin, Lac *see* Edward, Lake
Idîni *52 B2* Trarza, W Mauritania
Idlib *96 B3* Idlib, NW Syria
Idre *63 B5* Dalarna, C Sweden
Iecava *84 C3* Bauska, S Latvia
Ieper *65 A6 Fr.* Ypres. West-Vlaanderen, W Belgium
Ierápetra *83 D8* Kríti, Greece, E Mediterranean Sea
Ierisós *see* Ierissós
Ierissós *82 C4 var.* Ierisós. Kentrikí Makedonía, N Greece
Iferouâne *53 G2* Agadez, N Niger
Ifôghas, Adrar des *53 E2 var.* Adrar des Iforas. *mountain range* NE Mali
Iforas, Adrar des *see* Ifôghas, Adrar des
Igarka *92 D3* Krasnoyarskiy Kray, N Russian Federation
Igaunija *see* Estonia
Iglau/Iglawa/Igława *see* Jihlava
Iglesias *75 A5* Sardegna, Italy, C Mediterranean Sea
Igloolik *15 G2* Nunavut, N Canada
Igoumenítsa *82 A4* Ípeiros, W Greece
Iguaçu, Rio *41 E4 Sp.* Río Iguazú. *river* Argentina/Brazil
Iguaçu, Saltos do *41 E4 Sp.* Cataratas del Iguazú; *prev.* Victoria Falls. *waterfall* Argentina/Brazil
Iguala *29 E4 var.* Iguala de la Independencia. Guerrero, S Mexico
Iguala de la Independencia *see* Iguala
Iguazú, Cataratas del *see* Iguaçu, Saltos do
Iguazú, Río *see* Iguaçu, Rio
Iguid, Erg *see* Iguîdi, 'Erg
Iguîdi, 'Erg *48 C3 var.* Erg Iguid. *desert* Algeria/Mauritania
Ihavandhippolhu Atoll *110 A3 var.* Ihavandiffulu Atoll. *atoll* N Maldives
Ihavandiffulu Atoll *see* Ihavandhippolhu Atoll
Ihosy *57 F4* Fianarantsoa, S Madagascar
Iinnasuolu *see* Hinnøya
Iisalmi *62 E4 var.* Idensalmi. Itä-Suomi, C Finland
IJmuiden *64 C3* Noord-Holland, W Netherlands
IJssel *64 D3 var.* Yssel. *river* Netherlands
IJsselmeer *64 C2 prev.* Zuider Zee. *lake* N Netherlands
IJsselmuiden *64 D3* Overijssel, E Netherlands
Ijzer *65 A6 river* W Belgium
Ikaahuk *see* Sachs Harbour
Ikaluktutiak *see* Cambridge Bay
Ikaría *83 D6 var.* Kariot, Nicaria, Nikaria; *anc.* Icaria. *island* Dodekánisa, Greece, Aegean Sea
Ikela *55 D6* Equateur, C Dem. Rep. Congo
Iki *109 A7 island* SW Japan
Ilagan *117 E1* Luzon, N Philippines
Ilave *39 E4* Puno, S Peru
Iława *76 D3 Ger.* Deutsch-Eylau. Warmińsko-Mazurskie, NE Poland
Ilebo *55 C6 prev.* Port-Francqui. Kasai-Occidental, W Dem. Rep. Congo
Île-de-France *68 C3 cultural region* N France
Ilerda *see* Lleida
Ilfracombe *67 C7* SW England, United Kingdom
Ílhavo *70 B2* Aveiro, N Portugal
Iliamna Lake *14 C3 lake* Alaska, USA
Il'ichevsk *see* Illichivs'k, Ukraine
Ilici *see* Elche
Iligan *117 E2 off.* Iligan City. Mindanao, S Philippines
Iligan City *see* Iligan
Illapel *42 B4* Coquimbo, C Chile
Illichivs'k *87 E4 Rus.* Il'ichevsk. Odes'ka Oblast', SW Ukraine
Illicis *see* Elche
Illinois *18 A4 off.* State of Illinois, *also known as* Prairie State, Sucker State. *state* C USA
Illinois River *18 B4 river* Illinois, N USA
Illurco *see* Lorca
Illuro *see* Mataró
Ilo *39 E4* Moquegua, SW Peru
Iloilo *117 E2 off.* Iloilo City. Panay Island, C Philippines
Iloilo City *see* Iloilo
Ilorin *53 F4* Kwara, W Nigeria
Ilovlya *89 B6* Volgogradskaya Oblast', SW Russian Federation
Iluh *see* Batman
Il'yaly *see* Gurbansoltan Eje
Imatra *63 E5* Etelä-Suomi, SE Finland
Imbros *see* Gökçeada
Imishli *see* İmişli
İmişli *95 H3 Rus.* Imishli. C Azerbaijan
Imola *74 C3* Emilia-Romagna, N Italy
Imperatriz *41 F2* Maranhão, NE Brazil
Imperia *74 A3* Liguria, NW Italy
Impfondo *55 C5* Likouala, NE Congo
Imphāl *113 H3 state capital* Manipur, NE India
Imroz Adası *see* Gökçeada
Inagua Islands *see* Little Inagua
Inagua Islands *see* Great Inagua
Inarijärvi *62 D2 Lapp.* Aanaarjävri, *Swe.* Enareträsk. *lake* N Finland
Inău *see* Ineu
Inawashiro-ko *109 D5 var.* Inawasiro Ko. *lake* Honshū, C Japan
Inawasiro Ko *see* Inawashiro-ko
İncesu *94 D3* Kayseri, Turkey Asia
Incheon *107 E4 off.* Incheon Gwang-yeoksi, *prev.* Inch'ŏn, *Jap.* Jinsen; *prev.* Chemulpo. NW South Korea
Incheon-Gwang-yeoksi *see* Incheon
Inch'ŏn *see* Incheon
Incudine, Monte *69 E7 mountain* Corse, France, C Mediterranean Sea
Indefatigable Island *see* Santa Cruz, Isla
Independence *23 F4* Missouri, C USA
Independence Fjord *61 E1 fjord* N Greenland
Independence Island *see* Malden Island
Independence Mountains *24 C4 mountain range* Nevada, W USA
India *102 B3 off.* Republic of India, *var.* Indian Union, Union of India, *Hind.* Bhārat. *country* S Asia
India *see* Indija
Indiana *18 B4 off.* State of Indiana, *also known as* Hoosier State. *state* N USA
Indianapolis *18 C4 state capital* Indiana, N USA
Indian Church *30 C1* Orange Walk, N Belize
Indian Desert *see* Thar Desert
Indianola *23 F4* Iowa, C USA
Indian Union *see* India
India, Republic of *see* India
India, Union of *see* India
Indigirka *93 F2 river* NE Russian Federation
Inđija *78 D3 Hung.* India; *prev.* Indjija. Vojvodina, N Serbia
Indira Point *110 G3 headland* Andaman and Nicobar Island, India, NE Indian Ocean
Indjija *see* Inđija
Indomed Fracture Zone *119 B6 tectonic feature* SW Indian Ocean
Indonesia *116 B4 off.* Republic of Indonesia, *Ind.* Republik Indonesia; *prev.* Dutch East Indies, Netherlands East Indies, United States of Indonesia. *country* SE Asia
Indonesian Borneo *see* Kalimantan
Indonesia, Republic of *see* Indonesia
Indonesia, Republik *see* Indonesia
Indonesia, United States of *see* Indonesia
Indore *112 D4* Madhya Pradesh, C India
Indreville *see* Châteauroux
Indus *112 C2 Chin.* Yindu He; *prev.* Yin-tu Ho. *river* S Asia
Indus Cone *see* Indus Fan
Indus Fan *90 C5 var.* Indus Cone. *undersea fan* N Arabian Sea
Indus, Mouths of the *112 B4 delta* S Pakistan
İnebolu *94 C2* Kastamonu, N Turkey
Ineu *86 A4 Hung.* Borosjenő; *prev.* Inău. Arad, W Romania
Infiernillo, Presa del *29 E4 reservoir* S Mexico
Inglewood *24 D2* California, W USA
Ingolstadt *73 C6* Bayern, S Germany
Ingulets *see* Inhulets'
Inguri *see* Enguri
Inhambane *57 E4* Inhambane, SE Mozambique
Inhulets' *87 F3 Rus.* Ingulets. Dnipropetrovs'ka Oblast', E Ukraine
I-ning *see* Yining
Inis *see* Ennis
Inis Ceithleann *see* Enniskillen
Inn *73 C6 river* C Europe
Innaanganeq *60 C1 var.* Kap York. *headland* NW Greenland
Inner Hebrides *66 B4 island group* W Scotland, United Kingdom
Inner Islands *57 H1 var.* Central Group. *island group* NE Seychelles
Innisfail *126 D3* Queensland, NE Australia
Inniskilling *see* Enniskillen
Innsbruch *see* Innsbruck
Innsbruck *73 C7 var.* Innsbruch. Tirol, W Austria
Inoucdjouac *see* Inukjuak
Inowrazlaw *see* Inowrocław
Inowrocław *76 C3 Ger.* Hohensalza; *prev.* Inowrazlaw. Kujawski-pomorskie, C Poland
I-n-Sakane, 'Erg *53 E2 desert* N Mali
I-n-Salah *48 D3 var.* In Salah. C Algeria
Insterburg *see* Chernyakhovsk
Insula *see* Lille
Inta *88 E3* Respublika Komi, NW Russian Federation
Interamna *see* Teramo
Interamna Nahars *see* Terni
International Falls *23 F1* Minnesota, N USA
Inukjuak *16 D2 var.* Inoucdjouac; *prev.* Port Harrison. Québec, NE Canada
Inuuvik *see* Inuvik
Inuvik *14 D3 var.* Inuuvik. Northwest Territories, NW Canada
Invercargill *129 A7* Southland, South Island, New Zealand
Inverness *66 C3* N Scotland, United Kingdom
Investigator Ridge *119 D5 undersea ridge* E Indian Ocean
Investigator Strait *127 B7 strait* South Australia
Inyangani *56 D3 mountain* NE Zimbabwe
Ioánnina *82 A4 var.* Janina, Yannina. Ípeiros, W Greece
Iola *23 F5* Kansas, C USA
Ionia Basin *see* Ionian Basin
Ionian Basin *58 D5 var.* Ionia Basin. *undersea basin* Ionian Sea, C Mediterranean Sea
Ionian Islands *see* Iónia Nisiá
Ionian Sea *81 E3 Gk.* Iónio Pélagos, *It.* Mar Ionio. *sea* C Mediterranean Sea
Iónia Nisiá *83 A5 Eng.* Ionian Islands. *island group* W Greece
Ionio, Mar/Iónio Pélagos *see* Ionian Sea
Íos *83 D6 var.* Nio. *island* Kykládes, Greece, Aegean Sea
Ioulís *83 C6 prev.* Kéa. Tziá, Kykládes, Greece, Aegean Sea
Iowa *23 F3 off.* State of Iowa, *also known as* Hawkeye State. *state* C USA
Iowa City *23 G3* Iowa, C USA
Iowa Falls *23 G3* Iowa, C USA
Ipel' *77 C6 var.* Ipoly, *Ger.* Eipel. *river* Hungary/Slovakia
Ipiales *36 A4* Nariño, SW Colombia
Ipoh *116 B3* Perak, Peninsular Malaysia
Ipoly *see* Ipel'
Ippy *54 C4* Ouaka, C Central African Republic
Ipswich *127 E5* Queensland, E Australia
Ipswich *67 E6 hist.* Gipeswic. E England, United Kingdom
Iqaluit *15 H3 prev.* Frobisher Bay. *province capital* Baffin Island, Nunavut, NE Canada
Iquique *42 B1* Tarapacá, N Chile
Iquitos *38 C1* Loreto, N Peru
Irákleio *83 D7 var.* Herakleion, *Eng.* Candia; *prev.* Iráklion. Kríti, Greece, E Mediterranean Sea
Iráklion *see* Irákleio
Iran *98 C3 off.* Islamic Republic of Iran; *prev.* Persia. *country* SW Asia
Iranian Plateau *98 D3 var.* Plateau of Iran. *plateau* N Iran
Iran, Islamic Republic of *see* Iran
Iran, Plateau of *see* Iranian Plateau
Irapuato *29 E4* Guanajuato, C Mexico
Iraq *98 B3 off.* Republic of Iraq, *Ar.* 'Irāq. *country* SW Asia
'Irāq *see* Iraq
Iraq, Republic of *see* Iraq
Irbid *97 B5* Irbid, N Jordan
Irbīl *see* Arbīl
Ireland *67 A5 off.* Republic of Ireland, *Ir.* Éire. *country* NW Europe
Ireland *58 C3 Lat.* Hibernia. *island* Ireland/United Kingdom
Ireland, Republic of *see* Ireland
Irian *see* New Guinea
Irian Barat *see* Papua
Irian Jaya *see* Papua
Irian, Teluk *see* Cenderawasih, Teluk
Iringa *51 C7* Iringa, C Tanzania
Iriomote-jima *108 A4 island* Sakishima-shotō, SW Japan
Iriona *30 D2* Colón, NE Honduras
Irish Sea *67 C5 Ir.* Muir Éireann. *sea* C British Isles
Irkutsk *93 E4* Irkutskaya Oblast', S Russian Federation
Irminger Basin *see* Reykjanes Basin
Iroise *68 A3 sea* NW France
Iron Mountain *18 B2* Michigan, N USA
Ironwood *18 B1* Michigan, N USA
Irrawaddy *114 B2 var.* Ayeyarwady. *river* W Myanmar (Burma)
Irrawaddy, Mouths of the *115 A5 delta* SW Myanmar (Burma)
Irtish *see* Irtysh
Irtysh *92 C4 var.* Irtish, *Kaz.* Ertis. *river* C Asia
Irun *71 E1 Cast.* Irún. País Vasco, N Spain
Irún *see* Irun
Iruña *see* Pamplona
Isabela, Isla *38 A5 var.* Albemarle Island. *island* Galápagos Islands, Ecuador, E Pacific Ocean
Isaccea *86 D4* Tulcea, E Romania
Isachsen *15 F1* Ellef Ringnes Island, Nunavut, N Canada
Ísafjördur *61 E4* Vestfirdhir, NW Iceland
Isbarta *see* Isparta
Isca Damnoniorum *see* Exeter
Ise *109 C6* Mie, Honshū, SW Japan
Iseghem *see* Izegem
Isère *69 D5 river* E France
Isernia *75 D5 var.* Æsernia. Molise, C Italy
Ise-wan *109 C6 bay* S Japan
Isfahan *see* Eşfahān
Isha Baydhabo *see* Baydhabo
Ishigaki-jima *108 A4 island* Sakishima-shotō, SW Japan
Ishikari-wan *108 C2 bay* Hokkaidō, NE Japan
Ishim *92 C4* Tyumenskaya Oblast', C Russian Federation
Ishim *92 C4 Kaz.* Esil. *river* Kazakhstan/Russian Federation
Ishinomaki *108 D4 var.* Isinomaki. Miyagi, Honshū, C Japan
Ishkashim *see* Ishkoshim
Ishkoshim *101 F3 Rus.* Ishkashim. S Tajikistan
Isinomaki *see* Ishinomaki
Isiro *55 E5* Orientale, NE Dem. Rep. Congo
Iskăr *see* Iskŭr
İskenderun *94 D4 Eng.* Alexandretta. Hatay, S Turkey
İskenderun Körfezi *96 A2 Eng.* Gulf of Alexandretta. *gulf* S Turkey
Iskur *82 C2 var.* Iskăr. *river* NW Bulgaria
Yazovir Iskur *82 B2 prev.* Yazovir Stalin. *reservoir* W Bulgaria
Isla Cristina *70 C4* Andalucía, S Spain
Isla de León *see* San Fernando
Islāmābād *112 C1 country capital* (Pakistan) Federal Capital Territory Islāmābād, NE Pakistan
Island/Ísland *see* Iceland
Islay *66 B4 island* SW Scotland, United Kingdom
Isle *69 B5 river* W France
Isle of Man *67 B5 British Crown Dependency* NW Europe
Isles of Scilly *67 B8 island group* SW England, United Kingdom
Ismailia *see* Al Ismā'īlīya
Ismâ'ilîya *see* Al Ismā'īlīya
Ismid *see* İzmit
Isnā *50 B2 var.* Esna. SE Egypt
Isoka *56 D1* Northern, NE Zambia
Isparta *94 B4 var.* Isbarta. Isparta, SW Turkey
İspir *95 E3* Erzurum, NE Turkey
Israel *97 A7 off.* State of Israel, *var.* Medinat Israel, *Heb.* Yisrael, Yisra'el. *country* SW Asia
Israel, State of *see* Israel
Issa *see* Vis
Issiq Köl *see* Issyk-Kul', Ozero
Issoire *69 C5* Puy-de-Dôme, C France
Issyk-Kul' *see* Balykchy
Issyk-Kul', Ozero *101 G2 var.* Issiq Köl, *Kir.* Ysyk-Köl. *lake* E Kyrgyzstan
İstanbul *94 B2 Bul.* Tsarigrad, *Eng.* Istanbul, *prev.* Constantinople; *anc.* Byzantium. Istanbul, NW Turkey
İstanbul Boğazi *94 B2 var.* Bosporus Thracius, *Eng.* Bosphorus, Bosporus, *Turk.* Karadeniz Boğazi. *strait* NW Turkey
Istarska Županija *see* Istra
Istra *78 A3 off.* Istarska Županija. *province* NW Croatia
Istra *78 A3 Eng.* Istria, *Ger.* Istrien. *cultural region* NW Croatia
Istria/Istrien *see* Istra
Itabuna *41 G3* Bahia, E Brazil
Itagüí *36 B3* Antioquia, W Colombia
Itaipú, Represa de *41 E4 reservoir* Brazil/Paraguay
Itaituba *41 E2* Pará, NE Brazil
Italia/Italiana, Republica/Italian Republic, The *see* Italy
Italian Somaliland *see* Somalia
Italy *74 C3 off.* The Italian Republic, *It.* Italia, Repubblica Italiana. *country* S Europe
Iténez, Río *see* Guaporé, Rio
Ithaca *19 E3* New York, NE USA
It Hearrenfean *see* Heerenveen
Itoigawa *109 C5* Niigata, Honshū, C Japan
Itseqqortoormiit *see* Ittoqqortoormiit
Ittoqqortoormiit *61 E3 var.* Itseqqortoormiit, *Dan.* Scoresbysund, *Eng.* Scoresby Sound. Tunu, C Greenland
Iturup, Ostrov *108 E1 island* Kuril'skiye Ostrova, SE Russian Federation
Itzehoe *72 B2* Schleswig-Holstein, N Germany
Ivalo *62 D2 Lapp.* Avveel, Avvil. Lappi, N Finland
Ivanava *85 B7 Pol.* Janów, Janów Poleski, *Rus.* Ivanovo. Brestskaya Voblasts', SW Belarus
Ivangrad *see* Berane
Ivanhoe *127 C6* New South Wales, SE Australia
Ivano-Frankivs'k *86 C2 Ger.* Stanislau, *Pol.* Stanisławów, *Rus.* Ivano-Frankovsk; *prev.* Stanislav. Ivano-Frankivs'ka Oblast', W Ukraine
Ivano-Frankovsk *see* Ivano-Frankivs'k
Ivanovo *89 B5* Ivanovskaya Oblast', W Russian Federation
Ivanovo *see* Ivanava
Ivantsevichi/Ivatsevichi *see* Ivatsevichy
Ivatsevichy *85 B6 Pol.* Iwacewicze, *Rus.* Ivantsevichi, Ivatsevichi. Brestskaya Voblasts', SW Belarus
Ivigtut *see* Ivittuut
Ivittuut *60 B4 var.* Ivigtut. Kitaa, S Greenland
Iviza *see* Eivissa\Ibiza
Ivory Coast *see* Côte d'Ivoire
Ivujivik *16 D1* Québec, NE Canada
Iwacewicze *see* Ivatsevichy
Iwaki *109 D5* Fukushima, Honshū, N Japan
Iwakuni *109 B7* Yamaguchi, Honshū, SW Japan
Iwanai *108 C2* Hokkaidō, NE Japan
Iwate *108 D3* Iwate, Honshū, N Japan
Ixtapa *29 E5* Guerrero, S Mexico
Ixtepec *29 F5* Oaxaca, SE Mexico
Iyo-nada *109 B7 sea* S Japan
Izabal, Lago de *30 B2 prev.* Golfo Dulce. *lake* E Guatemala
Īzad Khvāst *98 D3* Fārs, C Iran
Izegem *65 A6 prev.* Iseghem. West-Vlaanderen, W Belgium
Izhevsk *89 D5 prev.* Ustinov. Udmurtskaya Respublika, NW Russian Federation
Izmail *see* Izmayil
Izmayil *86 D4 Rus.* Izmail. Odes'ka Oblast', SW Ukraine
İzmir *94 A3 prev.* Smyrna. İzmir, W Turkey
İzmit *94 B2 var.* Ismid; *anc.* Astacus. Kocaeli, NW Turkey
İznik Gölü *94 B3 lake* NW Turkey
Izu-hanto *109 D6 peninsula* Honshū, S Japan
Izu Shichito *see* Izu-shotō
Izu-shoto *109 D6 var.* Izu Shichito. *island group* S Japan
Izvor *82 B2* Pernik, W Bulgaria
Izyaslav *86 C2* Khmel'nyts'ka Oblast', W Ukraine
Izyum *87 G2* Kharkivs'ka Oblast', E Ukraine

J

Jabal ash Shifa *98 A4 desert* NW Saudi Arabia
Jabalpur *113 E4 prev.* Jubbulpore. Madhya Pradesh, C India
Jabbūl, Sabkhat al *96 B2 sabkha* NW Syria

Jablah – Kapsukas

Jablah *96 A3 var.* Jeble, *Fr.* Djéblé. Al Lādhiqīyah, W Syria
Jaca *71 F1* Aragón, NE Spain
Jacaltenango *30 A2* Huehuetenango, W Guatemala
Jackson *20 B2 state capital* Mississippi, S USA
Jackson *23 H5* Missouri, C USA
Jackson *20 C1* Tennessee, S USA
Jackson Head *129 A6 headland* South Island, New Zealand
Jacksonville *21 E3* Florida, SE USA
Jacksonville *18 B4* Illinois, N USA
Jacksonville *21 F1* North Carolina, SE USA
Jacksonville *27 G3* Texas, SW USA
Jacmel *32 D3 var.* Jaquemel. S Haiti
Jacob *see* Nkayi
Jacobābād *112 B3* Sind, SE Pakistan
Jadotville *see* Likasi
Jadransko More/Jadransko Morje *see* Adriatic Sea
Jaén *38 B2* Cajamarca, N Peru
Jaén *70 D4* Andalucía, SW Spain
Jaffna *110 D3* Northern Province, N Sri Lanka
Jagannath *see* Puri
Jagdalpur *113 E5* Chhattīsgarh, C India
Jagdaqi *105 G1* Nei Mongol Zizhiqu, N China
Jagodina *78 D4 prev.* Svetozarevo. Serbia, C Serbia
Jahra *see* Al Jahrā'
Jailolo *see* Halmahera, Pulau
Jaipur *112 D3 prev.* Jeypore. *state capital* Rājasthān, N India
Jaisalmer *112 C3* Rājasthān, NW India
Jajce *78 B3* Federacija Bosna I Hercegovina, W Bosnia and Herzegovina
Jakarta *116 C5 prev.* Djakarta, *Dut.* Batavia. *country capital* (Indonesia) Jawa, C Indonesia
Jakobstad *62 D4 Fin.* Pietarsaari. Länsi-Suomi, W Finland
Jakobstadt *see* Jēkabpils
Jalālābād *101 F4 var.* Jalalabad, Jelalabad. Nangarhār, E Afghanistan
Jalal-Abad *see* Dzhalal-Abad, Dzhalal-Abadskaya Oblast', Kyrgyzstan
Jalandhar *112 D2 prev.* Jullundur. Punjab, N India
Jalapa *30 D3* Nueva Segovia, NW Nicaragua
Jalpa *28 D4* Zacatecas, C Mexico
Jālū *49 G3 var.* Jūlā. NE Libya
Jaluit Atoll *122 D2 var.* Jālwōj. *atoll* Ralik Chain, S Marshall Islands
Jālwōj *see* Jaluit Atoll
Jamaame *51 D6 It.* Giamame; *prev.* Margherita. Jubbada Hoose, S Somalia
Jamaica *32 A4 country* W West Indies
Jamaica *34 A1 island* W West Indies
Jamaica Channel *32 D3 channel* Haiti/Jamaica
Jamālpur *113 F3* Bihār, NE India
Jambi *116 B4 var.* Telanaipura; *prev.* Djambi. Sumatera, W Indonesia
Jamdena *see* Yamdena, Pulau
James Bay *16 C3 bay* Ontario/Québec, E Canada
James River *23 E2 river* North Dakota/South Dakota, N USA
James River *19 E5 river* Virginia, NE USA
Jamestown *19 E3* New York, NE USA
Jamestown *23 E2* North Dakota, N USA
Jamestown *see* Holetown
Jammu *112 D2 prev.* Jummoo. *state capital* Jammu and Kashmir, NW India
Jammu and Kashmīr *112 D1 disputed region* India/Pakistan
Jāmnagar *112 C4 prev.* Navanagar. Gujarāt, W India
Jamshedpur *113 F4* Jhārkhand, NE India
Jamuna *see* Brahmaputra
Janaúba *41 F3* Minas Gerais, SE Brazil
Janesville *18 B3* Wisconsin, N USA
Janina *see* Ioánnina
Janischken *see* Joniškis
Jankovac *see* Jánoshalma
Jan Mayen *61 F4 Norwegian dependency* N Atlantic Ocean
Jánoshalma *77 C7 SCr.* Jankovac. Bács-Kiskun, S Hungary
Janów *see* Ivanava, Belarus
Janow/Janów *see* Jonava, Lithuania
Janów Poleski *see* Ivanava
Japan *108 C4 var.* Nippon, *Jap.* Nihon. *country* E Asia
Japan, Sea of *108 A4 var.* East Sea, *Rus.* Yapanskoye More. *sea* NW Pacific Ocean
Japan Trench *103 F1 trench* NW Pacific Ocean
Japen *see* Yapen, Pulau
Japiim *40 C2 var.* Máncio Lima. Acre, W Brazil
Japurá, Rio *40 C2 var.* Río Caquetá, Yapurá. *river* Brazil/Colombia
Japurá, Rio *see* Caquetá, Río
Jaqué *31 G5* Darién, SE Panama
Jaquemel *see* Jacmel
Jarablos *see* Jarābulus
Jarābulus *96 C2 var.* Jarablos, Jerablus, *Fr.* Djérablous. Ḩalab, N Syria
Jarbah, Jazīrat *see* Jerba, Île de
Jardines de la Reina, Archipiélago de los *32 B2 island group* C Cuba
Jarīd, Shaṭṭ al *see* Jerid, Chott el
Jarocin *76 C4* Wielkopolskie, C Poland
Jaroslau *see* Jarosław
Jarosław *77 E5 Ger.* Jaroslau, *Rus.* Yaroslav. Podkarpackie, SE Poland
Jarqo'rg'on *101 E3 Rus.* Dzharkurgan. Surkhondaryo Viloyati, S Uzbekistan
Jarvis Island *123 G2 US unincorporated territory* C Pacific Ocean
Jasło *77 D5* Podkarpackie, SE Poland
Jastrzębie-Zdrój *77 C5* Śląskie, S Poland
Jataí *41 E3* Goiás, C Brazil
Játiva *see* Xàtiva
Jauf *see* Al Jawf
Jaunpiebalga *84 D3* Gulbene, NE Latvia
Jaunpur *113 E3* Uttar Pradesh, N India
Java *130 A3* South Dakota, N USA
Javalambre *71 E3 mountain* E Spain
Java Sea *116 D4 Ind.* Laut Jawa. *sea* W Indonesia
Java Trench *102 D5 var.* Sunda Trench. *trench* E Indian Ocean
Jawa, Laut *see* Java Sea
Jawhar *51 D6 var.* Jowhar, *It.* Giohar. Shabeellaha Dhexe, S Somalia
Jaworów *see* Yavoriv
Jaya, Puncak *117 G4 prev.* Puntjak Carstensz, Puntjak Sukarno. *mountain* Papua, E Indonesia
Jayapura *117 H4 var.* Djajapura, *Dut.* Hollandia; *prev.* Kotabaru, Sukarnapura. Papua, E Indonesia
Jay Dairen *see* Dalian
Jayhawker State *see* Kansas
Jaz Murian, Hamun-e *98 E4 lake* SE Iran
Jebba *53 F4* Kwara, W Nigeria
Jebel, Bahr el *see* White Nile
Jeble *see* Jablah
Jedda *see* Jiddah
Jędrzejów *76 D4 Ger.* Endersdorf. Świętokrzyskie, C Poland
Jefferson City *23 G5 state capital* Missouri, C USA
Jega *53 F4* Kebbi, NW Nigeria
Jehol *see* Chengde
Jeju-do *107 E4 Jap.* Saishū; *prev.* Cheju-do, Quelpart. *island* S South Korea
Jeju Strait *107 E4 var.* Jeju-haehyŏp; *prev.* Cheju-Strait. *strait* S South Korea
Jeju-haehyŏp *see* Jeju Strait
Jēkabpils *84 D4 Ger.* Jakobstadt. Jēkabpils, S Latvia
Jelalabad *see* Jalālābād
Jelenia Góra *76 B4 Ger.* Hirschberg, Hirschberg im Riesengebirge, Hirschberg in Riesengebirge, Hirschberg in Schlesien. Dolnośląskie, SW Poland
Jelgava *84 C3 Ger.* Mitau. Jelgava, C Latvia
Jemappes *65 B6* Hainaut, S Belgium
Jember *116 D5 prev.* Djember. Jawa, C Indonesia
Jena *72 C4* Thüringen, C Germany
Jengish Chokusu *see* Tomür Feng
Jenin *97 A6* N West Bank
Jerablus *see* Jarābulus
Jerada *48 D2* NE Morocco
Jérémie *32 D3* SW Haiti
Jerez *see* Jeréz de la Frontera, Spain
Jerez de la Frontera *70 C5 var.* Jerez; *prev.* Xeres. Andalucía, SW Spain
Jerez de los Caballeros *70 C4* Extremadura, W Spain
Jericho *see* Arīḩā
Jerid, Chott el *49 E2 var.* Shaṭṭ al Jarīd. *salt lake* SW Tunisia
Jersey *67 D8 British Crown Dependency* Channel Islands, NW Europe
Jerusalem *81 H4 Ar.* Al Quds, Al Quds ash Sharīf, *Heb.* Yerushalayim; *anc.* Hierosolyma. *country capital* (Israel) Jerusalem, NE Israel
Jesenice *73 D7 Ger.* Assling. NW Slovenia
Jesselton *see* Kota Kinabalu
Jessore *113 G4* Khulna, W Bangladesh
Jesús María *42 C3* Córdoba, C Argentina
Jeypore *see* Jaipur, Rājasthān, India
Jhānsi *112 D3* Uttar Pradesh, N India
Jhārkhand *113 F4 cultural region* NE India
Jhelum *112 C2* Punjab, NE Pakistan
Ji *see* Hebei, China
Ji *see* Jilin, China
Jiangmen *106 C6* Guangdong, S China
Jiangsu *106 D4 var.* Chiang-su, Jiangsu Sheng, Kiangsu, Su. *province* E China
Jiangsu *see* Nanjing
Jiangsu Sheng *see* Jiangsu
Jiangxi *106 C6 var.* Chiang-hsi, Gan, Jiangxi Sheng, Kiangsi. *province* S China
Jiangxi Sheng *see* Jiangxi
Jiaxing *106 D5* Zhejiang, SE China
Jiayi *106 D6 prev.* Chiai; *var.* Chia-i, Chiayi, Kiayi, Jiayi, *Jap.* Kagi. C Taiwan
Jibuti *see* Djibouti
Jiddah *99 A5 Eng.* Jedda. Makkah, W Saudi Arabia
Jih-k'a-tse *see* Xigazê
Jihlava *77 B5 Ger.* Iglau, *Pol.* Igława. Vysocina, S Czech Republic
Jilib *51 D6 It.* Gelib. Jubbada Dhexe, S Somalia
Jilin *106 E3 var.* Chi-lin, Girin, Kirin; *prev.* Yungki, Yunki. Jilin, NE China
Jilin *106 E3 var.* Chi-lin, Girin, Ji, Jilin Sheng, Kirin. *province* NE China
Jilin Sheng *see* Jilin
Jilong *106 D6 prev.* Chilung; *var.* Keelung, *Jap.* Kirun, Kirun'; *prev. Sp.* Santissima Trinidad. N Taiwan
Jīma *51 C5 var.* Jimma, *It.* Gimma. Oromīya, C Ethiopia
Jimbolia *86 A4 Ger.* Hatzfeld, *Hung.* Zsombolya. Timiş, W Romania
Jiménez *28 D2* Chihuahua, N Mexico
Jimma *see* Jīma
Jimsar *104 C3* Xinjiang Uygur Zizhiqu, NW China
Jin *see* Shanxi
Jin *see* Tianjin Shi
Jinan *106 C4 var.* Chinan, Chi-nan, Tsinan. *province capital* Shandong, E China
Jingdezhen *106 C5* Jiangxi, S China
Jinghong *106 A6 var.* Yunjinghong. Yunnan, SW China
Jinhua *106 D5* Zhejiang, SE China
Jining *105 F3* Shandong, E China
Jinja *51 C6* S Uganda
Jinotega *30 D3* Jinotega, NW Nicaragua
Jinotepe *30 D3* Carazo, SW Nicaragua
Jinsen *see* Incheon
Jinzhong *106 C4 var.* Yuci. Shanxi, C China
Jinzhou *106 D3 var.* Chin-chou, Chinchow; *prev.* Chinhsien. Liaoning, NE China
Jirgalanta *see* Hovd
Jisr ash Shadadi *see* Ash Shadādah
Jiu *86 B5 Ger.* Schil, Schyl, *Hung.* Zsil, Zsily. *river* S Romania
Jiujiang *106 C5* Jiangxi, S China
Jixi *107 E2* Heilongjiang, NE China
Jīzān *99 B6 var.* Qīzān. Jīzān, SW Saudi Arabia
Jizzax *101 E2 Rus.* Dzhizak. Jizzax Viloyati, C Uzbekistan
João Belo *see* Xai-Xai
João Pessoa *41 G2 prev.* Paraíba. *state capital* Paraíba, E Brazil
Joazeiro *see* Juazeiro
Job'urg *see* Johannesburg
Jo-ch'iang *see* Ruoqiang
Jodhpur *112 C3* Rājasthān, NW India
Joensuu *63 E5* Itä-Suomi, SE Finland
Jōetsu *109 C5 var.* Zyôetu. Niigata, Honshū, C Japan
Jogjakarta *see* Yogyakarta
Johannesburg *56 D4 var.* Egoli, Erautini, Gauteng, *abbrev.* Job'urg. Gauteng, NE South Africa
Johannisburg *see* Pisz
John Day River *24 C3 river* Oregon, NW USA
John o'Groats *66 C2* N Scotland, United Kingdom
Johnston Atoll *121 E1 US unincorporated territory* C Pacific Ocean
Johor Baharu *see* Johor Bahru
Johor Bahru *116 B3 var.* Johor Baharu, Johore Bahru. Johor, Peninsular Malaysia
Johore Bahru *see* Johor Bahru
Johore Strait *116 A1 strait* Johor, Peninsular Malaysia, Malaysia/Singapore Asia Andaman Sea/ South China Sea
Joinvile *see* Joinville
Joinville *41 E4 var.* Joinvile. Santa Catarina, S Brazil
Jokkmokk *62 C3 Lapp.* Dálvvadis. Norrbotten, N Sweden
Jokyakarta *see* Yogyakarta
Joliet *18 B3* Illinois, N USA
Jonava *84 B4 Ger.* Janow, *Pol.* Janów. Kaunas, C Lithuania
Jonesboro *20 B1* Arkansas, C USA
Joniškis *84 C3 Ger.* Janischken. Šiaulai, N Lithuania
Jönköping *63 B7* Jönköping, S Sweden
Jonquière *17 E4* Québec, SE Canada
Joplin *23 F5* Missouri, C USA
Jordan *97 B6 off.* Hashemite Kingdom of Jordan, *Ar.* Al Mamlaka al Urdunīya al Hashemīyah, Al Urdunn; *prev.* Transjordan. *country* SW Asia
Jordan *97 B5 Ar.* Urdunn, *Heb.* HaYarden. *river* SW Asia
Jorhāt *113 H3* Assam, NE India
Jos *53 G4* Plateau, C Nigeria
Joseph Bonaparte Gulf *124 D2 gulf* N Australia
Jos Plateau *53 G4 plateau* C Nigeria
Jotunheimen *63 A5 mountain range* S Norway
Jounié *96 A4 var.* Junīyah. W Lebanon
Joure *64 D2 Fris.* De Jouwer. Friesland, N Netherlands
Joutseno *63 E5* Etelä-Suomi, SE Finland
Jowhar *see* Jawhar
J.Storm Thurmond Reservoir *see* Clark Hill Lake
Juan Aldama *28 D3* Zacatecas, C Mexico
Juan de Fuca, Strait of *24 A1 strait* Canada/USA
Juan Fernández, Islas *35 A6 Eng.* Juan Fernandez Islands. *island group* W Chile
Juan Fernandez Islands *see* Juan Fernández, Islas
Juazeiro *41 G2 prev.* Joazeiro. Bahia, E Brazil
Juazeiro do Norte *41 G2* Ceará, E Brazil
Juba *51 B5 var.* Jūbā. *country capital* (South Sudan) Bahr el Gabel, S South Sudan
Juba *51 D6 Amh.* Genalē Wenz, *It.* Guiba, *Som.* Ganaane, Webi Jubba. *river* Ethiopia/Somalia
Jubba, Webi *see* Juba
Jubbulpore *see* Jabalpur
Júcar *71 E3 var.* Jucar. *river* C Spain
Juchitán *29 F5 var.* Juchitán de Zaragosa. Oaxaca, SE Mexico
Juchitán de Zaragosa *see* Juchitán
Judayyidat Hāmir *98 B3* Al Anbār, S Iraq
Judenburg *73 D7* Steiermark, C Austria
Jugoslavija *see* Serbia
Juigalpa *30 D3* Chontales, S Nicaragua
Juiz de Fora *41 F4* Minas Gerais, SE Brazil
Jujuy *see* San Salvador de Jujuy
Jūlā *see* Jālū, Libya
Julia Beterrae *see* Béziers
Juliaca *39 E4* Puno, SE Peru
Juliana Top *37 G3 mountain* C Suriname
Julianehåb *see* Qaqortoq
Julio Briga *see* Bragança
Juliobriga *see* Logroño
Juliomagus *see* Angers
Jullundur *see* Jalandhar
Jumilla *71 E4* Murcia, SE Spain
Jummoo *see* Jammu
Jumna *see* Yamuna
Jumporn *see* Chumphon
Junction City *23 F4* Kansas, C USA
Juneau *14 D4 state capital* Alaska, USA
Junín *42 C4* Buenos Aires, E Argentina
Junīyah *see* Jounié
Junkseylon *see* Phuket
Jur *51 B5 river* C Sudan
Jura *68 D4 cultural region* E France
Jura *73 A7 var.* Jura Mountains. *mountain range* France/Switzerland
Jura *66 B4 island* SW Scotland, United Kingdom
Jura Mountains *see* Jura
Jurbarkas *84 B4 Ger.* Georgenburg, Jurburg. Tauragė, W Lithuania
Jurburg *see* Jurbarkas
Jūrmala *84 C3* Rīga, C Latvia
Juruá, Rio *40 C2 var.* Río Yuruá. *river* Brazil/Peru
Juruena, Rio *40 D3 river* W Brazil
Jutiapa *30 B2* Jutiapa, S Guatemala
Juticalpa *30 D2* Olancho, C Honduras
Jutland *63 A7 Den.* Jylland. *peninsula* W Denmark
Juvavum *see* Salzburg
Juventud, Isla de la *32 A2 var.* Isla de Pinos, *Eng.* Isle of Youth; *prev.* The Isle of the Pines. *island* W Cuba
Južna Morava *79 E5 Ger.* Südliche Morava. *river* SE Serbia
Jwaneng *56 C4* Southern, S Botswana
Jylland *see* Jutland
Jyrgalan *see* Dzhergalan
Jyväskylä *63 D5* Länsi-Suomi, C Finland

K

K2 *104 A4 Chin.* Qogir Feng, *Eng.* Mount Godwin Austen. *mountain* China/Pakistan
Kaafu Atoll *see* Male' Atoll
Kaaimanston *37 G3* Sipaliwini, N Suriname
Kaakhka *see* Kaka
Kaala *see* Caála
Kaamanen *62 D2 Lapp.* Gámas. Lappi, N Finland
Kaapstad *see* Cape Town
Kaaresuvanto *62 C3 Lapp.* Gárassavon. Lappi, N Finland
Kabale *51 B6* SW Uganda
Kabinda *55 D7* Kasai-Oriental, SE Dem. Rep. Congo
Kabinda *see* Cabinda
Kābol *see* Kābul
Kabompo *56 C2 river* W Zambia
Kābul *101 E4 prev.* Kābol. *country capital* (Afghanistan) Kābul, E Afghanistan
Kabul *101 E4 var.* Daryā-ye Kābul. *river* Afghanistan/Pakistan
Kābul, Daryā-ye *see* Kabul
Kabwe *56 D2* Central, C Zambia
Kachchh, Gulf of *112 B4 var.* Gulf of Cutch, Gulf of Kutch. *gulf* W India
Kachchh, Rann of *112 B4 var.* Rann of Kachh, Rann of Kutch. *salt marsh* India/Pakistan
Kachh, Rann of *see* Kachchh, Rann of
Kadan Kyun *115 B5 prev.* King Island. *island* Myeik Archipelago, S Myanmar (Burma)
Kadavu *123 E4 prev.* Kandavu. *island* S Fiji
Kadoma *56 D3 prev.* Gatooma. Mashonaland West, C Zimbabwe
Kadugli *50 B4* Southern Kordofan, S Sudan
Kaduna *53 G4* Kaduna, C Nigeria
Kadzhi-Say *101 G2 Kir.* Kajisay. Issyk-Kul'skaya Oblast', NE Kyrgyzstan
Kaédi *52 C3* Gorgol, S Mauritania
Kaffa *see* Feodosiya
Kafue *56 D2* Lusaka, SE Zambia
Kafue *56 C2 river* C Zambia
Kaga Bandoro *54 C4 prev.* Fort-Crampel. Nana-Grébizi, C Central African Republic
Kagan *see* Kogon
Kâghet *52 D1 var.* Karet. *physical region* N Mauritania
Kagi *see* Jiayi
Kagoshima *109 B8 var.* Kagosima. Kagoshima, Kyūshū, SW Japan
Kagoshima-wan *109 A8 bay* SW Japan
Kagosima *see* Kagoshima
Kagul *see* Cahul
Darya-ye Kahmard *101 E4 prev.* Darya-i-surkhab. *river* NE Afghanistan
Kahramanmaraş *94 D4 var.* Kahraman Maraş, Maraş, Marash. Kahramanmaraş, S Turkey
Kaiapoi *129 C6* Canterbury, South Island, New Zealand
Kaifeng *106 C4* Henan, C China
Kai, Kepulauan *117 F4 prev.* Kei Islands. *island group* Maluku, SE Indonesia
Kaikohe *128 C2* Northland, North Island, New Zealand
Kaikoura *129 C5* Canterbury, South Island, New Zealand
Kaikoura Peninsula *129 C5 peninsula* South Island, New Zealand
Kainji Lake *see* Kainji Reservoir
Kainji Reservoir *53 F4 var.* Kainji Lake. *reservoir* W Nigeria
Kaipara Harbour *128 C2 harbour* North Island, New Zealand
Kairouan *49 E2 var.* Al Qayrawān. E Tunisia
Kaisaria *see* Kayseri
Kaiserslautern *73 A5* Rheinland-Pfalz, SW Germany
Kaišiadorys *85 B5* Kaunas, S Lithuania
Kaitaia *128 C2* Northland, North Island, New Zealand
Kajaani *62 E4 Swe.* Kajana. Oulu, C Finland
Kajan *see* Kayan, Sungai
Kajana *see* Kajaani
Kajisay *see* Kadzhi-Say
Kaka *100 C2 Rus.* Kaakhka. Ahal Welaýaty, S Turkmenistan
Kake *14 D4* Kupreanof Island, Alaska, USA
Kakhovka *87 F4* Khersons'ka Oblast', S Ukraine
Kakhovs'ke Vodoskhovyshche *87 F4 Rus.* Kakhovskoye Vodokhranilishche. *reservoir* SE Ukraine
Kakhovskoye Vodokhranilishche *see* Kakhovs'ke Vodoskhovyshche
Kākināda *110 D1 prev.* Cocanada. Andhra Pradesh, E India
Kakshaal-Too, Khrebet *see* Kokshaal-Tau
Kaktovik *14 D2* Alaska, USA
Kalaallit Nunaat *see* Greenland
Kalahari Desert *56 B4 desert* Southern Africa
Kalaikhum *see* Qal'aikhum
Kalámai *see* Kalámata
Kalamariá *82 B4* Kentrikí Makedonía, N Greece
Kalamás *82 A4 var.* Thiamis; *prev.* Thýamis. *river* W Greece
Kalámata *83 B6 prev.* Kalámai. Pelopónnisos, S Greece
Kalamazoo *18 C3* Michigan, N USA
Kalambaka *see* Kalampáka
Kálamos *83 C5* Attikí, C Greece
Kalampáka *82 B4 var.* Kalambaka. Thessalía, C Greece
Kalanchak *87 F4* Khersons'ka Oblast', S Ukraine
Kalarash *see* Călăraşi
Kalasin *114 D4 var.* Muang Kalasin. Kalasin, E Thailand
Kālat *112 B2 var.* Kelat, Khelat. Baluchistān, SW Pakistan
Kalāt *see* Qalāt
Kalbarri *125 A5* Western Australia
Kalecik *94 C3* Ankara, N Turkey
Kalemie *55 E6 prev.* Albertville. Katanga, SE Dem. Rep. Congo
Kale Sultanie *see* Çanakkale
Kalgan *see* Zhangjiakou
Kalgoorlie *125 B6* Western Australia
Kalima *55 D6* Maniema, E Dem. Rep. Congo
Kalimantan *116 D4 Eng.* Indonesian Borneo. *geopolitical region* Borneo, C Indonesia
Kalinin *see* Tver'
Kaliningrad *84 A4* Kaliningradskaya Oblast', W Russian Federation
Kaliningrad *see* Kaliningradskaya Oblast'
Kaliningradskaya Oblast' *84 B4 var.* Kaliningrad. *province and enclave* W Russian Federation
Kalinkavichy *85 C7 Rus.* Kalinkovichi. Homyel'skaya Voblasts', SE Belarus
Kalinkovichi *see* Kalinkavichy
Kalisch/Kalish *see* Kalisz
Kalispell *22 B1* Montana, NW USA
Kalisz *76 C4 Ger.* Kalisch, *Rus.* Kalish; *anc.* Calisia. Wielkopolskie, C Poland
Kalix *62 D4* Norrbotten, N Sweden
Kalixälven *62 D3 river* N Sweden
Kallaste *84 E3 Ger.* Krasnogor. Tartumaa, SE Estonia
Kallavesi *63 E5 lake* SE Finland
Kalloní *83 D5* Lésvos, E Greece
Kalmar *63 C7 var.* Calmar. Kalmar, S Sweden
Kalmthout *65 C5* Antwerpen, N Belgium
Kalpáki *82 A4* Ípeiros, W Greece
Kalpeni Island *110 B3 island* Lakshadweep, India, N Indian Ocean
Kaltdorf *see* Pruszków
Kaluga *89 B5* Kaluzhskaya Oblast', W Russian Federation
Kalush *86 C2 Pol.* Kałusz. Ivano-Frankivs'ka Oblast', W Ukraine
Kałusz *see* Kalush
Kalutara *110 D4* Western Province, SW Sri Lanka
Kalvarija *85 B5 Pol.* Kalwaria. Marijampolė, S Lithuania
Kalwaria *see* Kalvarija
Kalyān *112 C5* Mahārāshtra, W India
Kálymnos *83 D6 var.* Kálimnos. *island* Dodekánisa, Greece, Aegean Sea
Kama *88 D4 river* NW Russian Federation
Kamarang *37 F3* W Guyana
Kambryk *see* Cambrai
Kamchatka *see* Kamchatka, Poluostrov
Kamchatka, Poluostrov *93 G3 Eng.* Kamchatka. *peninsula* E Russian Federation
Kamenets-Podol'skiy *see* Kam"yanets'-Podil's'kyy
Kamenka Dneprovskaya *see* Kam"yanka-Dniprovs'ka
Kamenskoye *see* Romaniv
Kamensk-Shakhtinskiy *89 B6* Rostovskaya Oblast', SW Russian Federation
Kamina *55 D7* Katanga, S Dem. Rep. Congo
Kamishli *see* Al Qāmishlī
Kamloops *15 E5* British Columbia, SW Canada
Kammu Seamount *130 C2 guyot* N Pacific Ocean
Kampala *51 B6 country capital* (Uganda) S Uganda
Kâmpóng Cham *115 D6 prev.* Kompong Cham. Kâmpóng Cham, C Cambodia
Kâmpóng Chhnăng *115 D6 prev.* Kompong. Kâmpóng Chhnăng, C Cambodia
Kâmpóng Saôm *see* Sihanoukville
Kâmpóng Spoe *115 D6 prev.* Kompong Speu. Kâmpóng Spœ, S Cambodia
Kâmpóng Thum *115 D5 prev.* Trâpeăng Vêng. Kâmpóng Thum, C Cambodia
Kâmpóng Trâbêk *115 D5 prev.* Phumĭ Kâmpóng Trâbêk, Phum Kompong Trabek. Kâmpóng Thum, C Cambodia
Kâmpôt *115 D6* Kâmpôt, SW Cambodia
Kampuchea *see* Cambodia
Kampuchea, Democratic *see* Cambodia
Kampuchea, People's Democratic Republic of *see* Cambodia
Kam"yanets'-Podil's'kyy *86 C3 Rus.* Kamenets-Podol'skiy. Khmel'nyts'ka Oblast', W Ukraine
Kam"yanka-Dniprovs'ka *87 F3 Rus.* Kamenka Dneprovskaya. Zaporiz'ka Oblast', SE Ukraine
Kamyshin *89 B6* Volgogradskaya Oblast', SW Russian Federation
Kanaky *see* New Caledonia
Kananga *55 D6 prev.* Luluabourg. Kasai-Occidental, S Dem. Rep. Congo
Kanara *see* Karnātaka
Kanash *89 C5* Chuvashskaya Respublika, W Russian Federation
Kanazawa *109 C5* Ishikawa, Honshū, SW Japan
Kanbe *114 B4* Yangon, SW Myanmar (Burma)
Kānchīpuram *110 C2 prev.* Conjeeveram. Tamil Nādu, SE India
Kandahār *101 E5 Per.* Qandahār. Kandahār, S Afghanistan
Kandalakša *see* Kandalaksha
Kandalaksha *88 B2 var.* Kandalakša, *Fin.* Kantalahti. Murmanskaya Oblast', NW Russian Federation
Kandangan *116 D4* Borneo, C Indonesia
Kandau *see* Kandava
Kandava *84 C3 Ger.* Kandau. Tukums, W Latvia
Kandavu *see* Kadavu
Kandi *53 F4* N Benin
Kandy *110 D3* Central Province, C Sri Lanka
Kane Fracture Zone *44 B4 fracture zone* NW Atlantic Ocean
Kāne'ohe *25 A8 var.* Kaneohe. O'ahu, Hawaii, USA, C Pacific Ocean
Kanestron, Akrotírio *see* Palioúri, Akrotírio
Kanëv *see* Kaniv
Kanevskoye Vodokhranilishche *see* Kanivs'ke Vodoskhovyshche
Kangān *see* Bandar-e Kangān
Kangaroo Island *127 A7 island* South Australia
Kangertittivaq *61 E4 Dan.* Scoresby Sund. *fjord* E Greenland
Kangikajik *61 E4 var.* Kap Brewster. *headland* E Greenland
Kaniv *87 E2 Rus.* Kanëv. Cherkas'ka Oblast', C Ukraine
Kanivs'ke Vodoskhovyshche *87 E2 Rus.* Kanevskoye Vodokhranilishche. *reservoir* C Ukraine
Kanjiža *78 D2 Ger.* Altkanischa, *Hung.* Magyarkanizsa, Ókanizsa; *prev.* Stara Kanjiža. Vojvodina, N Serbia
Kankaanpää *63 D5* Länsi-Suomi, SW Finland
Kankakee *18 B3* Illinois, N USA
Kankan *52 D4* E Guinea
Kannur *110 B2 var.* Cannanore. Kerala, SW India
Kano *53 G4* Kano, N Nigeria
Kānpur *113 E3 Eng.* Cawnpore. Uttar Pradesh, N India
Kansas *23 F5 off.* State of Kansas, *also known as* Jayhawker State, Sunflower State. *state* C USA
Kansas City *23 F4* Kansas, C USA
Kansas City *23 F4* Missouri, C USA
Kansas River *23 F5 river* Kansas, C USA
Kansk *93 E4* Krasnoyarskiy Kray, S Russian Federation
Kansu *see* Gansu
Kantalahti *see* Kandalaksha
Kántanos *83 C7* Kríti, Greece, E Mediterranean Sea
Kantemirovka *89 B6* Voronezhskaya Oblast', W Russian Federation
Kantipur *see* Kathmandu
Kanton *123 F3 var.* Abariringa, Canton Island; *prev.* Mary Island. *atoll* Phoenix Islands, C Kiribati
Kanye *56 C4* Southern, SE Botswana
Kaohsiung *see* Gaoxiong
Kaolack *52 B3 var.* Kaolak. W Senegal
Kaolak *see* Kaolack
Kaolan *see* Lanzhou
Kaoma *56 C2* Western, W Zambia
Kapelle *65 B5* Zeeland, SW Netherlands
Kapellen *65 C5* Antwerpen, N Belgium
Kapka, Massif du *54 C2 mountain range* E Chad
Kaplangky, Plato *see* Gaplaňgyr Platosy
Kapoeas *see* Kapuas, Sungai
Kapoeta *51 C5* E Equatoria, SE South Sudan
Kaposvár *77 C7* Somogy, SW Hungary
Kappeln *72 B2* Schleswig-Holstein, N Germany
Kapronca *see* Koprivnica
Kapstad *see* Cape Town
Kapsukas *see* Marijampolė

Kaptsevichy – Kirovakan

Kaptsevichy *85 C7 Rus.* Koptsevichi. Homyel'skaya Voblasts', SE Belarus
Kapuas, Sungai *116 C4 prev.* Kapoeas. *river* Borneo, C Indonesia
Kapuskasing *16 C4* Ontario, S Canada
Kapyl' *85 C6 Rus.* Kopyl'. Minskaya Voblasts', C Belarus
Kara-Balta *101 F2* Chuyskaya Oblast', N Kyrgyzstan
Karabil', Vozvyshennost' *see* Garabil Belentligi
Kara-Bogaz-Gol, Zaliv *see* Garabogaz Aylagy
Karabük *94 C2* Karabük, NW Turkey
Karāchi *112 B3* Sind, SE Pakistan
Karácsonkő *see* Piatra-Neamț
Karadeniz *see* Black Sea
Karadeniz Boğazi *see* İstanbul Boğazı
Karaferiye *see* Véroia
Karagandy *92 C4 prev.* Karaganda, *Kaz.* Qaraghandy. Karagandy, C Kazakhstan
Karaginskiy, Ostrov *93 H2 island* E Russian Federation
Karagumskiy Kanal *see* Garagum Kanaly
Karak *see* Al Karak
Kara-Kala *see* Magtymguly
Karakax *see* Moyu
Karakılısse *see* Ağrı
Karakol *101 G2 prev.* Przheval'sk. Issyk-Kul'skaya Oblast', NE Kyrgyzstan
Karakol *101 G2 var.* Karakolka. Issyk-Kul'skaya Oblast', NE Kyrgyzstan
Karakolka *see* Karakol
Karakoram Range *112 D1 mountain range* C Asia
Karaköse *see* Ağrı
Karakul' *see* Qarokŭl, Tajikistan
Kara Kum *see* Garagum
Kara Kum Canal/Karakumskiy Kanal *see* Garagum Kanaly
Karakumy, Peski *see* Garagum
Karamai *see* Karamay
Karaman *94 C4* Karaman, S Turkey
Karamay *104 B2 var.* Karamai, Kelamayi; *prev. Chin.* K'o-la-ma-i. Xinjiang Uygur Zizhiqu, NW China
Karamea Bight *129 B5 gulf* South Island, New Zealand
Karapelit *82 E1 Rom.* Stejarul. Dobrich, NE Bulgaria
Kara-Say *101 G2* Issyk-Kul'skaya Oblast', NE Kyrgyzstan
Karasburg *56 B4* Karas, S Namibia
Kara Sea *see* Karskoye More
Kara Strait *see* Karskiye Vorota, Proliv
Karatau *92 C5 Kaz.* Qarataū. Zhambyl, S Kazakhstan
Karavás *83 B7* Kýthira, S Greece
Karbalā' *98 B3 var.* Kerbala, Kerbela. Karbalā', S Iraq
Kardeljevo *see* Ploče
Kardhítsa *see* Kardítsa
Kardítsa *83 B5 var.* Kardhítsa. Thessalía, C Greece
Kärdla *84 C2 Ger.* Kertel. Hiiumaa, W Estonia
Karet *see* Kâgheț
Kargı *94 C2* Çorum, N Turkey
Kargilik *see* Yecheng
Kariba *56 D2* Mashonaland West, N Zimbabwe
Kariba, Lake *56 D3 reservoir* Zambia/Zimbabwe
Karibib *56 B3* Erongo, C Namibia
Karies *see* Karyés
Karigasniemi *62 D2 Lapp.* Garegegasnjárga. Lappi, N Finland
Karimata, Selat *116 C4 strait* W Indonesia
Karīmnagar *112 D5* Telangana, C India
Karin *50 D4* Sahil, N Somalia
Kariot *see* Ikaría
Káristos *see* Kárystos
Karkinits'ka Zatoka *87 E4 Rus.* Karkinitskiy Zaliv. *gulf* S Ukraine
Karkinitskiy Zaliv *see* Karkinits'ka Zatoka
Karkūk *see* Kirkūk
Karleby *see* Kokkola
Karl-Marx-Stadt *see* Chemnitz
Karlö *see* Hailuoto
Karlovac *78 B3 Ger.* Karlstadt, *Hung.* Károlyváros. Karlovac, C Croatia
Karlovy Vary *77 A5 Ger.* Karlsbad; *prev. Eng.* Carlsbad. Karlovarský Kraj, W Czech Republic
Karlsbad *see* Karlovy Vary
Karlsburg *see* Alba Iulia
Karlskrona *63 C7* Blekinge, S Sweden
Karlsruhe *73 B6 var.* Carlsruhe. Baden-Württemberg, SW Germany
Karlstad *63 B6* Värmland, C Sweden
Karlstadt *see* Karlovac
Karnāl *112 D2* Haryāna, N India
Karnātaka *110 C1 var.* Kanara; *prev.* Maisur, Mysore. *cultural region* W India
Karnobat *82 D2* Burgas, E Bulgaria
Karnul *see* Kurnool
Karol *see* Carei
Károly-Fehérvár *see* Alba Iulia
Károlyváros *see* Karlovac
Karpaten *see* Carpathian Mountains
Kárpathos *83 E7* Kárpathos, SE Greece
Kárpathos *83 E7 It.* Scarpanto; *anc.* Carpathos, Carpathus. *island* SE Greece
Karpaty *see* Carpathian Mountains
Karpenísi *83 B5 prev.* Karpenísion. Stereá Ellás, C Greece
Karpenísion *see* Karpenísi
Karpilovka *see* Aktsyabrski
Kars *95 F2 var.* Qars. Kars, NE Turkey
Karsau *see* Kārsava
Kārsava *84 D4 Ger.* Karsau; *prev. Rus.* Korsovka. Ludza, E Latvia
Karshi *see* Qarshi, Uzbekistan
Karskiye Vorota, Proliv *88 E2 Eng.* Kara Strait. *strait* N Russian Federation
Karskoye More *92 D2 Eng.* Kara Sea. *sea* Arctic Ocean
Karyés *82 C4 var.* Karies. Ágion Óros, N Greece
Kárystos *83 C6 var.* Káristos. Évvoia, C Greece
Kasai *55 C6 var.* Cassai, Kassai. *river* Angola/Dem. Rep. Congo
Kasaji *55 D7* Katanga, S Dem. Rep. Congo
Kasama *56 D1* Northern, N Zambia
Kasan *see* Koson
Kāsaragod *110 B2* Kerala, SW India
Kaschau *see* Košice
Kāshān *98 C3* Eşfahān, C Iran
Kashgar *see* Kashi
Kashi *104 A3 Chin.* Kaxgar, K'o-shih, *Uigh.* Kashgar. Xinjiang Uygur Zizhiqu, NW China
Kasi *see* Vārānasi
Kasongo *55 D6* Maniema, E Dem. Rep. Congo
Kasongo-Lunda *55 C7* Bandundu, SW Dem. Rep. Congo
Kásos *83 D7 island* S Greece
Kaspīy Mangy Oypaty *see* Caspian Depression
Kaspiysk *89 B8* Respublika Dagestan, SW Russian Federation
Kaspiyskoye More/Kaspiy Tengizi *see* Caspian Sea
Kassa *see* Košice
Kassai *see* Kasai
Kassala *50 C4* Kassala, E Sudan
Kassel *72 B4 prev.* Cassel. Hessen, C Germany
Kasserine *49 E2 var.* Al Qaşrayn. W Tunisia
Kastamonu *94 C2 var.* Castamoni, Kastamuni. Kastamonu, N Turkey
Kastamuni *see* Kastamonu
Kastaneá *82 B4* Kentrikí Makedonía, N Greece
Kastélli *see* Kíssamos
Kastoría *82 B4* Dytikí Makedonía, N Greece
Kástro *83 C6* Sífnos, Kykládes, Greece, Aegean Sea
Kastsyukovichy *85 E7 Rus.* Kostyukovichi. Mahilyowskaya Voblasts', E Belarus
Kastsyukowka *85 D7 Rus.* Kostyukovka. Homyel'skaya Voblasts', SE Belarus
Kasulu *51 B7* Kigoma, W Tanzania
Kasumiga-ura *109 D5 lake* Honshū, S Japan
Katahdin, Mount *19 G1 mountain* Maine, NE USA
Katalla *14 C3* Alaska, USA
Katana *see* Qaṭanā
Katanning *125 B7* Western Australia
Katawaz *see* Zarghūn Shahr
Katchall Island *111 F3 island* Nicobar Islands, India, NE Indian Ocean
Kateríni *82 B4* Kentrikí Makedonía, N Greece
Katha *114 B2* Sagaing, N Myanmar (Burma)
Katherine *126 A2* Northern Territory, N Australia
Kathmandu *102 C3 prev.* Kantipur. *country capital* (Nepal) Central, C Nepal
Katikati *128 D3* Bay of Plenty, North Island, New Zealand
Katima Mulilo *56 C3* Caprivi, NE Namibia
Katiola *52 D4* C Côte d'Ivoire
Káto Achaïa *83 B5 var.* Kato Ahaia, Káto Akhaía. Dytikí Ellás, S Greece
Kato Ahaia/Káto Akhaía *see* Káto Achaïa
Katoúna *83 A5* Dytikí Ellás, C Greece
Katowice *77 C5 Ger.* Kattowitz. Śląskie, S Poland
Katsina *53 G3* Katsina, N Nigeria
Kattakurgan *see* Kattaqo'rg'on
Kattaqo'rg'on *101 E2 Rus.* Kattakurgan. Samarqand Viloyati, C Uzbekistan
Kattavía *83 E7* Ródos, Dodekánisa, Greece, Aegean Sea
Kattegat *63 B7 Dan.* Kattegatt. *strait* N Europe
Kattegatt *see* Kattegat
Kattowitz *see* Katowice
Kaua'i *25 A7 var.* Kauai. *island* Hawai'ian Islands, Hawaii, USA, C Pacific Ocean
Kauai *see* Kaua'i
Kauen *see* Kaunas
Kaufbeuren *73 C6* Bayern, S Germany
Kaunas *84 B4 Ger.* Kauen, *Pol.* Kowno; *prev. Rus.* Kovno. Kaunas, C Lithuania
Kavadar *see* Kavadarci
Kavadarci *79 E6 Turk.* Kavadar. C Macedonia
Kavaja *see* Kavajë
Kavajë *79 C6 It.* Cavaia, Kavaja. Tiranë, W Albania
Kavakli *see* Topolovgrad
Kavála *82 C3 prev.* Kavála. Anatolikí Makedonía kai Thráki, NE Greece
Kāvali *110 D2* Andhra Pradesh, E India
Kaválla *see* Kavála
Kavango *see* Cubango/Okavango
Kavaratti Island *110 A3 island* Lakshadweep, Lakshadweep, SW India Asia N Indian Ocean
Kavarna *82 E2* Dobrich, NE Bulgaria
Kavengo *see* Cubango/Okavango
Kavir, Dasht-e *98 D3 var.* Great Salt Desert. *salt pan* N Iran
Kavkaz *see* Caucasus
Kawagoe *109 D5* Saitama, Honshū, S Japan
Kawasaki *108 A2* Kanagawa, Honshū, S Japan
Kawerau *128 E3* Bay of Plenty, North Island, New Zealand
Kaxgar *see* Kashi
Kaya *53 E3* C Burkina
Kayan *114 B4* Yangon, SW Burma (Myanmar)
Kayan, Sungai *116 D3 prev.* Kajan. *river* Borneo, C Indonesia
Kayes *52 C3* Kayes, W Mali
Kayseri *94 D3 var.* Kaisaria; *anc.* Caesarea Mazaca, Mazaca. Kayseri, C Turkey
Kazach'ye *93 F2* Respublika Sakha (Yakutiya), NE Russian Federation
Kazakhskaya SSR/Kazakh Soviet Socialist Republic *see* Kazakhstan
Kazakhstan *92 B4 off.* Republic of Kazakhstan, *var.* Kazakstan, *Kaz.* Qazaqstan, Qazaqstan Respublikasy; *prev.* Kazakh Soviet Socialist Republic, *Rus.* Kazakhskaya SSR. *country* C Asia
Kazakhstan, Republic of *see* Kazakhstan
Kazakh Uplands *92 C4 Eng.* Kazakh Uplands, Kirghiz Steppe, *Kaz.* Saryarqa. *uplands* C Kazakhstan
Kazakh Uplands *see* Kazakhskiy Melkosopochnik
Kazakstan *see* Kazakhstan
Kazan' *89 C5* Respublika Tatarstan, W Russian Federation
Kazandzhik *see* Bereket
Kazanlik *see* Kazanlŭk
Kazanlŭk *82 D2 prev.* Kazanlik. Stara Zagora, C Bulgaria
Kazatin *see* Kozyatyn
Kazbegi *see* Kazbek
Kazbek *95 F1 var.* Kazbegi, *Geor.* Mqinvartsveri. *mountain* N Georgia
Kāzerūn *98 D4* Fārs, S Iran
Kazi Magomed *see* Qazimämmäd
Kazvin *see* Qazvīn
Kéa *see* Tziá
Kéa *see* Ioulís
Kea, Mauna *25 B8 mountain* Hawaii, USA
Kéamu *see* Aneityum
Kearney *23 E4* Nebraska, C USA
Keban Barajı *95 E3 reservoir* C Turkey
Kebkabiya *50 A4* Northern Darfur, W Sudan
Kebnekaise *62 C3 mountain* N Sweden
Kecskemét *77 D7* Bács-Kiskun, C Hungary
Kediri *116 D5* Jawa, C Indonesia
Kędzierzyn-Kozle *77 C5 Ger.* Heydebrech. Opolskie, S Poland
Keelung *see* Jilong
Keetmanshoop *56 B4* Karas, S Namibia
Kefallinía *see* Kefalloniá
Kefalloniá *83 A5 var.* Kefallinía. *island* Iónia Nisiá, Greece, C Mediterranean Sea
Kefe *see* Feodosiya
Kegel *see* Keila
Kehl *73 A6* Baden-Württemberg, SW Germany
Kei Islands *see* Kai, Kepulauan
Keijō *see* Seoul
Keila *84 D2 Ger.* Kegel. Harjumaa, NW Estonia
Keïta *53 F3* Tahoua, C Niger
Keith *127 B7* South Australia
Kēk-Art *101 G2 prev.* Alaykel', Alay-Kuu. Oshskaya Oblast', SW Kyrgyzstan
Kékes *77 C6 mountain* N Hungary
Kelamayi *see* Karamay
Kelang *see* Klang
Kelat *see* Kālat
Kelifskiy Uzboy *see* Kelif Uzboýy
Kelif Uzboýy *100 D3 Rus.* Kelifskiy Uzboy. *salt marsh* E Turkmenistan
Kelkit Çayı *95 E3 river* N Turkey
Kelmė *84 B4* Šiauliai, C Lithuania
Kélo *54 B4* Tandjilé, SW Chad
Kelowna *15 E5* British Columbia, SW Canada
Kelso *24 B2* Washington, NW USA
Keltsy *see* Kielce
Keluang *116 B3 var.* Kluang. Johor, Peninsular Malaysia
Kem' *88 B3* Respublika Kareliya, NW Russian Federation
Kemah *95 E3* Erzincan, E Turkey
Kemaman *see* Cukai
Kemerovo *92 D4 prev.* Shcheglovsk. Kemerovskaya Oblast', C Russian Federation
Kemi *62 D4* Lappi, NW Finland
Kemijärvi *62 D3 Swe.* Kemiträsk. Lappi, N Finland
Kemijoki *62 D3 river* NW Finland
Kemin *101 G2 prev.* Bystrovka. Chuyskaya Oblast', N Kyrgyzstan
Kemins Island *see* Nikumaroro
Kemiträsk *see* Kemijärvi
Kemmuna *80 A5 var.* Comino. *island* C Malta
Kempele *62 D4* Oulu, C Finland
Kempten *73 B7* Bayern, S Germany
Kendal *67 D5* NW England, United Kingdom
Kendari *117 E4* Sulawesi, C Indonesia
Kenedy *27 G4* Texas, SW USA
Kenema *52 C4* SE Sierra Leone
Këneurgench *see* Köneürgenç
Kenge *55 C6* Bandundu, SW Dem. Rep. Congo
Kengtung *114 C3 prev.* Keng Tung. Shan State, E Myanmar (Burma)
Keng Tung *see* Kengtung
Kénitra *48 C2 prev.* Port-Lyautey. NW Morocco
Kennett *23 H5* Missouri, C USA
Kennewick *24 C2* Washington, NW USA
Kenora *16 A3* Ontario, S Canada
Kenosha *18 B3* Wisconsin, N USA
Kentau *92 B5* Yuzhnyy Kazakhstan, S Kazakhstan
Kentucky *18 C5 off.* Commonwealth of Kentucky, *also known as* Bluegrass State. *state* C USA
Kentucky Lake *18 B5 reservoir* Kentucky/Tennessee, S USA
Kentung *see* Keng Tung
Kenya *51 C6 off.* Republic of Kenya. *country* E Africa
Kenya, Mount *see* Kirinyaga
Kenya, Republic of *see* Kenya
Keokuk *23 G4* Iowa, C USA
Kéos *see* Tziá
Kępno *76 C4* Wielkopolskie, C Poland
Keppel Island *see* Niuatoputapu
Kerak *see* Al Karak
Kerala *110 C2 cultural region* S India
Kerasunt *see* Giresun
Keratéa *83 C6 var.* Keratea. Attikí, C Greece
Keratea *see* Keratéa
Kerbala/Kerbela *see* Karbalā'
Kerch *87 G5 Rus.* Kerch'. Respublika Krym, SE Ukraine
Kerch' *see* Kerch
Kerchens'ka Protska/Kerchenskiy Proliv *see* Kerch Strait
Kerch Strait *87 G4 var.* Bosporus Cimmerius, Enikale Strait, *Rus.* Kerchenskiy Proliv, *Ukr.* Kerchens'ka Protska. *strait* Black Sea/Sea of Azov
Keremitlik *see* Lyulyakovo
Kerguelen *119 C7 island* C French Southern and Antarctic Lands
Kerguelen Plateau *119 C7 undersea plateau* S Indian Ocean
Kerí *83 A6* Zákynthos, Iónia Nisiá, Greece, C Mediterranean Sea
Kerikeri *128 D2* Northland, North Island, New Zealand
Kerkenah, Îles de *80 D4 var.* Kerkenna Islands, *Ar.* Juzur Qarqannah. *island group* E Tunisia
Kerkenna Islands *see* Kerkenah, Îles de
Kerki *see* Atamyrat
Kérkira *see* Kérkyra
Kerkrade *65 D6* Limburg, SE Netherlands
Kerkuk *see* Kirkūk
Kérkyra *82 A4 var.* Kérkira, *Eng.* Corfu. Kérkyra, Iónia Nisiá, Greece, C Mediterranean Sea
Kermadec Islands *130 C4 island group* New Zealand, SW Pacific Ocean
Kermadec Trench *121 E4 trench* SW Pacific Ocean
Kermān *98 D3 var.* Kirman; *anc.* Carmana. Kermān, C Iran
Kermānshāh *98 C3 var.* Qahremānshahr; *prev.* Bākhtarān. Kermānshāhān, W Iran
Kerrville *27 F4* Texas, SW USA
Kertel *see* Kärdla
Kerulen *105 E2 Chin.* Herlen He, *Mong.* Herlen Gol. *river* China/Mongolia
Kerýneia *see* Girne
Kesennuma *108 D4* Miyagi, Honshū, C Japan
Keszthely *77 C7* Zala, SW Hungary
Ketchikan *14 D4* Revillagigedo Island, Alaska, USA
Kętrzyn *76 D2 Ger.* Rastenburg. Warmińsko-Mazurskie, NE Poland
Kettering *67 D6* C England, United Kingdom
Kettering *18 C4* Ohio, N USA
Keupriya *see* Primorsko
Keuruu *63 D5* Länsi-Suomi, C Finland
Keweenaw Peninsula *18 B1 peninsula* Michigan, N USA
Key Largo *21 F5* Key Largo, Florida, SE USA
Keystone State *see* Pennsylvania
Key West *21 E5* Florida Keys, Florida, SE USA
Kezdivásárhely *see* Târgu Secuiesc
Khabarovsk *93 G4* Khabarovskiy Kray, SE Russian Federation
Khachmas *see* Xaçmaz
Khadera *see* Hadera
Khairpur *112 B3* Sind, SE Pakistan
Khalīj as Suways *50 B2 var.* Suez, Gulf of. *gulf* NE Egypt
Khalkidhikí *see* Chalkidikí
Khalkís *see* Chalkída
Khambhat, Gulf of *112 C4 Eng.* Gulf of Cambay. *gulf* W India
Khamīs Mushayt *99 B6 var.* Hamīs Musait. 'Asīr, SW Saudi Arabia
Khānābād *101 E3* Kunduz, NE Afghanistan
Khān al Baghdādī *see* Al Baghdādī
Khandwa *112 D4* Madhya Pradesh, C India
Khanh Hung *see* Soc Trăng
Khaniá *see* Chaniá
Khanka, Lake *107 E2 var.* Hsing-K'ai Hu, Lake Hanka, *Chin.* Xingkai Hu, *Rus.* Ozero Khanka. *lake* China/Russian Federation
Khanka, Ozero *see* Khanka, Lake
Khankendi *see* Xankändi
Khanthabouli *114 D4 prev.* Savannakhét. Savannakhét, S Laos
Khanty-Mansiysk *92 C3 prev.* Ostyako-Voguls'k. Khanty-Mansiyskiy Avtonomnyy Okrug-Yugra, C Russian Federation
Khān Yūnis *97 A7 var.* Khān Yūnus. S Gaza Strip
Khān Yūnus *see* Khān Yūnis
Khanzi *see* Ghanzi
Kharagpur *113 F4* West Bengal, NE India
Kharbin *see* Harbin
Kharkiv *87 G2 Rus.* Khar'kov. Kharkivs'ka Oblast', NE Ukraine
Khar'kov *see* Kharkiv
Kharmanli *82 D3* Khaskovo, S Bulgaria
Khartoum *50 B4 var.* El Khartūm, Khartum. *country capital* (Sudan) Khartoum, C Sudan
Khartum *see* Khartoum
Khasavyurt *89 B8* Respublika Dagestan, SW Russian Federation
Khash, Dasht-e *100 D5 Eng.* Khash Desert. *desert* SW Afghanistan
Khash Desert *see* Khāsh, Dasht-e
Khashim Al Qirba/Khashm al Qirbah *see* Khashm el Girba
Khashm el Girba *50 C4 var.* Khashim Al Qirba, Khashm al Qirbah. Kassala, E Sudan
Khaskovo *82 D3* Khaskovo, S Bulgaria
Khaybar, Kowtal-e *see* Khyber Pass
Khaydarkan *101 F2 var.* Khaydarken. Batkenskaya Oblast', SW Kyrgyzstan
Khaydarken *see* Khaydarkan
Khazar, Baḩr-e/Khazar, Daryā-ye *see* Caspian Sea
Khelat *see* Kālat
Kherson *87 E4* Khersons'ka Oblast', S Ukraine
Kheta *93 E2 river* N Russian Federation
Khíos *see* Chíos
Khirbet el 'Aujā et Taḥtā *97 E7* E West Bank Asia
Khiva/Khiwa *see* Xiva
Khmel'nitskiy *see* Khmel 'nyts'kyy
Khmel 'nyts'kyy *86 C2 Rus.* Khmel'nitskiy; *prev.* Proskurov. Khmel'nyts'ka Oblast', W Ukraine
Khodasy *85 E6 Rus.* Khodosy. Mahilyowskaya Voblasts', E Belarus
Khodoriv *86 C2 Pol.* Chodorów, *Rus.* Khodorov. L'vivs'ka Oblast', NW Ukraine
Khodorov *see* Khodoriv
Khodosy *see* Khodasy
Khodzhent *see* Khujand
Khoi *see* Khvoy
Khojend *see* Khujand
Khokand *see* Qo'qon
Kholm *see* Khulm
Kholm *see* Chełm
Kholon *see* Holon
Khoms *see* Al Khums
Khong Sedone *see* Muang Khôngxédôn
Khon Kaen *114 D4 var.* Muang Khon Kaen. Khon Kaen, E Thailand
Khor *93 G4* Khabarovskiy Kray, SE Russian Federation
Khorat *see* Nakhon Ratchasima
Khorog *see* Khorugh
Khorugh *101 F3 Rus.* Khorog. S Tajikistan
Khōst *101 F4 prev.* Khowst. Khōst, E Afghanistan
Khotan *see* Hotan
Khouribga *48 B2* C Morocco
Khovd *see* Hovd
Khowst *see* Khōst
Khoy *see* Khvoy
Khoyniki *85 D8* Homyel'skaya Voblasts', SE Belarus
Khudzhand *see* Khujand
Khujand *101 E2 var.* Khodzhent, Khojend, *Rus.* Khudzhand; *prev.* Leninabad, *Taj.* Leninobod. N Tajikistan
Khulm *101 E3 var.* Kholm, Tashqurghan. Balkh, N Afghanistan
Khulna *113 G4* Khulna, SW Bangladesh
Khums *see* Al Khums
Khust *86 B3 var.* Husté, *Cz.* Chust, *Hung.* Huszt. Zakarpats'ka Oblast', W Ukraine
Khvoy *98 C2 var.* Khoi, Khoy. Āžarbāyjān-e Bākhtarī, NW Iran
Khyber Pass *112 C1 var.* Kowtal-e Khaybar. *pass* Afghanistan/Pakistan
Kiangmai *see* Chiang Mai
Kiang-ning *see* Nanjing
Kiangsi *see* Jiangxi
Kiangsu *see* Jiangsu
Kiáto *83 B6 prev.* Kiáton. Pelopónnisos, S Greece
Kiáton *see* Kiáto
Kiayi *see* Jiayi
Kibangou *55 B6* Niari, SW Congo
Kibombo *55 D6* Maniema, E Dem. Rep. Congo
Kıbrıs/Kıbrıs Cumhuriyeti *see* Cyprus
Kičevo *79 D6* SW FYR Macedonia
Kidderminster *67 D6* C England, United Kingdom
Kiel *72 B2* Schleswig-Holstein, N Germany
Kielce *76 D4 Rus.* Keltsy. Świętokrzyskie, C Poland
Kieler Bucht *72 B2 bay* N Germany
Kiev *see* Kyyiv
Kiev Reservoir *see* Kyyivs'ke Vodoskhovyshche
Kiffa *52 C3* Assaba, S Mauritania
Kigali *51 B6 country capital* (Rwanda) C Rwanda
Kigoma *51 B7* Kigoma, W Tanzania
Kihnu *84 C2 var.* Kihnu Saar, *Ger.* Kühnö. *island* SW Estonia
Kihnu Saar *see* Kihnu
Kii-suido *109 C7 strait* S Japan
Kikinda *78 D3 Ger.* Grosskikinda, *Hung.* Nagykikinda; *prev.* Velika Kikinda. Vojvodina, N Serbia
Kikládhes *see* Kykládes
Kikwit *55 C6* Bandundu, W Dem. Rep. Congo
Kilien Mountains *see* Qilian Shan
Kilimane *see* Quelimane
Kilimanjaro *47 E5 region* E Tanzania
Kilimanjaro *51 C7 var.* Uhuru Peak. *volcano* NE Tanzania
Kilingi-Nõmme *84 D3 Ger.* Kurkund. Pärnumaa, SW Estonia
Kilis *94 D4* Kilis, S Turkey
Kiliya *86 D4 Rom.* Chilia-Nouă. Odes'ka Oblast', SW Ukraine
Kilkenny *67 B6 Ir.* Cill Chainnigh. Kilkenny, S Ireland
Kilkís *82 B3* Kentrikí Makedonía, N Greece
Killarney *67 A6 Ir.* Cill Airne. Kerry, SW Ireland
Killeen *27 G3* Texas, SW USA
Kilmain *see* Quelimane
Kilmarnock *66 C4* W Scotland, United Kingdom
Kilwa *see* Kilwa Kivinje
Kilwa Kivinje *51 C7 var.* Kilwa. Lindi, SE Tanzania
Kimberley *56 C4* Northern Cape, C South Africa
Kimberley Plateau *124 C3 plateau* Western Australia
Kimch'aek *107 E3 prev.* Sŏngjin. E North Korea
Kími *see* Kými
Kinabalu, Gunung *116 D3 mountain* East Malaysia
Kindersley *15 F5* Saskatchewan, S Canada
Kindia *52 C4* Guinée-Maritime, SW Guinea
Kindley Field *20 A4 air base* E Bermuda
Kindu *55 D6 prev.* Kindu-Port-Empain. Maniema, C Dem. Rep. Congo
Kindu-Port-Empain *see* Kindu
Kineshma *89 C5* Ivanovskaya Oblast', W Russian Federation
King Charles Islands *see* Kong Karls Land
King Christian IX Land *see* Kong Christian IX Land
King Frederik VI Coast *see* Kong Frederik VI Kyst
King Frederik VIII Land *see* Kong Frederik VIII Land
King Island *127 B8 island* Tasmania, SE Australia
King Island *see* Kadan Kyun
Kingissepp *see* Kuressaare
Kingman *26 A1* Arizona, SW USA
Kingman Reef *123 E2 US territory* C Pacific Ocean
Kingsford Smith *126 E2* (Sydney) New South Wales, SE Australia
King's Lynn *67 E6 var.* Bishop's Lynn, Kings Lynn, Lynn, Lynn Regis. E England, United Kingdom
Kings Lynn *see* King's Lynn
King Sound *124 B3 sound* Western Australia
Kingsport *21 E1* Tennessee, S USA
Kingston *32 B5 country capital* (Jamaica) E Jamaica
Kingston *16 D5* Ontario, SE Canada
Kingston *19 F3* New York, NE USA
Kingston upon Hull *67 D5 var.* Hull. E England, United Kingdom
Kingston upon Thames *67 A8* SE England, United Kingdom
Kingstown *33 H4 country capital* (Saint Vincent and the Grenadines) Saint Vincent, Saint Vincent and the Grenadines
Kingstown *see* Dún Laoghaire
Kingsville *27 G5* Texas, SW USA
King William Island *15 F3 island* Nunavut, N Canada
Kinneret, Yam *see* Tiberias, Lake
Kinrooi *65 D5* Limburg, NE Belgium
Kinshasa *55 B6 prev.* Léopoldville. *country capital* (Dem. Rep. Congo) Kinshasa, W Dem. Rep. Congo
Kintyre *66 B4 peninsula* W Scotland, United Kingdom
Kinyeti *51 B5 mountain* S South Sudan
Kiparissía *see* Kyparissía
Kipili *51 B7* Rukwa, W Tanzania
Kipushi *55 D8* Katanga, SE Dem. Rep. Congo
Kirdzhali *see* Kŭrdzhali
Kirghizia *see* Kyrgyzstan
Kirghiz Range *101 F2 Rus.* Kirgizskiy Khrebet; *prev.* Alexander Range. *mountain range* Kazakhstan/Kyrgyzstan
Kirghiz SSR *see* Kyrgyzstan
Kirghiz Steppe *see* Kazakhskiy Melkosopochnik
Kirgizskaya SSR *see* Kyrgyzstan
Kirgizskiy Khrebet *see* Kirghiz Range
Kiriath-Arba *see* Hebron
Kiribati *123 F2 off.* Republic of Kiribati. *country* C Pacific Ocean
Kiribati, Republic of *see* Kiribati
Kırıkhan *94 D4* Hatay, S Turkey
Kırıkkale *94 C3 province* C Turkey
Kirin *see* Jilin
Kirinyaga *51 C6 prev.* Mount Kenya. *volcano* C Kenya
Kirishi *88 B4 var.* Kirisi. Leningradskaya Oblast', NW Russian Federation
Kirisi *see* Kirishi
Kiritimati *123 G2 prev.* Christmas Island. *atoll* Line Islands, E Kiribati
Kirkenes *62 E2 Fin.* Kirkkoniemi. Finnmark, N Norway
Kirk-Kilissa *see* Kırklareli
Kirkkoniemi *see* Kirkenes
Kirkland Lake *16 D4* Ontario, S Canada
Kırklareli *94 A2 prev.* Kirk-Kilissa. Kırklareli, NW Turkey
Kirkpatrick, Mount *132 B3 mountain* Antarctica
Kirksville *23 G4* Missouri, C USA
Kirkūk *98 B3 var.* Karkūk, Kerkuk. At Ta'mīn, N Iraq
Kirkwall *66 C2* NE Scotland, United Kingdom
Kirkwood *23 G4* Missouri, C USA
Kir Moab/Kir of Moab *see* Al Karak
Kirov *89 C5 prev.* Vyatka. Kirovskaya Oblast', NW Russian Federation
Kirovabad *see* Gäncä
Kirovakan *see* Vanadzor

Kirovo-Chepetsk – Kumul

Kirovo-Chepetsk 89 D5 Kirovskaya Oblast', NW Russian Federation
Kirovohrad 87 E3 Rus. Kirovograd; prev. Kirovo, Yelizavetgrad, Zinov'yevsk. Kirovohrads'ka Oblast', C Ukraine
Kirovo/Kirovograd see Kirovohrad
Kīrthar Range 112 B3 mountain range S Pakistan
Kiruna 62 C3 Lapp. Giron. Norrbotten, N Sweden
Kirun/Kirun' see Jilong
Kisalföld see Little Alföld
Kisangani 55 D5 prev. Stanleyville. Orientale, NE Dem. Rep. Congo
Kishinev see Chişinău
Kislovodsk 89 B7 Stavropol'skiy Kray, SW Russian Federation
Kismaayo 51 D6 var. Chisimayu, Kismayu, It. Chisimaio. Jubbada Hoose, S Somalia
Kismayu see Kismaayo
Kíssamos 83 C7 prev. Kastélli. Kríti, Greece, E Mediterranean Sea
Kissidougou 52 C4 Guinée-Forestière, S Guinea
Kissimmee, Lake 21 E4 lake Florida, SE USA
Kistna see Krishna
Kisumu 51 C6 prev. Port Florence. Nyanza, W Kenya
Kisvárda 77 E6 Ger. Kleinwardein. Szabolcs-Szatmár-Bereg, E Hungary
Kita 52 D3 Kayes, W Mali
Kitab see Kitob
Kitakyūshū 109 A7 var. Kitakyûsyû. Fukuoka, Kyūshū, SW Japan
Kitakyûsyû see Kitakyūshū
Kitami 108 D2 Hokkaidō, NE Japan
Kitchener 16 C5 Ontario, S Canada
Kíthnos see Kýthnos
Kitimat 14 D4 British Columbia, SW Canada
Kitinen 62 D3 river N Finland
Kitob 101 E3 Rus. Kitab. Qashqadaryo Viloyati, S Uzbekistan
Kitwe 56 D2 var. Kitwe-Nkana. Copperbelt, C Zambia
Kitwe-Nkana see Kitwe
Kitzbüheler Alpen 73 C7 mountain range W Austria
Kivalina 14 C2 Alaska, USA
Kivalo 62 D3 ridge C Finland
Kivertsi 86 C1 Pol. Kiwerce, Rus. Kivertsy. Volyns'ka Oblast', NW Ukraine
Kivertsy see Kivertsi
Kivu, Lac see Kivu, Lake
Kivu, Lake 55 E6 Fr. Lac Kivu. lake Rwanda/Dem. Rep. Congo
Kiwerce see Kivertsi
Kiyev see Kyyiv
Kiyevskoye Vodokhranilishche see Kyyivs'ke Vodoskhovyshche
Kızıl Irmak 94 C3 river C Turkey
Kizil Kum see Kyzyl Kum
Kizyl-Arvat see Serdar
Kjølen see Kölen
Kladno 77 A5 Středočeský, NW Czech Republic
Klagenfurt 73 D7 Slvn. Celovec. Kärnten, S Austria
Klaipėda 84 B3 Ger. Memel. Klaipeda, NW Lithuania
Klamath Falls 24 B4 Oregon, NW USA
Klamath Mountains 24 A4 mountain range California/Oregon, W USA
Klang 116 B3 var. Kelang; prev. Port Swettenham. Selangor, Peninsular Malaysia
Klarälven 63 B6 river Norway/Sweden
Klatovy 77 A5 Ger. Klattau. Plzeňský Kraj, W Czech Republic
Klattau see Klatovy
Klausenburg see Cluj-Napoca
Klazienaveen 64 E2 Drenthe, NE Netherlands
Kleines Ungarisches Tiefland see Little Alföld
Klein Karas 56 B4 Karas, S Namibia
Kleinwardein see Kisvárda
Kleisoúra 83 A5 Ípeiros, W Greece
Klerksdorp 56 D4 North-West, N South Africa
Klimavichy 85 E7 Rus. Klimovichi. Mahilyowskaya Voblasts', E Belarus
Klimovichi see Klimavichy
Klintsy 89 A5 Bryanskaya Oblast', W Russian Federation
Klisura 82 C2 Plovdiv, C Bulgaria
Ključ 78 B3 Federacija Bosna I Hercegovina, NW Bosnia and Herzegovina
Kłobuck 76 C4 Śląskie, S Poland
Klosters 73 B7 Graubünden, SE Switzerland
Kluang see Keluang
Kluczbork 76 C4 Ger. Kreuzburg, Kreuzburg in Oberschlesien. Opolskie, S Poland
Klyuchevskaya Sopka, Vulkan 93 H3 volcano E Russian Federation
Knin 78 B4 Šibenik-Knin, S Croatia
Knjaževac 78 E4 Serbia, E Serbia
Knokke-Heist 65 A5 West-Vlaanderen, NW Belgium
Knoxville 20 D1 Tennessee, S USA
Knud Rasmussen Land 60 D1 physical region N Greenland
Kobdo see Hovd
Kōbe 109 C6 Hyōgo, Honshū, SW Japan
København 63 B7 Eng. Copenhagen; anc. Hafnia. country capital (Denmark) Sjælland, København, E Denmark
Kobenni 52 D3 Hodh el Gharbi, S Mauritania
Koblenz 73 A5 prev. Coblenz, Fr. Coblence; anc. Confluentes. Rheinland-Pfalz, W Germany
Kobrin see Kobryn
Kobryn 85 A6 Rus. Kobrin. Brestskaya Voblasts', SW Belarus
Kobuleti 95 F2 prev. K'obulet'i. W Georgia
K'obulet'i see Kobuleti
Kočani 79 E6 NE FYR Macedonia
Kočevje 73 D8 Ger. Gottschee. S Slovenia
Koch Bihār 113 G3 West Bengal, NE India
Kochchi see Kochi
Kochi 110 C3 var. Cochin, Kochchi. Kerela, SW India
Kōchi 109 B7 var. Kôti. Kōchi, Shikoku, SW Japan
Kochiu see Gejiu
Kodiak 14 C3 Kodiak Island, Alaska, USA
Kodiak Island 14 C3 island Alaska, USA
Koedoes see Kudus
Koeln see Köln
Koepang see Kupang
Ko-erh-mu see Golmud
Koetai see Mahakam, Sungai
Koetaradja see Bandaaceh
Kōfu 109 D5 var. Kôhu. Yamanashi, Honshū, S Japan
Kogarah 126 E2 New South Wales, E Australia
Kogon 100 D2 Rus. Kagan. Buxoro Viloyati, C Uzbekistan
Kőhalom see Rupea
Kohīma 113 H3 state capital Nāgāland, E India
Koh I Noh see Büyükağrı Dağı
Kohtla-Järve 84 E2 Ida-Virumaa, NE Estonia
Kôhu see Kōfu
Kokand see Qo'qon
Kokchetav see Kokshetau
Kokkola 62 D4 Swe. Karleby; prev. Swe. Gamlakarleby. Länsi-Suomi, W Finland
Koko 53 F4 Kebbi, W Nigeria
Kokomo 18 C4 Indiana, N USA
Koko Nor see Qinghai, China
Koko Nor see Qinghai Hu, China
Kokrines 14 C2 Alaska, USA
Kokshaal-Tau 101 G2 Rus. Khrebet Kakshaal-Too. mountain range China/Kyrgyzstan
Kokshetau 92 C4 Kaz. Kökshetaū; prev. Kokchetav. Kokshetau, N Kazakhstan
Kökshetaū see Kokshetau
Koksijde 65 A5 West-Vlaanderen, W Belgium
Koksoak 16 D2 river Québec, E Canada
Kokstad 56 D5 KwaZulu/Natal, E South Africa
Kolaka 117 E4 Sulawesi, C Indonesia
K'o-la-ma-i see Karamay
Kola Peninsula see Kol'skiy Poluostrov
Kolari 62 D3 Lappi, NW Finland
Kolárovo 77 C6 Ger. Gutta; prev. Guta, Hung. Gúta. Nitriansky Kraj, SW Slovakia
Kolberg see Kołobrzeg
Kolda 52 C3 S Senegal
Kolding 63 A7 Vejle, C Denmark
Kölen 59 E1 Nor. Kjølen. mountain range Norway/Sweden
Kolguyev, Ostrov 88 C2 island NW Russian Federation
Kolhāpur 110 B1 Mahārāshtra, SW India
Kolhumadulu 110 A5 var. Thaa Atoll. atoll S Maldives
Kolín 77 B5 Ger. Kolin. Střední Čechy, C Czech Republic
Kolka 84 C2 Talsi, NW Latvia
Kolkasrags 84 C2 prev. Eng. Cape Domesnes. headland NW Latvia
Kolkata 113 G4 prev. Calcutta. West Bengal, N India
Kollam 110 C3 var. Quilon. Kerala, SW India
Kolmar see Colmar
Köln 72 A4 var. Koeln, Eng./Fr. Cologne, prev. Cöln; anc. Colonia Agrippina, Oppidum Ubiorum. Nordrhein-Westfalen, W Germany
Koło 76 C3 Wielkopolskie, C Poland
Kołobrzeg 76 B2 Ger. Kolberg. Zachodnio-pomorskie, NW Poland
Kolokani 52 D3 Koulikoro, W Mali
Kolomea see Kolomyya
Kolomna 89 B5 Moskovskaya Oblast', W Russian Federation
Kolomyya 86 C3 Ger. Kolomea. Ivano-Frankivs'ka Oblast', W Ukraine
Kolosjoki see Nikel'
Kolozsvár see Cluj-Napoca
Kolpa 78 A2 Ger. Kulpa, SCr. Kupa. river Croatia/Slovenia
Kolpino 88 B4 Leningradskaya Oblast', NW Russian Federation
Kólpos Mórfu see Güzelyurt Körfezi
Kol'skiy Poluostrov 88 C2 Eng. Kola Peninsula. peninsula NW Russian Federation
Kolwezi 55 D7 Katanga, S Dem. Rep. Congo
Kolyma 93 G2 river NE Russian Federation
Komatsu 109 C5 var. Komatu. Ishikawa, Honshū, SW Japan
Komatu see Komatsu
Kommunizm, Qullai see Ismoili Somonī, Qullai
Komoé 53 E4 var. Komoé Fleuve. river E Côte d'Ivoire
Komoé Fleuve see Komoé
Komotau see Chomutov
Komotiní 82 D3 var. Gümüljina, Turk. Gümülcine. Anatolikí Makedonía kai Thráki, NE Greece
Kompong see Kâmpóng Chhnăng
Kompong Cham see Kâmpóng Cham
Kompong Som see Sihanoukville
Kompong Speu see Kâmpóng Spoe
Komrat see Comrat
Komsomolets, Ostrov 93 E1 island Severnaya Zemlya, N Russian Federation
Komsomol'sk-na-Amure 93 G4 Khabarovskiy Kray, SE Russian Federation
Kondolovo 82 E3 Burgas, E Bulgaria
Kondopoga 88 B3 Respublika Kareliya, NW Russian Federation
Kondoz see Kunduz
Köneürgenç 100 C2 var. Köneürgench, Rus. Këneurgench; prev. Kunya-Urgench. Daşoguz Welaýaty, N Turkmenistan
Kong Christian IX Land 60 D4 Eng. King Christian IX Land. physical region SE Greenland
Kong Frederik IX Land 60 C3 physical region SW Greenland
Kong Frederik VIII Land 61 E2 Eng. King Frederik VIII Land. physical region NE Greenland
Kong Frederik VI Kyst 60 C4 Eng. King Frederik VI Coast. physical region SE Greenland
Kong Karls Land 61 G2 Eng. King Charles Islands. island group SE Svalbard
Kongo see Congo (river)
Kongolo 55 D6 Katanga, E Dem. Rep. Congo
Kongor 51 B5 Jonglei, E South Sudan
Kong Oscar Fjord 61 E3 fjord E Greenland
Kongsberg 63 B6 Buskerud, S Norway
Kông, Tônle 115 E5 var. Xê Kong. river Cambodia/Laos
Kong, Xê see Kông, Tônle
Königgrätz see Hradec Králové
Königshütte see Chorzów
Konin 76 C3 Ger. Kuhnau. Weilkopolskie, C Poland
Koninkrijk der Nederlanden see Netherlands
Konispol 79 C7 var. Konispoli. Vlorë, S Albania
Konispoli see Konispol
Kónitsa 82 A4 Ípeiros, W Greece
Konitz see Chojnice
Konjic 78 C4 Federacija Bosna I Hercegovina, S Bosnia and Herzegovina
Konosha 88 C4 Arkhangel'skaya Oblast', NW Russian Federation
Konotop 87 F1 Sums'ka Oblast', NE Ukraine
Konstantinovka see Kostyantynivka
Konstanz 73 B7 var. Constanz, Eng. Constance, hist. Kostnitz; anc. Constantia. Baden-Württemberg, S Germany
Konstanza see Constanţa
Konya 94 C4 var. Konieh, prev. Konia; anc. Iconium. Konya, C Turkey
Kopaonik 79 D5 mountain range S Serbia
Kopar see Koper
Koper 73 D8 It. Capodistria; prev. Kopar. SW Slovenia
Köpetdag Gershi 100 C3 mountain range Iran/Turkmenistan
Köpetdag Gershi/Kopetdag, Khrebet see Koppeh Dāgh
Koppeh Dāgh 98 D2 Rus. Khrebet Kopetdag, Turkm. Köpetdag Gershi. mountain range Iran/Turkmenistan
Kopreinitz see Koprivnica
Koprivnica 78 B2 Ger. Kopreinitz, Hung. Kaproncza. Koprivnica-Križevci, N Croatia
Köprülü see Veles
Koptsevichi see Kaptsevichy
Kopyl' see Kapyl'
Korat see Nakhon Ratchasima
Korat Plateau 114 D4 plateau E Thailand
Korba 113 E4 Chhattīsgarh, C India
Korça see Korçë
Korçë 79 D6 var. Korça, Gk. Korytsa, It. Corriza; prev. Koritsa. Korçë, SE Albania
Korčula 78 B4 It. Curzola; anc. Corcyra Nigra. island S Croatia
Korea Bay 105 G3 bay China/North Korea
Korea, Democratic People's Republic of see North Korea
Korea, Republic of see South Korea
Korea Strait 109 A7 Jap. Chōsen-kaikyō, Kor. Taehan-haehyŏp. channel Japan/South Korea
Korhogo 52 D4 N Côte d'Ivoire
Kórinthos 83 B6 anc. Corinthus Eng. Corinth. Pelopónnisos, S Greece
Korinthiakós Kólpos 83 B5 Eng. Gulf of Corinth; anc. Corinthiacus Sinus. gulf C Greece
Koritsa see Korçë
Kōriyama 109 D5 Fukushima, Honshū, C Japan
Korla 104 C3 Chin. K'u-erh-lo. Xinjiang Uygur Zizhiqu, NW China
Körmend 77 B7 Vas, W Hungary
Koróni 83 B6 Pelopónnisos, S Greece
Koror 122 A2 (Palau) Oreor, N Palau
Kőrös see Križevci
Korosten' 86 D1 Zhytomyrs'ka Oblast', NW Ukraine
Koro Toro 54 C2 Borkou-Ennedi-Tibesti, N Chad
Korsovka see Kārsava
Kortrijk 65 A6 Fr. Courtrai. West-Vlaanderen, W Belgium
Koryak Range 93 H2 var. Koryakskiy Khrebet, Eng. Koryak Range. mountain range NE Russian Federation
Koryak Range see Koryakskoye Nagor'ye
Koryakskiy Khrebet see Koryakskoye Nagor'ye
Koryazhma 88 C4 Arkhangel'skaya Oblast', NW Russian Federation
Korytsa see Korçë
Kos 83 E6 Kos, Dodekánisa, Greece, Aegean Sea
Kos 83 E6 It. Coo; anc. Cos. island Dodekánisa, Greece, Aegean Sea
Ko-saki 109 A7 headland Nagasaki, Tsushima, SW Japan
Kościan 76 B4 Ger. Kosten. Wielkopolskie, C Poland
Kościerzyna 76 C2 Pomorskie, NW Poland
Kosciusko, Mount see Kosciuszko, Mount
Kosciuszko, Mount 127 C7 prev. Mount Kosciusko. mountain New South Wales, SE Australia
K'o-shih see Kashi
Koshikijima-retto 109 A8 var. Kosikizima Rettô. island group SW Japan
Kōshū see Gwangju
Košice 77 D6 Ger. Kaschau, Hung. Kassa. Košický Kraj, E Slovakia
Kosikizima Rettô see Koshikijima-rettō
Köslin see Koszalin
Koson 101 E3 Rus. Kasan. Qashqadaryo Viloyati, S Uzbekistan
Kosovo 79 D5 prev. Autonomous Province of Kosovo and Metohija. country SE Europe
Kosovo and Metohija, Autonomous Province of see Kosovo
Kosovo Polje see Fushë Kosovë
Kosovska Mitrovica see Mitrovicë
Kosrae 122 C2 prev. Kusaie. island Caroline Islands, E Micronesia
Kossou, Lac de 52 D5 lake C Côte d'Ivoire
Kostanay 92 C4 var. Kustanay, Kaz. Qostanay. Kostanay, N Kazakhstan
Kosten see Kościan
Kostenets 82 C2 prev. Georgi Dimitrov. Sofiya, W Bulgaria
Kostnitz see Konstanz
Kostroma 88 B4 Kostromskaya Oblast', NW Russian Federation
Kostyantynivka 87 G3 Rus. Konstantinovka. Donets'ka Oblast', SE Ukraine
Kostyukovichi see Kastsyukovichy
Kostyukovka see Kastsyukowka
Koszalin 76 B2 Ger. Köslin. Zachodnio-pomorskie, NW Poland
Kota 112 D3 prev. Kotah. Rājasthān, N India
Kota Baharu see Kota Bharu
Kota Bahru see Kota Bharu
Kotabaru see Jayapura
Kota Bharu 116 B3 var. Kota Baharu, Kota Bahru. Kelantan, Peninsular Malaysia
Kotaboemi see Kotabumi
Kotabumi 116 B4 prev. Kotaboemi. Sumatera, W Indonesia
Kotah see Kota
Kota Kinabalu 116 D3 prev. Jesselton. Sabah, East Malaysia
Kotel'nyy, Ostrov 93 E2 island Novosibirskiye Ostrova, N Russian Federation
Kotka 63 E5 Etelä-Suomi, S Finland
Kotlas 88 C4 Arkhangel'skaya Oblast', NW Russian Federation
Kotonu see Cotonou
Kotor 79 C5 It. Cattaro. SW Montenegro
Kotovs'k 86 D3 Rus. Kotovsk. Odes'ka Oblast', SW Ukraine
Kotovsk see Hînceşti
Kottbus see Cottbus
Kotto 54 D4 river Central African Republic/Dem. Rep. Congo
Kotuy 93 E2 river N Russian Federation
Koudougou 53 E4 C Burkina
Koulamoutou 55 B6 Ogooué-Lolo, C Gabon
Koulikoro 52 D3 Koulikoro, SW Mali
Koumra 54 C4 Moyen-Chari, S Chad
Kourou 37 H3 N French Guiana
Kousseir see Al Quşayr
Kousséri 54 B3 prev. Fort-Foureau. Extrême-Nord, NE Cameroon
Koutiala 52 D4 Sikasso, S Mali
Kouvola 63 E5 Etelä-Suomi, S Finland
Kovel' 86 C1 Pol. Kowel. Volyns'ka Oblast', NW Ukraine
Kovno see Kaunas
Koweit see Kuwait
Kowel see Kovel'
Kowloon 106 A2 Hong Kong, S China
Kowno see Kaunas
Kozáni 82 B4 Dytikí Makedonía, N Greece
Kozara 78 B3 mountain range NW Bosnia and Herzegovina
Kozarska Dubica see Bosanska Dubica
Kozhikode 110 C2 var. Calicut. Kerala, SW India
Kozu-shima 109 D6 island E Japan
Kozyatyn 86 D2 Rus. Kazatin. Vinnyts'ka Oblast', C Ukraine
Kpalimé 53 E5 var. Palimé. SW Togo
Krâchéh 115 D6 prev. Kratie. Krâchéh, E Cambodia
Kragujevac 78 D4 Serbia, C Serbia
Krainburg see Kranj
Kra, Isthmus of 115 B6 isthmus Malaysia/Thailand
Krakau see Kraków
Kraków 77 D5 Eng. Cracow, Ger. Krakau; anc. Cracovia. Małopolskie, S Poland
Krâlănh 115 D5 Siĕmréab, NW Cambodia
Kralendijk 33 E5 dependent territory capital (Bonaire) Lesser Antilles, S Caribbean Sea
Kraljevo 78 D4 prev. Rankovićevo. Serbia, C Serbia
Kramators'k 87 G3 Rus. Kramatorsk. Donets'ka Oblast', SE Ukraine
Kramatorsk see Kramators'k
Kramfors 63 C5 Västernorrland, C Sweden
Kranéa see Kraniá
Kraniá 82 B4 var. Kranéa. Dytikí Makedonía, N Greece
Kranj 73 D7 Ger. Krainburg. NW Slovenia
Kranz see Zelenogradsk
Krāslava 84 D4 Krāslava, SE Latvia
Krasnaye 85 C5 Rus. Krasnoye. Minskaya Voblasts', C Belarus
Krasnoarmeysk 89 C6 Saratovskaya Oblast', W Russian Federation
Krasnodar 89 A7 prev. Ekaterinodar, Yekaterinodar. Krasnodarskiy Kray, SW Russian Federation
Krasnodon 87 H3 Luhans'ka Oblast', E Ukraine
Krasnogor see Kallaste
Krasnogvardeyskoye see Krasnohvardiys'ke
Krasnohvardiys'ke 87 F4 Rus. Krasnogvardeyskoye. Respublika Krym, S Ukraine
Krasnokamensk 93 F4 Chitinskaya Oblast', S Russian Federation
Krasnokamsk 89 D5 Permskaya Oblast', W Russian Federation
Krasnoperekops'k 87 F4 Rus. Krasnoperekopsk. Respublika Krym, S Ukraine
Krasnoperekopsk see Krasnoperekops'k
Krasnostav see Krasnystaw
Krasnovodsk see Türkmenbaşy
Krasnovodskiy Zaliv see Türkmenbasy Aylagy
Krasnovodsk Aylagy see Türkmenbasy Aylagy
Krasnoyarsk 92 D4 Krasnoyarskiy Kray, S Russian Federation
Krasnoye see Krasnaye
Krasnystaw 76 E4 Rus. Krasnostav. Lubelskie, SE Poland
Krasnyy Kut 89 C6 Saratovskaya Oblast', W Russian Federation
Krasnyy Luch 87 H3 prev. Krindachevka. Luhans'ka Oblast', E Ukraine
Kratie see Krâchéh
Krâvanh, Chuŏr Phnum 115 C6 Eng. Cardamom Mountains, Fr. Chaîne des Cardamomes. mountain range W Cambodia
Krefeld 72 A4 Nordrhein-Westfalen, W Germany
Kreisstadt see Krosno Odrzańskie
Kremenchug see Kremenchuk
Kremenchugskoye Vodokhranilishche/Kremenchuk Reservoir see Kremenchuts'ke Vodoskhovyshche
Kremenchuk 87 F2 Rus. Kremenchug. Poltavs'ka Oblast', NE Ukraine
Kremenchuk Reservoir 87 F2 Eng. Kremenchuk Reservoir, Rus. Kremenchugskoye Vodokhranilishche. reservoir C Ukraine
Kremenets' 86 C2 Pol. Krzemieniec, Rus. Kremenets. Ternopil's'ka Oblast', W Ukraine
Kremennaya see Kreminna
Kreminna 87 G2 Rus. Kremennaya. Luhans'ka Oblast', E Ukraine
Kresena see Kresna
Kresna 82 C3 var. Kresena. Blagoevgrad, SW Bulgaria
Kretikon Delagos see Kritikó Pélagos
Kretinga 84 B3 Ger. Krottingen. Klaipėda, NW Lithuania
Kreutz see Cristuru Secuiesc
Kreuz see Križevci, Croatia
Kreuz see Risti, Estonia
Kreuzburg/Kreuzburg in Oberschlesien see Kluczbork
Krichëv see Krychaw
Krievija see Russian Federation
Krindachevka see Krasnyy Luch
Krishna 110 C1 prev. Kistna. river C India
Krishnagiri 110 C2 Tamil Nādu, SE India
Kristiania see Oslo
Kristiansand 63 A6 var. Christiansand. Vest-Agder, S Norway
Kristianstad 63 B7 Skåne, S Sweden
Kristiansund 62 A4 var. Christiansund. Møre og Romsdal, S Norway
Kríti 83 C7 Eng. Crete. island Greece, Aegean Sea
Kritikó Pélagos 83 D7 var. Kretikon Delagos, Eng. Sea of Crete; anc. Mare Creticum. sea Greece, Aegean Sea
Krivoy Rog see Kryvyy Rih
Križevci 78 B2 Ger. Kreuz, Hung. Kőrös. Varaždin, NE Croatia
Krk 78 A3 It. Veglia; anc. Curieta. island NW Croatia
Kroatien see Croatia
Krolevets' 87 F1 Rus. Krolevets. Sums'ka Oblast', NE Ukraine
Krolevets see Krolevets'
Królewska Huta see Chorzów
Kronach 73 C5 Bayern, E Germany
Kronstadt see Braşov
Kroonstad 56 D4 Free State, C South Africa
Kropotkin 89 A7 Krasnodarskiy Kray, SW Russian Federation
Krosno 77 D5 Ger. Krossen. Podkarpackie, SE Poland
Krosno Odrzańskie 76 B3 Ger. Crossen, Kreisstadt. Lubuskie, W Poland
Krossen see Krosno
Krottingen see Kretinga
Krško 73 E8 Ger. Gurkfeld; prev. Videm-Krško. E Slovenia
Krugloye see Kruhlaye
Kruhlaye 85 D6 Rus. Krugloye. Mahilyowskaya Voblasts', E Belarus
Kruja see Krujë
Krujë 79 C6 var. Kruja, It. Croia. Durrës, C Albania
Krummau see Český Krumlov
Krung Thep, Ao 115 C5 var. Bight of Bangkok. bay S Thailand
Krung Thep Mahanakhon see Ao Krung Thep
Krupki 85 D6 Minskaya Voblasts', C Belarus
Krušné Hory see Erzgebirge
Krychaw 85 E7 Rus. Krichëv. Mahilyowskaya Voblasts', E Belarus
Kryms'ki Hory 87 F5 mountain range S Ukraine
Kryms'kyy Pivostriv 87 F5 Eng. Crimea. peninsula S Ukraine
Krynica 77 D5 Ger. Tannenhof. Małopolskie, S Poland
Kryve Ozero 87 E3 Odes'ka Oblast', SW Ukraine
Kryvyy Rih 87 F3 Rus. Krivoy Rog. Dnipropetrovs'ka Oblast', SE Ukraine
Krzemieniec see Kremenets'
Ksar al Kabir see Ksar-el-Kebir
Ksar al Soule see Er-Rachidia
Ksar-el-Kebir 48 C2 var. Alcázar, Ksar al Kabir, Ksar-el-Kébir, Ar. Al-Kasr al-Kebir, Al-Qsar al-Kbir, Sp. Alcazarquivir. NW Morocco
Ksar-el-Kébir see Ksar-el-Kebir
Kuala Dungun see Dungun
Kuala Lumpur 116 B3 country capital (Malaysia) Kuala Lumpur, Peninsular Malaysia
Kuala Terengganu 116 B3 var. Kuala Trengganu. Terengganu, Peninsular Malaysia
Kualatungkal 116 B4 Sumatera, W Indonesia
Kuang-chou see Guangzhou
Kuang-hsi see Guangxi Zhuangzu Zizhiqu
Kuang-tung see Guangdong
Kuang-yuan see Guangyuan
Kuantan 116 B3 Pahang, Peninsular Malaysia
Kuba see Quba
Kuban' 87 G5 var. Hypanis. river SW Russian Federation
Kubango see Cubango/Okavango
Kuching 116 C3 prev. Sarawak. Sarawak, East Malaysia
Kūchnay Darwāshān 100 D5 prev. Kūchnay Darweyshān. Helmand, S Afghanistan
Kūchnay Darweyshān see Kūchnay Darwāshān
Kuçova see Kuçovë
Kuçovë 79 C6 var. Kuçova; prev. Qyteti Stalin. Berat, C Albania
Kudara see Ghŭdara
Kudus 116 C5 prev. Koedoes. Jawa, C Indonesia
Kuei-lin see Guilin
Kuei-Yang/Kuei-yang see Guiyang
K'u-erh-lo see Korla
Kueyang see Guiyang
Kugaaruk 15 G3 prev. Pelly Bay. Nunavut, N Canada
Kugluktuk 31 E3 var. Qurlurtuuq; prev. Coppermine. Nunavut, NW Canada
Kuhmo 62 E4 Oulu, E Finland
Kuhnau see Konin
Kūhnö see Kihnu
Kuibyshev see Kuybyshevskoye Vodokhranilishche
Kuito 56 B2 Port. Silva Porto. Bié, C Angola
Kuji 108 D3 var. Kuzi. Iwate, Honshū, C Japan
Kukës 79 D5 var. Kukësi. Kukës, NE Albania
Kukësi see Kukës
Kukong see Shaoguan
Kukukhoto see Hohhot
Kula Kangri 113 G3 var. Kulhakangri. mountain Bhutan/China
Kuldīga 84 B3 Ger. Goldingen. Kuldīga, W Latvia
Kuldja see Yining
Kulhakangri see Kula Kangri
Kullorsuaq 60 D2 var. Kuvdlorssuak. Kitaa, C Greenland
Kulm see Chełmno
Kulmsee see Chełmża
Kŭlob 101 F3 Rus. Kulyab. SW Tajikistan
Kulpa see Kolpa
Kulu 94 C3 Konya, W Turkey
Kulunda Steppe 92 C4 Kaz. Qulyndy Zhazyghy, Rus. Kulundinskaya Ravnina. grassland Kazakhstan/Russian Federation
Kulundinskaya Ravnina see Kulunda Steppe
Kulyab see Kŭlob
Kum see Qom
Kuma 89 B7 river SW Russian Federation
Kumamoto 109 A7 Kumamoto, Kyūshū, SW Japan
Kumanova see Kumanovo
Kumanovo 79 E5 Turk. Kumanova. N Macedonia
Kumasi 53 E5 prev. Coomassie. C Ghana
Kumayri see Gyumri
Kumba 55 A5 Sud-Ouest, W Cameroon
Kumertau 89 D6 Respublika Bashkortostan, W Russian Federation
Kumillā see Comilla
Kumo 53 G4 Gombe, E Nigeria
Kumon Range 114 B2 mountain range N Myanmar (Burma)
Kumul see Hami

Luzern – Markounda

Luzern 73 B7 Fr. Lucerne, It. Lucerna. Luzern, C Switzerland
Luzon 117 E1 island N Philippines
Luzon Strait 103 E3 strait Philippines/Taiwan
L'viv 86 B2 Ger. Lemberg, Pol. Lwów, Rus. L'vov. L'vivs'ka Oblast', W Ukraine
L'vov see L'viv
Lwena see Luena
Lwów see L'viv
Lyakhavichy 85 B6 Rus. Lyakhovichi. Brestskaya Voblasts', SW Belarus
Lyakhovichi see Lyakhavichy
Lyallpur see Faisalābād
Lyangar see Langar
Lyck see Ełk
Lycksele 62 C4 Västerbotten, N Sweden
Lycopolis see Asyūṭ
Lyel'chytsy 85 C7 Rus. Lel'chitsy. Homyel'skaya Voblasts', SE Belarus
Lyepyel' 85 D5 Rus. Lepel'. Vitsyebskaya Voblasts', N Belarus
Lyme Bay 67 C7 bay S England, United Kingdom
Lynchburg 19 E5 Virginia, NE USA
Lynn see King's Lynn
Lynn Lake 15 F4 Manitoba, C Canada
Lynn Regis see King's Lynn
Lyon 69 D5 Eng. Lyons; anc. Lugdunum. Rhône, E France
Lyons see Lyon
Lyozna 85 E6 Rus. Liozno. Vitsyebskaya Voblasts', NE Belarus
Lypovets' 86 D2 Rus. Lipovets. Vinnyts'ka Oblast', C Ukraine
Lys see Leie
Lysychans'k 87 H3 Rus. Lisichansk. Luhans'ka Oblast', E Ukraine
Lyttelton 129 C6 South Island, New Zealand
Lyublin see Lublin
Lyubotin see Lyubotyn
Lyubotyn 87 G2 Rus. Lyubotin. Kharkivs'ka Oblast', E Ukraine
Lyulyakovo 82 E2 prev. Keremitlik. Burgas, E Bulgaria
Lyusina 85 B6 Rus. Lyusino. Brestskaya Voblasts', SW Belarus
Lyusino see Lyusina

M

Maale see Male'
Ma'ān 97 B7 Ma'ān, SW Jordan
Maardu 84 D2 Ger. Maart. Harjumaa, NW Estonia
Ma'aret-en-Nu'man see Ma'arrat an Nu'mān
Ma'arrat an Nu'mān 96 B3 var. Ma'aret-en-Nu'man, Fr. Maarret enn Naamâne. Idlib, NW Syria
Maarret enn Naamâne see Ma'arrat an Nu'mān
Maart see Maardu
Maas see Meuse
Maaseik 65 D5 prev. Maeseyck. Limburg, NE Belgium
Maastricht 65 D6 var. Maestricht; anc. Traiectum ad Mosam, Traiectum Tungorum. Limburg, SE Netherlands
Macao 107 C6 Port. Macau. Guangdong, SE China
Macapá 41 E1 state capital Amapá, N Brazil
Macarsca see Makarska
Macassar see Makassar
Macău see Makó, Hungary
Macau see Macao
MacCluer Gulf see Berau, Teluk
Macdonnell Ranges 124 D4 mountain range Northern Territory, C Australia
Macedonia see Macedonia, FYR
Macedonia, FYR 79 D6 off. the Former Yugoslav Republic of Macedonia, var. Macedonia, Mac. Makedonija, abbrev. FYR Macedonia, FYROM. country SE Europe
Macedonia, the Former Yugoslav Republic of see Macedonia, FYR
Maceió 41 G3 state capital Alagoas, E Brazil
Machachi 38 B1 Pichincha, C Ecuador
Machala 38 B2 El Oro, SW Ecuador
Machanga 57 E3 Sofala, E Mozambique
Machilīpatnam 110 D1 var. Bandar Masulipatnam. Andhra Pradesh, E India
Machiques 36 C2 Zulia, NW Venezuela
Macías Nguema Biyogo see Bioco, Isla de
Măcin 86 D5 Tulcea, SE Romania
Mackay 126 D4 Queensland, NE Australia
Mackay, Lake 124 C4 salt lake Northern Territory/Western Australia
Mackenzie 15 E3 river Northwest Territories, NW Canada
Mackenzie Bay 132 D3 bay Antarctica
Mackenzie Mountains 14 D3 mountain range Northwest Territories, NW Canada
Macleod, Lake 124 A4 lake Western Australia
Macomb 18 A4 Illinois, N USA
Macomer 75 A5 Sardegna, Italy, C Mediterranean Sea
Mâcon 69 D5 anc. Matisco, Matisco Ædourum. Saône-et-Loire, C France
Macon 20 D2 Georgia, SE USA
Macon 23 G4 Missouri, C USA
Macquarie Ridge 132 C5 undersea ridge SW Pacific Ocean
Macuspana 29 G4 Tabasco, SE Mexico
Ma'dabā 97 B6 var. Mādabā, Madeba; anc. Medeba. Ma'dabā, NW Jordan
Mādabā see Ma'dabā
Madagascar 57 F3 off. Democratic Republic of Madagascar, Malg. Madagasikara; prev. Malagasy Republic. country W Indian Ocean
Madagascar 57 F3 island W Indian Ocean
Madagascar Basin 47 E7 undersea basin W Indian Ocean
Madagascar, Democratic Republic of see Madagascar
Madagascar Plateau 47 E7 var. Madagascar Ridge, Madagascar Rise, Rus. Madagaskarskiy Khrebet. undersea plateau W Indian Ocean
Madagascar Rise/Madagascar Ridge see Madagascar Plateau
Madagasikara see Madagascar
Madagaskarskiy Khrebet see Madagascar Plateau
Madang 122 B3 Madang, N Papua New Guinea
Madanīyīn see Médenine
Mađarska see Hungary
Made 64 C4 Noord-Brabant, S Netherlands
Madeba see Ma'dabā
Madeira 48 A2 var. Ilha de Madeira. island Madeira, Portugal, NE Atlantic Ocean
Madeira, Ilha de see Madeira
Madeira Plain 44 C3 abyssal plain E Atlantic Ocean
Madeira, Rio 40 D2 var. Río Madera. river Bolivia/Brazil
Madeleine, Îles de la 17 F4 Eng. Magdalen Islands. island group Québec, E Canada
Madera 25 B6 California, W USA
Madera, Río see Madeira, Rio
Madhya Pradesh 113 E4 prev. Central Provinces and Berar. cultural region C India
Madīnat ath Thawrah 96 C2 var. Ath Thawrah. Ar Raqqah, N Syria
Madioen see Madiun
Madison 23 F3 South Dakota, N USA
Madison 18 B3 state capital Wisconsin, N USA
Madiun 116 D5 prev. Madioen. Jawa, C Indonesia
Madoera see Madura, Pulau
Madona 84 D4 Ger. Modohn. Madona, E Latvia
Madras see Chennai
Madras see Tamil Nādu
Madre de Dios, Río 39 E3 river Bolivia/Peru
Madre del Sur, Sierra 29 E5 mountain range S Mexico
Madre, Laguna 29 F3 lagoon NE Mexico
Madre, Laguna 27 G5 lagoon Texas, SW USA
Madre Occidental, Sierra 28 C3 var. Western Sierra Madre. mountain range C Mexico
Madre Oriental, Sierra 29 E3 var. Eastern Sierra Madre. mountain range C Mexico
Madre, Sierra 30 B2 var. Sierra de Soconusco. mountain range Guatemala/Mexico
Madrid 70 D3 country capital (Spain) Madrid, C Spain
Madura see Madurai
Madurai 110 C3 prev. Madura, Mathurai. Tamil Nādu, S India
Madura, Pulau 116 D5 prev. Madoera. island C Indonesia
Maebashi 109 D5 var. Maebasi, Mayebashi. Gunma, Honshū, S Japan
Maebasi see Maebashi
Mae Nam Khong see Mekong
Mae Nam Nan 114 C4 river NW Thailand
Mae Nam Yom 114 C4 river W Thailand
Maeseyck see Maaseik
Maestricht see Maastricht
Maéwo 122 D4 prev. Aurora. island C Vanuatu
Mafia 51 D7 island E Tanzania
Mafraq/Muḥāfaẓat al Mafraq see Al Mafraq
Magadan 93 G3 Magadanskaya Oblast', E Russian Federation
Magallanes see Punta Arenas
Magallanes, Estrecho de see Magellan, Strait of
Magangué 36 B2 Bolívar, N Colombia
Magdalena 39 F3 Beni, N Bolivia
Magdalena 28 B1 Sonora, NW Mexico
Magdalena, Isla 28 B3 island NW Mexico
Magdalena, Río 36 B2 river C Colombia
Magdalen Islands see Madeleine, Îles de la
Magdeburg 72 C4 Sachsen-Anhalt, C Germany
Magelang 116 C5 Jawa, C Indonesia
Magellan, Strait of 43 B8 Sp. Estrecho de Magallanes. strait Argentina/Chile
Magerøy see Magerøya
Magerøya 62 D1 var. Magerøy, Lapp. Máhkarávju. island N Norway
Maggiore, Lago see Maggiore, Lake
Maggiore, Lake 74 B1 It. Lago Maggiore. lake Italy/Switzerland
Maglaj 78 C3 Federacija Bosna I Hercegovina, N Bosnia and Herzegovina
Maglie 75 E6 Puglia, SE Italy
Magna 22 B4 Utah, W USA
Magnesia see Manisa
Magnitogorsk 92 B4 Chelyabinskaya Oblast', C Russian Federation
Magnolia State see Mississippi
Magta' Lahjar 52 C3 var. Magta Lahjar, Magta' Lahjar, Magtá Lahjar. Brakna, SW Mauritania
Magtymguly 100 C3 prev. Garrygala; Rus. Kara-gala. W Turkmenistan
Magway 114 A3 var. Magwe. Magway, W Myanmar (Burma)
Magyar-Becse see Bečej
Magyarkanizsa see Kanjiža
Magyarország see Hungary
Mahajanga 57 F2 var. Majunga. Mahajanga, NW Madagascar
Mahakam, Sungai 116 D4 var. Koetai, Kutai. river Borneo, C Indonesia
Mahalapye 56 D3 var. Mahalatswe. Central, SE Botswana
Mahalatswe see Mahalapye
Māhān 98 D3 Kermān, E Iran
Mahanādi 113 F4 river E India
Mahārāshtra 112 D5 cultural region W India
Mahbés see El Mahbas
Mahbūbnagar 112 D5 Telangana, C India
Mahdia 49 F2 var. Al Mahdīyah, Mehdia. NE Tunisia
Mahé 57 H1 island Inner Islands, NE Seychelles
Mahia Peninsula 128 E4 peninsula North Island, New Zealand
Mahilyow 85 D6 Rus. Mogilëv. Mahilyowskaya Voblasts', E Belarus
Máhkarávju see Magerøya
Mahmūd-e 'Erāqī see Maḥmūd-e Rāqī
Maḥmūd-e Rāqī 101 E4 var. Mahmūd-e 'Erāqī. Kāpīsā, NE Afghanistan
Mahón see Maó
Mähren see Moravia
Mährisch-Weisskirchen see Hranice
Maicao 36 C1 La Guajira, N Colombia
Mai Ceu/Mai Chio see Maych'ew
Maīdān Shahr 101 E4 prev. Meydān Shahr. Vardak, E Afghanistan
Maidstone 67 E7 SE England, United Kingdom
Maiduguri 53 H4 Borno, NE Nigeria
Mailand see Milano
Maimanah 100 D3 var. Meymaneh, Maymana. Fāryāb, NW Afghanistan
Main 73 B5 river C Germany
Mai-Ndombe, Lac 55 C6 prev. Lac Léopold II. lake W Dem. Rep. Congo
Maine 19 G2 off. State of Maine, also known as Lumber State, Pine Tree State. state NE USA
Maine 68 B3 cultural region NW France
Maine, Gulf of 19 H2 gulf NE USA
Mainland 66 C2 island N Scotland, United Kingdom
Mainland 66 D1 island NE Scotland, United Kingdom
Mainz 73 B5 Fr. Mayence. Rheinland-Pfalz, SW Germany
Maio 52 A3 var. Mayo. island Ilhas de Sotavento, SE Cape Verde
Maisur see Mysore, India
Maisur see Karnātaka, India
Maizhokunggar 104 C5 Xizang Zizhiqu, W China
Majorca see Mallorca
Mājro see Majuro Atoll
Majunga see Mahajanga
Majuro Atoll 122 D2 var. Mājro. atoll Ratak Chain, SE Marshall Islands
Makale see Mek'elē
Makarov Basin 133 B3 undersea basin Arctic Ocean
Makarska 78 B4 It. Macarsca. Split-Dalmacija, SE Croatia
Makasar see Makassar
Makasar, Selat see Makassar Straits
Makassar 117 E4 var. Macassar, Makasar; prev. Ujungpandang. Sulawesi, C Indonesia
Makassar Straits 116 D4 Ind. Makasar Selat. strait C Indonesia
Makay 57 F3 var. Massif du Makay. mountain range SW Madagascar
Makay, Massif du see Makay
Makedonija see Macedonia, FYR
Makeni 52 C4 C Sierra Leone
Makeyevka see Makiyivka
Makhachkala 92 A4 prev. Petrovsk-Port. Respublika Dagestan, SW Russian Federation
Makin 122 D2 prev. Pitt Island. atoll Tungaru, W Kiribati
Makira see San Cristobal
Makiyivka 87 G3 Rus. Makeyevka; prev. Dmitriyevsk. Donets'ka Oblast', E Ukraine
Makkah 99 A5 Eng. Mecca. Makkah, W Saudi Arabia
Makkovik 17 F2 Newfoundland and Labrador, NE Canada
Makó 77 D7 Rom. Macău. Csongrád, SE Hungary
Makoua 55 B5 Cuvette, C Congo
Makran Coast 98 E4 coastal region SE Iran
Makrany 85 A6 Rus. Mokrany. Brestskaya Voblasts', SW Belarus
Mākū 98 B2 Āžarbāyjān-e Gharbī, NW Iran
Makurdi 53 G4 Benue, C Nigeria
Mala see Malaita, Solomon Islands
Malabār Coast 110 B3 coast SW India
Malabo 55 A5 prev. Santa Isabel. country capital (Equatorial Guinea) Isla de Bioco, NW Equatorial Guinea
Malaca see Málaga
Malacca, Strait of 116 B3 Ind. Selat Malaka. strait Indonesia/Malaysia
Malacka see Malacky
Malacky 77 C6 Hung. Malacka. Bratislavský Kraj, W Slovakia
Maladzyechna 85 C5 Pol. Molodeczno, Rus. Molodechno. Minskaya Voblasts', C Belarus
Málaga 70 D5 anc. Malaca. Andalucía, S Spain
Malagarasi River 51 B7 river W Tanzania Africa
Malagasy Republic see Madagascar
Malaita 122 C3 var. Mala. island N Solomon Islands
Malakal 51 B5 Upper Nile, NE South Sudan
Malakula see Malekula
Malang 116 D5 Jawa, C Indonesia
Malange see Malanje
Malanje 56 B1 var. Malange. Malanje, NW Angola
Mälaren 63 C6 lake C Sweden
Malatya 95 E4 anc. Melitene. Malatya, SE Turkey
Mala Vyska 87 E3 Rus. Malaya Viska. Kirovohrads'ka Oblast', S Ukraine
Malawi 57 E1 off. Republic of Malaŵi; prev. Nyasaland, Nyasaland Protectorate. country S Africa
Malawi, Lake see Nyasa, Lake
Malaŵi, Republic of see Malawi
Malaya Viska see Mala Vyska
Malay Peninsula 102 D4 peninsula Malaysia/Thailand
Malaysia 116 B3 off. Malaysia, var. Federation of Malaysia; prev. the separate territories of Federation of Malaya, Sarawak and Sabah (North Borneo) and Singapore. country SE Asia
Malaysia, Federation of see Malaysia
Malbork 76 C2 Ger. Marienburg, Marienburg in Westpreussen. Pomorskie, N Poland
Malchin 72 C3 Mecklenburg-Vorpommern, N Germany
Malden 23 H5 Missouri, C USA
Malden Island 123 G3 prev. Independence Island. atoll E Kiribati
Maldives 110 A4 off. Maldivian Divehi, Republic of Maldives. country N Indian Ocean
Maldives, Republic of see Maldives
Maldivian Divehi see Maldives
Male' 110 B4 Div. Maale. country capital (Maldives) Male' Atoll, C Maldives
Male' Atoll 110 B4 var. Kaafu Atoll. atoll C Maldives
Malekula 122 D4 var. Malakula; prev. Mallicolo. island W Vanuatu
Malesína 83 C5 Stereá Ellás, E Greece
Malheur Lake 24 C3 lake Oregon, NW USA
Mali 53 E3 off. Republic of Mali, Fr. République du Mali; prev. French Sudan, Sudanese Republic. country W Africa
Malik, Wadi al see Milk, Wadi el
Mali Kyun 115 B5 var. Tavoy Island. island Myeik Archipelago, S Myanmar (Burma)
Malin see Malyn
Malindi 51 D7 Coast, SE Kenya
Malines see Mechelen
Mali, Republic of see Mali
Mali, République du see Mali
Malkiye see Al Mālikīyah
Malko Tŭrnovo 82 E3 Burgas, E Bulgaria
Mallaig 66 B3 N Scotland, United Kingdom
Mallawī 50 B2 var. Mallawi. C Egypt
Mallawi see Mallawī
Mallicolo see Malekula
Mallorca 71 G3 Eng. Majorca; anc. Baleares Major. island Islas Baleares, Spain, W Mediterranean Sea
Malmberget 62 C3 Lapp. Malmivaara. Norrbotten, N Sweden
Malmédy 65 D6 Liège, E Belgium
Malmivaara see Malmberget
Malmö 63 B7 Skåne, S Sweden
Maloelap see Maloelap Atoll
Maloelap Atoll 122 D1 var. Maloelap. atoll E Marshall Islands
Małopolska, Wyżyna 76 D4 plateau S Poland
Malozemel'skaya Tundra 88 D3 physical region NW Russian Federation
Malta 84 D4 Rēzekne, SE Latvia
Malta 22 C1 Montana, NW USA
Malta 75 C8 off. Republic of Malta. country C Mediterranean Sea
Malta 75 C8 island Malta, C Mediterranean Sea
Malta, Canale di see Malta Channel
Malta Channel 75 C8 It. Canale di Malta. strait Italy/Malta
Malta, Republic of see Malta
Maluku 117 F4 Dut. Molukken, Eng. Moluccas; prev. Spice Islands. island group E Indonesia
Maluku, Laut see Molucca Sea
Malung 63 B6 Dalarna, C Sweden
Malventum see Benevento
Malvina, Isla Gran see West Falkland
Malvinas, Islas see Falkland Islands
Malyn 86 D2 Rus. Malin. Zhytomyrs'ka Oblast', N Ukraine
Malyy Kavkaz see Lesser Caucasus
Mamberamo, Sungai 117 H4 river Papua, E Indonesia
Mambij see Manbij
Mamonovo 84 A4 Ger. Heiligenbeil. Kaliningradskaya Oblast', W Russian Federation
Mamoré, Rio 39 F3 river Bolivia/Brazil
Mamou 52 C4 W Guinea
Mamoudzou 57 F2 dependent territory capital (Mayotte) C Mayotte
Mamuno 56 C3 Ghanzi, W Botswana
Manacor 71 G3 Mallorca, Spain, W Mediterranean Sea
Manado 117 F3 prev. Menado. Sulawesi, C Indonesia
Managua 30 D3 country capital (Nicaragua) Managua, W Nicaragua
Managua, Lake 30 C3 var. Xolotlán. lake W Nicaragua
Manakara 57 G4 Fianarantsoa, SE Madagascar
Manama see Al Manāmah
Mananjary 57 G3 Fianarantsoa, SE Madagascar
Manáos see Manaus
Manapouri, Lake 129 A7 lake South Island, New Zealand
Manar see Mannar
Manas, Gora 101 E2 mountain Kyrgyzstan/Uzbekistan
Manaus 40 D2 prev. Manáos. state capital Amazonas, NW Brazil
Manavgat 94 B4 Antalya, SW Turkey
Manbij 96 C2 var. Mambij, Fr. Membidj. Ḥalab, N Syria
Manchester 67 D5 Lat. Mancunium. NW England, United Kingdom
Manchester 19 G3 New Hampshire, NE USA
Man-chou-li see Manzhouli
Manchurian Plain 103 E1 plain NE China
Máncio Lima see Japiim
Mancunium see Manchester
Mand see Mand, Rūd-e
Mandalay 114 B3 Mandalay, C Myanmar (Burma)
Mandan 23 E2 North Dakota, N USA
Mandeville 32 B5 C Jamaica
Mándra 83 C6 Attikí, C Greece
Rud-e Mand 98 D4 var. Mand. river S Iran
Mandurah 125 A6 Western Australia
Manduria 75 E5 Puglia, SE Italy
Mandya 110 C2 Karnātaka, C India
Manfredonia 75 D5 Puglia, SE Italy
Mangai 55 C6 Bandundu, W Dem. Rep. Congo
Mangaia 123 G5 island group S Cook Islands
Mangalia 86 D5 anc. Callatis. Constanţa, SE Romania
Mangalmé 54 C3 Guéra, SE Chad
Mangalore 110 B2 Karnātaka, W India
Mangaung see Bloemfontein
Mango see Sansanné-Mango, Togo
Mangoky 57 F3 river W Madagascar
Manhattan 23 F4 Kansas, C USA
Manicouagan, Réservoir 16 D3 lake Québec, E Canada
Manihiki 123 G4 atoll N Cook Islands
Manihiki Plateau 121 E3 undersea plateau C Pacific Ocean
Maniitsoq 60 C3 var. Manîtsoq, Dan. Sukkertoppen. Kitaa, S Greenland
Manila 117 E1 off. City of Manila. country capital (Philippines) Luzon, N Philippines
Manila, City of see Manila
Manisa 94 A3 var. Manissa, prev. Saruhan; anc. Magnesia. Manisa, W Turkey
Manissa see Manisa
Manitoba 15 F5 province S Canada
Manitoba, Lake 15 F5 lake Manitoba, S Canada
Manitoulin Island 16 C4 island Ontario, S Canada
Manîtsoq see Maniitsoq
Manizales 36 B3 Caldas, W Colombia
Manjimup 125 A7 Western Australia
Mankato 23 F3 Minnesota, N USA
Manlleu 71 G2 Cataluña, NE Spain
Manly 126 E1 Iowa, C USA
Manmād 112 C5 Mahārāshtra, W India
Mannar 110 C3 var. Manar. Northern Province, NW Sri Lanka
Mannar, Gulf of 110 C3 gulf India/Sri Lanka
Mannheim 73 B5 Baden-Württemberg, SW Germany
Manokwari 117 G4 New Guinea, E Indonesia
Manono 55 E7 Shaba, SE Dem. Rep. Congo
Manosque 69 D6 Alpes-de-Haute-Provence, SE France
Manra 123 F3 prev. Sydney Island. atoll Phoenix Islands, C Kiribati
Mansa 56 D2 prev. Fort Rosebery. Luapula, N Zambia
Mansel Island 15 G3 island Nunavut, NE Canada
Mansfield 18 D4 Ohio, N USA
Manta 38 A2 Manabí, W Ecuador
Manteca 25 B6 California, W USA
Mantoue see Mantova
Mantova 74 B2 Eng. Mantua, Fr. Mantoue. Lombardia, NW Italy
Mantua see Mantova
Manuae 123 G4 island S Cook Islands
Manukau see Manurewa
Manurewa 128 D3 var. Manukau. Auckland, North Island, New Zealand
Manzanares 71 E3 Castilla-La Mancha, C Spain
Manzanillo 32 C3 Granma, E Cuba
Manzanillo 28 D4 Colima, SW Mexico
Manzhouli 105 F1 var. Man-chou-li. Nei Mongol Zizhiqu, N China
Mao 54 B3 Kanem, W Chad
Maó 71 H3 Cast. Mahón, Eng. Port Mahon; anc. Portus Magonis. Menorca, Spain, W Mediterranean Sea
Maoke, Pegunungan 117 H4 Dut. Sneeuw-gebergte, Eng. Snow Mountains. mountain range Papua, E Indonesia
Maoming 106 C6 Guangdong, S China
Mapmaker Seamounts 103 H2 seamount range N Pacific Ocean
Maputo 56 D4 prev. Lourenço Marques. country capital (Mozambique) Maputo, S Mozambique
Marabá 41 F2 Pará, NE Brazil
Maracaibo 36 C1 Zulia, NW Venezuela
Maracaibo, Gulf of see Venezuela, Golfo de
Maracaibo, Lago de 36 C2 var. Lake Maracaibo. inlet NW Venezuela
Maracaibo, Lake see Maracaibo, Lago de
Maracay 36 D2 Aragua, N Venezuela
Marada see Marādah
Marādah 49 G3 var. Marada. N Libya
Maradi 53 G3 Maradi, S Niger
Maragha see Marāgheh
Marāgheh 98 C2 var. Maragha. Āžarbāyjān-e Khāvarī, NW Iran
Marajó, Baía de 41 F1 bay N Brazil
Marajó, Ilha de 41 E1 island N Brazil
Marakesh see Marrakech
Maramba see Livingstone
Maranhão 41 F2 off. Estado do Maranhão. state E Brazil
Maranhão 41 F2 off. Estado do Maranhão. region E Brazil
Maranhão, Estado do see Maranhão
Marañón, Río 38 B2 river N Peru
Marathon 16 C4 Ontario, S Canada
Marathón see Marathónas
Marathónas 83 C5 prev. Marathón. Attikí, C Greece
Märäzä 95 H2 Rus. Maraza. E Azerbaijan
Maraza see Märäzä
Marbella 70 D5 Andalucía, S Spain
Marble Bar 124 B4 Western Australia
Marburg see Marburg an der Lahn, Germany
Marburg see Maribor, Slovenia
Marburg an der Lahn 72 B4 hist. Marburg. Hessen, W Germany
March see Morava
Marche 74 C3 Eng. Marches. region C Italy
Marche 69 C5 cultural region C France
Marche-en-Famenne 65 C7 Luxembourg, SE Belgium
Marchena, Isla 38 B5 var. Bindloe Island. island Galápagos Islands, Ecuador, E Pacific Ocean
Marches see Marche
Mar Chiquita, Laguna 42 C3 lake C Argentina
Marcounda see Markounda
Mardān 112 C1 North-West Frontier Province, N Pakistan
Mar del Plata 43 D5 Buenos Aires, E Argentina
Mardin 95 E4 Mardin, SE Turkey
Maré 122 D5 island Îles Loyauté, E New Caledonia
Marea Neagră see Black Sea
Mareeba 126 D3 Queensland, NE Australia
Marek see Dupnitsa
Marganets see Marhanets'
Margarita, Isla de 37 E1 island N Venezuela
Margate 67 E7 prev. Mergate. SE England, United Kingdom
Margherita see Jamaame
Margherita, Lake 51 C5 Eng. Lake Margherita, It. Abbaia. lake SW Ethiopia
Margherita, Lake see Ābaya Hāyk'
Marghita 86 B3 Hung. Margitta. Bihor, NW Romania
Margitta see Marghita
Marhanets' 87 F3 Rus. Marganets. Dnipropetrovs'ka Oblast', E Ukraine
María Cleofas, Isla 28 C4 island C Mexico
Maria Island 127 C8 island Tasmania, SE Australia
María Madre, Isla 28 C4 island C Mexico
María Magdalena, Isla 28 C4 island C Mexico
Mariana Trench 103 G4 trench W Pacific Ocean
Mariánské Lázně 77 A5 Ger. Marienbad. Karlovarský Kraj, W Czech Republic
Marías, Islas 28 C4 island group C Mexico
Maria-Theresiopel see Subotica
Maribor 73 E7 Ger. Marburg. NE Slovenia
Marica see Maritsa
Maridi 51 B5 W Equatoria, S South Sudan
Marie Byrd Land 132 A3 physical region Antarctica
Marie-Galante 33 G4 var. Ceyre to the Caribs. island SE Guadeloupe
Marienbad see Mariánské Lázně
Marienburg see Alūksne, Latvia
Marienburg see Malbork, Poland
Marienburg in Westpreussen see Malbork
Marienhausen see Viļaka
Mariental 56 B4 Hardap, SW Namibia
Marienwerder see Kwidzyń
Mariestad 63 B6 Västra Götaland, S Sweden
Marietta 20 D2 Georgia, SE USA
Marijampolė 84 B4 prev. Kapsukas. Marijampolė, S Lithuania
Marília 41 E4 São Paulo, S Brazil
Marín 70 B1 Galicia, NW Spain
Mar'ina Gorka see Mar"ina Horka
Mar"ina Horka 85 C6 Rus. Mar'ina Gorka. Minskaya Voblasts', C Belarus
Maringá 41 E4 Paraná, S Brazil
Marion 23 G3 Iowa, C USA
Marion 18 D4 Ohio, N USA
Marion, Lake 21 E2 reservoir South Carolina, SE USA
Mariscal Estigarribia 42 D2 Boquerón, NW Paraguay
Maritsa 82 D3 var. Marica, Gk. Évros, Turk. Meriç; anc. Hebrus. river SW Europe
Maritzburg see Pietermaritzburg
Mariupol' 87 G4 prev. Zhdanov. Donets'ka Oblast', SE Ukraine
Marka 51 D6 var. Merca. Shabeellaha Hoose, S Somalia
Markham, Mount 132 B4 mountain Antarctica
Markounda 54 C4 var. Marcounda. Ouham, NW Central African Republic

Marktredwitz – Minnesota

Marktredwitz *73 C5* Bayern, E Germany
Marlborough *126 D4* Queensland, E Australia
Marmanda *see* Marmande
Marmande *69 B5 anc.* Marmanda. Lot-et-Garonne, SW France
Sea of Marmara *94 A2 Eng.* Sea of Marmara. *sea* NW Turkey
Marmara, Sea of *see* Marmara Denizi
Marmaris *94 A4* Muğla, SW Turkey
Marne *68 C3 cultural region* N France
Marne *68 D3 river* N France
Maro *54 C4* Moyen-Chari, S Chad
Maroantsetra *57 G2* Toamasina, NE Madagascar
Maromokotro *57 G2 mountain* N Madagascar
Maroni *37 G3 Dut.* Marowijne. *river* French Guiana/Suriname
Maroshevíz *see* Toplița
Marosludas *see* Luduș
Marosvásárhely *see* Târgu Mureș
Marotiri *121 F4 var.* Îlots de Bass, Morotiri. *island group* Îles Australes, SW French Polynesia
Maroua *54 B3* Extrême-Nord, N Cameroon
Marowijne *see* Maroni
Marquesas Fracture Zone *131 E3 fracture zone* E Pacific Ocean
Marquette *18 B1* Michigan, N USA
Marrakech *48 C2 var.* Marakesh, *Eng.* Marrakesh; *prev.* Morocco. W Morocco
Marrakesh *see* Marrakech
Marrawah *127 C8* Tasmania, SE Australia
Marree *127 B5* South Australia
Marsá al Burayqah *49 G3 var.* Al Burayqah. N Libya
Marsabit *51 C6* Eastern, N Kenya
Marsala *75 B7 anc.* Lilybaeum. Sicilia, Italy, C Mediterranean Sea
Marsberg *72 B4* Nordrhein-Westfalen, W Germany
Marseille *69 D6 Eng.* Marseilles; *anc.* Massilia. Bouches-du-Rhône, SE France
Marseilles *see* Marseille
Marshall *23 F2* Minnesota, N USA
Marshall *27 H2* Texas, SW USA
Marshall Islands *122 C1 off.* Republic of the Marshall Islands. *country* W Pacific Ocean
Marshall Islands, Republic of the *see* Marshall Islands
Marshall Seamounts *103 H3 seamount range* SW Pacific Ocean
Marsh Harbour *32 C1* Great Abaco, W Bahamas
Martaban *see* Mottama
Martha's Vineyard *19 G3 island* Massachusetts, NE USA
Martigues *69 D6* Bouches-du-Rhône, SE France
Martin *77 C5 Ger.* Sankt Martin, *Hung.* Turócszentmárton; *prev.* Turčianský Svätý Martin. Žilinský Kraj, N Slovakia
Martinique *33 G4 French overseas department* E West Indies
Martinique Channel *see* Martinique Passage
Martinique Passage *33 G4 var.* Dominica Channel, Martinique Channel. *channel* Dominica/Martinique
Marton *128 D4* Manawatu-Wanganui, North Island, New Zealand
Martos *70 D4* Andalucía, S Spain
Marungu *55 E7 mountain range* SE Dem. Rep. Congo
Mary *100 D3 prev.* Merv. Mary Welaýaty, S Turkmenistan
Maryborough *127 D4* Queensland, E Australia
Maryborough *see* Port Laoise
Mary Island *see* Kanton
Maryland *19 E5 off.* State of Maryland, *also known as* America in Miniature, Cockade State, Free State, Old Line State. *state* NE USA
Maryland, State of *see* Maryland
Maryville *23 F4* Missouri, C USA
Maryville *20 D1* Tennessee, S USA
Masai Steppe *51 C7 grassland* NW Tanzania
Masaka *51 B6* SW Uganda
Masallı *95 H3 Rus.* Masally. S Azerbaijan
Masally *see* Masallı
Masasi *51 C8* Mtwara, SE Tanzania
Masawa/Massawa *see* Mits'iwa
Masaya *30 D3* Masaya, W Nicaragua
Mascarene Basin *119 B5 undersea basin* W Indian Ocean
Mascarene Islands *57 H4 island group* W Indian Ocean
Mascarene Plain *119 B5 abyssal plain* W Indian Ocean
Mascarene Plateau *119 B5 undersea plateau* W Indian Ocean
Maseru *56 D4 country capital* (Lesotho) W Lesotho
Mas-ha *97 D7* W West Bank Asia
Mashhad *98 E2 var.* Meshed. Khorāsān-Razavī, NE Iran
Masindi *51 B6* W Uganda
Masīra *see* Maşīrah, Jazīrat
Masira, Gulf of *see* Maşīrah, Khalīj
Maşīrah, Jazīrat *99 E5 var.* Masīra. *island* E Oman
Maşīrah, Khalīj *99 E5 var.* Gulf of Masira. *bay* E Oman
Masis *see* Büyükağrı Dağı
Maskat *see* Masqaţ
Mason City *23 F3* Iowa, C USA
Masqaţ *99 E5 var.* Maskat, *Eng.* Muscat. *country capital* (Oman) NE Oman
Massa *74 B3* Toscana, C Italy
Massachusetts *19 G3 off.* Commonwealth of Massachusetts, *also known as* Bay State, Old Bay State, Old Colony State. *state* NE USA
Massenya *54 B3* Chari-Baguirmi, SW Chad
Massif Central *69 C5 plateau* C France
Massilia *see* Marseille
Massoukou *see* Franceville
Mastanli *see* Momchilgrad
Masterton *129 D5* Wellington, North Island, New Zealand
Masty *85 B5 Rus.* Mosty. Hrodzyenskaya Voblasts', W Belarus
Masuda *109 B6* Shimane, Honshū, SW Japan
Masuku *see* Franceville
Masvingo *56 D3 prev.* Fort Victoria, Nyanda, Victoria. Masvingo, SE Zimbabwe
Maşyāf *96 B3 Fr.* Misiaf. Ḩamāh, C Syria
Matadi *55 B6* Bas-Congo, W Dem. Rep. Congo
Matagalpa *30 D3* Matagalpa, C Nicaragua
Matale *110 D3* Central Province, C Sri Lanka
Matam *52 C3* NE Senegal
Matamata *128 D3* Waikato, North Island, New Zealand
Matamoros *28 D3* Coahuila, NE Mexico
Matamoros *29 E2* Tamaulipas, C Mexico
Matane *17 E4* Québec, SE Canada
Matanzas *32 B2* Matanzas, NW Cuba
Matara *110 D4* Southern Province, S Sri Lanka
Mataram *116 D5* Pulau Lombok, C Indonesia
Mataró *71 G2 anc.* Illuro. Cataluña, E Spain
Mataura *129 B7* Southland, South Island, New Zealand
Mataura *129 B7 river* South Island, New Zealand
Mata Uta *see* Matā'utu
Matā'utu *123 E4 var.* Mata Uta. *dependent territory capital* (Wallis and Futuna) Île Uvea, Wallis and Futuna
Matera *75 E5* Basilicata, S Italy
Mathurai *see* Madurai
Matianus *see* Orūmīyeh, Daryācheh-ye
Matías Romero *29 F5* Oaxaca, SE Mexico
Matisco/Matisco Ædourum *see* Mâcon
Mato Grosso *41 E3 off.* Estado de Mato Grosso; *prev.* Matto Grosso. *state* W Brazil
Mato Grosso *41 E3 off.* Estado de Mato Grosso; *prev.* Matto Grosso. *region* W Brazil
Mato Grosso do Sul *41 E4 off.* Estado de Mato Grosso do Sul. *region* S Brazil
Mato Grosso do Sul *41 E4 off.* Estado de Mato Grosso do Sul. *state* S Brazil
Mato Grosso do Sul, Estado de *see* Mato Grosso do Sul
Mato Grosso, Estado de *see* Mato Grosso
Mato Grosso, Planalto de *34 C4 plateau* C Brazil
Matosinhos *70 B2 prev.* Matozinhos. Porto, NW Portugal
Matozinhos *see* Matosinhos
Matsue *109 B6 var.* Matsuye, Matue. Shimane, Honshū, SW Japan
Matsumoto *109 C5 var.* Matumoto. Nagano, Honshū, S Japan
Matsuyama *109 B7 var.* Matuyama. Ehime, Shikoku, SW Japan
Matsuye *see* Matsue
Matterhorn *73 A8 It.* Monte Cervino. *mountain* Italy/Switzerland
Matthews Ridge *37 F2* N Guyana
Matthew Town *32 D2* Great Inagua, S Bahamas
Matto Grosso *see* Mato Grosso
Matucana *38 C4* Lima, W Peru
Matue *see* Matsue
Matumoto *see* Matsumoto
Maturín *37 E2* Monagas, NE Venezuela
Matuyama *see* Matsuyama
Mau *113 E3 var.* Maunāth Bhanjan. Uttar Pradesh, N India
Maui *25 B8 island* Hawaii, USA, C Pacific Ocean
Maun *56 C3* North-West, C Botswana
Maunāth Bhanjan *see* Mau
Mauren *72 E1* NE Liechtenstein Europe
Maurice *see* Mauritius
Mauritania *52 C2 off.* Islamic Republic of Mauritania, *Ar.* Mūrītānīyah. *country* W Africa
Mauritania, Islamic Republic of *see* Mauritania
Mauritius *57 H3 off.* Republic of Mauritius, *Fr.* Maurice. *country* W Indian Ocean
Mauritius *119 B5 island* W Indian Ocean
Mauritius, Republic of *see* Mauritius
Mawlamyaing *see* Mawlamyine
Mawlamyine *114 B4 var.* Mawlamyaing, Moulmein. Mon State, S Myanmar (Burma)
Mawson *132 D2 Australian research station* Antarctica
Mayadin *see* Al Mayādīn
Mayaguana *32 D2 island* SE Bahamas
Mayaguana Passage *32 D2 passage* SE Bahamas
Mayagüez *33 F3* W Puerto Rico
Mayamey *98 D2* Semnān, N Iran
Maya Mountains *30 B2 Sp.* Montañas Mayas. *mountain range* Belize/Guatemala
Mayas, Montañas *see* Maya Mountains
Maych'ew *50 C4 var.* Mai Chio, *It.* Mai Ceu. Tigray, N Ethiopia
Mayebashi *see* Maebashi
Mayence *see* Mainz
Mayfield *129 B6* Canterbury, South Island, New Zealand
Maykop *89 A7* Respublika Adygeya, SW Russian Federation
Maymana *see* Maīmanah
Maymyo *see* Pyin-Oo-Lwin
Mayo *see* Maio
Mayor Island *128 D3 island* NE New Zealand
Mayor Pablo Lagerenza *see* Capitán Pablo Lagerenza
Mayotte *57 F2 French territorial collectivity* E Africa
May Pen *32 B5* C Jamaica
Mayyit, Al Baḩr al *see* Dead Sea
Mazabuka *56 D2* Southern, S Zambia
Mazaca *see* Kayseri
Mazagan *see* El-Jadida
Mazār-e Sharīf *101 E3 var.* Mazār-i Sharif. Balkh, N Afghanistan
Mazār-i Sharif *see* Mazār-e Sharīf
Mazatlán *28 C3* Sinaloa, C Mexico
Mažeikiai *84 B3* Telšiai, NW Lithuania
Mazirbe *84 C2* Talsi, NW Latvia
Mazra'a *see* Al Mazra'ah
Mazury *76 D3 physical region* NE Poland
Mazyr *85 C7 Rus.* Mozyr'. Homyel'skaya Voblasts', SE Belarus
Mbabane *56 D4 country capital* (Swaziland) NW Swaziland
Mbacké *see* Mbaké
Mbaïki *55 C5 var.* M'Baiki. Lobaye, SW Central African Republic
M'Baiki *see* Mbaïki
Mbaké *52 B3 var.* Mbacké. W Senegal
Mbala *56 D1 prev.* Abercorn. Northern, NE Zambia
Mbale *51 C6* E Uganda
Mbandaka *55 C5 prev.* Coquilhatville. Equateur, NW Dem. Rep. Congo
M'Banza Congo *56 B1 var.* Mbanza Congo; *prev.* São Salvador, São Salvador do Congo. Dem. Rep. Congo, NW Angola
Mbanza-Ngungu *55 B6* Bas-Congo, W Dem. Rep. Congo
Mbarara *51 B6* SW Uganda
Mbé *54 B4* Nord, N Cameroon
Mbeya *51 C7* Mbeya, SW Tanzania
Mbomou/M'Bomu/Mbomu *see* Bomu
Mbour *52 B3* W Senegal
Mbuji-Mayi *55 D7 prev.* Bakwanga. Kasai-Oriental, S Dem. Rep. Congo
McAlester *27 G2* Oklahoma, C USA
McAllen *27 G5* Texas, SW USA
McCamey *27 E3* Texas, SW USA
M'Clintock Channel *15 F2 channel* Nunavut, N Canada
McComb *20 B3* Mississippi, S USA
McCook *23 E4* Nebraska, C USA
McKean Island *123 E3 island* Phoenix Islands, C Kiribati
Mount McKinley *14 C3 var.* Denali. *mountain* Alaska, USA
McKinley Park *14 C3* Alaska, USA
McMinnville *24 B3* Oregon, NW USA
McMurdo *132 B4 US research station* Antarctica
McPherson *23 E5* Kansas, C USA
McPherson *see* Fort McPherson
Mdantsane *56 D5* Eastern Cape, SE South Africa
Mead, Lake *25 D6 reservoir* Arizona/Nevada, W USA
Mecca *see* Makkah
Mechelen *65 C5 Eng.* Mechlin, *Fr.* Malines. Antwerpen, C Belgium
Mechlin *see* Mechelen
Mecklenburger Bucht *72 C2 bay* N Germany
Mecsek *77 C7 mountain range* SW Hungary
Medan *116 B3* Sumatera, E Indonesia
Medeba *see* Ma'dabā
Medellín *36 B3* Antioquia, NW Colombia
Médenine *49 F2 var.* Madanīyīn. SE Tunisia
Medeshamstede *see* Peterborough
Medford *24 B4* Oregon, NW USA
Medgidia *86 D5* Constanța, SE Romania
Medgyes *see* Mediaș
Mediaș *86 B4 Ger.* Mediasch, *Hung.* Medgyes. Sibiu, C Romania
Mediasch *see* Mediaș
Medicine Hat *15 F5* Alberta, SW Canada
Medina *see* Al Madīnah
Medinaceli *71 E2* Castilla-León, N Spain
Medina del Campo *70 D2* Castilla-León, N Spain
Medinat Israel *see* Israel
Mediolanum *see* Saintes, France
Mediolanum *see* Milano, Italy
Mediomatrica *see* Metz
Mediterranean Sea *80 D3 Fr.* Mer Méditerranée. *sea* Africa/Asia/Europe
Méditerranée, Mer *see* Mediterranean Sea
Médoc *69 B5 cultural region* SW France
Medvezh'yegorsk *88 B3* Respublika Kareliya, NW Russian Federation
Meekatharra *125 B5* Western Australia
Meemu Atoll *see* Mulakatholhu
Meerssen *65 D6 var.* Mersen. Limburg, SE Netherlands
Meerut *112 D2* Uttar Pradesh, N India
Megáli Préspa, Límni *see* Prespa, Lake
Meghālaya *91 G3 cultural region* NE India
Mehdia *see* Mahdia
Meheso *see* Mī'ēso
Me Hka *see* Nmai Hka
Mehrīz *98 D3* Yazd, C Iran
Mehtar Lām *101 F4 var.* Mehtarlām, Meterlam, Methariam, Metharlam. Laghmān, E Afghanistan
Mehtarlām *see* Mehtar Lām
Meiktila *114 B3* Mandalay, C Myanmar (Burma)
Méjico *see* Mexico
Mejillones *42 B2* Antofagasta, N Chile
Mek'elē *50 C4 var.* Makale. Tigray, N Ethiopia
Mékhé *52 B3* NW Senegal
Mekong *102 D3 var.* Lan-ts'ang Chiang, *Cam.* Mékôngk, *Chin.* Lancang Jiang, *Lao.* Mènam Khong, *Th.* Mae Nam Khong, *Tib.* Dza Chu, *Vtn.* Sông Tiền Giang. *river* SE Asia
Mékôngk *see* Mekong
Mekong, Mouths of the *115 E6 delta* S Vietnam
Melaka *116 B3 var.* Malacca. Melaka, Peninsular Malaysia
Melaka, Selat *see* Malacca, Strait of
Melanesia *122 D3 island group* W Pacific Ocean
Melanesian Basin *120 C2 undersea basin* W Pacific Ocean
Melbourne *127 C7 state capital* Victoria, SE Australia
Melbourne *21 E4* Florida, SE USA
Meleda *see* Mljet
Melghir, Chott *49 E2 var.* Chott Melrhir. *salt lake* E Algeria
Melilla *58 B5 anc.* Rusaddir, Russadir. Melilla, Spain, N Africa
Melilla *48 D2 enclave* Spain, N Africa
Melita *15 F5* Manitoba, S Canada
Melita *see* Mljet
Melitene *see* Malatya
Melitopol' *87 F4* Zaporiz'ka Oblast', SE Ukraine
Melle *65 B5* Oost-Vlaanderen, NW Belgium
Mellerud *63 B6* Västra Götaland, S Sweden
Mellieħa *80 B5* E Malta
Mellizo Sur, Cerro *43 A7 mountain* S Chile
Melo *42 E4* Cerro Largo, NE Uruguay
Melodunum *see* Melun
Melrhir, Chott *see* Melghir, Chott
Melsungen *72 B4* Hessen, C Germany
Melun *68 C3 anc.* Melodunum. Seine-et-Marne, N France
Melville Bay/Melville Bugt *see* Qimusseriarsuaq
Melville Island *124 D2 island* Northern Territory, N Australia
Melville Island *15 E2 island* Parry Islands, Northwest Territories, NW Canada
Melville, Lake *17 F2 lake* Newfoundland and Labrador, E Canada
Melville Peninsula *15 G3 peninsula* Nunavut, NE Canada
Melville Sound *see* Viscount Melville Sound
Membidj *see* Manbij
Memel *see* Neman, NE Europe
Memel *see* Klaipeda, Lithuania
Memmingen *73 B6* Bayern, S Germany
Memphis *20 C1* Tennessee, S USA
Menaam *see* Menaldum
Menado *see* Manado
Ménaka *53 F3* Goa, E Mali
Menaldum *64 D1 Fris.* Menaam. Friesland, N Netherlands
Mènam Khong *see* Mekong
Mendaña Fracture Zone *131 F4 fracture zone* E Pacific Ocean
Mende *69 C5 anc.* Mimatum. Lozère, S France
Mendeleyev Ridge *133 B2 undersea ridge* Arctic Ocean
Mendocino Fracture Zone *130 D2 fracture zone* NE Pacific Ocean
Mendoza *42 B4* Mendoza, W Argentina
Menemen *94 A3* İzmir, W Turkey
Menengiyn Tal *105 F2 plain* E Mongolia
Menongue *56 B2 var.* Vila Serpa Pinto, *Port.* Serpa Pinto. Cuando Cubango, C Angola
Menorca *71 H3 Eng.* Minorca; *anc.* Balearis Minor. *island* Islas Baleares, Spain, W Mediterranean Sea
Mentawai, Kepulauan *116 A4 island group* W Indonesia
Meppel *64 D2* Drenthe, NE Netherlands
Meran *see* Merano
Merano *74 C1 Ger.* Meran. Trentino-Alto Adige, N Italy
Merca *see* Marka
Mercedes *42 D3* Corrientes, NE Argentina
Mercedes *42 D4* Soriano, SW Uruguay
Meredith, Lake *27 E1 reservoir* Texas, SW USA
Merefa *87 G2* Kharkivs'ka Oblast', E Ukraine
Mergate *see* Margate
Mergui *see* Myeik
Mergui Archipelago *see* Myeik Archipelago
Mérida *29 H3* Yucatán, SW Mexico
Mérida *70 C4 anc.* Augusta Emerita. Extremadura, W Spain
Mérida *36 C2* Mérida, W Venezuela
Meridian *20 C2* Mississippi, S USA
Mérignac *69 B5* Gironde, SW France
Merín, Laguna *see* Mirim Lagoon
Merkulovichi *see* Myerkulavichy
Merowe *50 B3* Northern, N Sudan
Merredin *125 B6* Western Australia
Mersen *see* Meerssen
Mersey *67 D5 river* NW England, United Kingdom
Mersin *94 C4 var.* İçel. İçel, S Turkey
Mērsrags *84 C3* Talsi, NW Latvia
Meru *51 C6* Eastern, C Kenya
Merv *see* Mary
Merzifon *94 D2* Amasya, N Turkey
Merzig *73 A5* Saarland, SW Germany
Mesa *26 B2* Arizona, SW USA
Meseritz *see* Międzyrzecz
Meshed *see* Mashhad
Mesopotamia *35 C5 var.* Mesopotamia Argentina. *physical region* NE Argentina
Mesopotamia Argentina *see* Mesopotamia
Messalo, Rio *57 E2 var.* Mualo. *river* NE Mozambique
Messana/Messene *see* Messina
Messina *see* Musina
Messina, Strait of *see* Messina, Stretto di
Messina, Stretto di *75 D7 Eng.* Strait of Messina. *strait* SW Italy
Messíni *83 B6* Pelopónnisos, S Greece
Mesta *see* Néstos
Mestghanem *see* Mostaganem
Mestia *95 F1 var.* Mestiya. N Georgia
Mestiya *see* Mestia
Mestre *74 C2* Veneto, NE Italy
Metairie *20 B3* Louisiana, S USA
Metán *42 C2* Salta, N Argentina
Metapán *30 B2* Santa Ana, NW El Salvador
Meta, Río *36 D3 river* Colombia/Venezuela
Meterlam *see* Mehtar Lām
Methariam/Metharlam *see* Mehtar Lām
Metis *see* Metz
Metković *78 B4* Dubrovnik-Neretva, SE Croatia
Métsovo *82 B4 prev.* Métsovon. Ipeiros, C Greece
Métsovon *see* Métsovo
Metz *68 D3 anc.* Divodurum Mediomatricum, Mediomatrica, Metis. Moselle, NE France
Meulaboh *116 A3* Sumatera, W Indonesia
Meuse *65 C6 Dut.* Maas. *river* W Europe
Mexcala, Río *see* Balsas, Río
Mexicali *28 A1* Baja California Norte, NW Mexico
Mexicanos, Estados Unidos *see* Mexico
México *29 E4 var.* Ciudad de México, *Eng.* Mexico City. *country capital* (Mexico) México, C Mexico
Mexico *23 G4* Missouri, C USA
Mexico *28 C3 off.* United Mexican States, *var.* Méjico, México, *Sp.* Estados Unidos Mexicanos. *country* N Central America
México *see* Mexico
Mexico City *see* México
México, Golfo de *see* Mexico, Gulf of
Mexico, Gulf of *29 F2 Sp.* Golfo de México. *gulf* W Atlantic Ocean
Meyadine *see* Al Mayādīn
Meydān Shahr *see* Maīdān Shahr
Meymaneh *see* Maīmanah
Mezen' *88 D3 river* NW Russian Federation
Mezőtúr *77 D7* Jász-Nagykun-Szolnok, E Hungary
Mġarr *80 A5* Gozo, N Malta
Miadzioł Nowy *see* Myadzyel
Miahuatlán *29 F5 var.* Miahuatlán de Porfirio Díaz. Oaxaca, SE Mexico
Miahuatlán de Porfirio Díaz *see* Miahuatlán
Miami *21 F5* Florida, SE USA
Miami *27 G1* Oklahoma, C USA
Miami Beach *21 F5* Florida, SE USA
Mīāneh *98 C2 var.* Miyāneh. Āžarbāyjān-e Sharqī, NW Iran
Mianyang *106 B5* Sichuan, C China
Miastko *76 C2 Ger.* Rummelsburg in Pommern. Pomorskie, N Poland
Mi Chai *see* Nong Khai
Michalovce *77 E5 Ger.* Grossmichel, *Hung.* Nagymihály. Košický Kraj, E Slovakia
Michigan *18 C1 off.* State of Michigan, *also known as* Great Lakes State, Lake State, Wolverine State. *state* N USA
Michigan, Lake *18 C2 lake* N USA
Michurin *see* Tsarevo
Michurinsk *89 B5* Tambovskaya Oblast', W Russian Federation
Micoud *33 F2* SE Saint Lucia
Micronesia *122 B1 off.* Federated States of Micronesia. *country* W Pacific Ocean
Micronesia *122 C1 island group* W Pacific Ocean
Micronesia, Federated States of *see* Micronesia
Mid-Atlantic Cordillera *see* Mid-Atlantic Ridge
Mid-Atlantic Ridge *44 C3 var.* Mid-Atlantic Cordillera, Mid-Atlantic Rise, Mid-Atlantic Swell. *undersea ridge* Atlantic Ocean
Mid-Atlantic Rise/Mid-Atlantic Swell *see* Mid-Atlantic Ridge
Middelburg *65 B5* Zeeland, SW Netherlands
Middelharnis *64 B4* Zuid-Holland, SW Netherlands
Middelkerke *65 A5* West-Vlaanderen, W Belgium
Middle America Trench *13 B7 trench* E Pacific Ocean
Middle Andaman *111 F2 island* Andaman Islands, India, NE Indian Ocean
Middle Atlas *48 C2 Eng.* Middle Atlas. *mountain range* N Morocco
Middle Atlas *see* Moyen Atlas
Middleburg Island *see* 'Eua
Middle Congo *see* Congo (Republic of)
Middlesboro *18 C5* Kentucky, S USA
Middlesbrough *67 D5* N England, United Kingdom
Middletown *19 F4* New Jersey, NE USA
Middletown *19 F3* New York, NE USA
Mid-Indian Basin *119 C5 undersea basin* N Indian Ocean
Mid-Indian Ridge *119 C5 var.* Central Indian Ridge. *undersea ridge* C Indian Ocean
Midland *16 D5* Ontario, S Canada
Midland *18 C3* Michigan, N USA
Midland *27 E3* Texas, SW USA
Mid-Pacific Mountains *130 C2 var.* Mid-Pacific Seamounts. *seamount range* NW Pacific Ocean
Mid-Pacific Seamounts *see* Mid-Pacific Mountains
Midway Islands *130 D2 US territory* C Pacific Ocean
Miechów *77 D5* Małopolskie, S Poland
Międzyrzec Podlaski *76 E3* Lubelskie, E Poland
Międzyrzecz *76 B3 Ger.* Meseritz. Lubuskie, W Poland
Mielec *77 D5* Podkarpackie, SE Poland
Miercurea-Ciuc *86 C4 Ger.* Szeklerburg, *Hung.* Csíkszereda. Harghita, C Romania
Mieres del Camín *70 D1 var.* Mieres del Camino. Asturias, NW Spain
Mieres del Camino *see* Mieres del Camín
Mī'ēso *51 D5 var.* Meheso, Miesso. Oromīya, C Ethiopia
Miesso *see* Mī'ēso
Mifrats Hefa *97 A5 Eng.* Bay of Haifa; *prev.* MifrazHefa. *bay* N Israel
Miguel Asua *28 D3 var.* Miguel Auza. Zacatecas, C Mexico
Miguel Auza *see* Miguel Asua
Mijdrecht *64 C3* Utrecht, C Netherlands
Mikashevichi *see* Mikashevichy
Mikashevichy *85 C7 Pol.* Mikaszewicze, *Rus.* Mikashevichi. Brestskaya Voblasts', SW Belarus
Mikaszewicze *see* Mikashevichy
Mikhaylovgrad *see* Montana
Mikhaylovka *89 B6* Volgogradskaya Oblast', SW Russian Federation
Míkonos *see* Mýkonos
Mikre *82 C2* Lovech, N Bulgaria
Mikun' *88 D4* Respublika Komi, NW Russian Federation
Mikuni-sanmyaku *109 D5 mountain range* Honshū, N Japan Asia
Mikura-jima *109 D6 island* E Japan
Milagro *38 B2* Guayas, SW Ecuador
Milan *see* Milano
Milange *57 E2* Zambézia, NE Mozambique
Milano *74 B2 Eng.* Milan, *Ger.* Mailand; *anc.* Mediolanum. Lombardia, N Italy
Milas *94 A4* Muğla, SW Turkey
Milashavichy *85 C7 Rus.* Milashevichi. Homyel'skaya Voblasts', SE Belarus
Milashevichi *see* Milashavichy
Mildura *127 C6* Victoria, SE Australia
Mile *see* Mili Atoll
Miles *127 D5* Queensland, E Australia
Miles City *22 C2* Montana, NW USA
Milford *see* Milford Haven
Milford Haven *67 C6 prev.* Milford. SW Wales, United Kingdom
Milford Sound *129 A6* Southland, South Island, New Zealand
Milford Sound *129 A6 inlet* South Island, New Zealand
Milḩ, Baḩr al *see* Razāzah, Buḩayrat ar
Mili Atoll *122 D2 var.* Mile. *atoll* Ratak Chain, SE Marshall Islands
Mil'kovo *93 H3* Kamchatskaya Oblast', E Russian Federation
Milk River *15 E5* Alberta, SW Canada
Milk River *22 C1 river* Montana, NW USA
Milk, Wadi el *66 B4 var.* Wadi al Malik. *river* C Sudan
Milledgeville *21 E2* Georgia, SE USA
Mille Lacs Lake *23 F2 lake* Minnesota, N USA
Millennium Island *121 F3 prev.* Caroline Island, Thornton Island. *atoll* Line Islands, E Kiribati
Millerovo *89 B6* Rostovskaya Oblast', SW Russian Federation
Mílos *83 C7 island* Kykládes, Greece, Aegean Sea
Milton *129 B7* Otago, South Island, New Zealand
Milton Keynes *67 D6* SE England, United Kingdom
Milwaukee *18 B3* Wisconsin, N USA
Mimatum *see* Mende
Min *see* Fujian
Mīnā' Qābūs *118 B3* NE Oman
Minas Gerais *41 F3 off.* Estado de Minas Gerais. *state* E Brazil
Minas Gerais *41 F3 off.* Estado de Minas Gerais. *region* E Brazil
Minas Gerais, Estado de *see* Minas Gerais
Minatitlán *29 F4* Veracruz-Llave, E Mexico
Minbu *114 A3* Magway, W Myanmar (Burma)
Minch, The *66 B3 var.* North Minch. *strait* NW Scotland, United Kingdom
Mindanao *117 F2 island* S Philippines
Mindanao Sea *see* Bohol Sea
Mindelheim *73 C6* Bayern, S Germany
Mindello *see* Mindelo
Mindelo *52 A2 var.* Mindello; *prev.* Porto Grande. São Vicente, N Cape Verde
Minden *72 B4 anc.* Minthun. Nordrhein-Westfalen, NW Germany
Mindoro *117 E2 island* N Philippines
Mindoro Strait *117 E2 strait* W Philippines
Mineral Wells *27 F2* Texas, SW USA
Mingäçevir *95 G2 Rus.* Mingechaur, Mingechevir. C Azerbaijan
Mingechaur/Mingechevir *see* Mingäçevir
Mingora *see* Saidu Sharif
Minho *70 B2 former province* N Portugal
Minho *70 B2 Sp.* Miño. *river* Portugal/Spain
Minho, Rio *see* Miño
Minicoy Island *110 B3 island* SW India
Minius *see* Miño
Minna *53 G4* Niger, C Nigeria
Minneapolis *23 F2* Minnesota, N USA
Minnesota *23 F2 off.* State of Minnesota, *also known as* Gopher State, New England of the West, North Star State. *state* N USA

Miño – Mužlja

Miño *70 B2 var.* Mino, Minius, *Port.* Rio Minho. *river* Portugal/Spain
Miño *see* Minho, Rio
Minorca *see* Menorca
Minot *23 E1* North Dakota, N USA
Minsk *85 C6 country capital* (Belarus) Minskaya Voblasts', C Belarus
Minskaya Wzvyshsha *85 C6 mountain range* C Belarus
Mińsk Mazowiecki *76 D3 var.* Nowo-Minsk. Mazowieckie, C Poland
Minthun *see* Minden
Minto, Lac *16 D2 lake* Québec, C Canada
Minya *see* Al Minyā
Miraflores *28 C3* Baja California Sur, NW Mexico
Miranda de Ebro *71 E1* La Rioja, N Spain
Mirgorod *see* Myrhorod
Miri *116 D3* Sarawak, East Malaysia
Mirim Lagoon *41 E5 var.* Lake Mirim, *Sp.* Laguna Merín. *lagoon* Brazil/Uruguay
Mirim, Lake *see* Mirim Lagoon
Mírina *see* Mýrina
Mīrjāveh *98 E4* Sīstān va Balūchestān, SE Iran
Mirny *132 C3 Russian research station* Antarctica
Mirnyy *93 F3* Respublika Sakha (Yakutiya), NE Russian Federation
Mīrpur Khās *112 B3* Sind, SE Pakistan
Mirtoan Sea *see* Mirtóo Pélagos
Mirtóo Pélagos *83 C6 Eng.* Mirtoan Sea; *anc.* Myrtoum Mare. *sea* S Greece
Misiaf *see* Maşyāf
Miskito Coast *see* Mosquito Coast
Miskitos, Cayos *31 E2 island group* NE Nicaragua
Miskolc *77 D6* Borsod-Abaúj-Zemplén, NE Hungary
Misool, Pulau *117 F4 island* Maluku, E Indonesia
Mişrātah *49 F2 var.* Misurata. NW Libya
Mission *27 G5* Texas, SW USA
Mississippi *20 B2 off.* State of Mississippi, *also known as* Bayou State, Magnolia State. *state* SE USA
Mississippi Delta *20 B4 delta* Louisiana, S USA
Mississippi River *13 C6 river* C USA
Missoula *22 B1* Montana, NW USA
Missouri *23 F5 off.* State of Missouri, *also known as* Bullion State, Show Me State. *state* C USA
Missouri River *23 E3 river* C USA
Mistassini, Lac *16 D3 lake* Québec, SE Canada
Mistelbach an der Zaya *73 E6* Niederösterreich, NE Austria
Misti, Volcán *39 E4 volcano* S Peru
Misurata *see* Mişrātah
Mitau *see* Jelgava
Mitchell *127 D5* Queensland, E Australia
Mitchell *23 E3* South Dakota, N USA
Mitchell, Mount *21 E1 mountain* North Carolina, SE USA
Mitchell River *126 C2 river* Queensland, NE Australia
Mi Tho *see* My Tho
Mitilíni *see* Mytilíni
Mito *109 D5* Ibaraki, Honshū, S Japan
Mitrovica *see* Mitrovicë
Mitrovica/Mitrowitz *see* Sremska Mitrovica, Serbia
Mitrovicë *79 D5 Serb.* Mitrovica, *prev.* Kosovska Mitrovica, Titova Mitrovica. N Kosovo
Mits'iwa *50 C4 var.* Masawa, Massawa. E Eritrea
Mitspe Ramon *97 A7 prev.* Mizpe Ramon. Southern, S Israel
Mittelstadt *see* Baia Sprie
Mitú *36 C4* Vaupés, SE Colombia
Mitumba, Chaîne des/Mitumba Range *see* Mitumba, Monts
Mitumba Monts *55 E7 var.* Chaîne des Mitumba, Mitumba Range. *mountain range* E Dem. Rep. Congo
Mieru Wantipa, Lake *55 E7 lake* N Zambia
Miyake-jima *109 D6 island* Sakishima-shotō, SW Japan
Miyako *108 D4* Iwate, Honshū, C Japan
Miyakonojō *109 B8 var.* Miyakonzyô. Miyazaki, Kyūshū, SW Japan
Miyakonzyô *see* Miyakonojō
Miyāneh *see* Miāneh
Miyazaki *109 B8* Miyazaki, Kyūshū, SW Japan
Mizil *86 C5* Prahova, SE Romania
Miziya *82 C1* Vratsa, NW Bulgaria
Mizpe Ramon *see* Mitspe Ramon
Mjøsa *63 B6 var.* Mjøsen. *lake* S Norway
Mjøsen *see* Mjøsa
Mladenovac *78 D4* Serbia, C Serbia
Mława *76 D3* Mazowieckie, C Poland
Mljet *79 B5 It.* Meleda; *anc.* Melita. *island* S Croatia
Mmabatho *56 C4* North-West, N South Africa
Moab *22 B5* Utah, W USA
Moa Island *126 C1 island* Queensland, NE Australia
Moanda *55 B6 var.* Mouanda. Haut-Ogooué, SE Gabon
Moba *55 E7* Katanga, E Dem. Rep. Congo
Mobay *see* Montego Bay
Mobaye *55 C5* Basse-Kotto, S Central African Republic
Moberly *23 G4* Missouri, C USA
Mobile *20 C3* Alabama, S USA
Mobutu Sese Seko, Lac *see* Albert, Lake
Moçâmedes *see* Namibe
Mochudi *56 C4* Kgatleng, SE Botswana
Mocímboa da Praia *57 F2 var.* Vila de Mocímboa da Praia. Cabo Delgado, N Mozambique
Môco *56 B2 var.* Morro de Môco. *mountain* W Angola
Mocoa *36 A4* Putumayo, SW Colombia
Môco, Morro de *see* Môco
Mocuba *57 E3* Zambézia, NE Mozambique
Modena *74 B3 anc.* Mutina. Emilia-Romagna, N Italy
Modesto *25 B6* California, W USA
Modica *75 C7 anc.* Motyca. Sicilia, Italy, C Mediterranean Sea
Modimolle *56 D4 prev.* Nylstroom. Limpopo, NE South Africa
Modohn *see* Madona
Modriča *78 C3* Republika Srpska, N Bosnia and Herzegovina
Moe *127 C7* Victoria, SE Australia
Möen *see* Møn, Denmark
Moero, Lac *see* Mweru, Lake
Moeskroen *see* Mouscron
Mogadiscio/Mogadishu *see* Muqdisho
Mogador *see* Essaouira
Mogilëv *see* Mahilyow
Mogilev-Podol'skiy *see* Mohyliv-Podil's'kyy
Mogilno *76 C3* Kujawsko-pomorskie, C Poland
Moḥammadābād-e Rīgān *98 E4* Kermān, SE Iran
Mohammedia *48 C2 prev.* Fédala. NW Morocco
Mohave, Lake *25 D7 reservoir* Arizona/Nevada, W USA
Mohawk River *19 F3 river* New York, NE USA
Mohéli *see* Mwali
Mohns Ridge *61 F3 undersea ridge* Greenland Sea/Norwegian Sea
Moho *39 E4* Puno, SE Peru
Mohoro *51 C7* Pwani, E Tanzania
Mohyliv-Podil's'kyy *86 D3 Rus.* Mogilev-Podol'skiy. Vinnyts'ka Oblast', C Ukraine
Moi *63 A6* Rogaland, S Norway
Moili *see* Mwali
Mo i Rana *62 C3* Nordland, C Norway
Mõisaküla *84 D3 Ger.* Moiseküll. Viljandimaa, S Estonia
Moiseküll *see* Mõisaküla
Moissac *69 B6* Tarn-et-Garonne, S France
Mojácar *71 E5* Andalucía, S Spain
Mojave Desert *25 D7 plain* California, W USA
Mokrany *see* Makrany
Moktama *see* Mottama
Mol *65 C5 prev.* Moll. Antwerpen, N Belgium
Moldavia *see* Moldova
Moldavian SSR/Moldavskaya SSR *see* Moldova
Molde *63 A5* Møre og Romsdal, S Norway
Moldotau, Khrebet *see* Moldo-Too, Khrebet
Moldo-Too, Khrebet *101 G2 prev.* Khrebet Moldotau. *mountain range* C Kyrgyzstan
Moldova *86 D3 off.* Republic of Moldova, *var.* Moldavia; *prev.* Moldavian SSR, *Rus.* Moldavskaya SSR. *country* SE Europe
Moldova Nouă *86 A4 Ger.* Neumoldowa, *Hung.* Újmoldova. Caraș-Severin, SW Romania
Moldova, Republic of *see* Moldova
Moldoveanul *see* Vârful Moldoveanu
Molfetta *75 E5* Puglia, SE Italy
Moll *see* Mol
Mollendo *39 E4* Arequipa, SW Peru
Mölndal *63 B7* Västra Götaland, S Sweden
Molochans'k *87 G4 Rus.* Molochansk. Zaporiz'ka Oblast', SE Ukraine
Molodechno/Molodeczno *see* Maladzyechna
Molodezhnaya *132 C2 Russian research station* Antarctica
Moloka'i *25 B8 var.* Molokai. *island* Hawai'ian Islands, Hawaii, USA
Molokai Fracture Zone *131 E2 tectonic feature* NE Pacific Ocean
Molopo *56 C4 seasonal river* Botswana/South Africa
Mólos *83 B5* Stereá Ellás, C Greece
Molotov *see* Severodvinsk, Arkhangel'skaya Oblast', Russian Federation
Molotov *see* Perm', Permskaya Oblast', Russian Federation
Moluccas *see* Maluku
Molucca Sea *117 F4 Ind.* Laut Maluku. *sea* E Indonesia
Molukken *see* Maluku
Mombasa *51 D7* Coast, SE Kenya
Mombetsu *see* Monbetsu
Momchilgrad *82 D3 prev.* Mastanli. Kŭrdzhali, S Bulgaria
Møn *63 B8 prev.* Möen. *island* SE Denmark
Mona, Canal de la *see* Mona Passage
Monaco *69 C7 var.* Monaco-Ville; *anc.* Monoecus. *country capital* (Monaco) S Monaco
Monaco *69 E6 off.* Principality of Monaco. *country* W Europe
Monaco *see* München
Monaco, Port de *69 C8 bay* S Monaco W Mediterranean Sea
Monaco, Principality of *see* Monaco
Monaco-Ville *see* Monaco
Monahans *27 E3* Texas, SW USA
Mona, Isla *33 E3 island* W Puerto Rico
Mona Passage *33 E3 Sp.* Canal de la Mona. *channel* Dominican Republic/Puerto Rico
Monastir *see* Bitola
Monbetsu *108 D2 var.* Mombetsu, Monbetu. Hokkaidō, NE Japan
Monbetu *see* Monbetsu
Moncalieri *74 A2* Piemonte, NW Italy
Monchegorsk *88 C2* Murmanskaya Oblast', NW Russian Federation
Monclova *28 D2* Coahuila, NE Mexico
Moncton *17 F4* New Brunswick, SE Canada
Mondovì *74 A2* Piemonte, NW Italy
Monfalcone *74 D2* Friuli-Venezia Giulia, NE Italy
Monforte de Lemos *70 C1* Galicia, NW Spain
Mongo *54 C3* Guéra, C Chad
Mongolia *104 C2 Mong.* Mongol Uls. *country* E Asia
Mongolia, Plateau of *102 D1 plateau* E Mongolia
Mongol Uls *see* Mongolia
Mongora *see* Saidu Sharif
Mongos, Chaîne des *see* Bongo, Massif des
Mongu *56 C2* Western, W Zambia
Monkchester *see* Newcastle upon Tyne
Monkey Bay *57 E2* Southern, SE Malawi
Monkey River *see* Monkey River Town
Monkey River Town *30 C2 var.* Monkey River. Toledo, SE Belize
Monoecus *see* Monaco
Mono Lake *25 C6 lake* California, W USA
Monostor *see* Beli Manastir
Monóvar *71 F4 Cat.* Monòver. País Valenciano, E Spain
Monover *see* Monovar
Monroe *20 B2* Louisiana, S USA
Monrovia *52 C5 country capital* (Liberia) W Liberia
Mons *65 B6 Dut.* Bergen. Hainaut, S Belgium
Monselice *74 C2* Veneto, NE Italy
Montana *82 C2 prev.* Ferdinand, Mikhaylovgrad. Montana, NW Bulgaria
Montana *22 B1 off.* State of Montana, *also known as* Mountain State, Treasure State. *state* NW USA
Montargis *68 C4* Loiret, C France
Montauban *69 B6* Tarn-et-Garonne, S France
Montbéliard *68 D4* Doubs, E France
Mont Cenis, Col du *69 D5 pass* E France
Mont-de-Marsan *69 B6* Landes, SW France
Monteagudo *39 G4* Chuquisaca, S Bolivia
Montecarlo *69 C8* Misiones, NE Argentina
Monte Caseros *42 D3* Corrientes, NE Argentina
Monte Cristi *32 D3 var.* San Fernando de Monte Cristi. NW Dominican Republic
Monte Croce Carnico, Passo di *see* Plöcken Pass
Montegiardino *74 E2* SE San Marino
Montego Bay *32 A4 var.* Mobay. W Jamaica
Montélimar *69 D5 anc.* Acunum Acusio, Montilium Adhemari. Drôme, E France
Montemorelos *29 E3* Nuevo León, NE Mexico
Montenegro *79 C5 Serb.* Crna Gora. *country* SW Europe
Monte Patria *42 B3* Coquimbo, N Chile
Monterey *25 B6* California, W USA
Monterey *see* Monterrey
Monterey Bay *25 A6 bay* California, W USA
Montería *36 B2* Córdoba, NW Colombia
Montero *39 G4* Santa Cruz, C Bolivia
Monterrey *29 E3 var.* Monterey. Nuevo León, NE Mexico
Montes Claros *41 F3* Minas Gerais, SE Brazil
Montevideo *42 D4 country capital* (Uruguay) Montevideo, S Uruguay
Montevideo *23 F2* Minnesota, N USA
Montgenèvre, Col de *69 D5 pass* France/Italy
Montgomery *20 D2 state capital* Alabama, S USA
Montgomery *see* Sāhīwāl
Monthey *73 A7* Valais, SW Switzerland
Montilium Adhemari *see* Montélimar
Montluçon *68 C4* Allier, C France
Montoro *70 D4* Andalucía, S Spain
Montpelier *19 G2 state capital* Vermont, NE USA
Montpellier *69 C6* Hérault, S France
Montréal *17 E4 Eng.* Montreal. Québec, SE Canada
Montrose *66 D3* E Scotland, United Kingdom
Montrose *22 C5* Colorado, C USA
Montserrat *33 G3 var.* Emerald Isle. *UK dependent territory* E West Indies
Monywa *114 B3* Sagaing, C Myanmar (Burma)
Monza *74 B2* Lombardia, N Italy
Monze *56 D2* Southern, S Zambia
Monzón *71 F2* Aragón, NE Spain
Moonie *127 D5* Queensland, E Australia
Moon-Sund *see* Väinameri
Moora *125 A6* Western Australia
Moore *27 G1* Oklahoma, C USA
Moore, Lake *125 B6 lake* Western Australia
Moorhead *23 F2* Minnesota, N USA
Moose *16 C3 river* Ontario, S Canada
Moosehead Lake *19 G1 lake* Maine, NE USA
Moosonee *16 C3* Ontario, SE Canada
Mopti *53 E3* Mopti, C Mali
Moquegua *39 E4* Moquegua, SE Peru
Mora *63 C5* Dalarna, C Sweden
Morales *30 C2* Izabal, E Guatemala
Morant Bay *32 B5* E Jamaica
Moratalla *71 E4* Murcia, SE Spain
Morava *77 C5 var.* March. *river* C Europe
Morava *see* Moravia, Czech Republic
Morava *see* Velika Morava, Serbia
Moravia *77 B5 Cz.* Morava, *Ger.* Mähren. *cultural region* E Czech Republic
Moray Firth *66 C3 inlet* N Scotland, United Kingdom
Morea *see* Pelopónnisos
Moreau River *22 D2 river* South Dakota, N USA
Moree *127 D5* New South Wales, SE Australia
Morelia *29 E4* Michoacán, S Mexico
Morena, Sierra *70 C4 mountain range* S Spain
Moreni *86 C5* Dâmbovița, S Romania
Morgan City *20 B3* Louisiana, S USA
Morghab, Darya-ye *100 D3 Rus.* Murgab, Murghab, *Turkm.* Murgap, Murgap Deryasy. *river* Afghanistan/Turkmenistan
Morioka *108 D4* Iwate, Honshū, C Japan
Morlaix *68 A3* Finistère, NW France
Mormon State *see* Utah
Mornington Abyssal Plain *45 A7 abyssal plain* SE Pacific Ocean
Mornington Island *126 B2 island* Wellesley Islands, Queensland, N Australia
Morocco *48 B3 off.* Kingdom of Morocco, *Ar.* Al Mamlakah. *country* N Africa
Morocco *see* Marrakech
Morocco, Kingdom of *see* Morocco
Morogoro *51 C7* Morogoro, E Tanzania
Moro Gulf *117 E3 gulf* S Philippines
Morón *32 C2* Ciego de Ávila, C Cuba
Mörön *104 D2* Hövsgöl, N Mongolia
Morondava *57 F3* Toliara, W Madagascar
Moroni *57 F2 country capital* (Comoros) Grande Comore, NW Comoros
Morotai, Pulau *117 F3 island* Maluku, E Indonesia
Morotiri *see* Marotiri
Morphou *see* Güzelyurt
Morrinsville *128 D3* Waikato, North Island, New Zealand
Morris *23 F2* Minnesota, N USA
Morris Jesup, Kap *61 E1 headland* N Greenland
Morvan *68 D4 physical region* C France
Moscow *24 C2* Idaho, NW USA
Moscow *see* Moskva
Mosel *73 A5 Fr.* Moselle. *river* W Europe
Mosel *see* Moselle
Moselle *65 E8 Ger.* Mosel. *river* W Europe
Moselle *see* Mosel
Mosgiel *129 B7* Otago, South Island, New Zealand
Moshi *51 C7* Kilimanjaro, NE Tanzania
Mosjøen *62 B4* Nordland, C Norway
Moskovskiy *see* Moskva
Moskva *89 B5 Eng.* Moscow. *country capital* (Russian Federation) Gorod Moskva, W Russian Federation
Moskva *101 E3 Rus.* Moskovskiy; *prev.* Chubek. SW Tajikistan
Moson and Magyaróvár *see* Mosonmagyaróvár
Mosonmagyaróvár *77 C6 Ger.* Wieselburg-Ungarisch-Altenburg; *prev.* Moson and Magyaróvár, *Ger.* Wieselburg and Ungarisch-Altenburg. Győr-Moson-Sopron, NW Hungary
Mosquito Coast *31 E3 var.* Miskito Coast, *Eng.* Mosquito Coast. *coastal region* E Nicaragua
Mosquito Coast *see* La Mosquitia
Mosquito Gulf *31 F4 Eng.* Mosquito Gulf. *gulf* N Panama
Mosquito Gulf *see* Mosquitos, Golfo de los
Moss *63 B6* Østfold, S Norway
Mossâmedes *see* Namibe
Mosselbaai *56 C5 var.* Mosselbai, *Eng.* Mossel Bay. Western Cape, SW South Africa
Mosselbai/Mossel Bay *see* Mosselbaai
Mossendjo *55 B6* Niari, SW Congo
Mossoró *41 G2* Rio Grande do Norte, NE Brazil
Most *76 A4 Ger.* Brüx. Ústecký Kraj, NW Czech Republic
Mosta *80 B5 var.* Musta. C Malta
Mostaganem *48 D2 var.* Mestghanem. NW Algeria
Mostar *78 C4* Federacija Bosna I Hercegovina, S Bosnia and Herzegovina
Mosty *see* Masty
Mosul *see* Al Mawşil
Mota del Cuervo *71 E3* Castilla-La Mancha, C Spain
Motagua, Río *30 B2 river* Guatemala/Honduras
Mother of Presidents/Mother of States *see* Virginia
Motril *70 D5* Andalucía, S Spain
Motru *86 B4* Gorj, SW Romania
Mottama *114 B4 prev.* Martaban; *var.* Moktama. Mon State, S Myanmar (Burma)
Motueka *129 C5* Tasman, South Island, New Zealand
Motul *29 H3 var.* Motul de Felipe Carrillo Puerto. Yucatán, SE Mexico
Motul de Felipe Carrillo Puerto *see* Motul
Motyca *see* Modica
Mouanda *see* Moanda
Mouhoun *see* Black Volta
Mouila *55 A6* Ngounié, C Gabon
Moukden *see* Shenyang
Mould Bay *15 E2* Prince Patrick Island, Northwest Territories, N Canada
Moulins *68 C4* Allier, C France
Moulmein *see* Mawlamyine
Moundou *54 B4* Logone-Occidental, SW Chad
Moŭng Roessei *115 D5* Bătdâmbâng, W Cambodia
Moun Hou *see* Black Volta
Mountain Home *20 B1* Arkansas, C USA
Mountain State *see* Montana
Mountain State *see* West Virginia
Mount Cook *129 B6* Canterbury, South Island, New Zealand
Mount Desert Island *19 H2 island* Maine, NE USA
Mount Gambier *127 B7* South Australia
Mount Isa *126 B3* Queensland, C Australia
Mount Magnet *125 B5* Western Australia
Mount Pleasant *23 G4* Iowa, C USA
Mount Pleasant *18 C3* Michigan, N USA
Mount Vernon *18 B5* Illinois, N USA
Mount Vernon *24 B1* Washington, NW USA
Mourdi, Dépression du *54 C2 desert lowland* Chad/Sudan
Mouscron *65 A6 Dut.* Moeskroen. Hainaut, W Belgium
Mouse River *see* Souris River
Moussoro *54 B3* Kanem, W Chad
Moyen-Congo *see* Congo (Republic of)
Mo'ynoq *100 C1 Rus.* Muynak. Qoraqalpog'iston Respublikasi, NW Uzbekistan
Moyobamba *38 B2* San Martín, NW Peru
Moyu *104 B3 var.* Karakax. Xinjiang Uygur Zizhiqu, NW China
Moyynkum, Peski *101 F1 Kaz.* Moyynqum. *desert* S Kazakhstan
Moyynqum *see* Moyynkum, Peski
Mozambika, Lakandranon' i *see* Mozambique Channel
Mozambique *57 E3 off.* Republic of Mozambique; *prev.* People's Republic of Mozambique, Portuguese East Africa. *country* S Africa
Mozambique Basin *see* Natal Basin
Mozambique, Canal de *see* Mozambique Channel
Mozambique Channel *57 E3 Fr.* Canal de Mozambique, *Mal.* Lakandranon' i Mozambika. *strait* W Indian Ocean
Mozambique, People's Republic of *see* Mozambique
Mozambique Plateau *47 D7 var.* Mozambique Rise. *undersea plateau* SW Indian Ocean
Mozambique, Republic of *see* Mozambique
Mozambique Rise *see* Mozambique Plateau
Mozyr' *see* Mazyr
Mpama *55 B6 river* C Congo
Mpika *56 D2* Northern, NE Zambia
Mqinvartsveri *see* Kazbek
Mrągowo *76 D2 Ger.* Sensburg. Warmińsko-Mazurskie, NE Poland
Mthatha *56 D5 prev.* Umtata. Eastern Cape, SE South Africa
Mtkvari *see* Kura
Mtwara *51 D8* Mtwara, SE Tanzania
Mualo *see* Messalo, Rio
Muang Chiang Rai *see* Chiang Rai
Muang Kalasin *see* Kalasin
Muang Khammouan *see* Thakhèk
Muang Không *115 D5* Champasak, S Laos
Muang Khôngxédôn *115 D5 var.* Khong Sedone. Salavan, S Laos
Muang Khon Kaen *see* Khon Kaen
Muang Lampang *see* Lampang
Muang Loei *see* Loei
Muang Lom Sak *see* Lom Sak
Muang Nakhon Sawan *see* Nakhon Sawan
Muang Namo *114 C3* Oudômxai, N Laos
Muang Nan *see* Nan
Muang Phalan *114 D4 var.* Muang Phalane. Savannakhét, S Laos
Muang Phalane *see* Muang Phalan
Muang Phayao *see* Phayao
Muang Phitsanulok *see* Phitsanulok
Muang Phrae *see* Phrae
Muang Roi Et *see* Roi Et
Muang Sakon Nakhon *see* Sakon Nakhon
Muang Samut Prakan *see* Samut Prakan
Muang Sing *114 C3* Louang Namtha, N Laos
Muang Ubon *see* Ubon Ratchathani
Muang Xaignabouri *see* Xaignabouli
Muar *116 B3 var.* Bandar Maharani. Johor, Peninsular Malaysia
Mucojo *57 F2* Cabo Delgado, N Mozambique
Mudanjiang *107 E3 var.* Mu-tan-chiang. Heilongjiang, NE China
Mudon *115 B5* Mon State, S Myanmar (Burma)
Muenchen *see* München
Muenster *see* Münster
Mufulira *56 D2* Copperbelt, C Zambia
Mughla *see* Muğla
Muğla *94 A4 var.* Mughla. Muğla, SW Turkey
Mūḩ, Sabkhat al *96 C3 lake* C Syria
Muhu Väin *see* Väinameri
Muisne *38 A1* Esmeraldas, NW Ecuador
Mukacheve *86 B3 Hung.* Munkács, *Rus.* Mukachevo. Zakarpats'ka Oblast', W Ukraine
Mukachevo *see* Mukacheve
Mukalla *see* Al Mukallā
Mukden *see* Shenyang
Mula *71 E4* Murcia, SE Spain
Mulakatholhu *110 B4 var.* Meemu Atoll, Mulaku Atoll. *atoll* C Maldives
Mulaku Atoll *see* Mulakatholhu
Muleshoe *27 E2* Texas, SW USA
Mulhacén *71 E5 var.* Cerro de Mulhacén. *mountain* S Spain
Mulhacén, Cerro de *see* Mulhacén
Mülhausen *see* Mulhouse
Mülheim *73 A6 var.* Mulheim an der Ruhr. Nordrhein-Westfalen, W Germany
Mulheim an der Ruhr *see* Mülheim
Mulhouse *68 E4 Ger.* Mülhausen. Haut-Rhin, NE France
Müller-gerbergte *see* Muller, Pegunungan
Muller, Pegunungan *116 D4 Dut.* Müller-gerbergte. *mountain range* Borneo, C Indonesia
Mull, Isle of *66 B4 island* W Scotland, United Kingdom
Mulongo *55 D7* Katanga, SE Dem. Rep. Congo
Multān *112 C2* Punjab, E Pakistan
Mumbai *112 C5 prev.* Bombay. *state capital* Mahārāshtra, W India
Munamägi *see* Suur Munamägi
Münchberg *73 C5* Bayern, E Germany
München *73 C6 var.* Muenchen, *Eng.* Munich, *It.* Monaco. Bayern, SE Germany
Muncie *18 C4* Indiana, N USA
Mungbere *55 E5* Orientale, NE Dem. Rep. Congo
Mu Nggava *see* Rennell
Munich *see* München
Munkács *see* Mukacheve
Münster *72 A4 var.* Muenster, Münster in Westfalen. Nordrhein-Westfalen, W Germany
Munster *67 A6 Ir.* Cúige Mumhan. *cultural region* S Ireland
Münster in Westfalen *see* Münster
Muong Xiang Ngeun *114 C4 var.* Xieng Ngeun. Louangphabang, N Laos
Muonio *62 D3* Lappi, N Finland
Muonioälv/Muoniojoki *see* Muonionjoki
Muonionjoki *62 D3 var.* Muoniojoki, *Swe.* Muonioälv. *river* Finland/Sweden
Muqāţ *97 C5* Al Mafraq, E Jordan
Muqdisho *51 D6 Eng.* Mogadishu, *It.* Mogadiscio. *country capital* (Somalia) Banaadir, S Somalia
Mur *73 E7 SCr.* Mura. *river* C Europe
Mura *see* Mur
Muradiye *95 F3* Van, E Turkey
Murapara *see* Murupara
Murata *74 E2* S San Marino
Murchison River *125 A5 river* Western Australia
Murcia *71 F4* Murcia, SE Spain
Murcia *71 E4 autonomous community* SE Spain
Mureș *86 A4 river* Hungary/Romania
Murfreesboro *20 D1* Tennessee, S USA
Murgab *see* Morghāb, Daryā-yeb
Murgap *100 D3* Mary Welaýaty, S Turkmenistan
Murgap *see* Morghāb, Daryā-ye
Murgap Deryasy *see* Morghāb, Daryā-ye
Murghab *see* Morghāb, Daryā-ye
Murghob *101 F3 Rus.* Murgab. SE Tajikistan
Murgon *127 E5* Queensland, E Australia
Mūrītānīyah *see* Mauritania
Müritz *72 C3 var.* Müritzee. *lake* NE Germany
Müritzee *see* Müritz
Murmansk *88 C2* Murmanskaya Oblast', NW Russian Federation
Murmashi *88 C2* Murmanskaya Oblast', NW Russian Federation
Murom *89 B5* Vladimirskaya Oblast', W Russian Federation
Muroran *108 D3* Hokkaidō, NE Japan
Muros *70 B1* Galicia, NW Spain
Murray Fracture Zone *131 E2 fracture zone* NE Pacific Ocean
Murray Range *see* Murray Ridge
Murray Ridge *90 C5 var.* Murray Range. *undersea ridge* N Arabian Sea
Murray River *127 B6 river* SE Australia
Murrumbidgee River *127 C6 river* New South Wales, SE Australia
Murska Sobota *73 E7 Ger.* Olsnitz. NE Slovenia
Murupara *128 E3 var.* Murapara. Bay of Plenty, North Island, New Zealand
Murviedro *see* Sagunto
Murwāra *113 E4* Madhya Pradesh, N India
Murwillumbah *127 E5* New South Wales, SE Australia
Murzuq, Edeyin *see* Murzuq, Idhān
Murzuq, Idhan *49 F4 var.* Edeyin Murzuq. *desert* SW Libya
Mürzzuschlag *73 E7* Steiermark, E Austria
Muş *95 F3 var.* Mush. Muş, E Turkey
Musa, Gebel *50 C2 var.* Gebel Mûsa. *mountain* NE Egypt
Mûsa, Gebel *see* Mûsá, Jabal
Musala *82 B3 mountain* W Bulgaria
Muscat *see* Masqaţ
Muscat and Oman *see* Oman
Muscatine *23 G3* Iowa, C USA
Musgrave Ranges *125 D5 mountain range* South Australia
Musina *75 D7 var.* Messana, Messene; *anc.* Zancle. Sicilia, Italy, C Mediterranean Sea
Musina *56 D3 prev.* Messina. Limpopo, NE South Africa
Muskegon *18 C3* Michigan, N USA
Muskogean *see* Tallahassee
Muskogee *27 G1* Oklahoma, C USA
Musoma *51 C6* Mara, N Tanzania
Musta *see* Mosta
Mustafa-Pasha *see* Svilengrad
Musters, Lago *43 B6 lake* S Argentina
Muswellbrook *127 D6* New South Wales, SE Australia
Mut *94 C4* İçel, S Turkey
Mu-tan-chiang *see* Mudanjiang
Mutare *56 D3 var.* Mutari; *prev.* Umtali. Manicaland, E Zimbabwe
Mutari *see* Mutare
Mutina *see* Modena
Mutsu-wan *108 D3 bay* N Japan
Muttonbird Islands *129 A8 island group* SW New Zealand
Mu Us Shadi *105 E3 var.* Ordos Desert; *prev.* Mu Us Shamo. *desert* N China
Mu Us Shamo *see* Mu Us Shadi
Muy Muy *30 D3* Matagalpa, C Nicaragua
Muynak *see* Mo'ynoq
Mužlja *78 D3 Hung.* Felsőmuzslya; *prev.* Gornja Mužlja. Vojvodina, N Serbia

Mwali *57 F2 var.* Moili, Fr. Mohéli. *island* S Comoros

Mwanza *51 B6* Mwanza, NW Tanzania

Mweka *55 C6* Kasai-Occidental, C Dem. Rep. Congo

Mwene-Ditu *55 D7* Kasai-Oriental, S Dem. Rep. Congo

Mweru, Lake *55 D7 var.* Lac Moero. *lake* Dem. Rep. Congo/Zambia

Myadel' *see* Myadzyel

Myadzyel *85 C5 Pol.* Miadzioł Nowy, *Rus.* Myadel'. Minskaya Voblasts', N Belarus

Myanaung *114 B4* Ayeyarwady, SW Myanmar (Burma)

Myanmar *114 A3 off.* Union of Myanmar; *var.* Burma, Myanmar. *country* SE Asia

Myaungmya *114 A4* Ayeyarwady, SW Myanmar (Burma)

Myaydo *see* Aunglan

Myeik *115 B6 var.* Mergui. Tanintharyi, S Myanmar (Burma)

Myeik Archipelago *115 B6 prev.* Mergui Archipelago. *island group* S Myanmar (Burma)

Myerkulavichy *85 D7 Rus.* Merkulovichi. Homyel'skaya Voblasts', SE Belarus

Myingyan *114 B3* Mandalay, C Myanmar (Burma)

Myitkyina *114 B2* Kachin State, N Myanmar (Burma)

Mykolayiv *87 E4 Rus.* Nikolayev. Mykolayivs'ka Oblast', S Ukraine

Mykonos *83 D6 var.* Míkonos. *island* Kykládes, Greece, Aegean Sea

Myrhorod *87 F2 Rus.* Mirgorod. Poltavs'ka Oblast', NE Ukraine

Mýrina *82 D4 var.* Mírina. Límnos, SE Greece

Myrtle Beach *21 F2* South Carolina, SE USA

Mýrtos *83 D8* Kríti, Greece, E Mediterranean Sea

Myrtoum Mare *see* Mirtóo Pélagos

Myślibórz *76 B3* Zachodnio-pomorskie, NW Poland

Mysore *110 C2 var.* Maisur. Karnátaka, W India

Mysore *see* Karnátaka

My Tho *115 E6 var.* Mi Tho. Tiên Giang, S Vietnam

Mytilene *see* Mytilíni

Mytilíni *83 D5 var.* Mitilíni; *anc.* Mytilene. Lésvos, E Greece

Mzuzu *57 E2* Northern, N Malawi

N

Naberezhnyye Chelny *89 D5 prev.* Brezhnev. Respublika Tatarstan, W Russian Federation

Nablus *97 A6 var.* Nábulus, *Heb.* Shekhem; *anc.* Neapolis, *Bibl.* Shechem. N West Bank

Nacala *57 F2* Nampula, NE Mozambique

Na-Ch'ii *see* Nagqu

Nada *see* Danzhou

Nadi *123 E4 prev.* Nandi. Viti Levu, W Fiji

Nadur *80 A5* Gozo, N Malta

Nadvirna *86 C3 Pol.* Nadwórna, *Rus.* Nadvornaya. Ivano-Frankivs'ka Oblast', W Ukraine

Nadvoitsy *88 B3* Respublika Kareliya, NW Russian Federation

Nadvornaya/Nadwórna *see* Nadvirna

Nadym *92 C3* Yamalo-Nenetskiy Avtonomnyy Okrug, N Russian Federation

Náfpaktos *83 B5 var.* Návpaktos. Dytikí Ellás, C Greece

Náfplio *83 B6 prev.* Návplion. Pelopónnisos, S Greece

Naga *117 E2 off.* Naga City; *prev.* Nueva Caceres. Luzon, N Philippines

Naga City *see* Naga

Nagano *109 C5* Nagano, Honshú, S Japan

Nagaoka *109 D5* Niigata, Honshú, C Japan

Nagara Pathom *see* Nakhon Pathom

Nagara Sridharmaraj *see* Nakhon Si Thammarat

Nagara Svarga *see* Nakhon Sawan

Nagasaki *109 A7* Nagasaki, Kyúshú, SW Japan

Nagato *109 A7* Yamaguchi, Honshú, SW Japan

Nágercoil *110 C3* Tamil Nádu, SE India

Nagorno-Karabakh *95 G3 var.* Nagorno-Karabakhskaya Avtonomnaya Oblast, *Arm.* Lerrnayin Gharabakh, *Az.* Dağliq Qarabağ, *Rus.* Nagornyy Karabakh. *former autonomous region* SW Azerbaijan

Nagorno- Karabakhskaya Avtonomnaya Oblast *see* Nagorno-Karabakh

Nagornyy Karabakh *see* Nagorno-Karabakh

Nagoya *109 C6* Aichi, Honshú, SW Japan

Nágpur *112 D4* Mahárashtra, C India

Nagqu *104 C5 Chin.* Na-Ch'ii; *prev.* Hei-ho. Xizang Zizhiqu, W China

Nagybánya *see* Baia Mare

Nagybecskerek *see* Zrenjanin

Nagydisznód *see* Cisnădie

Nagyenyed *see* Aiud

Nagykálló *77 E6* Szabolcs-Szatmár-Bereg, E Hungary

Nagykanizsa *77 C7 Ger.* Grosskanizsa. Zala, SW Hungary

Nagykároly *see* Carei

Nagykikinda *see* Kikinda

Nagykörös *77 D7* Pest, C Hungary

Nagymihály *see* Michalovce

Nagysurány *see* Šurany

Nagyszalonta *see* Salonta

Nagyszeben *see* Sibiu

Nagyszentmiklós *see* Sânnicolau Mare

Nagyszöllös *see* Vynohradiv

Nagyszombat *see* Trnava

Nagytapolcsány *see* Topoľčany

Nagyvárad *see* Oradea

Naha *109 A3* Okinawa, Okinawa, SW Japan

Nahariya *97 A5 prev.* Nahariyya. Northern, N Israel

Nahariyya *see* Nahariya

Nahuel Huapí, Lago *43 B5 lake* W Argentina

Nain *17 F2* Newfoundland and Labrador, NE Canada

Nairobi *51 C6 country capital* (Kenya) Nairobi Area, S Kenya

Nairobi *51 C6* Nairobi Area, S Kenya

Naissus *see* Niš

Najaf *see* An Najaf

Najima *see* Fukuoka

Najin *107 E3* NE North Korea

Najrán *99 B6 var.* Abá as Su'úd. Najrán, S Saudi Arabia

Nakambé *see* White Volta

Nakamura *109 B7 var.* Shimanto. Kóchi, Shikoku, SW Japan

Nakatsugawa *109 C6 var.* Nakatsugawa. Gifu, Honshú, SW Japan

Nakatugawa *see* Nakatsugawa

Nakhichevan' *see* Naxçivan

Nakhodka *93 G5* Primorskiy Kray, SE Russian Federation

Nakhon Pathom *115 C5 var.* Nagara Pathom, Nakorn Pathom. Nakhon Pathom, W Thailand

Nakhon Ratchasima *115 C5 var.* Khorat, Korat. Nakhon Ratchasima, E Thailand

Nakhon Sawan *115 C5 var.* Muang Nakhon Sawan, Nagara Svarga. Nakhon Sawan, W Thailand

Nakhon Si Thammarat *115 C7 var.* Nagara Sridharmaraj, Nakhon Sithammarah. Nakhon Si Thammarat, SW Thailand

Nakhon Sithammaraj *see* Nakhon Si Thammarat

Nakorn Pathom *see* Nakhon Pathom

Nakuru *51 C6* Rift Valley, SW Kenya

Nal'chik *89 B8* Kabardino-Balkarskaya Respublika, SW Russian Federation

Nálút *49 F2* NW Libya

Namakan Lake *18 A1 lake* Canada/USA

Namangan *101 F2* Namangan Viloyati, E Uzbekistan

Nambala *56 D2* Central, C Zambia

Nam Co *104 C5 lake* W China

Nam Dinh *114 D3* Nam Ha, N Vietnam

Namib Desert *56 B3 desert* W Namibia

Namibe *56 A2 Port.* Moçámedes, Mossámedes. Namibe, SW Angola

Namibia *56 B3 off.* Republic of Namibia, *var.* South West Africa, *Afr.* Suidwes-Afrika, *Ger.* Deutsch-Südwestafrika; *prev.* German Southwest Africa, South-West Africa. *country* S Africa

Namibia, Republic of *see* Namibia

Namnetes *see* Nantes

Namo *see* Namu Atoll

Nam Ou *114 C3 river* N Laos

Nampa *24 D3* Idaho, NW USA

Nampula *57 E2* Nampula, NE Mozambique

Namsos *62 B4* Nord-Trøndelag, C Norway

Nam Tha *114 C4 river* N Laos

Namu Atoll *122 D2 var.* Namo. *atoll* Ralik Chain, C Marshall Islands

Namur *65 C6 Dut.* Namen. Namur, SE Belgium

Namyit Island *106 C8 island* S Spratly Islands

Nan *114 C4 var.* Muang Nan. Nan, NW Thailand

Nanaimo *14 D5* Vancouver Island, British Columbia, SW Canada

Nanchang *106 C5 var.* Nan-ch'ang, Nanch'ang-hsien. *province capital* Jiangxi, S China

Nan-ch'ang *see* Nanchang

Nanch'ang-hsien *see* Nanchang

Nancy *68 D3* Meurthe-et-Moselle, NE France

Nandaime *30 D3* Granada, SW Nicaragua

Nänded *112 D5* Mahárashtra, C India

Nandi *see* Nadi

Nándorhegy *see* Oţelu Roşu

Nandyál *110 C1* Andhra Pradesh, E India

Nan Hai *see* South China Sea

Naniwa *see* Ósaka

Nanjing *106 D5 var.* Nan-ching, Nanking; *prev.* Chianning, Chian-ning, Kiang-ning, Jiangsu. *province capital* Jiangsu, E China

Nanking *see* Nanjing

Nanning *106 B6 var.* Nan-ning; *prev.* Yung-ning. Guangxi Zhuangzu Zizhiqu, S China

Nan-ning *see* Nanning

Nanortalik *60 C5* Kitaa, S Greenland

Nanpan Jiang *114 D2 river* S China

Nanping *106 D6 var.* Nan-p'ing; *prev.* Yenping. Fujian, SE China

Nan-p'ing *see* Nanping

Nansei-shotó *108 A2 Eng.* Ryukyu Islands. *island group* SW Japan

Nansei Syotó Trench *see* Ryukyu Trench

Nansen Basin *133 C4 undersea basin* Arctic Ocean

Nansen Cordillera *133 B3 var.* Arctic Mid Oceanic Ridge, Nansen Ridge. *seamount range* Arctic Ocean

Nansen Ridge *see* Nansen Cordillera

Nansha Qundao *see* Spratly Islands

Nanterre *68 D1* Hauts-de-Seine, N France

Nantes *68 B4 Bret.* Naoned; *anc.* Condivincum, Namnetes. Loire-Atlantique, NW France

Nantucket Island *19 G3 island* Massachusetts, NE USA

Nanumaga *123 E3 var.* Nanumanga. *atoll* NW Tuvalu

Nanumanga *see* Nanumaga

Nanumea Atoll *123 E3 atoll* NW Tuvalu

Nanyang *106 C5 var.* Nan-yang. Henan, C China

Nan-yang *see* Nanyang

Naoned *see* Nantes

Napa *25 B6* California, W USA

Napier *128 E4* Hawke's Bay, North Island, New Zealand

Naples *21 E5* Florida, SE USA

Naples *see* Napoli

Napo *38 C1 river* NE Ecuador

Napoléon-Vendée *see* la Roche-sur-Yon

Napoli *75 C5 Eng.* Naples, *Ger.* Neapel; *anc.* Neapolis. Campania, S Italy

Napo, Río *38 C1 river* Ecuador/Peru

Naracoorte *127 B7* South Australia

Naradhivas *see* Narathiwat

Narathiwat *115 C7 var.* Naradhivas. Narathiwat, SW Thailand

Narbada *see* Narmada

Narbo Martius *see* Narbonne

Narbonne *69 C6 anc.* Narbo Martius. Aude, S France

Narborough Island *see* Fernandina, Isla

Nares Abyssal Plain *see* Nares Plain

Nares Plain *13 E6 var.* Nares Abyssal Plain. *abyssal plain* NW Atlantic Ocean

Nares Strædet *see* Nares Strait

Nares Strait *60 D1 Dan.* Nares Strædet. *strait* Canada/Greenland

Narew *76 E3 river* E Poland

Narmada *102 B3 var.* Narbada. *river* C India

Narova *see* Narva

Narovlya *85 C8 Rus.* Narovlya. Homyel'skaya Voblasts', SE Belarus

Närpes *63 D5 Fin.* Närpiö. Länsi-Suomi, W Finland

Närpiö *see* Närpes

Narrabri *127 D6* New South Wales, SE Australia

Narrogin *125 B6* Western Australia

Narva *84 E2* Ida-Virumaa, NE Estonia

Narva *84 E2 prev.* Narova. *river* Estonia/Russian Federation

Narva Bay *84 E2 Est.* Narva Laht, *Ger.* Narwa-Bucht, *Rus.* Narvskiy Zaliv. *bay* Estonia/Russian Federation

Narva Laht *see* Narva Bay

Narva Reservoir *84 E2 Est.* Narva Veehoidla, *Rus.* Narvskoye Vodokhranilishche. *reservoir* Estonia/Russian Federation

Narva Veehoidla *see* Narva Reservoir

Narvik *62 C3* Nordland, C Norway

Narvskiy Zaliv *see* Narva Bay

Narvskoye Vodokhranilishche *see* Narva Reservoir

Narwa-Bucht *see* Narva Bay

Nar'yan-Mar *88 D3 prev.* Beloshchel'ye, Dzerzhinskiy. Nenetskiy Avtonomnyy Okrug, NW Russian Federation

Naryn *101 G2* Narynskaya Oblast', C Kyrgyzstan

Nassau *32 C1 country capital* (The Bahamas) New Providence, N Bahamas

Nasser, Lake *50 B3 var.* Buhayrat Nasir, Buḥayrat Náşir, Buḩeiret Náşir. *lake* Egypt/Sudan

Naszód *see* Násáud

Nata *56 C3* Central, NE Botswana

Natal *41 G2 state capital* Rio Grande do Norte, E Brazil

Natal Basin *119 A6 var.* Mozambique Basin. *undersea basin* W Indian Ocean

Natanya *see* Netanya

Natchez *20 B3* Mississippi, S USA

Natchitoches *20 A2* Louisiana, S USA

Nathanya *see* Netanya

Natitingou *53 F4* NW Benin

Natsrat *see* Natzrat

Natuna Islands *see* Natuna, Kepulauan

Natuna, Kepulauan *102 D4 var.* Natuna Islands. *island group* W Indonesia

Naturaliste Plateau *119 E6 undersea plateau* E Indian Ocean

Natzrat *97 A5 var.* Natsrat, *Ar.* En Nazira, *Eng.* Nazareth; *prev.* Nazerat. Northern, N Israel

Naugard *see* Nowogard

Naujamiestis *84 C4* Panevėžys, C Lithuania

Nauru *122 D2 off.* Republic of Nauru; *prev.* Pleasant Island. *country* W Pacific Ocean

Nauru, Republic of *see* Nauru

Nauta *38 C2* Loreto, N Peru

Navahrudak *85 C6 Pol.* Nowogródek, *Rus.* Novogrudok. Hrodzyenskaya Voblasts', W Belarus

Navanagar *see* Jámnagar

Navapolatsk *85 D5 Rus.* Novopolotsk. Vitsyebskaya Voblasts', N Belarus

Navarra *71 E2 Eng./Fr.* Navarre. *autonomous community* N Spain

Navarre *see* Navarra

Navassa Island *32 C3 US unincorporated territory* C West Indies

Navoi *see* Navoiy

Navoiy *101 E2 Rus.* Navoi. Navoiy Viloyati, C Uzbekistan

Navojoa *28 C2* Sonora, NW Mexico

Navolat *see* Navolato

Navolato *28 C3 var.* Navolat. Sinaloa, C Mexico

Návpaktos *see* Náfpaktos

Návplion *see* Náfplio

Nawabashah *see* Nawábsháh

Nawábsháh *112 B3 var.* Nawabashah. Sind, S Pakistan

Naxçivan *95 G3 Rus.* Nakhichevan'. SW Azerbaijan

Náxos *83 D6 var.* Naxos. Náxos, Kykládes, Greece, Aegean Sea

Náxos *83 D6 island* Kykládes, Greece, Aegean Sea

Nayoro *108 D2* Hokkaidó, NE Japan

Nay Pyi Taw *114 B4 country capital* Myanmar (Burma)/ Mandalay, C Myanmar (Burma)

Nazareth *see* Natzrat

Nazca *38 D4* Ica, S Peru

Nazca Ridge *35 A5 undersea ridge* E Pacific Ocean

Naze *108 B3 var.* Nase. Kagoshima, Amami-óshima, SW Japan

Nazerat *see* Natzrat

Nazilli *94 A4* Aydin, SW Turkey

Nazrēt *51 C5 var.* Adama, Hadama. Oromíya, C Ethiopia

Nebaj *30 B2* Quiché, W Guatemala

Nebitdag *see* Balkanabat

Neblina, Pico da *40 C1 mountain* NW Brazil

Nebraska *22 D4 off.* State of Nebraska, *also known as* Blackwater State, Cornhusker State, Tree Planters State. *state* C USA

Nebraska City *23 F4* Nebraska, C USA

Neches River *27 H3 river* Texas, SW USA

Neckar *73 B6 river* SW Germany

Necochea *43 D5* Buenos Aires, E Argentina

Nederland *see* Netherlands

Neder Rhine *64 D4 Eng.* Lower Rhine. *river* C Netherlands

Nederweert *65 D5* Limburg, SE Netherlands

Neede *64 E3* Gelderland, E Netherlands

Neerpelt *65 D5* Limburg, NE Belgium

Neftekamsk *89 D5* Respublika Bashkortostan, W Russian Federation

Neftezavodsk *see* Seýdi

Negara Brunei Darussalam *see* Brunei

Negêlê *51 D5 var.* Negelli, *It.* Neghelli. Oromíya, C Ethiopia

Negelli *see* Negêlê

Negev *see* HaNegev

Neghelli *see* Negêlê

Negomane *57 E2 var.* Negomano. Cabo Delgado, N Mozambique

Negomano *see* Negomane

Negombo *110 C3* Western Province, SW Sri Lanka

Negotin *78 E4* Serbia, E Serbia

Negra, Punto *38 A3 headland* NW Peru

Negreşti *see* Negreşti-Oaş

Negreşti-Oaş *86 B3 Hung.* Avasfelsöfalu; *prev.* Negreşti. Satu Mare, NE Romania

Negro, Río *43 C5 river* E Argentina

Negro, Río *40 D1 river* N South America

Negro, Río *42 D4 river* Brazil/Uruguay

Negros *117 E2 island* C Philippines

Nehbandán *98 E3* Khorásán, E Iran

Neijiang *106 B5* Sichuan, C China

Neiva *36 B3* Huila, S Colombia

Nellore *110 D2* Andhra Pradesh, E India

Nelson *129 C5* Nelson, South Island, New Zealand

Nelson *15 G4 river* Manitoba, C Canada

Néma *52 D3* Hodh ech Chargui, SE Mauritania

Neman *84 A4 Bel.* Nyoman, *Ger.* Memel, *Lith.* Nemunas, *Pol.* Niemen. *river* NE Europe

Neman *84 A4 Bel.* Ragnit. Kaliningradskaya Oblast', W Russian Federation

Nemausus *see* Nîmes

Neméa *83 B6* Pelopónnisos, S Greece

Nemetocenna *see* Arras

Nemours *68 C3* Seine-et-Marne, N France

Nemunas *see* Neman

Nemuro *108 E2* Hokkaidó, NE Japan

Neóchóri *83 A5* Dytikí Ellás, C Greece

Nepal *113 E3 off.* Nepal. *country* S Asia

Nepal *see* Nepal

Nereta *84 C4* Aizkraukle, S Latvia

Neretva *78 C4 river* Bosnia and Herzegovina/ Croatia

Neris *85 C6 Pol.* Wilia, *Pol.; prev. Pol.* Wilja. *river* Belarus/Lithuania

Neris *see* Viliya

Nerva *70 C4* Andalucía, S Spain

Neryungri *93 F4* Respublika Sakha (Yakutiya), NE Russian Federation

Neskaupstaður *61 E5* Austurland, E Iceland

Ness, Loch *66 C3 lake* N Scotland, United Kingdom

Nesterov *see* Zhovkva

Néstos *82 C3 Bul.* Mesta, *Turk.* Kara Su. *river* Bulgaria/Greece

Nesvizh *see* Nyasvizh

Netanya *97 A6 var.* Natanya, Nathanya. Central, C Israel

Netherlands *64 C3 off.* Kingdom of the Netherlands, *var.* Holland, *Dut.* Koninkrijk der Nederlanden, Nederland. *country* NW Europe

Netherlands East Indies *see* Indonesia

Netherlands Guiana *see* Suriname

Netherlands, Kingdom of the *see* Netherlands

Netherlands New Guinea *see* Papua

Nettilling Lake *15 G3 lake* Baffin Island, Nunavut, N Canada

Netze *see* Noteć

Neu Amerika *see* Puławy

Neubrandenburg *72 D3* Mecklenburg-Vorpommern, NE Germany

Neuchâtel *73 A7 Ger.* Neuenburg. Neuchâtel, W Switzerland

Neuchâtel, Lac de *73 A7 Ger.* Neuenburger See. *lake* W Switzerland

Neuenburg *see* Neuchâtel

Neuenburger See *see* Neuchâtel, Lac de

Neufchâteau *65 D8* Luxembourg, SE Belgium

Neugradisk *see* Nova Gradiška

Neuhof *see* Zgierz

Neukuhren *see* Pionerskiy

Neumarkt *see* Târgu Secuiesc, Covasna, Romania

Neumarkt *see* Târgu Mureş

Neumoldowa *see* Moldova Nouă

Neumünster *72 B2* Schleswig-Holstein, N Germany

Neunkirchen *73 A5* Saarland, SW Germany

Neuquén *43 B5* Neuquén, SE Argentina

Neuruppin *72 C3* Brandenburg, NE Germany

Neusalz an der Oder *see* Nowa Sól

Neu Sandec *see* Nowy Sącz

Neusatz *see* Novi Sad

Neusiedler See *73 E6 Hung.* Fertő. *lake* Austria/Hungary

Neusohl *see* Banská Bystrica

Neustadt *see* Baia Mare, Maramureş, Romania

Neustadt an der Haardt *see* Neustadt an der Weinstrasse

Neustadt an der Weinstrasse *73 B5 prev.* Neustadt an der Haardt, *hist.* Niewenstat; *anc.* Nova Civitas. Rheinland-Pfalz, SW Germany

Neustadtl *see* Novo mesto

Neustettin *see* Szczecinek

Neustrelitz *72 D3* Mecklenburg-Vorpommern, NE Germany

Neutra *see* Nitra

Neu-Ulm *73 B6* Bayern, S Germany

Neuwied *73 A5* Rheinland-Pfalz, W Germany

Neuzen *see* Terneuzen

Nevada *25 C5 off.* State of Nevada, *also known as* Battle Born State, Sagebrush State, Silver State. *state* W USA

Nevada, Sierra *70 C4 mountain range* S Spain

Nevada, Sierra *25 C6 mountain range* W USA

Nevers *68 C4 anc.* Noviodunum. Nièvre, C France

Neves *54 E2* São Tomé, S Sao Tome and Principe, Africa

Nevinnomyssk *89 B7* Stavropol'skiy Kray, SW Russian Federation

Nevşehir *94 C3 var.* Nevsehir. Nevşehir, C Turkey

Newala *51 C8* Mtwara, SE Tanzania

New Albany *18 C5* Indiana, N USA

New Amsterdam *37 G3* E Guyana

Newark *19 F4* New Jersey, NE USA

New Bedford *19 G3* Massachusetts, NE USA

Newberg *24 B3* Oregon, NW USA

New Bern *21 F1* North Carolina, SE USA

New Braunfels *27 G4* Texas, SW USA

Newbridge *67 B6 Ir.* An Droichead Nua. Kildare, C Ireland

New Britain *122 B3 island* E Papua New Guinea

New Brunswick *17 F4 Fr.* Nouveau-Brunswick. *province* SE Canada

New Caledonia *122 D4 var.* Kanaky, Fr. Nouvelle-Calédonie. *French overseas territory* SW Pacific Ocean

New Caledonia *122 C5 island* SW Pacific Ocean

New Caledonia Basin *120 C4 undersea basin* W Pacific Ocean

Newcastle *127 D6* New South Wales, SE Australia

Newcastle *see* Newcastle upon Tyne

Newcastle upon Tyne *66 D4 var.* Newcastle, *hist.* Monkchester, *Lat.* Pons Aelii. NE England, United Kingdom

New Delhi *112 D3 country capital* (India) Delhi, N India

New England of the West *see* Minnesota

Newfoundland *17 G3 Fr.* Terre-Neuve. *island* Newfoundland and Labrador, SE Canada

Newfoundland and Labrador *17 F2 Fr.* Terre Neuve. *province* E Canada

Newfoundland Basin *44 B3 undersea feature* NW Atlantic Ocean

New Georgia Islands *122 C3 island group* NW Solomon Islands

New Glasgow *17 F4* Nova Scotia, SE Canada

New Goa *see* Panaji

New Guinea *122 A3 Dut.* Nieuw Guinea, *Ind.* Irian. *island* Indonesia/Papua New Guinea

New Hampshire *19 F2 off.* State of New Hampshire, *also known as* Granite State. *state* NE USA

New Haven *19 G3* Connecticut, NE USA

New Hebrides *see* Vanuatu

New Iberia *20 B3* Louisiana, S USA

New Ireland *122 C3 island* NE Papua New Guinea

New Jersey *19 F4 off.* State of New Jersey, *also known as* The Garden State. *state* NE USA

Newman *124 B4* Western Australia

Newmarket *67 E6 E* England, United Kingdom

New Mexico *26 C2 off.* State of New Mexico, *also known as* Land of Enchantment, Sunshine State. *state* SW USA

New Orleans *20 B3* Louisiana, S USA

New Plymouth *128 C4* Taranaki, North Island, New Zealand

Newport *67 C7* SE Wales, United Kingdom

Newport *18 C4* Kentucky, S USA

Newport *19 G2* Vermont, NE USA

Newport News *19 F5* Virginia, NE USA

New Providence *32 C1 island* N Bahamas

Newquay *67 C7* SW England, United Kingdom

Newry *67 B5 Ir.* An tIúr. SE Northern Ireland, United Kingdom

New Sarum *see* Salisbury

New Siberian Islands *see* Novosibirskiye Ostrova

New South Wales *127 C6 state* SE Australia

Newton *23 G3* Iowa, C USA

Newton *23 F5* Kansas, C USA

Newtownabbey *67 B5 Ir.* Baile na Mainistreach. E Northern Ireland, United Kingdom

New Ulm *23 F2* Minnesota, N USA

New York *19 F4* New York, NE USA

New York *19 F3 state* NE USA

New Zealand *128 A4 country* SW Pacific Ocean

Neyveli *110 C2* Tamil Nádu, SE India

Nezhin *see* Nizhyn

Ngangze Co *104 B5 lake* W China

Ngaoundéré *54 B4 var.* N'Gaoundéré. Adamaoua, N Cameroon

N'Gaoundéré *see* Ngaoundéré

Ngazidja *57 F2 var.* Grande-Comore. *island* NW Comoros

Ngerulmud *56 D4 country capital* (Palau) NW Pacific Ocean

N'Giva *56 B3 var.* Ondjiva, *Port.* Vila Pereira de Eça. Cunene, S Angola

Ngo *55 B6* Plateaux, SE Congo

Ngoko *55 B5 river* Cameroon/Congo

Ngourti *53 H3* Diffa, E Niger

Nguigmi *53 H3 var.* N'Guigmi. Diffa, SE Niger

N'Guigmi *see* Nguigmi

N'Gunza *see* Sumbe

Nguru *53 G3* Yobe, NE Nigeria

Nha Trang *115 E6* Khanh Hoa, S Vietnam

Niagara Falls *16 D5* Ontario, S Canada

Niagara Falls *19 E3* New York, NE USA

Niagara Falls *18 D3 waterfall* Canada/USA

Niamey *53 F3 country capital* (Niger) Niamey, SW Niger

Niangay, Lac *53 E3 lake* E Mali

Nia-Nia *55 E5* Orientale, N Dem. Rep. Congo

Nias, Pulau *116 A3 island* W Indonesia

Nicaea *see* Nice

Nicaragua *30 D3 off.* Republic of Nicaragua. *country* Central America

Nicaragua, Lago de *30 D4 var.* Cocibolca, Gran Lago, *Eng.* Lake Nicaragua. *lake* S Nicaragua

Nicaragua, Lake *see* Nicaragua, Lago de

Nicaragua, Republic of *see* Nicaragua

Nicaria *see* Ikaría

Nice *69 D6 It.* Nizza; *anc.* Nicaea. Alpes-Maritimes, SE France

Nicephorium *see* Ar Raqqah

Nicholas II Land *see* Severnaya Zemlya

Nicholls Town *32 C1* Andros Island, NW Bahamas

Nicobar Islands *102 B4 island group* India, E Indian Ocean

Nicosia *80 C5 Gk.* Lefkosía, *Turk.* Lefkoşa. *country capital* (Cyprus) C Cyprus

Nicoya *30 D4* Guanacaste, W Costa Rica

Nicoya, Golfo de *30 D5 gulf* W Costa Rica

Nicoya, Península de *30 D4 peninsula* NW Costa Rica

Nida *84 A3 Ger.* Nidden. Klaipėda, SW Lithuania

Nidaros *see* Trondheim

Nidden *see* Nida

Nidzica *76 D3 Ger.* Niedenburg. Warmińsko-Mazurskie, NE Poland

Niedenburg – Oldenburg in Holstein

Niedenburg *see* Nidzica
Niedere Tauern *77 A6 mountain range* C Austria
Niemen *see* Neman
Nieśwież *see* Nyasvizh
Nieuw Amsterdam *37 G3* Commewijne, NE Suriname
Nieuw-Bergen *64 D4* Limburg, SE Netherlands
Nieuwegein *64 C4* Utrecht, C Netherlands
Nieuw Guinea *see* New Guinea
Nieuw Nickerie *37 G3* Nickerie, NW Suriname
Niewenstat *see* Neustadt an der Weinstrasse
Niğde *94 C4* Niğde, C Turkey
Niger *53 F3 off.* Republic of Niger. *country* W Africa
Niger *53 F4 river* W Africa
Nigeria *53 F4 off.* Federal Republic of Nigeria. *country* W Africa
Nigeria, Federal Republic of *see* Nigeria
Niger, Mouths of the *53 F5 delta* S Nigeria
Niger, Republic of *see* Niger
Nihon *see* Japan
Niigata *109 D5* Niigata, Honshū, C Japan
Niihama *109 B7* Ehime, Shikoku, SW Japan
Ni'ihau *25 A7 var.* Niihau. *island* Hawaii, USA, C Pacific Ocean
Nii-jima *109 D6 island* E Japan
Nijkerk *64 D3* Gelderland, C Netherlands
Nijlen *65 C5* Antwerpen, N Belgium
Nijmegen *64 D4 Ger.* Nimwegen; *anc.* Noviomagus. Gelderland, SE Netherlands
Nikaria *see* Ikaría
Nikel' *88 C2 Finn.* Kolosjoki. Murmanskaya Oblast', NW Russian Federation
Nikiniki *117 E5* Timor, S Indonesia
Niklasmarkt *see* Gheorgheni
Nikolainkaupunki *see* Vaasa
Nikolayev *see* Mykolayiv
Nikol'sk *see* Ussuriysk
Nikol'sk-Ussuriyskiy *see* Ussuriysk
Nikopol' *87 F3* Dnipropetrovs'ka Oblast', SE Ukraine
Nikšić *79 C5* C Montenegro
Nikumaroro *123 E3* ; *prev.* Gardner Island. *atoll* Phoenix Islands, C Kiribati
Nikunau *123 E3 var.* Nukunau; *prev.* Byron Island. *atoll* Tungaru, W Kiribati
Nile *50 B2 former province* NW Uganda
Nile *46 D3 Ar.* Nahr an Nīl. *river* N Africa
Nile Delta *50 B1 delta* N Egypt
Nīl, Nahr an *see* Nile
Nîmes *69 C6 anc.* Nemausus, Nismes. Gard, S France
Nimwegen *see* Nijmegen
Nine Degree Channel *110 B3 channel* India/ Maldives
Ninetyeast Ridge *119 D5 undersea feature* E Indian Ocean
Ninety Mile Beach *128 C1 beach* North Island, New Zealand
Ningbo *106 D5 var.* Ning-po, Yin-hsien; *prev.* Ninghsien. Zhejiang, SE China
Ning-hsia *see* Ningxia
Ninghsien *see* Ningbo
Ning-po *see* Ningbo
Ningsia/Ningsia Hui/Ningsia Hui Autonomous Region *see* Ningxia
Ningxia *106 B4 off.* Ningxia Huizu Zizhiqu, *var.* Ning-hsia, Ningsia, *Eng.* Ningsia Hui, Ningsia Hui Autonomous Region. *autonomous region* N China
Ningxia Huizu Zizhiqu *see* Ningxia
Nio *see* Íos
Niobrara River *23 E3 river* Nebraska/Wyoming, C USA
Nioro *52 D3 var.* Nioro du Sahel. Kayes, W Mali
Nioro du Sahel *see* Nioro
Niort *68 B4* Deux-Sèvres, W France
Nipigon *16 B4* Ontario, S Canada
Nipigon, Lake *16 B3 lake* Ontario, S Canada
Nippon *see* Japan
Niš *79 E5 Eng.* Nish, *Ger.* Nisch; *anc.* Naissus. Serbia, SE Serbia
Nişab *98 B4* Al Ḥudūd ash Shamālīyah, N Saudi Arabia
Nisch/Nish *see* Niš
Nisibin *see* Nusaybin
Nisiros *see* Nísyros
Nisko *76 E4* Podkrapackie, SE Poland
Nismes *see* Nîmes
Nistru *see* Dniester
Nísyros *83 E7 var.* Nisiros. *island* Dodekánisa, Greece, Aegean Sea
Nitra *77 C6 Ger.* Neutra, *Hung.* Nyitra. Nitriansky Kraj, SW Slovakia
Nitra *77 C6 Ger.* Neutra, *Hung.* Nyitra. *river* W Slovakia
Niuatobutabu *see* Niuatoputapu
Niuatoputapu *123 E4 var.* Niuatobutabu; *prev.* Keppel Island. *island* N Tonga
Niue *123 F4 self-governing territory in free association with New Zealand* S Pacific Ocean
Niulakita *123 E3 var.* Nurakita. *atoll* S Tuvalu
Niutao *123 E3 atoll* NW Tuvalu
Nivernais *68 C4 cultural region* C France
Nizāmābād *112 D5* Telangana, C India
Nizhnegorskiy *see* Nyzhn'ohirs'kyy
Nizhnekamsk *89 C5* Respublika Tatarstan, W Russian Federation
Nizhnevartovsk *92 D3* Khanty-Mansiyskiy Avtonomnyy Okrug-Yugra, C Russian Federation
Nizhniy Novgorod *89 C5 prev.* Gor'kiy. Nizhegorodskaya Oblast', W Russian Federation
Nizhniy Odes *88 D4* Respublika Komi, NW Russian Federation
Nizhnyaya Tunguska *93 E3 Eng.* Lower Tunguska. *river* N Russian Federation
Nizhyn *87 E1 Rus.* Nezhin. Chernihivs'ka Oblast', NE Ukraine
Nizza *see* Nice
Njombe *51 C8* Iringa, S Tanzania
Nkayi *55 B6 prev.* Jacob. Bouenza, S Congo
Nkongsamba *54 A4 var.* N'Kongsamba. Littoral, W Cameroon
N'Kongsamba *see* Nkongsamba
Nmai Hka *114 B2 var.* Me Hka. *river* N Myanmar (Burma)
Nobeoka *109 B7* Miyazaki, Kyūshū, SW Japan
Noboribetsu *108 D3 var.* Noboribetu. Hokkaidō, NE Japan
Noboribetu *see* Noboribetsu
Nogales *28 B1* Sonora, NW Mexico
Nogales *26 B3* Arizona, SW USA
Nogal Valley *see* Dooxo Nugaaleed
Noire, Rivi`ere *see* Black River
Nokia *63 D5* Länsi-Suomi, W Finland
Nokou *54 B3* Kanem, W Chad
Nola *55 B5* Sangha-Mbaéré, SW Central African Republic
Nolinsk *89 C5* Kirovskaya Oblast', NW Russian Federation
Nongkaya *see* Nong Khai
Nong Khai *114 C4 var.* Mi Chai, Nongkaya. Nong Khai, E Thailand
Nonouti *122 D2 prev.* Sydenham Island. *atoll* Tungaru, W Kiribati
Noord-Beveland *64 B4 var.* North Beveland. *island* SW Netherlands
Noordwijk aan Zee *64 C3* Zuid-Holland, W Netherlands
Noordzee *see* North Sea
Nora *63 C6* Örebro, C Sweden
Norak *101 E3 Rus.* Nurek. W Tajikistan
Nord *61 F1* Avannaarsua, N Greenland
Nordaustlandet *61 G1 island* NE Svalbard
Norden *72 A3* Niedersachsen, NW Germany
Norderstedt *72 B3* Schleswig-Holstein, N Germany
Nordfriesische Inseln *see* North Frisian Islands
Nordhausen *72 C4* Thüringen, C Germany
Nordhorn *72 A3* Niedersachsen, NW Germany
Nord, Mer du *see* North Sea
Nord-Ouest, Territoires du *see* Northwest Territories
Nordsee/Nordsjøen/Nordsøen *see* North Sea
Norfolk *23 E3* Nebraska, C USA
Norfolk *19 F5* Virginia, NE USA
Norfolk Island *120 D4 Australian external territory* SW Pacific Ocean
Norfolk Ridge *120 D4 undersea feature* W Pacific Ocean
Norge *see* Norway
Norias *27 G5* Texas, SW USA
Noril'sk *92 D3* Taymyrskiy (Dolgano-Nenetskiy) Avtonomnyy Okrug, N Russian Federation
Norman *27 G1* Oklahoma, C USA
Normandes, Îles *see* Channel Islands
Normandie *68 B3 Eng.* Normandy. *cultural region* N France
Normandy *see* Normandie
Normanton *126 C3* Queensland, NE Australia
Norrköping *63 C6* Östergötland, S Sweden
Norrtälje *63 C6* Stockholm, C Sweden
Norseman *125 B6* Western Australia
Norske Havet *see* Norwegian Sea
North Albanian Alps *79 C5 Alb.* Bjeshkët e Namuna, *SCr.* Prokletije. *mountain range* SE Europe
Northallerton *67 D5* N England, United Kingdom
Northam *125 A6* Western Australia
North America *12 continent*
Northampton *67 D6* C England, United Kingdom
North Andaman *111 F2 island* Andaman Islands, India, NE Indian Ocean
North Australian Basin *119 E5 Fr.* Bassin Nord de l' Australie. *undersea feature* E Indian Ocean
North Bay *16 D4* Ontario, S Canada
North Beveland *see* Noord-Beveland
North Borneo *see* Sabah
North Cape *128 C1 headland* North Island, New Zealand
North Cape *62 D1 Eng.* North Cape. *headland* N Norway
North Cape *see* Nordkapp
North Carolina *21 E1 off.* State of North Carolina, *also known as* Old North State, Tar Heel State, Turpentine State. *state* SE USA
North Channel *18 D2 lake channel* Canada/USA
North Charleston *21 F2* South Carolina, SE USA
North Dakota *22 D2 off.* State of North Dakota, *also known as* Flickertail State, Peace Garden State, Sioux State. *state* N USA
North Devon Island *see* Devon Island
North East Frontier Agency/North East Frontier Agency of Assam *see* Arunāchal Pradesh
Northeast Providence Channel *32 C1 channel* N Bahamas
Northeim *72 B4* Niedersachsen, C Germany
Northern Cook Islands *123 F4 island group* N Cook Islands
Northern Dvina *see* Severnaya Dvina
Northern Ireland *66 B4 var.* The Six Counties. *cultural region* Northern Ireland, United Kingdom
Northern Mariana Islands *120 B1 US commonwealth territory* W Pacific Ocean
Northern Rhodesia *see* Zambia
Northern Sporades *see* Vóreies Sporádes
Northern Territory *122 A5 territory* N Australia
North European Plain *59 E3 plain* N Europe
Northfield *23 F2* Minnesota, N USA
North Fiji Basin *120 D3 undersea feature* N Coral Sea
North Frisian Islands *72 B2 var.* Nordfriesische Inseln. *island group* N Germany
North Huvadhu Atoll *110 B5 var.* Gaafu Alifu Atoll. *atoll* S Maldives
North Island *128 B2 island* N New Zealand
North Korea *107 E3 off.* Democratic People's Republic of Korea, *Kor.* Chosŏn-minjujuŭi-inmin-kanghwaguk. *country* E Asia
North Little Rock *20 B1* Arkansas, C USA
North Minch *see* Minch, The
North Mole *71 G4 harbour wall* NW Gibraltar Europe
North Platte *23 E4* Nebraska, C USA
North Platte River *22 D4 river* C USA
North Pole *133 B3 pole* Arctic Ocean
North Saskatchewan *15 F5 river* Alberta/ Saskatchewan, S Canada
North Sea *58 D3 Dan.* Nordsøen, *Dut.* Noordzee, *Fr.* Mer du Nord, *Ger.* Nordsee, *Nor.* Nordsjøen; *prev.* German Ocean, *Lat.* Mare Germanicum. *sea* NW Europe
North Siberian Lowland *93 E2 var.* North Siberian Plain, *Eng.* North Siberian Lowland. *lowlands* N Russian Federation
North Siberian Lowland/North Siberian Plain *see* Severo-Sibirskaya Nizmennost'
North Star State *see* Minnesota
North Taranaki Bight *128 C3 gulf* North Island, New Zealand
North Uist *66 B3 island* NW Scotland, United Kingdom
Northwest Atlantic Mid-Ocean Canyon *12 E4 undersea feature* N Atlantic Ocean
North West Highlands *66 C3 mountain range* N Scotland, United Kingdom
Northwest Pacific Basin *91 G4 undersea feature* NW Pacific Ocean
Northwest Territories *15 E3 Fr.* Territoires du Nord-Ouest. *territory* NW Canada
Northwind Plain *133 B2 undersea feature* Arctic Ocean
Norton Sound *14 C2 inlet* Alaska, USA
Norway *63 A5 off.* Kingdom of Norway, *Nor.* Norge. *country* N Europe
Norway, Kingdom of *see* Norway
Norwegian Basin *61 F4 undersea feature* NW Norwegian Sea
Norwegian Sea *61 F4 var.* Norske Havet. *sea* NE Atlantic Ocean
Norwich *67 E6* E England, United Kingdom
Nösen *see* Bistrița
Noshiro *108 D4 var.* Nosiro; *prev.* Noshirominato. Akita, Honshū, C Japan
Noshirominato/Nosiro *see* Noshiro
Nosivka *87 E1 Rus.* Nosovka. Chernihivs'ka Oblast', NE Ukraine
Nosop *56 C4 river* E Namibia
Nosovka *see* Nosivka
Noşratābād *98 E3* Sīstān va Balūchestān, E Iran
Noteć *76 C3 Ger.* Netze. *river* NW Poland
Nóties Sporádes *see* Dodekánisa
Nottingham *67 D6* C England, United Kingdom
Nouâdhibou *52 B2 prev.* Port-Étienne. Dakhlet Nouâdhibou, W Mauritania
Nouakchott *52 B2 country capital* (Mauritania) Nouakchott District, SW Mauritania
Nouméa *122 C5 dependent territory capital* (New Caledonia) Province Sud, S New Caledonia
Nouveau-Brunswick *see* New Brunswick
Nouvelle-Calédonie *see* New Caledonia
Nouvelle Écosse *see* Nova Scotia
Nova Civitas *see* Neustadt an der Weinstrasse
Nova Gorica *73 D8* W Slovenia
Nova Gradiška *78 C3 Ger.* Neugradisk, *Hung.* Újgradiska. Brod-Posavina, NE Croatia
Nova Iguaçu *41 F4* Rio de Janeiro, SE Brazil
Nova Lisboa *see* Huambo
Novara *74 B2 anc.* Novaria. Piemonte, NW Italy
Novaria *see* Novara
Nova Scotia *17 F4 Fr.* Nouvelle Écosse. *province* SE Canada
Nova Scotia *13 E5 physical region* SE Canada
Novaya Sibir', Ostrov *93 F1 island* Novosibirskiye Ostrova, NE Russian Federation
Novaya Zemlya *88 D1 island group* N Russian Federation
Novaya Zemlya Trough *see* East Novaya Zemlya Trough
Novgorod *see* Velikiy Novgorod
Novi Grad *see* Bosanski Novi
Novi Iskŭr *82 C2* Sofiya-Grad, W Bulgaria
Noviodunum *see* Nevers, Nièvre, France
Noviomagus *see* Lisieux, Calvados, France
Noviomagus *see* Nijmegen, Netherlands
Novi Pazar *79 D5 Turk.* Yenipazar. Serbia, S Serbia
Novi Sad *78 D3 Ger.* Neusatz, *Hung.* Újvidék. Vojvodina, N Serbia
Novoazovs'k *87 G4 Rus.* Novoazovsk. Donets'ka Oblast', E Ukraine
Novocheboksarsk *89 C5* Chuvashskaya Respublika, W Russian Federation
Novocherkassk *89 B7* Rostovskaya Oblast', SW Russian Federation
Novodvinsk *88 C3* Arkhangel'skaya Oblast', NW Russian Federation
Novograd-Volynskiy *see* Novohrad-Volyns'kyy
Novogrudok *see* Navahrudak
Novohrad-Volyns'kyy *86 D2 Rus.* Novograd-Volynskiy. Zhytomyrs'ka Oblast', N Ukraine
Novokazalinsk *see* Ayteke Bi
Novokuznetsk *92 D4 prev.* Stalinsk. Kemerovskaya Oblast', S Russian Federation
Novolazarevskaya *132 C2 Russian research station* Antarctica
Novo mesto *73 E8 Ger.* Rudolfswert; *prev. Ger.* Neustadtl. SE Slovenia
Novomoskovsk *89 B5* Tul'skaya Oblast', W Russian Federation
Novomoskovs'k *87 F3 Rus.* Novomoskovsk. Dnipropetrovs'ka Oblast', E Ukraine
Novopolotsk *see* Navapolatsk
Novoradomsk *see* Radomsko
Novo Redondo *see* Sumbe
Novorossiysk *89 A7* Krasnodarskiy Kray, SW Russian Federation
Novoshakhtinsk *89 B6* Rostovskaya Oblast', SW Russian Federation
Novosibirsk *92 D4* Novosibirskaya Oblast', C Russian Federation
Novosibirskiye Ostrova *93 F1 Eng.* New Siberian Islands. *island group* N Russian Federation
Novotroitsk *89 D6* Orenburgskaya Oblast', W Russian Federation
Novotroitskoye *see* Novotroyits'ke, Ukraine
Novotroyits'ke *87 F4 Rus.* Novotroitskoye. Khersons'ka Oblast', S Ukraine
Novo-Urgench *see* Urganch
Novovolyns'k *86 C1 Rus.* Novovolynsk. Volyns'ka Oblast', NW Ukraine
Novy Dvor *85 B6 Rus.* Novyy Dvor. Hrodzyenskaya Voblasts', W Belarus
Novyy Bug *see* Novyy Buh
Novyy Buh *87 E3 Rus.* Novyy Bug. Mykolayivs'ka Oblast', S Ukraine
Novyy Dvor *see* Novy Dvor
Novyy Margilan *see* Farg'ona
Novyy Uzen' *see* Zhanaozen
Nowa Sól *76 B4 var.* Nowasól, *Ger.* Neusalz an der Oder. Lubuskie, W Poland
Nowasól *see* Nowa Sól
Nowogard *76 B2 var.* Nowógard, *Ger.* Naugard. Zachodnio-pomorskie, NW Poland
Nowogródek *see* Navahrudak
Nowo-Minsk *see* Mińsk Mazowiecki
Nowy Dwór Mazowiecki *76 D3* Mazowieckie, C Poland
Nowy Sącz *77 D5 Ger.* Neu Sandec. Małopolskie, S Poland
Nowy Tomyśl *76 B3 var.* Nowy Tomysl. Wielkopolskie, C Poland
Nowy Tomysl *see* Nowy Tomyśl
Noyon *68 C3* Oise, N France
Nsanje *57 E3* Southern, S Malawi
Nsawam *53 E5* SE Ghana
Ntomba, Lac *55 C6 var.* Lac Tumba. *lake* NW Dem. Rep. Congo
Nubian Desert *50 B3 desert* NE Sudan
Nu Chiang *see* Salween
Nu'eima *97 E7* E West Bank Asia
Nueva Caceres *see* Naga
Nueva Gerona *32 B2* Isla de la Juventud, S Cuba
Nueva Rosita *28 D2* Coahuila, NE Mexico
Nuevitas *32 C2* Camagüey, E Cuba
Nuevo, Bajo *31 G1 island* NW Colombia South America
Nuevo Casas Grandes *28 C1* Chihuahua, N Mexico
Nuevo, Golfo *43 C6 gulf* S Argentina
Nuevo Laredo *29 E2* Tamaulipas, NE Mexico
Nui Atoll *123 E3 atoll* W Tuvalu
Nu Jiang *see* Salween
Nûk *see* Nuuk
Nukha *see* Şäki
Nuku'alofa *123 E5 country capital* (Tonga) Tongatapu, S Tonga
Nukufetau Atoll *123 E3 atoll* C Tuvalu
Nukulaelae Atoll *123 E3 var.* Nukulailai. *atoll* E Tuvalu
Nukulailai *see* Nukulaelae Atoll
Nukunau *see* Nikunau
Nukunonu Atoll *123 E3 island* C Tokelau
Nukus *100 C2* Qoraqalpog'iston Respublikasi, W Uzbekistan
Nullarbor Plain *125 C6 plateau* South Australia/ Western Australia
Nunap Isua *60 C5 var.* Uummannarsuaq, *Dan.* Kap Farvel, *Eng.* Cape Farewell. *cape* S Greenland
Nunavut *15 F3 territory* N Canada
Nuneaton *67 D6* C England, United Kingdom
Nunivak Island *14 B2 island* Alaska, USA
Nunspeet *64 D3* Gelderland, E Netherlands
Nuoro *75 A5* Sardegna, Italy, C Mediterranean Sea
Nuquí *36 A3* Chocó, W Colombia
Nurakita *see* Niulakita
Nurata *see* Nurota
Nurek *see* Norak
Nuremberg *see* Nürnberg
Nurmes *62 E4* Itä-Suomi, E Finland
Nürnberg *73 C5 Eng.* Nuremberg. Bayern, S Germany
Nurota *101 E2 Rus.* Nurata. Navoiy Viloyati, C Uzbekistan
Nusaybin *95 F4 var.* Nisibin. Manisa, SE Turkey
Nussdorf *see* Năsăud
Nutmeg State *see* Connecticut
Nuuk *60 C4 var.* Nûk, *Dan.* Godthaab, Godthåb. *dependent territory capital* (Greenland) Kitaa, SW Greenland
Nyagan' *92 C3* Khanty-Mansiyskiy Avtonomnyy Okrug-Yugra, N Russian Federation
Nyainqentanglha Shan *104 C5 mountain range* W China
Nyala *50 A4* Southern Darfur, W Sudan
Nyamapanda *56 D3* Mashonaland East, NE Zimbabwe
Nyamtumbo *51 C8* Ruvuma, S Tanzania
Nyanda *see* Masvingo
Nyandoma *88 C4* Arkhangel'skaya Oblast', NW Russian Federation
Nyantakara *51 B7* Kagera, NW Tanzania
Nyasa, Lake *57 E2 var.* Lake Malawi; *prev.* Lago Nyassa. *lake* E Africa
Nyasaland/Nyasaland Protectorate *see* Malawi
Nyassa, Lago *see* Nyasa, Lake
Nyasvizh *85 C6 Pol.* Nieśwież, *Rus.* Nesvizh. Minskaya Voblasts', C Belarus
Nyaunglebin *114 B4* Bago, SW Myanmar (Burma)
Nyeri *51 C6* Central, C Kenya
Nyíregyháza *77 D6* Szabolcs-Szatmár-Bereg, NE Hungary
Nyitra *see* Nitra
Nykøbing *63 B8* Storstrøm, SE Denmark
Nyköping *63 C6* Södermanland, S Sweden
Nylstroom *see* Modimolle
Nyngan *127 D6* New South Wales, SE Australia
Nyoman *see* Neman
Nyurba *93 F3* Respublika Sakha (Yakutiya), NE Russian Federation
Nyzhn'ohirs'kyy *87 F4 Rus.* Nizhnegorskiy. Respublika Krym, S Ukraine
NZ *see* New Zealand
Nzega *51 C7* Tabora, C Tanzania
Nzérékoré *52 D4* SE Guinea
Nzwani *see* Anjouan

O

Oa'hu *25 A7 var.* Oahu. *island* Hawai'ian Islands, Hawaii, USA
Oak Harbor *24 B1* Washington, NW USA
Oakland *25 B6* California, W USA
Oamaru *129 B7* Otago, South Island, New Zealand
Oaxaca *29 F5 var.* Oaxaca de Juárez; *prev.* Antequera. Oaxaca, SE Mexico
Oaxaca de Juárez *see* Oaxaca
Ob' *90 C2 river* C Russian Federation
Obal' *85 D5 Rus.* Obol'. Vitsyebskaya Voblasts', N Belarus
Oban *66 C4* W Scotland, United Kingdom
Oban *see* Halfmoon Bay
Obando *see* Puerto Inírida
Obdorsk *see* Salekhard
Óbecse *see* Bečej
Obeliai *84 C4* Panevėžys, NE Lithuania
Oberhollabrunn *see* Tulln
Ob', Gulf of *see* Obskaya Guba
Obidovichi *see* Abidavichy
Obihiro *108 D2* Hokkaidō, NE Japan
Obo *54 D4* Haut-Mbomou, E Central African Republic
Obock *50 D4* E Djibouti
Obol' *see* Obal'
Oborniki *76 C3* Wielkolpolskie, W Poland
Obrovo *see* Abrova
Obskaya Guba *92 D3 Eng.* Gulf of Ob. *gulf* N Russian Federation
Ob' Tablemount *119 B7 undersea feature* S Indian Ocean
Ocala *21 E4* Florida, SE USA
Ocaña *36 B2* Norte de Santander, N Colombia
Ocaña *70 D3* Castilla-La Mancha, C Spain
O Carballiño *70 C1 Cast.* Carballiño. Galicia, NW Spain
Occidental, Cordillera *36 B2 mountain range* W South America
Occidental, Cordillera *39 E4 mountain range* Bolivia/Chile
Ocean Falls *14 D5* British Columbia, SW Canada
Ocean Island *see* Banaba
Oceanside *25 C8* California, W USA
Ocean State *see* Rhode Island
Ochakiv *87 E4 Rus.* Ochakov. Mykolayivs'ka Oblast', S Ukraine
Ochakov *see* Ochakiv
Ochamchire *95 E2 prev.* Och'amch'ire. W Georgia
Och'amch'ire *see* Ochamchire
Ocho Rios *32 B4* C Jamaica
Ochrida *see* Ohrid
Ochrida, Lake *see* Ohrid, Lake
Ocotal *30 D3* Nueva Segovia, NW Nicaragua
Ocozocuautla *29 G5* Chiapas, SE Mexico
October Revolution Island *see* Oktyabr'skoy Revolyutsii, Ostrov
Ocú *31 F5* Herrera, S Panama
Ōdate *108 D3* Akita, Honshū, C Japan
Oddur *see* Xuddur
Ödemiş *94 A4* İzmir, SW Turkey
Ödenburg *see* Sopron
Odenpäh *see* Otepää
Odense *63 B7* Fyn, C Denmark
Oder *76 B3 Cz./Pol.* Odra. *river* C Europe
Oderhaff *see* Szczeciński, Zalew
Odesa *87 E4 Rus.* Odessa. Odes'ka Oblast', SW Ukraine
Odessa *27 E3* Texas, SW USA
Odessa *see* Odesa
Odessus *see* Varna
Odienné *52 D4* NW Côte d'Ivoire
Odisha *113 F4 prev.* Orissa. *cultural region* NE India
Ŏdôngk *115 D6* Kâmpóng Spoe, S Cambodia
Odoorn *64 E2* Drenthe, NE Netherlands
Odra *see* Oder
Oesel *see* Saaremaa
Of *95 E2* Trabzon, NE Turkey
Ofanto *75 D5 river* S Italy
Offenbach *73 B5 var.* Offenbach am Main. Hessen, W Germany
Offenbach am Main *see* Offenbach
Offenburg *73 B6* Baden-Württemberg, SW Germany
Ogaadeen *see* Ogaden
Ogaden *51 D5 Som.* Ogaadeen. *plateau* Ethiopia/ Somalia
Ōgaki *109 C6* Gifu, Honshū, SW Japan
Ogallala *22 D4* Nebraska, C USA
Ogbomosho *53 F4 var.* Ogmoboso. Oyo, W Nigeria
Ogden *22 B4* Utah, W USA
Ogdensburg *19 F2* New York, NE USA
Ogmoboso *see* Ogbomosho
Ogulin *78 A3* Karlovac, NW Croatia
Ohio *18 C4 off.* State of Ohio, *also known as* Buckeye State. *state* N USA
Ohio River *18 C4 river* N USA
Ohlau *see* Oława
Ohri *see* Ohrid
Ohrid *79 D6 Turk.* Ochrida, Ohri. SW FYR Macedonia
Ohrid, Lake *79 D6 var.* Lake Ochrida, *Alb.* Liqeni i Ohrit, *Mac.* Ohridsko Ezero. *lake* Albania/ FYR Macedonia
Ohridsko Ezero/Ohrit, Liqeni i *see* Ohrid, Lake
Ohura *128 D3* Manawatu-Wanganui, North Island, New Zealand
Oirschot *65 C5* Noord-Brabant, S Netherlands
Oise *68 C3 river* N France
Oistins *33 G2* S Barbados
Ōita *109 B7* Ōita, Kyūshū, SW Japan
Ojinaga *28 D2* Chihuahua, N Mexico
Ojos del Salado, Cerro *42 B3 mountain* W Argentina
Okaihau *128 C2* Northland, North Island, New Zealand
Ókanizsa *see* Kanjiža
Okāra *112 C2* Punjab, E Pakistan
Okavanggo *see* Cubango/Okavango
Okavango *see* Cubango
Okavango Delta *56 C3 wetland* N Botswana
Okayama *109 B6* Okayama, Honshū, SW Japan
Okazaki *109 C6* Aichi, Honshū, C Japan
Okeechobee, Lake *21 E4 lake* Florida, SE USA
Okefenokee Swamp *21 E3 wetland* Georgia, SE USA
Okhotsk *93 G3* Khabarovskiy Kray, E Russian Federation
Okhotsk, Sea of *91 F3 sea* NW Pacific Ocean
Okhtyrka *87 F2 Rus.* Akhtyrka. Sums'ka Oblast', NE Ukraine
Oki-guntō *see* Oki-shotō
Okinawa *108 A3 island* SW Japan
Okinawa-shoto *108 A3 island group* Nansei-shotō, SW Japan Asia
Oki-shoto *109 B6 var.* Oki-guntō. *island group* SW Japan
Oklahoma *27 F2 off.* State of Oklahoma, *also known as* The Sooner State. *state* C USA
Oklahoma City *27 G1 state capital* Oklahoma, C USA
Okmulgee *27 G1* Oklahoma, C USA
Oko, Wadi *50 C3 river* NE Sudan
Oktyabr'skiy *89 D6* Volgogradskaya Oblast', SW Russian Federation
Oktyabr'skiy *see* Aktsyabrski
Oktyabr'skoy Revolyutsii, Ostrov *93 E2 Eng.* October Revolution Island. *island* Severnaya Zemlya, N Russian Federation
Okulovka *88 B4 var.* Okulovka. Novgorodskaya Oblast', W Russian Federation
Okulovka *see* Okulovka
Okushiri-to *108 C3 var.* Okusiri Tô. *island* NE Japan
Okusiri Tô *see* Okushiri-tō
Oláh-Toplicza *see* Toplița
Öland *63 C7 island* S Sweden
Olavarría *43 D5* Buenos Aires, E Argentina
Oława *76 C4 Ger.* Ohlau. Dolnośląskie, SW Poland
Olbia *75 A5 prev.* Terranova Pausania. Sardegna, Italy, C Mediterranean Sea
Old Bay State/Old Colony State *see* Massachusetts
Old Dominion *see* Virginia
Oldebroek *64 D3* Gelderland, E Netherlands
Oldenburg *72 B3* Niedersachsen, NW Germany
Oldenburg *72 C2 var.* Oldenburg in Holstein. Schleswig-Holstein, N Germany
Oldenburg in Holstein *see* Oldenburg

Oldenzaal – Pará

Oldenzaal 64 E3 Overijssel, E Netherlands
Old Harbour 32 B5 C Jamaica
Old Line State see Maryland
Old North State see North Carolina
Olëkma 93 F4 river C Russian Federation
Olëkminsk 93 F3 Respublika Sakha (Yakutiya), NE Russian Federation
Oleksandrivka 87 E3 Rus. Aleksandrovka. Kirovohrads'ka Oblast', C Ukraine
Oleksandriya 87 F3 Rus. Aleksandriya. Kirovohrads'ka Oblast', C Ukraine
Olenegorsk 88 C2 Murmanskaya Oblast', NW Russian Federation
Olenëk 93 E3 Respublika Sakha (Yakutiya), NE Russian Federation
Olenëk 93 E3 river NE Russian Federation
Oléron, Île d' 69 A5 island W France
Olevs'k 86 D1 Rus. Olevsk. Zhytomyrs'ka Oblast', N Ukraine
Olevsk see Olevs'k
Ölgiy 104 C2 Bayan-Ölgiy, W Mongolia
Olhão 70 B5 Faro, S Portugal
Olifa 56 B3 Kunene, NW Namibia
Ólimbos see Ólympos
Olimpo see Fuerte Olimpo
Olisipo see Lisboa
Olita see Alytus
Oliva 71 F4 País Valenciano, E Spain
Olivet 68 C4 Loiret, C France
Olmaliq 101 E2 Rus. Almalyk. Toshkent Viloyati, E Uzbekistan
Olmütz see Olomouc
Olomouc 77 C5 Ger. Olmütz, Pol. Ołomuniec. Olomoucký Kraj, E Czech Republic
Ołomuniec see Olomouc
Olonets 88 B3 Respublika Kareliya, NW Russian Federation
Olovyannaya 93 F4 Chitinskaya Oblast', S Russian Federation
Olpe 72 B4 Nordrhein-Westfalen, W Germany
Olshanka see Vil'shanka
Olsnitz see Murska Sobota
Olsztyn 76 D2 Ger. Allenstein. Warmińsko-Mazurskie, N Poland
Olt 86 B5 var. Oltul, Ger. Alt. river S Romania
Oltenița 86 C5 prev. Eng. Oltenitsa; anc. Constantiola. Călărași, SE Romania
Oltenitsa see Oltenița
Oltul see Olt
Olvera 70 D5 Andalucía, S Spain
Ol'viopol' see Pervomays'k
Olympia 24 B2 state capital Washington, NW USA
Olympic Mountains 24 A2 mountain range Washington, NW USA
Olympus, Mount 82 B4 var. Ólimbos, Eng. Mount Olympus. mountain N Greece
Omagh 67 B5 Ir. An Ómaigh. W Northern Ireland, United Kingdom
Omaha 23 F4 Nebraska, C USA
Oman 99 D6 off. Sultanate of Oman, Ar. Salṭanat 'Umān; prev. Muscat and Oman. country SW Asia
Oman, Gulf of 98 E4 Ar. Khalīj 'Umān. gulf N Arabian Sea
Oman, Sultanate of see Oman
Omboué 55 A6 Ogooué-Maritime, W Gabon
Omdurman 50 B4 var. Umm Durmān. Khartoum, C Sudan
Ometepe, Isla de 30 D4 island S Nicaragua
Ommen 64 E3 Overijssel, E Netherlands
Omsk 92 C4 Omskaya Oblast', C Russian Federation
Ōmuta 109 A7 Fukuoka, Kyūshū, SW Japan
Onda 71 F3 País Valenciano, E Spain
Ondjiva see N'Giva
Öndörhaan 105 E2 var. Undur Khan; prev. Tsetsen Khan. Hentiy, E Mongolia
Onega 88 C3 Arkhangel'skaya Oblast', NW Russian Federation
Onega 88 B4 river NW Russian Federation
Onega, Lake see Onezhskoye Ozero
Onex 73 A7 Genève, SW Switzerland
Onezhskoye Ozero 88 B4 Eng. Lake Onega. lake NW Russian Federation
Ongole 110 D1 Andhra Pradesh, E India
Onitsha 53 G5 Anambra, S Nigeria
Onon Gol 105 E2 river N Mongolia
Onslow 124 A4 Western Australia
Onslow Bay 21 F1 bay North Carolina, E USA
Ontario 16 B3 province S Canada
Ontario, Lake 19 E3 lake Canada/USA
Onteniente see Ontinyent
Ontinyent 71 F4 var. Onteniente. País Valenciano, E Spain
Ontong Java Rise 103 H4 undersea feature W Pacific Ocean
Onuba see Huelva
Oodeypore see Udaipur
Oos-Londen see East London
Oostakker 65 B5 Oost-Vlaanderen, NW Belgium
Oostburg 65 B5 Zeeland, SW Netherlands
Oostende 65 A5 Eng. Ostend, Fr. Ostende. West-Vlaanderen, NW Belgium
Oosterbeek 64 D4 Gelderland, SE Netherlands
Oosterhout 64 C4 Noord-Brabant, S Netherlands
Opatija 78 A2 It. Abbazia. Primorje-Gorski Kotar, NW Croatia
Opava 77 C5 Ger. Troppau. Moravskoslezský Kraj, E Czech Republic
Opazova see Stara Pazova
Opelika 20 D2 Alabama, S USA
Opelousas 20 B3 Louisiana, S USA
Ophiusa see Formentera
Opmeer 64 C2 Noord-Holland, NW Netherlands
Opochka 88 A4 Pskovskaya Oblast', W Russian Federation
Opole 76 C4 Ger. Oppeln. Opolskie, S Poland
Oporto see Porto
Opotiki 128 E3 Bay of Plenty, North Island, New Zealand
Oppeln see Opole
Oppidum Ubiorum see Köln
Oqtosh 101 E2 Rus. Aktash. Samarqand Viloyati, C Uzbekistan
Oradea 86 B3 prev. Oradea Mare, Ger. Grosswardein, Hung. Nagyvárad. Bihor, NW Romania
Oradea Mare see Oradea
Orahovac see Rahovec
Oral see Ural'sk
Oran 48 D2 var. Ouahran, Wahran. NW Algeria
Orange 127 D6 New South Wales, SE Australia
Orange 69 D6 anc. Arausio. Vaucluse, SE France
Orangeburg 21 E2 South Carolina, SE USA
Orange Cone see Orange Fan
Orange Fan 47 C7 var. Orange Cone. undersea feature SW Indian Ocean
Orange Mouth/Orangemund see Oranjemund
Orange River 56 B4 Afr. Oranjerivier. river S Africa
Orange Walk 30 C1 Orange Walk, N Belize
Oranienburg 72 D3 Brandenburg, NE Germany
Oranjemund 56 B4 var. Orangemund; prev. Orange Mouth. Karas, SW Namibia
Oranjerivier see Orange River
Oranjestad 33 E5 dependent territory capital (Aruba) Lesser Antilles, S Caribbean Sea
Orany see Varėna
Orașul Stalin see Brașov
Oravicabánya see Oravița
Oravița 86 A4 Ger. Orawitza, Hung. Oravicabánya. Caraș-Severin, SW Romania
Orawitza see Oravița
Orbetello 74 B4 Toscana, C Italy
Orcadas 132 A1 Argentinian research station South Orkney Islands, Antarctica
Orchard Homes 22 B1 Montana, NW USA
Ordino 69 A8 Ordino, NW Andorra Europe
Ordos Desert see Mu Us Shadi
Ordu 94 D2 anc. Cotyora. Ordu, N Turkey
Ordzhonikidze 87 F3 Dnipropetrovs'ka Oblast', E Ukraine
Ordzhonikidze see Vladikavkaz, Russian Federation
Ordzhonikidze see Yenakiyeve, Ukraine
Orealla 37 G3 E Guyana
Örebro 63 C6 Örebro, C Sweden
Oregon 24 B3 off. State of Oregon, also known as Beaver State, Sunset State, Valentine State, Webfoot State. state NW USA
Oregon City 24 B3 Oregon, NW USA
Oregon, State of see Oregon
Orekhov see Orikhiv
Orël 89 B5 Orlovskaya Oblast', W Russian Federation
Orem 22 B4 Utah, W USA
Ore Mountains see Erzgebirge/Krušné Hory
Orenburg 89 D6 prev. Chkalov. Orenburgskaya Oblast', W Russian Federation
Orense see Ourense
Orestiáda 82 D3 prev. Orestiás. Anatolikí Makedonía kai Thráki, NE Greece
Orestiás see Orestiáda
Organ Peak 26 D3 mountain New Mexico, SW USA
Orgeyev see Orhei
Orhei 86 D3 var. Orheiu, Rus. Orgeyev. N Moldova
Orheiu see Orhei
Oriental, Cordillera 38 D3 mountain range Bolivia/Peru
Oriental, Cordillera 36 B3 mountain range C Colombia
Orihuela 71 F4 País Valenciano, E Spain
Orikhiv 87 G3 Rus. Orekhov. Zaporiz'ka Oblast', SE Ukraine
Orinoco, Río 37 E2 river Colombia/Venezuela
Orissa see Odisha
Orissaar see Orissaare
Orissaare 84 C2 Ger. Orissaar. Saaremaa, W Estonia
Oristano 75 A5 Sardegna, Italy, C Mediterranean Sea
Orito 36 A4 Putumayo, SW Colombia
Orizaba, Volcán Pico de 13 C7 var. Citlaltépetl. mountain S Mexico
Orkney see Orkney Islands
Orkney Islands 66 C2 var. Orkney, Orkneys. island group N Scotland, United Kingdom
Orkneys see Orkney Islands
Orlando 21 E4 Florida, SE USA
Orléanais 68 C4 cultural region C France
Orléans 68 C4 anc. Aurelianum. Loiret, C France
Orléansville see Chlef
Orly 68 E2 (Paris) Essonne, N France
Orlya 85 B6 Hrodzyenskaya Voblasts', W Belarus
Ormsö see Vormsi
Ormuz, Strait of see Hormuz, Strait of
Örnsköldsvik 63 C5 Västernorrland, C Sweden
Orolaunum see Arlon
Orol Dengizi see Aral Sea
Oromocto 17 F4 New Brunswick, SE Canada
Orona 123 F3 prev. Hull Island. atoll Phoenix Islands, C Kiribati
Oropeza see Cochabamba
Orosirá Rodhópis see Rhodope Mountains
Orpington 67 B8 United Kingdom
Orschowa see Orșova
Orsha 85 E6 Vitsyebskaya Voblasts', NE Belarus
Orsk 92 B4 Orenburgskaya Oblast', W Russian Federation
Orșova 86 A4 Ger. Orschowa, Hung. Orsova. Mehedinți, SW Romania
Ortelsburg see Szczytno
Orthez 69 B6 Pyrénées-Atlantiques, SW France
Ortona 74 D4 Abruzzo, C Italy
Oruba see Aruba
Orūmīyeh, Daryācheh-ye 99 C2 var. Matianus, Sha Hi, Urumi Yeh, Eng. Lake Urmia; prev. Daryācheh-ye Reẓā'īyeh. lake NW Iran
Oruro 39 F4 Oruro, W Bolivia
Oryokko see Yalu
Oss 64 D4 Noord-Brabant, S Netherlands
Ōsaka 109 C6 hist. Naniwa. Ōsaka, Honshū, SW Japan
Ōsaki see Furukawa
Osa, Península de 31 E5 peninsula S Costa Rica
Osborn Plateau 119 D5 undersea feature E Indian Ocean
Osca see Huesca
Ösel see Saaremaa
Osh 101 F2 Oshskaya Oblast', SW Kyrgyzstan
Oshawa 16 D5 Ontario, SE Canada
Oshikango 56 B3 Ohangwena, N Namibia
O-shima 109 D6 island S Japan
Oshkosh 18 B2 Wisconsin, N USA
Oshmyany see Ashmyany
Osiek see Osijek
Osijek 78 C3 prev. Osiek, Osjek, Ger. Esseg, Hung. Eszék. Osijek-Baranja, E Croatia
Osipenko see Berdyans'k
Osipovichi see Asipovichy
Osjek see Osijek
Oskaloosa 23 G4 Iowa, C USA
Oskarshamn 63 C7 Kalmar, S Sweden
Öskemen see Ust'-Kamenogorsk
Oskol 87 G2 Rus. Oskil. river Russian Federation/Ukraine
Oskil see Oskol
Oslo 63 B6 prev. Christiania, Kristiania. country capital (Norway) Oslo, S Norway
Osmaniye 94 D4 Osmaniye, S Turkey
Osnabrück 72 A3 Niedersachsen, NW Germany
Osogov Mountains 82 B3 var. Osogovske Planine, Osogovski Planina, Mac. Osogovski Planini. mountain range Bulgaria/FYR Macedonia
Osogovske Planine/Osogovski Planina/ Osogovski Planini see Osogov Mountains
Oșorhei see Târgu Mureș
Osorno 43 B5 Los Lagos, C Chile
Ossa, Serra d' 70 C4 mountain range SE Portugal
Ossora 93 H2 Koryakskiy Avtonomnyy Okrug, E Russian Federation
Ostee see Baltic Sea
Ostend/Ostende see Oostende
Oster 87 E1 Chernihivs'ka Oblast', N Ukraine
Östermyra see Seinäjoki
Osterode/Osterode in Ostpreussen see Ostróda
Österreich see Austria
Östersund 63 C5 Jämtland, C Sweden
Ostia Aterni see Pescara
Ostiglia 74 C2 Lombardia, N Italy
Ostrava 77 C5 Moravskoslezský Kraj, E Czech Republic
Ostróda 76 D3 Ger. Osterode, Osterode in Ostpreussen. Warmińsko-Mazurskie, NE Poland
Ostrołęka 76 D3 Ger. Wiesenhof, Rus. Ostrolenka. Mazowieckie, C Poland
Ostrolenka see Ostrołęka
Ostrov 88 A4 Latv. Austrava. Pskovskaya Oblast', W Russian Federation
Ostrovets see Ostrowiec Świętokrzyski
Ostrovnoy 88 C2 Murmanskaya Oblast', NW Russian Federation
Ostrów see Ostrów Wielkopolski
Ostrowiec see Ostrowiec Świętokrzyski
Ostrowiec Świętokrzyski 76 D4 var. Ostrowiec, Rus. Ostrovets. Świętokrzyskie, C Poland
Ostrów Mazowiecka 76 D3 var. Ostrów Mazowiecki. Mazowieckie, NE Poland
Ostrów Mazowiecki see Ostrów Mazowiecka
Ostrowo see Ostrów Wielkopolski
Ostrów Wielkopolski 76 C4 var. Ostrów, Ger. Ostrowo. Wielkopolskie, C Poland
Ostyako-Voguls'k see Khanty-Mansiysk
Osum see Osumit, Lumi i
Ōsumi-shotō 109 A8 island group Kagoshima, Nansei-shotō, SW Japan Asia East China Sea Pacific Ocean
Osumit, Lumi i 79 D7 var. Osum. river SE Albania
Osuna 70 D4 Andalucía, S Spain
Oswego 19 F2 New York, NE USA
Otago Peninsula 129 B7 peninsula South Island, New Zealand
Otaki 128 D4 Wellington, North Island, New Zealand
Otaru 108 C2 Hokkaidō, NE Japan
Otavalo 38 B1 Imbabura, N Ecuador
Otavi 56 B3 Otjozondjupa, N Namibia
Oțelu Roșu 86 B4 Ger. Ferdinandsberg, Hung. Nándorhgy. Caraș-Severin, SW Romania
Otepää 84 D3 Ger. Odenpäh. Valgamaa, SE Estonia
Oti 53 E4 river N Togo
Otira 129 C6 West Coast, South Island, New Zealand
Otjiwarongo 56 B3 Otjozondjupa, N Namibia
Otorohanga 128 D3 Waikato, North Island, New Zealand
Otranto, Canale d' see Otranto, Strait of
Otranto, Strait of 79 C6 It. Canale d'Otranto. strait Albania/Italy
Otrokovice 77 C5 Ger. Otrokowitz. Zlínský Kraj, E Czech Republic
Otrokowitz see Otrokovice
Ōtsu 109 C6 var. Ōtu. Shiga, Honshū, SW Japan
Ottawa 16 D5 country capital (Canada) Ontario, SE Canada
Ottawa 18 B3 Illinois, N USA
Ottawa 23 F5 Kansas, C USA
Ottawa 19 E2 Fr. Outaouais. river Ontario/Québec, SE Canada
Ottawa Islands 16 C1 island group Nunavut, C Canada
Ottignies 65 C6 Wallon Brabant, C Belgium
Ottumwa 23 G4 Iowa, C USA
Ōtu see Ōtsu
Ouachita Mountains 20 A1 mountain range Arkansas/Oklahoma, C USA
Ouachita River 20 B2 river Arkansas/Louisiana, C USA
Ouagadougou 53 E4 var. Wagadugu. country capital (Burkina) C Burkina
Ouahigouya 53 E3 NW Burkina
Ouahran see Oran
Oualâta 52 D3 var. Oualata. Hodh ech Chargui, SE Mauritania
Ouanary 37 H3 E French Guiana
Ouanda Djallé 54 D4 Vakaga, NE Central African Republic
Ouarâne 52 D2 desert C Mauritania
Ouargla 49 E2 var. Wargla. NE Algeria
Ouarzazate 48 C3 S Morocco
Oubangui 55 C5 Fr. Oubangui. river C Africa
Oubangui see Ubangi
Oubangui-Chari see Central African Republic
Oubangui-Chari, Territoire de l' see Central African Republic
Oudjda see Oujda
Ouessant, Île d' 68 A3 Eng. Ushant. island NW France
Ouésso 55 B5 Sangha, NW Congo
Oujda 48 D2 Ar. Oudjda, Ujda. NE Morocco
Oujeft 52 C2 Adrar, C Mauritania
Oulu 62 D4 Swe. Uleåborg. Oulu, C Finland
Oulujärvi 62 D4 Swe. Uleträsk. lake C Finland
Oulujoki 62 D4 Swe. Uleälv. river C Finland
Ounasjoki 62 D3 river N Finland
Ounianga Kébir 54 C2 Borkou-Ennedi-Tibesti, N Chad
Oup see Auob
Oupeye 65 D6 Liège, E Belgium
Our 65 D6 river NW Europe
Ourense 70 C1 Cast. Orense, Lat. Aurium. Galicia, NW Spain
Ourique 70 B4 Beja, S Portugal
Ours, Grand Lac de l' see Great Bear Lake
Ourthe 65 D7 river E Belgium
Ouse 67 D5 river N England, United Kingdom
Outaouais see Ottawa
Outer Hebrides 66 B3 var. Western Isles. island group NW Scotland, United Kingdom
Outer Islands 57 G1 island group SW Seychelles Africa W Indian Ocean
Outes 70 B1 Galicia, NW Spain
Ouvéa 122 D5 island Îles Loyauté, NE New Caledonia
Ouyen 127 C6 Victoria, SE Australia
Ovalle 42 B3 Coquimbo, N Chile
Ovar 70 B2 Aveiro, N Portugal
Overflakkee 64 B4 island SW Netherlands
Overijse 65 C6 Vlaams Brabant, C Belgium
Oviedo 70 C1 anc. Asturias. Asturias, NW Spain
Ovilava see Wels
Ovruch 86 D1 Zhytomyrs'ka Oblast', N Ukraine
Owando 55 B5 prev. Fort Rousset. Cuvette, C Congo
Owase 109 C6 Mie, Honshū, SW Japan
Owatonna 23 F3 Minnesota, N USA
Owen Fracture Zone 118 B4 tectonic feature W Arabian Sea
Owen, Mount 129 C5 mountain South Island, New Zealand
Owensboro 18 B5 Kentucky, S USA
Owen Stanley Range 122 B3 mountain range S Papua New Guinea
Owerri 53 G5 Imo, S Nigeria
Owo 53 F5 Ondo, SW Nigeria
Owyhee River 24 C4 river Idaho/Oregon, NW USA
Oxford 129 C6 Canterbury, South Island, New Zealand
Oxford 67 D6 Lat. Oxonia. S England, United Kingdom
Oxkutzcab 29 H4 Yucatán, SE Mexico
Oxnard 25 B7 California, W USA
Oxonia see Oxford
Oxus see Amu Darya
Oyama 109 D5 Tochigi, Honshū, S Japan
Oyem 55 B5 Woleu-Ntem, N Gabon
Oyo 55 B6 Cuvette, C Congo
Oyo 53 F4 Oyo, W Nigeria
Ozark 20 D3 Alabama, S USA
Ozark Plateau 23 G5 plain Arkansas/Missouri, C USA
Ozarks, Lake of the 23 F5 reservoir Missouri, C USA
Ozbourn Seamount 130 D4 undersea feature W Pacific Ocean
Ózd 77 D6 Borsod-Abaúj-Zemplén, NE Hungary
Ozieri 75 A5 Sardegna, Italy, C Mediterranean Sea

P

Paamiut 60 B4 var. Pâmiut, Dan. Frederikshåb. S Greenland
Pa-an see Hpa-an
Pabianice 76 C4 Łódzski, Poland
Pabna 113 G4 Rajshahi, W Bangladesh
Pacaraima, Sierra/Pacaraim, Serra see Pakaraima Mountains
Pachuca 29 E4 var. Pachuca de Soto. Hidalgo, C Mexico
Pachuca de Soto see Pachuca
Pacific-Antarctic Ridge 132 B5 undersea feature S Pacific Ocean
Pacific Ocean 130 D3 ocean
Padalung see Phatthalung
Padang 116 B4 Sumatera, W Indonesia
Paderborn 72 B4 Nordrhein-Westfalen, NW Germany
Padma see Brahmaputra
Padma see Ganges
Padova 74 C2 Eng. Padua; anc. Patavium. Veneto, NE Italy
Padre Island 27 G5 island Texas, SW USA
Padua see Padova
Paducah 18 B5 Kentucky, S USA
Paeroa 128 D3 Waikato, North Island, New Zealand
Páfos 80 C5 var. Paphos. W Cyprus
Pag 78 A3 It. Pago. island Zadar, C Croatia
Page 26 B1 Arizona, SW USA
Pago see Pag
Pago Pago 123 F4 dependent territory capital (American Samoa) Tutuila, W American Samoa
Pahiatua 128 D4 Manawatu-Wanganui, North Island, New Zealand
Pahsien see Chongqing
Paide 84 D2 Ger. Weissenstein. Järvamaa, N Estonia
Paihia 128 D2 Northland, North Island, New Zealand
Päijänne 63 D5 lake S Finland
Paine, Cerro 43 A7 mountain S Chile
Painted Desert 26 B1 desert Arizona, SW USA
Paisance see Piacenza
Paisley 66 C4 W Scotland, United Kingdom
País Valenciano 71 F3 var. Valencia, Cat. València; anc. Valentia. autonomous community NE Spain
País Vasco 71 E1 cultural region N Spain
Paita 38 B3 Piura, NW Peru
Pakanbaru see Pekanbaru
Pakaraima Mountains 37 E3 var. Serra Pacaraim, Sierra Pacaraima. mountain range N South America
Pakistan 112 A2 off. Islamic Republic of Pakistan, var. Islami Jamhuriya e Pakistan. country S Asia
Pakistan, Islamic Republic of see Pakistan
Pakistan, Islami Jamhuriya e see Pakistan
Paknam see Samut Prakan
Pakokku 114 A3 Magway, C Myanmar (Burma)
Pak Phanang 115 C7 var. Ban Pak Phanang. Nakhon Si Thammarat, SW Thailand
Pakruojis 84 C4 Šiaulai, N Lithuania
Paks 77 C7 Tolna, S Hungary
Paksé see Pakxé
Pakxé 115 D5 var. Paksé. Champasak, S Laos
Palafrugell 71 G2 Cataluña, NE Spain
Palagruža 79 B5 It. Pelagosa. island SW Croatia
Palaiá Epídavros 83 C6 Pelopónnisos, S Greece
Palaiseau 68 D2 Essonne, N France
Palamós 71 G2 Cataluña, NE Spain
Palamuse 84 E2 Ger. Sankt-Bartholomäi. Jõgevamaa, E Estonia
Palanka see Bačka Palanka
Pālanpur 112 C4 Gujarāt, W India
Palantia see Palencia
Palapye 56 D3 Central, SE Botswana
Palau 122 A2 var. Belau. country W Pacific Ocean
Palawan 117 E2 island W Philippines
Palawan Passage 116 D2 passage W Philippines
Paldiski 84 D2 prev. Baltiski, Eng. Baltic Port, Ger. Baltischport. Harjumaa, NW Estonia
Palembang 116 B4 Sumatera, W Indonesia
Palencia 70 D2 anc. Palantia, Pallantia. Castilla-León, NW Spain
Palerme see Palermo
Palermo 75 C7 Fr. Palerme; anc. Panhormus, Panormus. Sicilia, Italy, C Mediterranean Sea
Pāli 112 C3 Rājasthān, N India
Palikir 122 C2 country capital (Micronesia) Pohnpei, E Micronesia
Palimé see Kpalimé
Palioúri, Akrotírio 82 C4 var. Akrotírio Kanestron. headland N Greece
Palk Strait 110 C3 strait India/Sri Lanka
Pallantia see Palencia
Palliser, Cape 129 D5 headland North Island, New Zealand
Palma 71 G3 var. Palma de Mallorca. Mallorca, Spain, W Mediterranean Sea
Palma del Río 70 D4 Andalucía, S Spain
Palma de Mallorca see Palma
Palmar Sur 31 E5 Puntarenas, SE Costa Rica
Palma Soriano 32 C3 Santiago de Cuba, E Cuba
Palm Beach 126 E1 New South Wales, E Australia
Palmer 132 A2 US research station Antarctica
Palmer Land 132 A3 physical region Antarctica
Palmerston 123 F4 island S Cook Islands
Palmerston see Darwin
Palmerston North 128 D4 Manawatu-Wanganui, North Island, New Zealand
Palmetto State, The see South Carolina
Palmi 75 D7 Calabria, SW Italy
Palmira 36 B3 Valle del Cauca, W Colombia
Palm Springs 25 D7 California, W USA
Palmyra see Tudmur
Palmyra Atoll 123 G2 US privately owned unincorporated territory C Pacific Ocean
Palo Alto 25 B6 California, W USA
Paloe see Denpasar, Bali, C Indonesia
Paloe see Palu
Palu 117 E4 prev. Paloe. Sulawesi, C Indonesia
Pamiers 69 B6 Ariège, S France
Pamir 101 F3 var. Daryā-ye Pāmīr, Taj. Dar"yoi Pomir. river Afghanistan/Tajikistan
Pāmīr, Daryā-ye see Pamir
Pamir/Pāmīr, Daryā-ye see Pamirs
Pamirs 101 F3 Pash. Daryā-ye Pāmīr, Rus. Pamir. mountain range C Asia
Pâmiut see Paamiut
Pamlico Sound 21 G1 sound North Carolina, SE USA
Pampa 27 E1 Texas, SW USA
Pampa Aullagas, Lago see Poopó, Lago
Pampas 42 C4 plain C Argentina
Pampeluna see Pamplona
Pamplona 36 C2 Norte de Santander, N Colombia
Pamplona 71 E1 Basq. Iruña, prev. Pampeluna; anc. Pompaelo. Navarra, N Spain
Panaji 110 B1 var. Pangim, Panjim, New Goa. state capital Goa, W India
Panamá 31 G4 var. Ciudad de Panama, Eng. Panama City. country capital (Panama) Panamá, C Panama
Panama 31 G5 off. Republic of Panama. country Central America
Panama Basin 13 C8 undersea feature E Pacific Ocean
Panama Canal 31 F4 canal E Panama
Panama City 20 D3 Florida, SE USA
Panama City see Panamá
Panamá, Golfo de 31 G5 var. Gulf of Panama. gulf S Panama
Panama, Gulf of see Panamá, Golfo de
Panama, Isthmus of see Panamá, Istmo de
Panamá, Istmo de 31 G4 Eng. Isthmus of Panama; prev. Isthmus of Darien. isthmus E Panama
Panama, Republic of see Panama
Panay Island 117 E2 island C Philippines
Pančevo 78 D3 Ger. Pantschowa, Hung. Pancsova. Vojvodina, N Serbia
Pancsova see Pančevo
Paneas see Bāniyās
Panevėžys 84 C4 Panevėžys, C Lithuania
Pangim see Panaji
Pangkalpinang 116 C4 Pulau Bangka, W Indonesia
Pang-Nga see Phang-Nga
Panhormus see Palermo
Panjim see Panaji
Panopolis see Akhmīm
Pánormos 83 C7 Kríti, Greece, E Mediterranean Sea
Panormus see Palermo
Pantanal 41 E3 var. Pantanalmato-Grossense. swamp SW Brazil
Pantanalmato-Grossense see Pantanal
Pantelleria, Isola di 75 B7 island SW Italy
Pantschowa see Pančevo
Pánuco 29 E3 Veracruz-Llave, E Mexico
Pao-chi/Paoki see Baoji
Paola 80 B5 E Malta
Pao-shan see Baoshan
Pao-t'ou/Paotow see Baotou
Papagayo, Golfo de 30 C4 gulf NW Costa Rica
Papakura 128 D3 Auckland, North Island, New Zealand
Papantla 29 F4 var. Papantla de Olarte. Veracruz-Llave, E Mexico
Papantla de Olarte see Papantla
Papeete 123 H4 dependent territory capital (French Polynesia) Tahiti, W French Polynesia
Paphos see Páfos
Papilė 84 B3 Šiaulai, NW Lithuania
Papillion 23 F4 Nebraska, C USA
Papua 117 H4 var. Irian Barat, West Irian, West New Guinea, West Papua; prev. Dutch New Guinea, Irian Jaya, Netherlands New Guinea. province E Indonesia
Papua and New Guinea, Territory of see Papua New Guinea
Papua, Gulf of 122 B3 gulf S Papua New Guinea
Papua New Guinea 122 B3 off. Independent State of Papua New Guinea; prev. Territory of Papua and New Guinea. country NW Melanesia
Papua New Guinea, Independent State of see Papua New Guinea
Papuk 78 C3 mountain range NE Croatia
Pará 41 E2 off. Estado do Pará. state NE Brazil

Pokrovka – Qeshm

Pokrovka *see* Kyzyl-Suu
Pokrovs'ke *87 G3 Rus.* Pokrovskoye. Dnipropetrovs'ka Oblast', E Ukraine
Pokrovskoye *see* Pokrovs'ke
Pola *see* Pula
Pola de Lena *see* La Pola
Poland *76 B4 off.* Republic of Poland, *var.* Polish Republic, *Pol.* Polska, Rzeczpospolita Polska; *prev. Pol.* Polska Rzeczpospolita Ludowa, The Polish People's Republic. *country* C Europe
Poland, Republic of *see* Poland
Polatlı *94 C3* Ankara, C Turkey
Polatsk *85 D5 Rus.* Polotsk. Vitsyebskaya Voblasts', N Belarus
Pol-e Khomrī *see* Pul-e Khumrī
Poli *see* Pólis
Polikastro/Polýkastron *see* Polýkastro
Polikrayshte *see* Dolna Oryakhovitsa
Pólis *80 C5 var.* Poli. W Cyprus
Polish People's Republic, The *see* Poland
Polish Republic *see* Poland
Polkowice *76 B4 Ger.* Heerwegen. Dolnośląskie, W Poland
Pollença *71 G3* Mallorca, Spain, W Mediterranean Sea
Pologi *see* Polohy
Polohy *87 G3 Rus.* Pologi. Zaporiz'ka Oblast', SE Ukraine
Polokwane *56 D4 prev.* Pietersburg. Limpopo, NE South Africa
Polonne *86 D2 Rus.* Polonnoye. Khmel'nyts'ka Oblast', NW Ukraine
Polonnoye *see* Polonne
Polotsk *see* Polatsk
Polska/Polska, Rzeczpospolita/Polska Rzeczpospolita Ludowa *see* Poland
Polski Trŭmbesh *82 D2 prev.* Polsko Kosovo. Ruse, N Bulgaria
Polsko Kosovo *see* Polski Trŭmbesh
Poltava *87 F2* Poltavs'ka Oblast', NE Ukraine
Poltoratsk *see* Aşgabat
Põlva *84 E3 Ger.* Põlwe. Põlvamaa, SE Estonia
Põlwe *see* Põlva
Polyarnyy *88 C2* Murmanskaya Oblast', NW Russian Federation
Polýkastro *82 B3 var.* Polikastro; *prev.* Políkastron. Kentrikí Makedonía, N Greece
Polynesia *121 F4 island group* C Pacific Ocean
Pomeranian Bay *72 D2 Ger.* Pommersche Bucht, *Pol.* Zatoka Pomorska. *bay* Germany/Poland
Pomir, Dar"yoi *see* Pamir/Pāmīr, Daryā-ye
Pommersche Bucht *see* Pomeranian Bay
Pomorska, Zatoka *see* Pomeranian Bay
Pomorskiy Proliv *88 D2 strait* NW Russian Federation
Po, Mouth of the *74 C2 var.* Bocche del Po. *river* NE Italy
Pompaelo *see* Pamplona
Pompano Beach *21 F5* Florida, SE USA
Ponape Ascension Island *see* Pohnpei
Ponca City *27 G1* Oklahoma, C USA
Ponce *33 F3* C Puerto Rico
Pondicherry *110 C2 var.* Puducheri, *Fr.* Pondichéry. Pondicherry, SE India
Ponferrada *70 C1* Castilla-León, NW Spain
Poniatowa *76 E4* Lubelskie, E Poland
Pons Aelii *see* Newcastle upon Tyne
Pons Vetus *see* Pontevedra
Ponta Delgada *70 B5* São Miguel, Azores, Portugal, NE Atlantic Ocean
Ponta Grossa *41 E4* Paraná, S Brazil
Pontarlier *68 D4* Doubs, E France
Ponteareas *70 B2* Galicia, NW Spain
Ponte da Barca *70 B2* Viana do Castelo, N Portugal
Pontevedra *70 B1 anc.* Pons Vetus. Galicia, NW Spain
Pontiac *18 D3* Michigan, N USA
Pontianak *116 C4* Borneo, C Indonesia
Pontisarae *see* Pontoise
Pontivy *68 A3* Morbihan, NW France
Pontoise *68 C3 anc.* Briva Isarae, Cergy-Pontoise, Pontisarae. Val-d'Oise, N France
Ponziane Island *75 C5 island* C Italy
Poole *67 D7* S England, United Kingdom
Poona *see* Pune
Poopó, Lago *39 F4 var.* Lago Pampa Aullagas. *lake* W Bolivia
Popayán *36 B4* Cauca, SW Colombia
Poperinge *65 A6* West-Vlaanderen, W Belgium
Poplar Bluff *23 G5* Missouri, C USA
Popocatépetl *29 E4 volcano* S Mexico
Popper *see* Poprad
Poprad *77 D5 Ger.* Deutschendorf, *Hung.* Poprád. Prešovský Kraj, E Slovakia
Poprád *77 D5 Ger.* Popper, *Hung.* Poprád. *river* Poland/Slovakia
Porbandar *112 B4* Gujarāt, W India
Porcupine Plain *58 B3 undersea feature* E Atlantic Ocean
Pordenone *74 C2 anc.* Portenau. Friuli-Venezia Giulia, NE Italy
Poreč *78 A2 It.* Parenzo. Istra, NW Croatia
Porech'ye *see* Parechcha
Pori *63 D5 Swe.* Björneborg. Länsi-Suomi, SW Finland
Porirua *129 D5* Wellington, North Island, New Zealand
Porkhov *88 A4* Pskovskaya Oblast', W Russian Federation
Porlamar *37 E1* Nueva Esparta, NE Venezuela
Póros *83 C6* Póros, S Greece
Póros *83 A5* Kefallinía, Iónia Nisiá, Greece, C Mediterranean Sea
Pors *see* Porsangenfijorden
Porsangenfjorden *62 D2 Lapp.* Pors. *fjord* N Norway
Porsgrunn *63 B6* Telemark, S Norway
Portachuelo *39 G4* Santa Cruz, C Bolivia
Portadown *67 B5 Ir.* Port An Dúnáin. S Northern Ireland, United Kingdom
Portalegre *70 C3 anc.* Ammaia, Amoea. Portalegre, E Portugal
Port Alexander *14 D4* Baranof Island, Alaska, USA
Port Alfred *56 D5* Eastern Cape, S South Africa
Port Amelia *see* Pemba
Port An Dúnáin *see* Portadown
Port Angeles *24 B1* Washington, NW USA
Port Antonio *32 B5* NE Jamaica
Port Arthur *27 H4* Texas, SW USA
Port Augusta *127 B6* South Australia
Port-au-Prince *32 D3 country capital* (Haiti) C Haiti
Port Blair *111 F2* Andaman and Nicobar Islands, SE India
Port Charlotte *21 E4* Florida, SE USA
Port Darwin *see* Darwin
Port d'Envalira *69 B8* E Andorra Europe
Port Douglas *126 D3* Queensland, NE Australia
Port Elizabeth *56 C5* Eastern Cape, S South Africa
Portenau *see* Pordenone
Porterville *25 C7* California, W USA
Port-Étienne *see* Nouâdhibou
Port Florence *see* Kisumu
Port-Francqui *see* Ilebo
Port-Gentil *55 A6* Ogooué-Maritime, W Gabon
Port Harcourt *53 G5* Rivers, S Nigeria
Port Hardy *14 D5* Vancouver Island, British Columbia, SW Canada
Port Harrison *see* Inukjuak
Port Hedland *124 B4* Western Australia
Port Huron *18 D3* Michigan, N USA
Portimão *70 B4 var.* Vila Nova de Portimão. Faro, S Portugal
Port Jackson *126 E1 harbour* New South Wales, E Australia
Portland *127 B7* Victoria, SE Australia
Portland *19 G2* Maine, NE USA
Portland *24 B3* Oregon, NW USA
Portland *27 G4* Texas, SW USA
Portland Bight *32 B5 bay* S Jamaica
Portlaoighise *see* Port Laoise
Port Laoise *67 B6 var.* Portlaoise, *Ir.* Portlaoighise; *prev.* Maryborough. C Ireland
Portlaoise *see* Port Laoise
Port Lavaca *27 G4* Texas, SW USA
Port Lincoln *127 A6* South Australia
Port Louis *57 H3 country capital* (Mauritius) NW Mauritius
Port-Lyautey *see* Kénitra
Port Macquarie *127 E6* New South Wales, SE Australia
Port Mahon *see* Mahón
Portmore *32 B5* C Jamaica
Port Moresby *122 B3 country capital* (Papua New Guinea) Central/National Capital District, SW Papua New Guinea
Port Natal *see* Durban
Porto *70 B2 Eng.* Oporto; *anc.* Portus Cale. Porto, NW Portugal
Porto Alegre *41 E5 var.* Pôrto Alegre. *state capital* Rio Grande do Sul, S Brazil
Porto Alegre *54 E2* São Tomé, S Sao Tome and Principe, Africa
Porto Alexandre *see* Tombua
Porto Amélia *see* Pemba
Porto Bello *see* Portobelo
Portobelo *31 G4 var.* Porto Bello, Puerto Bello. Colón, N Panama
Port O'Connor *27 G4* Texas, SW USA
Porto Edda *see* Sarandë
Portoferraio *74 B4* Toscana, C Italy
Port-of-Spain *33 H5 country capital* (Trinidad and Tobago) Trinidad, Trinidad and Tobago
Porto Grande *see* Mindelo
Portogruaro *74 C2* Veneto, NE Italy
Porto-Novo *53 F5 country capital* (Benin) S Benin
Porto Rico *see* Puerto Rico
Porto Santo *48 A2 var.* Ilha do Porto Santo. *island* Madeira, Portugal, NE Atlantic Ocean
Porto Santo, Ilha do *see* Porto Santo
Porto Torres *75 A5* Sardegna, Italy, C Mediterranean Sea
Porto Velho *40 D2 var.* Velho. *state capital* Rondônia, W Brazil
Portoviejo *38 A2 var.* Puertoviejo. Manabí, W Ecuador
Port Pirie *127 B6* South Australia
Port Rex *see* East London
Port Said *see* Būr Sa'īd
Portsmouth *67 D7* S England, United Kingdom
Portsmouth *19 G3* New Hampshire, NE USA
Portsmouth *18 D4* Ohio, N USA
Portsmouth *19 F5* Virginia, NE USA
Port Stanley *see* Stanley
Port Sudan *50 C3* Red Sea, NE Sudan
Port Swettenham *see* Klang/Pelabuhan Klang
Port Talbot *67 C7* S Wales, United Kingdom
Portugal *70 B3 off.* Portuguese Republic. *country* SW Europe
Portuguese East Africa *see* Mozambique
Portuguese Guinea *see* Guinea-Bissau
Portuguese Republic *see* Portugal
Portuguese Timor *see* East Timor
Portuguese West Africa *see* Angola
Portus Cale *see* Porto
Portus Magnus *see* Almería
Portus Magonis *see* Mahón
Port-Vila *122 D4 var.* Vila. *country capital* (Vanuatu) Éfaté, C Vanuatu
Porvenir *39 E3* Pando, NW Bolivia
Porvenir *43 B8* Magallanes, S Chile
Porvoo *63 E6 Swe.* Borgå. Etelä-Suomi, S Finland
Porzecze *see* Parechcha
Posadas *42 D3* Misiones, NE Argentina
Poschega *see* Požega
Posen *see* Poznań
Posnania *see* Poznań
Postavy/Postawy *see* Pastavy
Posterholt *65 D5* Limburg, SE Netherlands
Postojna *73 D8 Ger.* Adelsberg, *It.* Postumia. SW Slovenia
Postumia *see* Postojna
Pöstyén *see* Piešt'any
Potamós *83 C7* Antikýthira, S Greece
Potentia *see* Potenza
Potenza *75 D5 anc.* Potentia. Basilicata, S Italy
Poti *95 F2 prev.* P'ot'i. W Georgia
P'ot'i *see* Poti
Potiskum *53 G4* Yobe, NE Nigeria
Potomac River *19 E5 river* NE USA
Potosí *39 F4* Potosí, S Bolivia
Potsdam *72 D3* Brandenburg, NE Germany
Potwar Plateau *112 C2 plateau* NE Pakistan
Poŭthĭsăt *115 D6 prev.* Pursat. Poŭthĭsăt, W Cambodia
Po, Valle del *see* Po Valley
Po Valley *74 C2 It.* Valle del Po. *valley* N Italy
Považská Bystrica *77 C5 Ger.* Waagbistritz, *Hung.* Vágbeszterce. Trenčiansky Kraj, W Slovakia
Poverty Bay *128 E4 inlet* North Island, New Zealand
Póvoa de Varzim *70 B2* Porto, NW Portugal
Powder River *22 D2 river* Montana/Wyoming, NW USA
Powell *22 C2* Wyoming, C USA
Powell, Lake *22 B5 lake* Utah, W USA
Požarevac *78 D4 Ger.* Passarowitz. Serbia, NE Serbia
Poza Rica *29 F4 var.* Poza Rica de Hidalgo. Veracruz-Llave, E Mexico
Poza Rica de Hidalgo *see* Poza Rica
Požega *78 D4 prev.* Slavonska Požega, *Ger.* Poschega, *Hung.* Pozsega. Požega-Slavonija, NE Croatia
Požega *78 D4* Serbia
Poznań *76 C3 Ger.* Posen, Posnania. Wielkolpolskie, C Poland
Pozoblanco *70 D4* Andalucía, S Spain
Pozsega *see* Požega
Pozsony *see* Bratislava
Pozzallo *75 C8* Sicilia, Italy, C Mediterranean Sea
Prachatice *77 A5 Ger.* Prachatitz. Jihočeský Kraj, S Czech Republic
Prachatitz *see* Prachatice
Prado del Ganso *see* Goose Green
Prae *see* Phrae
Prag/Praga/Prague *see* Praha
Praha *77 A5 Eng.* Prague, *Ger.* Prag, *Pol.* Praga. *country capital* (Czech Republic) Středočeský Kraj, NW Czech Republic
Praia *52 A3 country capital* (Cape Verde) Santiago, S Cape Verde
Prairie State *see* Illinois
Prathet Thai *see* Thailand
Prato *74 B3* Toscana, C Italy
Pratt *23 E5* Kansas, C USA
Prattville *20 D2* Alabama, S USA
Pravda *see* Glavinitsa
Pravia *70 C1* Asturias, N Spain
Preny *see* Prienai
Prenzlau *72 D3* Brandenburg, NE Germany
Prerau *see* Přerov
Přerov *77 C5 Ger.* Prerau. Olomoucký Kraj, E Czech Republic
Presov *see* Prešov
Prescott *26 B2* Arizona, SW USA
Preševo *79 D5* Serbia, SE Serbia
Presidente Epitácio *41 E4* São Paulo, S Brazil
Presidente Stroessner *see* Ciudad del Este
Prešov *77 D5 var.* Preschau, *Ger.* Eperies, *Hung.* Eperjes. Prešovský Kraj, E Slovakia
Prespa, Lake *79 D6 Alb.* Liqen i Prespës, *Gk.* Límni Megáli Préspa, Limni Prespa, *Mac.* Prespansko Ezero, *Serb.* Prespansko Jezero. *lake* SE Europe
Prespa, Limni/Prespansko Ezero/Prespansko Jezero/Prespës, Liqen i *see* Prespa, Lake
Presque Isle *19 H1* Maine, NE USA
Pressburg *see* Bratislava
Preston *67 D5* NW England, United Kingdom
Prestwick *66 C4* W Scotland, United Kingdom
Pretoria *56 D4 var.* Epitoli. *country capital* (South Africa-administrative capital) Gauteng, NE South Africa
Preussisch Eylau *see* Bagrationovsk
Preußisch Holland *see* Pasłęk
Preussisch-Stargard *see* Starogard Gdański
Préveza *83 A5* Ípeiros, W Greece
Pribilof Islands *14 A3 island group* Alaska, USA
Priboj *78 C4* Serbia, W Serbia
Price *22 B4* Utah, W USA
Prichard *20 C3* Alabama, S USA
Priekulė *84 B3 Ger.* Prökuls. Klaipėda, W Lithuania
Prienai *85 B5 Pol.* Preny. Kaunas, S Lithuania
Prieska *56 C4* Northern Cape, C South Africa
Prijedor *78 B3* Republika Srpska, NW Bosnia and Herzegovina
Prijepolje *78 D4* Serbia, W Serbia
Prikaspiyskaya Nizmennost' *see* Caspian Depression
Prilep *79 D6 Turk.* Perlepe. S FYR Macedonia
Priluki *see* Pryluky
Primorsk *84 A4 Ger.* Fischhausen. Kaliningradskaya Oblast', W Russian Federation
Primorsko *82 E2 prev.* Keupriya. Burgas, E Bulgaria
Primorsk/Primorskoye *see* Prymors'k
Prince Albert *15 F5* Saskatchewan, S Canada
Prince Edward Island *17 F4 Fr.* Île-du Prince-Édouard. *province* SE Canada
Prince Edward Islands *47 E8 island group* S South Africa
Prince George *15 E5* British Columbia, SW Canada
Prince of Wales Island *126 B1 island* Queensland, E Australia
Prince of Wales Island *15 F2 island* Queen Elizabeth Islands, Nunavut, NW Canada
Prince of Wales Island *see* Pinang, Pulau
Prince Patrick Island *15 E2 island* Parry Islands, Northwest Territories, NW Canada
Prince Rupert *14 D4* British Columbia, SW Canada
Prince's Island *see* Príncipe
Princess Charlotte Bay *126 C2 bay* Queensland, NE Australia
Princess Elizabeth Land *132 C3 physical region* Antarctica
Príncipe *55 A5 var.* Príncipe Island, *Eng.* Prince's Island. *island* N Sao Tome and Principe
Príncipe Island *see* Príncipe
Prinzapolka *31 E3* Región Autónoma Atlántico Norte, NE Nicaragua
Pripet *85 C7 Bel.* Prypyats', *Ukr.* Pryp"yat'. *river* Belarus/Ukraine
Pripet Marshes *85 B7 wetland* Belarus/Ukraine
Prishtinë *79 D5 Eng.* Pristina, *Serb.* Priština. C Kosovo
Pristina *see* Prishtinë
Priština *see* Prishtinë
Privas *69 D5* Ardèche, E France
Privolzhskaya Vozvyshennost' *59 G3 var.* Volga Uplands. *mountain range* W Russian Federation
Prizren *79 D5* S Kosovo
Probolinggo *116 D5* Jawa, C Indonesia
Probstberg *see* Wyszków
Progreso *29 H3* Yucatán, SE Mexico
Prokhladnyy *89 B8* Kabardino-Balkarskaya Respublika, SW Russian Federation
Prokletije *see* North Albanian Alps
Prökuls *see* Priekulė
Prokuplje *79 D5* Serbia, SE Serbia
Prome *see* Pyay
Promyshlennyy *88 E3* Respublika Komi, NW Russian Federation
Prościejów *see* Prostějov
Proskurov *see* Khmel 'nyts'kyy
Prossnitz *see* Prostějov
Prostějov *77 C5 Ger.* Prossnitz, *Pol.* Prościejów. Olomoucký Kraj, E Czech Republic
Provence *69 D6 cultural region* SE France
Providence *19 G3 state capital* Rhode Island, NE USA
Providence *see* Fort Providence
Providencia, Isla de *31 F3 island* NW Colombia, Caribbean Sea
Provideniya *133 B1* Chukotskiy Avtonomnyy Okrug, NE Russian Federation
Provo *22 B4* Utah, W USA
Prudhoe Bay *14 D2* Alaska, USA
Prusa *see* Bursa
Pruszków *76 D3 Ger.* Kaltdorf. Mazowieckie, C Poland
Prut *86 D4 Ger.* Pruth. *river* E Europe
Pruth *see* Prut
Pružana *see* Pruzhany
Pruzhany *85 B6 Pol.* Prużana. Brestskaya Voblasts', SW Belarus
Prychornomor'ska Nyzovyna *see* Black Sea Lowland
Prydniprovs'ka Nyzovyna/Prydnyaprowskaya Nizina *see* Dnieper Lowland
Prydz Bay *132 D3 bay* Antarctica
Pryluky *87 E2 Rus.* Priluki. Chernihivs'ka Oblast', NE Ukraine
Prymors'k *87 G4 Rus.* Primorsk; *prev.* Primorskoye. Zaporiz'ka Oblast', SE Ukraine
Pryp"yat'/Prypyats' *see* Pripet
Przemyśl *77 E5 Rus.* Peremyshl. Podkarpackie, C Poland
Przheval'sk *see* Karakol
Psará *83 D5 island* E Greece
Psel *87 F2 Rus.* Psël. *river* Russian Federation/Ukraine
Psël *see* Psel
Pskov *92 B2 Ger.* Pleskau, *Latv.* Pleskava. Pskovskaya Oblast', W Russian Federation
Pskov, Lake *84 E3 Est.* Pihkva Järv, *Ger.* Pleskauer See, *Rus.* Pskovskoye Ozero. *lake* Estonia/Russian Federation
Pskovskoye Ozero *see* Pskov, Lake
Ptich' *see* Ptsich
Ptsich *85 C7 Rus.* Ptich'. Homyel'skaya Voblasts', SE Belarus
Ptsich *85 C7 Rus.* Ptich'. *river* SE Belarus
Ptuj *73 E7 Ger.* Pettau; *anc.* Poetovio. NE Slovenia
Pucallpa *38 C3* Ucayali, C Peru
Puck *76 C2* Pomorskie, N Poland
Pudasjärvi *62 D4* Oulu, C Finland
Puebla *29 F4 var.* Puebla de Zaragoza. Puebla, S Mexico
Puebla de Zaragoza *see* Puebla
Pueblo *22 D5* Colorado, C USA
Puerto Acosta *39 E4* La Paz, W Bolivia
Puerto Aisén *43 B6* Aisén, S Chile
Puerto Ángel *29 F5* Oaxaca, SE Mexico
Puerto Argentino *see* Stanley
Puerto Ayacucho *36 D3* Amazonas, SW Venezuela
Puerto Baquerizo Moreno *38 B5 var.* Baquerizo Moreno. Galápagos Islands, Ecuador, E Pacific Ocean
Puerto Barrios *30 C2* Izabal, E Guatemala
Puerto Bello *see* Portobelo
Puerto Berrío *36 B2* Antioquia, C Colombia
Puerto Cabello *36 D1* Carabobo, N Venezuela
Puerto Cabezas *31 E2 var.* Bilwi. Región Autónoma Atlántico Norte, NE Nicaragua
Puerto Carreño *36 D3* Vichada, E Colombia
Puerto Cortés *30 C2* Cortés, NW Honduras
Puerto Cumarebo *36 C1* Falcón, N Venezuela
Puerto Deseado *43 C7* Santa Cruz, SE Argentina
Puerto Escondido *29 F5* Oaxaca, SE Mexico
Puerto Francisco de Orellana *38 B1 var.* Coca. NE Ecuador
Puerto Gallegos *see* Río Gallegos
Puerto Inírida *36 D3 var.* Obando. Guainía, E Colombia
Puerto La Cruz *37 E1* Anzoátegui, NE Venezuela
Puerto Lempira *31 E2* Gracias a Dios, E Honduras
Puerto Limón *see* Limón
Puertollano *70 D4* Castilla-La Mancha, C Spain
Puerto López *36 C1* La Guajira, N Colombia
Puerto Maldonado *39 E3* Madre de Dios, E Peru
Puerto México *see* Coatzacoalcos
Puerto Montt *43 B5* Los Lagos, C Chile
Puerto Natales *43 B7* Magallanes, S Chile
Puerto Obaldía *31 H5* Kuna Yala, NE Panama
Puerto Plata *33 E3 var.* San Felipe de Puerto Plata. N Dominican Republic
Puerto Presidente Stroessner *see* Ciudad del Este
Puerto Princesa *117 E2 off.* Puerto Princesa City. Palawan, W Philippines
Puerto Princesa City *see* Puerto Princesa
Puerto Príncipe *see* Camagüey
Puerto Rico *33 F3 off.* Commonwealth of Puerto Rico; *prev.* Porto Rico. *US commonwealth territory* C West Indies
Puerto Rico *34 B1 island* C West Indies
Puerto Rico, Commonwealth of *see* Puerto Rico
Puerto Rico Trench *34 B1 trench* NE Caribbean Sea
Puerto San José *see* San José
Puerto San Julián *43 B7 var.* San Julián. Santa Cruz, SE Argentina
Puerto Suárez *39 H4* Santa Cruz, E Bolivia
Puerto Vallarta *28 D4* Jalisco, SW Mexico
Puerto Varas *43 B5* Los Lagos, C Chile
Puerto Viejo *31 E4* Heredia, NE Costa Rica
Puertoviejo *see* Portoviejo
Puget Sound *24 B1 sound* Washington, NW USA
Puglia *75 E5 var.* Le Puglie, *Eng.* Apulia. *region* SE Italy
Pukaki, Lake *129 B6 lake* South Island, New Zealand
Pukekohe *128 D3* Auckland, North Island, New Zealand
Puket *see* Phuket
Pukhavichy *85 C6 Rus.* Pukhovichi. Minskaya Voblasts', C Belarus
Pukhovichi *see* Pukhavichy
Pula *78 A3 It.* Pola; *prev.* Pulj. Istra, NW Croatia
Pulaski *18 D5* Virginia, NE USA
Puławy *76 D4 Ger.* Neu Amerika. Lubelskie, E Poland
Pul-e Khumrī *101 E4 prev.* Pol-e Khomrī, *var.* Pul-i-Khumri. Baghlān, NE Afghanistan
Pul-i-Khumri *see* Pul-e Khumrī
Pulj *see* Pula
Pullman *24 C2* Washington, NW USA
Pułtusk *76 D3* Mazowieckie, C Poland
Puná, Isla *38 A2 island* SW Ecuador
Pune *112 C5 prev.* Poona. Mahārāshtra, W India
Punjab *112 C2 prev.* West Punjab, Western Punjab. *province* E Pakistan
Puno *39 E4* Puno, SE Peru
Punta Alta *43 C5* Buenos Aires, E Argentina
Punta Arenas *43 B8 prev.* Magallanes. Magallanes, S Chile
Punta Gorda *30 C2* Toledo, SE Belize
Punta Gorda *31 E4* Región Autónoma Atlántico Sur, SE Nicaragua
Puntarenas *30 D4* Puntarenas, W Costa Rica
Punto Fijo *36 C1* Falcón, N Venezuela
Pupuya, Nevado *39 E4 mountain* W Bolivia
Puri *113 F5 var.* Jagannath. Odisha, E India
Puriramya *see* Buriram
Purmerend *64 C3* Noord-Holland, C Netherlands
Pursat *see* Poŭthĭsăt, Poŭthĭsăt, W Cambodia
Purus, Rio *40 C2 var.* Río Purús. *river* Brazil/Peru
Pusan *see* Busan
Pushkino *see* Biläsuvar
Püspökladány *77 D6* Hajdú-Bihar, E Hungary
Putorana, Gory/Putorana Mountains *see* Putorana, Plato
Putorana Mountains *93 E3 var.* Gory Putorana, *Eng.* Putorana Mountains. *mountain range* N Russian Federation
Putrajaya *116 B3 administrative capital* (Malaysia) Kuala Lumpur, Peninsular Malaysia
Puttalam *110 C3* North Western Province, W Sri Lanka
Puttgarden *72 C2* Schleswig-Holstein, N Germany
Putumayo, Río *36 B5 var.* Içá, Rio. *river* NW South America
Puurmani *84 D2 Ger.* Talkhof. Jõgevamaa, E Estonia
Pyatigorsk *89 B7* Stavropol'skiy Kray, SW Russian Federation
Pyatikhatki *see* P"yatykhatky
P"yatykhatky *87 F3 Rus.* Pyatikhatki. Dnipropetrovs'ka Oblast', E Ukraine
Pyay *114 B4 var.* Prome, Pye. Bago, C Myanmar (Burma)
Pye *see* Pyay
Pyetrykaw *85 C7 Rus.* Petrikov. Homyel'skaya Voblasts', SE Belarus
Pyinmana *114 B4 country capital* (Myanmar (Burma)) Mandalay, C Myanmar (Burma)
Pyin-Oo-Lwin *114 B3 var.* Maymyo. Mandalay, C Myanmar (Burma)
Pýlos *83 B6 var.* Pilos. Pelopónnisos, S Greece
P'yŏngyang *107 E3 var.* P'yŏngyang-si, *Eng.* Pyongyang. *country capital* (North Korea) SW North Korea
P'yŏngyang-si *see* P'yŏngyang
Pyramid Lake *25 C5 lake* Nevada, W USA
Pyrenaei Montes *see* Pyrenees
Pyrenees *80 B2 Fr.* Pyrénées, *Sp.* Pirineos; *anc.* Pyrenaei Montes. *mountain range* SW Europe
Pýrgos *83 B6 var.* Pírgos. Dytikí Ellás, S Greece
Pyritz *see* Pyrzyce
Pyryatyn *87 E2 Rus.* Piryatin. Poltavs'ka Oblast', NE Ukraine
Pyrzyce *76 B3 Ger.* Pyritz. Zachodnio-pomorskie, NW Poland
Pyu *see* Phyu
Pyuntaza *114 B4* Bago, SW Myanmar (Burma)

Q

Qā' al Jafr *97 C7 lake* S Jordan
Qaanaaq *60 D1 var.* Qânâq, *Dan.* Thule. Avannaarsua, N Greenland
Qabātiya *97 E6* N West Bank Asia
Qābis *see* Gabès
Qābis, Khalīj *see* Gabès, Golfe de
Qacentina *see* Constantine
Qafṣah *see* Gafsa
Qagan Us *see* Dulan
Qahremānshahr *see* Kermānshāh
Qaidam Pendi *104 C4 basin* C China
Qal'aikhum *101 F3 Rus.* Kalaikhum. S Tajikistan
Qalāt *101 E5 Per.* Kalāt. Zābol, S Afghanistan
Qal'at Bīshah *99 B5* 'Asīr, SW Saudi Arabia
Qalqiliya *see* Qalqīlya
Qalqīlya *97 D6 var.* Qalqiliya. Central, W West Bank Asia
Qamdo *104 D5* Xizang Zizhiqu, W China
Qamishly *see* Al Qāmishlī
Qânâq *see* Qaanaaq
Qaqortoq *60 C4 Dan.* Julianehåb. Kitaa, S Greenland
Qaraghandy/Qaraghandy Oblysy *see* Karagandy
Qara Qum *see* Garagum
Qarataū *see* Karatau, Zhambyl, Kazakhstan
Qarkilik *see* Ruoqiang
Qarokŭl *101 F3 Rus.* Karakul'. E Tajikistan
Qarqannah, Juzur *see* Kerkenah, Îles de
Qars *see* Kars
Qarshi *101 E3 Rus.* Karshi; *prev.* Bek-Budi. Qashqadaryo Viloyati, S Uzbekistan
Qasigianguit *see* Qasigiannguit
Qasigiannguit *60 C3 var.* Qasigianguit, *Dan.* Christianshåb. Kitaa, C Greenland
Qaṣr al Farāfirah *50 B2 var.* Qasr Farâfra. W Egypt
Qasr Farâfra *see* Qaṣr al Farāfirah
Qaṭanā *97 B5 var.* Katana. Dimashq, S Syria
Qatar *98 C4 off.* State of Qatar, *Ar.* Dawlat Qaṭar. *country* SW Asia
Qatar, State of *see* Qatar
Qattara Depression *see* Qaṭṭārah, Munkhafaḍ al
Qaṭṭâra, Monkhafad el *see* Qaṭṭārah, Munkhafaḍ al
Qaṭṭārah, Munkhafaḍ al *50 A1 var.* Monkhafad el Qaṭṭâra, *Eng.* Qattara Depression. *desert* NW Egypt
Qausuittuq *see* Resolute
Qazaqstan/Qazaqstan Respublikasy *see* Kazakhstan
Qazimämmäd *95 H3 Rus.* Kazi Magomed. SE Azerbaijan
Qazris *see* Cáceres
Qazvīn *98 C2 var.* Kazvin. Qazvīn, N Iran
Qena *see* Qinā
Qeqertarssuaq *see* Qeqertarsuaq
Qeqertarsuaq *60 C3 var.* Qeqertarssuaq, *Dan.* Godhavn. Kitaa, S Greenland
Qeqertarsuaq *60 C3 island* W Greenland
Qeqertarsuup Tunua *60 C3 Dan.* Disko Bugt. *inlet* W Greenland
Qerveh *see* Qorveh
Qeshm *98 D4 var.* Jazīreh-ye Qeshm, Qeshm Island. *island* S Iran

R

Románvásár – San Cristóbal de Las Casas

Románvásár see Roman
Rome 20 D2 Georgia, SE USA
Rome see Roma
Romînia see Romania
Romny 87 F2 Sums'ka Oblast', NE Ukraine
Rømø 63 A7 Ger. Röm. island SW Denmark
Roncador, Serra do 34 D4 mountain range C Brazil
Ronda 70 D5 Andalucía, S Spain
Rondônia 40 D3 off. Estado de Rondônia; prev. Território de Rondônia. state W Brazil
Rondônia 40 D3 off. Estado de Rondônia; prev. Território de Rondônia. region W Brazil
Rondônia, Estado de see Rondônia
Rondônia, Território de see Rondônia
Rondonópolis 41 E3 Mato Grosso, W Brazil
Rongelap Atoll 122 D1 var. Rönlap. atoll Ralik Chain, NW Marshall Islands
Rõngu 84 D3 Ger. Ringen. Tartumaa, SE Estonia
Rönlap see Rongelap Atoll
Rønne 63 B8 Bornholm, E Denmark
Ronne Ice Shelf 132 A3 ice shelf Antarctica
Roosendaal 65 C5 Noord-Brabant, S Netherlands
Roosevelt Island 132 B4 island Antarctica
Roraima 40 D1 off. Estado de Roraima; prev. Território de Rio Branco, Território de Roraima. region N Brazil
Roraima 40 D1 off. Estado de Roraima; prev. Território de Rio Branco, Território de Roraima. state N Brazil
Roraima, Estado de see Roraima
Roraima, Mount 37 E3 mountain N South America
Roraima, Território de see Roraima
Røros 63 B5 Sør-Trøndelag, S Norway
Ross 129 B6 West Coast, South Island, New Zealand
Rosa, Lake 32 D2 lake Great Inagua, S Bahamas
Rosario 42 D4 Santa Fe, C Argentina
Rosario 42 D2 San Pedro, C Paraguay
Rosario see Rosarito
Rosarito 28 A1 var. Rosario. Baja California Norte, NW Mexico
Roscianum see Rossano
Roscommon 18 C2 Michigan, N USA
Roseau 33 G4 prev. Charlotte Town. country capital (Dominica) SW Dominica
Roseburg 24 B4 Oregon, NW USA
Rosenau see Rožňava
Rosenberg 27 G4 Texas, SW USA
Rosenberg see Ružomberok, Slovakia
Rosengarten 72 B3 Niedersachsen, N Germany
Rosenheim 73 C6 Bayern, S Germany
Rosia 71 H5 W Gibraltar Europe
Rosia Bay 71 H5 bay SW Gibraltar Europe W Mediterranean Sea Atlantic Ocean
Roşiori de Vede 86 B5 Teleorman, S Romania
Rositten see Rēzekne
Roslavl' 89 A5 Smolenskaya Oblast', W Russian Federation
Rosmalen 64 C4 Noord-Brabant, S Netherlands
Rossano 75 E6 anc. Roscianum. Calabria, SW Italy
Ross Ice Shelf 132 B4 ice shelf Antarctica
Rossiyskaya Federatsiya see Russian Federation
Rosso 52 B3 Trarza, SW Mauritania
Rossosh' 89 B6 Voronezhskaya Oblast', W Russian Federation
Ross Sea 132 B4 sea Antarctica
Rostak see Ar Rustāq
Rostock 72 C2 Mecklenburg-Vorpommern, NE Germany
Rostov see Rostov-na-Donu
Rostov-na-Donu 89 B7 var. Rostov, Eng. Rostov-on-Don. Rostovskaya Oblast', SW Russian Federation
Rostov-on-Don see Rostov-na-Donu
Roswell 26 D2 New Mexico, SW USA
Rota 122 B1 island S Northern Mariana Islands
Rotcher Island see Tamana
Rothera 132 A2 UK research station Antarctica
Rotomagus see Rouen
Rotorua 128 D3 Bay of Plenty, North Island, New Zealand
Rotorua, Lake 128 D3 lake North Island, New Zealand
Rotterdam 64 C4 Zuid-Holland, SW Netherlands
Rottweil 73 B6 Baden-Württemberg, S Germany
Rotuma 123 E4 island NW Fiji Oceania S Pacific Ocean
Roubaix 68 C2 Nord, N France
Rouen 68 C3 anc. Rotomagus. Seine-Maritime, N France
Roulers see Roeselare
Roumania see Romania
Round Rock 27 G3 Texas, SW USA
Rourkela see Rāurkela
Rousselaere see Roeselare
Roussillon 69 C6 cultural region S France
Rouyn-Noranda 16 D4 Québec, SE Canada
Rovaniemi 62 D3 Lappi, N Finland
Rovigno see Rovinj
Rovigo 74 C2 Veneto, NE Italy
Rovinj 78 A3 It. Rovigno. Istra, NW Croatia
Rovno see Rivne
Rovuma, Rio 57 F2 var. Ruvuma. river Mozambique/Tanzania
Rovuma, Rio see Ruvuma
Równe see Rivne
Roxas City 117 E2 Panay Island, C Philippines
Royale, Isle 18 B1 island Michigan, N USA
Royan 69 B5 Charente-Maritime, W France
Rozdol'ne 87 F4 Rus. Razdolnoye. Respublika Krym, S Ukraine
Rožňava 77 D6 Ger. Rosenau, Hung. Rozsnyó. Košický Kraj, E Slovakia
Rózsahegy see Ružomberok
Rozsnyó see Râşnov, Romania
Rozsnyó see Rožňava, Slovakia
Ruanda see Rwanda
Ruapehu, Mount 128 D4 volcano North Island, New Zealand
Ruapuke Island 129 B8 island SW New Zealand
Ruatoria 128 E3 Gisborne, North Island, New Zealand
Ruawai 128 D2 Northland, North Island, New Zealand
Rubezhnoye see Rubizhne
Rubizhne 87 H3 Rus. Rubezhnoye. Luhans'ka Oblast', E Ukraine
Ruby Mountains 25 D5 mountain range Nevada, W USA
Rucava 84 B3 Liepāja, SW Latvia
Rudensk see Rudzyensk
Rūdiškės 85 B5 Vilnius, S Lithuania
Rudnik see Dolni Chiflik
Rudny see Rudnyy
Rudnyy 92 C4 var. Rudny. Kostanay, N Kazakhstan
Rudolf, Lake see Turkana, Lake
Rudolfswert see Novo mesto
Rudzyensk 85 C6 Rus. Rudensk. Minskaya Voblasts', C Belarus
Rufiji 51 C7 river E Tanzania
Rufino 42 C4 Santa Fe, C Argentina
Rugāji 84 D4 Balvi, E Latvia
Rügen 72 D2 headland NE Germany
Ruggell 72 E1 N Liechtenstein Europe
Ruhja see Rūjiena
Ruhnu 84 C2 var. Ruhnu Saar, Swe. Runö. island SW Estonia
Ruhnu Saar see Ruhnu
Rujen see Rūjiena
Rūjiena 84 D3 Est. Ruhja, Ger. Rujen. Valmiera, N Latvia
Rukwa, Lake 51 B7 lake SE Tanzania
Rum see Rhum
Ruma 78 D3 Vojvodina, N Serbia
Rumadiya see Ar Ramādī
Rumania/Rumänien see Romania
Rumbek 51 B5 El Buhayrat, C South Sudan
Rum Cay 32 D2 island C Bahamas
Rumia 76 C2 Pomorskie, N Poland
Rummah, Wādī ar see Rimah, Wādī ar
Rummelsburg in Pommern see Miastko
Rumuniya/Rumŭniya/Rumunjska see Romania
Runanga 129 B5 West Coast, South Island, New Zealand
Rundu 56 C3 var. Runtu. Okavango, NE Namibia
Runö see Ruhnu
Runtu see Rundu
Ruoqiang 104 C3 var. Jo-ch'iang, Uigh. Charkhlik, Charkhliq, Qarkilik. Xinjiang Uygur Zizhiqu, NW China
Rupea 86 C4 Ger. Reps, Hung. Kőhalom; prev. Cohalm. Braşov, C Romania
Rupel 65 B5 river N Belgium
Rupella see la Rochelle
Rupert, Rivière de 16 D3 river Québec, C Canada
Rusaddir see Melilla
Ruschuk/Rusçuk see Ruse
Ruse 82 D1 var. Ruschuk, Rustchuk, Turk. Rusçuk. Ruse, N Bulgaria
Russadir see Melilla
Russellville 20 A1 Arkansas, C USA
Russia see Russian Federation
Russian America see Alaska
Russian Federation 90 D2 off. Russian Federation, var. Russia, Latv. Krievija, Rus. Rossiyskaya Federatsiya. country Asia/Europe
Rustaq see Ar Rustāq
Rustavi 95 G2 prev. Rust'avi. SE Georgia
Rust'avi see Rustavi
Rustchuk see Ruse
Ruston 20 B2 Louisiana, S USA
Rutanzige, Lake see Edward, Lake
Rutba see Ar Ruṭbah
Rutlam see Ratlām
Rutland 19 F2 Vermont, NE USA
Rutog 104 A4 var. Rutög, Rutok. Xizang Zizhiqu, W China
Rutok see Rutog
Ruvuma see Rovuma, Rio
Ruwenzori 55 E5 mountain range Dem. Rep. Congo/Uganda
Ruzhany 85 B6 Brestskaya Voblasts', SW Belarus
Ružomberok 77 C5 Ger. Rosenberg, Hung. Rózsahegy. Žilinský Kraj, N Slovakia
Rwanda 51 B6 off. Rwandese Republic; prev. Ruanda. country C Africa
Rwandese Republic see Rwanda
Ryazan' 89 B5 Ryazanskaya Oblast', W Russian Federation
Rybach'ye see Balykchy
Rybinsk 88 B4 prev. Andropov. Yaroslavskaya Oblast', W Russian Federation
Rybnik 77 C5 Śląskie, S Poland
Rybnitsa see Rîbniţa
Ryde 126 E1 United Kingdom
Ryki 76 D4 Lubelskie, E Poland
Rykovo see Yenakiyeve
Rypin 76 C3 Kujawsko-pomorskie, C Poland
Ryssel see Lille
Rysy 77 C5 mountain S Poland
Ryukyu Islands see Nansei-shotō
Ryukyu Trench 103 F3 var. Nansei Syotō Trench. trench S East China Sea
Rzeszów 77 E5 Podkarpackie, SE Poland
Rzhev 88 B4 Tverskaya Oblast', W Russian Federation

S

Saale 72 C4 river C Germany
Saalfeld 73 C5 var. Saalfeld an der Saale. Thüringen, C Germany
Saalfeld an der Saale see Saalfeld
Saarbrücken 73 A6 Fr. Sarrebruck. Saarland, SW Germany
Sääre 84 C2 var. Sjar. Saaremaa, W Estonia
Saare see Saaremaa
Saaremaa 84 C2 Ger. Oesel, Ösel; prev. Saare. island W Estonia
Saariselkä 62 D2 Lapp. Suoločielgi. Lappi, N Finland
Sab' Ābār 96 C4 var. Sab'a Biyar, Sa'b Bi'ār. Ḥimṣ, C Syria
Sab'a Biyar see Sab' Ābār
Šabac 78 D3 Serbia, W Serbia
Sabadell 71 G2 Cataluña, E Spain
Sabah 116 D3 prev. British North Borneo, North Borneo. state East Malaysia
Sabanalarga 36 B1 Atlántico, N Colombia
Sabaneta 36 C1 Falcón, N Venezuela
Sabaria see Szombathely
Sab'atayn, Ramlat as 99 C6 desert C Yemen
Sabaya 39 F4 Oruro, S Bolivia
Sa'b Bi'ār see Sab' Ābār
Saberi, Hamun-e 100 C5 var. Daryācheh-ye Hāmun, Daryācheh-ye Sīstān. lake Afghanistan/Iran
Sabhā 49 F3 C Libya
Sabi see Save
Sabinas 29 E2 Coahuila, NE Mexico
Sabinas Hidalgo 29 E2 Nuevo León, NE Mexico
Sabine River 27 H3 river Louisiana/Texas, SW USA
Sabkha see As Sabkhah
Sable, Cape 21 E5 headland Florida, SE USA
Sable Island 17 G4 island Nova Scotia, SE Canada
Şabyā 99 B6 Jīzān, SW Saudi Arabia
Sabzawar see Sabzevār
Sabzevār 98 D2 var. Sabzawar. Khorāsān-Razavī, NE Iran
Sachsen 72 D4 Eng. Saxony, Fr. Saxe. state E Germany
Sachs Harbour 15 E2 var. Ikaahuk. Banks Island, Northwest Territories, N Canada
Sächsisch-Reen/Sächsisch-Regen see Reghin
Sacramento 25 B5 state capital California, W USA
Sacramento Mountains 26 D2 mountain range New Mexico, SW USA
Sacramento River 25 B5 river California, W USA
Sacramento Valley 25 B5 valley California, W USA
Sá da Bandeira see Lubango
Şa'dah 99 B6 NW Yemen
Sado 109 C5 var. Sadoga-shima. island C Japan
Sadoga-shima see Sado
Saena Julia see Siena
Safad see Tsefat
Şafāqis see Sfax
Şafāshahr 98 D3 var. Deh Bīd. Fārs, C Iran
Safed see Tsefat
Säffle 63 B6 Värmland, C Sweden
Safford 26 C3 Arizona, SW USA
Safi 48 B2 W Morocco
Selseleh-ye Safīd Kuh 100 D4 Eng. Paropamisus Range. mountain range W Afghanistan
Sagaing 114 B3 Sagaing, C Myanmar (Burma)
Sagami-nada 109 D6 inlet SW Japan
Sagan see Żagań
Sāgar 112 D4 prev. Saugor. Madhya Pradesh, C India
Sagarmāthā see Everest, Mount
Sagebrush State see Nevada
Saghez see Saqqez
Saginaw 18 C3 Michigan, N USA
Saginaw Bay 18 D2 lake bay Michigan, N USA
Sagua la Grande 32 B2 Villa Clara, C Cuba
Sagunto 71 F3 Cat. Sagunt, Ar. Murviedro; anc. Saguntum. País Valenciano, E Spain
Sagunt/Saguntum see Sagunto
Sahara 46 B3 desert Libya/Algeria
Sahara el Gharbîya see Şaḩrā' al Gharbīyah
Saharan Atlas see Atlas Saharien
Sahel 52 D3 physical region C Africa
Sāḩilīyah, Jibāl as 96 B3 mountain range NW Syria
Sāhīwāl 112 C2 prev. Montgomery. Punjab, E Pakistan
Şaḩrā' al Gharbīyah 50 B2 var. Sahara el Gharbîya, Eng. Western Desert. desert C Egypt
Saïda 97 A5 var. Şaydā, Sayida; anc. Sidon. W Lebanon
Sa'īdābād see Sīrjān
Saidpur 113 G3 var. Syedpur. Rajshahi, NW Bangladesh
Saidu Sharif 112 C1 var. Mingora, Mongora. North-West Frontier Province, N Pakistan
Saigon see Hô Chi Minh
Saimaa 63 E5 lake SE Finland
St Albans 67 E6 anc. Verulamium. E England, United Kingdom
Saint Albans 18 D5 West Virginia, NE USA
St Andrews 66 C4 E Scotland, United Kingdom
Saint Anna Trough see Svyataya Anna Trough
St. Ann's Bay 32 B4 C Jamaica
St. Anthony 17 G3 Newfoundland and Labrador, SE Canada
Saint Augustine 21 E3 Florida, SE USA
St Austell 67 C7 SW England, United Kingdom
St.Botolph's Town see Boston
St-Brieuc 68 A3 Côtes d'Armor, NW France
St. Catharines 16 D5 Ontario, S Canada
St-Chamond 69 D5 Loire, E France
Saint Christopher and Nevis, Federation of see Saint Kitts and Nevis
Saint Christopher-Nevis see Saint Kitts and Nevis
Saint Clair, Lake 18 D3 var. Lac à L'Eau Claire. lake Canada/USA
St-Claude 69 D5 anc. Condate. Jura, E France
Saint Cloud 23 F2 Minnesota, N USA
Saint Croix 33 F3 island S Virgin Islands (US)
Saint Croix River 18 A2 river Minnesota/Wisconsin, N USA
St David's Island 20 B5 island E Bermuda
St-Denis 57 G4 dependent territory capital (Réunion) NW Réunion
St-Dié 68 E4 Vosges, NE France
St-Egrève 69 D5 Isère, E France
Sainte Marie, Cap see Vohimena, Tanjona
Saintes 69 B5 anc. Mediolanum. Charente-Maritime, W France
St-Étienne 69 D5 Loire, E France
St-Flour 69 C5 Cantal, C France
St-Gall/Saint Gall/St. Gallen see Sankt Gallen
St-Gaudens 69 B6 Haute-Garonne, S France
Saint George 127 D5 Queensland, E Australia
St George 20 B4 N Bermuda
Saint George 22 A5 Utah, W USA
St. George's 33 G5 country capital (Grenada) SW Grenada
St-Georges 17 E4 Québec, SE Canada
St-Georges 37 H3 E French Guiana
Saint George's Channel 67 B6 channel Ireland/Wales, United Kingdom
St George's Island 20 B4 island E Bermuda
Saint Helena 47 B6 UK dependent territory C Atlantic Ocean
St Helier 67 D8 dependent territory capital (Jersey) S Jersey, Channel Islands
St.Iago de la Vega see Spanish Town
Saint Ignace 18 C2 Michigan, N USA
St-Jean, Lac 17 E4 lake Québec, SE Canada
Saint Joe River 24 D2 river Idaho, NW USA
St. John 17 F4 New Brunswick, SE Canada
Saint-John see Saint John
Saint John 19 H1 Fr. Saint-John. river Canada/USA
St John's 33 G3 country capital (Antigua and Barbuda) Antigua, Antigua and Barbuda
St. John's 17 H3 province capital Newfoundland and Labrador, E Canada
Saint Joseph 23 F4 Missouri, C USA
St Julian's see San Giljan
St Kilda 66 A3 island NW Scotland, United Kingdom
Saint Kitts and Nevis 33 F3 off. Federation of Saint Christopher and Nevis, var. Saint Christopher-Nevis. country E West Indies
St-Laurent see St-Laurent-du-Maroni
St-Laurent-du-Maroni 37 H3 var. St-Laurent. NW French Guiana
St-Laurent, Fleuve see St. Lawrence
St. Lawrence 17 E4 Fr. Fleuve St-Laurent. river Canada/USA
St. Lawrence, Gulf of 17 F3 gulf NW Atlantic Ocean
Saint Lawrence Island 14 B2 island Alaska, USA
St-Lô 68 B3 anc. Briovera, Laudus. Manche, N France
St-Louis 68 E4 Haut-Rhin, NE France
Saint Louis 52 B3 NW Senegal
Saint Louis 23 G4 Missouri, C USA
Saint Lucia 33 E1 country SE West Indies
Saint Lucia Channel 33 H4 channel Martinique/Saint Lucia
St-Malo 68 B3 Ille-et-Vilaine, NW France
St-Malo, Golfe de 68 A3 gulf NW France
Saint Martin see Sint Maarten
St.Matthew's Island see Zadetkyi Kyun
St.Matthias Group 122 B3 island group NE Papua New Guinea
St. Moritz 73 B7 Ger. Sankt Moritz, Rmsch. San Murezzan. Graubünden, SE Switzerland
St-Nazaire 68 A4 Loire-Atlantique, NW France
Saint Nicholas see São Nicolau
Saint-Nicolas see Sint-Niklaas
St-Omer 68 C2 Pas-de-Calais, N France
Saint Paul 23 F2 state capital Minnesota, N USA
St-Paul, Île 119 C6 var. St.Paul Island. island Île St-Paul, NE French Southern and Antarctic Lands Antarctica Indian Ocea
St.Paul Island see St-Paul, Île
St Peter Port 67 D8 dependent territory capital (Guernsey) C Guernsey, Channel Islands
Saint Petersburg 21 E4 Florida, SE USA
Saint Petersburg see Sankt-Peterburg
St-Pierre and Miquelon 17 G4 Fr. Îles St-Pierre et Miquelon. French territorial collectivity NE North America
St-Quentin 68 C3 Aisne, N France
Saint Thomas see São Tomé, Sao Tome and Principe
Saint Thomas see Charlotte Amalie, Virgin Islands (US)
Saint Ubes see Setúbal
Saint Vincent 33 G4 island N Saint Vincent and the Grenadines
Saint Vincent see São Vicente
Saint Vincent and the Grenadines 33 H4 country SE West Indies
Saint Vincent, Cape see São Vicente, Cabo de
Saint Vincent Passage 33 H4 passage Saint Lucia/Saint Vincent and the Grenadines
Saint Yves see Setúbal
Saipan 120 B1 island/country capital (Northern Mariana Islands) S Northern Mariana Islands
Saishū see Jeju-do
Sajama, Nevado 39 F4 mountain W Bolivia
Sajószentpéter 77 D6 Borsod-Abaúj-Zemplén, NE Hungary
Sakākah 98 B4 Al Jawf, NW Saudi Arabia
Sakakawea, Lake 22 D1 reservoir North Dakota, N USA
Sak'art'velo see Georgia
Sakata 108 D4 Yamagata, Honshū, C Japan
Sakhalin 93 G4 var. Sakhalin. island SE Russian Federation
Sakhalin see Sakhalin, Ostrov
Sakhon Nakhon see Sakon Nakhon
Şäki 95 G2 Rus. Sheki; prev. Nukha. NW Azerbaijan
Saki see Saky
Sakishima-shoto 108 A3 var. Sakisima Syotô. island group SW Japan
Sakisima Syotô see Sakishima-shotō
Sakiz see Saqqez
Sakiz-Adasi see Chíos
Sakon Nakhon 114 D4 var. Muang Sakon Nakhon, Sakhon Nakhon. Sakon Nakhon, E Thailand
Saky 87 F5 Rus. Saki. Respublika Krym, S Ukraine
Sal 52 A3 island Ilhas de Barlavento, NE Cape Verde
Sala 63 C6 Västmanland, C Sweden
Salacgrīva 84 C3 Est. Salatsi. Limbaži, N Latvia
Sala Consilina 75 D5 Campania, S Italy
Salado, Río 40 D5 river E Argentina
Salado, Río 42 C3 river C Argentina
Şalālah 99 D6 SW Oman
Salamá 30 B2 Baja Verapaz, C Guatemala
Salamanca 42 B4 Coquimbo, C Chile
Salamanca 70 D2 anc. Helmantica, Salmantica. Castilla-León, NW Spain
Salamīyah 96 B3 var. As Salamīyah. Ḥamāh, W Syria
Salang see Phuket
Salantai 84 B3 Klaipėda, NW Lithuania
Salatsi see Salacgrīva
Salavan 115 D5 var. Saravan, Saravane. Salavan, S Laos
Salavat 89 D6 Respublika Bashkortostan, W Russian Federation
Sala y Gomez 131 F4 island Chile, E Pacific Ocean
Sala y Gomez Fracture Zone see Sala y Gomez Ridge
Sala y Gomez Ridge 131 G4 var. Sala y Gomez Fracture Zone. fracture zone SE Pacific Ocean
Salazar see N'Dalatando
Šalčininkai 85 C5 Vilnius, SE Lithuania
Salduba see Zaragoza
Saldus 84 B3 Ger. Frauenburg. Saldus, W Latvia
Sale 127 C7 Victoria, SE Australia
Salé 48 C2 NW Morocco
Salekhard 92 D3 prev. Obdorsk. Yamalo-Nenetskiy Avtonomnyy Okrug, N Russian Federation
Salem 110 C2 Tamil Nādu, SE India
Salem 24 B3 state capital Oregon, NW USA
Salerno 75 D5 anc. Salernum. Campania, S Italy
Salerno, Gulf of 75 C5 Eng. Gulf of Salerno. gulf S Italy
Salerno, Gulf of see Salerno, Golfo di
Salernum see Salerno
Salihorsk 85 C7 Rus. Soligorsk. Minskaya Voblasts', S Belarus
Salima 57 E2 Central, C Malawi
Salina 23 E5 Kansas, C USA
Salina Cruz 29 F5 Oaxaca, SE Mexico
Salinas 38 A2 Guayas, W Ecuador
Salinas 25 B6 California, W USA
Salisbury 67 D7 var. New Sarum. S England, United Kingdom
Salisbury see Harare
Sállan see Sørøya
Sallyana see Şalyān
Salmantica see Salamanca
Salmon River 24 D3 river Idaho, NW USA
Salmon River Mountains 24 D3 mountain range Idaho, NW USA
Salo 63 D6 Länsi-Suomi, SW Finland
Salon-de-Provence 69 D6 Bouches-du-Rhône, SE France
Salonica/Salonika see Thessaloníki
Salonta 86 A3 Hung. Nagyszalonta. Bihor, W Romania
Sal'sk 89 B7 Rostovskaya Oblast', SW Russian Federation
Salt see As Salt
Salta 42 C2 Salta, NW Argentina
Saltash 67 C7 SW England, United Kingdom
Saltillo 29 E3 Coahuila, NE Mexico
Salt Lake City 22 B4 state capital Utah, W USA
Salto 42 D4 Salto, N Uruguay
Salton Sea 25 D8 lake California, W USA
Salvador 41 G3 prev. São Salvador. state capital Bahia, E Brazil
Salween 102 C2 Bur. Thanlwin, Chin. Nu Chiang, Nu Jiang. river SE Asia
Şalyān 113 E3 var. Sallyana. Mid Western, W Nepal
Salzburg 73 D6 anc. Juvavum. Salzburg, N Austria
Salzgitter 72 C4 prev. Watenstedt-Salzgitter. Niedersachsen, C Germany
Salzwedel 72 C3 Sachsen-Anhalt, N Germany
Šamac see Bosanski Šamac
Samakhixai see Attapu
Samalayuca 28 C1 Chihuahua, N Mexico
Samar 117 F2 island C Philippines
Samara 92 B3 prev. Kuybyshev. Samarskaya Oblast', W Russian Federation
Samarang see Semarang
Samarinda 116 D4 Borneo, C Indonesia
Samarkand see Samarqand
Samarkandski/Samarkandskoye see Temirtau
Samarobriva see Amiens
Samarqand 101 E2 Rus. Samarkand. Samarqand Viloyati, C Uzbekistan
Samawa see As Samāwah
Sambalpur 113 F4 Odisha, E India
Sambava 57 G2 Antsiranana, NE Madagascar
Sambir 86 B2 Rus. Sambor. L'vivs'ka Oblast', NW Ukraine
Sambor see Sambir
Sambre 68 D2 river Belgium/France
Samfya 56 D2 Luapula, N Zambia
Saminatal 72 E2 valley Austria/Liechtenstein Europe
Samnān see Semnān
Sam Neua see Xam Nua
Samoa 123 E4 off. Independent State of Western Samoa, var. Sāmoa; prev. Western Samoa. country W Polynesia
Sāmoa see Samoa
Samoa Basin 121 E3 undersea basin W Pacific Ocean
Samobor 78 A2 Zagreb, N Croatia
Sámos 83 E6 prev. Limín Vathéos. Sámos, Dodekánisa, Greece, Aegean Sea
Sámos 83 D6 island Dodekánisa, Greece, Aegean Sea
Samothrace see Samothráki
Samothráki 82 D4 Samothráki, NE Greece
Samothraki 82 C4 anc. Samothrace. island NE Greece
Sampit 116 C4 Borneo, C Indonesia
Sâmraông 115 D5 prev. Phumĭ Sâmraông, Phum Samrong. Siĕmréab, NW Cambodia
Samsun 94 D2 anc. Amisus. Samsun, N Turkey
Samtredia 95 F2 W Georgia
Samui, Ko 115 C6 island SW Thailand
Samut Prakan 115 C5 var. Muang Samut Prakan, Paknam. Samut Prakan, C Thailand
San 52 D3 Ségou, C Mali
San 77 E5 river SE Poland
Şan'ā' 99 B6 Eng. Sana. country capital (Yemen) W Yemen
Sana 78 B3 river NW Bosnia and Herzegovina
Sanae 132 B2 South African research station Antarctica
Sanaga 55 B5 river C Cameroon
San Ambrosio, Isla 35 A5 Eng. San Ambrosio Island. island W Chile
San Ambrosio Island see San Ambrosio, Isla
Sanandaj 98 C3 prev. Sinneh. Kordestān, W Iran
San Andrés, Isla de 31 F3 island NW Colombia, Caribbean Sea
San Andrés Tuxtla 29 F4 var. Tuxtla. Veracruz-Llave, E Mexico
San Angelo 27 F3 Texas, SW USA
San Antonio 30 B2 Toledo, S Belize
San Antonio 42 B4 Valparaíso, C Chile
San Antonio 27 F4 Texas, SW USA
San Antonio Oeste 43 C5 Río Negro, E Argentina
San Antonio River 27 G4 river Texas, SW USA
Sanāw 99 C6 var. Sanaw. NE Yemen
San Benedicto, Isla 28 B4 island W Mexico
San Benito 30 B1 Petén, N Guatemala
San Benito 27 G5 Texas, SW USA
San Bernardino 25 C7 California, W USA
San Blas 28 C3 Sinaloa, C Mexico
San Blas, Cape 20 D3 headland Florida, SE USA
San Blas, Cordillera de 31 G4 mountain range NE Panama
San Carlos 30 D4 Río San Juan, S Nicaragua
San Carlos 26 B2 Arizona, SW USA
San Carlos see Quesada, Costa Rica
San Carlos de Ancud see Ancud
San Carlos de Bariloche 43 B5 Río Negro, SW Argentina
San Carlos del Zulia 36 C2 Zulia, W Venezuela
San Clemente Island 25 B8 island Channel Islands, California, W USA
San Cristóbal 36 C2 Táchira, W Venezuela
San Cristobal 122 C4 var. Makira. island SE Solomon Islands
San Cristóbal see San Cristóbal de Las Casas
San Cristóbal de Las Casas 29 G5 var. San Cristóbal. Chiapas, SE Mexico

San Cristóbal, Isla – Sembé

San Cristóbal, Isla 38 B5 var. Chatham Island. island Galápagos Islands, Ecuador, E Pacific Ocean
Sancti Spíritus 32 B2 Sancti Spíritus, C Cuba
Sandakan 116 D3 Sabah, East Malaysia
Sandalwood Island see Sumba, Pulau
Sandanski 82 C3 prev. Sveti Vrach. Blagoevgrad, SW Bulgaria
Sanday 66 D2 island NE Scotland, United Kingdom
Sanders 26 C2 Arizona, SW USA
Sand Hills 22 D3 mountain range Nebraska, C USA
San Diego 25 C8 California, W USA
Sandnes 63 A6 Rogaland, S Norway
Sandomierz 76 D4 Rus. Sandomir. Świętokrzyskie, C Poland
Sandomir see Sandomierz
Sandoway see Thandwe
Sandpoint 24 C1 Idaho, NW USA
Sand Springs 27 G1 Oklahoma, C USA
Sandusky 18 D3 Ohio, N USA
Sandvika 63 A6 Akershus, S Norway
Sandviken 63 C6 Gävleborg, C Sweden
Sandwich Island see Efate
Sandwich Islands see Hawaiian Islands
Sandy Bay 71 H5 Saskatchewan, C Canada
Sandy City 22 B4 Utah, W USA
Sandy Lake 16 B3 lake Ontario, C Canada
San Esteban 30 D2 Olancho, C Honduras
San Eugenio/San Eugenio del Cuareim see Artigas
San Felipe 36 D1 Yaracuy, NW Venezuela
San Felipe de Puerto Plata see Puerto Plata
San Félix, Isla 35 A5 Eng. San Felix Island. island W Chile
San Felix Island see San Félix, Isla
San Fernando 70 C5 prev. Isla de León. Andalucía, S Spain
San Fernando 33 H5 Trinidad, Trinidad and Tobago
San Fernando 24 D1 California, W USA
San Fernando 36 D2 var. San Fernando de Apure. Apure, C Venezuela
San Fernando de Apure see San Fernando
San Fernando del Valle de Catamarca 42 C3 var. Catamarca. Catamarca, NW Argentina
San Fernando de Monte Cristi see Monte Cristi
San Francisco 25 B6 California, W USA
San Francisco del Oro 28 C2 Chihuahua, N Mexico
San Francisco de Macorís 33 E3 C Dominican Republic
San Fructuoso see Tacuarembó
San Gabriel 38 B1 Carchi, N Ecuador
San Gabriel Mountains 24 E1 mountain range California, USA
Sangihe, Kepulauan see Sangir, Kepulauan
San Giljan 80 B5 var. St Julian's. N Malta
Sangir, Kepulauan 117 F3 var. Kepulauan Sangihe. island group N Indonesia
Sāngli 110 B1 Mahārāshtra, W India
Sangmélima 55 B5 Sud, S Cameroon
Sangre de Cristo Mountains 26 D1 mountain range Colorado/New Mexico, C USA
San Ignacio 30 B1 prev. Cayo, El Cayo. Cayo, W Belize
San Ignacio 39 F3 Beni, N Bolivia
San Ignacio 28 B2 Baja California Sur, NW Mexico
San Joaquin Valley 25 B7 valley California, W USA
San Jorge, Golfo 43 C6 var. Gulf of San Jorge. gulf S Argentina
San Jorge, Gulf of see San Jorge, Golfo
San José 31 E4 country capital (Costa Rica) San José, C Costa Rica
San José 39 G3 var. San José de Chiquitos. Santa Cruz, E Bolivia
San José 30 B3 var. Puerto San José. Escuintla, S Guatemala
San Jose 25 B6 California, W USA
San José see San José del Guaviare, Colombia
San José de Chiquitos see San José
San José de Cúcuta see Cúcuta
San José del Guaviare 36 C4 var. San José. Guaviare, S Colombia
San Juan 42 B4 San Juan, W Argentina
San Juan 33 F3 dependent territory capital (Puerto Rico) NE Puerto Rico
San Juan see San Juan de los Morros
San Juan Bautista 42 D3 Misiones, S Paraguay
San Juan Bautista see Villahermosa
San Juan Bautista Tuxtepec see Tuxtepec
San Juan de Alicante see Sant Joan d'Alacant
San Juan del Norte 31 E4 var. Greytown. Río San Juan, SE Nicaragua
San Juan de los Morros 36 D2 var. San Juan. Guárico, N Venezuela
San Juanito, Isla 28 C4 island C Mexico
San Juan Mountains 26 D1 mountain range Colorado, C USA
San Juan, Río 31 E4 river Costa Rica/Nicaragua
San Juan River 26 C1 river Colorado/Utah, W USA
San Julián see Puerto San Julián
Sankt-Bartholomäi see Palamuse
Sankt Gallen 73 B7 var. St. Gallen, Eng. Saint Gall, Fr. St-Gall. Sankt Gallen, NE Switzerland
Sankt-Georgen see Sfântu Gheorghe
Sankt-Jakobi see Pärnu-Jaagupi, Pärnumaa, Estonia
Sankt Martin see Martin
Sankt Moritz see St. Moritz
Sankt-Peterburg 88 B4 prev. Leningrad, Petrograd, Eng. Saint Petersburg, Fin. Pietari. Leningradskaya Oblast', NW Russian Federation
Sankt Pölten 73 E6 Niederösterreich, N Austria
Sankt Veit am Flaum see Rijeka
Sankuru 55 D6 river C Dem. Rep. Congo
Şanlıurfa 95 E4 prev. Sanli Urfa, Urfa; anc. Edessa. şanlıurfa, S Turkey
Sanli Urfa see Şanlıurfa
San Lorenzo 39 G5 Tarija, S Bolivia
San Lorenzo 38 A1 Esmeraldas, N Ecuador
San Lorenzo, Isla 38 C4 island W Peru
Sanlúcar de Barrameda 70 C5 Andalucía, S Spain
San Luis 42 C4 San Luis, C Argentina
San Luis 30 B2 Petén, NE Guatemala
San Luis see San Luis Río Colorado
San Luis Obispo 25 B7 California, W USA
San Luis Potosí 29 E3 San Luis Potosí, C Mexico
San Luis Río Colorado 28 A1 var. San Luis Río Colorado. Sonora, NW Mexico
San Marcos 30 A2 San Marcos, W Guatemala
San Marcos 27 G4 Texas, SW USA
San Marcos de Arica see Arica
San Marino 74 E1 country capital (San Marino) C San Marino
San Marino 74 D1 off. Republic of San Marino. country S Europe
San Marino, Republic of see San Marino
San Martín 132 A2 Argentinian research station Antarctica
San Mateo 37 E2 Anzoátegui, NE Venezuela
San Matías 39 H3 Santa Cruz, E Bolivia
San Matías, Golfo 43 C5 var. Gulf of San Matías. gulf E Argentina
San Matías, Gulf of see San Matías, Golfo
Sanmenxia 106 C4 var. Shan Xian. Henan, C China
Sânmiclăuş Mare see Sânnicolau Mare
San Miguel 30 C3 San Miguel, SE El Salvador
San Miguel 28 D2 Coahuila, N Mexico
San Miguel de Ibarra see Ibarra
San Miguel de Tucumán 42 C3 var. Tucumán. Tucumán, N Argentina
San Miguelito 31 G4 Panamá, C Panama
San Miguel, Río 39 G3 river E Bolivia
San Murezzan see St. Moritz
Sannār see Sennar
Sânnicolaul-Mare see Sânnicolau Mare
Sânnicolau Mare 86 A4 var. Sânnicolaul-Mare, Hung. Nagyszentmiklós; prev. Sânmiclăuş Mare, Sînnicolau Mare. Timiş, W Romania
Sanok 77 E5 Podkarpackie, SE Poland
San Pablo 39 F5 Potosí, S Bolivia
San Pedro 30 C1 Corozal, NE Belize
San-Pédro 52 D5 S Côte d'Ivoire
San Pedro 28 D3 var. San Pedro de las Colonias. Coahuila, NE Mexico
San Pedro de la Cueva 28 C2 Sonora, NW Mexico
San Pedro de las Colonias see San Pedro
San Pedro de Lloc 38 B3 La Libertad, NW Peru
San Pedro Mártir, Sierra 28 A1 mountain range NW Mexico
San Pedro Sula 30 C2 Cortés, NW Honduras
San Rafael 42 B4 Mendoza, W Argentina
San Rafael Mountains 25 C7 mountain range California, W USA
San Ramón de la Nueva Orán 42 C2 Salta, N Argentina
San Remo 74 A3 Liguria, NW Italy
San Salvador 30 B3 country capital (El Salvador) San Salvador, SW El Salvador
San Salvador 32 D2 prev. Watlings Island. island E Bahamas
San Salvador de Jujuy 42 C2 var. Jujuy. Jujuy, N Argentina
San Salvador, Isla 38 A4 island Ecuador
Sansanné-Mango 53 E4 var. Mango. N Togo
San Sebastián 71 E1 País Vasco, N Spain see also Donostia
Sansepolcro 74 C3 Toscana, C Italy
San Severo 75 D5 Puglia, SE Italy
Santa Ana 39 F3 Beni, N Bolivia
Santa Ana 30 B3 Santa Ana, NW El Salvador
Santa Ana 24 D2 California, W USA
Santa Ana de Coro see Coro
Santa Ana Mountains 24 E2 mountain range California, W USA
Santa Barbara 28 C2 Chihuahua, N Mexico
Santa Barbara 25 C7 California, W USA
Santa Cataliña de Armada 70 B1 var. Santa Comba. Galicia, NW Spain
Santa Catalina Island 25 B8 island Channel Islands, California, W USA
Santa Catarina 41 E5 off. Estado de Santa Catarina. region S Brazil
Santa Catarina 41 E5 off. Estado de Santa Catarina. state S Brazil
Santa Catarina, Estado de see Santa Catarina
Santa Clara 32 B2 Villa Clara, C Cuba
Santa Clarita 24 D1 California, USA
Santa Comba see Santa Cataliña de Armada
Santa Cruz 54 E2 São Tomé, S Sao Tome and Principe, Africa
Santa Cruz 25 B6 California, W USA
Santa Cruz 39 G4 department E Bolivia
Santa Cruz Barillas see Barillas
Santa Cruz del Quiché 30 B2 Quiché, W Guatemala
Santa Cruz de Tenerife 48 A3 Tenerife, Islas Canarias, Spain, NE Atlantic Ocean
Santa Cruz, Isla 38 B5 var. Indefatigable Island, Isla Chávez. island Galápagos Islands, Ecuador, E Pacific Ocean
Santa Cruz Islands 122 D3 island group E Solomon Islands
Santa Cruz, Río 43 B7 river S Argentina
Santa Elena 30 B1 Cayo, W Belize
Santa Fe 42 C4 Santa Fe, C Argentina
Santa Fe 26 D1 state capital New Mexico, SW USA
Santa Fe see Bogotá
Santa Fe de Bogotá see Bogotá
Santa Genoveva 28 B3 mountain NW Mexico
Santa Isabel 122 C3 var. Bughotu. island N Solomon Islands
Santa Isabel see Malabo
Santa Lucia Range 25 B7 mountain range California, W USA
Santa Margarita, Isla 28 B3 island NW Mexico
Santa Maria 41 E5 Rio Grande do Sul, S Brazil
Santa Maria 25 B7 California, USA
Santa Maria 70 A5 island Azores, Portugal, NE Atlantic Ocean
Santa María del Buen Aire see Buenos Aires
Santa María, Isla 38 A5 var. Isla Floreana, Charles Island. island Galápagos Islands, Ecuador, E Pacific Ocean
Santa Marta 36 B1 Magdalena, N Colombia
Santa Maura see Lefkáda
Santa Monica 24 D1 California, W USA
Santana 54 E2 São Tomé, C Sao Tome and Principe, Africa
Santander 70 D1 Cantabria, N Spain
Santarém 41 E2 Pará, N Brazil
Santarém 70 B3 anc. Scalabis. Santarém, W Portugal
Santa Rosa 42 C4 La Pampa, C Argentina
Santa Rosa see Santa Rosa de Copán
Santa Rosa de Copán 30 C2 var. Santa Rosa. Copán, W Honduras
Santa Rosa Island 25 B8 island California, W USA
Santa Uxía de Ribeira 70 B1 var. Ribeira. NW Spain
Sant Carles de la Rápida see Sant Carles de la Ràpita
Sant Carles de la Ràpita 71 F3 var. Sant Carles de la Rápida. Cataluña, NE Spain
Santiago 42 B4 var. Gran Santiago. country capital (Chile) Santiago, C Chile
Santiago 33 E3 var. Santiago de los Caballeros. N Dominican Republic
Santiago 31 F5 Veraguas, S Panama
Santiago see Santiago de Compostela
Santiago 52 A3 var. São Tiago. island Ilhas de Sotavento, S Cape Verde
Santiago see Santiago de Cuba, Cuba
Santiago de Compostela 70 B1 var. Santiago, Eng. Compostella; anc. Campus Stellae. Galicia, NW Spain
Santiago de Cuba 32 C3 var. Santiago. Santiago de Cuba, E Cuba
Santiago de Guayaquil see Guayaquil
Santiago del Estero 42 C3 Santiago del Estero, C Argentina
Santiago de los Caballeros see Santiago, Dominican Republic
Santiago de los Caballeros see Ciudad de Guatemala, Guatemala
Santiago Pinotepa Nacional see Pinotepa Nacional
Santiago, Río 38 B2 river N Peru
Santi Quaranta see Sarandë
Santissima Trinidad see Jilong
San Joan d'Alacant 71 F4 Cast. San Juan de Alicante. País Valenciano, E Spain
Sant Julià de Lòria 69 A8 Sant Julià de Lòria, SW Andorra Europe
Santo see Espiritu Santo
Santo Antão 52 A2 island Ilhas de Barlavento, N Cape Verde
Santo António 54 E1 Príncipe, N Sao Tome and Principe, Africa
Santo Domingo 33 D3 prev. Ciudad Trujillo. country capital (Dominican Republic) SE Dominican Republic
Santo Domingo de los Colorados 38 B1 Pichincha, NW Ecuador
Santo Domingo Tehuantepec see Tehuantepec
Santorini see Thíra
Santorini 83 D7 island Kykládes, Greece, Aegean Sea
San Tomé de Guayana see Ciudad Guayana
Santos 41 F4 São Paulo, S Brazil
Santos Plateau 35 D5 undersea plateau SW Atlantic Ocean
Santo Tomé 42 D3 Corrientes, NE Argentina
Santo Tomé de Guayana see Ciudad Guayana
San Valentín, Cerro 43 A6 mountain S Chile
San Vicente 30 C3 San Vicente, C El Salvador
São Francisco, Rio 41 F3 river E Brazil
Sao Hill 51 C7 Iringa, S Tanzania
São João da Madeira 70 B2 Aveiro, N Portugal
São Jorge 70 A5 island Azores, Portugal, NE Atlantic Ocean
São Luís 41 F2 state capital Maranhão, NE Brazil
São Mandol see São Manuel, Rio
São Manuel, Rio 41 E3 var. São Mandol, Teles Pirés. river C Brazil
São Marcos, Baía de 41 F1 bay N Brazil
São Miguel 70 A5 island Azores, Portugal, NE Atlantic Ocean
Saona, Isla 33 E3 island SE Dominican Republic
Saône 69 D5 river E France
São Nicolau 52 A3 Eng. Saint Nicholas. island Ilhas de Barlavento, N Cape Verde
São Paulo 41 E4 state capital São Paulo, S Brazil
São Paulo 41 E4 off. Estado de São Paulo. state S Brazil
São Paulo 41 E4 off. Estado de São Paulo. region S Brazil
São Paulo de Loanda see Luanda
São Paulo, Estado de see São Paulo
São Pedro do Rio Grande do Sul see Rio Grande
São Roque, Cabo de 41 G2 headland E Brazil
São Salvador see Salvador, Brazil
São Salvador/São Salvador do Congo see M'Banza Congo, Angola
São Tiago see Santiago
São Tomé 55 A5 country capital (Sao Tome and Principe) São Tomé, S Sao Tome and Principe
São Tomé 54 E2 Eng. Saint Thomas. island S Sao Tome and Principe
Sao Tome and Principe 54 D1 off. Democratic Republic of Sao Tome and Principe. country E Atlantic Ocean
Sao Tome and Principe, Democratic Republic of see Sao Tome and Principe
São Tomé, Pico de 54 D2 mountain São Tomé, C Sao Tome and Principe, Africa
São Vicente 52 A3 Eng. Saint Vincent. island Ilhas de Barlavento, N Cape Verde
São Vicente, Cabo de 70 B5 Eng. Cape Saint Vincent, Port. Cabode São Vicente. cape S Portugal
São Vicente, Cabo de see São Vicente, Cabo de
Sápai see Sápes
Sapele 53 F5 Delta, S Nigeria
Sápes 82 D3 var. Sápai. Anatolikí Makedonía kai Thráki, NE Greece
Sapir 97 B7 prev. Sappir. Southern, S Israel
Sa Pobla 71 G3 Mallorca, Spain, W Mediterranean Sea
Sappir see Sapir
Sapporo 108 D2 Hokkaidō, NE Japan
Sapri 75 D6 Campania, S Italy
Sapulpa 27 G1 Oklahoma, C USA
Saqqez 98 C2 var. Saghez, Sakiz, Saqqiz. Kordestān, NW Iran
Saqqiz see Saqqez
Sara Buri 115 C5 var. Saraburi. Saraburi, C Thailand
Saraburi see Sara Buri
Saragossa see Zaragoza
Saragt see Sarahs
Saraguro 38 B2 Loja, S Ecuador
Sarahs 100 D3 var. Saragt, Rus. Serakhs. Ahal Welaýaty, S Turkmenistan
Sarajevo 78 C4 country capital (Bosnia and Herzegovina) Federacija Bosna I Hercegovina, SE Bosnia and Herzegovina
Sarakhs 98 E2 Khorāsān-Razavī, NE Iran
Saraktash 89 D6 Orenburgskaya Oblast', W Russian Federation
Saran' 92 C4 Kaz. Saran. Karagandy, C Kazakhstan
Saranda see Sarandë
Sarandë 79 C7 var. Saranda, It. Porto Edda; prev. Santi Quaranta. Vlorë, S Albania
Saransk 89 C5 Respublika Mordoviya, W Russian Federation
Sarasota 21 E4 Florida, SE USA
Saratov 92 B3 Saratovskaya Oblast', W Russian Federation
Saravan/Saravane see Salavan
Sarawak 116 D3 state East Malaysia
Sarawak see Kuching
Sarcelles 68 D1 Val-d'Oise, Île-de-France, N France Europe
Sardegna 75 A5 Eng. Sardinia. island Italy, C Mediterranean Sea
Sardinia see Sardegna
Sarera, Teluk see Cenderawasih, Teluk
Sargasso Sea 44 B4 sea W Atlantic Ocean
Sargodha 112 C2 Punjab, NE Pakistan
Sarh 54 C4 prev. Fort-Archambault. Moyen-Chari, S Chad
Sārī 98 D2 var. Sari, Sāri. Māzandarān, N Iran
Saría 83 E7 island SE Greece
Sarıkamış 95 F3 Kars, NE Turkey
Sarikol Range 101 G3 Rus. Sarykol'skiy Khrebet. mountain range China/Tajikistan
Sark 67 D8 Fr. Sercq. island Channel Islands
Şarkışla 94 D3 Sivas, C Turkey
Sarmiento 43 B6 Chubut, S Argentina
Sarnia 16 C5 Ontario, S Canada
Sarny 86 C1 Rivnens'ka Oblast', NW Ukraine
Sarochyna 85 D5 Rus. Sorochino. Vitsyebskaya Voblasts', N Belarus
Sarov 89 C5 prev. Sarova. Respublika Mordoviya, SW Russian Federation
Sarova see Sarov
Sarpsborg 63 B6 Østfold, S Norway
Sarrebruck see Saarbrücken
Sartène 69 E7 Corse, France, C Mediterranean Sea
Sarthe 68 B4 river N France
Sárti 82 C4 Kentrikí Makedonía, N Greece
Saruhan see Manisa
Saryarqa see Kazakhskiy Melkosopochnik
Sarykol'skiy Khrebet see Sarikol Range
Sary-Tash 101 F2 Oshskaya Oblast', SW Kyrgyzstan
Saryyesik-Atyrau, Peski 101 G1 desert E Kazakhstan
Sasebo 109 A7 Nagasaki, Kyūshū, SW Japan
Saskatchewan 15 F5 province SW Canada
Saskatchewan 15 F5 river Manitoba/ Saskatchewan, C Canada
Saskatoon 15 F5 Saskatchewan, S Canada
Sasovo 89 B5 Ryazanskaya Oblast', W Russian Federation
Sassandra 52 D5 S Côte d'Ivoire
Sassandra 52 D5 var. Ibo, Sassandra Fleuve. river S Côte d'Ivoire
Sassandra Fleuve see Sassandra
Sassari 75 A5 Sardegna, Italy, C Mediterranean Sea
Sassenheim 64 C3 Zuid-Holland, W Netherlands
Sassnitz 72 D2 Mecklenburg-Vorpommern, NE Germany
Sathmar see Satu Mare
Sátoraljaújhely 77 D6 Borsod-Abaúj-Zemplén, NE Hungary
Satpura Range 112 D4 mountain range C India
Satsuma-Sendai 109 A8 Kagoshima, Kyūshū, SW Japan
Satsunan-shoto 108 A3 island group Nansei-shotō, SW Japan Asia
Sattanen 62 D3 Lappi, NE Finland
Satu Mare 86 B3 Ger. Sathmar, Hung. Szatmárrnémeti. Satu Mare, NW Romania
Sau see Sava
Saudi Arabia 99 B5 off. Kingdom of Saudi Arabia, Al 'Arabīyah as Su'ūdīyah, Ar. Al Mamlakah al 'Arabīyah as Su'ūdīyah. country SW Asia
Saudi Arabia, Kingdom of see Saudi Arabia
Sauer see Sûre
Saugor see Sāgar
Saulkrasti 84 C3 Rīga, C Latvia
Sault Sainte Marie 16 C4 Michigan, N USA
Sault Sainte Marie 18 C1 Michigan, N USA
Sault Ste. Marie 16 C4 Ontario, S Canada
Saumur 68 B4 Maine-et-Loire, NW France
Saurimo 56 C1 Port. Henrique de Carvalho, Vila Henrique de Carvalho. Lunda Sul, NE Angola
Sava 85 E6 Mahilyowskaya Voblasts', E Belarus
Sava 78 B3 Eng. Save, Ger. Sau, Hung. Száva. river SE Europe
Savá 30 D2 Colón, N Honduras
Savai'i 123 E4 island NW Samoa
Savannah 21 E2 Georgia, SE USA
Savannah River 21 E2 river Georgia/South Carolina, SE USA
Savannakhét see Khanthabouli
Savanna-La-Mar 32 A5 W Jamaica
Savaria see SzombathelySave see Sava
Save, Rio 57 E3 var. Sabi. river Mozambique/ Zimbabwe
Saverne 68 E3 var. Zabern; anc. Tres Tabernae. Bas-Rhin, NE France
Savigliano 74 A2 Piemonte, NW Italy
Savigsivik see Savissivik
Savinski see Savinskiy
Savinskiy 88 C3 var. Savinski. Arkhangel'skaya Oblast', NW Russian Federation
Savissivik 60 D1 var. Savigsivik. Avannaarsua, N Greenland
Savoie 69 D5 cultural region E France
Savona 74 A2 Liguria, NW Italy
Savu Sea 117 E5 Ind. Laut Sawu. sea S Indonesia
Sawakin see Suakin
Sawdirí see Sodiri
Sawhāj 50 B2 var. Sawhāj var. Sohâg, Suliag. C Egypt
Sawhāj see Sawhāj
Şawqirah 99 D6 var. Suqrah. S Oman
Sawu, Laut see Savu Sea
Saxe see Sachsen
Saxony see Sachsen
Sayaboury see Xaignabouli
Sayanskiy Khrebet 90 D3 mountain range S Russian Federation
Saýat 100 D3 Rus. Sayat. Lebap Welaýaty, E Turkmenistan
Sayaxché 30 B2 Petén, N Guatemala
Şaydā/Sayida see Saida
Sayḩūt 99 D6 E Yemen
Saynshand 105 E2 Dornogovĭ, SE Mongolia
Sayre 19 E3 Pennsylvania, NE USA
Say'ūn 99 C6 var. Saywūn. C Yemen
Saywūn see Say'ūn
Scalabis see Santarém
Scandinavia 44 D2 geophysical region NW Europe
Scarborough 67 D5 N England, United Kingdom
Scarpanto see Kárpathos
Scebeli see Shebeli
Schaan 72 E1 W Liechtenstein Europe
Schaerbeek 65 C6 Brussels, C Belgium
Schaffhausen 73 B7 Fr. Schaffhouse. Schaffhausen, N Switzerland
Schaffhouse see Schaffhausen
Schagen 64 C2 Noord-Holland, NW Netherlands
Schaulen see Šiauliai
Schebschi Mountains see Shebshi Mountains
Scheessel 72 B3 Niedersachsen, NW Germany
Schefferville 17 E2 Québec, E Canada
Schelde see Scheldt
Scheldt 65 B5 Dut. Schelde, Fr. Escaut. river W Europe
Schell Creek Range 25 D5 mountain range Nevada, W USA
Schenectady 19 F3 New York, NE USA
Schertz 27 G4 Texas, SW USA
Schiermonnikoog 64 D1 Fris. Skiermûntseach. island Waddeneilanden, N Netherlands
Schijndel 64 D4 Noord-Brabant, S Netherlands
Schil see Jiu
Schiltigheim 68 E3 Bas-Rhin, NE France
Schivelbein see Świdwin
Schleswig 72 B2 Schleswig-Holstein, N Germany
Schleswig-Holstein 72 B2 state N Germany
Schlettstadt see Sélestat
Schlochau see Człuchów
Schneekoppe see Sněžka
Schneidemühl see Piła
Schoden see Skuodas
Schönebeck 72 C4 Sachsen-Anhalt, C Germany
Schönlanke see Trzcianka
Schooten see Schoten
Schoten 65 C5 var. Schooten. Antwerpen, N Belgium
Schouwen 64 B4 island SW Netherlands
Schwabenalb see Schwäbische Alb
Schwäbische Alb 73 B6 var. Schwabenalb, Eng. Swabian Jura. mountain range S Germany
Schwandorf 73 C5 Bayern, SE Germany
Schwaz 73 C7 Tirol, W Austria
Schweidnitz see Świdnica
Schweinfurt 73 B5 Bayern, SE Germany
Schweiz see Switzerland
Schwerin 72 C3 Mecklenburg-Vorpommern, N Germany
Schwertberg see Świecie
Schwiebus see Świebodzin
Schwyz 73 B7 var. Schwiz. Schwyz, C Switzerland
Schyl see Jiu
Scio see Chíos
Scoresby Sound/Scoresbysund see Ittoqqortoormiit
Scoresby Sund see Kangertittivaq
Scotia Sea 35 C8 sea SW Atlantic Ocean
Scotland 66 C3 cultural region Scotland, U K
Scott Base 132 B4 NZ research station Antarctica
Scott Island 132 B5 island Antarctica
Scottsbluff 22 D3 Nebraska, C USA
Scottsboro 20 D1 Alabama, S USA
Scottsdale 26 B2 Arizona, SW USA
Scranton 19 F3 Pennsylvania, NE USA
Scrobesbyrig' see Shrewsbury
Scupi see Skopje
Scutari see Shkodër
Scutari, Lake 79 C5 Alb. Liqeni i Shkodrës, SCr. Skadarsko Jezero. lake Albania/Montenegro
Scyros see Skýros
Searcy 20 B1 Arkansas, C USA
Seattle 24 B2 Washington, NW USA
Sébaco 30 D3 Matagalpa, W Nicaragua
Sebaste/Sebastia see Sivas
Sebastían Vizcaíno, Bahía 28 A2 bay NW Mexico
Sebastopol see Sevastopol'
Sebenico see Šibenik
Sechura, Bahía de 38 A3 bay NW Peru
Secunderābād 112 D5 var. Sikandarabad. Telangana, C India
Sedan 68 D3 Ardennes, N France
Seddon 129 D5 Marlborough, South Island, New Zealand
Seddonville 129 C5 West Coast, South Island, New Zealand
Sédhiou 52 B3 SW Senegal
Sedlez see Siedlce
Sedona 26 B2 Arizona, SW USA
Sedunum see Sion
Seeland see Sjælland
Seenu Atoll see Addu Atoll
Seesen 72 B4 Niedersachsen, C Germany
Segestica see Sisak
Segezha 88 B3 Respublika Kareliya, NW Russian Federation
Seghedin see Szeged
Segna see Senj
Segodunum see Rodez
Ségou 52 D3 var. Segu. Ségou, C Mali
Segovia 70 D2 Castilla-León, C Spain
Segoviao Wangkí see Coco, Río
Segu see Ségou
Séguédine 53 H2 Agadez, NE Niger
Seguin 27 G4 Texas, SW USA
Segura 71 E4 river S Spain
Seinäjoki 63 D5 Swe. Östermyra. Länsi-Suomi, W Finland
Seine 68 D1 river N France
Seine, Baie de la 68 B3 bay N France
Sejong City 106 E4 administrative capital (South Korea) C South Korea
Sekondi see Sekondi-Takoradi
Sekondi-Takoradi 53 E5 var. Sekondi. S Ghana
Selānik see Thessaloníki
Selenga 105 E1 Mong. Selenge Mörön. river Mongolia/Russian Federation
Selenge Mörön see Selenga
Sélestat 68 E4 Ger. Schlettstadt. Bas-Rhin, NE France
Seleucia see Silifke
Selfoss 61 E5 Sudhurland, SW Iceland
Sélibabi 52 C3 var. Sélibaby. Guidimaka, S Mauritania
Sélibaby see Sélibabi
Selma 25 C6 California, W USA
Selway River 24 D2 river Idaho, NW USA North America
Selwyn Range 126 B3 mountain range Queensland, C Australia
Selzaete see Zelzate
Semarang 116 C5 var. Samarang. Jawa, C Indonesia
Sembé 55 B5 Sangha, NW Congo

Semendria – Slavyansk

Semendria *see* Smederevo
Semey *92 D4 prev.* Semipalatinsk. Vostochnyy Kazakhstan, E Kazakhstan
Semezhevo *see* Syemyezhava
Seminole *27 E3* Texas, SW USA
Seminole, Lake *20 D3 reservoir* Florida/Georgia, SE USA
Semipalatinsk *see* Semey
Semnān *98 D3 var.* Samnān. Semnān, N Iran
Semois *65 C8 river* SE Belgium
Sendai *108 D4* Miyagi, Honshū, C Japan
Sendai-wan *108 D4 bay* E Japan
Senec *77 C6 Ger.* Wartberg, *Hung.* Szenc; *prev.* Szempcz. Bratislavský Kraj, W Slovakia
Senegal *52 B3 off.* Republic of Senegal, *Fr.* Sénégal. *country* W Africa
Senegal *52 C3 Fr.* Sénégal. *river* W Africa
Senegal, Republic of *see* Senegal
Senftenberg *72 D4* Brandenburg, E Germany
Senia *see* Senj
Senica *77 C6 Ger.* Senitz, *Hung.* Szenice. Trnavský Kraj, W Slovakia
Seniça *see* Sjenica
Senitz *see* Senica
Senj *78 A3 Ger.* Zengg, *It.* Segna; *anc.* Senia. Lika-Senj, NW Croatia
Senja *62 C2 prev.* Senjen. *island* N Norway
Senjen *see* Senja
Senkaku-shoto *108 A3 island group* SW Japan
Senlis *68 C3* Oise, N France
Sennar *50 C4 var.* Sannār. Sinnar, C Sudan
Senones *see* Sens
Sens *68 C3 anc.* Agendicum, Senones. Yonne, C France
Sensburg *see* Mrągowo
Sên, Stœng *115 D5 river* C Cambodia
Senta *78 D3 Hung.* Zenta. Vojvodina, N Serbia
Seo de Urgel *see* La Seu d'Urgel
Seoul *107 E4 off.* Seoul Teukbyeolsi, *prev.* Sŏul, *Jap.* Keijō; *prev.* Kyŏngsŏng. *country capital* (South Korea) NW South Korea
Seoul Teukbyeolsi *see* Seoul
Şepşi-Sângeorz/Sepsiszentgyőrgy *see* Sfântu Gheorghe
Sept-Îles *17 E3* Québec, SE Canada
Seraing *65 D6* Liège, E Belgium
Serakhs *see* Sarahs
Seram, Laut *117 F4 Eng.* Ceram Sea. *sea* E Indonesia
Pulau Seram *117 F4 var.* Serang, *Eng.* Ceram. *island* Maluku, E Indonesia
Serang *116 C5* Jawa, C Indonesia
Serang *see* Seram, Pulau
Serasan, Selat *116 C3 strait* Indonesia/Malaysia
Serbia *78 D4 off.* Federal Republic of Serbia; *prev.* Yugoslavia, *SCr.* Jugoslavija. *country* SE Europe
Serbia, Federal Republic of *see* Serbia
Sercq *see* Sark
Serdar *100 C2 prev. Rus.* Gyzyrlabat, Kizyl-Arvat. Balkan Welaýaty, W Turkmenistan
Serdica *see* Sofiya
Serdobol' *see* Sortavala
Serenje *56 D2* Central, E Zambia
Seres *see* Sérres
Seret/Sereth *see* Siret
Serhetabat *100 D4 prev. Rus.* Gushgy, Kushka. Mary Welaýaty, S Turkmenistan
Sérifos *83 C6 anc.* Seriphos. *island* Kykládes, Greece, Aegean Sea
Seriphos *see* Sérifos
Serov *92 C3* Sverdlovskaya Oblast', C Russian Federation
Serowe *56 D3* Central, SE Botswana
Serpa Pinto *see* Menongue
Serpent's Mouth, The *37 F2 Sp.* Boca de la Serpiente. *strait* Trinidad and Tobago/Venezuela
Serpiente, Boca de la *see* Serpent's Mouth, The
Serpukhov *89 B5* Moskovskaya Oblast', W Russian Federation
Sérrai *see* Sérres
Serrana, Cayo de *31 F2 island group* NW Colombia South America
Serranilla, Cayo de *31 F2 island group* NW Colombia South America Caribbean Sea
Serravalle *74 E1* N San Marino
Sérres *82 C3 var.* Seres; *prev.* Sérrai. Kentrikí Makedonía, NE Greece
Sesdlets *see* Siedlce
Sesto San Giovanni *74 B2* Lombardia, N Italy
Sesvete *78 B2* Zagreb, N Croatia
Setabis *see* Xàtiva
Sète *69 C6 prev.* Cette. Hérault, S France
Setesdal *63 A6 valley* S Norway
Sétif *49 E2 var.* Stif. N Algeria
Setté Cama *55 A6* Ogooué-Maritime, SW Gabon
Setúbal *70 B4 Eng.* Saint Ubes, Saint Yves. Setúbal, W Portugal
Setúbal, Baía de *70 B4 bay* W Portugal
Seul, Lac *16 B3 lake* Ontario, S Canada
Sevan *95 G2* C Armenia
Sevana Lich *95 G3 Eng.* Lake Sevan, *Rus.* Ozero Sevan. *lake* E Armenia
Sevan, Lake/Sevan, Ozero *see* Sevana Lich
Sevastopol' *87 F5 Eng.* Sebastopol. Respublika Krym, S Ukraine
Severn *16 B2 river* Ontario, S Canada
Severn *67 D6 Wel.* Hafren. *river* England/Wales, United Kingdom
Severnaya Dvina *88 C4 var.* Northern Dvina. *river* NW Russian Federation
Severnaya Zemlya *93 E2 var.* Nicholas II Land. *island group* N Russian Federation
Severnyy *88 E3* Respublika Komi, NW Russian Federation
Severodonetsk *see* Syeverodonets'k
Severodvinsk *88 C3 prev.* Molotov, Sudostroy. Arkhangel'skaya Oblast', NW Russian Federation
Severomorsk *88 C2* Murmanskaya Oblast', NW Russian Federation
Seversk *92 D4* Tomskaya Oblast', C Russian Federation
Sevier Lake *22 A4 lake* Utah, W USA
Sevilla *70 C4 Eng.* Seville; *anc.* Hispalis. Andalucía, SW Spain
Seville *see* Sevilla
Sevlievo *82 D2* Gabrovo, N Bulgaria
Sevluš/Sevlyush *see* Vynohradiv
Seward's Folly *see* Alaska
Seychelles *57 G1 off.* Republic of Seychelles. *country* W Indian Ocean
Seychelles, Republic of *see* Seychelles
Seyðisfjörður *61 E5* Austurland, E Iceland
Seýdi *100 D2 Rus.* Seydi; *prev.* Neftezavodsk. Lebap Welaýaty, E Turkmenistan
Seyhan *see* Adana
Sfákia *see* Chóra Sfakíon
Sfântu Gheorghe *86 C4 Ger.* Sankt-Georgen, *Hung.* Sepsiszentgyőrgy; *prev.* Şepşi-Sângeorz, Sfîntu Gheorghe. Covasna, C Romania
Sfax *49 F2 Ar.* Şafāqis. E Tunisia
Sfîntu Gheorghe *see* Sfântu Gheorghe
's-Gravenhage *64 B4 var.* Den Haag, *Eng.* The Hague, *Fr.* La Haye. *country capital* (Netherlands-seat of government) Zuid-Holland, W Netherlands
's-Gravenzande *64 B4* Zuid-Holland, W Netherlands
Shaan/Shaanxi Sheng *see* Shaanxi
Shaanxi *106 B5 var.* Shaan, Shaanxi Sheng, Shan-hsi, Shenshi, Shensi. *province* C China
Shabani *see* Zvishavane
Shabeelle, Webi *see* Shebeli
Shache *104 A3 var.* Yarkant. Xinjiang Uygur Zizhiqu, NW China
Shacheng *see* Huailai
Shackleton Ice Shelf *132 D3 ice shelf* Antarctica
Shadādī *see* Ash Shadādah
Shāhābād *see* Eslāmābād
Sha Hi *see* Orūmīyeh, Daryācheh-ye
Shahjahanabad *see* Delhi
Shahr-e Kord *98 C3 var.* Shahr Kord. Chahār Maḩall va Bakhtīārī, C Iran
Shahr Kord *see* Shahr-e Kord
Shāhrūd *98 D2 prev.* Emāmrūd, Emāmshahr. Semnān, N Iran
Shalkar *92 B4 var.* Chelkar. Aktyubinsk, W Kazakhstan
Shām, Bādiyat ash *see* Syrian Desert
Shana *see* Kuril'sk
Shandī *see* Shendi
Shandong *106 D4 var.* Lu, Shandong Sheng, Shantung. *province* E China
Shandong Sheng *see* Shandong
Shanghai *106 D5 var.* Shang-hai. Shanghai Shi, E China
Shangrao *106 D5* Jiangxi, S China
Shan-hsi *see* Shaanxi, China
Shan-hsi *see* Shanxi, China
Shannon *67 A6 Ir.* An tSionainn. *river* W Ireland
Shan Plateau *114 B3 plateau* E Myanmar (Burma)
Shansi *see* Shanxi
Shantar Islands *see* Shantarskiye Ostrova
Shantarskiye Ostrova *93 G3 Eng.* Shantar Islands. *island group* E Russian Federation
Shantou *106 D6 var.* Shan-t'ou, Swatow. Guangdong, S China
Shan-t'ou *see* Shantou
Shantung *see* Shandong
Shanxi *106 C4 var.* Jin, Shan-hsi, Shansi, Shanxi Sheng. *province* C China
Shan Xian *see* Sanmenxia
Shanxi Sheng *see* Shanxi
Shaoguan *106 C6 var.* Shao-kuan, *Cant.* Kukong; *prev.* Ch'u-chiang. Guangdong, S China
Shao-kuan *see* Shaoguan
Shaqrā' *98 B4* Ar Riyāḑ, C Saudi Arabia
Shaqrā *see* Shuqrah
Shar *92 D5 var.* Charsk. Vostochnyy Kazakhstan, E Kazakhstan
Shari *108 D2* Hokkaidō, NE Japan
Shari *see* Chari
Sharjah *see* Ash Shāriqah
Shark Bay *125 A5 bay* Western Australia
Sharqī, Al Jabal ash/Sharqi, Jebel esh *see* Anti-Lebanon
Shashe *56 D3 var.* Shashi. *river* Botswana/Zimbabwe
Shashi *see* Shashe
Shatskiy Rise *103 G1 undersea rise* N Pacific Ocean
Shawnee *27 G1* Oklahoma, C USA
Shaykh, Jabal ash *see* Hermon, Mount
Shchadryn *85 D7 Rus.* Shchedrin. Homyel'skaya Voblasts', SE Belarus
Shchedrin *see* Shchadryn
Shcheglovsk *see* Kemerovo
Shchëkino *89 B5* Tul'skaya Oblast', W Russian Federation
Shchors *87 E1* Chernihivs'ka Oblast', N Ukraine
Shchuchin *see* Shchuchyn
Shchuchinsk *92 C4 prev.* Shchuchye. Akmola, N Kazakhstan
Shchuchye *see* Shchuchinsk
Shchuchyn *85 B5 Pol.* Szczuczyn Nowogródzki, *Rus.* Shchuchin. Hrodzyenskaya Voblasts', W Belarus
Shebekino *89 A6* Belgorodskaya Oblast', W Russian Federation
Shebelē Wenz, Wabē *see* Shebeli
Shebeli *51 D5 Amh.* Wabē Shebelē Wenz, *It.* Scebeli, *Som.* Webi Shabeelle. *river* Ethiopia/Somalia
Sheberghān *see* Shibirghān
Sheboygan *18 B2* Wisconsin, N USA
Shebshi Mountains *54 A4 var.* Schebschi Mountains. *mountain range* E Nigeria
Shechem *see* Nablus
Shedadi *see* Ash Shadādah
Sheffield *67 D5* N England, United Kingdom
Shekhem *see* Nablus
Sheki *see* Şäki
Shelby *22 B1* Montana, NW USA
Sheldon *23 F3* Iowa, C USA
Shelekhov Gulf *see* Shelikhova, Zaliv
Shelikhova, Zaliv *93 G2 Eng.* Shelekhov Gulf. *gulf* E Russian Federation
Shendi *50 C4 var.* Shandī. River Nile, NE Sudan
Shengking *see* Liaoning
Shenking *see* Liaoning
Shenshi/Shensi *see* Shaanxi
Shenyang *106 D3 Chin.* Shen-yang, *Eng.* Moukden, Mukden; *prev.* Fengtien. *province capital* Liaoning, NE China
Shen-yang *see* Shenyang
Shepetivka *86 D2 Rus.* Shepetovka. Khmel'nyts'ka Oblast', NW Ukraine
Shepetovka *see* Shepetivka
Shepparton *127 C7* Victoria, SE Australia
Sherbrooke *17 E4* Québec, SE Canada
Shereik *50 C3* River Nile, N Sudan
Sheridan *22 C2* Wyoming, C USA
Sherman *27 G2* Texas, SW USA
's-Hertogenbosch *64 C4 Fr.* Bois-le-Duc, *Ger.* Herzogenbusch. Noord-Brabant, S Netherlands
Shetland Islands *66 D1 island group* NE Scotland, United Kingdom
Shevchenko *see* Aktau
Shiberghān/Shiberghan *see* Shibirghān
Shibirghān *101 E3 var.* Sheberghān, Shiberghan, Shiberghān. Jowzjān, N Afghanistan
Shibetsu *108 D2 var.* Sibetu. Hokkaidō, NE Japan
Shibh Jazirat Sīnā' *see* Sinai
Shibushi-wan *109 B8 bay* SW Japan
Shigatse *see* Xigazê
Shih-chia-chuang/Shihmen *see* Shijiazhuang
Shihezi *104 C2* Xinjiang Uygur Zizhiqu, NW China
Shiichi *see* Shyichy
Shijiazhuang *106 C4 var.* Shih-chia-chuang; *prev.* Shihmen. *province capital* Hebei, E China
Shikārpur *112 B3* Sind, S Pakistan
Shikoku *109 C7 var.* Sikoku. *island* SW Japan
Shikoku Basin *103 F2 var.* Sikoku Basin. *undersea basin* N Philippine Sea
Shikotan, Ostrov *108 E2 Jap.* Shikotan-tō. *island* NE Russian Federation
Shikotan-tō *see* Shikotan, Ostrov
Shilabo *51 D5* Sumalē, E Ethiopia
Shiliguri *113 F3 prev.* Siliguri. West Bengal, NE India
Shilka *93 F4 river* S Russian Federation
Shillong *113 G3 state capital* Meghālaya, NE India
Shimanto *see* Nakamura
Shimbir Berris *see* Shimbiris
Shimbiris *50 E4 var.* Shimbir Berris. *mountain* N Somalia
Shimoga *110 C2* Karnātaka, W India
Shimonoseki *109 A7 var.* Simonoseki, *hist.* Akamagaseki, Bakan. Yamaguchi, Honshū, SW Japan
Shinano-gawa *109 C5 var.* Sinano Gawa. *river* Honshū, C Japan
Shīndand *100 D4 prev.* Shīndand. Herāt, W Afghanistan
Shīndand *see* Shīndand
Shingū *109 C6 var.* Singû. Wakayama, Honshū, SW Japan
Shinjō *108 D4 var.* Sinzyô. Yamagata, Honshū, C Japan
Shinyanga *51 C7* Shinyanga, NW Tanzania
Shiprock *26 C1* New Mexico, SW USA
Shīrāz *98 D4 var.* Shīrāz. Fārs, S Iran
Shishchitsy *see* Shyshchytsy
Shivpuri *112 D3* Madhya Pradesh, C India
Shizugawa *108 D4* Miyagi, Honshū, NE Japan
Shizuoka *109 D6 var.* Sizuoka. Shizuoka, Honshū, S Japan
Shklov *see* Shklow
Shklow *85 D6 Rus.* Shklov. Mahilyowskaya Voblasts', E Belarus
Shkodër *79 C5 var.* Shkodra, *It.* Scutari, *SCr.* Skadar. Shkodër, NW Albania
Shkodra *see* Shkodër
Shkodrës, Liqeni i *see* Scutari, Lake
Shkumbinit, Lumi i *79 C6 var.* Shkumbi, Shkumbin. *river* C Albania
Shkumbi/Shkumbin *see* Shkumbinit, Lumi i
Sholāpur *see* Solāpur
Shostka *87 F1* Sums'ka Oblast', NE Ukraine
Show Low *26 B2* Arizona, SW USA
Show Me State *see* Missouri
Shpola *87 E3* Cherkas'ka Oblast', N Ukraine
Shqipëria/Shqipërisë, Republika e *see* Albania
Shreveport *20 A2* Louisiana, S USA
Shrewsbury *67 D6 hist.* Scrobesbyrig'. W England, United Kingdom
Shu *92 C5 Kaz.* Shū. Zhambyl, SE Kazakhstan
Shuang-liao *see* Liaoyuan
Shū, Kazakhstan *see* Shu
Shumagin Islands *14 B3 island group* Alaska, USA
Shumen *82 D2* Shumen, NE Bulgaria
Shumilina *85 E5 Rus.* Shumilino. Vitsyebskaya Voblasts', NE Belarus
Shumilino *see* Shumilina
Shunsen *see* Chuncheon
Shuqrah *99 B7 var.* Shaqrā. SW Yemen
Shwebo *114 B3* Sagaing, C Myanmar (Burma)
Shyichy *85 C7 Rus.* Shiichi. Homyel'skaya Voblasts', SE Belarus
Shymkent *92 B5 prev.* Chimkent. Yuzhnyy Kazakhstan, S Kazakhstan
Shyshchytsy *85 C6 Rus.* Shishchitsy. Minskaya Voblasts', C Belarus
Siam *see* Thailand
Siam, Gulf of *see* Thailand, Gulf of
Sian *see* Xi'an
Siang *see* Brahmaputra
Siangtan *see* Xiangtan
Šiauliai *84 B4 Ger.* Schaulen. Šiauliai, N Lithuania
Siazan' *see* Siyäzän
Sibay *89 D6* Respublika Bashkortostan, W Russian Federation
Šibenik *78 B4 It.* Sebenico. Šibenik-Knin, S Croatia
Siberia *see* Sibir'
Siberoet *see* Siberut, Pulau
Siberut, Pulau *116 A4 prev.* Siberoet. *island* Kepulauan Mentawai, W Indonesia
Sibi *112 B2* Baluchistān, SW Pakistan
Sibir' *93 E3 var.* Siberia. *physical region* NE Russian Federation
Sibiti *55 B6* Lékoumou, S Congo
Sibiu *86 B4 Ger.* Hermannstadt, *Hung.* Nagyszeben. Sibiu, C Romania
Sibolga *116 B3* Sumatera, W Indonesia
Sibu *116 D3* Sarawak, East Malaysia
Sibut *54 C4 prev.* Fort-Sibut. Kémo, S Central African Republic
Sibuyan Sea *117 E2 sea* W Pacific Ocean
Sichon *115 C6 var.* Ban Sichon, Si Chon. Nakhon Si Thammarat, SW Thailand
Si Chon *see* Sichon
Sichuan *106 B5 var.* Chuan, Sichuan Sheng, Ssu-ch'uan, Szechuan, Szechwan. *province* C China
Sichuan Pendi *106 B5 basin* C China
Sichuan Sheng *see* Sichuan
Sicilian Channel *see* Sicily, Strait of
Sicily *75 C7 Eng.* Sicily; *anc.* Trinacria. *island* Italy, C Mediterranean Sea
Sicily, Strait of *75 B7 var.* Sicilian Channel. *strait* C Mediterranean Sea
Sicuani *38 D4* Cusco, S Peru
Sidári *82 A4* Kérkyra, Iónia Nisiá, Greece, C Mediterranean Sea
Sidas *116 C4* Borneo, C Indonesia
Siderno *75 D7* Calabria, SW Italy
Sidhirókastron *see* Sidirókastro
Sîdi Barrâni *50 A1* NW Egypt
Sidi Bel Abbès *48 D2 var.* Sidi bel Abbès, Sidi-Bel-Abbès. NW Algeria
Sidirókastro *82 C3 prev.* Sidhirókastron. Kentrikí Makedonía, NE Greece
Sidley, Mount *132 B4 mountain* Antarctica
Sidney *22 D1* Montana, NW USA
Sidney *22 D4* Nebraska, C USA
Sidney *18 C4* Ohio, N USA
Sidon *see* Saïda
Sidra *see* Surt
Sidra/Sidra, Gulf of *see* Surt, Khalīj, N Libya
Siebenbürgen *see* Transylvania
Siedlce *76 E3 Ger.* Sedlez, *Rus.* Sesdlets. Mazowieckie, C Poland
Siegen *72 B4* Nordrhein-Westfalen, W Germany
Siemiatycze *76 E3* Podlaskie, NE Poland
Siena *74 B3 Fr.* Sienne; *anc.* Saena Julia. Toscana, C Italy
Sienne *see* Siena
Sieradz *76 C4* Sieradz, C Poland
Sierpc *76 D3* Mazowieckie, C Poland
Sierra Leone *52 C4 off.* Republic of Sierra Leone. *country* W Africa
Sierra Leone Basin *44 C4 undersea basin* E Atlantic Ocean
Sierra Leone, Republic of *see* Sierra Leone
Sierra Leone Ridge *see* Sierra Leone Rise
Sierra Leone Rise *44 C4 var.* Sierra Leone Ridge, Sierra Leone Schwelle. *undersea rise* E Atlantic Ocean
Sierra Leone Schwelle *see* Sierra Leone Rise
Sierra Vista *26 B3* Arizona, SW USA
Sífnos *83 C6 anc.* Siphnos. *island* Kykládes, Greece, Aegean Sea
Sigli *116 A3* Sumatera, W Indonesia
Siglufjörður *61 E4* Nordhurland Vestra, N Iceland
Signal Peak *26 A2 mountain* Arizona, SW USA
Signan *see* Xi'an
Signy *132 A2 UK research station* South Orkney Islands, Antarctica
Siguatepeque *30 C2* Comayagua, W Honduras
Siguiri *52 D4* NE Guinea
Sihanoukville *115 D6 var.* Kâmpóng Saôm; *prev.* Kompong Som. Kâmpóng Saôm, SW Cambodia
Siilinjärvi *62 E4* Itä-Suomi, C Finland
Siirt *95 F4 var.* Sert; *anc.* Tigranocerta. Siirt, SE Turkey
Sikandarabad *see* Secunderābād
Sikasso *52 D4* Sikasso, S Mali
Sikeston *23 H5* Missouri, C USA
Sikhote-Alin', Khrebet *93 G4 mountain range* SE Russian Federation
Siking *see* Xi'an
Siklós *77 C7* Baranya, SW Hungary
Sikoku *see* Shikoku
Sikoku Basin *see* Shikoku Basin
Šilalė *84 B4* Tauragė, W Lithuania
Silchar *113 G3* Assam, NE India
Silesia *76 B4 physical region* SW Poland
Silifke *94 C4 anc.* Seleucia. İçel, S Turkey
Siliguri *see* Shiliguri
Siling Co *104 C5 lake* W China
Silinhot *see* Xilinhot
Silistra *82 E1 var.* Silistria; *anc.* Durostorum. Silistra, NE Bulgaria
Silistria *see* Silistra
Sillamäe *84 E2 Ger.* Sillamäggi. Ida-Virumaa, NE Estonia
Sillamäggi *see* Sillamäe
Sillein *see* Žilina
Šilutė *84 B4 Ger.* Heydekrug. Klaipėda, W Lithuania
Silvan *95 E4* Dıyarbakır, SE Turkey
Silva Porto *see* Kuito
Silver State *see* Colorado
Silver State *see* Nevada
Simanichy *85 C7 Rus.* Simonichi. Homyel'skaya Voblasts', SE Belarus
Simav *94 B3* Kütahya, W Turkey
Simav Çayı *94 A3 river* NW Turkey
Simbirsk *see* Ul'yanovsk
Simeto *75 C7 river* Sicilia, Italy, C Mediterranean Sea
Simeulue, Pulau *116 A3 island* NW Indonesia
Simferopol' *87 F5* Respublika Krym, S Ukraine
Simitla *82 C3* Blagoevgrad, SW Bulgaria
Şimlăul Silvaniei/Şimleul Silvaniei *see* Şimleu Silvaniei
Şimleu Silvaniei *86 B3 Hung.* Szilágysomlyó; *prev.* Şimlăul Silvaniei, Şimleul Silvaniei. Sălaj, NW Romania
Simonichi *see* Simanichy
Simonoseki *see* Shimonoseki
Simpelveld *65 D6* Limburg, SE Netherlands
Simplon Pass *73 B8 pass* S Switzerland
Simpson *see* Fort Simpson
Simpson Desert *126 B4 desert* Northern Territory/South Australia
Sinai *50 C2 var.* Sinai Peninsula, *Ar.* Shibh Jazirat Sīnā', Sīnā. *physical region* NE Egypt
Sinaia *86 C4* Prahova, SE Romania
Sinano Gawa *see* Shinano-gawa
Sīnā/Sinai Peninsula *see* Sinai
Sincelejo *36 B2* Sucre, NW Colombia
Sind *112 B3 var.* Sindh. *province* SE Pakistan
Sindelfingen *73 B6* Baden-Württemberg, SW Germany
Sindh *see* Sind
Sindi *84 D2 Ger.* Zintenhof. Pärnumaa, SW Estonia
Sines *70 B4* Setúbal, S Portugal
Singan *see* Xi'an
Singapore *116 B3 country capital* (Singapore) S Singapore
Singapore *116 A1 off.* Republic of Singapore. *country* SE Asia
Singapore, Republic of *see* Singapore
Singen *73 B6* Baden-Württemberg, S Germany
Singida *51 C7* Singida, C Tanzania
Singkang *117 E4* Sulawesi, C Indonesia
Singkawang *116 C3* Borneo, C Indonesia
Singora *see* Songkhla
Singû *see* Shingū
Sining *see* Xining
Siniscola *75 A5* Sardegna, Italy, C Mediterranean Sea
Sinj *78 B4* Split-Dalmacija, SE Croatia
Sinkiang/Sinkiang Uighur Autonomous Region *see* Xinjiang Uygur Zizhiqu
Sinnamarie *see* Sinnamary
Sinnamary *37 H3 var.* Sinnamarie. N French Guiana
Sinneh *see* Sanandaj
Sînnicolau Mare *see* Sânnicolau Mare
Sinoe, Lacul *see* Sinoie, Lacul
Sinoie, Lacul *86 D5 prev.* Lacul Sinoe. *lagoon* SE Romania
Sinop *94 D2 anc.* Sinope. Sinop, N Turkey
Sinope *see* Sinop
Sinsheim *73 B6* Baden-Württemberg, SW Germany
Sint Maarten *33 G3 Eng.* Saint Martin. *island* Lesser Antilles
Sint-Michielsgestel *64 C4* Noord-Brabant, S Netherlands
Sîn-Miclăuş *see* Gheorgheni
Sint-Niklaas *65 B5 Fr.* Saint-Nicolas. Oost-Vlaanderen, N Belgium
Sint-Pieters-Leeuw *65 B6* Vlaams Brabant, C Belgium
Sintra *70 B3 prev.* Cintra. Lisboa, W Portugal
Sinujiif *51 E5* Nugaal, NE Somalia
Sinus Aelaniticus *see* Aqaba, Gulf of
Sinus Gallicus *see* Lion, Golfe du
Sinyang *see* Xinyang
Sinzyô *see* Shinjō
Sion *73 A7 Ger.* Sitten; *anc.* Sedunum. Valais, SW Switzerland
Sioux City *23 F3* Iowa, C USA
Sioux Falls *23 F3* South Dakota, N USA
Sioux State *see* North Dakota
Siphnos *see* Sífnos
Siping *106 D3 var.* Ssu-p'ing, Szeping; *prev.* Ssu-p'ing-chieh. Jilin, NE China
Siple, Mount *132 A4 mountain* Siple Island, Antarctica
Siquirres *31 E4* Limón, E Costa Rica
Siracusa *75 D7 Eng.* Syracuse. Sicilia, Italy, C Mediterranean Sea
Sir Edward Pellew Group *126 B2 island group* Northern Territory, NE Australia
Siret *86 C3 var.* Siretul, *Ger.* Sereth, *Rus.* Seret. *river* Romania/Ukraine
Siretul *see* Siret
Siria *see* Syria
Sirikit Reservoir *114 C4 lake* N Thailand
Sīrjān *98 D4 prev.* Sa'īdābād. Kermān, S Iran
Sirna *see* Sýrna
Şırnak *95 F4* Şırnak, SE Turkey
Síros *see* Sýros
Sirte *see* Surt
Sirti, Gulf of *see* Surt, Khalīj
Sisak *78 B3 var.* Siscia, *Ger.* Sissek, *Hung.* Sziszek; *anc.* Segestica. Sisak-Moslavina, C Croatia
Siscia *see* Sisak
Sisimiut *60 C3 var.* Holsteinborg, Holsteinsborg, Holstenborg, Holstensborg. Kitaa, S Greenland
Sissek *see* Sisak
Sīstān, Daryācheh-ye *see* Şāberī, Hāmūn-e
Sitaş Cristuru *see* Cristuru Secuiesc
Siteía *83 D8 var.* Sitía. Kríti, Greece, E Mediterranean Sea
Sitges *71 G2* Cataluña, NE Spain
Sitía *see* Siteía
Sittang *see* Sittoung
Sittard *65 D5* Limburg, SE Netherlands
Sitten *see* Sion
Sittoung *114 B4 var.* Sittang. *river* S Myanmar (Burma)
Sittwe *114 A3 var.* Akyab. Rakhine State, W Myanmar (Burma)
Siuna *30 D3* Región Autónoma Atlántico Norte, NE Nicaragua
Siut *see* Asyūţ
Sivas *94 D3 anc.* Sebastia, Sebaste. Sivas, C Turkey
Siverek *95 E4* Şanlıurfa, S Turkey
Siwa *see* Sīwah
Sīwah *50 A2 var.* Siwa. NW Egypt
Six Counties, The *see* Northern Ireland
Six-Fours-les-Plages *69 D6* Var, SE France
Siyäzän *95 H2 Rus.* Siazan'. NE Azerbaijan
Sjar *see* Sääre
Sjenica *79 D5 Turk.* Seniça. Serbia, SW Serbia
Skadar *see* Shkodër
Skadarsko Jezero *see* Scutari, Lake
Skagerak *see* Skagerrak
Skagerrak *63 A6 var.* Skagerak. *channel* N Europe
Skagit River *24 B1 river* Washington, NW USA
Skalka *62 C3 lake* N Sweden
Skarżysko-Kamienna *76 D4* Świętokrzyskie, C Poland
Skaudvilė *84 B4* Tauragė, SW Lithuania
Skegness *67 E6* E England, United Kingdom
Skellefteå *62 D4* Västerbotten, N Sweden
Skellefteälven *62 C4 river* N Sweden
Ski *63 B6* Akershus, S Norway
Skíathos *83 C5* Skíathos, Vóreies Sporádes, Greece, Aegean Sea
Skidal' *85 B5 Rus.* Skidel'. Hrodzyenskaya Voblasts', W Belarus
Skidel' *see* Skidal'
Skiermûntseach *see* Schiermonnikoog
Skierniewice *76 D3* Łódzkie, C Poland
Skiftet *84 C1 strait* Finland Atlantic Ocean Baltic Sea Gulf of Bothnia/Gulf of Finland
Skíros *see* Skýros
Skópelos *83 C5* Skópelos, Vóreies Sporádes, Greece, Aegean Sea
Skopje *79 D6 var.* Üsküb, *Turk.* Üsküp, *prev.* Skoplje; *anc.* Scupi. *country capital* (FYR Macedonia) N FYR Macedonia
Skoplje *see* Skopje
Skovorodino *93 F4* Amurskaya Oblast', SE Russian Federation
Skudnesfjorden *63 A6 fjord* S Norway
Skuodas *84 B3 Ger.* Schoden, *Pol.* Szkudy. Klaipėda, NW Lithuania
Skye, Isle of *66 B3 island* NW Scotland, United Kingdom
Skylge *see* Terschelling
Skýros *83 C5 var.* Skíros. Skńyros, Vóreies Sporádes, Greece, Aegean Sea
Skýros *83 C5 var.* Skíros; *anc.* Scyros. *island* Vóreies Sporádes, Greece, Aegean Sea
Slagelse *63 B7* Vestsjælland, E Denmark
Slatina *78 C3 Hung.* Szlatina; *prev.* Podravska Slatina. Virovitica-Podravina, NE Croatia
Slatina *86 B5* Olt, S Romania
Slavgorod *see* Slawharad
Slavonski Brod *78 C3 Ger.* Brod, *Hung.* Bród; *prev.* Brod, Brod na Savi. Brod-Posavina, NE Croatia
Slavuta *86 C2* Khmel'nyts'ka Oblast', NW Ukraine
Slavyansk *see* Slov"yans'k

Slawharad - Sulu, Laut

Slawharad *85 E7 Rus.* Slavgorod. Mahilyowskaya Voblasts', E Belarus
Sławno *76 C2* Zachodnio-pomorskie, NW Poland
Slēmānī *see* As Sulaymānīyah
Sliema *80 B5* N Malta
Sligo *67 A5 Ir.* Sligeach. Sligo, NW Ireland
Sliven *82 D2 var.* Slivno. Sliven, C Bulgaria
Slivnitsa *82 B2* Sofiya, W Bulgaria
Slivno *see* Sliven
Slobozia *86 C5* Ialomiţa, SE Romania
Slonim *85 B6 Pol.* Słonim. Hrodzyenskaya Voblasts', W Belarus
Słonim *see* Slonim
Slovakia *77 C6 off.* Slovenská Republika, *Ger.* Slowakei, *Hung.* Szlovákia, *Slvk.* Slovensko. *country* C Europe
Slovak Ore Mountains *see* Slovenské rudohorie
Slovenia *73 D8 off.* Republic of Slovenia, *Ger.* Slowenien, *Slvn.* Slovenija. *country* SE Europe
Slovenia, Republic of *see* Slovenia
Slovenija *see* Slovenia
Slovenská Republika *see* Slovakia
Slovenské rudohorie *77 D6 Eng.* Slovak Ore Mountains, *Ger.* Slowakisches Erzgebirge, Ungarisches Erzgebirge. *mountain range* C Slovakia
Slovensko *see* Slovakia
Slov"yans'k *87 G3 Rus.* Slavyansk. Donets'ka Oblast', E Ukraine
Slowakei *see* Slovakia
Slowakisches Erzgebirge *see* Slovenské rudohorie
Slowenien *see* Slovenia
Słubice *76 B3 Ger.* Frankfurt. Lubuskie, W Poland
Sluch *86 D1 river* NW Ukraine
Słupsk *76 C2 Ger.* Stolp. Pomorskie, N Poland
Slutsk *85 C6* Minskaya Voblasts', S Belarus
Smallwood Reservoir *17 F2 lake* Newfoundland and Labrador, S Canada
Smara *48 B3 var.* Es Semara. N Western Sahara
Smarhon' *85 C5 Pol.* Smorgonie, *Rus.* Smorgon'. Hrodzyenskaya Voblasts', W Belarus
Smederevo *78 D4 Ger.* Semendria. Serbia, N Serbia
Smederevska Palanka *78 D4* Serbia, C Serbia
Smela *see* Smila
Smila *87 E2 Rus.* Smela. Cherkas'ka Oblast', C Ukraine
Smilten *see* Smiltene
Smiltene *84 D3 Ger.* Smilten. Valka, N Latvia
Smøla *62 A4 island* W Norway
Smolensk *89 A5* Smolenskaya Oblast', W Russian Federation
Smorgon'/Smorgonie *see* Smarhon'
Smyrna *see* İzmir
Snake *12 B4 river* Yukon Territory, NW Canada
Snake River *24 C3 river* NW USA
Snake River Plain *24 D4 plain* Idaho, NW USA
Sneek *64 D2* Friesland, N Netherlands
Sneeuw-gebergte *see* Maoke, Pegunungan
Sněžka *76 B4 Ger.* Schneekoppe, *Pol.* Śnieżka. *mountain* N Czech Republic/Poland
Śniardwy, Jezioro *76 D2 Ger.* Spirdingsee. *lake* NE Poland
Sniečkus *see* Visaginas
Śnieżka *see* Sněžka
Snina *77 E5 Hung.* Szinna. Prešovský Kraj, E Slovakia
Snowdonia *67 C6 mountain range* NW Wales, United Kingdom
Snowdrift *see* Łutselk'e
Snow Mountains *see* Maoke, Pegunungan
Snyder *27 F3* Texas, SW USA
Sobradinho, Barragem de *see* Sobradinho, Represa de
Sobradinho, Represa de *41 F2 var.* Barragem de Sobradinho. *reservoir* E Brazil
Sochi *89 A7* Krasnodarskiy Kray, SW Russian Federation
Société, Îles de la/Society Islands *see* Société, Archipel de la
Society Islands *123 G4 var.* Archipel de Tahiti, Îles de la Société, *Eng.* Society Islands. *island group* W French Polynesia
Soconusco, Sierra de *see* Madre, Sierra
Socorro *26 D2* New Mexico, SW USA
Socorro, Isla *28 B5 island* W Mexico
Socotra *see* Suquṭrā
Soc Trăng *115 D6 var.* Khanh Hung. Soc Trăng, S Vietnam
Socuéllamos *71 E3* Castilla-La Mancha, C Spain
Sodankylä *62 D3* Lappi, N Finland
Sodari *see* Sodiri
Söderhamn *63 C5* Gävleborg, C Sweden
Södertälje *63 C6* Stockholm, C Sweden
Sodiri *50 B4 var.* Sawdirī, Sodari. Northern Kordofan, C Sudan
Soekaboemi *see* Sukabumi
Soemba *see* Sumba, Pulau
Soengaipenoeh *see* Sungaipenuh
Soerabaja *see* Surabaya
Soerakarta *see* Surakarta
Sofia *see* Sofiya
Sofiya *82 C2 var.* Sophia, *Eng.* Sofia, *Lat.* Serdica. *country capital* (Bulgaria) (Bulgaria) Sofiya-Grad, W Bulgaria
Sogamoso *36 B3* Boyacá, C Colombia
Sognefjorden *63 A5 fjord* NE North Sea
Sohâg *see* Sawhāj
Sohar *see* Şuḩār
Sohm Plain *44 B3 abyssal plain* NW Atlantic Ocean
Sohrau *see* Żory
Sokal' *86 C2 Rus.* Sokal. L'vivs'ka Oblast', NW Ukraine
Söke *94 A4* Aydın, SW Turkey
Sokhumi *95 E1 Rus.* Sukhumi. NW Georgia
Sokodé *53 F4* C Togo
Sokol *88 C4* Vologodskaya Oblast', NW Russian Federation
Sokółka *76 E3* Podlaskie, NE Poland
Sokolov *77 A5 Ger.* Falkenau an der Eger; *prev.* Falknov nad Ohří. Karlovarský Kraj, W Czech Republic
Sokone *52 B3* W Senegal
Sokoto *53 F3* Sokoto, NW Nigeria
Sokoto *53 F4 river* NW Nigeria
Sokotra *see* Suquṭrā
Solāpur *102 B3 var.* Sholāpur. Mahārāshtra, W India
Solca *86 C3 Ger.* Solka. Suceava, N Romania
Sol, Costa del *70 D5 coastal region* S Spain
Soldeu *69 B7* NE Andorra Europe
Solec Kujawski *76 C3* Kujawsko-pomorskie, C Poland
Soledad *36 B1* Atlántico, N Colombia
Isla Soledad *see* East Falkland
Soligorsk *see* Salihorsk
Solikamsk *92 C3* Permskaya Oblast', NW Russian Federation
Sol'-Iletsk *89 D6* Orenburgskaya Oblast', W Russian Federation
Solingen *72 A4* Nordrhein-Westfalen, W Germany
Solka *see* Solca
Sollentuna *63 C6* Stockholm, C Sweden
Solo *see* Surakarta
Solok *116 B4* Sumatera, W Indonesia
Solomon Islands *122 C3 prev.* British Solomon Islands Protectorate. *country* W Solomon Islands N Melanesia W Pacific Ocean
Solomon Islands *122 C3 island group* Papua New Guinea/Solomon Islands
Solomon Sea *122 B3 sea* W Pacific Ocean
Soltau *72 B3* Niedersachsen, NW Germany
Sol'tsy *88 A4* Novgorodskaya Oblast', W Russian Federation
Solun *see* Thessaloníki
Solwezi *56 D2* North Western, NW Zambia
Sōma *108 D4* Fukushima, Honshū, C Japan
Somalia *51 D5 off.* Somali Democratic Republic, *Som.* Jamuuriyada Demuqraadiga Soomaaliyeed, Soomaaliya; *prev.* Italian Somaliland, Somaliland Protectorate. *country* E Africa
Somali Basin *47 E5 undersea basin* W Indian Ocean
Somali Democratic Republic *see* Somalia
Somaliland *51 D5 disputed territory* N Somalia
Somaliland Protectorate *see* Somalia
Sombor *78 C3 Hung.* Zombor. Vojvodina, NW Serbia
Someren *65 D5* Noord-Brabant, SE Netherlands
Somerset *20 A5 var.* Somerset Village. W Bermuda
Somerset *18 C5* Kentucky, S USA
Somerset Island *20 A5 island* W Bermuda
Somerset Island *15 F2 island* Queen Elizabeth Islands, Nunavut, NW Canada
Somerset Village *see* Somerset
Somers Islands *see* Bermuda
Somerton *26 A2* Arizona, SW USA
Someş *86 B3 river* Hungary/Romania Europe
Somme *68 C2 river* N France
Sommerfeld *see* Lubsko
Somotillo *30 C3* Chinandega, NW Nicaragua
Somoto *30 D3* Madriz, NW Nicaragua
Songea *51 C8* Ruvuma, S Tanzania
Sŏngjin *see* Kimch'aek
Songkhla *115 C7 var.* Songkla, *Mal.* Singora. Songkhla, SW Thailand
Songkla *see* Songkhla
Sonoran Desert *26 A3 var.* Desierto de Altar. *desert* Mexico/USA
Sonsonate *30 B3* Sonsonate, W El Salvador
Soochow *see* Suzhou
Soomaaliya/Soomaaliyeed, Jamuuriyada Demuqraadiga *see* Somalia
Soome Laht *see* Finland, Gulf of
Sop Hao *114 D3* Houaphan, N Laos
Sophia *see* Sofiya
Sopianae *see* Pécs
Sopot *76 C2 Ger.* Zoppot. Pomorskie, N Poland
Sopron *77 B6 Ger.* Ödenburg. Győr-Moson-Sopron, NW Hungary
Sorau/Sorau in der Niederlausitz *see* Żary
Sorgues *69 D6* Vaucluse, SE France
Sorgun *94 D3* Yozgat, C Turkey
Soria *71 E2* Castilla-León, N Spain
Soroca *86 D3 Rus.* Soroki. N Moldova
Sorochino *see* Sarochyna
Soroki *see* Soroca
Sorong *117 F4* Papua, E Indonesia
Sørøy *see* Sørøya
Sørøya *62 C2 var.* Sørøy, *Lapp.* Sállan. *island* N Norway
Sortavala *88 B3 prev.* Serdobol'. Respublika Kareliya, NW Russian Federation
Sotavento, Ilhas de *52 A3 var.* Leeward Islands. *island group* S Cape Verde
Sotkamo *62 E4* Oulu, C Finland
Souanké *55 B5* Sangha, NW Congo
Soueida *see* As Suwaydā'
Soufli *82 D3 prev.* Souflíon. Anatolikí Makedonía kai Thráki, NE Greece
Souflíon *see* Soufli
Soufrière *33 F2* W Saint Lucia
Soukhné *see* As Sukhnah
Sŏul *see* Seoul
Soûr *97 A5 var.* Şūr; *anc.* Tyre. SW Lebanon
Souris River *23 E1 var.* Mouse River. *river* Canada/USA
Soúrpi *83 B5* Thessalía, C Greece
Sousse *49 F2 var.* Sūsah. NE Tunisia
South Africa *56 C4 off.* Republic of South Africa, *Afr.* Suid-Afrika. *country* S Africa
South Africa, Republic of *see* South Africa
South America *34 continent*
Southampton *67 D7 hist.* Hamwih, *Lat.* Clausentum. S England, United Kingdom
Southampton Island *15 G3 island* Nunavut, NE Canada
South Andaman *111 F2 island* Andaman Islands, India, NE Indian Ocean
South Australia *127 A5 state* S Australia
South Australian Basin *120 B5 undersea basin* SW Indian Ocean
South Bend *18 C3* Indiana, N USA
South Beveland *see* Zuid-Beveland
South Bruny Island *127 C8 island* Tasmania, SE Australia
South Carolina *21 E2 off.* State of South Carolina, *also known as* The Palmetto State. *state* SE USA
South Carpathians *see* Carpaţii Meridionalii
South China Basin *103 E4 undersea basin* SE South China Sea
South China Sea *103 E4 Chin.* Nan Hai, *Ind.* Laut Cina Selatan, *Vtn.* Biên Đông. *sea* SE Asia
South Dakota *22 D2 off.* State of South Dakota, *also known as* The Coyote State, Sunshine State. *state* N USA
Southeast Indian Ridge *119 D7 undersea ridge* Indian Ocean/Pacific Ocean
Southeast Pacific Basin *131 E5 var.* Belling Hausen Mulde. *undersea basin* SE Pacific Ocean
South East Point *127 C7 headland* Victoria, S Australia
Southend-on-Sea *67 E6* E England, United Kingdom
Southern Alps *129 B6 mountain range* South Island, New Zealand
Southern Cook Islands *123 F4 island group* S Cook Islands
Southern Cross *125 B6* Western Australia
Southern Indian Lake *15 F4 lake* Manitoba, C Canada
Southern Ocean *45 B7 ocean* Atlantic Ocean/Indian Ocean/Pacific Ocean
Southern Uplands *66 C4 mountain range* S Scotland, United Kingdom
South Fiji Basin *120 D4 undersea basin* S Pacific Ocean
South Geomagnetic Pole *132 B3 pole* Antarctica
South Georgia *35 D8 island* South Georgia and the South Sandwich Islands, SW Atlantic Ocean
South Goulburn Island *124 E2 island* Northern Territory, N Australia
South Huvadhu Atoll *110 A5 atoll* S Maldives
South Indian Basin *119 D7 undersea basin* Indian Ocean/Pacific Ocean
South Island *129 C6 island* S New Zealand
South Korea *107 E4 off.* Republic of Korea, *Kor.* Taehan Min'guk. *country* E Asia
South Lake Tahoe *25 C5* California, W USA
South Orkney Islands *132 A2 island group* Antarctica
South Ossetia *95 F2 former autonomous region* SW Georgia
South Pacific Basin *see* Southwest Pacific Basin
South Platte River *22 D4 river* Colorado/Nebraska, C USA
South Pole *132 B3 pole* Antarctica
South Sandwich Islands *35 D8 island group* SW Atlantic Ocean
South Sandwich Trench *35 E8 trench* SW Atlantic Ocean
South Shetland Islands *132 A2 island group* Antarctica
South Shields *66 D4* NE England, United Kingdom
South Sioux City *23 F3* Nebraska, C USA
South Sudan *50 B5 off.* Republic of South Sudan, *country* N Africa
South Taranaki Bight *128 C4 bight* SE Tasman Sea
South Tasmania Plateau *see* Tasman Plateau
South Uist *66 B3 island* NW Scotland, United Kingdom
South-West Africa/South West Africa *see* Namibia
South West Cape *129 A8 headland* Stewart Island, New Zealand
Southwest Indian Ocean Ridge *see* Southwest Indian Ridge
Southwest Indian Ridge *119 B6 var.* Southwest Indian Ocean Ridge. *undersea ridge* SW Indian Ocean
Southwest Pacific Basin *121 E4 var.* South Pacific Basin. *undersea basin* SE Pacific Ocean
Sovereign Base Area *80 C5 uk military installation* S Cyprus
Soweto *56 D4* Gauteng, NE South Africa
Sōya-kaikyō *see* La Pérouse Strait
Spain *70 D3 off.* Kingdom of Spain, *Sp.* España; *anc.* Hispania, Iberia, *Lat.* Hispana. *country* SW Europe
Spain, Kingdom of *see* Spain
Spalato *see* Split
Spanish Town *32 B5 hist.* St.Iago de la Vega. C Jamaica
Sparks *25 C5* Nevada, W USA
Sparta *see* Spárti
Spartanburg *21 E1* South Carolina, SE USA
Spárti *83 B6 Eng.* Sparta. Pelopónnisos, S Greece
Spearfish *22 D2* South Dakota, N USA
Speightstown *33 G1* NW Barbados
Spencer *23 F3* Iowa, C USA
Spencer Gulf *127 B6 gulf* South Australia
Spey *66 C3 river* NE Scotland, United Kingdom
Spice Islands *see* Maluku
Spiess Seamount *45 C7 seamount* S Atlantic Ocean
Spijkenisse *64 B4* Zuid-Holland, SW Netherlands
Spíli *83 C8* Kríti, Greece, E Mediterranean Sea
Spin Būldak *101 E5* Kandahār, S Afghanistan
Spirdingsee *see* Śniardwy, Jezioro
Spitsbergen *61 F2 island* NW Svalbard
Split *78 B4 It.* Spalato. Split-Dalmacija, S Croatia
Špoģi *84 D4* Daugvapils, SE Latvia
Spokane *24 C2* Washington, NW USA
Spratly Islands *116 B2 Chin.* Nansha Qundao. *disputed territory* SE Asia
Spree *72 D4 river* E Germany
Springfield *18 B4 state capital* Illinois, N USA
Springfield *19 G3* Massachusetts, NE USA
Springfield *23 G5* Missouri, C USA
Springfield *18 C4* Ohio, N USA
Springfield *24 B3* Oregon, NW USA
Spring Garden *37 F2* NE Guyana
Spring Hill *21 E4* Florida, SE USA
Springs Junction *129 C5* West Coast, South Island, New Zealand
Springsure *126 D4* Queensland, E Australia
Sprottau *see* Szprotawa
Spruce Knob *19 E4 mountain* West Virginia, NE USA
Srbinje *see* Foča
Srbobran *78 D3 var.* Bácsszenttamás, *Hung.* Szenttamás. Vojvodina, N Serbia
Srebrenica *78 C4* Republika Srpska, E Bosnia and Herzegovina
Sredets *82 D2 prev.* Syulemeshlii. Stara Zagora, C Bulgaria
Sredets *82 E2 prev.* Grudovo. Burgas, E Bulgaria
Srednerusskaya Vozvyshennost' *87 G1 Eng.* Central Russian Upland. *mountain range* W Russian Federation
Sremska Mitrovica *78 C3 prev.* Mitrovica, *Ger.* Mitrowitz. Vojvodina, NW Serbia
Srepok, Sông *see* Srêpôk, Tônle
Srêpôk, Tônle *115 E5 var.* Sông Srepok. *river* Cambodia/Vietnam
Sri Aman *116 C3* Sarawak, East Malaysia
Sri Jayawardanapura *see* Sri Jayewardenapura Kotte
Sri Jayewardenapura Kotte *110 D3 administrative capital* (Sri Lanka) *var.* Sri Jayawardanapura. Western Province, W Sri Lanka
Srīkākulam *113 F5* Andhra Pradesh, E India
Sri Lanka *110 D3 off.* Democratic Socialist Republic of Sri Lanka; *prev.* Ceylon. *country* S Asia
Sri Lanka, Democratic Socialist Republic of *see* Sri Lanka
Srinagarind Reservoir *115 C5 lake* W Thailand
Srpska, Republika *78 B3 republic* Bosnia and Herzegovina
Ssu-ch'uan *see* Sichuan
Ssu-p'ing/Ssu-p'ing-chieh *see* Siping
Stabroek *65 B5* Antwerpen, N Belgium
Stade *72 B3* Niedersachsen, NW Germany
Stadskanaal *64 E2* Groningen, NE Netherlands
Stafford *67 D6* C England, United Kingdom
Staicele *84 D3* Limbaži, N Latvia
Ştaierdorf-Anina *see* Anina
Stájerlakanina *see* Anina
Stakhanov *87 H3* Luhans'ka Oblast', E Ukraine
Stalin *see* Varna
Stalinabad *see* Dushanbe
Stalingrad *see* Volgograd
Stalino *see* Donets'k
Stalinobod *see* Dushanbe
Stalinsk *see* Novokuznetsk
Stalinski Zaliv *see* Varnenski Zaliv
Stalin, Yazovir *see* Iskŭr, Yazovir
Stalowa Wola *76 E4* Podkarpackie, SE Poland
Stamford *19 F3* Connecticut, NE USA
Stampalia *see* Astypálaia
Stanislau *see* Ivano-Frankivs'k
Stanislav *see* Ivano-Frankivs'k
Stanisławów *see* Ivano-Frankivs'k
Stanke Dimitrov *see* Dupnitsa
Stanley *43 D7 var.* Port Stanley, Puerto Argentino. *dependent territory capital* (Falkland Islands) East Falkland, Falkland Islands
Stanleyville *see* Kisangani
Stann Creek *see* Dangriga
Stanovoy Khrebet *91 E3 mountain range* SE Russian Federation
Stanthorpe *127 D5* Queensland, E Australia
Staphorst *64 D2* Overijssel, E Netherlands
Starachowice *76 D4* Świętokrzyskie, C Poland
Stara Kanjiža *see* Kanjiža
Stara Pazova *78 D3 Ger.* Altpasua, *Hung.* Ópazova. Vojvodina, N Serbia
Stara Planina *see* Balkan Mountains
Stara Zagora *82 D2 Lat.* Augusta Trajana. Stara Zagora, C Bulgaria
Starbuck Island *123 G3 prev.* Volunteer Island. *island* E Kiribati
Stargard in Pommern *see* Stargard Szczeciński
Stargard Szczeciński *76 B3 Ger.* Stargard in Pommern. Zachodnio-pomorskie, NW Poland
Stari Bečej *see* Bečej
Starobel'sk *see* Starobil's'k
Starobil's'k *87 H2 Rus.* Starobel'sk. Luhans'ka Oblast', E Ukraine
Starobin *85 C7 var.* Starobyn. Minskaya Voblasts', S Belarus
Starobyn *see* Starobin
Starogard Gdański *76 C2 Ger.* Preussisch-Stargard. Pomorskie, N Poland
Starokonstantinov *see* Starokostyantyniv
Starokostyantyniv *86 D2 Rus.* Starokonstantinov. Khmel'nyts'ka Oblast', NW Ukraine
Starominskaya *89 A7* Krasnodarskiy Kray, SW Russian Federation
Staryya Darohi *85 C6 Rus.* Staryye Dorogi. Minskaya Voblasts', S Belarus
Staryye Dorogi *see* Staryya Darohi
Staryy Oskol *89 B6* Belgorodskaya Oblast', W Russian Federation
State College *19 E4* Pennsylvania, NE USA
Staten Island *see* Estados, Isla de los
Statesboro *21 E2* Georgia, SE USA
States, The *see* United States of America
Staunton *19 E5* Virginia, NE USA
Stavanger *63 A6* Rogaland, S Norway
Stavers Island *see* Vostok Island
Stavropol' *89 B7 prev.* Voroshilovsk. Stavropol'skiy Kray, SW Russian Federation
Stavropol' *see* Tol'yatti
Steamboat Springs *22 C4* Colorado, C USA
Steenwijk *64 D2* Overijssel, N Netherlands
Steier *see* Steyr
Steierdorf/Steierdorf-Anina *see* Anina
Steinamanger *see* Szombathely
Steinkjer *62 B4* Nord-Trøndelag, C Norway
Stejarul *see* Karapelit
Stendal *72 C3* Sachsen-Anhalt, C Germany
Stepanakert *see* Xankändi
Stephenville *27 F3* Texas, SW USA
Sterling *22 D4* Colorado, C USA
Sterling *18 B3* Illinois, N USA
Sterlitamak *92 B3* Respublika Bashkortostan, W Russian Federation
Stettin *see* Szczecin
Stettiner Haff *see* Szczeciński, Zalew
Stevenage *67 E6* E England, United Kingdom
Stevens Point *18 B2* Wisconsin, N USA
Stewart Island *129 A8 island* S New Zealand
Steyerlak-Anina *see* Anina
Steyr *73 D6 var.* Steier. Oberösterreich, N Austria
St. Helena Bay *56 B5 bay* SW South Africa
Stif *see* Sétif
Stillwater *27 G1* Oklahoma, C USA
Štip *79 E6* E FYR Macedonia
Stirling *66 C4* C Scotland, United Kingdom
Stjørdalshalsen *62 B4* Nord-Trøndelag, C Norway
St-Maur-des-Fossés *68 E2* Val-de-Marne, Île-de-France, N France Europe
Stockach *73 B6* Baden-Württemberg, S Germany
Stockholm *63 C6 country capital* (Sweden) Stockholm, C Sweden
Stockmannshof *see* Pļaviņas
Stockton *25 B6* California, W USA
Stockton Plateau *27 E4 plain* Texas, SW USA
Stœng Trêng *115 D5 prev.* Stung Treng. Stœng Trêng, NE Cambodia
Stoke *see* Stoke-on-Trent
Stoke-on-Trent *67 D6 var.* Stoke. C England, United Kingdom
Stolbce *see* Stowbtsy
Stolbtsy *see* Stowbtsy
Stolp *see* Słupsk
Stolpmünde *see* Ustka
Stómio *82 B4* Thessalía, C Greece
Store Bælt *see* Storebælt
Storebelt *see* Storebælt
Støren *63 B5* Sør-Trøndelag, S Norway
Storfjorden *61 G2 fjord* S Norway
Storhammer *see* Hamar
Stornoway *66 B2* NW Scotland, United Kingdom
Storsjön *63 B5 lake* C Sweden
Storuman *62 C4* Västerbotten, N Sweden
Storuman *62 C4 lake* N Sweden
Stowbtsy *85 C6 Pol.* Stołbce, *Rus.* Stolbtsy. Minskaya Voblasts', C Belarus
Strabane *67 B5 Ir.* An Srath Bán. W Northern Ireland, United Kingdom
Strakonice *77 A5 Ger.* Strakonitz. Jihočeský Kraj, S Czech Republic
Strakonitz *see* Strakonice
Stralsund *72 D2* Mecklenburg-Vorpommern, NE Germany
Stranraer *67 C5* S Scotland, United Kingdom
Strasbourg *68 E3 Ger.* Strassburg; *anc.* Argentoratum. Bas-Rhin, NE France
Strășeni *86 D3 var.* Strasheny. C Moldova
Strasheny *see* Strășeni
Strassburg *see* Strasbourg, France
Strassburg *see* Aiud, Romania
Stratford *128 D4* Taranaki, North Island, New Zealand
Strathfield *126 E2* New South Wales, E Australia
Straubing *73 C6* Bayern, SE Germany
Strehaia *86 B5* Mehedinți, SW Romania
Strelka *92 D4* Krasnoyarskiy Kray, C Russian Federation
Strigonium *see* Esztergom
Strofilia *see* Strofyliá
Strofyliá *83 C5 var.* Strofilia. Évvoia, C Greece
Stromboli *75 D6 island* Isole Eolie, S Italy
Stromeferry *66 C3* N Scotland, United Kingdom
Strömstad *63 B6* Västra Götaland, S Sweden
Strömsund *62 C4* Jämtland, C Sweden
Struga *79 D6* SW FYR Macedonia
Struma *see* Strymónas
Strumica *79 E6* E FYR Macedonia
Strumyani *82 C3* Blagoevgrad, SW Bulgaria
Strymónas *82 C3 Bul.* Struma. *river* Bulgaria/Greece
Stryy *86 B2* L'vivs'ka Oblast', NW Ukraine
Studholme *129 B6* Canterbury, South Island, New Zealand
Stuhlweissenberg *see* Székesfehérvár
Stung Treng *see* Stœng Trêng
Sturgis *22 D3* South Dakota, N USA
Stuttgart *73 B6* Baden-Württemberg, SW Germany
Stykkishólmur *61 E4* Vesturland, W Iceland
Styr *86 C1 Rus.* Styr'. *river* Belarus/Ukraine
Su *see* Jiangsu
Suakin *50 C3 var.* Sawakin. Red Sea, NE Sudan
Subačius *84 C4* Panevėžys, NE Lithuania
Subaykhān *96 E3* Dayr az Zawr, E Syria
Subotica *78 D2 Ger.* Maria-Theresiopel, *Hung.* Szabadka. Vojvodina, N Serbia
Suceava *86 C3 Ger.* Suczawa, *Hung.* Szucsava. Suceava, NE Romania
Su-chou *see* Suzhou
Suchow *see* Suzhou, Jiangsu, China
Suchow *see* Xuzhou, Jiangsu, China
Sucker State *see* Illinois
Sucre *39 F4 hist.* Chuquisaca, La Plata. *country capital* (Bolivia-legal capital) Chuquisaca, S Bolivia
Suczawa *see* Suceava
Sudan *50 A4 off.* Republic of Sudan, *Ar.* Jumhuriyat as-Sudan; *prev.* Anglo-Egyptian Sudan. *country* N Africa
Sudanese Republic *see* Mali
Sudan, Jumhuriyat as- *see* Sudan
Sudan, Republic of *see* Sudan
Sudbury *16 C4* Ontario, S Canada
Sudd *51 B5 swamp region* N South Sudan
Sudeten *76 B4 var.* Sudetes, Sudetic Mountains, *Cz./Pol.* Sudety. *mountain range* Czech Republic/Poland
Sudetes/Sudetic Mountains/Sudety *see* Sudeten
Südkarpaten *see* Carpații Meridionalii
Südliche Morava *see* Južna Morava
Sudong, Pulau *116 A2 island* SW Singapore Asia
Sudostroy *see* Severodvinsk
Sue *51 B5 river* S South Sudan
Sueca *71 F3* País Valenciano, E Spain
Sue Wood Bay *20 B5 bay* W Bermuda North America W Atlantic Ocean
Suez *50 B1 Ar.* As Suways, El Suweis. NE Egypt
Suez Canal *50 B1 Ar.* Qanāt as Suways. *canal* NE Egypt
Suez, Gulf of *see* Khalīj as Suways
Suğla Gölü *94 C4 lake* SW Turkey
Şuḩār *99 D5 var.* Sohar. N Oman
Sühbaatar *105 E1* Selenge, N Mongolia
Suhl *73 C5* Thüringen, C Germany
Suicheng *see* Suixi
Suid-Afrika *see* South Africa
Suidwes-Afrika *see* Namibia
Suixi *106 C6 var.* Suicheng. Guangdong, S China
Sujāwal *112 B3* Sind, SE Pakistan
Sukabumi *116 C5 prev.* Soekaboemi. Jawa, C Indonesia
Sukagawa *109 D5* Fukushima, Honshū, C Japan
Sukarnapura *see* Jayapura
Sukarno, Puntjak *see* Jaya, Puncak
Sukhne *see* As Sukhnah
Sukhona *88 C4 var.* Tot'ma. *river* NW Russian Federation
Sukhumi *see* Sokhumi
Sukkertoppen *see* Maniitsoq
Sukkur *112 B3* Sind, SE Pakistan
Sukumo *109 B7* Kōchi, Shikoku, SW Japan
Sulaimaniya *see* As Sulaymānīyah
Sulaiman Range *112 C2 mountain range* C Pakistan
Sula, Kepulauan *117 E4 island group* C Indonesia
Sulawesi *117 E4 Eng.* Celebes. *island* C Indonesia
Sulawesi, Laut *see* Celebes Sea
Sulechów *76 B3 Ger.* Züllichau. Lubuskie, W Poland
Suliag *see* Sawhāj
Sullana *38 B2* Piura, NW Peru
Sullivan Island *see* Lanbi Kyun
Sulphur Springs *27 G2* Texas, SW USA
Sultānābād *see* Arāk
Sulu Archipelago *117 E3 island group* SW Philippines
Sülüktü *see* Sulyukta
Sulu, Laut *see* Sulu Sea

Sulu Sea – Tehuantepec, Gulf of

Sulu Sea *117 E2 var.* Laut Sulu. *sea* SW Philippines
Sulyukta *101 E2 Kir.* Sülüktü. Batkenskaya Oblast', SW Kyrgyzstan
Sumatera *115 B8 Eng.* Sumatra. *island* W Indonesia
Sumatra *see* Sumatera
Šumava *see* Bohemian Forest
Sumba, Pulau *117 E5 Eng.* Sandalwood Island; *prev.* Soemba. *island* Nusa Tenggara, C Indonesia
Sumba, Selat *117 E5 strait* Nusa Tenggara, S Indonesia
Sumbawanga *51 B7* Rukwa, W Tanzania
Sumbe *56 B2 var.* N'Gunza, *Port.* Novo Redondo. Cuanza Sul, W Angola
Sumeih *51 B5* Southern Darfur, S Sudan
Sumgait *see* Sumqayıt, Azerbaijan
Summer Lake *24 B4 lake* Oregon, NW USA
Summit *71 H5* Alaska, USA
Sumqayıt *95 H2 Rus.* Sumgait. E Azerbaijan
Sumy *87 F2* Sums'ka Oblast', NE Ukraine
Sunbury *127 C7* Victoria, SE Australia
Sunda Islands *see* Greater Sunda Islands
Sunda, Selat *116 B5 strait* Jawa/Sumatera, SW Indonesia
Sunda Trench *see* Java Trench
Sunderland *66 D4 var.* Wearmouth. NE England, United Kingdom
Sundsvall *63 C5* Västernorrland, C Sweden
Sunflower State *see* Kansas
Sungaipenuh *116 B4 prev.* Soengaipenoeh. Sumatera, W Indonesia
Sunnyvale *25 A6* California, W USA
Sunset State *see* Oregon
Sunshine State *see* Florida
Sunshine State *see* New Mexico
Sunshine State *see* South Dakota
Suntar *93 F3* Respublika Sakha (Yakutiya), NE Russian Federation
Sunyani *53 E5* W Ghana
Suoločielgi *see* Saariselkä
Suomenlahti *see* Finland, Gulf of
Suomen Tasavalta/Suomi *see* Finland
Suomussalmi *62 E4* Oulu, E Finland
Suŏng *115 D6* Kâmpóng Cham, C Cambodia
Suoyarvi *88 B3* Respublika Kareliya, NW Russian Federation
Supe *38 C3* Lima, W Peru
Supérieur, Lac *see* Superior, Lake
Superior *18 A1* Wisconsin, N USA
Superior, Lake *18 B1 Fr.* Lac Supérieur. *lake* Canada/USA
Suqrah *see* Şawqirah
Suquṭrā *99 C7 var.* Sokotra, *Eng.* Socotra. *island* SE Yemen
Şūr *99 E5* NE Oman
Şūr *see* Soûr
Surabaja *see* Surabaya
Surabaya *116 D5 prev.* Surabaja, Soerabaja. Jawa, C Indonesia
Surakarta *116 C5 Eng.* Solo; *prev.* Soerakarta. Jawa, S Indonesia
Šurany *77 C6 Hung.* Nagysurány. Nitriansky Kraj, SW Slovakia
Sūrat *112 C4* Gujarāt, W India
Suratdhani *see* Surat Thani
Surat Thani *115 C6 var.* Suratdhani. Surat Thani, SW Thailand
Surazh *85 E5* Vitsyebskaya Voblasts', NE Belarus
Surdulica *79 E5* Serbia, SE Serbia
Sûre *65 D7 var.* Sauer. *river* W Europe
Surendranagar *112 C4* Gujarāt, W India
Surfers Paradise *127 E5* Queensland, E Australia
Surgut *92 D3* Khanty-Mansiyskiy Avtonomnyy Okrug-Yugra, C Russian Federation
Surin *115 D5* Surin, E Thailand
Surinam *see* Suriname
Suriname *37 G3 off.* Republic of Suriname, *var.* Surinam; *prev.* Dutch Guiana, Netherlands Guiana. *country* N South America
Suriname, Republic of *see* Suriname
Sūrīya/Sūrīyah, Al-Jumhūrīyah al-'Arabīyah as- *see* Syria
Surkhab, Darya-i- *see* Kahmard, Daryā-ye
Surkhob *101 F3 river* C Tajikistan
Surt *49 G2 var.* Sidra, Sirte. N Libya
Surt, Khalīj *49 F2 Eng.* Gulf of Sidra, Gulf of Sirti, Sidra. *gulf* N Libya
Surtsey *61 E5 island* S Iceland
Suruga-wan *109 D6 bay* SE Japan
Susa *74 A2* Piemonte, NE Italy
Sūsah *see* Sousse
Susanville *25 B5* California, W USA
Susitna *14 C3* Alaska, USA
Susteren *65 D5* Limburg, SE Netherlands
Susuman *93 G3* Magadanskaya Oblast', E Russian Federation
Sutlej *112 C2 river* India/Pakistan
Suur Munamägi *84 D3 var.* Munamägi, *Ger.* Eier-Berg. *mountain* SE Estonia
Suur Väin *84 C2 Ger.* Grosser Sund. *strait* W Estonia
Suva *123 E4 country capital* (Fiji) Viti Levu, W Fiji
Suvalkai/Suvalki *see* Suwałki
Suvorovo *82 E2 prev.* Vetrino. Varna, E Bulgaria
Suwałki *76 E2 Lith.* Suvalkai, *Rus.* Suvalki. Podlaskie, NE Poland
Şuwār *see* Aş Şuwār
Suways, Qanāt as *see* Suez Canal
Suweida *see* As Suwaydā'
Suzhou *106 D5 var.* Soochow, Su-chou, Suchow; *prev.* Wuhsien. Jiangsu, E China
Svalbard *61 E1 Norwegian dependency* Arctic Ocean
Svartisen *62 C3 glacier* C Norway
Svay Riěng *115 D6* Svay Riěng, S Cambodia
Sveg *63 B5* Jämtland, C Sweden
Svenstavik *63 C5* Jämtland, C Sweden
Sverdlovsk *see* Yekaterinburg
Sverige *see* Sweden
Sveti Vrach *see* Sandanski
Svetlogorsk *see* Svyetlahorsk
Svetlograd *89 B7* Stavropol'skiy Kray, SW Russian Federation
Svetlovodsk *see* Svitlovods'k
Svetozarevo *see* Jagodina
Svilengrad *82 D3 prev.* Mustafa-Pasha. Khaskovo, S Bulgaria
Svitlovods'k *87 F3 Rus.* Svetlovodsk. Kirovohrads'ka Oblast', C Ukraine
Svizzera *see* Switzerland
Svobodnyy *93 G4* Amurskaya Oblast', SE Russian Federation
Svyataya Anna Trough *133 C4 var.* Saint Anna Trough. *trough* N Kara Sea
Svyetlahorsk *85 D7 Rus.* Svetlogorsk. Homyel'skaya Voblasts', SE Belarus
Swabian Jura *see* Schwäbische Alb
Swakopmund *56 B3* Erongo, W Namibia
Swan Islands *31 E1 island group* NE Honduras North America
Swansea *67 C7 Wel.* Abertawe. S Wales, United Kingdom
Swarzędz *76 C3* Poznań, W Poland
Swatow *see* Shantou
Swaziland *56 D4 off.* Kingdom of Swaziland. *country* S Africa
Swaziland, Kingdom of *see* Swaziland
Sweden *62 B4 off.* Kingdom of Sweden, *Swe.* Sverige. *country* N Europe
Sweden, Kingdom of *see* Sweden
Sweetwater *27 F3* Texas, SW USA
Świdnica *76 B4 Ger.* Schweidnitz. Wałbrzych, SW Poland
Świdwin *76 B2 Ger.* Schivelbein. Zachodnio-pomorskie, NW Poland
Świebodzice *76 B4 Ger.* Freiburg in Schlesien, Swiebodzice. Wałbrzych, SW Poland
Świebodzin *76 B3 Ger.* Schwiebus. Lubuskie, W Poland
Świecie *76 C3 Ger.* Schwertberg. Kujawsko-pomorskie, C Poland
Swindon *67 D7* S England, United Kingdom
Swinemünde *see* Świnoujście
Świnoujście *76 B2 Ger.* Swinemünde. Zachodnio-pomorskie, NW Poland
Swiss Confederation *see* Switzerland
Switzerland *73 A7 off.* Swiss Confederation, *Fr.* La Suisse, *Ger.* Schweiz, *It.* Svizzera; *anc.* Helvetia. *country* C Europe
Sycaminum *see* Hefa
Sydenham Island *see* Nonouti
Sydney *126 D1 state capital* New South Wales, SE Australia
Sydney *17 G4* Cape Breton Island, Nova Scotia, SE Canada
Sydney Island *see* Manra
Syedpur *see* Saidpur
Syemyezhava *85 C6 Rus.* Semezhevo. Minskaya Voblasts', C Belarus
Syene *see* Aswān
Syeverodonets'k *87 H3 Rus.* Severodonetsk. Luhans'ka Oblast', E Ukraine
Syktyvkar *88 D4 prev.* Ust'-Sysol'sk. Respublika Komi, NW Russian Federation
Sylhet *113 G3* Sylhet, NE Bangladesh
Synel'nykove *87 G3* Dnipropetrovs'ka Oblast', E Ukraine
Syowa *132 C2 Japanese research station* Antarctica
Syracuse *19 E3* New York, NE USA
Syracuse *see* Siracusa
Syrdar'ya *92 B4* Sirdaryo Viloyati, E Uzbekistan
Syria *96 B3 off.* Syrian Arab Republic, *var.* Siria, Syrie, *Ar.* Al-Jumhūrīyah al-'Arabīyah as-Sūrīyah, Sūrīya. *country* SW Asia
Syrian Arab Republic *see* Syria
Syrian Desert *97 D5 Ar.* Al Hamad, Bādiyat ash Shām. *desert* SW Asia
Syrie *see* Syria
Sýrna *83 E7 var.* Sirna. *island* Kykládes, Greece, Aegean Sea
Sýros *83 C6 var.* Síros. *island* Kykládes, Greece, Aegean Sea
Syulemeshlii *see* Sredets
Syvash, Zaliv *see* Syvash, Zatoka
Syvash, Zatoka *87 F4 Rus.* Zaliv Syvash. *inlet* S Ukraine
Syzran' *89 C6* Samarskaya Oblast', W Russian Federation
Szabadka *see* Subotica
Szamotuły *76 B3* Poznań, W Poland
Szászrégen *see* Reghin
Szatmárrnémeti *see* Satu Mare
Száva *see* Sava
Szczecin *76 B3 Eng./Ger.* Stettin. Zachodnio-pomorskie, NW Poland
Szczecinek *76 B2 Ger.* Neustettin. Zachodnio-pomorskie, NW Poland
Szczeciński, Zalew *76 A2 var.* Stettiner Haff, *Ger.* Oderhaff. *bay* Germany/Poland
Szczuczyn Nowogródzki *see* Shchuchyn
Szczytno *76 D3 Ger.* Ortelsburg. Warmińsko-Mazurskie, NE Poland
Szechuan/Szechwan *see* Sichuan
Szeged *77 D7 Ger.* Szegedin, *Rom.* Seghedin. Csongrád, SE Hungary
Szegedin *see* Szeged
Székelykeresztúr *see* Cristuru Secuiesc
Székesfehérvár *77 C6 Ger.* Stuhlweissenberg; *anc.* Alba Regia. Fejér, W Hungary
Szeklerburg *see* Miercurea-Ciuc
Szekler Neumarkt *see* Târgu Secuiesc
Szekszárd *77 C7* Tolna, S Hungary
Szempcz/Szenc *see* Senec
Szenice *see* Senica
Szenttamás *see* Srbobran
Szeping *see* Siping
Szilágysomlyó *see* Şimleu Silvaniei
Szinna *see* Snina
Sziszek *see* Sisak
Szitás-Keresztúr *see* Cristuru Secuiesc
Szkudy *see* Skuodas
Szlatina *see* Slatina
Szlovákia *see* Slovakia
Szolnok *77 D6* Jász-Nagykun-Szolnok, C Hungary
Szombathely *77 B6 Ger.* Steinamanger; *anc.* Sabaria, Savaria. Vas, W Hungary
Szprotawa *76 B4 Ger.* Sprottau. Lubuskie, W Poland
Sztálinváros *see* Dunaújváros
Szucsava *see* Suceava

T

Tabariya, Bahrat *see* Tiberias, Lake
Table Rock Lake *27 G1 reservoir* Arkansas/Missouri, C USA
Tábor *77 B5* Jihočeský Kraj, S Czech Republic
Tabora *51 B7* Tabora, W Tanzania
Tabrīz *98 C2 var.* Tebriz; *anc.* Tauris. Āžarbāyjān-e Sharqī, NW Iran
Tabuaeran *123 G2 prev.* Fanning Island. *atoll* Line Islands, E Kiribati
Tabūk *98 A4* Tabūk, NW Saudi Arabia
Täby *63 C6* Stockholm, C Sweden
Tachau *see* Tachov
Tachov *77 A5 Ger.* Tachau. Plveňský Kraj, W Czech Republic
Tacloban *117 F2 off.* Tacloban City. Leyte, C Philippines
Tacloban City *see* Tacloban
Tacna *39 E4* Tacna, SE Peru
Tacoma *24 B2* Washington, NW USA
Tacuarembó *42 D4 prev.* San Fructuoso. Tacuarembó, C Uruguay
Tademaït, Plateau du *48 D3 plateau* C Algeria
Tadmor/Tadmur *see* Tudmur
Tādpatri *110 C2* Andhra Pradesh, E India
Tadzhikistan *see* Tajikistan
Taegu *see* Daegu
Taehan-haehyŏp *see* Korea Strait
Taehan Min'guk *see* South Korea
Taejŏn *see* Daejeon
Tafassâsset, Ténéré du *53 G2 desert* N Niger
Tafila/Ţafīlah, Muḩāfaẓat aṭ *see* Aṭ Ţafīlah
Taganrog *89 A7* Rostovskaya Oblast', SW Russian Federation
Taganrog, Gulf of *87 G4 Rus.* Taganrogskiy Zaliv, *Ukr.* Tahanroz'ka Zatoka. *gulf* Russian Federation/Ukraine
Taganrogskiy Zaliv *see* Taganrog, Gulf of
Taguatinga *41 F3* Tocantins, C Brazil
Tagus *70 C3 Port.* Rio Tejo, *Sp.* Río Tajo. *river* Portugal/Spain
Tagus Plain *58 A4 abyssal plain* E Atlantic Ocean
Tahanroz'ka Zatoka *see* Taganrog, Gulf of
Tahat *49 E4 mountain* SE Algeria
Tahiti *123 H4 island* Îles du Vent, W French Polynesia
Tahiti, Archipel de *see* Société, Archipel de la
Tahlequah *27 G1* Oklahoma, C USA
Tahoe, Lake *25 B5 lake* California/Nevada, W USA
Tahoua *53 F3* Tahoua, W Niger
Taibei *106 D6 var.* T'aipei; *Jap.* Taihoku; *prev.* Daihoku. *country capital* (Taiwan) N Taiwan
Taichū *see* Taizhong
T'aichung *see* Tiazhong
Taiden *see* Daejeon
Taieri *129 B7 river* South Island, New Zealand
Taihape *128 D4* Manawatu-Wanganui, North Island, New Zealand
Taihoku *see* Taibei
Taikyū *see* Daegu
Tailem Bend *127 B7* South Australia
T'ainan *see* Tainan
Tainan *106 D6 prev.* T'ainan, Dainan. S Taiwan
T'aipei *see* Taibei
Taiping *116 B3* Perak, Peninsular Malaysia
Taiwan *106 D6 off.* Republic of China, *var.* Formosa, Formo'sa. *country* E Asia
Taiwan *see* Tiazhong
T'aiwan Haihsia/Taiwan Haixia *see* Taiwan Strait
Taiwan Strait *106 D6 var.* Formosa Strait, *Chin.* T'aiwan Haihsia, Taiwan Haixia. *strait* China/Taiwan
Taiyuan *106 C4 var.* T'ai-yuan, T'ai-yüan; *prev.* Yangku. *province capital* Shanxi, C China
T'ai-yuan/T'ai-yüan *see* Taiyuan
Taizhong *106 D6 Jap.* Taichū; *prev.* T'aichung, Taiwan. C Taiwan
Ta'izz *99 B7* SW Yemen
Tajikistan *101 E3 off.* Republic of Tajikistan, *Rus.* Tadzhikistan, *Taj.* Jumhurii Tojikiston; *prev.* Tajik S.S.R. *country* C Asia
Tajikistan, Republic of *see* Tajikistan
Tajik S.S.R *see* Tajikistan
Tajo, Río *see* Tagus
Tak *114 C4 var.* Rahaeng. Tak, W Thailand
Takao *see* Gaoxiong
Takaoka *109 C5* Toyama, Honshū, SW Japan
Takapuna *128 D2* Auckland, North Island, New Zealand
Takeshima *see* Liancourt Rocks
Takhiatash *see* Taxiatosh
Takhtakupyr *see* Taxtako'pir
Takikawa *108 D2* Hokkaidō, NE Japan
Takla Makan Desert *see* Taklimakan Shamo
Taklimakan Shamo *104 B3 Eng.* Takla Makan Desert. *desert* NW China
Takow *see* Gaoxiong
Takutea *123 G4 island* S Cook Islands
Talabriga *see* Aveiro, Portugal
Talabriga *see* Talavera de la Reina, Spain
Talachyn *85 D6 Rus.* Tolochin. Vitsyebskaya Voblasts', NE Belarus
Talamanca, Cordillera de *31 E5 mountain range* S Costa Rica
Talara *38 B2* Piura, NW Peru
Talas *101 F2* Talasskaya Oblast', NW Kyrgyzstan
Talaud, Kepulauan *117 F3 island group* E Indonesia
Talavera de la Reina *70 D3 anc.* Caesarobriga, Talabriga. Castilla-La Mancha, C Spain
Talca *42 B4* Maule, C Chile
Talcahuano *43 B5* Bío Bío, C Chile
Taldykorgan *92 C5 Kaz.* Taldyqorghan; *prev.* Taldy-Kurgan. Taldykorgan, SE Kazakhstan
Taldy-Kurgan/Taldyqorghan *see* Taldykorgan
Ta-lien *see* Dalian
Taliq-an *see* Tāloqān
Tal'ka *85 C6* Minskaya Voblasts', C Belarus
Talkhof *see* Puurmani
Tallahassee *20 D3 prev.* Muskogean. *state capital* Florida, SE USA
Tall al Abyaḍ *see* At Tall al Abyaḍ
Tallin *see* Tallinn
Tallinn *84 D2 Ger.* Reval, *Rus.* Tallin; *prev.* Revel. *country capital* (Estonia) Harjumaa, NW Estonia
Tall Kalakh *96 B4 var.* Tell Kalakh. Ḩimş, C Syria
Tallulah *20 B2* Louisiana, S USA
Talnakh *92 D3* Taymyrskiy (Dolgano-Nenetskiy) Avtonomnyy Okrug, N Russian Federation
Tal'ne *87 E3 Rus.* Tal'noye. Cherkas'ka Oblast', C Ukraine
Tal'noye *see* Tal'ne
Taloga *27 F1* Oklahoma, C USA
Tāloqān *101 E3 var.* Taliq-an. Takhār, NE Afghanistan
Talsen *see* Talsi
Talsi *84 C3 Ger.* Talsen. Talsi, NW Latvia
Taltal *42 B2* Antofagasta, N Chile
Talvik *62 D2* Finnmark, N Norway
Tamabo, Banjaran *116 D3 mountain range* East Malaysia
Tamale *53 E4* C Ghana
Tamana *123 E3 prev.* Rotcher Island. *atoll* Tungaru, W Kiribati
Tamanrasset *49 E4 var.* Tamenghest. S Algeria
Tamar *67 C7 river* SW England, United Kingdom
Tamar *see* Tudmur
Tamatave *see* Toamasina
Tamazunchale *29 E4* San Luis Potosí, C Mexico
Tambacounda *52 C3* SE Senegal
Tambov *89 B6* Tambovskaya Oblast', W Russian Federation
Tambura *51 B5* W Equatoria, SW South Sudan
Tâmchekket *see* Tâmchekkeţ
Tâmchekkeţ *52 C3 var.* Tamchaket. Hodh el Gharbi, S Mauritania
Tamenghest *see* Tamanrasset
Tamil Nādu *110 C3 prev.* Madras. *cultural region* SE India
Tam Ky *115 E5* Quang Nam-ƒa Nãng, C Vietnam
Tammerfors *see* Tampere
Tampa *21 E4* Florida, SE USA
Tampa Bay *21 E4 bay* Florida, SE USA
Tampere *63 D5 Swe.* Tammerfors. Länsi-Suomi, W Finland
Tampico *29 E3* Tamaulipas, C Mexico
Tamworth *127 D6* New South Wales, SE Australia
Tanabe *109 C7* Wakayama, Honshū, SW Japan
Tana Bru *62 D2* Finnmark, N Norway
T'ana Hāyk' *50 C4 var.* Lake Tana. *lake* NW Ethiopia
Tanais *see* Don
Tana, Lake *see* T'ana Hāyk'
Tanami Desert *124 D3 desert* Northern Territory, N Australia
Tananarive *see* Antananarivo
Ţăndărei *86 D5* Ialomiţa, SE Romania
Tandil *43 D5* Buenos Aires, E Argentina
Tandjoengkarang *see* Bandar Lampung
Tanega-shima *109 B8 island* Nansei-shotō, SW Japan
Tanen Taunggyi *see* Tane Range
Tane Range *114 B4 Bur.* Tanen Taunggyi. *mountain range* W Thailand
Tanezrouft *48 D4 desert* Algeria/Mali
Ţanf, Jabal aţ *96 D4 mountain* SE Syria
Tanga *51 C7* Tanga, E Tanzania
Tanganyika and Zanzibar *see* Tanzania
Tanganyika, Lake *51 B7 lake* E Africa
Tanger *48 C2 var.* Tangiers, Tangier, *Fr./Ger.* Tangerk, *Sp.* Tánger; *anc.* Tingis. NW Morocco
Tangerk *see* Tanger
Tanggula Shan *104 C4 mountain* W China
Tangier *see* Tanger
Tangiers *see* Tanger
Tangra Yumco *104 B5 var.* Tangro Tso. *lake* W China
Tangro Tso *see* Tangra Yumco
Tangshan *106 D3 var.* T'ang-shan. Hebei, E China
T'ang-shan *see* Tangshan
Tanimbar, Kepulauan *117 F5 island group* Maluku, E Indonesia
Tanintharyi *115 B6 prev.* Tenasserim. S Myanmar (Burma)
Tanjungkarang/Tanjungkarang-Telukbetung *see* Bandar Lampung
Tanna *122 D4 island* S Vanuatu
Tannenhof *see* Krynica
Tan-Tan *48 B3* SW Morocco
Tan-tung *see* Dandong
Tanzania *51 C7 off.* United Republic of Tanzania, *Swa.* Jamhuri ya Muungano wa Tanzania; *prev.* German East Africa, Tanganyika and Zanzibar. *country* E Africa
Tanzania, Jamhuri ya Muungano wa *see* Tanzania
Tanzania, United Republic of *see* Tanzania
Taoudenit *see* Taoudenni
Taoudenni *53 E2 var.* Taoudenit. Tombouctou, N Mali
Tapa *84 E2 Ger.* Taps. Lääne-Virumaa, NE Estonia
Tapachula *29 G5* Chiapas, SE Mexico
Tapaiu *see* Gvardeysk
Tapajós, Rio *41 E2 var.* Tapajóz. *river* NW Brazil
Tapajóz *see* Tapajós, Rio
Taps *see* Tapa
Ţarābulus *49 F2 var.* Ţarābulus al Gharb, *Eng.* Tripoli. *country capital* (Libya) NW Libya
Ţarābulus al Gharb *see* Ţarābulus
Ţarābulus/Ţarābulus ash Shām *see* Tripoli
Taraclia *86 D4 Rus.* Tarakilya. S Moldova
Tarakilya *see* Taraclia
Taranaki, Mount *128 C4 var.* Egmont. *volcano* North Island, New Zealand
Tarancón *71 E3* Castilla-La Mancha, C Spain
Taranto *75 E5 var.* Tarentum. Puglia, SE Italy
Taranto, Gulf of *75 E6 Eng.* Gulf of Taranto. *gulf* S Italy
Taranto, Gulf of *see* Taranto, Golfo di
Tarapoto *38 C2* San Martín, N Peru
Tarare *69 D5* Rhône, E France
Tarascon *69 D6* Bouches-du-Rhône, SE France
Tarawa *122 D2 atoll* Tungaru, W Kiribati
Taraz *92 C5 prev.* Aulie Ata, Auliye-Ata, Dzhambul, Zhambyl. Zhambyl, S Kazakhstan
Tarazona *71 E2* Aragón, NE Spain
Tarbes *69 B6 anc.* Bigorra. Hautes-Pyrénées, S France
Tarcoola *127 A6* South Australia
Taree *127 D6* New South Wales, SE Australia
Tarentum *see* Taranto
Târgovişte *86 C5 prev.* Tîrgovişte. Dâmboviţa, S Romania
Targu Jiu *86 B4 prev.* Tîrgu Jiu. Gorj, W Romania
Târgul-Neamţ *see* Târgu-Neamţ
Târgul-Săcuiesc *see* Târgu Secuiesc
Târgu Mureş *86 B4 prev.* Oşorhei, Tîrgu Mures, *Ger.* Neumarkt, *Hung.* Marosvásárhely. Mureş, C Romania
Târgu-Neamţ *86 C3 var.* Târgul-Neamţ; *prev.* Tîrgu-Neamţ. Neamţ, NE Romania
Târgu Ocna *86 C4 Hung.* Aknavásár; *prev.* Tîrgu Ocna. Bacău, E Romania
Târgu Secuiesc *86 C4 Ger.* Neumarkt, Szekler Neumarkt, *Hung.* Kezdivásárhely; *prev.* Chezdi-Oşorheiu, Târgul-Săcuiesc, Tîrgu Secuiesc. Covasna, E Romania
Tar Heel State *see* North Carolina
Tarija *39 G5* Tarija, S Bolivia
Tarīm *99 C6* C Yemen
Tarim Basin *see* Tarim Pendi
Tarim Pendi *102 C2 Eng.* Tarim Basin. *basin* NW China
Tarim He *104 B3 river* NW China
Tarma *38 C3* Junín, C Peru
Tarn *69 C6 cultural region* S France
Tarn *69 C6 river* S France
Tarnobrzeg *76 D4* Podkarpackie, SE Poland
Tarnopol *see* Ternopil'
Tarnów *77 D5* Małopolskie, S Poland
Tarraco *see* Tarragona
Tarragona *71 G2 anc.* Tarraco. Cataluña, E Spain
Tarrasa *see* Terrassa
Tàrrega *71 F2 var.* Tarrega. Cataluña, NE Spain
Tarsatica *see* Rijeka
Tarsus *94 C4* İçel, S Turkey
Tartous/Tartouss *see* Ţarţūs
Tartu *84 D3 Ger.* Dorpat; *prev. Rus.* Yurev, Yury'ev. Tartumaa, SE Estonia
Ţarţūs *96 A3 off.* Muḩāfaẓat Ţarţūs, *var.* Tartous, Tartus. *governorate* W Syria
Ţarţūs, Muḩāfaẓat *see* Ţarţūs
Ta Ru Tao, Ko *115 B7 island* S Thailand Asia
Tarvisio *74 D2* Friuli-Venezia Giulia, NE Italy
Tarvisium *see* Treviso
Tashauz *see* Daşoguz
Tashi Chho Dzong *see* Thimphu
Tashkent *see* Toshkent
Tash-Kömür *see* Tash-Kumyr
Tash-Kumyr *101 F2 Kir.* Tash-Kömür. Dzhalal-Abadskaya Oblast', W Kyrgyzstan
Tashqurghan *see* Khulm
Tasikmalaja *see* Tasikmalaya
Tasikmalaya *116 C5 prev.* Tasikmalaja. Jawa, C Indonesia
Tasman Basin *120 C5 var.* East Australian Basin. *undersea basin* S Tasman Sea
Tasman Bay *129 C5 inlet* South Island, New Zealand
Tasmania *127 B8 prev.* Van Diemen's Land. *state* SE Australia
Tasmania *130 B4 island* SE Australia
Tasman Plateau *120 C5 var.* South Tasmania Plateau. *undersea plateau* SW Tasman Sea
Tasman Sea *120 C5 sea* SW Pacific Ocean
Tassili-n-Ajjer *49 E4 plateau* E Algeria
Tatabánya *77 C6* Komárom-Esztergom, NW Hungary
Tatar Pazardzhik *see* Pazardzhik
Tathlīth *99 B5* 'Asīr, S Saudi Arabia
Tatra Mountains *77 D5 Ger.* Tatra, *Hung.* Tátra, *Pol./Slvk.* Tatry. *mountain range* Poland/Slovakia
Tatra/Tátra *see* Tatra Mountains
Tatry *see* Tatra Mountains
Ta-t'ung/Tatung *see* Datong
Tatvan *95 F3* Bitlis, SE Turkey
Ta'ū *123 F4 var.* Tau. *island* Manua Islands, E American Samoa
Taukum, Peski *101 G1 desert* SE Kazakhstan
Taumarunui *128 D4* Manawatu-Wanganui, North Island, New Zealand
Taungdwingyi *114 B3* Magway, C Myanmar (Burma)
Taunggyi *114 B3* Shan State, C Myanmar (Burma)
Taungoo *114 B4* Bago, C Myanmar (Burma)
Taunton *67 C7* SW England, United Kingdom
Taupo *128 D3* Waikato, North Island, New Zealand
Taupo, Lake *128 D3 lake* North Island, New Zealand
Tauragė *84 B4 Ger.* Tauroggen. Tauragė, SW Lithuania
Tauranga *128 D3* Bay of Plenty, North Island, New Zealand
Tauris *see* Tabrīz
Tauroggen *see* Tauragė
Taurus Mountains *see* Toros Dağları
Tavas *94 B4* Denizli, SW Turkey
Tavastehus *see* Hämeenlinna
Tavira *70 C5* Faro, S Portugal
Tavoy *see* Dawei
Tavoy Island *see* Mali Kyun
Ta Waewae Bay *129 A7 bay* South Island, New Zealand
Tawakoni, Lake *27 G2 reservoir* Texas, SW USA
Tawau *116 D3* Sabah, East Malaysia
Ţawkar *see* Tokar
Tawzar *see* Tozeur
Taxco *29 E4 var.* Taxco de Alarcón. Guerrero, S Mexico
Taxco de Alarcón *see* Taxco
Taxiatosh *100 C2 Rus.* Takhiatash. Qoraqalpog'iston Respublikasi, W Uzbekistan
Taxtako'pir *100 D1 Rus.* Takhtakupyr. Qoraqalpog'iston Respublikasi, NW Uzbekistan
Tay *66 C3 river* C Scotland, United Kingdom
Taylor *27 G3* Texas, SW USA
Taymā' *98 A4* Tabūk, NW Saudi Arabia
Taymyr, Ozero *93 E2 lake* N Russian Federation
Taymyr, Poluostrov *93 E2 peninsula* N Russian Federation
Taz *92 D3 river* N Russian Federation
Tbilisi *95 G2 var.* T'bilisi, *Eng.* Tiflis. *country capital* (Georgia) SE Georgia
T'bilisi *see* **Tbilisi**
Tchad *see* Chad
Tchad, Lac *see* Chad, Lake
Tchien *see* Zwedru
Tchongking *see* Chongqing
Tczew *76 C2 Ger.* Dirschau. Pomorskie, N Poland
Te Anau *129 A7* Southland, South Island, New Zealand
Te Anau, Lake *129 A7 lake* South Island, New Zealand
Teapa *29 G4* Tabasco, SE Mexico
Teate *see* Chieti
Tebingtinggi *116 B3* Sumatera, N Indonesia
Tebriz *see* Tabrīz
Techirghiol *86 D5* Constanţa, SE Romania
Tecomán *28 D4* Colima, SW Mexico
Tecpan *29 E5 var.* Tecpan de Galeana. Guerrero, S Mexico
Tecpan de Galeana *see* Tecpan
Tecuci *86 C4* Galaţi, E Romania
Tedzhen *see* Harīrūd/Tejen
Tedzhen *see* Tejen
Tees *67 D5 river* N England, United Kingdom
Tefé *40 D2* Amazonas, N Brazil
Tegal *116 C4* Jawa, C Indonesia
Tegelen *65 D5* Limburg, SE Netherlands
Tegucigalpa *30 C3 country capital* (Honduras) Francisco Morazán, SW Honduras
Teheran *see* Tehrān
Tehrān *98 C3 var.* Teheran. *country capital* (Iran) Tehrān, N Iran
Tehuacán *29 F4* Puebla, S Mexico
Tehuantepec *29 F5 var.* Santo Domingo Tehuantepec. Oaxaca, SE Mexico
Tehuantepec, Golfo de *29 F5 var.* Gulf of Tehuantepec. *gulf* S Mexico
Tehuantepec, Gulf of *see* Tehuantepec, Golfo de

Tehuantepec, Isthmus of - Trbovlje

Tehuantepec, Isthmus of see Tehuantepec, Istmo de
Tehuantepec, Istmo de 29 F5 var. Isthmus of Tehuantepec. isthmus SE Mexico
Tejen 100 C3 Rus. Tedzhen. Ahal Welaýaty, S Turkmenistan
Tejen see Harīrūd
Tejo, Rio see Tagus
Te Kao 128 C1 Northland, North Island, New Zealand
Tekax 29 H4 var. Tekax de Álvaro Obregón. Yucatán, SE Mexico
Tekax de Álvaro Obregón see Tekax
Tekeli 92 C5 Almaty, SE Kazakhstan
Tekirdağ 94 A2 It. Rodosto; anc. Bisanthe, Raidestos, Rhaedestus. Tekirdağ, NW Turkey
Te Kuiti 128 D3 Waikato, North Island, New Zealand
Tela 30 C2 Atlántida, NW Honduras
Telanaipura see Jambi
Telangana 112 D5 cultural region SE India
Tel Aviv-Jaffa see Tel Aviv-Yafo
Tel Aviv-Yafo 97 A6 var. Tel Aviv-Jaffa. Tel Aviv, C Israel
Teles Pirés see São Manuel, Rio
Telish 82 C2 prev. Azizie. Pleven, N Bulgaria
Tell Abiad/Tell Abyad see At Tall al Abyaḑ
Tell Kalakh see Tall Kalakh
Tell Shedadi see Ash Shadādah
Tel'man/Tel'mansk see Gubadag
Teloekbetoeng see Bandar Lampung
Telo Martius see Toulon
Telschen see Telšiai
Telšiai 84 B3 Ger. Telschen. Telšiai, NW Lithuania
Telukbetung see Bandar Lampung
Temerin 78 D3 Vojvodina, N Serbia
Temeschburg/Temeschwar see Timişoara
Temesvár/Temeswar see Timişoara
Temirtau 92 C4 prev. Samarkandski, Samarkandskoye. Karagandy, C Kazakhstan
Tempio Pausania 75 A5 Sardegna, Italy, C Mediterranean Sea
Temple 27 G3 Texas, SW USA
Temuco 43 B5 Araucanía, C Chile
Temuka 129 B6 Canterbury, South Island, New Zealand
Tenasserim see Tanintharyi
Ténenkou 52 D3 Mopti, C Mali
Ténéré 53 G3 physical region C Niger
Tenerife 48 A3 island Islas Canarias, Spain, NE Atlantic Ocean
Tengger Shamo 105 E3 desert N China
Tengréla 52 D4 var. Tingréla. N Côte d'Ivoire
Tenkodogo 53 E4 S Burkina
Tennant Creek 126 A3 Northern Territory, C Australia
Tennessee 20 C1 off. State of Tennessee, also known as The Volunteer State. state SE USA
Tennessee River 20 C1 river S USA
Tenos see Tínos
Tepelena see Tepelenë
Tepelenë 79 C7 var. Tepelena, It. Tepeleni. Gjirokastër, S Albania
Tepeleni see Tepelenë
Tepic 28 D4 Nayarit, C Mexico
Teplice 76 A4 Ger. Teplitz; prev. Teplice-Šanov, Teplitz-Schönau. Ústecký Kraj, NW Czech Republic
Teplice-Šanov/Teplitz/Teplitz-Schönau see Teplice
Tequila 28 D4 Jalisco, SW Mexico
Teraina 123 G2 prev. Washington Island. atoll Line Islands, E Kiribati
Teramo 74 C4 anc. Interamna. Abruzzi, C Italy
Tercan 95 E3 Erzincan, NE Turkey
Terceira 70 A5 var. Ilha Terceira. island Azores, Portugal, NE Atlantic Ocean
Terceira, Ilha see Terceira
Terekhovka see Tsyerakhowka
Teresina 41 F2 var. Therezina. state capital Piauí, NE Brazil
Termez see Termiz
Termia see Kýthnos
Términos, Laguna de 29 G4 lagoon SE Mexico
Termiz 101 E3 Rus. Termez. Surkhondaryo Viloyati, S Uzbekistan
Termoli 74 D4 Molise, C Italy
Terneuzen 65 B5 var. Neuzen. Zeeland, SW Netherlands
Terni 74 C4 anc. Interamna Nahars. Umbria, C Italy
Ternopil' 86 C2 Pol. Tarnopol, Rus. Ternopol'. Ternopil's'ka Oblast', W Ukraine
Ternopol' see Ternopil'
Terracina 75 C5 Lazio, C Italy
Terranova di Sicilia see Gela
Terranova Pausania see Olbia
Terrassa 71 G2 Cast. Tarrasa. Cataluña, E Spain
Terre Adélie 132 C4 physical region Antarctica
Terre Haute 18 B4 Indiana, N USA
Terre Neuve see Newfoundland and Labrador
Terschelling 64 C1 Fris. Skylge. island Waddeneilanden, N Netherlands
Teruel 71 F3 anc. Turba. Aragón, E Spain
Tervel 82 E1 prev. Kurtbunar, Rom. Curtbunar. Dobrich, NE Bulgaria
Tervueren see Tervuren
Tervuren 65 C6 var. Tervueren. Vlaams Brabant, C Belgium
Teseney 50 C4 var. Tessenei. W Eritrea
Tessalit 53 E2 Kidal, NE Mali
Tessaoua 53 G3 Maradi, S Niger
Tessenderlo 65 C5 Limburg, NE Belgium
Tessenei see Teseney
Testigos, Islas los 37 E1 island group N Venezuela
Tete 57 E2 Tete, NW Mozambique
Teterow 72 C3 Mecklenburg-Vorpommern, NE Germany
Tétouan 48 C2 var. Tetouan, Tetuán. N Morocco
Tetovo 79 D5 Razgrad, N Bulgaria
Tetschen see Děčín
Tetuán see Tétouan
Teverya see Tverya
Te Waewae Bay 129 A7 bay South Island, New Zealand
Texarkana 20 A2 Arkansas, C USA
Texarkana 27 H2 Texas, SW USA
Texas 27 F3 off. State of Texas, also known as Lone Star State. state S USA
Texas City 27 H4 Texas, SW USA
Texel 64 C2 island Waddeneilanden, NW Netherlands
Texoma, Lake 27 G2 reservoir Oklahoma/Texas, C USA
Teziutlán 29 F4 Puebla, S Mexico
Thaa Atoll see Kolhumadulu
Thai, Ao see Thailand, Gulf of
Thai Binh 114 D3 Thai Binh, N Vietnam
Thailand 115 C5 off. Kingdom of Thailand, Th. Prathet Thai; prev. Siam. country SE Asia
Thailand, Gulf of 115 C6 var. Gulf of Siam, Th. Ao Thai, Vtn. Vinh Thai Lan. gulf SE Asia
Thailand, Kingdom of see Thailand
Thai Lan, Vinh see Thailand, Gulf of
Thai Nguyên 114 D3 Bắc Thai, N Vietnam
Thakhèk 114 D4 var. Muang Khammouan. Khammouan, C Laos
Thamarīd see Thamarīt
Thamarīt 99 D6 var. Thamarīd, Thumrayt. SW Oman
Thames 128 D3 Waikato, North Island, New Zealand
Thames 67 B8 river S England, United Kingdom
Thandwe 114 A4 var. Sandoway. Rakhine State, W Myanmar (Burma)
Thanh Hoa 114 D3 Thanh Hoa, N Vietnam
Thanintari Taungdan see Bilauktaung Range
Thanlwin see Salween
Thar Desert 112 C3 var. Great Indian Desert, Indian Desert. desert India/Pakistan
Tharthar, Buhayrat ath 98 B3 lake C Iraq
Thásos 82 C4 Thásos, E Greece
Thásos 82 C4 island E Greece
Thaton 114 B4 Mon State, S Myanmar (Burma)
Thayetmyo 114 A4 Magway, C Myanmar (Burma)
The Crane 33 H2 var. Crane. S Barbados
The Dalles 24 B3 Oregon, NW USA
The Flatts Village see Flatts Village
The Hague see 's-Gravenhage
Theodosia see Feodosiya
The Pas 15 F5 Manitoba, C Canada
Therezina see Teresina
Thérma 83 D6 Ikaría, Dodekánisa, Greece, Aegean Sea
Thermaic Gulf/Thermaicus Sinus see Thermaïkós Kólpos
Thermaïkós Kólpos 82 B4 Eng. Thermaic Gulf; anc. Thermaicus Sinus. gulf N Greece
Thermiá see Kýthnos
Thérmo 83 B5 Dytikí Elláda, C Greece
The Rock 71 H4 New South Wales, SE Australia
The Sooner State see Oklahoma
Thessaloníki 82 C3 Eng. Salonica, Salonika, SCr. Solun, Turk. Selânik. Kentrikí Makedonía, N Greece
The Valley 33 G3 dependent territory capital (Anguilla) E Anguilla
The Village 27 G1 Oklahoma, C USA
The Volunteer State see Tennessee
Thiamis see Kalamás
Thian Shan see Tien Shan
Thibet see Xizang Zizhiqu
Thief River Falls 23 F1 Minnesota, N USA
Thienen see Tienen
Thiers 69 C5 Puy-de-Dôme, C France
Thiès 52 B3 W Senegal
Thikombia see Cikobia
Thimbu see Thimphu
Thimphu 113 G3 var. Thimbu; prev. Tashi Chho Dzong. country capital (Bhutan) W Bhutan
Thionville 68 D3 Ger. Diedenhofen. Moselle, NE France
Thíra 83 D7 var. Santorini. Kykládes, Greece, Aegean Sea
Thiruvananthapuram 110 C3 var. Trivandrum, Tiruvantapuram. state capital Kerala, SW India
Thitu Island 106 C8 island NW Spratly Islands
Tholen 64 B4 island SW Netherlands
Thomasville 20 D3 Georgia, SE USA
Thompson 15 F4 Manitoba, C Canada
Thonon-les-Bains 69 D5 Haute-Savoie, E France
Thorenburg see Turda
Thorlákshöfn 61 E5 Sudhurland, SW Iceland
Thorn see Toruń
Thornton Island see Millennium Island
Thorshavn see Tórshavn
Thospitis see Van Gölü
Thouars 68 B4 Deux-Sèvres, W France
Thoune see Thun
Thracian Sea 82 D4 Gk. Thrakikó Pélagos; anc. Thracium Mare. sea Greece/Turkey
Thracium Mare/Thrakikó Pélagos see Thracian Sea
Three Gorges Reservoir 107 C5 reservoir C China
Three Kings Islands 128 C1 island group N New Zealand
Thrissur 110 C3 var. Trichūr. Kerala, SW India
Thuin 65 B7 Hainaut, S Belgium
Thule see Qaanaaq
Thumrayt see Thamarīt
Thun 73 A7 Fr. Thoune. Bern, W Switzerland
Thunder Bay 16 B4 Ontario, S Canada
Thuner See 73 A7 lake C Switzerland
Thung Song 115 C7 var. Cha Mai. Nakhon Si Thammarat, SW Thailand
Thurso 66 C2 N Scotland, United Kingdom
Thýamis see Kalamás
Tianjin 106 D4 var. Tientsin. Tianjin Shi, E China
Tianjin see Tianjin Shi
Tianjin Shi 106 D4 var. Jin, Tianjin, T'ien-ching, Tientsin. municipality E China
Tian Shan see Tien Shan
Tianshui 106 B4 Gansu, C China
Tiba see Chiba
Tiber 74 C4 Eng. Tiber. river C Italy
Tiber see Tevere, Italy
Tiber see Tivoli, Italy
Tiberias see Tverya
Tiberias, Lake 97 B5 var. Chinnereth, Sea of Bahr Tabariya, Sea of Galilee, Ar. Bahrat Tabariya, Heb. Yam Kinneret. lake N Israel
Tibesti 54 C2 var. Tibesti Massif, Ar. Tibistī. mountain range N Africa
Tibesti Massif see Tibesti
Tibet see Xizang Zizhiqu
Tibetan Autonomous Region see Xizang Zizhiqu
Tibet, Plateau of see Qingzang Gaoyuan
Tībistī see Tibesti
Tibnī see At Tibnī
Tiburón, Isla 28 B2 var. Isla del Tiburón. island NW Mexico
Tiburón, Isla del see Tiburón, Isla
Tichau see Tychy
Tichit 52 D2 var. Tichitt. Tagant, C Mauritania
Tichitt see Tichit
Ticinum see Pavia
Ticul 29 H3 Yucatán, SE Mexico
Tidjikdja see Tidjikja
Tidjikja 52 C2 var. Tidjikdja; prev. Fort-Cappolani. Tagant, C Mauritania
T'ien-ching see Tianjin Shi
Tienen 65 C6 var. Thienen, Fr. Tirlemont. Vlaams Brabant, C Belgium
Tiên Giang, Sông see Mekong
Tien Shan 104 B3 Chin. Thian Shan, Tian Shan, T'ien Shan, Rus. Tyan'-Shan'. mountain range C Asia
Tientsin see Tianjin
Tierp 63 C6 Uppsala, C Sweden
Tierra del Fuego 43 B8 island Argentina/Chile
Tiflis see T'bilisi
Tifton 20 D3 Georgia, SE USA
Tifu 117 F4 Pulau Buru, E Indonesia
Tighina 86 D4 Rus. Bendery; prev. Bender. E Moldova
Tigranocerta see Siirt
Tigris 98 B2 Ar. Dijlah, Turk. Dicle. river Iraq/Turkey
Tiguentourine 49 E3 E Algeria
Ti-hua/Tihwa see Ürümqi
Tijuana 28 A1 Baja California Norte, NW Mexico
Tikhoretsk 89 A7 Krasnodarskiy Kray, SW Russian Federation
Tikhvin 88 B4 Leningradskaya Oblast', NW Russian Federation
Tiki Basin 121 G3 undersea basin S Pacific Ocean
Tikinsso 52 C4 river C Guinea
Tiksi 93 F2 Respublika Sakha (Yakutiya), NE Russian Federation
Tilburg 64 C4 Noord-Brabant, S Netherlands
Tilimsen see Tlemcen
Tilio Martius see Toulon
Tillabéri 53 F3 var. Tillabéry. Tillabéri, W Niger
Tillabéry see Tillabéri
Tílos 83 E7 island Dodekánisa, Greece, Aegean Sea
Timan Ridge see Timanskiy Kryazh
Timanskiy Kryazh 88 D3 Eng. Timan Ridge. ridge NW Russian Federation
Timaru 129 B6 Canterbury, South Island, New Zealand
Timbaki/Timbákion see Tympáki
Timbedgha 52 D3 var. Timbédra. Hodh ech Chargui, SE Mauritania
Timbédra see Timbedgha
Timbuktu see Tombouctou
Timiş 86 A4 county SW Romania
Timişoara 86 A4 Ger. Temeschwar, Temeswar, Hung. Temesvár; prev. Temeschburg. Timiş, W Romania
Timmins 16 C4 Ontario, S Canada
Timor 103 F5 island Nusa Tenggara, C Indonesia
Timor Sea 103 F5 sea E Indian Ocean
Timor Timur see East Timor
Timor Trench see Timor Trough
Timor Trough 103 F5 var. Timor Trench. trough NE Timor Sea
Timrå 63 C5 Västernorrland, C Sweden
Tindouf 48 C3 W Algeria
Tineo 70 C1 Asturias, N Spain
Tingis see Tanger
Tingo María 38 C3 Huánuco, C Peru
Tingréla see Tengréla
Tinhosa Grande 54 E2 island N Sao Tome and Principe, Africa, E Atlantic Ocean
Tinhosa Pequena 54 E1 island N Sao Tome and Principe, Africa, E Atlantic Ocean
Tinian 122 B1 island S Northern Mariana Islands
Tínos 83 D6 Tínos, Kykládes, Greece, Aegean Sea
Tínos 83 D6 anc. Tenos. island Kykládes, Greece, Aegean Sea
Tip 79 E6 Papua, E Indonesia
Tipitapa 30 D3 Managua, W Nicaragua
Tip Top Mountain 16 C4 mountain Ontario, S Canada
Tirana see Tiranë
Tiranë 79 C6 var. Tirana. country capital (Albania) Tiranë, C Albania
Tiraspol 86 D4 Rus. Tiraspol'. E Moldova
Tiraspol' see Tiraspol
Tiree 66 B3 island W Scotland, United Kingdom
Tîrgovişte see Târgovişte
Tîrgu Jiu see Targu Jiu
Tîrgu Mures see Târgu Mureş
Tîrgu-Neamţ see Târgu-Neamţ
Tîrgu Ocna see Târgu Ocna
Tîrgu Secuiesc see Târgu Secuiesc
Tirlemont see Tienen
Tírnavos see Týrnavos
Tirnovo see Veliko Tŭrnovo
Tirol 73 C7 off. Land Tirol, var. Tyrol, It. Tirolo. state W Austria
Tirol, Land see Tirol
Tirolo see Tirol
Tirreno, Mare see Tyrrhenian Sea
Tiruchchirāppalli 110 C3 prev. Trichinopoly. Tamil Nādu, SE India
Tiruppattūr 110 C2 Tamil Nādu, SE India
Tiruvantapuram see Thiruvananthapuram
Tisa see Tisza
Tisza 81 F1 Ger. Theiss, Rom./Slvn./SCr. Tisa, Rus. Tissa, Ukr. Tysa. river SE Europe
Tiszakécske 77 D7 Bács-Kiskun, C Hungary
Titano, Monte 74 E1 mountain C San Marino
Titicaca, Lake 39 E4 lake Bolivia/Peru
Titograd see Podgorica
Titose see Chitose
Titova Mitrovica see Mitrovicë
Titovo Užice see Užice
Titu 86 C5 Dâmboviţa, S Romania
Titule 55 D5 Orientale, N Dem. Rep. Congo
Tiverton 67 C7 SW England, United Kingdom
Tivoli 74 C4 anc. Tiber. Lazio, C Italy
Tizimín 29 H3 Yucatán, SE Mexico
Tizi Ouzou 49 E1 var. Tizi-Ouzou. N Algeria
Tizi-Ouzou see Tizi Ouzou
Tiznit 48 B3 SW Morocco
Tjilatjap see Cilacap
Tjirebon see Cirebon
Tlaquepaque 28 D4 Jalisco, C Mexico
Tlascala see Tlaxcala
Tlaxcala 29 F4 var. Tlascala, Tlaxcala de Xicohténcatl. Tlaxcala, C Mexico
Tlaxcala de Xicohténcatl see Tlaxcala
Tlemcen 48 D2 var. Tilimsen, Tlemsen. NW Algeria
Tlemsen see Tlemcen
Toamasina 57 G3 var. Tamatave. Toamasina, E Madagascar
Toba, Danau 116 B3 lake Sumatera, W Indonesia
Tobago 33 H5 island NE Trinidad and Tobago
Toba Kakar Range 112 B2 mountain range NW Pakistan
Tobol 92 C4 Kaz. Tobyl. river Kazakhstan/Russian Federation
Tobol'sk 92 C3 Tyumenskaya Oblast', C Russian Federation
Tobruch/Tobruk see Ţubruq
Tobyl see Tobol
Tocantins 41 E3 off. Estado do Tocantins. state C Brazil
Tocantins 41 E3 off. Estado do Tocantins. region C Brazil
Tocantins, Estado do see Tocantins
Tocantins, Rio 41 F2 river N Brazil
Tocoa 30 D2 Colón, N Honduras
Tocopilla 42 B2 Antofagasta, N Chile
Todi 74 C4 Umbria, C Italy
Todos os Santos, Baía de 41 G3 bay E Brazil
Toetoes Bay 129 B8 bay South Island, New Zealand
Tofua 123 E4 island Ha'apai Group, C Tonga
Togo 53 E4 off. Togolese Republic; prev. French Togoland. country W Africa
Togolese Republic see Togo
Tojikiston, Jumhurii see Tajikistan
Tokanui 129 B7 Southland, South Island, New Zealand
Tokar 50 C3 var. Ţawkar. Red Sea, NE Sudan
Tokat 94 D3 Tokat, N Turkey
Tokelau 123 E3 NZ overseas territory W Polynesia
Tőketerebes see Trebišov
Tokio see Tōkyō
Tokmak 101 G2 Kir. Tokmok. Chuyskaya Oblast', N Kyrgyzstan
Tokmak 87 G4 var. Velykyy Tokmak. Zaporiz'ka Oblast', SE Ukraine
Tokmok see Tokmak
Tokoroa 128 D3 Waikato, North Island, New Zealand
Tokounou 52 C4 C Guinea
Tokushima 109 C6 var. Tokusima. Tokushima, Shikoku, SW Japan
Tōkyō 108 A1 var. Tokio. country capital (Japan) Tōkyō, Honshū, S Japan
Tōkyō-wan 108 A2 bay S Japan
Tolbukhin see Dobrich
Toledo 70 D3 anc. Toletum. Castilla-La Mancha, C Spain
Toledo 18 D3 Ohio, N USA
Toledo Bend Reservoir 27 G3 reservoir Louisiana/Texas, SW USA
Toletum see Toledo
Toliara 57 F4 var. Toliary; prev. Tuléar. Toliara, SW Madagascar
Toliary see Toliara
Tolmein see Tolmin
Tolmin 73 D7 Ger. Tolmein, It. Tolmino. W Slovenia
Tolmino see Tolmin
Tolna 77 C7 Ger. Tolnau. Tolna, S Hungary
Tolnau see Tolna
Tolochin see Talachyn
Tolosa 71 E1 País Vasco, N Spain
Tolosa see Toulouse
Toluca 29 E4 var. Toluca de Lerdo. México, S Mexico
Toluca de Lerdo see Toluca
Tol'yatti 89 C6 prev. Stavropol'. Samarskaya Oblast', W Russian Federation
Tomah 18 B2 Wisconsin, N USA
Tomakomai 108 D2 Hokkaidō, NE Japan
Tomar 70 B3 Santarém, W Portugal
Tomaschow see Tomaszów Mazowiecki
Tomaschow see Tomaszów Lubelski
Tomaszów see Tomaszów Mazowiecki
Tomaszów Lubelski 76 E4 Ger. Tomaschow. Lubelskie, E Poland
Tomaszów Mazowiecka see Tomaszów Mazowiecki
Tomaszów Mazowiecki 76 D4 var. Tomaszów Mazowiecka; prev. Tomaszów, Ger. Tomaschow. Łódzkie, C Poland
Tombigbee River 20 C3 river Alabama/Mississippi, S USA
Tombouctou 53 E3 Eng. Timbuktu. Tombouctou, N Mali
Tombua 56 A2 Port. Porto Alexandre. Namibe, SW Angola
Tomelloso 71 E3 Castilla-La Mancha, C Spain
Tomini, Gulf of see Teluk Tomini
Tomini, Teluk 117 E4 Eng. Gulf of Tomini; prev. Teluk Gorontalo. bay Sulawesi, C Indonesia
Tomsk 92 D4 Tomskaya Oblast', C Russian Federation
Tomür Feng 104 B3 pre. Pik Pobedy; Kyrg. Jengish Chokusu. mountain China/Kyrgyzstan
Tonezh see Tonyezh
Tonga 123 E4 off. Kingdom of Tonga, var. Friendly Islands. country SW Pacific Ocean
Tonga, Kingdom of see Tonga
Tongatapu 123 E5 island Tongatapu Group, S Tonga
Tongatapu Group 123 E5 island group S Tonga
Tonga Trench 121 E3 trench S Pacific Ocean
Tongchuan 106 C4 Shaanxi, C China
Tongeren 65 D6 Fr. Tongres. Limburg, NE Belgium
Tongking, Gulf of see Tonkin, Gulf of
Tongliao 105 G2 Nei Mongol Zizhiqu, N China
Tongres see Tongeren
Tongshan see Xuzhou, Jiangsu, China
Tongtian He 104 C4 river C China
Tonj 51 B5 Warab, C South Sudan
Tonkin, Gulf of 106 B7 var. Tongking, Gulf of, Chin. Beibu Wan, Vtn. Vinh Bắc Bộ. gulf China/Vietnam
Tônlé Sap 115 D5 Eng. Great Lake. lake W Cambodia
Tonopah 25 C6 Nevada, W USA
Tonyezh 85 C7 Rus. Tonezh. Homyel'skaya Voblasts', SE Belarus
Tooele 22 B4 Utah, W USA
Toowoomba 127 E5 Queensland, E Australia
Topeka 23 F4 state capital Kansas, C USA
Toplicza see Topliţa
Topliţa 86 C3 Ger. Toplitz, Hung. Marosheviz; prev. Topliţa Română, Hung. Oláh-Toplicza, Toplicza. Harghita, C Romania
Topliţa Română/Töplitz see Topliţa
Topoľčany 77 C6 Hung. Nagytapolcsány. Nitriansky Kraj, W Slovakia
Topolovgrad 82 D3 prev. Kavakli. Khaskovo, S Bulgaria
Topolya see Bačka Topola
Top Springs Roadhouse 124 E3 Northern Territory, N Australia
Torda see Turda
Torez 87 H3 Donets'ka Oblast', SE Ukraine
Torgau 72 D4 Sachsen, E Germany
Torhout 65 A5 West-Vlaanderen, W Belgium
Torino 74 A2 Eng. Turin. Piemonte, NW Italy
Tornacum see Tournai
Torneå see Tornio
Torneträsk 62 C3 lake N Sweden
Tornio 62 D4 Swe. Torneå. Lappi, NW Finland
Tornionjoki 62 D3 river Finland/Sweden
Toro 70 D2 Castilla-León, N Spain
Toronto 16 D5 province capital Ontario, S Canada
Toros Dağları 94 C4 Eng. Taurus Mountains. mountain range S Turkey
Torquay 67 C7 SW England, United Kingdom
Torrance 24 D2 California, W USA
Torre, Alto da 70 B3 mountain C Portugal
Torre del Greco 75 D5 Campania, S Italy
Torrejón de Ardoz 71 E3 Madrid, C Spain
Torrelavega 70 D1 Cantabria, N Spain
Torrens, Lake 127 A6 salt lake South Australia
Torrent 71 F3 Cas. Torrente, var. Torrent de l'Horta. País Valenciano, E Spain
Torrent de l'Horta/Torrente see Torrent
Torreón 28 D3 Coahuila, NE Mexico
Torres Strait 126 C1 strait Australia/Papua New Guinea
Torres Vedras 70 B3 Lisboa, C Portugal
Torrington 22 D3 Wyoming, C USA
Tórshavn 61 F5 Dan. Thorshavn. Dependent territory capital Faroe Islands
To'rtkok'l 100 D2 var. Türtkül, Rus. Turtkul'; prev. Petroaleksandrovsk. Qoraqalpog'iston Respublikasi, W Uzbekistan
Tortoise Islands see Galápagos Islands
Tortosa 71 F2 anc. Dertosa. Cataluña, E Spain
Tortue, Montagne 37 H3 mountain range C French Guiana
Tortuga, Isla see La Tortuga, Isla
Toruń 76 C3 Ger. Thorn. Toruń, Kujawsko-pomorskie, C Poland
Tõrva 84 D3 Ger. Törwa. Valgamaa, S Estonia
Törwa see Tõrva
Torzhok 88 B4 Tverskaya Oblast', W Russian Federation
Tosa-wan 109 B7 bay SW Japan
Toscana 74 B3 Eng. Tuscany. region C Italy
Toscano, Archipelago 74 B4 Eng. Tuscan Archipelago. island group C Italy
Toshkent 101 E2 Eng./Rus. Tashkent. country capital (Uzbekistan) Toshkent Viloyati, E Uzbekistan
Totana 71 E4 Murcia, SE Spain
Tot'ma see Sukhona
Totness 37 G3 Coronie, N Suriname
Tottori 109 B6 Tottori, Honshū, SW Japan
Touâjil 52 C2 Tiris Zemmour, N Mauritania
Touggourt 49 E2 NE Algeria
Toukoto 52 C3 Kayes, W Mali
Toul 68 D3 Meurthe-et-Moselle, NE France
Toulon 69 D6 anc. Telo Martius, Tilio Martius. Var, SE France
Toulouse 69 B6 anc. Tolosa. Haute-Garonne, S France
Toungoo see Taungoo
Touraine 68 B4 cultural region C France
Tourane see Đà Nẵng
Tourcoing 68 C2 Nord, N France
Tournai 65 A6 var. Tournay, Dut. Doornik; anc. Tornacum. Hainaut, SW Belgium
Tournay see Tournai
Tours 68 B4 anc. Caesarodunum, Turoni. Indre-et-Loire, C France
Tovarkovskiy 89 B5 Tul'skaya Oblast', W Russian Federation
Tower Island see Genovesa, Isla
Townsville 126 D3 Queensland, NE Australia
Towoeti Meer see Towuti, Danau
Towraghoudī 100 D4 Herāt, NW Afghanistan
Towson 19 F4 Maryland, NE USA
Towuti, Danau 117 E4 Dut. Towoeti Meer. lake Sulawesi, C Indonesia
Toyama 109 C5 Toyama, Honshū, SW Japan
Toyama-wan 109 B5 bay W Japan
Toyohara see Yuzhno-Sakhalinsk
Toyota 109 C6 Aichi, Honshū, SW Japan
Tozeur 49 E2 var. Tawzar. W Tunisia
Trâblous see Tripoli
Trabzon 95 E2 Eng. Trebizond; anc. Trapezus. Trabzon, NE Turkey
Traiectum ad Mosam/Traiectum Tungorum see Maastricht
Traiskirchen 73 E6 Niederösterreich, NE Austria
Trajani Portus see Civitavecchia
Trajectum ad Rhenum see Utrecht
Trakai 85 C5 Ger. Traken, Pol. Troki. Vilnius, SE Lithuania
Traken see Trakai
Tralee 67 A6 Ir. Trá Lí. SW Ireland
Trá Lí see Tralee
Tralles Aydin see Aydın
Trang 115 C7 Trang, S Thailand
Transantarctic Mountains 132 B3 mountain range Antarctica
Transilvania see Transylvania
Transilvaniei, Alpi see Carpaţii Meridionalii
Transjordan see Jordan
Transnistria 86 D3 cultural region NE Moldova
Transsylvanische Alpen/Transylvanian Alps see Carpaţii Meridionalii
Transylvania 86 B4 Eng. Ardeal, Transilvania, Ger. Siebenbürgen, Hung. Erdély. cultural region NW Romania
Trapani 75 B7 anc. Drepanum. Sicilia, Italy, C Mediterranean Sea
Trâpeăng Vêng see Kâmpóng Thum
Trapezus see Trabzon
Traralgon 127 C7 Victoria, SE Australia
Trasimenischersee see Trasimeno, Lago
Trasimeno, Lago 74 C4 Eng. Lake of Perugia, Ger. Trasimenischersee. lake C Italy
Traù see Trogir
Traverse City 18 C2 Michigan, N USA
Tra Vinh 115 D6 var. Phu Vinh. Tra Vinh, S Vietnam
Travis, Lake 27 F3 reservoir Texas, SW USA
Travnik 78 C4 Federacija Bosna I Hercegovina, C Bosnia and Herzegovina
Trbovlje 73 E7 Ger. Trifail. C Slovenia

Treasure State – Uulu

Treasure State *see* Montana
Třebíč *77 B5 Ger.* Trebitsch. Vysočina, C Czech Republic
Trebinje *79 C5* Republika Srpska, S Bosnia and Herzegovina
Trebišov *77 D6 Hung.* Tőketerebes. Košický Kraj, E Slovakia
Trebitsch *see* Třebíč
Trebnitz *see* Trzebnica
Tree Planters State *see* Nebraska
Trélazé *68 B4* Maine-et-Loire, NW France
Trelew *43 C6* Chubut, SE Argentina
Tremelo *65 C5* Vlaams Brabant, C Belgium
Trenčín *77 C5 Ger.* Trentschin, *Hung.* Trencsén. Trenčiansky Kraj, W Slovakia
Trencsén *see* Trenčín
Trengganu, Kuala *see* Kuala Terengganu
Trenque Lauquen *42 C4* Buenos Aires, E Argentina
Trent *see* Trento
Trento *74 C2 Eng.* Trent, *Ger.* Trient; *anc.* Tridentum. Trentino-Alto Adige, N Italy
Trenton *19 F4 state capital* New Jersey, NE USA
Trentschin *see* Trenčín
Tres Arroyos *43 D5* Buenos Aires, E Argentina
Treskavica *78 C4 mountain range* SE Bosnia and Herzegovina
Tres Tabernae *see* Saverne
Treves/Trèves *see* Trier
Treviso *74 C2 anc.* Tarvisium. Veneto, NE Italy
Trichinopoly *see* Tiruchchirāppalli
Trichūr *see* Thrissur
Tridentum/Trient *see* Trento
Trier *73 A5 Eng.* Treves, *Fr.* Trèves; *anc.* Augusta Treverorum. Rheinland-Pfalz, SW Germany
Triesen *72 E2* SW Liechtenstein Europe
Triesenberg *72 E2* SW Liechtenstein
Trieste *74 D2 Slvn.* Trst. Friuli-Venezia Giulia, NE Italy
Trifail *see* Trbovlje
Tríkala *82 B4 prev.* Trikkala. Thessalía, C Greece
Trikkala *see* Tríkala
Trimontium *see* Plovdiv
Trinacria *see* Sicilia
Trincomalee *110 D3 var.* Trinkomali. Eastern Province, NE Sri Lanka
Trindade, Ilha da *45 C5 island* Brazil, W Atlantic Ocean
Trinidad *39 F3* Beni, N Bolivia
Trinidad *42 D4* Flores, S Uruguay
Trinidad *22 D5* Colorado, C USA
Trinidad *33 H5 island* C Trinidad and Tobago
Trinidad and Tobago *33 H5 off.* Republic of Trinidad and Tobago. *country* SE West Indies
Trinidad and Tobago, Republic of *see* Trinidad and Tobago
Trinité, Montagnes de la *37 H3 mountain range* C French Guiana
Trinity River *27 G3 river* Texas, SW USA
Trinkomali *see* Trincomalee
Trípoli *83 B6 prev.* Trípolis. Pelopónnisos, S Greece
Tripoli *96 B4 var.* Tarābulus, Ţarābulus ash Shām, Trâblous; *anc.* Tripolis. N Lebanon
Tripoli *see* Ţarābulus
Trípolis *see* Trípoli, Greece
Tripolis *see* Tripoli, Lebanon
Tristan da Cunha *47 B7 UK dependent territory* SE Atlantic Ocean
Triton Island *106 B7 island* S Paracel Islands
Trivandrum *see* **Thiruvananthapuram**
Trnava *77 C6 Ger.* Tyrnau, *Hung.* Nagyszombat. Trnavský Kraj, W Slovakia
Trnovo *see* Veliko Tŭrnovo
Trogir *78 B4 It.* Traù. Split-Dalmacija, S Croatia
Troglav *78 B4 mountain* Bosnia and Herzegovina/Croatia
Trois-Rivières *17 E4* Québec, SE Canada
Troki *see* Trakai
Trollhättan *63 B6* Västra Götaland, S Sweden
Tromsø *62 C2 Fin.* Tromssa. Troms, N Norway
Tromssa *see* Tromsø
Trondheim *62 B4 Ger.* Drontheim; *prev.* Nidaros, Trondhjem. Sør-Trøndelag, S Norway
Trondheimsfjorden *62 B4 fjord* S Norway
Trondhjem *see* Trondheim
Troódos *80 C5 var.* Troodos Mountains. *mountain range* C Cyprus
Troodos Mountains *see* Troódos
Troppau *see* Opava
Troy *20 D3* Alabama, S USA
Troy *19 F3* New York, NE USA
Troyan *82 C2* Lovech, N Bulgaria
Troyes *68 D3 anc.* Augustobona Tricassium. Aube, N France
Trst *see* Trieste
Trstenik *78 E4* Serbia, C Serbia
Trucial States *see* United Arab Emirates
Trujillo *30 D2* Colón, NE Honduras
Trujillo *38 B3* La Libertad, NW Peru
Trujillo *70 C3* Extremadura, W Spain
Truk Islands *see* Chuuk Islands
Trŭn *82 B2* Pernik, W Bulgaria
Truro *17 F4* Nova Scotia, SE Canada
Truro *67 C7* SW England, United Kingdom
Trzcianka *76 B3 Ger.* Schönlanke. Pila, Wielkopolskie, C Poland
Trzebnica *76 C4 Ger.* Trebnitz. Dolnośląskie, SW Poland
Tsalka *95 F2* S Georgia Asia
Tsamkong *see* Zhanjiang
Tsangpo *see* Brahmaputra
Tsarevo *82 E2 prev.* Michurin. Burgas, E Bulgaria
Tsarigrad *see* İstanbul
Tsaritsyn *see* Volgograd
Tschakathurn *see* Čakovec
Tschaslau *see* Čáslav
Tschenstochau *see* Częstochowa
Tsefat *97 B5 var.* Safed, *Ar.* Safad; *prev.* Ẕefat. Northern, N Israel
Tselinograd *see* Astana
Tsetsen Khan *see* Öndörhaan
Tsetserleg *104 D2* Arhangay, C Mongolia
Tshela *55 B6* Bas-Congo, W Dem. Rep. Congo
Tshikapa *55 C7* Kasai-Occidental, SW Dem. Rep. Congo
Tshuapa *55 D6 river* C Dem. Rep. Congo
Tsinan *see* Jinan
Tsing Hai *see* Qinghai Hu, China
Tsinghai *see* Qinghai, China
Tsingtao/Tsingtau *see* Qingdao
Tsinkiang *see* Quanzhou
Tsintao *see* Qingdao
Tsitsihar *see* Qiqihar
Tsu *109 C6 var.* Tu. Mie, Honshū, SW Japan
Tsugaru-kaikyo *108 C3 strait* N Japan
Tsumeb *56 B3* Otjikoto, N Namibia
Tsuruga *109 C6 var.* Turuga. Fukui, Honshū, SW Japan
Tsuruoka *108 D4 var.* Turuoka. Yamagata, Honshū, C Japan
Tsushima *109 A7 var.* Tsushima-tō, Tusima. *island group* SW Japan
Tsushima-tō *see* Tsushima
Tsyerakhowka *85 D8 Rus.* Terekhovka. Homyel'skaya Voblasts', SE Belarus
Tsyurupinsk *see* Tsyurupyns'k
Tsyurupyns'k *87 E4 Rus.* Tsyurupinsk. Khersons'ka Oblast', S Ukraine
Tu *see* Tsu
Tuamotu, Archipel des *see* Tuamotu, Îles
Tuamotu Fracture Zone *121 H3 fracture zone* E Pacific Ocean
Tuamotu, Îles *123 H4 var.* Archipel des Tuamotu, Dangerous Archipelago, Tuamotu Islands. *island group* N French Polynesia
Tuamotu Islands *see* Tuamotu, Îles
Tuapi *31 E2* Región Autónoma Atlántico Norte, NE Nicaragua
Tuapse *89 A7* Krasnodarskiy Kray, SW Russian Federation
Tuba City *26 B1* Arizona, SW USA
Tubbergen *64 E3* Overijssel, E Netherlands
Tubeke *see* Tubize
Tubize *65 B6 Dut.* Tubeke. Walloon Brabant, C Belgium
Tubmanburg *52 C5* NW Liberia
Ţubruq *49 H2 Eng.* Tobruk, *It.* Tobruch. NE Libya
Tubuai, Îles/Tubuai Islands *see* Australes, Îles
Tucker's Town *20 B5* E Bermuda
Tuckum *see* Tukums
Tucson *26 B3* Arizona, SW USA
Tucumán *see* San Miguel de Tucumán
Tucumcari *27 E2* New Mexico, SW USA
Tucupita *37 E2* Delta Amacuro, NE Venezuela
Tucuruí, Represa de *41 F2 reservoir* NE Brazil
Tudela *71 E2 Basq.* Tutera; *anc.* Tutela. Navarra, N Spain
Tudmur *96 C3 var.* Tadmur, Tamar, *Gk.* Palmyra, *Bibl.* Tadmor. Ḥimş, C Syria
Tuguegarao *117 E1* Luzon, N Philippines
Tuktoyaktuk *15 E3* Northwest Territories, NW Canada
Tukums *84 C3 Ger.* Tuckum. Tukums, W Latvia
Tula *89 B5* Tul'skaya Oblast', W Russian Federation
Tulancingo *29 E4* Hidalgo, C Mexico
Tulare Lake Bed *25 C7 salt flat* California, W USA
Tulcán *38 B1* Carchi, N Ecuador
Tulcea *86 D5* Tulcea, E Romania
Tul'chin *see* Tul'chyn
Tul'chyn *86 D3 Rus.* Tul'chin. Vinnyts'ka Oblast', C Ukraine
Tuléar *see* Toliara
Tulia *27 E2* Texas, SW USA
Tūlkarm *97 D7* West Bank, Israel
Tulle *69 C5 anc.* Tutela. Corrèze, C France
Tulln *73 E6 var.* Oberhollabrunn. Niederösterreich, NE Austria
Tully *126 D3* Queensland, NE Australia
Tulsa *27 G1* Oklahoma, C USA
Tuluá *36 B3* Valle del Cauca, W Colombia
Tulun *93 E4* Irkutskaya Oblast', S Russian Federation
Tumaco *36 A4* Nariño, SW Colombia
Tumba, Lac *see* Ntomba, Lac
Tumbes *38 A2* Tumbes, NW Peru
Tumkūr *110 C2* Karnātaka, W India
Tumuc-Humac Mountains *41 E1 var.* Serra Tumucumaque. *mountain range* N South America
Tumucumaque, Serra *see* Tumuc-Humac Mountains
Tunca Nehri *see* Tundzha
Tunduru *51 C8* Ruvuma, S Tanzania
Tundzha *82 D3 Turk.* Tunca Nehri. *river* Bulgaria/Turkey
Tungabhadra Reservoir *110 C2 lake* S India
Tungaru *123 E2 prev.* Gilbert Islands. *island group* W Kiribati
T'ung-shan *see* Xuzhou
Tungsten *14 D4* Northwest Territories, W Canada
Tung-t'ing Hu *see* Dongting Hu
Tunis *49 E1 var.* Tūnis. *country capital* (Tunisia) NE Tunisia
Tunis, Golfe de *80 D3 Ar.* Khalīj Tūnis. *gulf* NE Tunisia
Tunisia *49 F2 off.* Tunisian Republic, *Ar.* Al Jumhūrīyah at Tūnisīyah, *Fr.* République Tunisienne. *country* N Africa
Tunisian Republic *see* Tunisia
Tunisienne, République *see* Tunisia
Tūnisīyah, Al Jumhūrīyah at *see* Tunisia
Tūnis, Khalīj *see* Tunis, Golfe de
Tunja *36 B3* Boyacá, C Colombia
Tuong Buong *see* Tương Đương
Tương Đương *114 D4 var.* Tuong Buong. Nghê An, N Vietnam
Tüp *see* Tyup
Tupelo *20 C2* Mississippi, S USA
Tupiza *39 G5* Potosí, S Bolivia
Turabah *99 B5* Makkah, W Saudi Arabia
Turangi *128 D4* Waikato, North Island, New Zealand
Turan Lowland *100 C2 var.* Turan Plain, *Kaz.* Turan Oypaty, *Rus.* Turanskaya Nizmennost', *Turk.* Turan Pesligi, *Uzb.* Turan Pasttekisligi. *plain* C Asia
Turan Oypaty/Turan Pesligi/Turan Plain/ Turanskaya Nizmennost' *see* Turan Lowland
Turan Pasttekisligi *see* Turan Lowland
Ţurayf *98 A3* Al Ḥudūd ash Shamālīyah, NW Saudi Arabia
Turba *see* Teruel
Turbat *112 A3* Baluchistān, SW Pakistan
Turčianský Svätý Martin *see* Martin
Turda *86 B4 Ger.* Thorenburg, *Hung.* Torda. Cluj, NW Romania
Turek *76 C3* Wielkopolskie, C Poland
Turfan *see* Turpan
Turin *see* Torino
Turkana, Lake *51 C6 var.* Lake Rudolf. *lake* N Kenya
Turkestan *see* Turkistan
Turkey *94 B3 off.* Republic of Turkey, *Turk.* Türkiye Cumhuriyeti. *country* SW Asia
Turkey, Republic of *see* Turkey
Turkish Republic of Northern Cyprus *80 D5 disputed territory* Cyprus
Turkistan *92 B5 prev.* Turkestan. Yuzhnyy Kazakhstan, S Kazakhstan
Türkiye Cumhuriyeti *see* Turkey
Türkmenabat *100 D3 prev. Rus.* Chardzhev, Chardzhou, Chardzhui, Lenin-Turkmenski, *Turkm.* Chärjew. Lebap Welaýaty, E Turkmenistan
Türkmen Aylagy *100 B2 Rus.* Turkmenskiy Zaliv. *lake gulf* W Turkmenistan
Turkmenbashi *see* Türkmenbaşy
Türkmenbaşy *100 B2 Rus.* Turkmenbashi; *prev.* Krasnovodsk. Balkan Welaýaty, W Turkmenistan
Türkmenbasy Aylagy *100 A2 prev. Rus.* Krasnovodskiy Zaliv, *Turkm.* Krasnowodsk Aylagy. *lake Gulf* W Turkmenistan
Turkmenistan *100 B2* ; *prev.* Turkmenskaya Soviet Socialist Republic. *country* C Asia
Turkmenskaya Soviet Socialist Republic *see* Turkmenistan
Turkmenskiy Zaliv *see* Türkmen Aylagy
Turks and Caicos Islands *33 E2 UK dependent territory* N West Indies
Turku *63 D6 Swe.* Åbo. Länsi-Suomi, SW Finland
Turlock *25 B6* California, W USA
Turnagain, Cape *128 D4 headland* North Island, New Zealand
Turnau *see* Turnov
Turnhout *65 C5* Antwerpen, N Belgium
Turnov *76 B4 Ger.* Turnau. Liberecký Kraj, N Czech Republic
Tŭrnovo *see* Veliko Tŭrnovo
Turnu Măgurele *86 B5 var.* Turnu-Măgurele. Teleorman, S Romania
Turnu Severin *see* Drobeta-Turnu Severin
Turócszentmárton *see* Martin
Turoni *see* Tours
Turpan *104 C3 var.* Turfan. Xinjiang Uygur Zizhiqu, NW China
Turpan Depression *see* Turpan Pendi
Turpan Pendi *104 C3 Eng.* Turpan Depression. *depression* NW China
Turpentine State *see* North Carolina
Tŭrtkŭl/Turtkul' *see* To'rtkok'l
Turuga *see* Tsuruga
Turuoka *see* Tsuruoka
Tuscaloosa *20 C2* Alabama, S USA
Tuscan Archipelago *see* Toscano, Archipelago
Tuscany *see* Toscana
Tusima *see* Tsushima
Tutela *see* Tulle, France
Tutela *see* Tudela, Spain
Tutera *see* Tudela
Tuticorin *110 C3* Tamil Nādu, SE India
Tutrakan *82 D1* Silistra, NE Bulgaria
Tutuila *123 F4 island* W American Samoa
Tuvalu *123 E3 prev.* Ellice Islands. *country* SW Pacific Ocean
Tuwayq, Jabal *99 C5 mountain range* C Saudi Arabia
Tuxpan *28 D4* Jalisco, C Mexico
Tuxpán *29 F4 var.* Tuxpán de Rodríguez Cano. Veracruz-Llave, E Mexico
Tuxpán de Rodríguez Cano *see* Tuxpán
Tuxtepec *29 F4 var.* San Juan Bautista Tuxtepec. Oaxaca, S Mexico
Tuxtla *29 G5 var.* Tuxtla Gutiérrez. Chiapas, SE Mexico
Tuxtla *see* San Andrés Tuxtla
Tuxtla Gutiérrez *see* Tuxtla
Tuy Hoa *115 E5* Phu Yên, S Vietnam
Tuz, Lake *94 C3 lake* C Turkey
Tver' *88 B4 prev.* Kalinin. Tverskaya Oblast', W Russian Federation
Tverya *97 B5 var.* Tiberias; *prev.* Teverya. Northern, N Israel
Twin Falls *24 D4* Idaho, NW USA
Tyan'-Shan' *see* Tien Shan
Tychy *77 D5 Ger.* Tichau. Śląskie, S Poland
Tyler *27 G3* Texas, SW USA
Tylos *see* Bahrain
Tympáki *83 C8 var.* Timbaki; *prev.* Timbákion. Kríti, Greece, E Mediterranean Sea
Tynda *93 F4* Amurskaya Oblast', SE Russian Federation
Tyne *66 D4 river* N England, United Kingdom
Tyôsi *see* Chōshi
Tyras *see* Dniester
Tyre *see* Soûr
Tyrnau *see* Trnava
Týrnavos *82 B4 var.* Tírnavos. Thessalía, C Greece
Tyrol *see* Tirol
Tyros *see* Bahrain
Tyrrhenian Sea *75 B6 It.* Mare Tirreno. *sea* N Mediterranean Sea
Tyumen' *92 C3* Tyumenskaya Oblast', C Russian Federation
Tyup *101 G2 Kir.* Tüp. Issyk-Kul'skaya Oblast', NE Kyrgyzstan
Tywyn *67 C6* W Wales, United Kingdom
Tzekung *see* Zigong
Tziá *83 C6 prev.* Kéa, Kéos; *anc.* Ceos. *island* Kykládes, Greece, Aegean Sea

U

Uaco Cungo *56 B1* C Angola
UAE *see* United Arab Emirates
Uanle Uen *see* Wanlaweyn
Uaupés, Rio *see* Vaupés, Río
Ubangi-Shari *see* Central African Republic
Ube *109 B7* Yamaguchi, Honshū, SW Japan
Ubeda *71 E4* Andalucía, S Spain
Uberaba *41 F4* Minas Gerais, SE Brazil
Uberlândia *41 F4* Minas Gerais, SE Brazil
Ubol Rajadhani/Ubol Ratchathani *see* Ubon Ratchathani
Ubon Ratchathani *115 D5 var.* Muang Ubon, Ubol Rajadhani, Ubol Ratchathani, Udon Ratchathani. Ubon Ratchathani, E Thailand
Ubrique *70 D5* Andalucía, S Spain
Ubsu-Nur, Ozero *see* Uvs Nuur
Ucayali, Río *38 D3 river* C Peru
Uchiura-wan *108 D3 bay* NW Pacific Ocean
Uchkuduk *see* Uchquduq
Uchquduq *100 D2 Rus.* Uchkuduk. Navoiy Viloyati, N Uzbekistan
Uchtagan Gumy/Uchtagan, Peski *see* Uçtagan Gumy
Uçtagan Gumy *100 C2 var.* Uchtagan Gumy, *Rus.* Peski Uchtagan. *desert* NW Turkmenistan
Udaipur *112 C3 prev.* Oodeypore. Rājasthān, N India
Uddevalla *63 B6* Västra Götaland, S Sweden
Udine *74 D2 anc.* Utina. Friuli-Venezia Giulia, NE Italy
Udintsev Fracture Zone *132 A5 tectonic feature* S Pacific Ocean
Udipi *see* Udupi
Udon Ratchathani *see* Ubon Ratchathani
Udon Thani *114 C4 var.* Ban Mak Khaeng, Udorndhani. Udon Thani, N Thailand
Udorndhani *see* Udon Thani
Udupi *110 B2 var.* Udipi. Karnātaka, SW India
Uele *55 D5 var.* Welle. *river* NE Dem. Rep. Congo
Uelzen *72 C3* Niedersachsen, N Germany
Ufa *89 D6* Respublika Bashkortostan, W Russian Federation
Ugāle *84 C2* Ventspils, NW Latvia
Uganda *51 B6 off.* Republic of Uganda. *country* E Africa
Uganda, Republic of *see* Uganda
Uhorshchyna *see* Hungary
Uhuru Peak *see* Kilimanjaro
Uíge *56 B1 Port.* Carmona, Vila Marechal Carmona. Uíge, NW Angola
Uinta Mountains *22 B4 mountain range* Utah, W USA
Uitenhage *56 C5* Eastern Cape, S South Africa
Uithoorn *64 C3* Noord-Holland, C Netherlands
Ujda *see* Oujda
Ujelang Atoll *122 C1 var.* Wujlān. *atoll* Ralik Chain, W Marshall Islands
Újgradiska *see* Nova Gradiška
Újmoldova *see* Moldova Nouă
Ujungpandang *see* Makassar
Ujung Salang *see* Phuket
Újvidék *see* Novi Sad
UK *see* United Kingdom
Ukhta *92 C3* Respublika Komi, NW Russian Federation
Ukiah *25 B5* California, W USA
Ukmergė *84 C4 Pol.* Wiłkomierz. Vilnius, C Lithuania
Ukraine *86 C2 off.* Ukraine, *Rus.* Ukraina, *Ukr.* Ukrayina; *prev.* Ukrainian Soviet Socialist Republic, Ukrainskay S.S.R. *country* SE Europe
Ukraine *see* Ukraine
Ukrainian Soviet Socialist Republic *see* Ukraine
Ukrainskay S.S.R/Ukrayina *see* Ukraine
Ulaanbaatar *105 E2 Eng.* Ulan Bator; *prev.* Urga. *country capital* (Mongolia) Töv, C Mongolia
Ulaangom *104 C2* Uvs, NW Mongolia
Ulan Bator *see* Ulaanbaatar
Ulanhad *see* Chifeng
Ulan-Ude *93 E4 prev.* Verkhneudinsk. Respublika Buryatiya, S Russian Federation
Uleälv *see* Oulujoki
Uleträsk *see* Oulujärvi
Ulft *64 E4* Gelderland, E Netherlands
Ullapool *66 C3* N Scotland, United Kingdom
Ulm *73 B6* Baden-Württemberg, S Germany
Ulsan *107 E4 Jap.* Urusan. SE South Korea
Ulster *67 B5 province* Northern Ireland, United Kingdom/Ireland
Ulungur Hu *104 B2 lake* NW China
Uluṟu *125 D5 var.* Ayers Rock. *monolith* Northern Territory, C Australia
Ulyanivka *87 E3 Rus.* Ul'yanovka. Kirovohrads'ka Oblast', C Ukraine
Ul'yanovka *see* Ulyanivka
Ul'yanovsk *89 C5 prev.* Simbirsk. Ul'yanovskaya Oblast', W Russian Federation
Umán *29 H3* Yucatán, SE Mexico
Uman' *87 E3 Rus.* Uman. Cherkas'ka Oblast', C Ukraine
Uman *see* Uman'
Umanak/Umanaq *see* Uummannaq
'Umān, Khalīj *see* Oman, Gulf of
'Umān, Salṭanat *see* Oman
Umbrian-Machigian Mountains *see* Umbro-Marchigiano, Appennino
Umbro-Marchigiano, Appennino *74 C3 Eng.* Umbrian-Machigian Mountains. *mountain range* C Italy
Umeå *62 C4* Västerbotten, N Sweden
Umeälven *62 C4 river* N Sweden
Umiat *14 D2* Alaska, USA
Umm Buru *50 A4* Western Darfur, W Sudan
Umm Durmān *see* Omdurman
Umm Ruwaba *50 C4 var.* Umm Ruwābah, Um Ruwāba. Northern Kordofan, C Sudan
Umm Ruwābah *see* Umm Ruwaba
Umnak Island *14 A3 island* Aleutian Islands, Alaska, USA
Um Ruwāba *see* Umm Ruwaba
Umtali *see* Mutare
Umtata *see* Mthatha
Una *78 B3 river* Bosnia and Herzegovina/Croatia
Unac *78 B3 river* W Bosnia and Herzegovina
Unalaska Island *14 A3 island* Aleutian Islands, Alaska, USA
'Unayzah *98 B4 var.* Anaiza. Al Qaşīm, C Saudi Arabia
Unci *see* Almería
Uncía *39 F4* Potosí, C Bolivia
Uncompahgre Peak *22 B5 mountain* Colorado, C USA
Undur Khan *see* Öndörhaan
Ungaria *see* Hungary
Ungarisches Erzgebirge *see* Slovenské rudohorie
Ungarn *see* Hungary
Ungava Bay *17 E1 bay* Québec, E Canada
Ungava Peninsula *see* Ungava, Péninsule d'
Ungava, Péninsule d' *16 D1 Eng.* Ungava Peninsula. *peninsula* Québec, SE Canada
Ungeny *see* Ungheni
Ungheni *86 D3 Rus.* Ungeny. W Moldova
Unguja *see* Zanzibar
Üngüz Angyrsyndaky Garagum *100 C2 Rus.* Zaunguzskiye Garagumy. *desert* N Turkmenistan
Ungvár *see* Uzhhorod
Unimak Island *14 B3 island* Aleutian Islands, Alaska, USA
Union *21 E1* South Carolina, SE USA
Union City *20 C1* Tennessee, S USA
Union of Myanmar *see* Myanmar
United Arab Emirates *99 C5 Ar.* Al Imārāt al 'Arabīyah al Muttaḥidah, *abbrev.* UAE; *prev.* Trucial States. *country* SW Asia
United Arab Republic *see* Egypt
United Kingdom *67 B5 off.* United Kingdom of Great Britain and Northern Ireland, *abbrev.* UK. *country* NW Europe
United Kingdom of Great Britain and Northern Ireland *see* United Kingdom
United Mexican States *see* Mexico
United Provinces *see* Uttar Pradesh
United States of America *13 B5 off.* United States of America, *var.* America, The States, *abbrev.* U.S., USA. *country* North America
Unst *66 D1 island* NE Scotland, United Kingdom
Ünye *94 D2* Ordu, W Turkey
Upala *30 D4* Alajuela, NW Costa Rica
Upata *37 E2* Bolívar, E Venezuela
Upemba, Lac *55 D7 lake* SE Dem. Rep. Congo
Upernavik *60 C2 var.* Upernivik. Kitaa, C Greenland
Upernivik *see* Upernavik
Upington *56 C4* Northern Cape, W South Africa
'Upolu *123 F4 island* SE Samoa
Upper Klamath Lake *24 A4 lake* Oregon, NW USA
Upper Lough Erne *67 A5 lake* SW Northern Ireland, United Kingdom
Upper Red Lake *23 F1 lake* Minnesota, N USA
Upper Volta *see* Burkina
Uppsala *63 C6* Uppsala, C Sweden
Uqsuqtuuq *see* Gjoa Haven
Ural *90 B3 Kaz.* Zhayyk. *river* Kazakhstan/Russian Federation
Ural Mountains *see* Ural'skiye Gory
Ural'sk *92 B3 Kaz.* Oral. Zapadnyy Kazakhstan, NW Kazakhstan
Ural'skiye Gory *92 C3 var.* Ural'skiy Khrebet, *Eng.* Ural Mountains. *mountain range* Kazakhstan/Russian Federation
Ural'skiy Khrebet *see* Ural'skiye Gory
Uraricoera *40 D1* Roraima, N Brazil
Ura-Tyube *see* Ŭroteppa
Urbandale *23 F3* Iowa, C USA
Urdunn *see* Jordan
Uren' *89 C5* Nizhegorodskaya Oblast', W Russian Federation
Urga *see* Ulaanbaatar
Urganch *100 D2 Rus.* Urgench; *prev.* Novo-Urgench. Xorazm Viloyati, W Uzbekistan
Urgench *see* Urganch
Urgut *101 E3* Samarqand Viloyati, C Uzbekistan
Urmia, Lake *see* Orūmīyeh, Daryācheh-ye
Uroševac *see* Ferizaj
Ŭroteppa *101 E2 Rus.* Ura-Tyube. NW Tajikistan
Uruapan *29 E4 var.* Uruapan del Progreso. Michoacán, SW Mexico
Uruapan del Progreso *see* Uruapan
Uruguai, Rio *see* Uruguay
Uruguay *42 D4 off.* Oriental Republic of Uruguay; *prev.* La Banda Oriental. *country* E South America
Uruguay *42 D3 var.* Rio Uruguai, Río Uruguay. *river* E South America
Uruguay, Oriental Republic of *see* Uruguay
Uruguay, Río *see* Uruguay
Urumchi *see* Ürümqi
Urumi Yeh *see* Orūmīyeh, Daryācheh-ye
Ürümqi *104 C3 var.* Tihwa, Urumchi, Urumqi, Urumtsi, Wu-lu-k'o-mu-shi, Wu-lu-mu-ch'i; *prev.* Ti-hua. Xinjiang Uygur Zizhiqu, NW China
Urumtsi *see* Ürümqi
Urundi *see* Burundi
Urup, Ostrov *93 H4 island* Kuril'skiye Ostrova, SE Russian Federation
Urusan *see* Ulsan
Urziceni *86 C5* Ialomiţa, SE Romania
Usa *88 E3 river* NW Russian Federation
Uşak *94 B3 prev.* Ushak. Uşak, W Turkey
Ushak *see* Uşak
Ushant *see* Ouessant, Île d'
Ushuaia *43 B8* Tierra del Fuego, S Argentina
Usinsk *88 E3* Respublika Komi, NW Russian Federation
Üsküb/Üsküp *see* Skopje
Usmas Ezers *84 B3 lake* NW Latvia
Usol'ye-Sibirskoye *93 E4* Irkutskaya Oblast', C Russian Federation
Ussel *69 C5* Corrèze, C France
Ussuriysk *93 G5 prev.* Nikol'sk, Nikol'sk-Ussuriyskiy, Voroshilov. Primorskiy Kray, SE Russian Federation
Ustica *75 B6 island* S Italy
Ust'-Ilimsk *93 E4* Irkutskaya Oblast', C Russian Federation
Ústí nad Labem *76 A4 Ger.* Aussig. Ústecký Kraj, NW Czech Republic
Ustinov *see* Izhevsk
Ustka *76 C2 Ger.* Stolpmünde. Pomorskie, N Poland
Ust'-Kamchatsk *93 H2* Kamchatskaya Oblast', E Russian Federation
Ust'-Kamenogorsk *92 D5 Kaz.* Öskemen. Vostochnyy Kazakhstan, E Kazakhstan
Ust'-Kut *93 E4* Irkutskaya Oblast', C Russian Federation
Ust'-Olenëk *93 E3* Respublika Sakha (Yakutiya), NE Russian Federation
Ustrzyki Dolne *77 E5* Podkarpackie, SE Poland
Ust'-Sysol'sk *see* Syktyvkar
Ust Urt *see* Ustyurt Plateau
Ustyurt Plateau *100 B1 var.* Ust Urt, *Uzb.* Ustyurt Platosi. *plateau* Kazakhstan/Uzbekistan
Ustyurt Platosi *see* Ustyurt Plateau
Usulután *30 C3* Usulután, SE El Salvador
Usumacinta, Río *30 B1 river* Guatemala/Mexico
Usumbura *see* Bujumbura
U.S./USA *see* United States of America
Utah *22 B4 off.* State of Utah, *also known as* Beehive State, Mormon State. *state* W USA
Utah Lake *22 B4 lake* Utah, W USA
Utena *84 C4* Utena, E Lithuania
Utica *19 F3* New York, NE USA
Utina *see* Udine
Utrecht *64 C4 Lat.* Trajectum ad Rhenum. Utrecht, C Netherlands
Utsunomiya *109 D5 var.* Utunomiya. Tochigi, Honshū, S Japan
Uttarakhand *113 E2 cultural region* N India
Uttar Pradesh *113 E3 prev.* United Provinces, United Provinces of Agra and Oudh. *cultural region* N India
Utunomiya *see* Utsunomiya
Uulu *84 D2* Pärnumaa, SW Estonia

Uummannaq – Vyerkhnyadzvinsk

Uummannaq *60 C3 var.* Umanak, Umanaq. Kitaa, C Greenland
Uummannarsuaq *see* Nunap Isua
Uvalde *27 F4* Texas, SW USA
Uvaravichy *85 D7 Rus.* Uvarovichi. Homyel'skaya Voblasts', SE Belarus
Uvarovichi *see* Uvaravichy
Uvea, Île *123 E4 island* N Wallis and Futuna
Uvs Nuur *104 C1 var.* Ozero Ubsu-Nur. *lake* Mongolia/Russian Federation
'Uwaynāt, Jabal al *66 A3 var.* Jebel Uweinat. *mountain* Libya/Sudan
Uweinat, Jebel *see* 'Uwaynāt, Jabal al
Uyo *53 G5* Akwa Ibom, S Nigeria
Uyuni *39 F5* Potosí, W Bolivia
Uzbekistan *100 D2 off.* Republic of Uzbekistan. *country* C Asia
Uzbekistan, Republic of *see* Uzbekistan
Uzhgorod *see* Uzhhorod
Uzhhorod *86 B2 Rus.* Uzhgorod; *prev.* Ungvár. Zakarpats'ka Oblast', W Ukraine
Užice *78 D4 prev.* Titovo Užice. Serbia, W Serbia

V

Vaal *56 D4 river* C South Africa
Vaals *65 D6* Limburg, SE Netherlands
Vaasa *63 D5 Swe.* Vasa; *prev.* Nikolainkaupunki. Länsi-Suomi, W Finland
Vaassen *64 D3* Gelderland, E Netherlands
Vác *77 C6 Ger.* Waitzen. Pest, N Hungary
Vadodara *112 C4 prev.* Baroda. Gujarāt, W India
Vaduz *72 E2 country capital* (Liechtenstein) W Liechtenstein
Vág *see* Váh
Vágbeszterce *see* Považská Bystrica
Váh *77 C5 Ger.* Waag, *Hung.* Vág. *river* W Slovakia
Váhtjer *see* Gällivare
Väinameri *84 C2 prev.* Muhu Väin, *Ger.* Moon-Sund. *sea* E Baltic Sea
Vajdahunyad *see* Hunedoara
Valachia *see* Wallachia
Valday *88 B4* Novgorodskaya Oblast', W Russian Federation
Valdecañas, Embalse de *70 D3 reservoir* W Spain
Valdepeñas *71 E4* Castilla-La Mancha, C Spain
Valdez *14 C3* Alaska, USA
Valdia *see* Weldiya
Valdivia *43 B5* Los Lagos, C Chile
Val-d'Or *16 D4* Québec, SE Canada
Valdosta *21 E3* Georgia, SE USA
Valence *69 D5 anc.* Valentia, Valentia Julia, Ventia. Drôme, E France
Valencia *71 F3* País Valenciano, E Spain
Valencia *24 D1* California, USA
Valencia *36 D1* Carabobo, N Venezuela
Valencia, Gulf of *71 F3 var.* Gulf of Valencia. *gulf* E Spain
Valencia, Gulf of *see* Valencia, Golfo de
Valencia/València *see* País Valenciano
Valenciennes *68 D2* Nord, N France
Valentia *see* Valence, France
Valentia *see* País Valenciano
Valentia Julia *see* Valence
Valentine State *see* Oregon
Valera *36 C2* Trujillo, NW Venezuela
Valetta *see* Valletta
Valga *84 D3 Ger.* Walk, *Latv.* Valka. Valgamaa, S Estonia
Valira *69 A8 river* Andorra/Spain Europe
Valjevo *78 C4* Serbia, W Serbia
Valjok *see* Válljohka
Valka *84 D3 Ger.* Walk. Valka, N Latvia
Valka *see* Valga
Valkenswaard *65 D5* Noord-Brabant, S Netherlands
Valladolid *29 H3* Yucatán, SE Mexico
Valladolid *70 D2* Castilla-León, NW Spain
Vall D'Uxó *see* La Vall d'Uixó
Valle de La Pascua *36 D2* Guárico, N Venezuela
Valledupar *36 B1* Cesar, N Colombia
Vallejo *25 B6* California, W USA
Vallenar *42 B3* Atacama, N Chile
Valletta *75 C8 prev.* Valetta. *country capital* (Malta) E Malta
Valley City *23 E2* North Dakota, N USA
Válljohka *62 D2 var.* Valjok. Finnmark, N Norway
Valls *71 G2* Cataluña, NE Spain
Valmiera *84 D3 Est.* Volmari, *Ger.* Wolmar. Valmiera, N Latvia
Valona *see* Vlorë
Valozhyn *85 C5 Pol.* Wołożyn, *Rus.* Volozhin. Minskaya Voblasts', C Belarus
Valparaíso *42 B4* Valparaíso, C Chile
Valparaiso *18 C3* Indiana, N USA
Valverde del Camino *70 C4* Andalucía, S Spain
Van *95 F3* Van, E Turkey
Vanadzor *95 F2 prev.* Kirovakan. N Armenia
Vancouver *14 D5* British Columbia, SW Canada
Vancouver *24 B3* Washington, NW USA
Vancouver Island *14 D5 island* British Columbia, SW Canada
Vanda *see* Vantaa
Van Diemen Gulf *124 D2 gulf* Northern Territory, N Australia
Van Diemen's Land *see* Tasmania
Vaner, Lake *see* Vänern
Vänern *63 B6 Eng.* Lake Vaner; *prev.* Lake Vener. *lake* S Sweden
Vangaindrano *57 G4* Fianarantsoa, SE Madagascar
Van Gölü *95 F3 Eng.* Lake Van; *anc.* Thospitis. *salt lake* E Turkey
Van Horn *26 D3* Texas, SW USA
Van, Lake *see* Van Gölü
Vannes *68 A3 anc.* Dariorigum. Morbihan, NW France
Vantaa *63 D6 Swe.* Vanda. Etelä-Suomi, S Finland
Vanua Levu *123 E4 island* N Fiji
Vanuatu *122 C4 off.* Republic of Vanuatu; *prev.* New Hebrides. *country* SW Pacific Ocean
Vanuatu, Republic of *see* Vanuatu
Van Wert *18 C4* Ohio, N USA
Vapincum *see* Gap
Varakļāni *84 D4* Madona, C Latvia
Vārānasi *113 E3 prev.* Banaras, Benares, *hist.* Kasi. Uttar Pradesh, N India
Varangerfjorden *62 E2 Lapp.* Várjjatvuotna. *fjord* N Norway
Varangerhalvøya *62 D2 Lapp.* Várnjárga. *peninsula* N Norway
Varannó *see* Vranov nad Topl'ou
Varasd *see* Varaždin
Varaždin *78 B2 Ger.* Warasdin, *Hung.* Varasd. Varaždin, N Croatia
Varberg *63 B7* Halland, S Sweden
Vardar *79 E6 Gk.* Axiós. *river* FYR Macedonia/Greece
Varde *63 A7* Ribe, W Denmark
Vareia *see* Logroño
Varėna *85 B5 Pol.* Orany. Alytus, S Lithuania
Varese *74 B2* Lombardia, N Italy
Vârful Moldoveanu *86 B4 var.* Moldoveanul; *prev.* Vîrful Moldoveanu. *mountain* C Romania
Várjjatvuotna *see* Varangerfjorden
Varkaus *63 E5* Itä-Suomi, C Finland
Varna *82 E2 prev.* Stalin; *anc.* Odessus. Varna, E Bulgaria
Varnenski Zaliv *82 E2 prev.* Stalinski Zaliv. *bay* E Bulgaria
Várnjárga *see* Varangerhalvøya
Varshava *see* Warszawa
Vasa *see* Vaasa
Vasilikí *83 A5* Lefkáda, Iónia Nisiá, Greece, C Mediterranean Sea
Vasilishki *85 B5 Pol.* Wasiliszki. Hrodzyenskaya Voblasts', W Belarus
Vasil'kov *see* Vasyl'kiv
Vaslui *86 D4* Vaslui, C Romania
Västerås *63 C6* Västmanland, C Sweden
Vasyl'kiv *87 E2 var.* Vasil'kov. Kyyivs'ka Oblast', N Ukraine
Vaté *see* Efate
Vatican City *75 A7 off.* Vatican City. *country* S Europe
Vatnajökull *61 E5 glacier* SE Iceland
Vatter, Lake *see* Vättern
Vättern *63 B6 Eng.* Lake Vatter; *prev.* Lake Vetter. *lake* S Sweden
Vaughn *26 D2* New Mexico, SW USA
Vaupés, Río *36 C4 var.* Rio Uaupés. *river* Brazil/Colombia
Vava'u Group *123 E4 island group* N Tonga
Vavuniya *110 D3* Northern Province, N Sri Lanka
Vawkavysk *85 B6 Pol.* Wołkowysk, *Rus.* Volkovysk. Hrodzyenskaya Voblasts', W Belarus
Växjö *63 C7 var.* Vexiö. Kronoberg, S Sweden
Vaygach, Ostrov *88 E2 island* NW Russian Federation
Veendam *64 E2* Groningen, NE Netherlands
Veenendaal *64 D4* Utrecht, C Netherlands
Vega *62 B4 island* C Norway
Veglia *see* Krk
Veisiejai *85 B5* Alytus, S Lithuania
Vejer de la Frontera *70 C5* Andalucía, S Spain
Veldhoven *65 D5* Noord-Brabant, S Netherlands
Velebit *78 A3 mountain range* C Croatia
Velenje *73 E7 Ger.* Wöllan. N Slovenia
Veles *79 E6 Turk.* Köprülü. C FYR Macedonia
Velho *see* Porto Velho
Velika Kikinda *see* Kikinda
Velika Morava *78 D4 var.* Glavn'a Morava, Morava, *Ger.* Grosse Morava. *river* C Serbia
Velikaya *91 G2 river* NE Russian Federation
Veliki Bečkerek *see* Zrenjanin
Velikiye Luki *88 A4* Pskovskaya Oblast', W Russian Federation
Velikiy Novgorod *88 B4 prev.* Novgorod. Novgorodskaya Oblast', W Russian Federation
Veliko Tŭrnovo *82 D2 prev.* Tirnovo, Trnovo, Tŭrnovo. Veliko Tŭrnovo, N Bulgaria
Velingrad *82 C3* Pazardzhik, C Bulgaria
Vel'ký Krtíš *77 D6* Banskobystrický Kraj, C Slovakia
Vellore *110 C2* Tamil Nādu, SE India
Velobriga *see* Viana do Castelo
Velsen *see* Velsen-Noord
Velsen-Noord *64 C3 var.* Velsen. Noord-Holland, W Netherlands
Vel'sk *88 C4 var.* Velsk. Arkhangel'skaya Oblast', NW Russian Federation
Velvendós *see* Velvéntos
Velvéntos *82 B4 var.* Velvendós. C Greece
Velykyy Tokmak *see* Tokmak
Vendôme *68 C4* Loir-et-Cher, C France
Venedig *see* Venezia
Vener, Lake *see* Vänern
Venetia *see* Venezia
Venezia *74 C2 Eng.* Venice, *Fr.* Venise, *Ger.* Venedig; *anc.* Venetia. Veneto, NE Italy
Venezia, Golfo di *see* Venice, Gulf of
Venezuela *36 D2 off.* Republic of Venezuela; *prev.* Estados Unidos de Venezuela, United States of Venezuela. *country* N South America
Venezuela, Estados Unidos de *see* Venezuela
Venezuela, Gulf of *36 C1 Eng.* Gulf of Maracaibo, Gulf of Venezuela. *gulf* NW Venezuela
Venezuela, Gulf of *see* Venezuela, Golfo de
Venezuelan Basin *34 B1 undersea basin* E Caribbean Sea
Venezuela, Republic of *see* Venezuela
Venezuela, United States of *see* Venezuela
Venice *20 C4* Louisiana, S USA
Venice *see* Venezia
Venice, Gulf of *74 C2 It.* Golfo di Venezia, *Slvn.* Beneški Zaliv. *gulf* N Adriatic Sea
Venise *see* Venezia
Venlo *65 D5 prev.* Venloo. Limburg, SE Netherlands
Venloo *see* Venlo
Venta *84 B3 Ger.* Windau. *river* Latvia/Lithuania
Venta Belgarum *see* Winchester
Ventia *see* Valence
Ventimiglia *74 A3* Liguria, NW Italy
Ventspils *84 B2 Ger.* Windau. Ventspils, NW Latvia
Vera *42 D3* Santa Fe, C Argentina
Veracruz *29 F4 var.* Veracruz Llave. Veracruz-Llave, E Mexico
Veracruz Llave *see* Veracruz
Vercellae *see* Vercelli
Vercelli *74 A2 anc.* Vercellae. Piemonte, NW Italy
Verdal *see* Verdalsøra
Verdalsøra *62 B4 var.* Verdal. Nord-Trøndelag, C Norway
Verde, Cabo *see* Cape Verde
Verde, Costa *70 D1 coastal region* N Spain
Verden *72 B3* Niedersachsen, NW Germany
Veria *see* Véroia
Verkhnedvinsk *see* Vyerkhnyadzvinsk
Verkhneudinsk *see* Ulan-Ude
Verkhoyanskiy Khrebet *93 F3 mountain range* NE Russian Federation
Vermillion *23 F3* South Dakota, N USA
Vermont *19 F2 off.* State of Vermont, *also known as* Green Mountain State. *state* NE USA
Vernal *22 B4* Utah, W USA
Vernon *27 F2* Texas, SW USA
Verőcze *see* Virovitica
Véroia *82 B4 var.* Veria, Vérroia, *Turk.* Karaferiye. Kentrikí Makedonía, N Greece
Verona *74 C2* Veneto, NE Italy
Vérroia *see* Véroia
Versailles *68 D1* Yvelines, N France
Versecz *see* Vršac
Verulamium *see* St Albans
Verviers *65 D6* Liège, E Belgium
Vesdre *65 D6 river* E Belgium
Veselinovo *82 D2* Shumen, NE Bulgaria
Vesontio *see* Besançon
Vesoul *68 D4 anc.* Vesulium, Vesulum. Haute-Saône, E France
Vesterålen *62 B2 island group* N Norway
Vestfjorden *62 C3 fjord* C Norway
Vestmannaeyjar *61 E5* Sudhurland, S Iceland
Vesulium/Vesulum *see* Vesoul
Vesuna *see* Périgueux
Vesuvio *75 D5 Eng.* Vesuvius. *volcano* S Italy
Vesuvius *see* Vesuvio
Veszprém *77 C7 Ger.* Veszprim. Veszprém, W Hungary
Veszprim *see* Veszprém
Vetrino *see* Suvorovo
Vetrino *see* Vyetryna
Vetter, Lake *see* Vättern
Veurne *65 A5 var.* Furnes. West-Vlaanderen, W Belgium
Vexiö *see* Växjö
Viacha *39 F4* La Paz, W Bolivia
Viana de Castelo *see* Viana do Castelo
Viana do Castelo *70 B2 var.* Viana de Castelo; *anc.* Velobriga. Viana do Castelo, NW Portugal
Vianen *64 C4* Utrecht, C Netherlands
Viangchan *114 C4 Eng./Fr.* Vientiane. *country capital* (Laos) C Laos
Viangphoukha *114 C3 var.* Vieng Pou Kha. Louang Namtha, N Laos
Viareggio *74 B3* Toscana, C Italy
Viborg *63 A7* Viborg, NW Denmark
Vic *71 G2 var.* Vich; *anc.* Ausa, Vicus Ausonensis. Cataluña, NE Spain
Vicentia *see* Vicenza
Vicenza *74 C2 anc.* Vicentia. Veneto, NE Italy
Vich *see* Vic
Vichy *69 C5* Allier, C France
Vicksburg *20 B2* Mississippi, S USA
Victoria *57 H1 country capital* (Seychelles) Mahé, SW Seychelles
Victoria *14 D5 province capital* Vancouver Island, British Columbia, SW Canada
Victoria *80 A5 var.* Rabat. Gozo, NW Malta
Victoria *27 G4* Texas, SW USA
Victoria *127 C7 state* SE Australia
Victoria *see* Masvingo, Zimbabwe
Victoria Bank *see* Vitória Seamount
Victoria de Durango *see* Durango
Victoria de las Tunas *see* Las Tunas
Victoria Falls *56 C3* Matabeleland North, W Zimbabwe
Victoria Falls *56 C2 waterfall* Zambia/Zimbabwe
Victoria Falls *see* Iguaçu, Saltos do
Victoria Island *15 F3 island* Northwest Territories/Nunavut, NW Canada
Victoria, Lake *51 B6 var.* Victoria Nyanza. *lake* E Africa
Victoria Land *132 C4 physical region* Antarctica
Victoria Nyanza *see* Victoria, Lake
Victoria River *124 D3 river* Northern Territory, N Australia
Victorville *25 C7* California, W USA
Vicus Ausonensis *see* Vic
Vicus Elbii *see* Viterbo
Vidalia *21 E2* Georgia, SE USA
Videm-Krško *see* Krško
Vídeň *see* Wien
Vidin *82 B1 anc.* Bononia. Vidin, NW Bulgaria
Vidzy *85 C5* Vitsyebskaya Voblasts', NW Belarus
Viedma *43 C5* Río Negro, E Argentina
Vieja, Sierra *26 D3 mountain range* Texas, SW USA
Vieng Pou Kha *see* Viangphoukha
Vienna *see* Wien, Austria
Vienna *see* Vienne, France
Vienne *69 D5 anc.* Vienna. Isère, E France
Vienne *68 B4 river* W France
Vientiane *see* Viangchan
Vientos, Paso de los *see* Windward Passage
Vierzon *68 C4* Cher, C France
Viesīte *84 C4 Ger.* Eckengraf. Jēkabpils, S Latvia
Vietnam *114 D4 off.* Socialist Republic of Vietnam, *Vtn.* Công Hoa Xa Hôi Chu Nghia Viêt Nam. *country* SE Asia
Vietnam, Socialist Republic of *see* Vietnam
Vietri *see* Viêt Tri
Viêt Tri *114 D3 var.* Vietri. Vinh Phu, N Vietnam
Vieux Fort *33 F2* S Saint Lucia
Vigo *70 B2* Galicia, NW Spain
Viipuri *see* Vyborg
Vijayawāda *110 D1 prev.* Bezwada. Andhra Pradesh, SE India
Vila *see* Port-Vila
Vila Artur de Paiva *see* Cubango
Vila da Ponte *see* Cubango
Vila de João Belo *see* Xai-Xai
Vila de Mocímboa da Praia *see* Mocímboa da Praia
Vila do Conde *70 B2* Porto, NW Portugal
Vila do Zumbo *56 D2 prev.* Vila do Zumbu, Zumbo. Tete, NW Mozambique
Vila do Zumbu *see* Vila do Zumbo
Vilafranca del Penedès *71 G2 var.* Villafranca del Panadés. Cataluña, NE Spain
Vila General Machado *see* Camacupa
Vila Henrique de Carvalho *see* Saurimo
Viļaka *84 D4 Ger.* Marienhausen. Balvi, NE Latvia
Vilalba *70 C1* Galicia, NW Spain
Vila Marechal Carmona *see* Uíge
Vila Nova de Gaia *70 B2* Porto, NW Portugal
Vila Nova de Portimão *see* Portimão
Vila Pereira de Eça *see* N'Giva
Vila Real *70 C2 var.* Vila Rial. Vila Real, N Portugal
Vila Rial *see* Vila Real
Vila Robert Williams *see* Caála
Vila Salazar *see* N'Dalatando
Vila Serpa Pinto *see* Menongue
Vileyka *see* Vilyeyka
Vilhelmina *62 C4* Västerbotten, N Sweden
Vilhena *40 D3* Rondônia, W Brazil
Vília *83 C5* Attikí, C Greece
Viliya *85 C5 Lith.* Neris. *river* W Belarus
Viliya *see* Neris
Viljandi *84 D2 Ger.* Fellin. Viljandimaa, S Estonia
Vilkaviškis *84 B4 Pol.* Wyłkowyszki. Marijampolė, SW Lithuania
Villa Acuña *28 D2 var.* Ciudad Acuña. Coahuila, NE Mexico
Villa Bella *39 F2* Beni, N Bolivia
Villacarrillo *71 E4* Andalucía, S Spain
Villa Cecilia *see* Ciudad Madero
Villach *73 D7 Slvn.* Beljak. Kärnten, S Austria
Villacidro *75 A5* Sardegna, Italy, C Mediterranean Sea
Villa Concepción *see* Concepción
Villa del Pilar *see* Pilar
Villafranca de los Barros *70 C4* Extremadura, W Spain
Villafranca del Panadés *see* Vilafranca del Penedès
Villahermosa *29 G4 prev.* San Juan Bautista. Tabasco, SE Mexico
Villajoyosa *71 F4 Cat.* La Vila Joíosa. País Valenciano, E Spain
Villa María *42 C4* Córdoba, C Argentina
Villa Martín *39 F5* Potosí, SW Bolivia
Villa Mercedes *42 C4* San Juan, Argentina
Villanueva *28 D3* Zacatecas, C Mexico
Villanueva de la Serena *70 C3* Extremadura, W Spain
Villanueva de los Infantes *71 E4* Castilla-La Mancha, C Spain
Villarrica *42 D2* Guairá, SE Paraguay
Villavicencio *36 B3* Meta, C Colombia
Villaviciosa *70 D1* Asturias, N Spain
Villazón *39 G5* Potosí, S Bolivia
Villena *71 F4* País Valenciano, E Spain
Villeurbanne *69 D5* Rhône, E France
Villingen-Schwenningen *73 B6* Baden-Württemberg, S Germany
Villmanstrand *see* Lappeenranta
Vilna *see* Vilnius
Vilnius *85 C5 Pol.* Wilno, *Ger.* Wilna; *prev. Rus.* Vilna. *country capital* (Lithuania) Vilnius, SE Lithuania
Vil'shanka *87 E3 Rus.* Olshanka. Kirovohrads'ka Oblast', C Ukraine
Vilvoorde *65 C6 Fr.* Vilvorde. Vlaams Brabant, C Belgium
Vilvorde *see* Vilvoorde
Vilyeyka *85 C5 Pol.* Wilejka, *Rus.* Vileyka. Minskaya Voblasts', NW Belarus
Vilyuy *93 F3 river* NE Russian Federation
Viña del Mar *42 B4* Valparaíso, C Chile
Vinaròs *71 F3* País Valenciano, E Spain
Vincennes *18 B4* Indiana, N USA
Vindhya Mountains *see* Vindhya Range
Vindhya Range *112 D4 var.* Vindhya Mountains. *mountain range* N India
Vindobona *see* Wien
Vineland *19 F4* New Jersey, NE USA
Vinh *114 D4* Nghê An, N Vietnam
Vinh Loi *see* Bac Liêu
Vinishte *82 C2* Montana, NW Bulgaria
Vinita *27 G1* Oklahoma, C USA
Vinkovci *78 C3 Ger.* Winkowitz, *Hung.* Vinkovcze. Vukovar-Srijem, E Croatia
Vinkovcze *see* Vinkovci
Vinnitsa *see* Vinnytsya
Vinnytsya *86 D2 Rus.* Vinnitsa. Vinnyts'ka Oblast', C Ukraine
Vinogradov *see* Vynohradiv
Vinson Massif *132 A3 mountain* Antarctica
Viranşehir *95 E4* Şanlıurfa, SE Turkey
Vîrful Moldoveanu *see* Vârful Moldoveanu
Virginia *23 G1* Minnesota, N USA
Virginia *19 E5 off.* Commonwealth of Virginia, *also known as* Mother of Presidents, Mother of States, Old Dominion. *state* NE USA
Virginia Beach *19 F5* Virginia, NE USA
Virgin Islands *see* British Virgin Islands
Virgin Islands (US) *33 F3 var.* Virgin Islands of the United States; *prev.* Danish West Indies. *US unincorporated territory* E West Indies
Virgin Islands of the United States *see* Virgin Islands (US)
Virôchey *115 E5* Rôtânôkiri, NE Cambodia
Virovitica *78 C2 Ger.* Virovititz, *Hung.* Verőcze; *prev. Ger.* Werowitz. Virovitica-Podravina, NE Croatia
Virovititz *see* Virovitica
Virton *65 D8* Luxembourg, SE Belgium
Virtsu *84 D2 Ger.* Werder. Läänemaa, W Estonia
Vis *78 B4 It.* Lissa; *anc.* Issa. *island* S Croatia
Vis *see* Fish
Visaginas *84 C4 prev.* Sniečkus. Utena, E Lithuania
Visākhapatnam *113 E5 var.* Vishakhapatnam. Andhra Pradesh, SE India
Visalia *25 C6* California, W USA
Visby *63 C7 Ger.* Wisby. Gotland, SE Sweden
Viscount Melville Sound *15 F2 prev.* Melville Sound. *sound* Northwest Territories, N Canada
Visé *65 D6* Liège, E Belgium
Viseu *70 C2 prev.* Vizeu. Viseu, N Portugal
Vishakhapatnam *see* Visākhapatnam
Vislinskiy Zaliv *see* Vistula Lagoon
Visoko *78 C4* Federacija Bosna I Hercegovina, C Bosnia and Herzegovina
Visttasjohka *62 D3 river* N Sweden
Vistula *76 C2 Eng.* Vistula, *Ger.* Weichsel. *river* C Poland
Vistula *see* Wisła
Vistula Lagoon *76 C2 Ger.* Frisches Haff, *Pol.* Zalew Wiślany, *Rus.* Vislinskiy Zaliv. *lagoon* Poland/Russian Federation
Vitebsk *see* Vitsyebsk
Viterbo *74 C4 anc.* Vicus Elbii. Lazio, C Italy
Viti *see* Fiji
Viti Levu *123 E4 island* W Fiji
Vitim *93 F4 river* C Russian Federation
Vitória *41 F4 state capital* Espírito Santo, SE Brazil
Vitoria *see* Vitoria-Gasteiz
Vitoria Bank *see* Vitória Seamount
Vitória da Conquista *41 F3* Bahia, E Brazil
Vitoria-Gasteiz *71 E1 var.* Vitoria, *Eng.* Vittoria. País Vasco, N Spain
Vitória Seamount *45 B5 var.* Victoria Bank, Vitoria Bank. *seamount* C Atlantic Ocean
Vitré *68 B3* Ille-et-Vilaine, NW France
Vitsyebsk *85 E5 Rus.* Vitebsk. Vitsyebskaya Voblasts', NE Belarus
Vittoria *75 C7* Sicilia, Italy, C Mediterranean Sea
Vittoria *see* Vitoria-Gasteiz
Vizcaya, Golfo de *see* Biscay, Bay of
Vizianagaram *113 E5 var.* Vizianagram. Andhra Pradesh, E India
Vizianagram *see* Vizianagaram
Vjosës, Lumi i *79 C7 var.* Vijosa, Vijosë, *Gk.* Aóos. *river* Albania/Greece
Vlaanderen *see* Flanders
Vlaardingen *64 B4* Zuid-Holland, SW Netherlands
Vladikavkaz *89 B8 prev.* Dzaudzhikau, Ordzhonikidze. Respublika Severnaya Osetiya, SW Russian Federation
Vladimir *89 B5* Vladimirskaya Oblast', W Russian Federation
Vladimirovka *see* Yuzhno-Sakhalinsk
Vladimir-Volynskiy *see* Volodymyr-Volyns'kyy
Vladivostok *93 G5* Primorskiy Kray, SE Russian Federation
Vlagtwedde *64 E2* Groningen, NE Netherlands
Vlasotince *79 E5* Serbia, SE Serbia
Vlieland *64 C1 Fris.* Flylân. *island* Waddeneilanden, N Netherlands
Vlijmen *64 C4* Noord-Brabant, S Netherlands
Vlissingen *65 B5 Eng.* Flushing, *Fr.* Flessingue. Zeeland, SW Netherlands
Vlodava *see* Włodawa
Vlonë/Vlora *see* Vlorë
Vlorë *79 C7 prev.* Vlonë, *It.* Valona, Vlora. Vlorë, SW Albania
Vlotslavsk *see* Włocławek
Vöcklabruck *73 D6* Oberösterreich, NW Austria
Vogelkop *see* Doberai, Jazirah
Vohimena, Tanjona *57 F4 Fr.* Cap Sainte Marie. *headland* S Madagascar
Voiron *69 D5* Isère, E France
Vojvodina *78 D3 Ger.* Wojwodina. Vojvodina, N Serbia
Volga *89 B7 river* NW Russian Federation
Volga Uplands *see* Privolzhskaya Vozvyshennost'
Volgodonsk *89 B7* Rostovskaya Oblast', SW Russian Federation
Volgograd *89 B7 prev.* Stalingrad, Tsaritsyn. Volgogradskaya Oblast', SW Russian Federation
Volkhov *88 B4* Leningradskaya Oblast', NW Russian Federation
Volkovysk *see* Vawkavysk
Volmari *see* Valmiera
Volnovakha *87 G3* Donets'ka Oblast', SE Ukraine
Volodymyr-Volyns'kyy *86 C1 Pol.* Włodzimierz, *Rus.* Vladimir-Volynskiy. Volyns'ka Oblast', NW Ukraine
Vologda *88 B4* Vologodskaya Oblast', W Russian Federation
Vólos *83 B5* Thessalía, C Greece
Volozhin *see* Valozhyn
Vol'sk *89 C6* Saratovskaya Oblast', W Russian Federation
Volta *53 E5 river* SE Ghana
Volta Blanche *see* White Volta
Volta, Lake *53 E5 reservoir* SE Ghana
Volta Noire *see* Black Volta
Volturno *75 D5 river* S Italy
Volunteer Island *see* Starbuck Island
Volzhskiy *89 B6* Volgogradskaya Oblast', SW Russian Federation
Võnnu *84 E3 Ger.* Wendau. Tartumaa, SE Estonia
Voorst *64 D3* Gelderland, E Netherlands
Voranava *85 C5 Pol.* Werenów, *Rus.* Voronovo. Hrodzyenskaya Voblasts', W Belarus
Vorderrhein *73 B7 river* SE Switzerland
Vóreies Sporádes *83 C5 var.* Vóreioi Sporádes, Vórioi Sporádhes, *Eng.* Northern Sporades. *island group* E Greece
Vóreioi Sporádes *see* Vóreies Sporádes
Vórioi Sporádhes *see* Vóreies Sporádes
Vorkuta *92 C2* Respublika Komi, NW Russian Federation
Vormsi *84 C2 var.* Vormsi Saar, *Ger.* Worms, *Swed.* Ormsö. *island* W Estonia
Vormsi Saar *see* Vormsi
Voronezh *89 B6* Voronezhskaya Oblast', W Russian Federation
Voronovo *see* Voranava
Voroshilov *see* Ussuriysk
Voroshilovgrad *see* Luhans'k, Ukraine
Voroshilovsk *see* Stavropol', Russian Federation
Võru *84 D3 Ger.* Werro. Võrumaa, SE Estonia
Vosges *68 E4 mountain range* NE France
Vostochno-Sibirskoye More *93 F1 Eng.* East Siberian Sea. *sea* Arctic Ocean
Vostochnyy Sayan *93 E4 Mong.* Dzüün Soyonï Nuruu, *Eng.* Eastern Sayans. *mountain range* Mongolia/Russian Federation
Vostock Island *see* Vostok Island
Vostok *132 C3 Russian research station* Antarctica
Vostok Island *123 G3 var.* Vostock Island; *prev.* Stavers Island. *island* Line Islands, SE Kiribati
Voznesens'k *87 E3 Rus.* Voznesensk. Mykolayivs'ka Oblast', S Ukraine
Vranje *79 E5* Serbia, SE Serbia
Vranov *see* Vranov nad Topl'ou
Vranov nad Topl'ou *77 D5 var.* Vranov, *Hung.* Varannó. Prešovský Kraj, E Slovakia
Vratsa *82 C2* Vratsa, NW Bulgaria
Vrbas *78 C3* Vojvodina, NW Serbia
Vrbas *78 C3 river* N Bosnia and Herzegovina
Vršac *78 E3 Ger.* Werschetz, *Hung.* Versecz. Vojvodina, NE Serbia
Vsetín *77 C5 Ger.* Wsetin. Zlínský Kraj, E Czech Republic
Vučitrn *see* Vushtrri
Vukovar *78 C3 Hung.* Vukovár. Vukovar-Srijem, E Croatia
Vulcano, Isola *75 C7 island* Isole Eolie, S Italy
Vung Tau *115 E6 prev. Fr.* Cape Saint Jacques, Cap Saint-Jacques. Ba Ria-Vung Tau, S Vietnam
Vushtrri *79 D5 Serb.* Vučitrn. N Kosovo
Vyatka *89 C5 river* NW Russian Federation
Vyatka *see* Kirov
Vyborg *88 B3 Fin.* Viipuri. Leningradskaya Oblast', NW Russian Federation
Vyerkhnyadzvinsk *85 D5 Rus.* Verkhnedvinsk. Vitsyebskaya Voblasts', N Belarus

Vyetryna – Yalpug, Ozero

Vyetryna *85 D5 Rus.* Vetrino. Vitsyebskaya Voblasts', N Belarus
Vynohradiv *86 B3 Cz.* Sevluš, *Hung.* Nagyszöllös, *Rus.* Vinogradov; *prev.* Sevlyush. Zakarpats'ka Oblast', W Ukraine

W

Wa *53 E4* NW Ghana
Waag *see* Váh
Waagbistritz *see* Považská Bystrica
Waal *64 C4 river* S Netherlands
Wabash *18 C4* Indiana, N USA
Wabash River *18 B5 river* N USA
Waco *27 G3* Texas, SW USA
Wad Al-Hajarah *see* Guadalajara
Waddān *49 F3* NW Libya
Waddeneilanden *64 C1 Eng.* West Frisian Islands. *island group* N Netherlands
Waddenzee *64 C1 var.* Wadden Zee. *sea* SE North Sea
Wadden Zee *see* Waddenzee
Waddington, Mount *14 D5 mountain* British Columbia, SW Canada
Wādī as Sīr *97 B6 var.* Wadi es Sir. 'Ammān, NW Jordan
Wadi es Sir *see* Wādī as Sīr
Wadi Halfa *50 B3 var.* Wādī Ḥalfā'. Northern, N Sudan
Wādī Mūsā *97 B7 var.* Petra. Ma'ān, S Jordan
Wad Madanī *see* Wad Medani
Wad Medani *50 C4 var.* Wad Madanī. Gezira, C Sudan
Waflia *117 F4* Pulau Buru, E Indonesia
Wagadugu *see* Ouagadougou
Wagga Wagga *127 C7* New South Wales, SE Australia
Wagin *125 B7* Western Australia
Wāh *112 C1* Punjab, NE Pakistan
Wahai *117 F4* Pulau Seram, E Indonesia
Wahaybah, Ramlat Al *see* Wahībah, Ramlat Āl
Wahiawā *25 A8 var.* Wahiawa. O'ahu, Hawaii, USA, C Pacific Ocean
Wahībah, Ramlat Ahl *see* Wahībah, Ramlat Āl
Wahībah Sands *99 E5 var.* Ramlat Ahl Wahībah, Ramlat Al Wahaybah, *Eng.* Wahībah Sands. *desert* N Oman
Wahībah Sands *see* Wahībah, Ramlat Āl
Wahpeton *23 F2* North Dakota, N USA
Wahran *see* Oran
Waiau *129 A7 river* South Island, New Zealand
Waigeo, Pulau *117 G4 island* Maluku, E Indonesia
Waikaremoana, Lake *128 E4 lake* North Island, New Zealand
Wailuku *25 B8* Maui, Hawaii, USA, C Pacific Ocean
Waimate *129 B6* Canterbury, South Island, New Zealand
Waiouru *128 D4* Manawatu-Wanganui, North Island, New Zealand
Waipara *129 C6* Canterbury, South Island, New Zealand
Waipawa *128 E4* Hawke's Bay, North Island, New Zealand
Waipukurau *128 D4* Hawke's Bay, North Island, New Zealand
Wairau *129 C5 river* South Island, New Zealand
Wairoa *128 E4* Hawke's Bay, North Island, New Zealand
Wairoa *128 D2 river* North Island, New Zealand
Waitaki *129 B6 river* South Island, New Zealand
Waitara *128 D4* Taranaki, North Island, New Zealand
Waitzen *see* Vác
Waiuku *128 D3* Auckland, North Island, New Zealand
Wakasa-wan *109 C6 bay* C Japan
Wakatipu, Lake *129 A7 lake* South Island, New Zealand
Wakayama *109 C6* Wakayama, Honshū, SW Japan
Wake Island *130 C2 US unincorporated territory* NW Pacific Ocean
Wake Island *120 D1 atoll* NW Pacific Ocean
Wakkanai *108 C1* Hokkaidō, NE Japan
Walachei/Walachia *see* Wallachia
Wałbrzych *76 B4 Ger.* Waldenburg, Waldenburg in Schlesien. Dolnośląskie, SW Poland
Walcourt *65 C7* Namur, S Belgium
Wałcz *76 B3 Ger.* Deutsch Krone. Zachodnio-pomorskie, NW Poland
Waldenburg/Waldenburg in Schlesien *see* Wałbrzych
Waldia *see* Weldiya
Wales *14 C2* Alaska, USA
Wales *67 C6 Wel.* Cymru. *cultural region* Wales, United Kingdom
Walgett *127 D5* New South Wales, SE Australia
Walk *see* Valga, Estonia
Walk *see* Valka, Latvia
Walker Lake *25 C5 lake* Nevada, W USA
Wallachia *86 B5 var.* Walachia, *Ger.* Walachei, *Rom.* Valachia. *cultural region* S Romania
Walla Walla *24 C2* Washington, NW USA
Wallenthal *see* Hațeg
Wallis and Futuna *123 E4 Fr.* Territoire de Wallis et Futuna. *French overseas territory* C Pacific Ocean
Wallis et Futuna, Territoire de *see* Wallis and Futuna
Walnut Ridge *20 B1* Arkansas, C USA
Waltenberg *see* Zalău
Walthamstow *67 B7* Waltham Forest, SE England, United Kingdom
Walvisbaai *see* Walvis Bay
Walvis Bay *56 A4 Afr.* Walvisbaai. Erongo, NW Namibia
Walvish Ridge *see* Walvis Ridge
Walvis Ridge *47 B7 var.* Walvish Ridge. *undersea ridge* E Atlantic Ocean
Wan *see* Anhui
Wanaka *129 B6* Otago, South Island, New Zealand
Wanaka, Lake *129 A6 lake* South Island, New Zealand
Wanchuan *see* Zhangjiakou
Wandel Sea *61 E1 sea* Arctic Ocean
Wandsworth *67 A8* Wandsworth, SE England, United Kingdom
Wanganui *128 D4* Manawatu-Wanganui, North Island, New Zealand
Wangaratta *127 C7* Victoria, SE Australia
Wankie *see* Hwange
Wanki, Río *see* Coco, Río
Wanlaweyn *51 D6 var.* Wanle Weyn, *It.* Uanle Uen. Shabeellaha Hoose, SW Somalia
Wanle Weyn *see* Wanlaweyn
Wanxian *see* Wanzhou
Wanzhou *106 B5 var.* Wanxian. Chongqing Shi, C China
Warangal *113 E5* Telangana, C India
Warasdin *see* Varaždin
Warburg *72 B4* Nordrhein-Westfalen, W Germany
Ware *15 E4* British Columbia, W Canada
Waremme *65 C6* Liège, E Belgium
Waren *72 C3* Mecklenburg-Vorpommern, NE Germany
Wargla *see* Ouargla
Warkworth *128 D2* Auckland, North Island, New Zealand
Warnemünde *72 C2* Mecklenburg-Vorpommern, NE Germany
Warner *27 G1* Oklahoma, C USA
Warnes *39 G4* Santa Cruz, C Bolivia
Warrego River *127 C5 seasonal river* New South Wales/Queensland, E Australia
Warren *18 D3* Michigan, N USA
Warren *18 D3* Ohio, N USA
Warren *19 E3* Pennsylvania, NE USA
Warri *53 F5* Delta, S Nigeria
Warrnambool *127 B7* Victoria, SE Australia
Warsaw/Warschau *see* Warszawa
Warszawa *76 D3 Eng.* Warsaw, *Ger.* Warschau, *Rus.* Varshava. *country capital* (Poland) Mazowieckie, C Poland
Warta *76 B3 Ger.* Warthe. *river* W Poland
Wartberg *see* Senec
Warthe *see* Warta
Warwick *127 E5* Queensland, E Australia
Washington *22 A2* NE England, United Kingdom
Washington D.C. *19 E4 country capital* (USA) District of Columbia, NE USA
Washington Island *see* Teraina
Washington, Mount *19 G2 mountain* New Hampshire, NE USA
Wash, The *67 E6 inlet* E England, United Kingdom
Wasiliszki *see* Vasilishki
Waspam *31 E2 var.* Waspán. Región Autónoma Atlántico Norte, NE Nicaragua
Waspán *see* Waspam
Watampone *117 E4 var.* Bone. Sulawesi, C Indonesia
Watenstedt-Salzgitter *see* Salzgitter
Waterbury *19 F3* Connecticut, NE USA
Waterford *67 B6 Ir.* Port Láirge. Waterford, S Ireland
Waterloo *23 G3* Iowa, C USA
Watertown *19 F2* New York, NE USA
Watertown *23 F2* South Dakota, N USA
Waterville *19 G2* Maine, NE USA
Watford *67 A7* E England, United Kingdom
Watlings Island *see* San Salvador
Watsa *55 E5* Orientale, NE Dem. Rep. Congo
Watts Bar Lake *20 D1 reservoir* Tennessee, S USA
Wau *51 B5 var.* Wāw. Western Bahr el Ghazal, C South Sudan
Waukegan *18 B3* Illinois, N USA
Waukesha *18 B3* Wisconsin, N USA
Wausau *18 B2* Wisconsin, N USA
Waverly *23 G3* Iowa, C USA
Wavre *65 C6* Walloon Brabant, C Belgium
Wāw *see* Wau
Wawa *16 C4* Ontario, S Canada
Waycross *21 E3* Georgia, SE USA
Wearmouth *see* Sunderland
Webfoot State *see* Oregon
Webster City *23 F3* Iowa, C USA
Weddell Plain *132 A2 abyssal plain* SW Atlantic Ocean
Weddell Sea *132 A2 sea* SW Atlantic Ocean
Weener *72 A3* Niedersachsen, NW Germany
Weert *65 D5* Limburg, SE Netherlands
Weesp *64 C3* Noord-Holland, C Netherlands
Węgorzewo *76 D2 Ger.* Angerburg. Warmińsko-Mazurskie, NE Poland
Weichsel *see* Wisła
Weimar *72 C4* Thüringen, C Germany
Weissenburg *see* Alba Iulia, Romania
Weissenburg in Bayern *73 C6* Bayern, SE Germany
Weissenstein *see* Paide
Weisskirchen *see* Bela Crkva
Weisswampach *65 D7* Diekirch, N Luxembourg
Wejherowo *76 C2* Pomorskie, NW Poland
Welchman Hall *33 G1* C Barbados
Weldiya *50 C4 var.* Waldia, *It.* Valdia. Āmara, N Ethiopia
Welkom *56 D4* Free State, C South Africa
Welle *see* Uele
Wellesley Islands *126 B2 island group* Queensland, N Australia
Wellington *129 D5 country capital* (New Zealand) Wellington, North Island, New Zealand
Wellington *23 F5* Kansas, C USA
Wellington *see* Wellington, Isla
Wellington, Isla *43 A7 var.* Wellington. *island* S Chile
Wells *24 D4* Nevada, W USA
Wellsford *128 D2* Auckland, North Island, New Zealand
Wells, Lake *125 C5 lake* Western Australia
Wels *73 D6 anc.* Ovilava. Oberösterreich, N Austria
Wembley *67 A8* Alberta, W Canada
Wemmel *65 B6* Vlaams Brabant, C Belgium
Wenatchee *24 B2* Washington, NW USA
Wenchi *53 E4* W Ghana
Wen-chou/Wenchow *see* Wenzhou
Wendau *see* Võnnu
Wenden *see* Cēsis
Wenzhou *106 D5 var.* Wen-chou, Wenchow. Zhejiang, SE China
Werda *56 C4* Kgalagadi, S Botswana
Werder *see* Virtsu
Werenów *see* Voranava
Werkendam *64 C4* Noord-Brabant, S Netherlands
Werowitz *see* Virovitica
Werro *see* Võru
Werschetz *see* Vršac
Wesenberg *see* Rakvere
Weser *72 B3 river* NW Germany
Wessel Islands *126 B1 island group* Northern Territory, N Australia
West Antarctica *132 A3 var.* Lesser Antarctica. *physical region* Antarctica
West Australian Basin *see* Wharton Basin
West Bank *97 A6 disputed region* SW Asia
West Bend *18 B3* Wisconsin, N USA
West Bengal *113 F4 cultural region* NE India
West Cape *129 A7 headland* South Island, New Zealand
West Des Moines *23 F3* Iowa, C USA
Westerland *72 B2* Schleswig-Holstein, N Germany
Western Australia *124 B4 state* W Australia
Western Bug *see* Bug
Western Carpathians *77 E7 mountain range* W Romania Europe
Western Desert *see* Şaḩrā' al Gharbīyah
Western Dvina *63 E7 Bel.* Dzvina, *Ger.* Düna, *Latv.* Daugava, *Rus.* Zapadnaya Dvina. *river* W Europe
Western Ghats *112 C5 mountain range* SW India
Western Isles *see* Outer Hebrides
Western Punjab *see* Punjab
Western Sahara *48 B3 disputed territory* N Africa
Western Samoa *see* Samoa
Western Samoa, Independent State of *see* Samoa
Western Sayans *see* Zapadnyy Sayan
Western Scheldt *see* Westerschelde
Western Sierra Madre *see* Madre Occidental, Sierra
Westerschelde *65 B5 Eng.* Western Scheldt; *prev.* Honte. *inlet* S North Sea
West Falkland *43 C7 var.* Gran Malvina, Isla Gran Malvina. *island* W Falkland Islands
West Fargo *23 F2* North Dakota, N USA
West Frisian Islands *see* Waddeneilanden
West Irian *see* Papua
Westliche Morava *see* Zapadna Morava
West Mariana Basin *120 B1 var.* Perece Vela Basin. *undersea feature* W Pacific Ocean
West Memphis *20 B1* Arkansas, C USA
West New Guinea *see* Papua
Weston-super-Mare *67 D7* SW England, United Kingdom
West Palm Beach *21 F4* Florida, SE USA
West Papua *see* Papua
Westport *129 C5* West Coast, South Island, New Zealand
West Punjab *see* Punjab
West River *see* Xi Jiang
West Siberian Plain *see* Zapadno-Sibirskaya Ravnina
West Virginia *18 D4 off.* State of West Virginia, *also known as* Mountain State. *state* NE USA
Wetar, Pulau *117 F5 island* Kepulauan Damar, E Indonesia
Wetzlar *73 B5* Hessen, W Germany
Wevok *14 C2 var.* Wewuk. Alaska, USA
Wewuk *see* Wevok
Wexford *67 B6 Ir.* Loch Garman. SE Ireland
Weyburn *15 F5* Saskatchewan, S Canada
Weymouth *67 D7* S England, United Kingdom
Wezep *64 D3* Gelderland, E Netherlands
Whakatane *128 E3* Bay of Plenty, North Island, New Zealand
Whale Cove *15 G3* Nunavut, C Canada
Whangarei *128 D2* Northland, North Island, New Zealand
Wharton Basin *119 D5 var.* West Australian Basin. *undersea feature* E Indian Ocean
Whataroa *129 B6* West Coast, South Island, New Zealand
Wheatland *22 D3* Wyoming, C USA
Wheeler Peak *26 D1 mountain* New Mexico, SW USA
Wheeling *18 D4* West Virginia, NE USA
Whitby *67 D5* N England, United Kingdom
Whitefish *22 B1* Montana, NW USA
Whitehaven *67 C5* NW England, United Kingdom
Whitehorse *14 D4 territory capital* Yukon Territory, W Canada
White Nile *50 B4 Ar.* Al Baḩr al Abyaḑ, An Nīl al Abyaḑ, Bahr el Jebel. *river* C South Sudan
White River *22 D3 river* South Dakota, N USA
White Sea *see* Beloye More
White Volta *53 E4 var.* Nakambé, *Fr.* Volta Blanche. *river* Burkina/Ghana
Whitianga *128 D2* Waikato, North Island, New Zealand
Whitney, Mount *25 C6 mountain* California, W USA
Whitsunday Group *126 D3 island group* Queensland, E Australia
Whyalla *127 B6* South Australia
Wichita *23 F5* Kansas, C USA
Wichita Falls *27 F2* Texas, SW USA
Wichita River *27 F2 river* Texas, SW USA
Wickenburg *26 B2* Arizona, SW USA
Wicklow *67 B6 Ir.* Cill Mhantáin. *county* E Ireland
Wicklow Mountains *67 B6 Ir.* Sléibhte Chill Mhantáin. *mountain range* E Ireland
Wieliczka *77 D5* Małopolskie, S Poland
Wieluń *76 C4* Sieradz, C Poland
Wien *73 E6 Eng.* Vienna, *Hung.* Bécs, *Slvk.* Vídeň, *Slvn.* Dunaj; *anc.* Vindobona. *country capital* (Austria) Wien, NE Austria
Wiener Neustadt *73 E6* Niederösterreich, E Austria
Wierden *64 E3* Overijssel, E Netherlands
Wiesbaden *73 B5* Hessen, W Germany
Wieselburg and Ungarisch-Altenburg/ Wieselburg-Ungarisch-Altenburg *see* Mosonmagyaróvár
Wiesenhof *see* Ostrołęka
Wight, Isle of *67 D7 island* United Kingdom
Wigorna Ceaster *see* Worcester
Wijchen *64 D4* Gelderland, SE Netherlands
Wijk bij Duurstede *64 D4* Utrecht, C Netherlands
Wilcannia *127 C6* New South Wales, SE Australia
Wilejka *see* Vilyeyka
Wilhelm, Mount *122 B3 mountain* C Papua New Guinea
Wilhelm-Pieck-Stadt *see* Guben
Wilhelmshaven *72 B3* Niedersachsen, NW Germany
Wilia/Wilja *see* Neris
Wilkes Barre *19 F3* Pennsylvania, NE USA
Wilkes Land *132 C4 physical region* Antarctica
Wiłkomierz *see* Ukmergė
Willard *26 D2* New Mexico, SW USA
Willcox *26 C3* Arizona, SW USA
Willebroek *65 B5* Antwerpen, C Belgium
Willemstad *33 E5 dependent territory capital* (Curaçao) Lesser Antilles, S Caribbean Sea
Williston *22 D1* North Dakota, N USA
Wilmington *19 F4* Delaware, NE USA
Wilmington *21 F2* North Carolina, SE USA
Wilmington *18 C4* Ohio, N USA
Wilna/Wilno *see* Vilnius
Wilrijk *65 C5* Antwerpen, N Belgium
Winchester *67 D7 hist.* Wintanceaster, *Lat.* Venta Belgarum. S England, United Kingdom
Winchester *19 E4* Virginia, NE USA
Windau *see* Ventspils, Latvia
Windau *see* Venta, Latvia/Lithuania
Windhoek *56 B3 Ger.* Windhuk. *country capital* (Namibia) Khomas, C Namibia
Windhuk *see* Windhoek
Windorah *126 C4* Queensland, C Australia
Windsor *16 C5* Ontario, S Canada
Windsor *67 D7* S England, United Kingdom
Windsor *19 G3* Connecticut, NE USA
Windward Islands *33 H4 island group* E West Indies
Windward Islands *see* Barlavento, Ilhas de, Cape Verde
Windward Passage *32 D3 Sp.* Paso de los Vientos. *channel* Cuba/Haiti
Winisk *16 C2 river* Ontario, C Canada
Winkowitz *see* Vinkovci
Winnebago, Lake *18 B2 lake* Wisconsin, N USA
Winnemucca *25 C5* Nevada, W USA
Winnipeg *15 G5 province capital* Manitoba, S Canada
Winnipeg, Lake *15 G5 lake* Manitoba, C Canada
Winnipegosis, Lake *16 A3 lake* Manitoba, C Canada
Winona *23 G3* Minnesota, N USA
Winschoten *64 E2* Groningen, NE Netherlands
Winsen *72 B3* Niedersachsen, N Germany
Winston Salem *21 E1* North Carolina, SE USA
Winsum *64 D1* Groningen, NE Netherlands
Wintanceaster *see* Winchester
Winterswijk *64 E4* Gelderland, E Netherlands
Winterthur *73 B7* Zürich, NE Switzerland
Winton *126 C4* Queensland, E Australia
Winton *129 A7* Southland, South Island, New Zealand
Wisby *see* Visby
Wisconsin *18 A2 off.* State of Wisconsin, *also known as* Badger State. *state* N USA
Wisconsin Rapids *18 B2* Wisconsin, N USA
Wisconsin River *18 B3 river* Wisconsin, N USA
Wiślany, Zalew *see* Vistula Lagoon
Wismar *72 C2* Mecklenburg-Vorpommern, N Germany
Wittenberge *72 C3* Brandenburg, N Germany
Wittlich *73 A5* Rheinland-Pfalz, SW Germany
Wittstock *72 C3* Brandenburg, NE Germany
W. J. van Blommesteinmeer *37 G3 reservoir* E Suriname
Władysławowo *76 C2* Pomorskie, N Poland
Włocławek *76 C3 Ger./Rus.* Vlotslavsk. Kujawsko-pomorskie, C Poland
Włodawa *76 E4 Rus.* Vlodava. Lubelskie, SE Poland
Włodzimierz *see* Volodymyr-Volyns'kyy
Wlotzkasbaken *56 B3* Erongo, W Namibia
Wodonga *127 C7* Victoria, SE Australia
Wodzisław Śląski *77 C5 Ger.* Loslau. Śląskie, S Poland
Wojerecy *see* Hoyerswerda
Wōjjā *see* Wotje Atoll
Wojwodina *see* Vojvodina
Woking *67 D7* SE England, United Kingdom
Wolf, Isla *38 A4 island* Galápagos Islands, W Ecuador South America
Wolfsberg *73 D7* Kärnten, SE Austria
Wolfsburg *72 C3* Niedersachsen, N Germany
Wolgast *72 D2* Mecklenburg-Vorpommern, NE Germany
Wołkowysk *see* Vawkavysk
Wöllan *see* Velenje
Wollaston Lake *15 F4* Saskatchewan, C Canada
Wollongong *127 D6* New South Wales, SE Australia
Wolmar *see* Valmiera
Wołożyn *see* Valozhyn
Wolvega *64 D2 Fris.* Wolvegea. Friesland, N Netherlands
Wolvegea *see* Wolvega
Wolverhampton *67 D6* C England, United Kingdom
Wolverine State *see* Michigan
Wōnsan *107 E3* SE North Korea
Woodburn *24 B3* Oregon, NW USA
Woodland *25 B5* California, W USA
Woodruff *18 B2* Wisconsin, N USA
Woods, Lake of the *16 A3 Fr.* Lac des Bois. *lake* Canada/USA
Woodville *128 D4* Manawatu-Wanganui, North Island, New Zealand
Woodward *27 F1* Oklahoma, C USA
Worcester *56 C5* Western Cape, SW South Africa
Worcester *67 D6 hist.* Wigorna Ceaster. W England, United Kingdom
Worcester *19 G3* Massachusetts, NE USA
Workington *67 C5* NW England, United Kingdom
Worland *22 C3* Wyoming, C USA
Wormatia *see* Worms
Worms *73 B5 anc.* Augusta Vangionum, Borbetomagus, Wormatia. Rheinland-Pfalz, SW Germany
Worms *see* Vormsi
Worthington *23 F3* Minnesota, N USA
Wotje Atoll *122 D1 var.* Wōjjā. *atoll* Ratak Chain, E Marshall Islands
Woudrichem *64 C4* Noord-Brabant, S Netherlands
Wrangel Island *93 F1 Eng.* Wrangel Island. *island* NE Russian Federation
Wrangel Island *see* Vrangelya, Ostrov
Wrangel Plain *133 B2 undersea feature* Arctic Ocean
Wrocław *76 C4 Eng./Ger.* Breslau. Dolnośląskie, SW Poland
Września *76 C3* Wielkopolskie, C Poland
Wsetin *see* Vsetín
Wuchang *see* Wuhan
Wuday'ah *99 C6 spring/well* S Saudi Arabia
Wuhai *105 E3 var.* Haibowan. Nei Mongol Zizhiqu, N China
Wuhan *106 C5 var.* Han-kou, Han-k'ou, Hanyang, Wuchang, Wu-han; *prev.* Hankow. *province capital* Hubei, C China
Wu-han *see* Wuhan
Wuhsien *see* Suzhou
Wuhsi/Wu-his *see* Wuxi
Wuhu *106 D5 var.* Wu-na-mu. Anhui, E China
Wujlān *see* Ujelang Atoll
Wukari *53 G4* Taraba, E Nigeria
Wuliang Shan *106 A6 mountain range* SW China
Wu-lu-k'o-mu-shi/Wu-lu-mu-ch'i *see* Ürümqi
Wu-na-mu *see* Wuhu
Wuppertal *72 A4 prev.* Barmen-Elberfeld. Nordrhein-Westfalen, W Germany
Würzburg *73 B5* Bayern, SW Germany
Wusih *see* Wuxi
Wuxi *106 D5 var.* Wuhsi, Wu-hsi, Wusih. Jiangsu, E China
Wuyi Shan *103 E3 mountain range* SE China
Wye *67 C6 Wel.* Gwy. *river* England/Wales, United Kingdom
Wyłkowyszki *see* Vilkaviškis
Wyndham *124 D3* Western Australia
Wyoming *18 C3* Michigan, N USA
Wyoming *22 B3 off.* State of Wyoming, *also known as* Equality State. *state* C USA
Wyszków *76 D3 Ger.* Probstberg. Mazowieckie, NE Poland

X

Xaafuun, Raas *50 E4 var.* Ras Hafun. *cape* NE Somalia
Xaçmaz *95 H2 Rus.* Khachmas. N Azerbaijan
Xaignabouli *114 C4 prev.* Muang Xaignabouri, *Fr.* Sayaboury. Xaignabouli, N Laos
Xai-Xai *57 E4 prev.* João Belo, Vila de João Belo. Gaza, S Mozambique
Xalapa *29 F4* Veracruz-Llave, Mexico
Xam Nua *114 D3 var.* Sam Neua. Houaphan, N Laos
Xankändi *95 G3 Rus.* Khankendi; *prev.* Stepanakert. SW Azerbaijan
Xánthi *82 C3* Anatolikí Makedonía kai Thráki, NE Greece
Xàtiva *71 F3 Cas.* Xátiva; *anc.* Setabis, *var.* Jativa. País Valenciano, E Spain
Xauen *see* Chefchaouen
Xäzär Dänizi *see* Caspian Sea
Xeres *see* Jeréz de la Frontera
Xiaguan *see* Dali
Xiamen *106 D6 var.* Hsia-men; *prev.* Amoy. Fujian, SE China
Xi'an *106 C4 var.* Changan, Sian, Signan, Siking, Singan, Xian. *province capital* Shaanxi, C China
Xiang *see* Hunan
Xiangkhoang *see* Phônsaven
Xiangtan *106 C5 var.* Hsiang-t'an, Siangtan. Hunan, S China
Xiao Hinggan Ling *106 D2 Eng.* Lesser Khingan Range. *mountain range* NE China
Xichang *106 B5* Sichuan, C China
Xieng Khouang *see* Phônsaven
Xieng Ngeun *see* Muong Xiang Ngeun
Xigazê *104 C5 var.* Jih-k'a-tse, Shigatse, Xigaze. Xizang Zizhiqu, W China
Xi Jiang *102 D3 var.* Hsi Chiang, *Eng.* West River. *river* S China
Xilinhot *105 F2 var.* Silinhot. Nei Mongol Zizhiqu, N China
Xilokastro *see* Xylókastro
Xin *see* Xinjiang Uygur Zizhiqu
Xingkai Hu *see* Khanka, Lake
Xingu, Rio *41 E2 river* C Brazil
Xingxingxia *104 D3* Xinjiang Uygur Zizhiqu, NW China
Xining *105 E4 var.* Hsining, Hsi-ning, Sining. *province capital* Qinghai, C China
Xinjiang *see* Xinjiang Uygur Zizhiqu
Xinjiang Uygur Zizhiqu *104 B3 var.* Sinkiang, Sinkiang Uighur Autonomous Region, Xin, Xinjiang. *autonomous region* NW China
Xinpu *see* Lianyungang
Xinxiang *106 C4* Henan, C China
Xinyang *106 C5 var.* Hsin-yang, Sinyang. Henan, C China
Xinzo de Limia *70 C2* Galicia, NW Spain
Xiqing Shan *102 D2 mountain range* C China
Xiva *100 D2 Rus.* Khiva, Khiwa. Xorazm Viloyati, W Uzbekistan
Xixón *see* Gijón
Xizang *see* Xizang Zizhiqu
Xizang Gaoyuan *see* Qingzang Gaoyuan
Xizang Zizhiqu *104 B4 var.* Thibet, Tibetan Autonomous Region, Xizang, *Eng.* Tibet. *autonomous region* W China
Xolotlán *see* Managua, Lago de
Xucheng *see* Xuwen
Xuddur *51 D5 var.* Hudur, *It.* Oddur. Bakool, SW Somalia
Xuwen *106 C7 var.* Xucheng. Guangdong, S China
Xuzhou *106 D4 var.* Hsu-chou, Suchow, Tongshan; *prev.* T'ung-shan. Jiangsu, E China
Xylókastro *83 B5 var.* Xilokastro. Pelopónnisos, S Greece

Y

Ya'an *106 B5 var.* Yaan. Sichuan, C China
Yabēlo *51 C5* Oromīya, C Ethiopia
Yablis *31 E2* Región Autónoma Atlántico Norte, NE Nicaragua
Yablonovyy Khrebet *93 F4 mountain range* S Russian Federation
Yabrai Shan *105 E3 mountain range* NE China
Yafran *49 F2* NW Libya
Yaghan Basin *45 B7 undersea feature* SE Pacific Ocean
Yagotin *see* Yahotyn
Yahotyn *87 E2 Rus.* Yagotin. Kyyivs'ka Oblast', N Ukraine
Yahualica *28 D4* Jalisco, SW Mexico
Yakima *24 B2* Washington, NW USA
Yakima River *24 B2 river* Washington, NW USA
Yakoruda *82 C3* Blagoevgrad, SW Bulgaria
Yaku-shima *109 B8 island* Nansei-shotō, SW Japan
Yakutat *14 D4* Alaska, USA
Yakutsk *93 F3* Respublika Sakha (Yakutiya), NE Russian Federation
Yala *115 C7* Yala, SW Thailand
Yalizava *85 D6 Rus.* Yelizovo. Mahilyowskaya Voblasts', E Belarus
Yalong Jiang *106 A5 river* C China
Yalova *94 B3* Yalova, NW Turkey
Yalpug, Ozero *see* Yalpuh, Ozero